The Routledge International Handbook of Innovation Education

The Routledge International Handbook of Innovation Education is the international reference work on innovation education intended to open an entirely new direction in education. The overall goal of the handbook is to address the question of how to develop innovators in general and how to develop the innovative potential of today's young people with exceptional talents in science, technology, engineering, and maths (STEM) disciplines in particular. Today many governments around the world are interested in the development of STEM innovators.

This handbook provides the first and most comprehensive account available of what should be done in order to develop innovators and how to do it successfully. It includes chapters by leading specialists from around the world responsible for much of the current research in the fields of innovation, gifted education, scientific talent, science education, and high ability studies. Based on the latest research findings and expert opinion, this book goes beyond mere anecdotes to consider what science can tell us about the development of innovators.

By enlisting chapters from innovation experts, educators, psychologists, policy makers, and researchers in the field of management, *The Routledge International Handbook of Innovation Education* allows all of these scholars to speak to each other about how to develop innovators via innovation education, including such issues as:

- the nature of innovation education;
- its basis, main components, and content;
- its criteria and specificity in various domains and contexts;
- societal demands placed upon it.

This ground-breaking work will thus serve as the first authoritative resource on all aspects of theory, research, and practice of innovation education.

Larisa V. Shavinina, PhD, is a Professor at the Université du Québec en Outaouais, Canada. She has been an innovation consultant to the governments of Canada, the USA, Dubai, Singapore, and other countries, and has acted as editor-in-chief of two other field-defining handbooks.

The Routledge International Handbook Series

The Routledge International Handbook of English, Language and Literacy Teaching
Edited by Dominic Wyse, Richard Andrews and James Hoffman

The Routledge International Handbook of the Sociology of Education
Edited by Michael W. Apple, Stephen J. Ball and Luis Armand Gandin

The Routledge International Handbook of Higher Education
Edited by Malcolm Tight, Ka Ho Mok, Jeroen Huisman and Christopher C. Morpew

The Routledge International Companion to Multicultural Education
Edited by James A. Banks

The Routledge International Handbook of Creative Learning
Edited by Julian Sefton Green, Pat Thomson, Ken Jones and Liora Bresler

The Routledge International Handbook of Critical Education
Edited by Michael W. Apple, Wayne Au, and Luis Armando Gandin

The Routledge International Handbook of Lifelong Learning
Edited by Peter Jarvis

The Routledge International Handbook of Early Childhood Education
Edited by Tony Bertram, John Bennett, Philip Gammage and Christine Pascal

The Routledge International Handbook of Teacher and School Development
Edited by Christopher Day

The Routledge International Handbook of Education, Religion and Values
Edited by James Arthur and Terence Lovat

The Routledge International Handbook of Innovation Education

Edited by Larisa V. Shavinina

LONDON AND NEW YORK

First published 2013
by Routledge
2 Park Square, Milton Park, Abingdon, Oxon OX14 4RN

Simultaneously published in the USA and Canada
by Routledge
711 Third Avenue, New York, NY 10017

Routledge is an imprint of the Taylor & Francis Group, an informa business

© 2013 Larisa V. Shavinina

The right of the editor to be identified as the author of the editorial material, and of the authors for their individual chapters, has been asserted in accordance with sections 77 and 78 of the Copyright, Designs and Patents Act 1988.

All rights reserved. No part of this book may be reprinted or reproduced or utilized in any form or by any electronic, mechanical, or other means, now known or hereafter invented, including photocopying and recording, or in any information storage or retrieval system, without permission in writing from the publishers.

Trademark notice: Product or corporate names may be trademarks or registered trademarks, and are used only for identification and explanation without intent to infringe.

British Library Cataloguing in Publication Data
A catalogue record for this book is available from the British Library

Library of Congress Cataloging in Publication Data
The Routledge international handbook of innovation education/edited by Larisa V. Shavinina.
  pages cm
  1. Science–Study and teaching–Handbooks, manuals, etc. 2. Technology–Study and teaching–Handbooks, manuals, etc. 3. Engineering–Study and teaching–Handbooks, manuals, etc. 4. Mathematics–Study and teaching–Handbooks, manuals, etc. 5. Creative thinking–Study and teaching–Handbooks, manuals, etc. I. Shavinina, Larisa V., editor of compilation.
  Q181.R846 2013
  507.1–dc23
                                                                                    2012036137

ISBN: 978-0-415-68221-3 (hbk)
ISBN: 978-0-203-38714-6 (ebk)

Typeset in Bembo
by Wearset Ltd, Boldon, Tyne and Wear

Printed and bound in the United States of America by Publishers Graphics, LLC on sustainably sourced paper.

This handbook is dedicated to my wonderful children Alexander, Denis, and Maxim

Contents

Contributors xiii
Preface: universal readiness to develop innovators xxvi

PART I
Introduction 1

1 Innovation education: the emergence of a new discipline 3
  Larisa V. Shavinina

PART II
The nature of innovation education 15

2 Innovation education: defining the phenomenon 17
  Rósa Gunnarsdóttir

3 The fundamentals of innovation education 29
  Larisa V. Shavinina

4 How advances in gifted education contribute to innovation education, and vice versa 52
  David Yun Dai

5 Innovation education meets conceptual change research: conceptual analysis and instructional implications 68
  Stella Vosniadou and Panagiotis Kampylis

6 New brain-imaging studies indicate how prototyping is related to entrepreneurial giftedness and innovation education in children 79
  Larry R. Vandervert and Kimberly J. Vandervert-Weathers

7 How can scientific innovators–geniuses be developed?: the case of Albert Einstein 92
  Larisa V. Shavinina

PART III
Creativity as a foundation of innovation education — 109

8 From creativity education to innovation education: what will it take? 111
   Joyce Van Tassel-Baska

9 The three-ring conception of innovation and a triad of processes for developing creative productivity in young people 128
   Marcia A. B. Delcourt and Joseph S. Renzulli

10 New creative education: when creative thinking, entrepreneurial education, and innovative education come together 142
   Fangqi Xu

PART IV
Assessment and identification related issues of innovation education — 151

11 Torrance's innovator meter and the decline of creativity in America 153
   Kyung Hee Kim and Robert A. Pierce

12 Do not overlook innovators!: discussing the "silent" issues of the assessment of innovative abilities in today's children–tomorrow's innovators 168
   Larisa V. Shavinina

PART V
From advances in giftedness and gifted education to innovation education — 183

13 Innovation education: perspectives from research and practice in gifted education 185
   Lynn H. Fox

14 An application of the schoolwide enrichment model and high-end learning theory to innovation education 201
   Ruth E. Lyons and Sally M. Reis

15 Future problem solving as education for innovation 215
   Bonnie L. Cramond and Elizabeth C. Fairweather

16 The trajectory of early development of prominent innovators: entrepreneurial giftedness in childhood 227
   Larisa V. Shavinina

PART VI
The role of teachers, parents, and schools in the development of innovators   243

17 Educating wizards: developing talent through innovation education   245
     Sarah J. Noonan

18 Where did all great innovators come from?: lessons from early childhood and adolescent education of Nobel laureates in science   258
     Larisa V. Shavinina

19 Settings and pedagogy in innovation education   273
     Svanborg R. Jónsdóttir and Allyson Macdonald

20 Exploring innovative schools with preservice teachers   288
     Michael Kamen and Deborah Erickson Shepherd

PART VII
Research on mathematical talent and innovations in math education for developing innovators   301

21 The dynamic curriculum: a fresh view of teaching mathematics for inspiring innovation   303
     Mark Saul

22 School textbooks as a medium for the intellectual development of children during the mathematics teaching process   315
     Marina A. Kholodnaya and Emanuila G. Gelfman

23 The interfaces of innovation in mathematics and the arts   330
     Bharath Sriraman and Kristina Juter

24 NASA press releases and mission statements: exploring the mathematics behind the science   341
     Sten Odenwald

PART VIII
Innovations in science education for developing innovators   357

25 Innovation in science, technology, engineering, and mathematics (STEM) disciplines: implications for educational practices   359
     David F. Feldon, Melissa D. Hurst, Christopher A. Rates, and Jennifer Elliott

26 The importance of informal learning in science for innovation education 372
    Susan M. Stocklmayer and Bobby Cerini

27 Designing an innovative approach to engage students in learning science: the evolving case of hybridized writing 385
    Stephen M. Ritchie and Louisa Tomas

28 An integrated approach to the study of biology 396
    Sergei Danilov and Olga Danilova

29 Socioscientific innovation for the common good 404
    John Lawrence Bencze

PART IX
How does technology education contribute to innovation education? 417

30 The role and place of science and technology education in developing innovation education 419
    Alister Jones and Cathy Buntting

31 Nurturing innovation through online learning 430
    Patricia Wallace

32 E-learning as educational innovation in universities: two case studies 442
    Lorraine Carter and Vince Salyers

33 Developing an understanding of the pedagogy of using a Virtual Reality Learning Environment (VRLE) to support innovation education 456
    Gisli Thorsteinsson

PART X
Innovation management, entrepreneurship, and innovation education 471

34 Creating an innovative and entrepreneurial collegiate academic program 473
    Lynn A. Fish and Ji-Hee Kim

35 Educating the innovation managers of the Web 2.0 age: a problem-based learning approach to user innovation training programs 487
    Peter Keinz and Reinhard Prügl

36 What can innovation education learn from innovators with longstanding records of breakthrough innovations?   499
    Larisa V. Shavinina

37 The role of entrepreneurs' career solidarity toward innovation: an irreplaceable relationship in career capital pyramid   513
    Masaru Yamashita and Jin-ichiro Yamada

38 Modeling the firm: constructing an integrated entrepreneurship course for undergraduate engineers   527
    Pius Baschera, Fredrik Hacklin, Georg von Krogh, and Boris Battistini

39 Igniting the spark: utilization of positive emotions in developing radical innovators   534
    Birgitta Sandberg

40 Introducing the phenomenon of the "abortion" of new ideas and describing the impact of "saved" ideas and thus implemented innovations on the economy in the case of distinguished innovators   545
    Larisa V. Shavinina

PART XI
Policy implications, institutional, and government efforts in innovation education   555

41 Innovation education through science, technology, engineering and math (STEM) subjects: the UK experience   557
    Frank Banks

42 Policy on knowledge exchange, innovation and entrepreneurship   570
    Alice Frost

43 The worldwide interest in developing innovators: the case of the Center for Talented Youth (United States) and PERMATApintar (Malaysia)   583
    Julian Jones and Noriah Mohd. Ishak

44 How does Singapore foster the development of innovators?   590
    Chwee Geok Quek and Liang See Tan

PART XII
Conclusions 605

45 Overall perspectives on the future promise (and forward thrusts) of innovation education 607
Larry R. Vandervert

Index 619

Contributors

Frank Banks is a Professor of Teacher Education and Director for International Development in Teacher Education at The Open University (OU) in the United Kingdom. He is responsible for teacher professional development projects in India, Bangladesh, and Sub-Saharan Africa, funded by charities and UK Aid, and has directed the innovative online initial teacher education program for the University. Before joining the OU he worked as a school teacher of science, engineering, and technology in different secondary high schools in England and in Wales, including head of department, and as an elementary school advisory teacher in Powys, Wales. Banks' research interests are in the fields of teacher professional knowledge, and science and technology education. He has authored or edited 12 books/handbooks for teachers and over 100 academic papers; and acted as a consultant in the professional development of teachers to Egyptian, South African and Argentinean government agencies, UNESCO, and the World Bank.

Pius Baschera is a Professor in the Department of Management, Technology, and Economics at ETH Zurich, where he holds the Chair of Entrepreneurship. He is the chairman of Hilti Corporation and the Venture Incubator, a seed-stage venture capital fund in Switzerland. He is a member of the Board of Directors of Roche Group, Schindler Holding, Vorwerk & Co. KG, and Ardex GmbH. He holds a PhD in management and an MSc in mechanical engineering from ETH Zurich.

Boris Battistini is a Research Associate in the Department of Management, Technology, and Economics at ETH Zurich. He is a project leader of the Corporate Venturing Research Initiative, undertaken in cooperation with Bain & Company. His research focuses on innovation, strategic investments, and venture capital, and his work has appeared in journals such as *Long Range Planning* and *Research-Technology Management*. He is a graduate of King's College London and holds an MSc and an MRes from the University of London.

John Lawrence Bencze, PhD, is an Associate Professor in Science Education at the Ontario Institute for Studies in Education, University of Toronto, where he teaches in the graduate studies and teacher education programs. In addition to his PhD (University Toronto, 1995) and BEd (Queen's University, 1977) in education, he holds BSc (Queen's University, 1974) and MSc (Queen's University, 1977) degrees in biology. Prior to his work as a professor, he worked as a teacher of science in elementary and secondary schools and as a science education consultant in Ontario, Canada. His teaching and research emphasize history, philosophy, and sociology of science and technology, along with student-led, research-informed, socio-political actions to address personal, social, and environmental issues associated with fields of science and technology.

Contributors

Cathy Buntting is an experienced researcher with the Wilf Malcolm Institute of Educational Research at the University of Waikato in New Zealand. She has a Master's degree in biochemistry and a PhD in science education. Her research interests are in science and technology education, including biotechnology education and the teaching of ethical and futures thinking skills.

Lorraine Carter, PhD, is the Academic Director of the Centre for Flexible Teaching and Learning at Nipissing University in North Bay, Canada. She conducts research in the health education field (cultural safety, health care in rural and northern settings, telemedicine, interprofessional care) and the broader field of post-secondary education (the scholarship of teaching and learning, use of social media in education, meaningful e-learning in universities). Carter holds an Honours BA in English (University of Western Ontario), MA in English (University of Western Ontario), BEd (University of Toronto), and PhD in Educational Studies (University of Windsor). Carter also has extensive teaching, research, and administrative experience in technology-supported education through Laurentian University, the Ontario Telemedicine Network, and the Northern Ontario School of Medicine. She is an active conference presenter and has won several awards for her leadership in enabling access to education through technology-mediated strategies.

Elizabeth (Bobby) Cerini is a science communication practitioner, consultant, and award-winning film-maker. Her PhD research at the Australian National University examines the roles and experiences of science heroes within contemporary society, and investigates whole of life engagement with science. Cerini is a previous recipient of a Melbourne University Alumni Award (1995), a British Council Chevening Scholarship (1998), and a Scinema Award (2008). Since 1998, she has been responsible for creating science engagement programs for a wide variety of organizations and audiences in the UK, Europe, Australia, and New Zealand. This includes overseeing events and programs during the UK's Science Year and Planet Science campaigns, hosting interactive workshops and programs for live and broadcast audiences, and designing science communication initiatives such as public exhibitions, online resources, and education activities. Cerini is the founding director of the Science and Factual Filmmakers' Network, an online community supporting emerging filmmakers.

Bonnie L. Cramond, PhD, is the Director of the Torrance Center for Creativity and Talent Development at the University of Georgia, and a Professor of Educational Psychology there. She has been a member of the Board of Directors of the National Association for Gifted Children, editor of the *Journal of Secondary Gifted Education*, and is on the review board for several journals. A former school teacher, she is a survivor of parenting two gifted and creative people. An international and national speaker, she has published numerous articles and chapters, a book on creativity research, and teaches classes on giftedness and creativity. She is particularly interested in the identification and nurturance of creativity, especially among students considered at risk because of their different way of thinking, such as those misdiagnosed with ADHD, emotional problems, or those who drop out.

David Yun Dai is an Associate Professor of Educational Psychology and Methodology at University at Albany, State University of New York. He received his doctoral degree in psychology from Purdue University, and worked as a post-doctoral fellow at the National Research Center on the Gifted and Talented, University of Connecticut. His research interests include gifted education, talent development, inquiry-based learning, and intellectual development. Dai was the recipient of the Early Scholar Award in 2006 conferred by the National Association for Gifted Children, and he was a Fulbright Scholar to China during 2008–2009. He currently serves on the editorial boards of *Gifted Child Quarterly*, *Journal for the Education of the Gifted*, and *Roeper Review*. He has published seven books and over 70 journal articles, book chapters, encyclopedia entries, and book reviews in general psychology, educational psychology, and gifted education.

Sergei Danilov, PhD, is a Lecturer in Biology at the National Taras Shevschenko University in Kiev, Ukraine. He authored and co-authored more than 30 publications and participated in the development of the national conception of biology education in Ukraine and has been a member of the Evaluation Committee of the National Olympiad in Biology, Ukraine. Danilov is actively involved in the professional development of biology teachers.

Olga Danilova, PhD, is a Professor Emerita of Biology at the National Taras Shevschenko University in Kiev, Ukraine and has authored more than 150 publications. She participated in the development of the national conception of biology education in Ukraine and has been a member of the Evaluation Committee of the International Olympiads in Biology.

Marcia A.B. Delcourt is a Professor in the Department of Education and Educational Psychology at Western Connecticut State University, Danbury, Connecticut, where she coordinates a Doctor of Education in Instructional Leadership Program. She has been active in education for over 20 years as a teacher, graduate level program coordinator, curriculum and program consultant, and researcher. She has also been a principal investigator for several state and federal grants, including for the National Research Center on the Gifted and Talented. Her present research interests include developing inquiry skills in teachers and students, encouraging gifted and creative behavior, developing and analyzing talent development programs in schools, and encouraging teachers as researchers. She is presently working with educators and researchers through the government of India to identify and mentor children who have high ability in mathematics and science.

Jennifer Elliott is an Assistant Professor of Instructional Design and Technology at the University of Cincinnati in the College of Education, Criminal Justice and Human Services, whose work focuses on the research, design, and development of games and simulations and immersive learning environments for education. Her research includes the design and development of mobile augmented reality applications, theoretical framework development in immersive learning environments, the intersection of cognition, pedagogy, and technology, 3D visualization tools, and technology for universal design. Additionally she has published research on collaborative problem solving and participatory augmented reality, social media and professional development, and complex systems understanding with simulations. She teaches courses in game design, transdisciplinary collaboration and innovation, learning theory in online learning environments, and legal, social, and ethical issues in educational technology.

Deborah Erickson Shepherd is the reading and dyslexia specialist at Meridian School, an International Baccalaureate charter school she co-founded in Round Rock, Texas. She has 18 years of teaching experience both in traditional public and charter schools across several grade levels, including multi-age classes. She has served as a literacy coach and reading specialist for eight years. Deborah earned a BA from Trinity University in elementary education, special education, and education for the hearing impaired. She is currently working on her Master's in Elementary Education at Texas State University. Deborah co-taught the Innovative Schools class with Michael Kamen for two years at Southwestern University in Georgetown, Texas.

Elizabeth C. Fairweather is a part-time Assistant Professor in the Educational Psychology Department at the University of Georgia. In addition, she teaches middle school advanced/gifted language arts and social studies in Walton County, Florida. She is a veteran teacher with over 20 years of experience in the field of education and has earned certification from the National Board for Professional Teaching Standards. She has taught in public elementary, middle, and high schools as well as at the university level. She is particularly interested in facilitating creativity among children and adolescents especially through developing diagnostic measures of creative process. She has presented at state and national educational conferences, as well as published in educational journals and books.

Contributors

David F. Feldon is an Associate Professor of STEM education and educational psychology at the University of Virginia's Curry School of Education and the Associate Director of the Center for the Advanced Study of Teaching and Learning in Higher Education (CASTL-HE). Feldon's research examines two lines of inquiry that are distinct but mutually supportive. The first characterizes the cognitive components of expertise as they contribute to effective and innovative problem solving, as well as how they affect the quality of instruction that experts can provide. The second examines the development of research skills within STEM disciplines as a function of instruction and other educational support mechanisms. His scholarship has been published in various journals, including *Educational Psychologist*, *Educational Technology Research and Development*, *Journal of Research in Science Teaching*, *Journal of Higher Education*, and *Science*.

Lynn A. Fish is a Professor of Management at Canisius College in Buffalo, New York. She earned her BS, MS and PhD in Industrial Engineering and MBA from the University of Buffalo. MBA students have honored her three times as the Donald E. Calvert Distinguished Professor. Her strong background in industrial and production systems engineering provide her students with a real-world understanding of concepts. She has published several articles on her 'hands-on' methods in the prestigious *Decision Sciences Journal of Innovative Education*. Her research interests focus on supply chain, project management, and operations management. Fish has published over 60 articles in scholarly journals, conference proceedings, and professional publications, and has given presentations at numerous professional meetings in the US, Canada, and Europe.

Lynn H. Fox is an educational psychologist and the former Dean of the School of Education at American University where she taught courses in Educational Psychology, Testing and Measurement, Quantitative Research Methods and Math for Elementary School Teachers. Her PhD is in psychology from the Johns Hopkins University where she was a Professor of Education, an Assistant/Associate Director of the Study of Mathematically Precocious Youth from 1971 to 1974, and the Director of the Intellectually Gifted Child Study Group from 1974 to 1985. Throughout her career her focus has been on gender differences, gifted and 'at risk' learners, and testing and program evaluation. She is the author of numerous books, book chapters, and journal articles. She is currently an Associate Professor Emeritus at American University in Washington, DC, and is enjoying consulting and volunteer work in Venice, Florida.

Alice Frost took up the post of Head of Business and Community policy at the Higher Education Funding Council for England (HEFCE – the primary public funder of higher education in England) in April 2006. She has previously held a number of policy positions at the Council, including Head of Research Policy and Head of Learning and Teaching. Alice Frost has had a varied career in the UK in national public policy (including in the Department for Education, Cabinet Office/Office of Science and Technology, and House of Commons), higher education policy (Universities UK and the charitable sector), and in HE and the regions. Her areas of policy interest include science and technology policy and regional economic development. She studied politics at undergraduate and postgraduate levels at Oxford University.

Emanuila G. Gelfman, PhD, is a Professor and the Chair of the Division of Mathematics and Math Teaching at the Tomsk Pedagogical University in Tomsk, Russia. She is also the Director of the Research Center on the Intellectual Development of Personality. She authored more than 70 publications on how to teach mathematics in school and university. Her research interests focus on math teaching.

Contributors

Rósa Gunnarsdóttir, PhD, has been working with young inventors for nearly 20 years, both in Iceland and internationally. Her PhD thesis was published at the University of Leeds under the title 'Innovation education: defining the phenomenon'. There she mapped the teaching and learning processes in Innovation Education classes in a school in Iceland. She has been in the forefront of the development and implementation of the curriculum subject Innovation Education in Iceland. She has also been active in the emerging discussions of Innovation Education and Entrepreneurial Education in Europe. One of her current projects is working with the Mawhiba Institution in Saudi Arabia in developing an Innovation Education and Entrepreneurial Education track for their National Olympiad. She is now working as Director of Teaching Affairs and Registry at Reykjavík University.

Fredrik Hacklin is a Lecturer in the Department of Management, Technology and Economics at ETH Zurich, where he is heading research activities of the entrepreneurship group. Previously, he was an associate with Booz & Company and had visiting positions at Stanford University and Keio University. His research interests are in corporate innovation and strategic entrepreneurship. He holds a PhD in management from ETH Zurich and an MSc in computer science from KTH Stockholm.

Melissa D. Hurst is the Director of Career Development in the office of graduate studies and postdoctoral programs at the University of Virginia. She provides career and student development services designed to assist Master's and PhD students with the transition into the workplace. Previously she was a postdoctoral research associate in the Curry School of Education where she explored graduate student development among students in the STEM fields. She holds a PhD in educational psychology from the University of South Carolina.

Alister Jones is a Research Professor and the Deputy Vice-Chancellor at the University of Waikato, Hamilton, New Zealand. Prior to his current role, he was Dean of Education at the University of Waikato. He is a past director of the Wilf Malcolm Institute of Educational Research and the Centre for Science and Technology Education Research at the university. He has worked in a variety of university settings internationally to build research capability. He has research interests in curriculum, teaching, learning, and assessment, particularly in science education.

Julian Jones, PhD, is the Senior Director for International Development at the Johns Hopkins University Center for Talented Youth (CTY). He is helping build a worldwide network of exceptional pre-university students and organizations that serve them. The network will feature common admission and shared research in giftedness and gifted education. In an earlier career, he taught international relations at the University of Maryland and later directed the university's programs in Asia. Jones holds a PhD from the Fletcher School of Law and Diplomacy.

Svanborg R. Jónsdóttir is a lecturer in Arts and Creative Work at the School of Education at the University of Iceland with a PhD in pedagogy and education, and her doctoral studies were about Innovation Education in Icelandic Compulsory Schools. Her research interests are in creative work, school change and innovation, and entrepreneurial education.

Kristina Juter works as an Associate Professor at Kristianstad University and Lund University in Sweden. She has many years of experience in mathematics teacher education at all levels, but has mainly been working with students becoming secondary and upper secondary school teachers. Juter has also worked with regular mathematics courses at upper secondary level as well as university level.

Her research interests are concept formation in calculus at university level, development of teacher identity, and student achievements in mathematics. She is also interested in artistic applications of mathematics.

Michael Kamen is an Associate Professor and Education Department Chair at Southwestern University in Georgetown, Texas, where he has taught since 2002. Prior to accepting his current position, he taught for 11 years at Auburn University, where he was an Associate Professor and Elementary Education Program Coordinator. Kamen earned a BA in Elementary Education from SUNY at Stony Brook, an MS in Supervision and Administration from Bank Street College of Education, and a PhD in Science Education from the University of Texas at Austin. He has ten years of experience teaching in public and private schools at the elementary and middle school levels. His teaching and research interests include innovative schools, Vygotskian/cultural-historical activity theory, science education, math education, educational technology, creative drama, play, lesson study, environmental education, and the role of language in learning.

Panagiotis Kampylis joined the European Commission, Joint Research Centre, Institute for Prospective Technological Studies (Information Society Unit) in January 2012. He contributes—a Research Fellow—to projects on ICT for Learning, Creativity, and Innovation in Life Long Learning. Kampylis holds a PhD in Cognitive Sciences from the University of Jyväskylä (Finland) thanks to a scholarship from the Greek State Scholarships Foundation. The title of his PhD dissertation is 'Fostering creative thinking: the role of primary teachers'. He co-authored several scientific publications in the fields of creativity and innovation in education and training, arts education, and ICT-enhanced learning. He is co-founder and associate editor of the open-access *Hellenic Journal of Music, Education and Culture*.

Peter Keinz, PhD, is an Assistant Professor at the Institute for Entrepreneurship and Innovation at the WU—Vienna University of Economics and Business. In research, he mainly focuses on user innovation and open innovation. In particular, Keinz works on the development of methods and tools that help organizations to benefit from the creative potential of external individuals and/or collectives in the course of innovation processes. Furthermore, he is interested in the strategic implications for companies when pursuing open and user innovation strategies. In teaching, Keinz applies problem-based learning approaches. Transferring his scientific insights about open and user innovation into his course formats, he offers classes in which students conduct innovation management projects. Over the last seven years, Keinz has run more than 40 of these projects with partners like Magna, Siemens, CERN, and the European Space Agency.

Marina A. Kholodnaya, PhD, is a Professor of Psychology and the Director of the Laboratory of High Abilities at the Institute of Psychology of the Russian Academy of Sciences in Moscow, Russia. She authored more than 170 publications. Her research interests include the nature of human intelligence, cognitive styles, conceptual thinking, intellectual giftedness, and innovation education.

Ji-Hee Kim is an Associate Professor of Entrepreneurship and the Director of the Undergraduate Entrepreneurship Program at Canisius College in Buffalo, New York. Her research interests include innovative entrepreneurial education, family business, small business, economic development, and consumer decision making. She is an Advisory Board member of the Korea Small and Medium Business Administration, the Korean Federation of Small and Medium Business, and the Center of Korean Women Entrepreneurship for the 21st Century. She is also an Associate Director of the High-Tech Business Development and Research Institute and a regional coordinator for the Family Firm Institute.

Kim designed and taught several Entrepreneurship courses at Canisius including: Creativity, Innovation, and Entrepreneurship; Family Business and Entrepreneurship; Small Business Management and Planning; New Venture Creation and Contemporary Issues in Entrepreneurship.

Kyung Hee Kim is an Associate Professor of Educational Psychology at the College of William and Mary. She trains parents and teachers around the world to foster creativity in children. She is the elected chair for the Creativity Network of the National Association for Gifted Children. In July 2010, she opened a national and international dialogue with her study 'The Creativity Crisis'. Numerous organizations have recognized Kim's scholarly contributions including, in 2011, the Early Scholar Award from the National Association for Gifted Children (NAGC); in 2009, the Berlyne Award from the American Psychology Association, and the New Voice in Intelligence and Creativity Award from the University of Kansas and the Counseling Laboratory for the Exploration of Optimal States; in 2008, the Hollingworth Award from NAGC and the Ronald W. Collins Distinguished Faculty Research Award from Eastern Michigan University; and others. Kim was recently honored as the 2012 Torrance Lecturer.

Georg von Krogh is a Professor at ETH Zurich, where he holds the Chair of Strategic Management and Innovation. He has been appointed a member of the National Research Council (*Nationaler Forschungsrat*) of the Swiss National Science Foundation. Georg was also the Head of ETH Zurich's Department of Management, Technology, and Economics from 2008–2011. He specializes in competitive strategy, technological innovation, and knowledge management. He has conducted research in various industries including financial services, media, computer software and hardware, life-sciences, and consumer goods, and has co-authored books on strategic management, knowledge creation, innovation and organization, and management theory. His articles have been published in journals including *Management Science*, *MIS Quarterly*, *Organization Science*, *Research Policy*, *Strategic Management Journal*, and *Harvard Business Review*. He is a Senior Editor of *Organization Studies*, and an editorial board member of a number of journals.

Ruth E. Lyons is the Director of the Dr. Joseph S. Renzulli Gifted and Talented Academy in Hartford, Connecticut. During her 11 years in education she witnessed the profound power and impact gifted pedagogy has, which encouraged her to pursue a PhD in the field. Ruth is a doctoral candidate in the Educational Psychology Department at the University of Connecticut. Ruth received a Bachelor's of Fine Arts from Rochester Institute of Technology, a Master's of Science in Education from Fordham University, and a Certificate of Advanced Studies in Educational Leadership from the University of Maine. She has been awarded the 2010 NAGC A. Harry Passow Classroom Teacher Scholarship and the Connecticut Association of the Gifted 2011 Administrator of the Year. While at the University of Connecticut Ruth was the on-site coordinator for Confratute, the longest running summer institute on enrichment-based differentiated teaching.

Allyson Macdonald is a Professor in Educational Studies at the School of Education at the University of Iceland. She has an interest in science, technology and innovation, and in educational change. She has studied in South Africa, the USA, England, and Iceland.

Noriah Mohd. Ishak, PhD, is the Director of Pusat PERMATApintar Negara, a Malaysian National Gifted Center at the Universiti Kebangsaan Malaysia (UKM). She was involved in building the Malaysian gifted program which was targeted to all Malaysian children since its inception in 2007. She is actively doing research on giftedness and counseling for the gifted among Malaysian gifted students. In an earlier career, she taught counseling, psychology and research methods at the Faculty of Education, UKM, and served as the Deputy Dean at the faculty. Mohd. Ishak holds a PhD from the Western Michigan University.

Sarah J. Noonan, EdD, an Associate Professor in the Leadership, Policy and Administration Department at the University of St Thomas in Minneapolis, Minnesota, teaches graduate courses in leadership and organizational theory, intercultural communication, teacher learning, and professional development, and issues and challenges in executive leadership. Noonan previously served as a superintendent and assistant superintendent of schools, director of teaching and learning, state director of gifted education, and K-12 teacher before receiving her faculty appointment at the University of St Thomas in 2000. A passionate and innovative teacher, Noonan offers workshops on 'motivating pathways' for student learning, the power of story, peer reviews of teaching, and interest, novelty, and emotion in learning. Her publications include several books on leadership and articles on culturally sensitive pedagogy. Noonan earned a Doctor of Education in Educational Leadership from the University of Wyoming.

Sten Odenwald, PhD, is an astronomer at the NASA Goddard Spaceflight Center, and Senior Scientist with ADNET Corporation. He received his PhD from Harvard University in 1982, and has been involved with investigations of the star formation, galaxy evolution, and the nature of the cosmic infrared background. He is an active science popularizer and book author, and participates in many NASA programs in space science and math education. He has a number of websites promoting science education including 'The Astronomy Café' (www.astronomycafe.net) and 'Space Weather' (www.solarstorms.org). Currently, he is the Director of the Space Math @ NASA project which is NASA-based, 'STEM' education program that develops K-12 math problems featuring scientific discoveries from across NASA. SpaceMath achieved its five-millionth problem download by teachers and students in May, 2012 (http://spacemath.gsfc.nasa.gov).

Robert A. Pierce is a Lecturer in History at Christopher Newport University (CNU). His area of specialization is Italy in the late Middle Ages, and, among other publications, he authored *Pier Paolo Vergerio the Propagandist* (2003) and co-edited *Ritratti: La dimensione individuale nella storia* (2009). Pierce has received research grants from the Pro Helvetia Foundation, the Bibliographical Society of America, the Bibliographical Society (UK), and the American Philosophical Society. He is also active in the area of teacher education, k-12 curriculum, and assessment. He teaches in the MAT program at CNU, is the Social Studies faculty member for Project AERO, the curriculum and instruction arm of the Office of Overseas Schools of the US Department of State, and serves as a Chief Examiner for the International Baccalaureate. Pierce trains social studies teachers around the world in various social studies teaching methods, including ways to foster creativity.

Reinhard Prügl, PhD, is a Professor and holder of the Chair of Innovation, Technology and Entrepreneurship at Zeppelin University Friedrichshafen at Lake Constance, Germany. His research is anchored in the areas of user and open innovation, technological competence leveraging, family entrepreneurship, and business model innovation emphasizing aspects of search for innovation-related knowledge. He is particularly interested in gaining a deeper understanding of mechanisms and effects of different approaches toward a systematic search for innovation.

Chwee Geok Quek, PhD, received her Master's and Doctoral degrees from the College of William and Mary in Virginia, USA. She is a Principal Specialist in Gifted Education at the Gifted Education Branch, Curriculum Planning and Development Division, Ministry of Education, Singapore.

She spent a major part of her career in the field of gifted education. Her research interests include gender differences, the use of gifted education pedagogy to improve teaching and learning, effectiveness of professional development and factors impacting talent development.

Christopher A. Rates is currently a doctoral student in the STEM Education program at the Curry School of Education at the University of Virginia. He has a Master's degree from Curry's Educational Psychology and Applied Developmental Science program where he acquired a background in examining educational issues using mixed methods. His primary research interest focuses on the development of critical thinking skills in the context of STEM learning.

Sally M. Reis is the Vice Provost for Academic Affairs and a Board of Trustees Distinguished Professor at the University of Connecticut where she also serves as a Principal Investigator for the National Research Center on the Gifted and Talented. She was a teacher for 15 years, 11 of which were spent working with gifted students on the elementary, junior high, and high school levels. She has authored or co-authored over 250 articles, books, book chapters, monographs, and technical reports. Her research interests are related to special populations of gifted and talented students, including students with learning disabilities, gifted females, and diverse groups of talented students. She is also interested in extensions of the Schoolwide Enrichment Model for both gifted and talented students and as a way to expand offerings and provide general enrichment to identify talents and potentials in students who have not been previously identified as gifted.

Joseph S. Renzulli is a Distinguished Professor of Educational Psychology at the University of Connecticut and the Director of the National Research Center on the Gifted and Talented. His research has focused on strength-based assessment, the identification and development of creativity and giftedness in young people, and models for personalized learning. A focus of his work has been on applying the pedagogy of gifted education to the improvement of learning for all students. His most recent work is a computer-based assessment of student strengths and a teacher-planning tool integrated with an Internet-based search engine that matches highly challenging enrichment activities and resources to individual student profiles and teacher selected curricular topics. The American Psychological Association named Renzulli among the 25 most influential psychologists in the world and in 2009 he received the Harold W. McGraw, Jr. Award for Innovation In Education, considered by many to be 'the Nobel' for educators.

Stephen M. Ritchie is a Professor of Science Education at Queensland University of Technology, Brisbane, Australia. His research has focused mostly on classroom issues that relate to teaching and learning science. He currently conducts research on the emotional engagement of students in science classes as they become more scientifically literate. He is also interested in the emotional experiences of beginning science teachers and the quality of pre-service science teacher education. He is the Editor-in-Chief (2008–2012) of *Research in Science Education*.

Vince Salyers, PhD, is the Associate Dean, Faculty of Health and Community Studies at Mount Royal University in Calgary, Alberta, Canada. He is responsible for initiatives in the areas of curriculum development, international education, research, e-learning, and partnerships. Salyers holds a BA in Psychology and a Master of Science in Nursing from San Francisco State University and an EdD in Curriculum and Instruction from the University of San Francisco. The integration and utilization of technology and e-learning strategies into program curricula is his passion. His doctoral research focused on the effectiveness of web-enhanced instruction in teaching psychomotor nursing skills and he continues to develop his web-enhanced and e-learning research. Salyers is currently involved in an international, multi-institution study aimed at helping students and faculty to identify their needs and systematically implement support strategies for integrating e-learning technologies into their learning and teaching activities in effective, meaningful, and sustainable ways.

Birgitta Sandberg, PhD, is a Post-Doctoral Researcher at the Department of Marketing and International Business, Turku School of Economics, University of Turku. She coordinates the Global Innovation Management Master's Degree Programme and teaches various courses. Her main research interests include the development and marketing of radical innovations, entrepreneurial innovators, and the role of emotions in innovation processes. Her publications include a book entitled *Managing and Marketing Radical Innovations: Marketing New Technology* (2008, Routledge), and articles in, for example, *Journal of Business Research, Creativity and Innovation Management, Technovation*, and *European Journal of Innovation Management*.

Mark Saul is the Director of the Center for Mathematical Talent at the Couarnt Institute of Mathematical Sciences, New York University. He grew up in the Bronx and graduated from Columbia University and New York University. He then spent 35 years in and around New York City, teaching mathematics from grades 3 through 12. More recently, he has been consultant in gifted education for the John Templeton Foundation and program director for the National Science Foundation. He is a 1984 recipient of the Presidential Award for Mathematics and Science Teaching. Saul directed the Research Science Institute at MIT and in Shanghai. Internationally, he has given talks and led workshops in 20 countries. He served as the President of the American Regions Mathematics League, mathematics field editor of *Quantum* (the English-language version of the Russian journal *Kvant*), and is an Associate Editor of the AMS *Notices*. His publications include numerous articles, books, and translations.

Larisa V. Shavinina, PhD, is a Professor at the Université du Québec en Outaouais, Canada. Her research focuses on the nature of giftedness, the child prodigy phenomenon, scientific talent in the case of Nobel laureates, entrepreneurial giftedness, managerial talent, new assessment procedures for the identification of the gifted, high intellectual and creative educational multimedia technologies (HICEMTs) aimed at developing potential abilities of gifted and talented individuals. Over the years Dr. Shavinina's research has expanded to encompass innovation. Her bestselling *International Handbook on Innovation* (1171 pages; Elsevier, 2003), was the first and only book of its type, and is considered the beginning of innovation science. She introduced innovation education as a new direction in gifted education. Innovation is an important element in Dr. Shavinina's research on giftedness and economy. Her *International Handbook on Giftedness* (1539 pages; Springer, 2009) sets a new standard for the field.

Bharath Sriraman is a Professor of Mathematics in the Department of Mathematical Sciences and an Adjunct Professor of Central/SW Asian Studies at the University of Montana, where he occasionally offers courses in Indo-Iranian studies and languages. He lives in Montana byway of the merchant marine. He received his BS in mathematics from the University of Alaska, and his MS and PhD in mathematical sciences from the Northern Illinois University. He maintains an active interest in educational philosophy, interdisciplinarity (math–science–arts), history and philosophy of mathematics and science, creativity; innovation and talent development; and political and social justice dimensions of education. He has published over 300 journal articles, commentaries, book chapters, edited books, and reviews in his areas of interest, and presented over 120 papers at international conferences, symposia, and invited colloquia. Sriraman is the founding editor of *The Mathematics Enthusiast*, as well as the founding series co-editor of *Advances in Mathematics Education* (Springer Berlin/Heidelberg).

Susan M. Stocklmayer is a Professor of Science Communication at the Australian National University, Canberra, and the Director of the Australian National Centre for the Public Awareness of Science, which is also the UNESCO Centre for Science Communication. The Centre has a full graduate and undergraduate program in Science Communication. Her major research concerns issues related to science learning at the interface between science and the public, and gender and multicultural issues. As part of the university's outreach she has presented science shows, lectures, and workshops on all five continents. She is the co-Editor in Chief of the *International Journal of Science Education Part B: Communication and Public Engagement* and was awarded the Order of Australia in 2004 for services to science communication.

Liang See Tan, received her PhD from Nanyang Technological University, Singapore and Master of Science in Education at Purdue University, United States. Before joining the National Institute of Education (NIE) as a lecturer, she was a high school teacher for 16 years. She fosters talent through mentorship programs in school. Currently, she is a research scientist at the Office of Education Research at NIE. Tan's research focuses on differentiated curriculum and instruction, talent development, and the relations between motivational beliefs, goal orientations, and student outcomes. She is the Principal Investigator and Co-Principal Investigator for more than $500,000 in competitive research funding from the Office of Educational Research (OER), National Institute of Education. These projects include the SEYLS (Singapore Early Years Longitudinal Study) project, Project ARTS (Arts Research on Teachers and Students: Pedagogies and Practices), and the Phase 2 of Project ARTS II (Arts Research for Students and Teachers 2).

Gisli Thorsteinsson, PhD, is an Associate Professor in Design and Craft in the School of Education at the University of Iceland. He holds a doctoral degree in Philosophy from Loughborough University in England, which focused on using a Virtual Reality Learning Environment for idea generation training in Innovation Education in Iceland. Thorsteinsson was the Chair of the Icelandic Craft Teachers Society in 1995–2005, the Chair for the Nordic Sloyd Teachers Society in 2001–2005, and on the board of NordFo, an Academic Research Society for Pedagogic Craft Education in the Nordic countries in 2000–2005. Thorsteinsson was involved in building up a new national curriculum for ICT and Technology Education in 1999. He has published numerous articles and several textbooks on Innovations Education and Design and Craft Education.

Louisa Tomas is a lecturer in Science Education at James Cook University, Townsville, Australia. Tomas' research primarily explores curricula and pedagogical approaches that engage diverse students in the learning of science and enhance the development of scientific literacy. She is currently researching school students' emotional engagement with socioscientific issues with her colleague and co-author, Stephen Ritchie, as well as Science and Sustainability Education in pre-service teacher education.

Larry R. Vandervert is widely published in the study of creativity, and innovation. In 2003 he contributed two chapters on innovation education to Larisa V. Shavinina's *International Handbook on Innovation*, including (1) the neurophysiology of innovation and (2) the book's concluding chapter, 'Research on Innovation at the Beginning of the 21st Century: What do we know about it?' Vandervert's works on innovation are extensions of his fundamental published contributions on the collaborative roles of the cerebral cortex and the cognitive functions of the cerebellum in creativity, in the neurophysiology of the child prodigy phenomenon, and in the neurophysiology of working memory in the evolution of language. These contributions (along with the foregoing reference citations) are described in Wikipedia's discussions of *creativity* (Sec. 8.1) and of *child prodigies*. He is currently a retired college Professor.

Contributors

Kimberly J. Vandervert-Weathers has a Master's Degree in Teaching from Whitworth University. During the preparation of her chapter contribution to this handbook, she taught at Seattle Hebrew Academy, where she also supervised Camp Invention summer camps. She is currently employed with Spokane Public Schools.

Joyce VanTassel-Baska is the Smith Professor Emerita at the College of William and Mary in Virginia where she developed a graduate program and a Research and Development Center in Gifted Education. Formerly, she initiated and directed the Center for Talent Development at Northwestern University. She has also served as the state Director of gifted programs for Illinois, as a regional Director of a Gifted Service Center in the Chicago area, as coordinator of gifted programs for the Toledo, Ohio public school system, and as a teacher of gifted high school students in English and Latin. VanTassel-Baska has published widely including 27 books and over 500 refereed journal articles, book chapters, and scholarly reports. Her major research interests are on the talent development process and effective curricular interventions with the gifted.

Stella Vosniadou is a Professor of Cognitive Psychology and the Chair of the Division of Cognitive Science in the Department of Philosophy and History of Science at the National and Kapodistrian University of Athens. She is also the Chair of the Interdisciplinary Graduate Program in Cognitive Science and the Director of the Cognitive Science Laboratory at the University of Athens. Her research interests are in the areas of learning and cognitive development. She is internationally known for her studies on conceptual change in the learning of science and mathematics. Amongst others, she is the Editor of the *International Handbook of Research on Conceptual Change* (Routledge, 2008). Vosniadou is a Fellow of the International Academy of Education and of AERA, current Chair of the Cognitive Science Society, and past Chair of the European Association for Research on Learning and Instruction (EARLI).

Patricia Wallace, PhD, is the Senior Director, CTY*Online* and Information Technology, at the Johns Hopkins University Center for Talented Youth. She heads the rapidly growing online programs for gifted youth in grades preK to 12, and is also principal investigator for Cogito.org, a website and online community for students interested in STEM fields. Her research and writing focus on the relationships among technology, learning, and human behavior, and she has published ten books, many articles, and several educational software programs. Books include *The Psychology of the Internet* and *The Internet in the Workplace*, both published by Cambridge University Press. She earned her PhD in psychology and also holds an MS in computer systems management. Before joining CTY in 2001, she held positions as a faculty member, CIO, and Head of a research center.

Fangqi Xu is a Professor of Management at the Kinki University in Japan. He obtained his PhD at the School of Knowledge Science of JAIST (Japan Advanced Institute of Science and Technology) under the supervision of Professors Ikujiro Nonaka and Susumu Kunifuji. Xu is the author of many papers, books, and book chapters including *The Comparative Research between Japanese and Chinese Enterprises* and *Liu Chuanzhi: The Founder of Lenovo*. He is the founder and the Director of the Institute for Creative Management and Innovation, Kinki University, Editor of *Kindai Management Review*, and the President of the Japan Creativity Society.

Jin-ichiro Yamada, PhD, is an Associate Professor of Entrepreneurship Strategy in School of Business at the Osaka City University, and the visiting research officer of National Science, Technology and Policy, Japan. He was a visiting fellow in Cranfield School of Management, UK, and a Visiting Professor in Chair Arts, Culture and Management in Europe, Bordeaux Business School, France. His main interest is entrepreneurship strategy and strategic management in creative industries and science-based industries. His current research work focuses on the multidimensional roles of entrepreneurship in start-ups, innovation, and industrial clustering.

Masaru Yamashita, PhD, is an Associate Professor of Business at the Aoyama Gakuin University, Tokyo, Japan, and a Visiting Scholar at the San Diego State University, USA. His research interests include organizational creativity in Japanese companies, especially in Japanese creative industry. He studies what organizational process makes an organizational creativity, and focuses on social capital and human resource development.

Preface
Universal readiness to develop innovators

> If you publish ... you are trying to create something that is original, that stands out from the crowd. ... Above all, you want to create something you are proud of.
>
> (Richard Branson, 2002, p. 57)

Today, innovation is the cornerstone of economic prosperity, scientific discovery, technological invention, and cultural vibrancy. As the Premier Minister of Québec Jean Charest emphasized launching Québec's innovation strategy, 'Innovation has been critical to the economic development of modern societies ... Our prosperity in the future depends on it.' Governments around the world are launching innovation strategies and design innovation policies to make their economies the innovation-based economies and to transform their societies into innovative societies. At the same time, little is known about exactly what should be done and how it should be done. Because of this confusion, those strategies and policies are rarely successful. Financial investments in research and development (R & D) and new technologies are essential, but this is not the whole story. Despite the ever-increasing importance of innovation in society, one should acknowledge that innovation remains somewhat mysterious. During many years of my work on the bestselling *International Handbook on Innovation* (the first encyclopedia of this type that is considered the beginning of innovation science; published by Elsevier Science, 1171 pages, Shavinina, 2003), I found that something important was missing: a supply of new innovators and educational theories and techniques on how to develop them. This topic has never been studied. Echoing Minister Charest's statement, societies can progress today only through new levels of innovation and every effort should be made to develop potential innovators. *The Routledge International Handbook of Innovation Education* is, consequently, a timely and much needed endeavor of great importance to the whole world.

On August 24–25, 2009 the US National Science Board invited a group of experts for a panel discussion on how to develop the next generation of innovators in science, technology, engineering, and mathematics (STEM) disciplines. This event—with a subsequent report to Congress and President Obama—shows that the US government is interested in developing innovators. Governments of many other countries have also come to realize that they have to develop innovators in one way or another, but too often do not know how. Governments are ready to invest a great deal in the development of innovators, but they do not know what to do exactly or where to start. This is a new trend on a global scale.

There exists, therefore, a universal need in developing innovators. The whole world is thus ready to absorb the first comprehensive publication on how to develop innovators that this handbook offers. This is a unique historical moment in time that presents an unprecedented opportunity, and we have to act on it. This handbook aims exactly at addressing the issue of how to develop innovators by offering a new approach which is ever-gaining worldwide momentum and acceptance: *innovation education*. The handbook will have a potentially huge worldwide impact due to the above-described universal readiness (and hunger) to learn how to develop innovators. This emphasizes the urgency and importance of the handbook.

The Routledge International Handbook of Innovation Education is the very first handbook on innovation education. It opens an entirely novel direction in education. In my invited speech on how to develop innovators at the US National Science Board in August 2009, I emphasized that a radically new type of education is needed: innovation education. This is the beginning of a new field of scientific inquiry. If we are indeed interested in developing tomorrow's innovators who will be able to *implement* creative ideas into practice in the form of new products, services, or processes, and to make great discoveries and inventions, then the best way to do this is via innovation education. The overall goal of the handbook is to provide direct answers to the question of how to develop innovators.

Specifically, this handbook presents the first and most comprehensive account available of *what* should be done in order to develop innovators and *how* to do it successfully. With respect to *what*, the handbook offers a clear answer: innovation education. With respect to *how*, the handbook puts forward the existing research findings, which provide a wide range of perspectives leading to promising approaches.

There is a consensus among researchers today that innovation refers to the implementation of ideas into practice in the form of new products, processes, or services (Shavinina, 2003). Sometimes innovation and creativity are considered to be synonymous. This is not correct. Creativity refers to the generation of novel, original, and appropriate ideas. Innovation is essentially about the *implementation* of ideas. Creators generate new ideas, innovators implement them into practice. The handbook is mainly about how to develop innovators who can implement new ideas.

Many would agree that innovators are vital for the scientific, economic, technological, and societal advancements in the world. Many would also agree that human beings must progress in their study of innovators and especially to understand their development. However, there are only a relatively small number of researchers who really study innovators in this way. The number is getting even smaller if one looks at those scholars who address the issue of *how* to develop innovators. My goal was to put them together in a single handbook in order to present a comprehensive picture of the contemporary research on what should be done in order to develop tomorrow's innovators and how. The specific objectives of the handbook are twofold. The first is to lay out the conceptual basis of innovation education as a new, emerging scientific discipline. The second objective consists in presenting a wide range of best practices in innovation education worldwide. For the very first time, researchers from a variety of disciplines are brought together in one handbook to outline the foundations of innovation education and give practitioners sound advice on how to develop innovators. The handbook is thus international in scope reflecting American, Asian, and European knowledge and practices within overall global perspectives.

The target international audience for this handbook is broad and includes a wide range of specialists—both researchers and practitioners (mostly teachers) alike—in the areas of education, talent development and high ability, innovation, science education and gifted education, and policy making. Non-specialists (mainly parents) will also be interested in this handbook, and it will be useful in a wide range of undergraduate and graduate courses as supplementary reading. Because the coverage of the handbook is broad, it can be used as a reference on an as-needed basis for those who would like information about a particular topic, or read from cover to cover as a sourcebook. In addition, it can be used as a textbook in courses dealing with general education, innovation education, science education, and gifted education. In short, anyone interested in knowing the wide range of issues regarding innovation education will want to have easy access to this handbook.

The handbook hopes to accomplish at least four things for readers. First, the reader will obtain expert insight into what innovation education is all about. Indeed, some of the world's leading

Preface

specialists contributed their chapters. Second, the handbook presents many facets of innovation education and this breadth will allow the reader to acquire a comprehensive and panoramic picture of the nature of innovation within a single book. Third, based on this picture, the reader will develop an accurate sense of what spurs potential innovators toward their extraordinary achievements and exceptional performances leading to breakthrough innovations. The reader will also understand why some individuals become innovators and why others, while they could potentially become innovators, do not. Fourth, and perhaps most importantly, the reader (whether a student, teacher, researcher, or business professional) will be able to apply the ideas and findings in this handbook to critically consider how best to foster his or her own innovative abilities and innovative performance.

I wanted to do almost the impossible with this handbook: to cover every conceivable facet of innovation education as it is today. As readers proceed from one chapter to another, they will see that ultimately this desire became possible. In line with the Richard Branson quote above, I and the contributing authors are indeed proud of the final product: the handbook is exceptional in many ways (as addressed in the introduction).

Financial support of granting agencies for my research that led to the idea of this handbook deserves special mention. I am extremely grateful to the Social Sciences and Humanities Research Council of Canada (SSHRC) for their three-year grant *A Study of Early Childhood and Adolescent Education of Nobel Laureates and the Implications for Gifted and General Education: Developing Scientific Talent of Nobel Calibre* in 2007–2010. I wish to thank the Templeton Foundation and the Institute for Research and Policy on Acceleration of the Belin-Blank International Center for Gifted Education and Talent Development at the University of Iowa in the USA for their support of the project *The Role of Academic Acceleration in the Development of Scientific Talent: The Case of Nobel Laureates*.

I must also express my exceptional debt of gratitude to the *Fonds québécois de la recherche sur la société et la culture* (FQRSC) for their two grants under the Support for Innovative Projects program entitled *Introducing and Explaining the Phenomenon of Individual Innovation: When Intuition, Wisdom, Creativity, Managerial Talent, Entrepreneurial Giftedness, and Excellence Come Together* (in 2008–2010) and *Introducing the Phenomenon of the 'Abortion' of New Ideas and Estimating the Potential Impact of 'Killed' Ideas and Thus Lost Innovations on the Economy* (in 2010–2012). I benefited greatly from the FQRSC support of my radically innovative ideas and projects. The handbook is one of the outcomes of those remarkable projects.

There are many people to thank for helping this handbook come to fruition. Most important are the authors: I thank them very much for their willingness to undertake the difficult and challenging task of contributing chapters. I am particularly grateful to Professor Marina A. Kholodnaya, my former PhD supervisor, who to a great extent 'made' me a researcher, developing my perception of important scientific issues. She continually inspires me to take on innovative endeavors and to constantly move ahead.

I am very grateful to Larry Vandervert, Bruce Roberts, Michael Isichenko, and Evgueni Ponomarev for their thorough reviews of the chapters. Their promptness, as well as long-term, sustained commitments to the project and/or short-term emergency assistance, were highly appreciated. Although I must have driven him crazy at times, Larry tremendously inspired me during the work on the handbook and pushed my own innovative abilities to their limits.

I am especially grateful to my research assistants—Alla Vlasenko, David Lefebvre, and Eugenia Iurcu—for their successful handling of countless duties on this immense project in addition to their regular jobs and/or university assignments. I exceptionally appreciate their amazing ability to quickly grasp my intentions when 'time' constraints were tough and I may have given contradictory instructions. Somehow they had a rare talent to figure out what I indeed wanted to say and what was needed at each particular moment. Their ability to prioritize tasks and to figure out exactly what was required is remarkable. At times, I did not know how they managed to put up with me. They were simply excellent throughout months of the preparation of the handbook: I could not have done it better myself.

Special thanks to my editor at Routledge/Taylor & Francis publishing company—Bruce Roberts—who provided just the right blend of freedom, encouragement, patience, and guidance needed for successful completion of this immense project. He liberated me from many routine tasks that have emerged over the course of this project, and I am grateful to him. We have had a remarkable working relationship since the publication of my *International Handbook on Innovation* by Elsevier, when Bruce was there. Over the years Bruce tried to convince me to undertake new publishing projects with him; and he succeeded. I am happy to publish this new handbook with him. I could not have dreamed of a better publisher.

I must mention the exceptional professionalism of Mary Bailes, who translated two chapters from Russian into English. She provided extra effort and insight in her superb contributions, and, as the authors admitted, the final chapters in English looked even better than they did in their original Russian.

I am exceptionally grateful to my role-models of wisdom—Evgenia L. Yakovleva, Valentina Levyshkina, and Olga V. Danilova—for their unconditional support and incredibly high level of inspiration, which I always need working under strict deadlines. Multiple hours of phone conversations—leading to heightened emotional energy necessary for remarkable achievements and speedy handling of my numerous duties—are greatly appreciated. Their constant inspiration and 'push' kept me moving forward.

I also wish to acknowledge my debt of gratitude to my parents, Anna Shavinina and Vladimir Shavinin, who aroused a passionate intellectual curiosity and love for challenges in me. I express sincere appreciation to my aunt Elena Avdienko, who has been helping us by taking care of our little Maxim and cooking nice meals over the last six months of the project.

Finally, I owe my biggest debt of gratitude to my husband, Evgueni Ponomarev, and our 12-year-old, eight-year-old, and two-year-old sons, Alexander, Denis, and Maxim, respectively. In countless ways, Evgueni has been a true advisor, critic, and friend throughout the project. He provided the moral and technical support, and—more importantly—the time I needed to complete this project. He did so by performing a number of great tasks, from administering PC problems when I worked at nights to assuming the lion's share (and the lioness's, too) of child care for our kids and particularly for Maxim, who was just ten-months old when I started the project. Very simply, this is his handbook, too.

I also want to thank Alexander, Denis, and Maxim whose entry into the world taught me more about innovators and the need to develop their versatile innovative abilities, especially when they are at those unique sensitive periods, than have any other events in my life. They were patient with me, encouraged me, and curiously asked 'when will you eventually finish this handbook?' Thinking about the futures of my own children, I sincerely hope that educational systems of all countries around the world, as well as parents and other caregivers, will be able to develop unique innovative talents of each child on our planet.

References

Branson, R. (2002). *Losing My Virginity: The Autobiography*. London: Virgin Books.
Shavinina, L. V. (Ed.). (2003). *The International Handbook on Innovation*. Oxford, UK: Elsevier Science.

Larisa V. Shavinina
June 2012
Gatineau, Québec, Canada

Part I
Introduction

1

Innovation education
The emergence of a new discipline

Larisa V. Shavinina
UNIVERSITÉ DU QUÉBEC EN OUTAOUAIS, CANADA

Summary: This chapter provides a general introduction to the handbook thus creating a broad picture on what to expect in the chapters that follow. Specifically, it presents a short overview of a new, emerging field: innovation education. Its multifaceted and multidimensional nature is discussed via a short description of chapters included in this handbook. The uniqueness and novelty of the handbook are pointed out. The main contents of each chapter are summarized and approaches taken by chapter authors are briefly outlined.

Key words: Innovation education, innovation, new emerging field of innovation education, approaches to understanding innovation education.

> We have to find a new view of the world. … If you can find any other view of the world which agrees over the entire range where things have already been observed, but disagrees somewhere else, you have made a great discovery. It is very nearly impossible, but not quite.
> *(Richard Feynman, Nobel Laureate)*

This handbook is about a great discovery: innovation education, which refers to a wide range of educational interventions aimed at identifying, developing, and transforming child talent into adult innovation. These are those societal actions aimed at preparing children to become adult innovators. Such educational interventions should include, but should not be limited to, those elements discussed in the chapters included in this handbook. Its mission is an extraordinary one: to provide a comprehensive, deep insight into how to develop innovators. In order to fulfill this mission, the content of the handbook is unique in a number of ways.

First of all, it is the first handbook on innovation education, which marks the beginning of a new, emerging discipline: innovation education. The science of innovation education will originate from this handbook. This is the first book that directly and comprehensively addresses the above-described universal need in knowing how to develop innovators. The handbook will not only improve education in many ways; it has significant economic implications as well and greatly contributes to the resolution of important social problems. Taking into account today's increasing demand on innovative ideas, solutions, products, and services, the handbook will answer educators' and governments' concerns related to the best development of innovators.

Second, the handbook is devoted to the multidisciplinary and multifaceted nature of innovation education, its foundations and main components, the current state of research and practice, as well as future developments. The purpose of the handbook is to present the existing knowledge on this multidimensional phenomenon—innovation education—from the viewpoints of various scientific disciplines: mainly innovation, gifted education, science education, psychology of high ability and talent, entrepreneurship, innovation management, business, and technology.

This is thus a truly innovative and unique handbook. All these facets of its novelty convincingly demonstrate that the handbook is a much needed endeavor. It lays out the foundations of a new field: innovation education. The aim of bringing a wide range of experts together in this handbook was to present a comprehensive picture of contemporary research on and practice of innovation education.

In selecting chapter authors, I was especially interested in their contributions to the emerging discipline of innovation education and/or in their challenging ideas, which will advance the field in the near future. My deepest belief is that any handbook on any scientific topic should not only report the current findings in the field, but must also move forward that field by introducing thought-provoking novel ideas. In one way or another, each chapter in the handbook brings something new to our understanding of innovation education. This is one of the key merits of this handbook.

The chapter contributors take a number of different approaches, reflecting a variety of perspectives on innovation education and related concepts such as innovation, creativity, entrepreneurial giftedness, enterprise education, and others. Sometimes even the interpretations of a key concept of the handbook—innovation education—differ. However, this is quite normal at the current stage in the development of the field of innovation education, when it is just at its infancy. I will briefly mention the main ideas of chapters below. The descriptions of each chapter are intentionally short in order to entice readers to seek further details contained within chapters.

The handbook is divided into 12 (XII) parts. Part I comprises Chapter 1, *Innovation education: the emergence of a new discipline*, which sets the stage for understanding innovation education by providing a general introduction to a variety of issues discussed in the chapters that follow.

Parts II to XII, consisting of 44 chapters, represent distinctive, although definitely overlapping, approaches to innovation education. Specifically, Part II of the handbook describes the nature of innovation education, its basic mechanisms, and its various facets. This part includes six chapters.

Chapter 2, *Innovation education: defining the phenomenon*, by Rósa Gunnarsdóttir, describes the beginnings and achievements of innovation education in Iceland. This small country is probably the only in the world where innovation education is the mandatory subject in school curriculum. Dr. Gunnarsdóttir is one of a few pioneers of innovation education in Iceland. In her chapter she defines the phenomenon of innovation education and analyzes its developmental essence and conceptual basis.

In Chapter 3, *The fundamentals of innovation education*, Larisa V. Shavinina discusses the bases of innovation education, which are associated with the 11 interrelated components. These include: gifted education programs; science and technology education; new programs aiming to develop entrepreneurial giftedness that is closely related to innovation; programs for fostering children's metacognitive abilities or abilities to implement things: the so-called executive talent; new programs based on recent progress in the study of scientific talent of Nobel laureates; programs that incorporate the essentials of research on polymaths; new programs for nurturing applied wisdom and moral responsibility; programs aiming to develop managerial talent; the fundamentals of deadline management; the foundations of innovation science; and courage-related issues, because courage is mandatory for innovators.

Chapter 4, *How advances in gifted education contribute to innovation education, and vice versa* by David Yun Dai, examines how achievements in the field of gifted education and related areas contribute to shaping a new discipline of innovation education. Dr. Dai identifies various phases in the development of gifted education and describes its impact on innovation education. The author concludes his deep analysis by calling for a number of imperatives vital for innovation education such as the curriculum, pedagogical, capacity building, and assessment imperatives.

In Chapter 5, *Innovation education meets conceptual change research: conceptual analysis and instructional implications*, Stella Vosniadou and Panagiotis Kampylis discuss innovation in science and mathematics, namely the role of conceptual change in scientific and mathematical discovery. The authors argue that teaching science and mathematics from the perspective of a conceptual change develops the skills for the gradual but intentional restructuring of students' prior knowledge. This is one of the best ways to cultivate future innovators in STEM disciplines.

Chapter 6, *New brain-imaging studies indicate how prototyping is related to entrepreneurial giftedness and innovation education in children*, by Larry R. Vandervert and Kimberly J. Vandervert-Weathers, is the only chapter in the handbook, which offers an interesting insight into how an innovator's brain functions. The authors thus explain the neuro-scientific basis of innovation education. However, this is not the whole story. They also present a range of exciting educational options aimed at developing children's innovative abilities such as, for example, Camp Invention in the USA.

In Chapter 7, *How can scientific innovators–geniuses be developed?: the case of Albert Einstein*, Larisa V. Shavinina explicates the developmental foundation of individual innovation that is related to a child's sensitive periods—periods of heightened and selective responsiveness to everything that is going on around him or her. Specifically, the chapter focuses on the developmental basis of innovation in the case of Albert Einstein. It was found that a number of overlapping sensitive periods characterized a trajectory of Einstein's educational development and this is why his scientific genius emerged. The implications for innovation education are discussed.

Part III of the Handbook, *Creativity as a foundation of innovation education*, concentrates on creativity as the first step in innovation process and emphasizes a need to develop creative abilities as an essential facet of innovation education. This Part includes three chapters.

Chapter 8, *From creativity education to innovation education: what will it take?* by Joyce VanTassel-Baska, traces the development of creativity in giftedness research and links it to innovation education. The author highlights distinctions between creativity education and innovation education, as well as analyzes similarities, using examples of work in curriculum, instruction, and assessment to illustrate key points. Dr. VanTassel-Baska considers creativity as associated with new ideas and methodologies that break from the past, while innovation represents an attempt to synthesize past efforts and render them pragmatic in a given field of endeavor.

In Chapter 9, *The three-ring conception of innovation and a triad of processes for developing creative productivity in young people*, Marcia A. B. Delcourt and Joseph S. Renzulli discuss components of innovative process and types of educational services for promoting innovation in young people. These services include exposing students to areas of potential interest and task commitment, providing them with the methodological skills to pursue their interests in a professionally authentic manner, and providing the opportunities, resources, and encouragement to see their ideas through to fruition.

Chapter 10, *New creative education: when creative thinking, entrepreneurial education, and innovative education come together*, by Fangqi Xu, introduces the concept of a new creative education, which consists of creative thinking, entrepreneurial education and innovative education. The author analyzes each of these elements and shows that they cannot exist one without the other if one is really concerned with the development of innovators.

Part IV of the Handbook, *Assessment and identification related issues of innovation education*, is devoted to the exceptionally important concept in the field of innovation education: how to measure innovative abilities. It includes just two chapters, thus reflecting a current state of affair in this area: an almost complete absence of the assessment methods aimed at the identification of innovative talents.

In Chapter 11, *Torrance's innovator meter and the decline of creativity in America*, Kyung Hee Kim and Robert A. Pierce describe the Torrance Tests of Creative Thinking (TTCT)—Figural, which is the most widely accepted measure of adaptive creativity, innovative creativity, and creative personality. Research demonstrates that even as intelligence quotient (IQ) scores increased annually, TTCT scores continually decreased in the United States since 1990. A creative and innovative soul is part of what enabled America to ascend to world leader, but if the United States is no longer an environment that fosters creativity and innovation, will it continue to lead the world? The authors provide explanations for and implications of this trend.

Chapter 12, *Do not overlook innovators!: discussing the "silent" issues of the assessment of innovative abilities in today's children–tomorrow's innovators*, by Larisa V. Shavinina, is about a wide range of topics related to the identification of innovation talent. These include the theoretical basis of innovation, why the normal notion of assessment that applies to intelligence testing does not necessarily apply to innovation talent, and how the best measurements can be developed. The chapter also discusses what should be included in the comprehensive approach to the psychological assessment of innovative abilities. It is emphasized that innovative abilities can be identified in the form of what Dr. Vandervert calls "trajectory analysis" or "trajectory assessment."

Part V of the Handbook, *From advances in giftedness and gifted education to innovation education*, examines what innovation education can take from giftedness research and gifted education. In other words, it looks at those best practices in the area of high ability aim to develop innovation talent. It contains four chapters.

In Chapter 13, *Innovation education: perspectives from research and practice in gifted education*, Lynn H. Fox states that research and practice in Gifted and Talented Education can inform efforts to conceptualize and implement programs for innovation education. She reviews some of the major findings and conclusions from over 100 years of research and programming in gifted education. Successes and failures in ways to formulate definitions of giftedness and talent are instructive as educators seek to develop program models and strategies in innovation education. The author specifically looks at findings from 50 years of study of mathematically precocious students and efforts to foster interest and excellence in education for science, technology, engineering, and mathematics (STEM Education), which are particularly relevant for the emerging discipline of innovation education.

Chapter 14, *An application of the schoolwide enrichment model and high-end learning theory to innovation education*, by Ruth E. Lyons and Sally M. Reis, presents the Renzulli Academy, an innovative urban school, in which students were transformed from consumers of information to producers of information, inventors, historians, and scientists. By providing a detailed blueprint for total school innovation, schools that implement the Schoolwide Enrichment Model (SEM) can develop a unique program based on local resources, student populations, school leadership dynamics, and faculty strengths and creativity. The SEM includes numerous opportunities for differentiated instruction and an assessment of students' interests, learning styles, and product styles. In SEM schools, students access quality, challenging, and engaging curriculum and experience a pedagogy of enrichment and engagement that increases achievement and creates a setting where academic excellence and innovation is expected and creativity is celebrated in all students.

In Chapter 15, *Future problem solving as education for innovation*, Bonnie L. Cramond and Elizabeth C. Fairweather describe the International Future Problem Solving Program aimed at developing creative thinking and innovative dispositions in students. The brain child of Paul Torrance, this program helps to prepare a workforce for the future capable of innovation. The Future Problem Solving Program should, therefore, be an important component of innovation education.

Chapter 16, *The trajectory of early development of prominent innovators: entrepeneurial giftedness in childhood*, by Larisa V. Shavinina, concentrates on early development of gifted entrepreneurs–innovators. The cases of Richard Branson, Warren Buffett, Steven Case, Michael Dell, Bill Gates, and Sam Walton are discussed, which show that the trajectories of their entrepreneurial giftedness originated from childhood. It was found that early developmental path is common for these gifted entrepreneurs–innovators. The influence of early manifestations of entrepreneurial giftedness on the subsequent development of their innovative talent is also considered.

Part VI of the Handbook, *The role of teachers, parents, and schools in the development of innovators*, examines an exceptionally important role that great teachers, supporting parents, and excellent schools play in nurturing innovation talent. This part includes four chapters.

In Chapter 17, *Educating wizards: developing talent through innovation education*, Sarah J. Noonan offers an interesting perspective on how to make innovation education an interesting and enjoyable experience that leads to the development of innovation talent. Her new approach consists in the analysis of a case study of Harry Potter, an adolescent with unrealized special powers, as portrayed in *Harry Potter and the Sorcerer's Stone* (Rowling, 1997). The analysis highlights the author's idea that the educational conditions favoring the realization and use of a capacity for innovation should include (1) interactions with *expert teachers* in a variety of settings, (2) access to a *challenging curriculum*, (3) opportunities to receive *skilled coaching* to develop talents, and (4) *opportunities to satisfy individual needs and achieve developmental goals* in innovator-friendly environments.

Chapter 18, *Where did all great innovators come from?: lessons from early childhood and adolescent education of Nobel laureates in science*, by Larisa V. Shavinina, analyzes interesting findings from the project about early childhood and adolescent education of Nobel laureates in science. Specifically, it focuses on the exceptional role of parents and teachers in developing innovators-geniuses. The understanding of the principles involved in the educational development of Nobel laureates will allow teachers to accordingly improve, develop, modify, and transcend areas in the current curriculum in an attempt to cultivate scientific talent, of Nobel caliber, in future generations of innovators.

In Chapter 19, *Settings and pedagogy in innovation education*, Svanborg R. Jónsdóttir and Allyson Macdonald discuss findings from research on innovation and entrepreneurial education (IEE) in Icelandic compulsory schools. IEE is an emerging curriculum area in Iceland that aims to develop creativity and problem solving abilities in students. The authors analyze the interaction between IEE and the settings into which it is introduced. The distinguishing features of those settings are considered, which support or hinder the development of IEE and the types of pedagogy it requires.

Chapter 20, *Exploring innovative schools with preservice teachers*, by Michael Kamen and Deborah Erickson Shepherd, presents an elective course in which preservice teachers study innovative schools. Students examine various facets of such schools with a focus on the theoretical and philosophical assumptions that create the school's ethos. Many of these schools have a social justice mission. The class visits a number of very different schools, discussing the pedagogy of each school with teachers, administrators, and students themselves. This course prepares future teachers to be innovators in their classrooms rather than teachers focused on high-stakes standardized testing and external accountability.

Part VII of the Handbook, *Research on mathematical talent and innovations in math education for developing innovators*, is about nurturing innovative talents in mathematics. It consists of four chapters.

Chapter 21, *The dynamic curriculum: a fresh view of teaching mathematics for inspiring innovation*, by Mark Saul, is about an alternative educational model, in which curriculum is seen as dynamic and the teacher as responding interactively to the unfolding needs of the students. Examples from working classrooms, using mathematical content, are considered.

In Chapter 22, *School textbooks as a medium for the intellectual development of children during the mathematics teaching process*, Marina A. Kholodnaya and Emanuila G. Gelfman argue that one of the ways to develop children's intelligence and innovative talent is through the design of new school textbooks and educational materials of a type that meets the requirements of a psycho-didactic approach. As part of the "Mathematics, Psychology, Intelligence" (MPI) education project, the authors developed mathematics textbooks and educational materials (study books, practical work, workbooks for independent study, computer software) for middle school students within the framework of an "enrichment" teaching model. The goal of this model is the intellectual development of students through their mathematical education using specially constructed educational texts. Their specificity is related to the fact that, at the same time as conveying structures of formal mathematical knowledge, they also (1) support development of the basic components of students' mental experience (including cognitive, conceptual, metacognitive and intentional experience), and (2) create conditions for students to employ their own individual cognitive styles.

Chapter 23, *The interfaces of innovation in mathematics and the arts*, by Bharath Sriraman and Kristina Juter, is devoted to innovation in architecture and art with an emphasis on mathematical creativity. To this end, the work of Buckminster Fuller is considered. Architectural creation in society is tightly connected to geometry, topology, and other parts of mathematics. The authors view buildings and art as results of human thinking linking abstract mathematical representations and concrete physical structures. For such links to occur, inventors need to be able to work in interdisciplinary settings. Findings from mathematics education are described in the light of fostering creative innovators in mathematics related to the arts.

In Chapter 24, *NASA press releases and mission statements: exploring the mathematics behind the science*, Sten Odenwald presents a unique approach to teaching math: SpaceMath@NASA program. NASA press releases and mission statements contain a wealth of quantitative information and understated mathematics, which are used in this program aimed at stimulating children's interest in mathematics and science. Students are often curious about space topics, such as the search for extraterrestrial life, black holes, or space exploration. For this reason, press releases about discoveries in space make mathematics relevant and exciting to students beyond the mundane application problems so common in modern-day mathematics textbooks. This is a truly innovative program for teaching mathematics.

Part VIII of the Handbook, *Innovations in science education for developing innovators*, concentrates on domain-specific innovations in various scientific areas. It consists of five chapters.

Chapter 25, *Innovation in science, technology, engineering, and mathematics (STEM) disciplines: implications for educational practices*, by David F. Feldon, Melissa D. Hurst, Christopher A. Rates, and Jennifer Elliott, reviews research from education, psychology, and the sociology of science that inform educational practice. Findings from these fields shed light on how to develop innovators in science and engineering. Scientists endeavor to generate new knowledge about the universe, and engineers apply that knowledge to generate effective solutions for new problems. The authors bring interesting insights into how systematic approaches to science education at all levels of schooling can prepare students to enter their respective fields as innovators.

In Chapter 26, *The importance of informal learning in science for innovation education*, Susan M. Stocklmayer and Bobby Cerini compare the roles of formal and informal learning in nurturing the skills needed for innovative and creative thinking in science. Informal learning, whether it be at the hands of a mentor, in the environment of a science center, museum, or a science club, or through a demanding hobby which requires an understanding of science, can be a powerful change agent in fostering such skills. Formal education can, on the other hand, stifle innovative skills. Through specific case studies of science innovators, the authors demonstrate that there are common factors, which may be identified and explored in order to bring more opportunities for developing innovators into the formal science education area.

Chapter 27, *Designing an innovative approach to engage students in learning science: the evolving case of hybridized writing*, by Stephen M. Ritchie and Louisa Tomas, is about a pioneering online project aimed to engage students in learning science related to the socially relevant issue of biosecurity. In the project, students are required to merge narrative story lines with scientific information. This hybridized writing project called *BioStories* is innovative in that it is a new educational response to the problem of disengagement of students in science, and has proven to be effective in changing classroom practice and improving students' scientific literacy. The authors highlight the innovative design process of the project, emphasize its effectiveness, and identify future directions for further development and research.

In Chapter 28, *An integrated approach to the study of biology*, Sergei Danilov and Olga Danilova discuss a whole set of issues related to the design and use of an innovative set of electronic technologies for teaching biology. From the authors' point of view, the successful biology education should be directed toward: (1) the development of students' analytical and practical skills; (2) the promotion of healthy way of life; (3) a personal orientation to continuing education; (4) individualized checking of knowledge and self-checking; and (5) ecologization of education and creating the foundations for nature-conservation activity.

Chapter 29, *Socioscientific innovation for the common good*, by John Lawrence Bencze, raises a whole range of important issues for innovation education. His research shows that in many contexts throughout the world the ability to innovate has been increasingly limited to a few powerful people and groups. Financiers and corporations seem to be the main controllers of innovation—using it to encourage most citizens to function as unquestioning consumers of for-profit products and services. This hyper-consumerism leads to many problems, including increasing disparities between rich and poor and dramatic environmental degradation. As science and engineering play a significant role in innovation and capitalist activities, school science can make great contributions to generating societies that may effectively address potential social and environmental problems associated with extreme capitalism.

Part IX of the Handbook, *How does technology education contribute to innovation education?* brings to the attention a set of diverse issues related to the role of technology education in developing innovators. This part consists of four chapters.

In Chapter 30, *The role and place of science and technology education in developing innovation education*, Alister Jones and Cathy Buntting observe that little is known how science and technology education might contribute to the development of a future-focused innovative culture. The authors examine the role and place of science and technology education in developing innovative culture and thinking by providing a broad international scope of the field. They explore the connection between science and technology education and emergent approaches that focus on developing strong connections between the work of science and innovation and the science and technology classroom to enhance student engagement and an understanding of modern innovation. The place of futures thinking in science and technology education is also analyzed.

Chapter 31, *Nurturing innovation through online learning*, by Patricia Wallace, considers the evolution of online learning environments, their psychological, social, and cognitive characteristics, and how these environments contribute to innovation education. As the technologies advance, opportunities arise to tap the unique capabilities of these environments to foster critical thinking, creativity, and innovation. The author found that some characteristics of the online world—and, therefore, online education—may actually help create an environment that promotes attitudes and behaviors thought to be related to innovation, such as risk taking and willingness to challenge or defy norms. Dr. Wallace provides examples of online learning programs, which lead to the development of innovative abilities in children.

In Chapter 32, *E-learning as educational innovation in universities: two case studies*, Lorraine Carter and Vince Salyers present an innovative educational framework, ICARE, as a means of structuring e-learning opportunities. Research demonstrates that courses using the ICARE framework provide a valid alternative to more traditional face-to-face classroom formats, and remove some of the barriers usually associated with face-to-face, blended, and fully online learning environments. The authors discuss the developing quantitative and qualitative evidence which educators need to understand as they search for, adopt, and implement effective and innovative e-learning strategies.

Chapter 33, *Developing an understanding of the pedagogy of using a Virtual Reality Learning Environment (VRLE) to support innovation education*, by Gisli Thorsteinsson, is about a specific e-learning aimed at supporting ideation. The goal of Innovation Education (IE) as a new subject area in Icelandic schools is to teach students to identify needs and problems in their environment and to find solutions: a process of ideation. This activity has been classroom based, but now a unique VRLE technology has been created to support ideation. This technology supports online communications between students and teacher and enables them to develop drawings and descriptions of the solutions.

Part X of the Handbook, *Innovation management, entrepreneurship, and innovation education*, mainly discusses how entrepreneurial abilities, which are closely related to innovation, and innovation talents in business can be developed. It consists of seven chapters.

In Chapter 34, *Creating an innovative and entrepreneurial collegiate academic program*, Lynn A. Fish and Ji-Hee Kim describe a successful entrepreneurship program that emphasizes creativity, innovation, interpersonal skills, and entrepreneurial leadership.

The integrated curriculum provides students with a broad and necessary business background while developing entrepreneurial skills and insights. Students discover their innate entrepreneurial potential; develop processes, tools, and perspectives to capitalize on that potential; learn to identify and evaluate business opportunities; acquire capital and other resources; and to start, develop, grow, operate, and harvest a business. The thriving program continues to grow to other college academic areas, locally and internationally.

Chapter 35, *Educating the innovation managers of the Web 2.0 Age: a problem-based learning approach to user innovation training programs*, by Peter Keinz and Reinhard Prügl, concentrates on open innovation paradigm, which changes the role of a company in innovation process. Today, companies are facilitators rather than innovators themselves. It means that they are concerned with providing users and other external stakeholders with the know-how, skills, and means necessary to perform innovation-related tasks. The question then arises: how can business schools develop workforce for such companies? The authors present a training program focused on the idea of problem-based learning in which students get to know some of the essential user innovation methods (e.g., lead-user method as well as user community-based technological competence leveraging) and actually get involved in a real-life project.

In Chapter 36, *What can innovation education learn from innovators with longstanding records of breakthrough innovations?* Larisa V. Shavinina considers findings from the study of the phenomenon of individual innovation in the case of outstanding innovators and their implications for innovation education. In sharp contrast to the conventional wisdom of innovation science emphasizing that (1) innovation is a team sport, and (2) people are good either in generating ideas (i.e., creativity) or in their implementing into practice (i.e., innovation), just to mention a few dogmas, the author found that there is a rare group of individual innovators. They possess by a unique ability to both generate great ideas and to implement them into practice in the form of new products, services, and processes by putting into place all the necessary organizational, human, and "environmental" structures. This is the phenomenon of individual innovation.

Chapter 37, *The role of entrepreneurs' career solidarity toward innovation: an irreplaceable relationship in career capital pyramid*, by Masaru Yamashita and Jin-ichiro Yamada, brings to the attention the role of a single outstanding entrepreneur in innovation process in contrast to a team of entrepreneurs. The authors introduce the issue of "career solidarity," which requires accumulation of hierarchic career capital that is composed by the asset of knowing-whom, knowing-why, and knowing-how. One of the best practices in the case of the Japanese film industry is presented and explained in light of the career solidarity concept.

In Chapter 38, *Modeling the firm: constructing an integrated entrepreneurship course for undergraduate engineers*, Pius Baschera, Fredrik Hacklin, Georg von Krogh, and Boris Battistini examine the increasing managerial challenges of innovation and entrepreneurship for engineers and consider a course aimed to develop their managerial talents and entrepreneurial abilities. As the industry reality is becoming increasingly complex and boundary-spanning, tomorrow's engineers are required to take on integrative and interdisciplinary roles of coordination and management early on. Consequently, the authors argue that today's teaching programs need to respond to this trend by equipping engineers with a more holistic and integrated view of managerial concepts, allowing them to act as innovators and entrepreneurs in their fields.

Chapter 39, *Igniting the spark: utilization of positive emotions in developing radical innovators*, by Birgitta Sandberg, is the only contribution to the handbook that explores the important role of emotions in nurturing innovative talents. The author focuses on radical innovation, which is characterized by significant uncertainties, mixed emotions, and rapid changes. This presents a great challenge to the current education and requires the development of novel forms of teaching. Even though the role of the affective dimensions in learning is acknowledged, the influence of emotions in higher education is still a relatively unexplored field. Dr. Sandberg examines how positive emotions can be incorporated into university education targeted at developing radical innovators. Three positive emotions—joy, enthusiasm, and pride—are considered in some detail and their use in the development of radical innovators is assessed.

In Chapter 40, *Introducing the phenomenon of the "abortion" of new ideas and describing the impact of "saved" ideas and thus implemented innovations on the economy in the case of distinguished innovators*, Larisa V. Shavinina brings interesting insights to the topic of innovation education. She states that when people intentionally or unintentionally abandon their ideas without a wish to develop them any further and eventually implement them into practice, they thus abort potential innovations. This is what the phenomenon of the abortion of ideas is all about. The chapter concentrates on the well-known cases when individuals resisted abandoning their creative ideas and finally implemented them. These cases shed light on what today's children can learn from outstanding innovators in order to be able to save, develop, and implement their ideas into practice in the form of new products, processes, and services.

Part XI of the Handbook, *Policy implications, institutional, and government efforts in innovation education*, demonstrates what organizations and countries should—and should not—do in order to develop innovators. This part consists of four chapters.

Chapter 41, *Innovation education through science, technology, engineering, and math (STEM) subjects: the UK experience*, by Frank Banks, examines 30 years of different projects and initiatives in the United Kingdom intended to increase the relevance of the curriculum to life outside the school, to promote creativity and enterprise in young people, and to foster innovation through "minds-on as well as hands-on" teaching strategies. Dr. Banks analyzes both successful approaches and "lessons to be learned" in the cases of the Technical and Vocational Educational Initiative (TVEI) in the 1980s; the introduction in the 1990s of the manufacture of innovative products through the new national curriculum subject of Technology for *all* students aged 5–16 years; the Young Foresight Programme in the 2000s; and the recent different curriculum organization models to encourage STEM coordination in schools.

In Chapter 42, *Policy on knowledge exchange, innovation and entrepreneurship,* Alice Frost describes the recent development of policy and practice related to the contribution of universities to innovation and entrepreneurship in the UK. The author especially concentrates on the role of the Higher Education Funding Council for England (HEFCE) in innovation policy. Funding for knowledge exchange through HEFCE's HE Innovation Funding (HEIF) links innovation and enterprise agendas with academic and teaching developments, creating an ideal environment for innovation education. Knowledge exchange leads to universities and academics, which understand the commercial world and innovation, and which can support and inspire students—tomorrow's innovators.

Chapter 43, *The worldwide interest in developing innovators: the case of the Center for Talented Youth (United States) and PERMATApintar (Malaysia),* by Julian Jones and Noriah Mohd. Ishak, is about the organizational example in developing innovators. The CTY at the Johns Hopkins University in the USA identifies and nurtures high ability pre-university youth. It employs a talent search to discover bright student using above-grade-level tests. For those who qualify, the top 3%, it offers summer, online, and weekend advanced programs. Although CTY has received considerable academic attention worldwide, over its 33-year history, it is only in the last five years that international attention has broadened to include CTY's economic development potential as a developer of innovators. Malaysia was the first major adapter of the CTY approach. This chapter discusses Malaysian government and university interest in deepening the pool of youthful genius to encourage economic growth and innovation, and particularly the success to date of the Malaysian model called *PERMATApintar* ("Gifted Gems").

In Chapter 44, *How does Singapore foster the development of innovators?* Chwee Geok Quek and Liang See Tan consider what Singapore does in order to nurture innovators and becomes one of the most innovative countries in the world. Specifically, the authors discuss the initiatives introduced by the Ministry of Education in the K-16 system to develop a spirit of innovation and enterprise and thus creating an innovation culture. Descriptions of key programs for students, as well as innovations introduced by schools and tertiary institutions, are presented. The programs aim to develop students who are able to meet the challenges of global economic development and international competition, and to try new and untested routes toward innovation without fear of failure.

Part XII of the Handbook, *Conclusions,* contains a single chapter, *Overall perspectives on the future promise (and forward thrusts) of innovation education,* by Larry R. Vandervert, which serves to integrate the other chapters in the handbook and forecasts the future prospects for innovation education. This chapter points out common as well as unique features of the various facets of innovation education presented in the chapters of the handbook and suggests directions in which tomorrow's research, practice, and policy might lead us. Specifically, the author addresses the issues of (a) why is innovation education necessary in the first place? (b) does innovation education actually work? and (c) how do the innovative abilities of the gifted individual interpenetrate with the social frameworks of innovative teams? Answers to these questions draw upon both (1) the long history of human innovation and (2) the innovation concepts, methods, and programs discussed in the chapters in this handbook.

Therefore, the chapters of this handbook discussed the foundations of innovation education as a new, emerging field, analyzed its current state of research, presented a great variety of practical approaches to developing innovators, and described future prospects. Authors brought to the attention of readers a wide range of perspectives on innovation education and an array of teaching options aimed at the development of innovators. This handbook thus provides the first and the most comprehensive account available of what innovation education is and how to develop innovators, which will change the world. And—what is probably the most important—the handbook launched many new research directions in the emerging field of innovation education that even have not been mentioned elsewhere earlier. This handbook thus greatly advances innovation education worldwide by making a lot of great discoveries ... exactly as Richard Feynman recommended.

Reference

Rowling, J. K. (1997). *Harry Potter and the Sorcerer's Stone*. New York: Scholastic.

Part II
The nature of innovation education

2

Innovation education
Defining the phenomenon

Rósa Gunnarsdóttir
UNIVERSITY OF REYKJAVÍK, ICELAND

Summary: Innovation education (IE) was introduced into the Icelandic education system by enthusiastic teachers in 1991, as an effort to capture the creative resourcefulness of children. The momentum grew and in 1997 the pioneers of the IE movement were approached to develop a new curriculum subject, IE and use of knowledge. IE spread through the Icelandic system from teacher to teacher, in-service training and word of mouth. Experienced teachers reported that IE affected the children in an engaging and most surprising way. But still there was no definition of what this phenomenon was. The research presented in this chapter will depict a definition of IE as it presented itself in Iceland, and how IE can be discussed using different pedagogical and didactical theories. In short a translation of what teachers knew as good teaching, into theoretical abstraction of learning and teaching in IE.

Key words: Pedagogy, theory, practise, Innovation Education, linked case studies, definitions of key concepts.

Introduction

Nations are made of persons who are different and unique. These persons are capable of extraordinary things when minds meet and thoughts and ideas are put into actions and products. In Iceland, and surely in many other countries, innovation is viewed as one of the major contributors to the wealth and future of our nation. Some go so far to say that the most important resource a nation has is its human capital. But how can one harness that treasure? Or more to the point, how to cultivate it and securing it as a resource for future use?

At the start of the IE movement in Iceland in the 1990s this aim of national importance was never the main goal. The group of teachers and enthusiasts were looking at the individual and how important it is to allow each and every one to become the best they can possibly be. All of us had our own agenda and ideas about what we wanted to happen in IE, or, as it was known back then, as Innovation classes. But what bound us together was the *belief in the individual creativity and that everyone can, and should, be a part of creating and developing the human environment we call our culture and habitat*. This became our mantra and is still the main reason that we devote our time and effort to the cause.

The classes started out as afterschool provisions, that were run by a craft teacher assisted by two elderly inventors and supported by visionaries in the local authority in Reykjavík. The result of that was the first Young Inventors Competition (YIC) that was held in 1991. Gradually these afterschool classes became more frequent and bigger. The children, aged 8–12, became more active and inventive, so in 1993 the dawn of IE began in Foldaskóli in Reykjavík. A group of teachers, led by the craft teacher and later a science teacher in school, began to develop teaching materials that would support the children and themselves in their efforts of producing inventions and designs that allowed the children to feel that they were builders of their own lives and surroundings.

IE spread by word of mouth around Reykjavík and around the country. The founders were asked to set up in-service training for fellow teachers from all around. The teaching material was distributed and it evolved quickly in the hands of the practising teachers around the country into culture specific approaches that linked the students to the everyday life around them.

In 1997 the founders were approached by the Ministry of Education, Science and Culture to produce a suggestion for a new curriculum subject that would be called IE and the practical use of knowledge (in Icelandic, *Nýsköpun og hagnýting þekkingar*). In the National Curriculum that was published in 1999 this subject was present. It was a part of the Design and Technology section and had no allocation of time per se, but presented as an elective to the schools that were equipped and ready to take this on. Since then IE has been taught in around 10–25% of Icelandic primary and secondary schools as a subject and the Young Inventors Competition is still going strong.

IE has now been a part of the Icelandic culture for two decades. The story of the YIC is an integral part of the development of the curriculum subject Innovation Education (IE), both when one looks at who were the key actors in the development and also that the competition has served as a lifeline between the world of work or reality as some of us prefer to call it, and what is happening in schools. During this time more than 35,000 inventions have been entered into the competition, some of patentable quality and others, though too few, have been produced and marketed.

Innovation Education practical description

As a practitioner developed reaction to a pedagogical need, IE in Iceland has cultural specific agenda and appearance. When developing a way to *further the ability of the young to use and produce knowledge in creative way* the teachers acted more on insight and impulse in their work than from a specific theoretical perspective. Over the years the best definition of IE in schools was found by asking the children that took part. Their answer was simple: *in IE we invent things and we design new things*. In the teachers' minds the concept and understanding was developing further; IE was beginning to mean more than just a subject or tasks they did as a part of their jobs, it was becoming an ideology, a pedagogy or teaching method.

What would we see in an IE class?

IE classes in Iceland vary. Some schools have it as a part of their school curriculum, and may implement it in different ways. They might organise it as themed projects work for several days or weeks at a time, or as a part of the weekly timetable. Other schools might run afterschool classes of IE. Others encourage their students to be active in the YIC and use the pedagogical tools of IE as a regular part of their school day. What links all the activities suggested above is that in IE, the children focus on gathering and analysing needs in their own surroundings, come up with solutions to them, make models and even prototypes of them and present them in some way, in school or outside.

In short, they use a very simple ideation process that can be shown as *Needs – Solution – Product* to access their creative mind, start creative work and finish by presenting their ingenuity to the world and thus influencing themselves as well as the real world.

The ideation process describes the activity of the students rather well; however, if that is all that happens in IE lessons, the results are at best limited. This process is not enough though for IE to be effective. *Facilitation ideation for creativity* is a process in which the emphasis is on providing the child with the knowledge, skill and competence needed to realise the full potential of his or her idea and invention. If the work in schools focuses too much on the teaching of creative relevant skills and knowledge, the results are also very limited. For maximum effect, both of these must be set in parallel in the students' work.

Finding the balance

The IE movement produced a series of teaching materials for schools in the early 1990s, under the name *Innovation and Science*, and that material is still partly in use. The development of that took four years, resulting in materials that were aimed at 9–12 year-old students, though it has been used from pre-primary up through higher education. The material is organised as a sequence, starting with introducing *the language of the inventor* to the children in book one, and then *technology as a tool to make your inventions* work in book two and in book three the *concept enterprise is introduced* by asking the students to set up companies and produce items for a market set up for them. The fourth book addresses *societal needs and architecture*, and how the inventive and creative mind can be an actor and builder of societies. In short, *the teaching materials aim towards developing the language, skills and attitude of the young inventors*.

The results of the evaluation of piloting the teaching materials in the 1990s, reviled several interesting issues (Þorsteinsson & Gunnarsdóttir, 1996). Many of the teachers had expressed curiosity as to *why this subject had the influence that it had on the children*. After accumulating information, a pattern started to appear in the teachers' accounts. This pattern referred to the creative process present in the interactions between teacher, student and environment in IE episodes. This process was called in Icelandic *Sköpunarferill* or the process of facilitating creativity for ideation abbreviated to FCI mentioned earlier. Though this attempt to capture the learning and teaching practices in IE was helpful it still did not give an in-depth understanding of IE in action.

The study of IE in practice

During the summer of 1996, Gunnarsdóttir and Þorsteinsson explored feedback on the experiences that they had collected from the teachers who had piloted the Year 4 and Year 5 teaching materials in IE. They analysed what was actually happening in the IE sessions. One of the findings of that analysis was that the teachers expressed the opinion that their experience with the children in IE had affected their way of viewing children and learning, but few of them were able to articulate their understanding. Most of them described a 'gut feeling' that what they were doing felt 'right' in the cases when the planned lessons 'worked'. In other lessons 'it' just did not work. The teachers talked about issues such as they 'felt' that the students' identity had changed, and that they, the teachers, were starting to change their understanding and attitude towards children and teaching even after more than 20 years of successful teaching.

Such teachers' comments intrigued Þorsteinsson and Gunnarsdóttir who were heading the developmental project and thus by using exemplars from the participants, teachers and students, Þorsteinsson and Gunnarsdóttir set out to find what the 'it' was.

Viewing the action observed in the classroom as a process helped define contributing aspects of the 'it'.

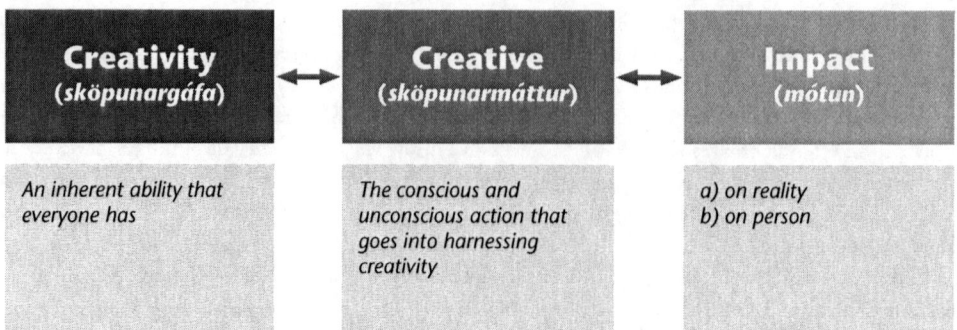

Figure 2.1 The ideation process as depicted in the first editions of the IE teaching materials.

While this version of the FCI process was forming, Gunnarsdóttir and Þorsteinsson tried to unpack what was happening in IE from the point of view of the child, looking for explanations for the transformation that they, as well as the piloting teachers, had observed in the children. The first thing that was noted was that *the children were creative*, and moreover *they were using their creativity*, and in some way developing their creative efficacy and *ability to produce ideas and inventions*.

The FCI as depicted in Figure 2.1, consists of three parts:

- *Creativity*. The talents, mental dowry and personal traits that the person was born with.
- *Creative efficacy*. The ability to use the creative element within and take the initiative to actively seek out needs and develop solutions to these needs. This is the part of the process by which the influences of the social and physical environment come into effect and is the way that the IE teaching materials and the teachers can 'reach' the students' creative work.
- *Impact*. The realisation of mental functions in the material world. Two main kinds of impact were identified: the impact that a product produced by the child has on the physical and social environment, and the impact that the production process has on the child. By working on a product the child accumulates knowledge and skills as a by-product of the effort made.

The understanding of the concept 'creativity' that had been adopted by the IE movement was that creativity is part of human nature, just as much as other personality traits. An understanding that everyone is creative and that it is desirable to make the best use of that ability had an influence on the way that the ethos in the classroom evolved. It was also important to realise that creativity included both the ability to make and to create, highlighting the subtle difference in these two concepts.

The concept of creativity has often been explained to new IE teachers as the inherent ability of man to evolve, to make choices and act on them, or in other words, the ability to originate ideas and take initiative. It is also the ability to make things by utilising the idea in question, using one's mental dowry and knowledge to take the initiative and make and do things. Creativity seems a characteristic of every human and is therefore not given to anybody, nor taught but is a part of the individual's personality. This is the paradox of IE. *How can you enable or develop an inherent characteristic if creativity cannot be taught?*

Defining the phenomenon of IE

The quest to find the 'it' the teachers described during the piloting had to be taken on as a sizeable research project. And it needed to explore suitable language for understanding the phenomenon as well as documentation of the interaction and activities that were related to learning and teaching IE. This was the aim of my PhD research project IE – *defining the phenomenon*, published in 2001.[1]

Nearing the classroom activities

It had been observed in the previous evaluation of IE that the children actively engage in IE both in school and at home, which suggests that these two niches must be included in the research design. By choosing to use the children and their experiences in each case as a link between the two matrices, and by gathering information from both social niches, a deeper and wider understanding of the factors that contribute to the participation in the social event IE could be gained.

The concept of social matrix was used in the research project in conjunction with the concept of social event to represent the two major components of social actions, participants and interactions. The social matrix is viewed as the grid of interactions that forms between individuals as they communicate, more to the point not only the individuals but all interactions that come to be in any given instant. The social matrix is built up of the individuals (the actors), the environment and time (the factors).

The individuals are all different in make and behaviour. Each person has a personality partly determined by genes and partly by the surroundings in which the person lives. An important point to make is that an individual or person exists both on intramental and intermental planes and brings complex understandings and aspects of personality, mostly subconsciously, to events.

The other factors that constitute the social matrix are the environment and time. The physical environment in which the interactions take place contributes to the way that the interactions develop, such as unconscious and conscious use of resources and time.

Time is an important concept in this argument. In order to understand the phenomenon of IE, the research design had to allow for time to be acknowledged as contributing to the formation of the social matrix, and time was an emerging attribute during the analysis of the social event.

> Understanding through time cannot simply be added on to existing perspectives and theories of social science. It requires a reconceptualisation of not merely social time but the very nature of 'the social'. Despite the inevitable continuity with existing perspectives, it alters the method and the vision, the methodology, and ontology.
>
> *(Adam, 1990, p. 8)*

In order to understand a social event that is taking place through time, the ontology and methodology of the research must recognise social time. This supports the way that the analysis and eventually the emerging theoretical representations of the findings are structured.

The social matrix is thus a grid defined by the physical environment and the presence of people in any given point in time. The interactions between the people and the environment are interlaced multidirectional connections that continually change through time: one can then say that the product of this ever changing mass of interconnections and changing individuals is a defined social event. Thus a social event is the collective product of intermental and intramental activities in a given place over a given time period, involving people and their physical environment.

The interactions taking place between the actors and factors in a social matrix may be understood differently by different individuals, as their previous experiences and understanding of the situation may differ, thus the messages that the individual internalises through the experience is tinted.

The above social matrices are made up of possible contributing factors and actors that make up the social event, co-constructed by interactions over time. The factors and actors in the matrices all

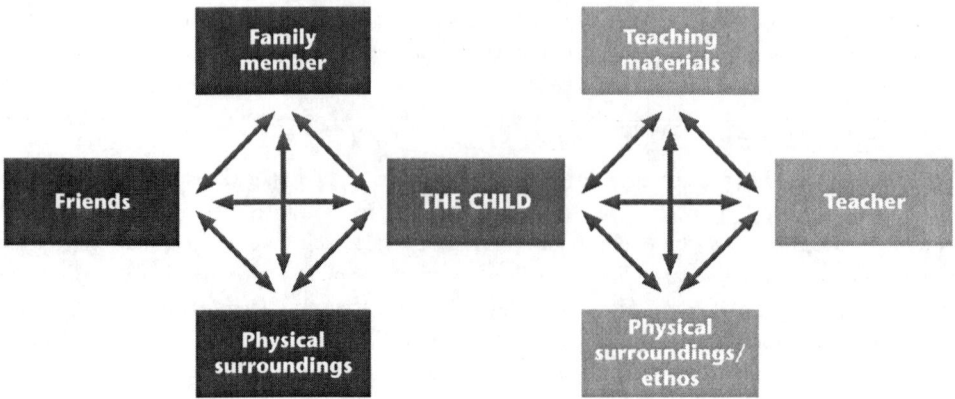

Figure 2.2 The Social Matrix active in innovation education.

contribute to the social event. Data about the social history of each actor had to be collected in order to be able to get a sense of the way the actors explicated the nature of IE. The teacher's professional background and personal set of values and norms, the child's wealth of experiences, values and norms, and the family and peer relations of the children were important in establishing an understanding of the participants' views of and their engagement in IE. The physical surroundings of the home/real life situation of the children were also important features of the social events as well as their physical manifestations of IE in a social context.

The other factors that contributed to a socially shared event of IE in the school context were the teaching material, which offered a structure to the interactions in the classroom and provided both context and opportunity; and the physical surroundings of the school, which restricted the lessons in time and space.

The social matrix is ever changing, never stationary, both in relation to time and space as well as in composition. The basic units of the matrix undertake constant changes that can be attributed to the social events that the individuals take part in. Everything that an individual experiences, changes him/her in some way. These changes can be slow and gradual over long periods of time, like when a person grows up in a certain culture with its norms, or very sudden, for example when a child learns to ride a bicycle.

The theoretical language that suits a discourse on IE

One of the major things needed for further development of IE was a discourse language that was suited to academic explanation of the pedagogical activities found in IE. Variety of fields can be coined as preferable contributors to the structure of that language. Following is a shortlist of contributors that best help explicate the phenomenon in academic terms.

In his short life Vygotsky developed a theory of the way an individual can reach higher levels of cognitive development through interaction and communication with others. Socio-culturalists view the mind as 'both constituted by and realised in the use of human culture' (Bruner, 1996, p. 1).

The part of Vygotsky's theories that touches on learning and teaching are of particular interest here as they emphasise the development of individuals in social contexts. An individual incorporates and acquires skills and knowledge that are embedded in the culture in which the individual grows up.

Vygotsky's perspective on learning is summarised as follows:

> Any function in the child's cultural development appears on the stage twice, or on two planes. First, it appears on the social plane and then on the psychological plane. First it appears between people as a interpsychological category and then within the child as a intrapsychological category. This is equally true concerning voluntary attention, logical memory, the formation of concepts, and the development of volition.
>
> *(Vygotsky, 1981, p. 163)*

A child or novice learner comes across knowledge, a concept or a skill in social surroundings where it is used in communication or interaction. The novice then internalises the novel concept or skill and over time, conversations become more 'egocentric', moving from the social to the individual until the novice starts to use 'egocentric' talk in the Vygotskian sense and in the end uses internal talk. The interactions move from the social plane to the psychological. Learning occurs then through appropriation of the cultural tools of the cultural language in use (i.e. meaning), how words and artefacts are used as well as the cultural goals of the society.

Vygotsky proposes that learning originates in social contexts and argues that knowledge is acquired through social interaction and the use of mediating tools such as language. He suggests that individuals learn by linguistic means and that more accomplished persons will teach the novice. Bruner expresses his understanding of Vygotsky's view of learning as regarding individual cognitive development subject to a dialectical interplay between nature and history, biology and culture, the lone intellect and society. Vygotsky believed that mind is transmitted across history by means of successive mental 'sharings' which pass ideas from those more able or advanced to those who are less able (Bruner, 1996).

An important concept here is mediation, particularly how information, skills and concepts are mediated in social contexts. Vygotsky states that higher mental functions (e.g., thinking, voluntary attention and logical memory) are mediated by the use of tools and signs, i.e. psychological tools such as language, 'various ways of counting; mnemonic techniques; algebraic symbol systems; works of art; writing; schemes, diagrams, maps, and mechanical drawings; all sorts of conventional signs' (1981, p. 137). There are two major properties of signs one must take into account when trying to understand what Vygotsky meant by tool use. First, when included in the process of behaviour the tool will affect the entire flow and structure of mental functions and, second, signs are by nature social, not organic or individual (1981, p. 137).

Vygotsky also recognised that signs and tools such as language are a product of history and society and that the social context in which the signs are used is significant.

The concept of tool use urges the consideration of where and how cultural knowledge is transferred to individuals. Vygotsky focused on the capability of children as learners. He identified a heuristic he called the Zone of Proximal Development, or ZPD. ZPD is defined as the difference in the ability an individual exhibits on tasks, with or without 'probing help' from a more capable individual. The idea of the ZPD is based on the premise that more expert others can support the learning of the less expert. The ZPD is the zone of development that occurs between initial assessment of the learner and assessment of the same learner on a similar task after she or he has received expert help.

Once the ZPD for the learner has been observed, it is then possible to use it to make judgements about a child as a learner; that is, the more easily the child moves through the ZPD the more likely one could describe the child as a quick learner. This is an important break away from an idea of measurement, which simply assesses a child's knowledge and understanding at a fixed point in time.

Since Vygotsky's ideas became publicly available in the western world, many have drawn upon his ideas in order to modify them into analytic tools or explanations of development and schooling (for example: Bruner, 1996; Wertsch, 1985, 1998; Edwards & Collison, 1996; Tharp & Gallimore,

1988). Cole (1985) developed a definition of Vygotsky's notion of the ZPD drawing on the work of one of Vygotsky's contemporaries, Leontév. Cole suggests that the ZPD is 'the structure of joint activity in any context where there are participants who exercise differential responsibilities by virtue of differential expertise' (1985, p. 155). Joint activity is central to the contemporary notion of the ZPD, as Cole points out (quoting Leontév):

> Human psychology is concerned with the activity of concrete individuals, which takes place either in a collective – i.e., jointly with other people – or in situation in which the subject deals directly with the surrounding world of objects – e.g., at the potter's wheel or the writer's desk ... if we removed human activity from the system of social relationships and social life it would not exist.
>
> *(Leontév, 1981, p. 46)*

Cole's analysis emphasises the role that activity has in mediating cognitive growth or learning alongside language and other methods of communication. Cole's work importantly allows us to see learning in the ZPD as a process of participation in cultural action.

Vygotsky argued that linguistic means of development of higher mental functions had supremacy over other means of mediation. Tharp and Gallimore, however, emphasise the importance of activity to learning: 'only when linguistic tools have been integrated with the tools of physical action can the potential of full cognitive development be reached' (1988, p. 44). However, Tharp and Gallimore's strategies do appear to emphasise the language-based role of the teacher's interactions while the learner is performing. Questioning, feeding back, instructing and cognitive structuring are all linguistic means of mediation, but for Tharp and Gallimore, modelling and contingency management are predominantly non-linguistic. Modelling means here the way that the teacher acts in class and how the students react to that. Bandura (1977) suggests that there are three ways that modelling can influence the observer (the student): by learning new responses, inhibition or disinhibition, and facilitation. Contingency management has more to do with the reward and punishment in the behaviourist sense of theorists such as Skinner (1938).

Building on these theoretical premises a part of the findings of the research project can be presented and interpreted.

Emergence of a revised version of the FCI

One of the findings of the PhD research project could be portrayed as a model of interactions in IE episodes. The model showed in Figure 2.3 is based on two processes that emerged as findings from the data analysis, first the process of Creative Relevant Skills and second the Ideation process.

Creative Relevant Skills (CRS) is a concept used for the activities and interactions regarding the introduction of new skills and knowledge by the teacher to the community of practice in the classroom. These could be identified in all of the seven cases in different measures and in different parts of the curriculum and of course with different participants, both teachers and students.

Ideation is a concept used to depict the activity of the child when working on an invention or design. This progress of activity was also identified in all of the seven cases in the research project, though they were all individually acted out and not always in the same order of activities. However a good foundation of a process of ideation could be found with all the children and in most cases more than one example was recorded.

Figure 2.3 attempts to bring together the two patterns of interaction that came up as results of the content analysis, creative relevant skills and ideation, ending up with a model of interaction patterns which, despite operating in opposite directions, are interrelated and mutually dependent.

The phenomenon of innovation

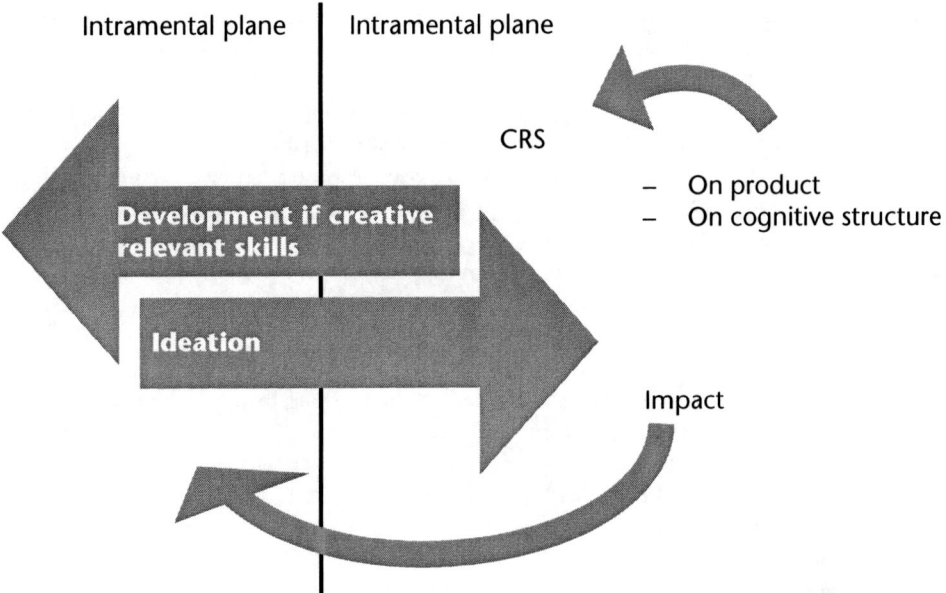

Figure 2.3 Representation of the Facilitation of Creativity for Ideation (FCI) process.

It has already been established that these two directions of activities and influences are based on findings of the research presented in the PhD thesis (Gunnarsdóttir, 2001). However the link between them needs further theoretical investigation.

The relation between ideation and CRS is depicted as causal. Ideation serves as motivation for the learning taking place within the CRS, or the structured agenda of the teaching material. Classroom actions in turn provide access to knowledge and skills, reinforcing the ideation process. In Figure 2.3 the CRS is situated as a part of the social event in the intermental plane. This by no means implies that the CRS is the only thing on the intermental plane that contributes to the facilitation taking place. Other aspects such as the influence of peers, physical surroundings and individual differences play a major part in the facilitation.

Ideation affects the development of CRS as a motivational influence in individual development in two ways. First of all, when the students work on an idea or make an idea public they need to use themselves as a resource, which in most cases means using knowledge that they have been taught in IE lessons. These include skills such as drawing three dimensionally or following the design process of looking for needs, finding solutions and making a product. This makes the students highly motivated when approaching the learning in IE lessons, as they see a use for the skills and knowledge presented, and regard that knowledge as valuable.

Ideation has proven to be effective as both a learning process as well as a creative process in IE. The students are exploring their understanding of reality when finding solutions to needs, identified by themselves, thus redefining and suggesting improvements using knowledge and skills already accumulated. The students are actively searching for 'bits and pieces' that they still lack in knowledge, skills and materials, whilst realising the solution that they are working on. The final step is producing new knowledge, both in the form of new understandings of their reality and in the form of a new product. The product that they are working on might not be new under the sun in the eyes of experts in the culture, but, to the students, their products are new and represent their revised understanding of reality and the way that they suggest that the surroundings can be modified and improved, and in that way new under their roof. This is why the two curved arrows are represented in Figure 2.3. The upper one represents the first option of influences of ideation on the CRS. The lower one represents the link between the ideation and the development of the individual cognitive structure that is the learning taking place while ideating.

The two features of facilitation in regards to the CRS and Ideation can be explained by the use of Vygotsky's model of ZPD. The CRS agenda and facilitation of cognitive structures represents a more traditional understanding of the ZPD as being defined and led by the teacher. Ideation is on the other hand defined and executed by the student in IE. This can be seen as the *locus of control* (Rotter, 1966) in the interactions in IE lessons fluctuating and being dependent on the tasks in place in lessons.

Figure 2.4 represents the way that the ZPD can be used to explain the interactions and the locus of control in IE episodes. During sessions structured by the teaching materials, stressing the CRS and the agenda that IE as a subject is promoted; the locus of control is situated more with the teacher rather than the student. The teacher reaches towards the student and the interactions are situated closer to the upper level of the ZPD, as depicted above. The interactions can therefore be understood as externally directed assistance.

By looking at the same figure but from the point of view of the ideation process, another aspect emerges. When the student is involved in ideation, the locus of control in the interactions is situated with the student. The student engages in what can be viewed as self-initiated and directed assistance. The student has identified that assistance is required for the idea that he or she is working on, in order to develop it further. The student then actively seeks out the necessary assistance, choosing from whom or what source the assistance is sought. These interactions can arguably be

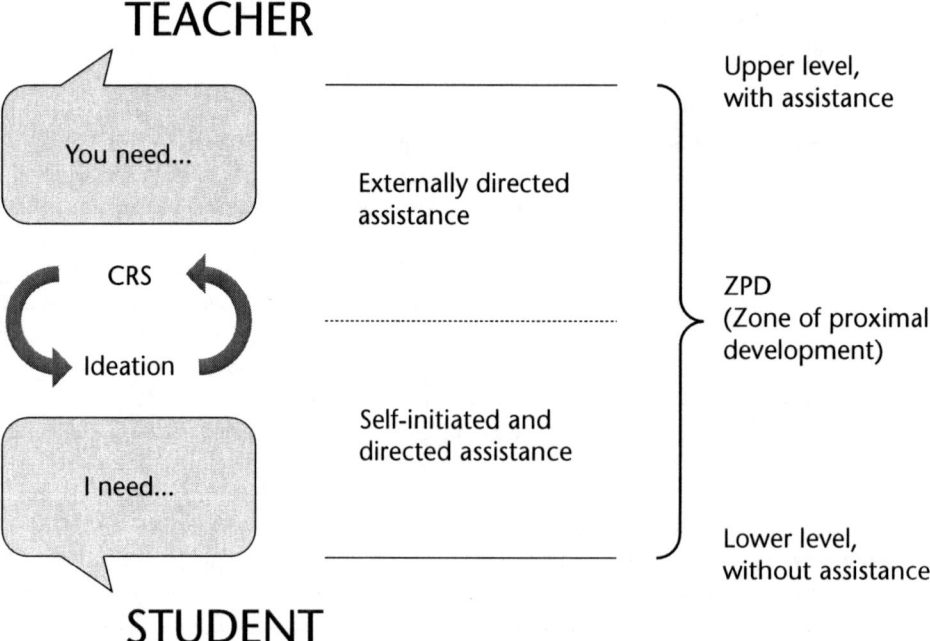

Figure 2.4 Interpretation of the interactions between teacher and learner in innovation education, using Vygotski's model of Zone of Proximal Development.

said to be situated closer to the lower level of the ZPD depicted above. The assistance that the student seeks is in closer proximity to the cognitive structure forming the idea that the student is working on. In other words, the child is presenting the idea to the source of assistance, usually the teacher or peer, and doing it in a way that makes sense to himself or herself. The child's identity and communicative skills tint the idea, making it easier for the child to select the information relevant to the problem at hand as he or she holds the definition of the problem. This makes it evident that the ownership of the idea and the locus of control in the interactions surrounding the child facilitate the child's learning and development of the idea. Whilst the student is using individuals and resources that were present in the niche that the ideation was carried out in, the student adds to the knowledge base (memory and higher mental functions, Vygotsky, 1978) and skills, in other words the student is learning.

So to sum it all up…

The main finding of the PhD research project is that IE can be defined in terms of the pedagogical heuristic or model shown as the refined FCI (Facilitation of Creativity for Ideation) process. The aims and agenda of the subject dictate the direction of school-based IE as investigated here. The counterbalance is the students' engagement in ideation, giving them control and requesting that they take initiative and responsibility over their tasks. This leaves us with a picture of the FCI as heuristic made up of two patterns of interactions connected by influences between the intramental and intermental. These influences are in essence bi-directional. Once this heuristic has been initiated, it continues (maybe not consciously all of the time), however the ideation process which the FCI supports is ongoing. Locking the FCI into a heuristic, which gains momentum with no obvious beginning or end, provides a foundation for further discussions about the locus of control of the interactions between persons involved in ideation or creative work.

The study of the emerging field of IE as it represents itself in Iceland is still young, though two new PhD projects are being published in 2011 and are sited in this handbook. Much has been done in Iceland to identify and implement creative work as a part of the education system.

In the laws on education, pre-primary up through upper secondary from 2008, a further confirmation of IE was adopted, by further emphasising creative work and in schools. In the summer of 2011 a general part of the new national curriculum for all school levels, bar university, was published. In that six issues were introduced as being core functions in the Icelandic education system. Those are literacy, democracy, equality, health and wellbeing, sustainability and creative work. Each of these concepts is to illuminate all work in school from pre-primary though upper secondary education. Within that remit IE has secured its place as a defined curriculum subject that addresses all of the core functions.

From one research project several issues emerged as critical to the future of IE. Not just as a part of the Icelandic National Curriculum but also as the nature of IE is becoming clearer the impact of the ideology in IE on the progression of the student's identity development.

Defining the phenomenon IE as it has developed in the field, can be said to be a step towards identifying a structure for educational change emerging from the needs of the children and the teaching professionals in the field, but not as a politically defined initiative structured to win votes. This statement highlights the need for further look at the nature of education and educational systems and re-evaluation of the content, context and processes used in the systems, and even questioning the fundamental assumptions that underpin education. All around the world, and especially in Europe, the call for new skills for jobs in the future, is getting louder. We all know that *the only constant* that the children we teach today will face in the future, *is change*. Teaching and learning by mediation and distribution of knowledge is not what our children need. They need to be addressed as creative individuals that have the capacity to unpick the mess that our generation and the generations before us got us into.

And the only way I see towards that goal is by exposing all children around the world to IE, which is rooted in their creative capacity and bound by their moral judgement.

Note

1 Research methods and rational as well as detailed analysis of the data gathered will not be discussed here but can be found in the thesis (see References).

References

Adam, B. (1990). *Time and Social Theory*. Cambridge: Polity Press.
Bandura, A. (1977). *Social Learning Theory*. Engelwood Cliffs, NJ: Prentice-Hall.
Bruner, J. (1996). *The Culture of Education*. Cambridge, MA: Harvard University Press.
Cole, M. (1985). The zone of proximal development. In J. V. Wertcsh (Ed.), *Culture, Communication and Cognition: Vygotskian Perspectives* (pp. 146–161). Cambridge: Cambridge University Press.
Edwards, A., & Collison, J. (1996). *Mentoring and Developing Practice in Primary Schools*. London: Open University Press.
Gunnarsdóttir, R. (2001). Unpublished PhD thesis. *Innovation Education: Defining the Phenomenon*. University of Leeds.
Rotter, J. B. (1966). Generalized expectancies for internal versus external control of reinforcement. *Psychological Monographs*, 18 (Whole No. 609).
Skinner, B. F. (1938). *The Behaviour of Organisms: An Experimental Analysis*. New York: Appelton-Century-Crofts.
Tharp, R. G., & Gallimore, R. (1988). *Rousing Minds to Life: Teaching, Learning, and Schooling in Social Context*. Cambridge: Cambridge University Press.
Þorsteinsson, G., & Gunnarsdóttir, R. (1996). *Frumkvæði: Sköpun* [*Initiative: Creativity*]. Reykjavík: Nýsköpun og náttúruvísindi.
Vygotsky, L. S. (1978). *Mind in Society: The Development of Higher Psychological Processes*, M. Cole, V. John-Steiner, S. Scribner & E. Sauberman (Eds.), Cambridge, MA: Harvard University Press.
Vygotsky, L. S. (1981). The genesis of higher mental functions. In J. V. Wertsch (Ed.), *The Concept of Activity in Soviet Psychology* (pp. 148–188). Armonk, NY: Sharpe.
Wertsch, J. V. (1985). *Culture, Communication and Cognition*. London: Cambridge University Press.
Wertsch, J. V. (1998). *Mind as Action*. Oxford: Oxford University Press.

3

The fundamentals of innovation education

Larisa V. Shavinina

UNIVERSITÉ DU QUÉBEC EN OUTAOUAIS, CANADA

Summary: If we want to develop innovative abilities of children and adolescents, then a special, new direction in education is needed: innovation education. This chapter describes the fundamentals of innovation education, which refers to a wide range of educational interventions aimed at identifying,[1] developing, and transforming child talent into adult innovation. These are those societal actions aimed at preparing children and adolescents to become adult innovators. Such educational interventions should include, but should not be limited to, the eleven interrelated elements. The chapter discusses each of them.

Key words: Innovation, innovation education, innovation science, innovation gap, individual differences principle of innovation, courage, deadline management.

Introduction

In contrast to the universal readiness to develop innovators pointed out in the Preface (Shavinina, this volume), researchers occasionally study how to develop innovators. From many chapters included in the best-selling *International Handbook on Innovation* (Shavinina, 2003a), only few of them discussed the development of innovators in science (Root-Bernstein, 2003; Root-Bernstein & Root-Bernstein, 2003; Shavinina, 2003b; Weisberg, 2003) or the development of innovative abilities via the stimulation of creativity (Clapham, 2003; Reis & Renzulli, 2003; Shavinina & Ponomarev, 2003). How to specifically develop innovators in science, technology, engineering, and mathematics is a relatively new direction in the interdisciplinary science of innovation.

The US government, who realized a need to develop innovators in science, technology, engineering, and math (STEM) disciplines, is behind a recent push for more research in this area. The National Science Board led the efforts of the US administration in this direction (NSF, 2009). Specifically, in August 2009 the Board invited a group of experts in the areas of giftedness, creativity, cognitive psychology, and science education for a panel discussion on how to develop the next generation of innovators in STEM disciplines (NSF, 2009). This event—with a subsequent report to Congress and President Obama—demonstrates that the US government is indeed interested in developing innovators.

For the most part, experts invited for a hearing in August 2009 emphasized policy issues related to the development of innovators (Subotnik et al., 2009), findings from high ability studies (Lubinski & Benbow, 2006; VanTassel-Baska & MacFarlane, 2009), and a need to radically change the existing practice of the development of scientists in sciences, engineering, and mathematics (Root-Bernstein, 2008). Shavinina (2009g) underlined the need to better study the high achievers in STEM disciplines (e.g., Nobel Prize winners in science and great inventors), whose innovation achievements are undeniable, as well as a need to focus on innovation education for everyone. Experts agree that the existing research base is rather thin (Benbow, 2009), that a comprehensive research plan is necessary, and that there is need for basic research in this area (Marett & Johnson, 2009).

If we want to actualize the innovative potential of today's children and develop their unique innovative talents, then we have to concentrate on innovation education. It is necessary to point out that innovation education is not about innovations in teaching mathematics, physics, biology, chemistry, and other disciplines. These are the so-called domain-specific innovations in each particular area of human endeavor (Shavinina, 2003a). They also help develop innovators. However, this is not what innovation education is all about. Innovation education refers to a wide range of educational interventions aimed at developing and transforming child talent into adult innovation. It means those societal actions aimed at helping children to become adult innovators. These educational interventions should include, but should not be limited to, the following eleven elements:

1. The programs for gifted and talented children existing in the area of gifted education, which seem to be effective, and may be useful for many other categories of learners (e.g., the Renzulli Enrichment Program, the Future Problem Solving, Creative Problem Solving, the Center for Talented Youth programs, the Belin-Blank International Center for Gifted Education and Talent Development programs, and other programs, just to mention a few: Brody, 2009; Cramond, 2009; Cramond et al., this volume; Jones et al., this volume; Lyons & Reis, this volume; Renzulli & Reis, 2009; VanTassel-Baska & MacFarlane, 2009; Wallace, this volume).
2. The best programs from science and technology education.
3. New programs aimed at the development of entrepreneurial giftedness that is closely related to innovation.
4. Programs for the development of children's metacognitive abilities. It means that special emphasis should be made on fostering their abilities to implement things: the so-called executive abilities. It is important for innovators, because innovation is about the implementation of ideas into practice.
5. New programs, which should be based on recent progress in the study of scientific talent of Nobel laureates.
6. Programs that would incorporate the essentials of research on polymaths.
7. New programs for the development of applied wisdom and moral responsibility.
8. Programs aimed at the development of managerial talent.
9. The basics of deadline management.
10. The foundations of innovation science: a general "know-what" and "know-how" about innovation, including the psychological basis of innovation, innovation gap, the individual differences principle of innovation, innovation management, just to mention a few.
11. Courage-related issues. For innovators to succeed, the courage is compulsory. It is sad that no one teaches today's children and adolescents to be courageous in pursuing their unique interests and implementing their novel ideas into practice in the form of new products, processes, or services.

The structure of innovation education can, therefore, be presented as eleven overlapping elements; each of them is described below.

The structure of innovation education

Gifted education programs

Successful programs for gifted and talented children are well described in the literature (Colangelo & Davis, 2003; Heller, Monks, Sternberg, & Subotnik, 2000; Shavinina, 2009a). They will not be, therefore, discussed in detail here. Nevertheless, it is important to emphasize that although gifted education programs are included as the first element in the structure of innovation education, they are not sufficient for developing innovators. It would be a huge mistake to suppose that gifted education itself will "produce" innovators. There are at least two reasons for this.

First, by definition, innovation is the implementation of ideas into practice in the form of new products, processes, and services. This is not about the generation of novel, original, and appropriate ideas, that is, creativity. People—and especially the gifted—are far better in generating new ideas than in their implementation into practice. The implementation of new ideas into practice requires many talents (e.g., an ability to meet deadlines), which are not normally taught in gifted education programs. Therefore, if we indeed want to develop innovators, then innovation education should incorporate such elements, which may help the gifted to cultivate talents necessary for becoming innovators (e.g., to foster their practical intuition and applied wisdom).

It should be pointed out that innovation education is not intended to replace gifted education. At the same time successful development of all talents of the gifted will eventually put them at the cutting edge in their areas of endeavors and, consequently, chances are pretty high that they will innovate. Similarly, if the ultimate goal of gifted education is to produce innovators, and the majority of students do not become innovators, then it probably means that programs are not quite successful.

Second, research on gifted entrepreneurs–innovators demonstrates that many of them could not be considered the gifted in accordance with the prevailing definitions of giftedness and practice of gifted education (Shavinina, 2008, 2009a). Entrepreneurial giftedness is a special type of high ability that significantly deviates from existing types of giftedness. The study of entrepreneurial giftedness is a relatively recent enterprise (Shavinina, 2006). This line of research shows that innovators do not necessarily come from the population of gifted children, are viewed as the gifted, and are enrolled in the existing gifted education programs (see also Chapter 16 of this volume, *The trajectory of early development of prominent innovators: entrepreneurial giftedness in childhood*). Based on this, one cannot, consequently, expect that 100% of future innovators will come from the population of today's gifted and talented children. In order to increase the chance of the gifted becoming innovators, innovation education should thus include other elements, which will be presented and discussed below.

Science and technology education

Today innovation is often associated with scientific discoveries and technological inventions implemented into practice. Tomorrow's innovators should be aware of scientific and technological breakthroughs, as well as to develop an advanced scientific thinking and a technological state of mind. As this element of innovation education is well presented in the chapters included in Parts VII, VIII, and IX in the handbook, it will not be considered here.

Entrepreneurial giftedness

The most important element of innovation education should be related to entrepreneurial giftedness.[2] This is because entrepreneurs put ideas into practice via the creation of new ventures, which, in turn, create employment and—in successful cases—lead to economic prosperity. The essence of entrepreneurship thus greatly, but not entirely, coincides with the nature of innovation. Innovation is about the implementation of ideas into practice in the form of *new* products, processes, and services, whereas entrepreneurship, per se, may or may not be about new ideas.[3] At the same time, keen entrepreneurs are necessary to carry innovations forward to market. This is why the research findings regarding entrepreneurial giftedness must be the cornerstone of innovation education.

Based on the existing research, the following facets of entrepreneurial giftedness should be incorporated in innovation education:

Early manifestations of entrepreneurial giftedness. From early childhood, gifted entrepreneurs demonstrated two types of characteristics—*specific* and *general ones*—which helped them to succeed in business (Shavinina, 2008). Specific characteristics refer to those abilities, skills, or personality traits that are directly related to *entrepreneurial* giftedness. For instance, the creation of ventures with money-making potential belongs to the specific manifestations of entrepreneurial giftedness. General manifestations refer to those abilities, skills, or personality traits that can also be useful in other types of giftedness and are not exclusively associated with entrepreneurial giftedness. For example, competitiveness as a personality trait is helpful in business and sports alike. The following interrelated yet different *specific* manifestations of entrepreneurial giftedness were identified:

- *Constantly generate ideas on how to make money* (i.e., creative abilities conducive to entrepreneurial giftedness).
- *Love to generate and implement real-life projects with at least a minimal financial reward.* This is the key characteristic of entrepreneurial giftedness that incorporates both creative and innovative abilities of entrepreneurs. This is because innovation is essentially about the implementation of creative ideas into practice in the form of new products, processes, or services.
- *Love doing real business plans with predicted financial outcomes.*
- *Work passionately and hard on executing their plans.* Young gifted entrepreneurs do everything necessary for their projects to succeed (e.g., they are able to convince other people to participate in the implementation of their ventures).
- *Wish to do "real" things that bring money and try to do whatever possible to cut unnecessary steps.* It is one of the manifestations of practical intelligence in young gifted entrepreneurs. Examples are abundant and they are discussed in Shavinina (2008).

The *general* manifestations of entrepreneurial giftedness include the following interrelated characteristics.

- *Perseverance to succeed: if I put my mind to something, I can do almost everything.* The best manifestation of gifted entrepreneurs' persistence is the fact that they do not give up after the first failed project(s). Failures do not stop them.
- *Optimism and "change the world" attitude.* Gifted entrepreneurs from early years believe in themselves and their ability to change the world by succeeding with their projects. They have a positive vision of the future and of every venture they initiate. Optimism helps gifted entrepreneurs succeed. This supports scientific findings demonstrating that optimists always outperform pessimists (Carver & Scheier, 2003).
- *Early exposure to challenges.* It is amazing how gifted entrepreneurs liked challenges from their early years, had a lot of exposure to them, and, as a result, the love of challenges became one of the distinguishing characteristics of talented entrepreneurs.

- *Competitiveness, excellence,* and *perfection*. As a consequence of the early exposure to challenges or an intensive involvement in sport activities, young gifted entrepreneurs possess competitive personalities. When they compete, they always try to be the best and win.
- *Neglect of academic subjects*. Probably because gifted entrepreneurs live in their own world of "real practical" projects (often with money-making potential), school subjects do not make much sense to them. Many do not do well at school and simply ignore academic subjects (e.g., Richard Branson). This is also because teachers often do not show the practical applications of those subjects; they just ask children to memorize a great deal of knowledge. This directly contradicts the essence of the practical mind of gifted entrepreneurs who are eager to do "real" things in real life; not in the classroom for their teachers. However, with respect to this characteristic, there are some exceptions. Bill Gates is one of them. For instance, he was doing well in the elementary school. Nevertheless, it is not clear whether Bill's success in academic subjects was determined by his intellectual abilities or by his extraordinary competitiveness. Jeff Bezos, who was enrolled in a gifted education program, is another exception.
- *Independence in thoughts and actions*. From early years gifted entrepreneurs are very independent in their thoughts and actions: authorities do not exist for them. For example, Mary Gates, in describing her son, Bill, has said that "he has pretty much done what he wanted since the age of eight" (Wallace & Erickson, 1992, p. 11).
- *Rule-breaking attitude*. As a result of their extreme independence, a rule-breaking attitude is another distinguishing characteristic of gifted entrepreneurs. This is why talented entrepreneurs are innovators: they are able to break all the existing rules of the game and introduce something new. This is how and why great innovations happen.

Therefore, this element of innovation education should convey a valuable knowledge about specific and general signs of entrepreneurial giftedness. It is interesting to point out that these signs manifested in childhood became strong characteristics of talented entrepreneurs in their adulthood (Shavinina, 2006).

Creative abilities of great entrepreneurs. Gifted entrepreneurs developed their own, highly individual methods and techniques for producing new, original, and appropriate ideas (Shavinina, 2009d). It is interesting to note that traditional creativity training does not contain anything similar. Creativity techniques and methods of great entrepreneurs will help to enhance abilities of tomorrow's innovators. For instance, I translated some of those methods into a series of practical techniques for my workshops on creativity for teachers and managers alike.

Furthermore, gifted entrepreneurs are polymaths (this concept will also be discussed below), that is, multi-talented individuals, because they are able to both generate ideas with money-making potential and to implement them into practice. It is a rare ability.[4] This unique ability of talented entrepreneurs has allowed me to talk about the phenomenon of *individual innovation*. Usually, innovation is a "team sport" in that it involves many individuals to implement ideas into practice. The case of individual innovators is a complete exception from this rule: they are able to put into action all the ingredients necessary for the implementation of their ideas (for a detailed account of this phenomenon see Chapter 36, *What can innovation education learn from innovators with longstanding records of breakthrough innovations?* included in this volume).

Metacognitive or executive abilities of gifted entrepreneurs. Innovation is essentially about the *implementation* of ideas into practice in the form of new products, processes, or services. Lessons from famous entrepreneurs on how they do it successfully and what helps them will be of immense importance to today's children. Translated into a set of special exercises, these "know-what" and "know-how" facets of the executive abilities of talented entrepreneurs will be an important part in developing innovative abilities. From early childhood gifted entrepreneurs tried to implement their ideas into practice. For instance, when Richard Branson was twelve years old, he decided to grow Christmas trees. Today's children should use every opportunity to implement their ideas into practice and parents and teachers should support them in their endeavors. For example, if a child wants to write a book, parents and/or teachers should help him in this endeavor and a real book should be eventually produced.

Motivation of gifted entrepreneurs. Many gifted entrepreneurs founded new companies not because they wanted to make money. They started businesses for variety of reasons, which were not related to financial rewards (Shavinina, 2009e). For instance, creativity has always been behind the entrepreneurial motivation of Richard Branson. Thus, he wrote in his autobiography that "I have never gone into any business purely to make money. ... A business has to exercise your creative instincts" (Branson, 2002, p. 57). In the case of Michael Dell it is excellence that is in the heart of his entrepreneurial motivation. When 18-year-old Michael almost left university and his father asked him what he was planning to do with his life, he replied: "I want to compete with IBM!" (Dell, 1999, p. 10). High abilities—for example, creativity or excellence—and achievement related issues are behind great entrepreneurs' wish to create new ventures. Children and adolescents should know that money was not the driving force for many renowned entrepreneurs.

Unique vision: unusual type of representations. Unique point of view or unique type of representations is the very essence of giftedness (Shavinina, 2009c). This is the basis of entrepreneurial giftedness as well. Talented entrepreneurs see, understand, and interpret everything in an unusual way. Thus, not many eighteen-year-olds could say with all the certitude that they "want to compete with IBM" and consider it as the goal of their life. Cases of the unique vision of great entrepreneurs and how many important business decisions were made based on their unusual points of view will contribute to the development of innovative abilities of today's children. Independently of high ability studies, leadership researchers[5] have also found that vision is a critical aspect of great leaders (for a more detailed account on a unique vision see Chapter 12, *Do not overlook innovators!: discussing the "silent" issues of the assessment of innovative abilities in today's children–tomorrow's innovators* included in this volume).

Specific extracognitive abilities: practical intuition. Gifted entrepreneurs are characterized by specific feelings, intentions and beliefs, preferences and values, as well as intuition (their definition will be presented below). Examples of these abilities (which can be found in Shavinina, 2009d) and the demonstration of their exceptional role in entrepreneurial giftedness along with exercises designed to develop them in children should be an essential part of innovation education. It is fascinating how many important business decisions in all areas of human endeavor were made based on intuition (Branson, 2002; Morita, 1987). This is especially appropriate for women-entrepreneurs who have more developed intuition in comparison with men (Kay Ash, 1996; Roddick, 2000).

Micro-social factors stimulating or inhibiting the development of entrepreneurial giftedness. Tomorrow's innovators should also be aware of the macro- and micro-social factors, which can facilitate or hinder the development of their entrepreneurial giftedness. There is a difference between micro-social and macro-social factors. The micro-social factors refer to the influence of such social institutions as family, school, university, and proximal social surrounding (e.g., childhood friends). The macro-social factors refer to those societal, cultural, and historical contexts in which individuals live (i.e., the contemporary Zeitgeist). Definitely, macro-social factors often operate through micro-social factors. Examples from the lives of gifted entrepreneurs are necessary for innovation education. Specifically, the information about the role of nuclear and extended family,[6] "significant others", and great contemporaries, as well as knowledge of how gifted entrepreneurs were able to turn negative micro-social influences into the beneficial ones are of great importance to today's children and adolescents (Shavinina, 2006).

To sum up this section, entrepreneurial giftedness should be a major element in the structure of innovation education. Any child will benefit from this, regardless of his or her actual abilities. It will also help us to identify many potentially talented young entrepreneurs and businessmen. At the moment entrepreneurial giftedness is a type of high ability that is not measured in any way. When talented children are selected for gifted education programs, the entrepreneurially gifted are usually overlooked and, therefore, lost.

Metacognition in action: ability to implement things

The development of children and adolescents' abilities to implement things—or their executive abilities—is an important element of innovation education, because innovation is mainly about the implementation of new ideas into practice. This is a unique ability and not everyone is capable of it.

A few years ago one management book became a bestseller on Amazon.com. Why? Because its authors demonstrated that many executives never execute. This is why the issue of the *innovation gap* is an important one. The concept of the innovation gap means that people have a lot of creative ideas, but they are not able to implement them due to various reasons. There are many barriers to innovation (Shavinina, 2007). Although a majority of people believe that innovation is a good thing, researchers found that obstacles are the norm, rather than the exception, on the way to innovation (Hadjimanolis, 2003). There exist human-related, technology-related, and policy-related barriers to innovation, just to mention the main, broad groups of barriers. A wide range of these obstacles make it difficult for potential innovators to implement their creative ideas into practice. Consequently, the topic of the innovation gap should be kept in mind when we are discussing the executive abilities of today's children. This also explains in part why innovation is still a relatively rare thing and why innovators' executive abilities—that is, individuals' abilities to both generate and implement great ideas—are unique ones and should be highly appreciated. Every effort should thus be made to develop them in today's children. And this is not an easy task.

Human abilities to implement things—or our executive abilities—are in fact our metacognitive abilities. Metacognition refers to one's own knowledge about one's own cognitive abilities, as well as guiding, monitoring, and executing of one's own mental processes (Brown, 1978, 1987; Flavell, 1976, 1979; Kholodnaya, 1997). Ann Brown (1994) and other researchers (Brown & Palincsar, 1989) working in the area of metacognition designed special educational programs aiming to develop children's metacognitive abilities. Innovation education can benefit from those programs by incorporating its best practices (Barfurth, Ritchie, Irving, & Shore, 2009).

Scientific talent: lessons learned from Nobel laureates

Tomorrow's innovators should learn interesting lessons from Nobel laureates in science. Scientific discoveries made by Nobel laureates are scientific innovations. The nature of innovation is cross-disciplinary, that is, in any field of human endeavor innovation begins from great ideas. If we are concerned with the development of children's scientific talents, then we should study the Nobel Prize winners and their high abilities in detail (Shavinina, 2003b, 2004). Why? This is because winning a Nobel Prize represents the pinnacle of accomplishment possible in one's field of expertise. This prize in science is associated with a rare, high degree of intellectually creative achievement that testifies to innovative minds of its recipients.

Today we know that during their early years, Nobel laureates in science encompassed a wide range of human abilities including the gifted (e.g., Marie Curie or Gertrude B. Elion), the gifted underachievers (e.g., Albert Einstein), and children without any special talents (e.g., Barbara McClintock). It is amazing that these different profiles of abilities—that manifested themselves in clearly divergent trajectories of talent development—eventually led to the same outcome: great scientific discoveries, which means that those who made them possessed exceptional intellectually creative abilities. At the very end, all the various trajectories of talent development led to the same result: astonishing scientific achievements. The study of (auto) biographical accounts on Nobel Prize winners shows that among many factors, which made it possible, their extracognitive abilities and their objectivization of cognition played an important role (Shavinina, 1996a, 1996b, 2003b, 2009b). Extracognitive abilities refers to the following interrelated elements, including:

- specific intellectually creative feelings: feelings of direction, harmony, beauty, and style, including senses of "important problems," "good" ideas, "correct" theories, elegant solutions; and feelings of "being right, being wrong, or having come across something important";
- specific intellectually creative beliefs and intentions (e.g., belief in elevated standards of performance);
- specific preferences and intellectual values (e.g., the "inevitable" choice of great mentors and internally developed standards of working); and
- intuition (Marton, Fensham, & Chaiklin, 1994; Root-Bernstein & Root-Bernstein, 2003).

The word "specific" embodies the uniqueness of these aspects in the structure of Nobel laureates' high abilities. For instance, Barbara McClintock pointed out that her "feeling for organism" was behind her breakthrough discoveries in genetics (Keller, 1984). The extracognitive abilities of Nobel laureates represent the highest level of the manifestation of human intellectual and creative potential, as well as predict outstanding scientific talent of Nobel caliber (Shavinina, 2003b, 2004). It is useful for future innovators to know how specific feelings, belief and intentions, preferences, and intuitive processes helped Nobel laureates to be innovative. In short, a special program for the gifted can be based on these research findings, which would explain the nature of extracognitive abilities, why they are needed to make innovation happen, and how everyone can develop them.

Lessons from Nobel laureates' childhood and adolescence are another important facet of learning from them. I am currently working on a project entitled *A Study of Early Childhood and Adolescent Education of Nobel Laureates and the Implications for Gifted and General Education: Developing Scientific Talent of Nobel Calibre*. The project studies early childhood and adolescent education of Nobel laureates in science—starting with the first laureate, who received his prize in 1901, and ending with the most recent laureates, who received their prizes in 2010 (N = 611). The goals of this research are twofold. The first is to discover the unique aspects of the early childhood and adolescent education of Nobel laureates, which contributed to their superior intellectual and creative development and had led to their excellent achievements in science. The second goal is related to translating the results of this investigation into larger-scale innovation, gifted, science, and general education worldwide. The discovery of the principles involved in the educational development of Nobel laureates will allow educators to accordingly improve, develop, modify, and transcend areas in the current curriculum in an attempt to cultivate outstanding scientific talent, of Nobel caliber, in today's children. This is why this research is an exceptionally timely and important endeavor and lessons that can be derived for gifted education are enormous (see Chapter 18, *Where did all great innovators come from?: lessons from early childhood and adolescent education of Nobel laureates in science*, this volume).

Another important and unexpected aspect of learning from Nobel laureates came from a gifted ten-year-old boy, Alexander, who became interested in this project. When he started to read books about Nobel Prize winners' childhood and adolescence education written for children, he discovered many interesting things. In spite of the fact that Alexander heard many times that Nobel laureates were excellent students, he found, for example, that Guglielmo Marconi was not a good boy at school. Nonetheless, it was *he* who pioneered radio communication. Albert Einstein was expelled from his school in Munich, Germany; however, *he* had one of the greatest minds of the 20th century. Fred Banting was not good at school either, but it was only *him* who was able to discover insulin and saved the lives of millions of diabetics worldwide. The father of Barbara McClintock told teachers to not give any homework to his children. Her dad believed that eight hours at school was more than enough for learning.

Alexander, therefore, originally thought that the Nobel laureates in science were great students. However, he has since found out that they were not. To be more precise, many of them were not the best in school. This made their stories so much more interesting to the boy. As the principal investigator on the above-mentioned project, I was intrigued by Alexander's research findings and suggested that other kids and adults might also be fascinated. As a result, he is currently working on a book about young Nobel laureates' educational stories that will be of interest to children, their parents, and teachers. The stories of Nobel laureates are very encouraging and motivating for today's children (Ponomarev, 2013).

It is interesting to note that in a time when researchers think of how to stimulate kids' interest in science, Alexander has found some useful ways to do this. When invited to present at a special hearing at the US National Science Board on how to develop the next generation of innovators in STEM disciplines with a subsequent report to Congress and President Obama in August 2009, I discussed Alexander's case. Specifically, his first degree of interest was his interest in why not all Nobel laureates in science were good boys or girls at school. Then this initial interest motivated him to learn more about Nobel Prize winners' discoveries. This was Alexander's second degree of interest. As a result, he is becoming seriously interested in science, which can be considered the third degree of interest. A deep involvement in science might be a desired final outcome and this is the highest degree of interest. Alexander's book on how "supposed" delinquent boys and girls in school still managed to make great scientific discoveries and became Nobel laureates is a "by-product" of his developing interest in science (Ponomarev, 2013).

Teachers can learn from today's talented children how they get interested in science and use those lessons. In a time when world-renowned specialists from a variety of disciplines try to suggest ways to stimulate children's interest in science, Alexander discovered a great way to motivate kids to learn more about science. By using these lessons, parents and teachers will foster scientific talents of Nobel caliber in today's children, which will enrich the world by subsequent scientific innovations. The development of scientific talents is, therefore, an important element of innovation education.

Polymathy or multiple giftedness

Polymathy should be another element of innovation education. The concept polymathy refers to the cases of multiply talented individuals who made significant major contributions to multiple domains (Root-Bernstein, 2003, 2009; Root-Bernstein & Root-Bernstein, 1999). Researchers studying the nature of expertise often state that specialization is a requirement for adult success, that skills and knowledge do not transfer across domains, and that the domain-dependence of creativity and giftedness makes general creativity and talent impossible (Ericsson, Nandagopal, & Roring, 2009). The absence of people who have made key contributions to multiple domains supposedly supports the specialization thesis. Root-Bernstein's (2009) study challenges all three legs of the specialization thesis. He had identified individuals who have made important contributions to several domains; thus demonstrating polymathy among creatively gifted adults. Today's children–future innovators will greatly benefit from the knowledge about the phenomenon of polymathy that must be included in innovation education programs. Parents and teachers should also encourage the gifted to develop their talents to the fullest extent in all possible areas of human endeavor.

Applied wisdom and moral responsibility

It is import to develop wisdom[7] in tomorrow's innovators because this is the most critical talent in the structure of individual innovation (see Chapter 36 of this volume, *What can innovation education learn from innovators with longstanding records of breakthrough innovations?*). The collapse of *Enron*, *Worldcom*, and other companies demonstrated that wisdom is a much needed, yet a relatively rare, human ability. "Mistakes will be made," repeats the Auditor General of Canada, Sheila Fraser. The most powerful way to prevent future mistakes of any kind is to rely on human wisdom. (Auto)biographical accounts on great innovators show that they have highly developed wisdom-related skills (Branson, 2002; Dell, 1999; Grove, 1987, 1996; Lowenstein, 1996; Morita, 1987). Wisdom and innovation are thus highly related. Societies can progress today only by innovating; therefore, every effort should be made to develop wisdom in today's gifted children—tomorrow's innovators. Wisdom is behind success of any human endeavor.

Shavinina and Medvid (2009) analyzed the wisdom-based performance of Richard Branson, a famous innovator and entrepreneur. This and similar cases (e.g., the business philosophy of Mary Kay Ash or the entrepreneurial approach of Anita Roddick), as well as practical techniques aimed at developing wisdom-related abilities of children, should be included in innovation education. Future innovators must understand the nature of wisdom and its important role in the innovative decision making process, as well as be aware of the basic traits of wise individuals, and know how wisdom-related performance can be enhanced.

Moral responsibility is closely related to wisdom. Prominent innovators are characterized by a high level of moral development (see Chapter 36 of this volume, *What can innovation education learn from innovators with longstanding records of breakthrough innovations?*). They feel personal responsibility for the future of our planet. Thus, Warren Buffet gave almost all his fortune to the Melinda and Bill Gates Foundation being convinced that they are doing great things in Africa and other poor regions of the world.

Richard Branson is another impressive example. He has always given a lot to charities and founded his own not-for-profit foundation, *Virgin Unite* (see www.virginunite.com). He often helps people who are in difficult situations. For instance, Richard flew on the *Virgin* plane to Iraq under fire in 1991 to release British and other foreign nationals. His recent publications are full of concerns about environment and sustainable development of the Earth (Branson, 2008), and about turning capitalism upside down—shifting the focus from profit alone to also caring for people and communities all around the world (Branson, 2011).

Today's children–tomorrow's innovators should be aware of such remarkable examples and well understand that end results of their innovative efforts may lead to incredible fortunes, which automatically implies extraordinary moral responsibility.

Developing managerial talent:[8] lessons from great managers

As noticed above, innovators are distinguished by an excellent ability to put into place all the organizational, human, and "environmental" structures necessary for implementing their ideas into practice in the form of new products, processes, or services. The human aspect is very important here. David Ogilvy, the legendary founder and CEO of the advertising agency *Ogilvy & Mather Worldwide* liked to repeat: "people are the only thing that matters, and the only thing you should think about, because when that part is right, everything else works" (Wademan, 2005, p. 36). Innovators should, therefore, be very good in managing people. This is why managerial talent is a critical ingredient of innovation talent. Research shows that many famous innovators were excellent in managing the human part of business. See, for example, Shavinina and Medvid (2009) for a special case study of Richard Branson as a great manager. Mary Kay Ash, Anita Roddick, Nelson Mandela, Akio Morita, Michael Dell, Andy Grove, just to mention a few, were also good in managing people. Lessons from these managers should be an important element of innovation education. The knowledge of their unique approaches to and methods of managing people will be an asset for today's children–tomorrow's innovators.

Another part of learning from great managers should be based on the Gallup organization's study of more than 80,000 best managers in the world (Buckingham & Coffman, 1990). The gifted will benefit from knowing how and why the best managers in the world break all the rules of conventional management wisdom in every facet of their professional activity. Practical cases with exercises will help develop children's managerial abilities. For example, future innovators should know that best managers hire for talent (not only for brainpower, experience, and willpower), because they are convinced that every role performed at excellence requires talent. When these managers set expectations for their employees, they define the right outcomes, not the right steps. It is up to employees to determine the appropriate steps. When motivating someone, great managers focus on human strengths, not on weaknesses. This is because they understand well that the greatest room for our enhancement is in the area of our strengths, not weaknesses. It means that best managers in the world even do not try to change or improve people. Instead, they concentrate on human strengths and work from there. When developing employees, great managers try to find the right fit between the individual's abilities and the job requirements; not just promote someone. Great managers also realized that they can manage people only by remote control. Real control is an illusion of mediocre managers. The best managers spend more time with their "stars"—the best, high performing employees. In brief, everything that great managers do is a complete deviation from conventional management wisdom. This is why the Gallup researchers concluded that the best managers around the globe break all the rules of conventional management wisdom (Buckingham & Coffman, 1990).

Adding these two parts of lessons from great managers to innovation education will allow educators to guarantee that tomorrow's innovators will really develop their managerial talents and thus will be well prepared to deal with a human part, the most important part of innovation.

Deadline management

A new and exciting research discipline has recently emerged within administrative sciences: deadline management. Every single project—including innovative one—has its beginning and end. The same is true for any innovation that often depends on timely efforts, namely: time needed to develop it, to introduce it to the market place, and so forth. In general, shorter period of time is better for innovation. It means that with respect to innovation those individuals and organizations, who are fast in the innovation process, will win. This is why knowledge about deadline management should be an essential aspect of innovation education. The fundamentals of deadline management include the understanding that deadlines bring out the best in people, a need to cultivate positive attitude to deadlines, as well as to develop a culture conducive to deadlines (Amabile, Hadly, & Kramer, 2002; Gersick, 1995).

The basics of innovation science

Innovation education will not be comprehensive without some general "know-what" and "know-how" about innovation science and practice such as: the tyranny of success—when winners often become losers—(i.e., when companies lose their innovative edge), conflicting organizational pressures (i.e., functioning efficiently today while innovating effectively for the future), problems associated with partial views of innovation, individual differences in innovation, multiple barriers to innovation, how innovation can help individuals and organizations increase performance, why it is important to work on many types of innovation simultaneously, and many others (Shavinina, 2003a).

Look at the individual differences principle of innovation (Shavinina, 2007). Because of individual differences among people and organizations, there is no one best "recipe" for making innovation happen, which would fit everyone. Something that works perfectly for one individual or organization may not work well for others and might lead to collapses in the third case. Even if one company does almost always come up with great solutions, which excellently work for them—say, *Dell Computer Inc.*—it does not follow at all that everyone else adopting those solutions will meet with the same degree of success. Copying may simply make the problem worse, if those things, which we try to copy, do not correspond to our individual differences. Our individual differences play an important role in success or failure of our efforts aimed at innovation. This is what "the individual differences principle of innovation" is all about (Shavinina, 2007). For example, Richard Branson's innovative approaches to creating a highly successful *Virgin* group and a unique brand name—as well as *Virgin*'s expansion through branded venture capital—are well known for many years (Branson, 2002). Nevertheless, how many entrepreneurs and businessmen did follow his example and did the same?[9]

The individual differences principle of innovation explains why innovation is anything but business as usual. "There are no reliable templates, rules, processes, or even measures of success. ... Innovation is not like most other business functions and activities" (Editorial, *Harvard Business Review*, 2002, p. 39). Innovation poses a constantly mutating puzzle, or set of puzzles. Individuals and organizations cannot afford not to play the game but the rules are anything but clear. There is not the comfort of a single "right" answer because the question keeps changing. Yet certain individuals and organizations are somehow able to come up with great ideas and implement them over and over again into practice. The theory of individual innovation explains why and how it happens (Shavinina & Sheeratan, 2003). The individual differences principle of innovation also suggests that innovation agenda or innovation strategy will vary significantly for different individuals and organizations.

Children are able to understand these complex issues in detail.

Courage: much needed and untrained talent

Research on famous innovators—such as Jeff Bezos, Richard Branson, Michael Dell, Thomas Edison, Mary Kay Ash, Akio Morita, Anita Roddick, Fred Smith, Sam Walton,[10] and many others—shows that they were and are very courageous people (Shavinina, 2006, 2008, this volume). Courage is much needed for future innovators. When they are about to implement great ideas into practice and thus introduce breakthrough innovations that have not existed before and nobody can predict the response from markets, they should not afraid to go ahead. Markets simply do not exist for the revolutionary innovations. It is innovators who have to create them. In order to reach that point, they should be courageous enough to convince everyone around them that the market for such and such particular product, process, or service will exist. Akio Morita is a great example.

He was a co-founder and the Chairman of Sony when he decided to develop Walkman. And he faced strong resistance. "Everybody gave me a hard time. It seemed as though nobody liked the idea" (Morita, 1987, p. 79). Sony's Walkman thus appeared despite very strong marketing input to suggest there was no demand for this kind of product. The marketing department at Sony even resisted using the word Walkman, explaining that it does not exist in English language and sounds very strange to English-speaking people. The incredible intuition of Akio Morita made Walkman possible. He pioneered the Walkman project and took personal responsibility for it. At the peak of a very strong resistance to the idea of the Walkman, he resolved the problem by threatening to leave the Chairman position if Sony did not sell 100,000 Walkman in the first six months. Sony sold many more in the first half a year, and Akio Morita was later awarded by the Royal Academy of the UK for his contributions to the development of English language (Morita, 1987). That is, he introduced the words *Sony* and *Walkman*. It is impossible to imagine modern English language without these words. He emphasized later that creativity "requires human thought, spontaneous intuition, and a lot of courage" (Morita, 1987, p. 83). It is sad that today's children are not taught to be courageous. This element of innovation education will thus fill in an apparent niche in general and gifted education by developing this much needed and untrained talent.

The "know-how" part of innovation education: high intellectual and creative educational multimedia technologies

Up to this point, various elements of innovation education were considered. This is the *know-what* part of innovation education. Its *know-how* part is about how to present these elements in a better possible way that would be the most productive for teaching the world's innovators. High intellectual and creative educational multimedia technologies (HICEMTs) can be a good carrier of the know-how part of innovation education. These technologies aim to develop an individual's intellectual and creative abilities in accordance with his or her time, speed, and learning needs.

Psychoeducational multimedia technologies (PMTs) and HICEMTs were first introduced in 1997 (Shavinina, 1997a). PMTs are multimedia technologies that base their five-part educational essence (discussed below) on fundamental psychological processes and phenomena (Shavinina, 1998a). HICEMTs constitute a special type of PMT whose general content is elaborated in accordance with underlying psychological mechanisms and states and whose special content is developed, structured, presented, and delivered according to the key principles of human intelligence and creativity (Shavinina, 2000).

The term *high* in HICEMTs refers to a significant saturation of the special content of these technologies through educational materials directed toward the actualization and development of human intellectual potential and creative abilities. HICEMTs emerge at the crossroads of many subfields of psychology (e.g., general, cognitive, developmental, educational, personality, media, cyber, and applied), education, and multimedia technology. At the moment HICEMTs represent an ideal, the realization of which will require the joint efforts of many psychologists, educators, technology specialists, and even venture capitalists. The next high peak of the development of high technology in, say, Silicon Valley will allow generous financial investments in the development of HICEMTs. Considerable support of public granting agencies will be required as well. Leaving for now the investment part for the future, this section briefly describes HICEMTs and their huge potential for educating tomorrow's innovators.

General characteristics of HICEMTs

The nature of HICEMTs can be described through a set of their general and specific characteristics. The general characteristics of HICEMTs include: (a) general psychological basis, (b) actualization of fundamental cognitive mechanisms, (c) new targets of educational and developmental influences, (d) better adaptation to an individual's psychological organization,[11] and (e) "psycho-edutainment" as an overall framework. Each of these five characteristics actually represents clusters of characteristics, which are highly interrelated within clusters.

General psychological basis

The general psychological basis means that HICEMTs are based on psychological processes and phenomena and particularly on the mechanisms of human intellectual and creative functioning. General psychological processes and phenomena include perception, attention, short-term and long-term memory, visual thinking, knowledge base, mental space, concept formation, analytical reasoning, metacognitive abilities, cognitive and learning styles, critical thinking, motivation, and many other psychological mechanisms. One of the strong arguments to place psychological foundations at the heart of HICEMTs is the fact that the educational process is always a psychological process. Learning, teaching, and training are based on fundamental psychological mechanisms.

For instance, an individual's knowledge base is an important psychological foundation for the process of knowledge transfer, which is usually considered a purpose of education. Researchers have found that knowledge base is a critical one in the successful functioning of human mind (Bjorklund & Schneider, 1996; Chi & Greeno, 1987; Chi & Koeske, 1983; Kholodnaya, 1997; Runco, 2006; Runco & Albert, 1990; Schneider, 1993; Shavinina & Kholodnaya, 1996; Sternberg, 1985, 1990). Psychologists demonstrated that domain-specific knowledge is crucial in highly intellectual performance and in the process of acquiring new knowledge, especially its quantity and quality. For example, an individual will not be able to solve any problem productively if he or she does not possess relevant prior knowledge (Chi & Greeno, 1987). The knowledge base can facilitate the use of particular learning strategies, generalize strategy use to related domains, or even diminish the need for strategy activation. Furthermore, a rich knowledge base can sometimes compensate for an overall lack of general cognitive abilities (Bjorklund & Schneider, 1996; Schneider, 1993). Despite all these findings, current educational multimedia products do not take into account the psychological nature of the human knowledge base, particularly with respect to its optimum functioning, which characterizes exceptional creative and intellectual performance that is the case of the gifted (Rabinowitz & Glaser, 1985; Shavinina & Kholodnaya, 1996; Shore & Kanevsky, 1993; Sternberg & Lubart, 1995) and innovators. Innovation is a special type of giftedness. Other psychological processes and phenomena and their appropriateness for the development of HICEMTs have been described elsewhere (Shavinina, 1997a, 1997b, 1998a).

Fundamental cognitive mechanisms

HICEMTs should be directed toward actualization of fundamental cognitive mechanisms, which play a significant role in an individual's intellectual and creative functioning. Broadly speaking, these technologies aim to actualize and develop an individual's cognitive abilities. This is the second main characteristic of HICEMTs. The following cognitive processes and phenomena are viewed as important in contemporary accounts of human intelligence and creativity (Brown, 1978, 1984; Flavell, 1976, 1979; Kholodnaya, 1997; Runco & Albert, 1990; Shavinina & Kholodnaya, 1996; Sternberg, 1984, 1985, 1988a, 1988b, 1990): conceptual structures (Case, 1995; Kholodnaya, 1983, 1997), knowledge base (Chi & Greeno, 1987; Chi & Koeske, 1983), mental space (Kholodnaya, 1997; Shavinina & Kholodnaya, 1996), cognitive strategies (Bjorklund & Schneider, 1996; Pressley, Borkowski, & Schneider, 1987), meta-cognitive processes (Borkowski, 1992; Borkowski & Peck, 1986; Brown, 1978, 1984; Campione & Brown, 1978; Flavell, 1976, 1979; Shore & Dover, 1987; Barfurth et al., 2009; Sternberg, 1985, 1990), specific intellectual intentions or extracognitive abilities (Shavinina, 1996b, 2003b, 2004; Shavinina & Ferrari, 2004), objectivization of cognition (Shavinina, 1996a), and intellectual and cognitive styles (Jones, 1997; Kholodnaya, 1997; Martinsen, 1997; Riding, 1997; Sternberg, 1985, 1987), just to mention a few. Other processes and phenomena should also be embedded in HICEMTs. These psychological mechanisms provide a necessary foundation for the further successful development of human creative and intellectual abilities, because development of an individual's mental potential must be based on already-actualized cognitive resources (Shavinina, 1998a).

The great French scientist Blaise Pascal said, "Chance favours the prepared Mind." To rephrase Pascal's words with respect to HICEMTs, it should be said that the prepared mind will gain much more advantage from specially elaborated psychoeducational multimedia technologies designed for the development of human mental abilities. Through repeated exposure to the fundamental processes and phenomena of the human cognitive system, HICEMTs will become "know-how" learning and training multimedia technologies by providing an underlying educational basis for the subsequent development of an innovator's mind.

New targets of educational and developmental influences

HICEMTs will change the conventional point of view about the targets of educational and developmental influences. The conventional wisdom of companies and developers of current educational multimedia technologies is to address their products to the abstract "user" as a whole. Designers are mostly unconcerned with where their products and services are directed. There is a general user (i.e., children, adolescents, or adults), and what is needed is to specify to what age category a given product is addressed. From a psychological point of view, it is unproductive to direct educational multimedia technologies to users in general, because people in general and children in particular are complex psychological systems with many hierarchical components, multidimensional variables, multifaceted parameters, and structural interrelations. To direct any educational influence to such complex systems as a whole is to decrease immediately the quality of education. As a result, the exact targets of the existing educational multimedia applications are missing, although they could be responsible for the higher productivity of the educational process. When analyzing educational multimedia technologies, one has always to ask him or herself: "To what exactly, in the structure of users' mind or personality, is the given educational multimedia technology directed?" It is not easy to answer this question, because the developers of educational multimedia usually do not ask it. The importance of asking the right questions was emphasized by Einstein (1949) and other brilliant minds of the 20th century.

For the most part, the main goal of current educational multimedia is "knowledge transfer." However, today such transfer is not fully realized, because successful knowledge transfer is a derivative of fundamental psychological mechanisms. The basic mental processes and phenomena should be viewed as the real targets of innovation education. The development of HICEMTs leads to a change in the traditional audience (i.e., targets) of available educational multimedia products from "users as a whole" to the underlying mechanisms of human intellectual and creative functioning. The primary objective of education has also changed: from realizing simple knowledge transfer to developing an individual's intellectual and creative abilities. Such changes will result in significant increases in the quality of education for today's children–tomorrow's innovators (Shavinina, 2000).

Better adaptation to individuals' psychological organization

The foregoing features of HICEMTs will allow them to be better adapted to an individual's psychological organization than current educational multimedia. HICEMTs can take into account numerous psychological characteristics of users, such as behavioral, developmental, emotional, motivational, personality, and social ones. HICEMTs should be directly built on the psychological specificity of users. For instance, any two educational multimedia courses for learning of a foreign language are certainly different for children and adults. It is clear that this difference is mainly connected to content and developers of educational multimedia, for the most part, are limited by content issues. Their way of thinking is as follows: "Children cannot understand some educational material, so we need to present them with more simple information." Nonetheless, this is not enough. Another difference between educational multimedia courses for children and adults concerns the psychological organization of the two groups of users. Adults have an internal motivation to study foreign languages, whereas children should have a strong, ongoing external motivation in addition to the internal one (or even instead of it; Shavinina & Ponomarev, 2003).

HICEMTs have the potential to generate the necessary conditions for the appearance and maintenance on the appropriate level of a child's motivation to learn, cognitive behavior, emotional involvement, and personal satisfaction with her or his gradual progress through the educational content. It does not matter by what means developers of educational multimedia can reach this goal (e.g., exciting scenarios, specific multimedia effects, user-friendly means, innovative learning methods, and so on; of course, their combination would be preferable). The bottom line is that HICEMTs should fit the internal psychological structures of human intellectual, creative, behavioral, cognitive, developmental, emotional, motivational, and other systems.

Psycho-edutainment

As was predicted earlier, the appearance of "psycho-edutainment"—a new area of the global educational multimedia market—is an inevitable event (Shavinina, 1998a). This innovative multidisciplinary multimedia field that emerges at the crossroads of existing multimedia fields (i.e., education, entertainment, and edutainment) and a new area—psychology—is the only scientifically viable framework for the development of HICEMTs. Taking into account that (a) education is, in its essence, a psychological process, (b) entertainment involves its own psychological mechanisms related to games, (c) play is a preferable and leading form of children's activity, and (d) fun and positive emotions accompany successful learning, as well as is highly desirable as one of the means to sustain children's curiosity, then one can conclude that HICEMTs cannot be developed other than through the synthesized regrouping of contemporary fields of multimedia (Shavinina, 2000). The nature of HICEMTs cannot be associated with one particular multimedia field; they may be created only in the space of "psycho-edutainment."

The five features described above provide the general characterization of HICEMTs. The other considerable portion of the features of HICEMTs relates to their more specific characteristics.

Specific characteristics of HICEMTs

Among specific characteristics of HICEMTs, the following can be distinguished: (a) "intellectual" content, (b) "creative" content, and (c) "intellectually creative edutainment." The first two characteristics deal with the specific content of HICEMTs, and they demonstrate "what" should be included in these technologies. Intellectually creative edutainment, on the other hand, represents a substantial part of the important characteristics related to the mode of presentation of this content. Therefore, intellectually creative edutainment describes "how" specific content might be embedded in HICEMTs.

Intellectual and creative contents cover many different, multidimensional, and interrelated aspects. However, it is not possible to consider all of them here, because they vary significantly and depend on developers. At the current level of the development of psychology, there are many different theories of human intelligence and creativity (Detterman, 1994; Kholodnaya, 1997; Miller, 1996; Runco, 2006; Runco & Albert, 1990; Shavinina, 1998b; Simonton, 1988; Sternberg, 1982, 1985, 1988b, 1990; Sternberg & Lubart, 1995) that determine developers' conceptions of creative and intellectual contents. A variety of the psychological approaches to the understanding of the nature of individual intelligence and creativity will strongly influence differences in the content of HICEMTs through developers' conceptions. These different approaches will predetermine what is intellectual and creative in HICEMTs.

One of the most promising approaches to developing the content of HICEMTs can be based on Kholodnaya's (1997) theory of individual intelligence. In other words, this theory might underlie the development of HICEMTs. Sternberg's triarchic theory of human intelligence and Gardner's multiple intelligences theory are examples of other theories that could also serve as foundations for the development of HICEMTs.

The joining of entertainment and education has led to the appearance of "edutainment," a rapidly developing multimedia field. Similarly, the combining of psychology and edutainment has led to the emergence of "psycho-edutainment," a promising new multimedia area. Likewise, the combination of intelligence and creativity with edutainment leads to "intellectually creative edutainment," a framework for the development of HICEMTs. HICEMTs will significantly transform "edutainment." Taken together, modern multimedia technology, educational games, and entertainment—built on the fundamental psychological processes and basic principles of human intellectual functioning and creative performance—form an "intellectually creative edutainment" that provides a real opportunity to develop HICEMTs.

How many HICEMTs: one, two, or more?

Definitely, a question might arise: how many HICEMTs can be developed? The answer is: an unlimited number can be developed. There are a few scientific reasons for this. First, as discussed above, the variety of psychological approaches and theories in the areas of human intelligence and creativity—which provide a foundation for developers' conceptions of creativity and intelligence—exclude at all a limited number of HICEMTs.

Second, the complex multidimensional nature of intellectual and creative abilities of today's children–tomorrow's innovators (toward the actualization and development of which HICEMTs are directed) excludes in principle a single educational multimedia technology that would be more productive than other technologies, because the development of innovative talent can be achieved through various psychoeducational methods. Different methods are suitable for different individuals in general and for innovators in particular, who have various, multiple profiles of innovative abilities.

Third, the transdisciplinary nature of HICEMTs (i.e., in contrast to traditional interdisciplinary and multidisciplinary approaches, "the complete fundamental-level merging of 'disciplinary' sources of knowledge is the focus," Vandervert, 2001) provides a basis for the development of a variety of new psycho-educational multimedia technologies (Vandervert, Shavinina, & Cornell, 2001). Finally, technological and economic factors also result in the exclusion of a limited quantity of HICEMTs.

Therefore, with their general and specific sets of characteristics, HICEMTs seem to be a good way to deliver all the above-mentioned elements of innovation education. This is a challenging task for the future. However, even today some of the new educational multimedia technologies incorporate certain general characteristics of HICEMTs (see Shavinina, 2009f, for examples and directions in which HICEMTs can be developed).

Innovation education for today's adults: the case of INNOCREX

So far innovation education for children and adolescents was discussed. However, it should be accessible to all members of any society. Everyone will benefit from innovation education: children, adolescents, and adults. Innovation can come from any person in society (Bessant, 2003). In this case the above-described elements of innovation education should be slightly modified. INNOCREX (www.innocrex.com), a special organization founded in 2005, is an example of such modification. Its mission is to develop adults' abilities via a series of seven one-day workshops on creativity (*From New Ideas to Success and Prosperity: Managing for Creativity*), innovation (*How to Realize the Innovative Potential of Your Organization: The Basics of Innovation Management*), excellence (*Managing for Excellence in Your Organization*), managerial talent (*On Managerial Talent: Skills Recognition and Development*), entrepreneurial giftedness (*Recognizing Entrepreneurial Giftedness*), practical intuition (*Developing and Using Practical Intuition*), and applied wisdom (*Applied Wisdom for Individuals and Organizations: Managing for Wisdom-Related Performance*).

The intention was to offer these workshops to all people working in the private, public, and not-for-profit organizations. For instance, teachers and managers alike find these workshops useful for their professional development. Specifically, they learn important practical tools (for example, intuition and wisdom), which help them to channel their talent, to motivate their students and employees to be more creative, to solve everyday problems in innovative ways, and to perform professional activities at the level of excellence.

Conclusion

This chapter described the basics of innovation education. Its structure consists of the following eleven elements:

- The programs available in the field of gifted education, which prove their effectiveness.
- The best programs from science and technology education.
- New programs aim to develop entrepreneurial giftedness, which is closely related to innovation.
- Programs for the development of children's metacognitive abilities or abilities to implement things: the so-called executive talent. It is essential, because innovation is about the implementation of ideas into practice.
- New programs based on recent progress in the study of scientific talent of Nobel laureates.
- Programs, which would incorporate the essentials of research on polymaths.

- New programs for the development of applied wisdom and moral responsibility.
- Programs aim to develop managerial talent.
- The fundamentals of deadline management.
- The foundations of innovation science: a general "know-what" and "know-how" about innovation, including the psychological basis of innovation, innovation gap, the individual differences principle of innovation, innovation management, and so on.
- Courage-related issues. Courage is mandatory for tomorrow's innovators.

HICEMTs seem to represent one of the best ways to deliver these components of innovation education. If educators are indeed concerned about the future of our world, then they have to focus on innovation education. Policy makers have an important role to play in order to help teachers and parents to achieve this goal. Its successful accomplishment will ultimately mean the fulfillment of one of the most important missions of any government: economic prosperity for all in the whole world.

Acknowledgments

The work presented herein was supported under the Support for Innovative Projects program (Grants AN-129135 and AN-143064) of the *Fonds québécois de la recherche sur la société et la culture* (FQRSC) and by the Social Sciences and Humanities Research Council (SSHRC) of Canada and the Templeton Foundation/the Institute for Research and Policy on Acceleration of the Belin-Blank International Center for Gifted Education and Talent Development of the University of Iowa. The findings and opinions expressed in the chapter do not reflect the positions or policies of these granting agencies. Special thanks to Larry Vandervert for his detailed feedback on early drafts of this chapter.

Notes

1. As the identification aspects are discussed in Shavinina's chapter on the assessment of innovative abilities in this volume, they are not, therefore, considered here.
2. *Entrepreneurial giftedness* refers to talented individuals who have succeeded in business by creating new ventures with at least a minimal financial reward (*fulfilled* entrepreneurial giftedness) or who demonstrated an exceptional potential ability to succeed (*prospective* entrepreneurial giftedness; Shavinina, 2009d).
3. The difference between entrepreneurship and innovation is based on the fact that entrepreneurs can implement any ideas into practice, not necessarily new ones (e.g., to open a pizzeria); while innovation is associated with the implementation of new ideas (e.g., to produce iPhones or iPods). In other words, one can distinguish between conventional entrepreneurs and entrepreneurs–innovators.
4. For the most part, people either produce ideas or implement them in life. The first group is called "creators" and the second group "innovators."
5. Leadership research traditionally belongs to the field of management and business studies, administrative sciences. It means that leadership scholars found the importance of vision in great leaders independently of giftedness researchers.
6. For definitions of these concepts please see Shavinina (2009d).
7. Wisdom refers to the application of intelligence, creativity, and experience as guided by values toward the achievement of a common good, through a balance among (a) intrapersonal, (b) interpersonal, and (c) extrapersonal interests, over the (a) short and (b) long term, to achieve a balance among (a) adaptation to existing environments, (b) shaping of existing environments, and (c) selection of new environments. A lack of wisdom is thus characterized by the faulty acquisition or application of intelligence, creativity, and experience knowledge as guided by values away from the achievement of a common good, through an imbalance among intrapersonal, interpersonal, and extrapersonal interests, of the short and long term, resulting in a failure in balance among adaptation to existing environments, shaping of existing environments, and selection of new environments. Foolishness is an extreme failure of wisdom (Sternberg, 1998).

8 Traditionally, managerial talent is associated with an individual's exceptional ability to deal with people, mainly to motivate them to achieve high performance (Buckingham & Coffman, 1990). Recently Shavinina and Medvid (2009) defined managerial talent as a combination of applied wisdom, practical intuition, excellence, entrepreneurial giftedness, creative abilities, and innovation.
9 Stelios Haji-Ioannou, a founder of *EasyJet*, is an obvious exception. *EasyJet* has been so successful that it now has a stock-market quote and Stelios Haji-Ioannou is developing new businesses, using the same brand, through his separate private venture-capital vehicle *easyGroup*.
10 The names of these innovators are mentioned in an alphabetical order.
11 Working on the development of a series of new textbooks on mathematics, which are largely based on her theory of human intelligence, Kholodnaya (1997) introduced the principles of new targets of educational influences and a need to better adapt learning and teaching material to children's psychological organization.

References

Amabile, T. M., Hadly, C. N., & Kramer, S. (2002). Creativity under the gun. *Harvard Business Review* (August), *80*(8), 52–61.
Barfurth, M. A., Ritchie, K. C., Irving, J. A., & Shore, B. M. (2009). A metacognitive portrait of gifted learners. In L. V. Shavinina (Ed.), *International Handbook on Giftedness* (pp. 397–420). Dordrecht: Springer Science.
Benbow, C. P. (2009). Fostering STEM innovators. In National Science Board, *Expert Panel Discussion on Preparing the Next Generation of STEM Innovators* (pp. 66–75). Arlington, VA: National Science Foundation.
Bessant, J. (2003). Challenges in innovation management. In L. V. Shavinina (Ed.), *The International Handbook on Innovation* (pp. 761–774). Oxford, UK: Elsevier Science.
Bjorklund, D. F., & Schneider, W. (1996). The interaction of knowledge, aptitude, and strategies in children's memory development. In H. W. Reese (Ed.), *Advances in Child Development and Behavior* (pp. 59–89). San Diego, CA: Academic Press.
Borkowski, J. G. (1992). Metacognitive theory: A framework for teaching literacy, writing, and math skills. *Journal of Learning Disabilities*, *25*(4), 253–257.
Borkowski, J. G., & Peck, V. A. (1986). Causes and consequences of metamemory in gifted children. In R. J. Sternberg & J. E. Davidson (Eds.), *Conceptions of Giftedness* (pp. 182–200). Cambridge, UK: Cambridge University Press.
Branson, R. (2002). *Losing My Virginity: The Autobiography*. London: Virgin Books.
Branson, R. (2008). *Screw It, Let's Do It*. London: Virgin Books.
Branson, R. (2011). *Screw Business as Usual*. New York: Portfolio/Penguin.
Brody, L. E. (2009). The John Hopkins talent search model for identifying and developing exceptional mathematical and verbal abilities. In L. V. Shavinina (Ed.), *International Handbook on Giftedness* (pp. 999–1016). Dordrecht: Springer Science.
Brown, A. L. (1978). Knowing when, where, and how to remember: A problem of metacognition. In R. Glaser (Ed.), *Advances in Instructional Psychology*, vol. 1 (pp. 77–165). Hillsdale, NJ: Erlbaum.
Brown, A. L. (1984). Metacognition, executive control, self-regulation, and other even more mysterious mechanisms. In F. E. Weinert & R. H. Kluwe (Eds.), *Metacognition, Motivation, and Learning* (pp. 60–108). Stuttgart, West Germany: Kuhlhammer.
Brown, A. L. (1987). Metacognition, executive control, self-regulation, and other even more mysterious mechanisms. In F. E. Weinert & R. H. Kluwe (Eds.), *Metacognition, Motivation, and Understanding* (pp. 64–116). Hillsdale, NJ: Erlbaum.
Brown, A. L. (1994). The advancement of learning. *Educational Researcher*, *23*(8), 4–12.
Brown, A. L., & Palincsar, A. S. (1989). Guided, cooperative learning and individual knowledge acquisition. In L. B. Resnick (Ed.), *Knowing, Learning, and Instruction* (pp. 76–108). Hillsdale, NJ: Erlbaum.
Buckingham, M., & Coffman, C. (1990). *First, Break All the Rules: What the World's Greatest Managers Do Differently*. New York: Simon & Schuster.
Campione, J. C., & Brown, A. L. (1978). Toward a theory of intelligence: Contributions from research with retarded children. *Intelligence*, *2*(3), 279–304.
Carver, C., & Scheier, M. (2003). Optimism. In S. J. Lopez & C. R. Snyder (Eds.), *Positive Psychological Assessment* (pp. 75–87). Washington, DC: American Psychological Association.
Case, R. (1995). The development of conceptual structures. In W. Damon, D. Kuhn, & R. Siegler (Eds.), *Handbook of Child Psychology* (pp. 745–800). New York: Wiley.

Chi, M. T. H., & Greeno, J. G. (1987). Cognitive research relevant to education. In J. A. Sechzer & S. M. Pfafflin (Eds.), *Psychology and Educational Policy* (pp. 39–57). New York: New York Academy of Sciences.

Chi, M. T. H., & Koeske, R. D. (1983). Network representation of a child's dinosaur knowledge. *Developmental Psychology, 19*(1), 29–39.

Clapham, M. M. (2003). The development of innovative ideas through creativity training. In L. V. Shavinina (Ed.), *The International Handbook on Innovation* (pp. 366–376). Oxford, UK: Elsevier Science.

Colangelo, N., & Davis, G. (2003). *Handbook of Gifted Education*. Boston: Allyn & Bacon.

Cramond, B. L. (2009). Future problem solving in gifted education. In L. V. Shavinina (Ed.), *International Handbook on Giftedness* (pp. 1143–1156). Dordrecht: Springer Science.

Dell, M. (1999). *Direct From Dell*. New York: Harper Business.

Detterman, D. K. (Ed.) (1994). *Current Topics in Human Intelligence: Vol. 4. Theories of Intelligence*. Norwood, NJ: Ablex.

Editorial. (2002). Inspiring innovation: Voices. *Harvard Business Review* (August), *80*(8), 39–49.

Einstein, A. (1949). Autobiographical notes. In P. A. Schlipp (Ed.), *Albert Einstein: Philosopher and Scientist* (pp. 3–49). New York: Library of Living Philosophers.

Ericsson, K. A., Nandagopal, K., & Roring, R. W. (2009). An expert performance approach to the study of giftedness. In L. V. Shavinina (Ed.), *International Handbook on Giftedness* (pp. 129–154). Dordrecht: Springer Science.

Flavell, J. H. (1976). Metacognitive aspects of problem solving. In L. B. Resnick (Ed.), *The Nature of Intelligence* (pp. 231–235). Hillside, NJ: Erlbaum.

Flavell, J. H. (1979). Metacognition and cognitive monitoring: A new area of cognitive-development inquiry. *American Psychologist, 34*(10), 906–911.

Gersick, C. J. (1995). Everything new under the gun. In C. M. Ford & D. A. Gioia (Eds.), *Creative Action in Organizations* (pp. 142–148). Thousand Oaks, CA: Sage Publications.

Grove, A. (1987). *One-on-One With Andy Grove*. New York: Penguin Books.

Grove, A. (1996). *Only the Paranoid Survive*. New York: Doubleday.

Hadjimanolis, A. (2003). The barriers approach to innovation. In L. V. Shavinina (Ed.), *The International Handbook on Innovation* (pp. 559–573). Oxford, UK: Elsevier Science.

Heller, K. A., Monks, F., Sternberg, R. J., & Subotnik, R. F. (2000). *The International Handbook of Giftedness and Talent*. Oxford, UK: Elsevier Science.

Jones, A. E. (1997). Reflection-impulsivity and wholist-analytic: Two fledglings?... or is R-I a cuckoo? *Educational Psychology, 17*(1–2), 65–77.

Kay Ash, M. (1996). *You Can Have It All: Lifetime Wisdom from America's Foremost Woman Entrepreneur*. New York: Doubleday.

Keller, E. F. (1984). *A Feeling for the Organism: The Life and Work of Barbara McClintock*. New York: Doubleday.

Kholodnaya, M. A. (1983). *The Integrated Structures of Conceptual Thinking*. Tomsk, Russia: Tomsk University Press.

Kholodnaya, M. A. (1997). *The Psychology of Intelligence*. Moscow: Science Press.

Kholodnaya, M. A. (2002). *The Psychology of Intelligence*, 2nd ed. St-Petersbourg: Peter.

Lowenstein, R. (1996). *Buffett: The Making of an American Capitalist*. New York: Main Street Books.

Lubinski, D., & Benbow, C. P. (2006). Study of mathematically precocious youth after 35 years. *Perspectives on Psychological Science, 1*(4), 123–152.

Marrett, C. B., & Johnson, P. (2009). Cultivating talent for innovation. In National Science Board, *Expert Panel Discussion on Preparing the Next Generation of STEM Innovators* (pp. 80–84). Arlington, VA: National Science Foundation.

Martinsen, O. (1997). The construct of cognitive style and its implications for creativity. *High Ability Studies, 8*(2), 135–158.

Marton, F., Fensham, P., & Chaiklin, S. (1994). A Nobel's eye view of scientific intuition. *International Journal of Science Education, 16*(4), 457–473.

Miller, A. (1996). *Insights of Genius: Visual Imagery and Creativity in Science and Art*. New York: Springer-Verlag.

Morita, A. (1987). *Made in Japan*. London: Collins.

National Science Board (NSF). (2009). *Expert Panel Discussion on Preparing the Next Generation of STEM Innovators*. Arlington, VA: National Science Foundation.

Ponomarev, A. E. (2013). *What Did I Learn from Nobel Laureates in Science?* In preparation.

Pressley, M., Borkowski, J. G., & Schneider, W. (1987). Cognitive strategies: Good strategy users coordinate metacognition and knowledge. In R. Vasta (Ed.), *Annals of Child Development*, vol. 4 (pp. 89–129). Greenwich, CT: JAI Press.

Rabinowitz, M., & Glaser, R. (1985). Cognitive structure and process in highly competent performance. In F. D. Horowitz & M. O'Brien (Eds.), *The Gifted and Talented: Developmental Perspectives* (pp. 75–97). Washington, DC: American Psychological Association.

Reis, S. M., & Renzulli, J. S. (2003). Developing high potentials for innovation in young people through the schoolwide enrichment model. In L. V. Shavinina (Ed.), *The International Handbook on Innovation* (pp. 333–346). Oxford, UK: Elsevier Science.

Renzulli, J. S., & Reis, S. M. (2009). A technology-based application of the schoolwide enrichment model and high-end learning theory. In L. V. Shavinina (Ed.), *International Handbook on Giftedness* (pp. 1203–1223). Dordrecht: Springer Science.

Riding, R. (1997). On the nature of cognitive style. *Educational Psychology, 17*(1–2), 29–49.

Roddick, A. (2000). *Business as Unusual*. London: Thorsons.

Root-Bernstein, R. (2003). The art of innovation. In L. V. Shavinina (Ed.), *The International Handbook on Innovation* (pp. 267–278). Oxford, UK: Elsevier Science.

Root-Bernstein, R. (2008). I don't know! (toward a curriculum of questioning). In W. Vitek and W. Jackson (Eds.), *Virtues of Ignorance: Complexity, Sustainability, and the Limits of Knowledge* (pp. 233–250). Salina, KA: University Press of Kentucky.

Root-Bernstein, R. (2009). Multiple giftedness in adults: The case of polymaths. In L. V. Shavinina (Ed.), *International Handbook on Giftedness* (pp. 853–872). Dordrecht: Springer Science.

Root-Bernstein, R., & Root-Bernstein, M. (1999). *Sparks of Genius*. Boston: Houghton Mifflin.

Root-Bernstein, R., & Root-Bernstein, M. (2003). Intuitive tools for innovative thinking. In L. V. Shavinina (Ed.), *The International Handbook on Innovation* (pp. 377–387). Oxford, UK: Elsevier Science.

Runco, M. A. (Ed.) (2006). *Creativity: Theories and Themes: Research, Development, and Practice*. London, UK: Elsevier.

Runco, M. A., & Albert, R. S. (Eds.) (1990). *Theories of Creativity*. Newbury Park, CA: Sage.

Schneider, W. (1993). Domain-specific knowledge and memory performance in children. *Educational Psychology Review, 5*(3), 257–273.

Shavinina, L. V. (1996a). The objectivization of cognition and intellectual giftedness. *High Ability Studies, 7*(1), 91–98.

Shavinina, L. V. (1996b). Specific intellectual intentions and creative giftedness. In A. J. Cropley & D. Dehn (Eds.), *Fostering the Growth of High Ability: European Pperspectives* (pp. 373–381). Norwood, NJ: Ablex.

Shavinina, L. V. (1997a, July). *Educational multimedia of "tomorrow": High intellectual and creative psychoeducational technologies* (Paper presented at the European Congress of Psychology, Dublin, Ireland).

Shavinina, L. V. (1997b, September). *High intellectual and creative technologies as an educational multimedia of the 21st century* (Paper presented at the European Open Classroom II Conference: School Education in the Information Society, Sisi, Crete, Greece).

Shavinina, L. (1998a). Interdisciplinary innovation: Psychoeducational multimedia technologies. *New Ideas in Psychology, 16*(3), 189–204.

Shavinina, L. V. (1998b). On Miller's insights of genius: What do we know about it? *Creativity Research Journal, 11*(2), 183–185.

Shavinina, L. V. (2000). High intellectual and creative educational multimedia technologies. *CyberPsychology & Behaviour, 3*(2), 1–8.

Shavinina, L. V. (2003a). *The International Handbook on Innovation*. Oxford, UK: Elsevier Science.

Shavinina, L. V. (2003b). Understanding scientific innovation: The case of Nobel laureates. In L. V. Shavinina (Ed.), *The International Handbook on Innovation* (pp. 445–457). Oxford, UK: Elsevier Science.

Shavinina, L. V. (2004). Explaining high abilities of Nobel laureates. *High Ability Studies, 15*(2), 243–254.

Shavinina, L. V. (2006). Micro-social factors in the development of entrepreneurial giftedness: The case of Richard Branson. *High Ability Studies, 17*(2), 225–235.

Shavinina, L. V. (2007). Comment l'innovation peut-elle accroître la performance organisationnelle? In L. Chaput (Ed.), *Modèles Contemporains en Gestion: Un Nouveau Paradigme, la Performance* (pp. 167–197). Le Delta I, Québec, Canada: Presses de l'Université du Québec.

Shavinina, L. V. (2008). Early signs of entrepreneurial giftedness. *Gifted and Talented International, 23*(2), 3–17.

Shavinina, L. V. (Ed.) (2009a). *International Handbook on Giftedness*. Dordrecht: Springer Science.

Shavinina, L. V. (2009b). Scientific talent: The case of Nobel laureates. In L. V. Shavinina (Ed.), *International Handbook on Giftedness* (pp. 649–669). Dordrecht: Springer Science.

Shavinina, L. V. (2009c). A unique type of representation is the essence of giftedness: Toward a cognitive-developmental theory of giftedness. In L. V. Shavinina (Ed.), *International Handbook on Giftedness* (pp. 231–257). Dordrecht: Springer.

Shavinina, L. V. (2009d). On entrepreneurial giftedness: Where did all great entrepreneurs come from? In L. V. Shavinina (Ed.), *International Handbook on Giftedness* (pp. 793–807). Dordrecht: Springer Science.

Shavinina, L. V. (2009e). On giftedness and economy: The impact of talented individuals on the global economy. In L. V. Shavinina (Ed.), *International Handbook on Giftedness* (pp. 925–944). Dordrecht: Springer Science.

Shavinina, L. V. (2009f). High intellectual and creative educational multimedia technologies for gifted education. In L. V. Shavinina (Ed.), *International Handbook on Giftedness* (pp. 1181–1202). Dordrecht: Springer Science.

Shavinina, L. V. (2009g). How to develop innovators: Lessons from Nobel laureates and great entrepreneurs. Innovation Education. An invited presentation at the US National Science Board *Expert panel discussion on preparing the next generation of STEM innovators*. Arlington, VA: National Science Foundation (24 August).

Shavinina, L. V., & Ferrari, M. (Eds.) (2004). *Beyond Knowledge*. Mahwah, NJ: Erlbaum Publishers.

Shavinina, L. V., & Kholodnaya, M. A. (1996). The cognitive experience as a psychological basis of intellectual giftedness. *Journal for the Education of the Gifted, 20*(1), 3–35.

Shavinina, L. V., & Medvid, M. (2009). Understanding managerial talent. In L. V. Shavinina (Ed.), *International Handbook on Giftedness* (pp. 839–851). Dordrecht: Springer Science.

Shavinina, L. V., & Ponomarev, E. A. (2003). Developing innovative ideas through high intellectual and creative educational multimedia technologies. In L. V. Shavinina (Ed.), *The International Handbook on Innovation* (pp. 401–418). Oxford, UK: Elsevier Science.

Shavinina, L. V., & Sheeratan, K. (2003). On the nature of individual innovation. In L. V. Shavinina (Ed.), *The International Handbook on Innovation* (pp. 31–43). Oxford, UK: Elsevier Science.

Shore, B. M., & Dover, A. C. (1987). Metacognition, intelligence and giftedness. *Gifted Child Quarterly, 31*, 37–39.

Shore, B. M., & Kanevsky, L. S. (1993). Thinking processes: Being and becoming gifted. In K. A. Heller, F. J. Mönks, & A. H. Passow (Eds.), *International Handbook of Research and Development of Giftedness and Talent* (pp. 133–147). Oxford, UK: Pergamon Press.

Simonton, D. K. (1988). *Scientific Genius: A Psychology of Science*. New York: Cambridge University Press.

Simonton, D. K. (2009). Gifts, talents, and their societal repercussions. In L. V. Shavinina (Ed.), *International Handbook on Giftedness* (pp. 905–915). Dordrecht: Springer Science.

Sternberg, R. J. (Ed.) (1982). *Handbook of Intelligence*. New York: Cambridge University Press.

Sternberg, R. J. (Ed.) (1984). *Mechanisms of Cognitive Development*. New York: Freeman.

Sternberg, R. J. (1985). *Beyond IQ: A Triarchic Theory of Human Intelligence*. New York: Cambridge University Press.

Sternberg, R. J. (Ed.) (1988a). *The Nature of Creativity*. New York: Cambridge University Press.

Sternberg, R. J. (1988b). A three-facet model of creativity. In R. J. Sternberg (Ed.), *The Nature of Creativity* (pp. 125–147). New York: Cambridge University Press.

Sternberg, R. J. (1990). *Metaphors of Mind: Conceptions of the Nature of Intelligence*. New York: Cambridge University Press.

Sternberg, R. J. (1998). *Handbook of Wisdom*. New York: Cambridge University Press.

Sternberg, R. J., & Lubart, T. (1995). *Defying the Crowd: Cultivating Creativity in a Culture of Conformity*. New York: Free Press.

Subotnik, R., Orland, M., Rayhack, K., Schuck, J., Edmiston, A., Earle, J., Crowe, E., Johnson, P., Carroll, T., Berch, D., & Fuchs, B. (2009). Identifying and developing talent in science, technology, engineering, and mathematics (STEM): An agenda for research, policy, and practice. In L. V. Shavinina (Ed.), *International Handbook on Giftedness* (pp. 104–117). Dordrecht: Springer Science.

Vandervert, L. R. (2001). A provocative view of how algorithms of the human brain will embed in cybereducation. In L. R. Vandervert, L. V. Shavinina, & R. Cornell (Eds.), *CyberEducation: The Future of Long Distance Learning* (pp. 41–62). Larchmont, NY: Liebert Publishers.

Vandervert, L. R., Shavinina, L. V., & Cornell, R. (Eds.) (2001). *CyberEducation: The Future of Long Distance Learning*. Larchmont, NY: Liebert Publishers.

VanTassel-Baska, J., & MacFarlane, B. (2009). Enhancing creativity in curriculum. In L. V. Shavinina (Ed.), *International Handbook on Giftedness* (pp. 1061–1083). Dordrecht: Springer Science.

Wademan, D. (2005). The best advice I ever got. *Harvard Business Review* (January), 35–44.

Wallace, J., & Erickson, J. (1992). *Hard Drive: Bill Gates and the Making of the Microsoft Empire*. New York: HarperCollins Publishers.

Weisberg, R. (2003). Case studies of innovation: Ordinary thinking, extraordinary outcomes. In L. V. Shavinina (Ed.), *The International Handbook on Innovation* (pp. 204–247). Oxford, UK: Elsevier Science.

4

How advances in gifted education contribute to innovation education, and vice versa

David Yun Dai

STATE UNIVERSITY OF NEW YORK AT ALBANY, USA

Summary: Gifted education can be seen as a frontier of innovation education, wherein the nurturing of creativity has been an educational priority. Parallel to the first two stages of research on creativity, with its emphasis on person and process, gifted education has focused on two aspects of innovation education: how to provide a good educational match for those who demonstrate unique creative potential, and how to nurture creativity through curricular and instructional designs. These explorations have proved highly meaningful for general innovation education. In recent years, creativity researchers have broadened their perspectives beyond person and process to encompass complex social-cognitive dynamics and synergistic power. Technological advances and availability of cyber resources also make it possible to design an education gearing toward developing personal creativity on all fronts of human endeavor. It is argued that gifted education can learn from these new movements in innovation education and broaden its education scope accordingly.

Key words: Gifted education, innovation education, creativity enhancement, the person, process, and context dimensions.

> To give a fair chance to potential creativity is a matter of life and death for any society.
> *(Arnold Toynbee, 1964)*

Introduction: links between innovation education and gifted education

It is not without reason that, in education settings, the term "creativity" is used more often (e.g., development of creative productivity, Renzulli, 2005; creativity enhancement, Beghetto & Kaufman, 2010), while in work settings, innovation is used more frequently (e.g., Estrin, 2009). For the former, the concern is more about developing desired characteristics conducive to mature creative expressions, and increasing the chances that youths will become more capable of producing novel and useful ideas and products in their adulthood. When the term "innovation" is used, the concern is more over facilitating creative products and services that have a direct social impact and practical consequences. Thus innovation is always context-specific, product-driven, while creativity could refer to more general conditions and characteristics conducive to development and expression of creativity. Defined this way, we might think of innovation education as having two phases: general preparation in K-12 education and direct application in college and beyond. Whether we use creativity education and innovation education interchangeably or make a distinction between the two is a matter of convention.

Gifted education has a natural connection to innovation education, given its goal of producing next generations of leaders who can lead; that is, who can innovate in varied and many ways on all fronts of human endeavor. At least three links can be established between gifted education and innovation education: theoretical, practical, and social ones, which almost guarantee the reciprocation of the two.

The first link is theoretical in nature. Gifted education is particularly keen on creativity as (a) an aptitude variable (personal characteristics, cognitive or affective, conducive to creative expressions, broadly defined), (b) a contextual variable (environmental structures and social conditions that induce or inhibit creativity), and (c) an educational outcome variable (what kind of educational experiences promotes creativity enhancement and development). Therefore, gifted education has a huge overlap with innovation education in seeking a deep understanding of how the creative potential can be unleashed, identified, carefully nurtured, and systematically developed.

The second link is more practical. Gifted education arose partly as a reaction to the traditional age-graded schooling, with a fixed curriculum for all, the transmission model of learning and pedagogy, and a uniformed evaluation system. Gifted education seeks a more individualized mode of education, responsive to individuals' strengths and interests, rather than adhering to a one-size-fits-all curriculum and teaching to the average. Pedagogically, it seeks a more productive mode of learning rather than mere regurgitation or passive absorption of information for testing purposes (Renzulli & Dai, 2001; Renzulli & De Wet, 2010). It can be argued that an emphasis on creative and critical thinking in classroom now advocated by leading educational scholars (e.g., Halpern, 2008; Resnick, 2010) as well as prominent advocacy groups (e.g., Partnership for 21st Century Skills, 2008) dates back to the early years of gifted education (e.g., DeHaan & Havighurst, 1957; Torrance, 1963).

Finally, the third link, the social one, refers to the fact that many leading researchers and scholars in gifted education have also been educational psychologists and active researchers on creativity (e.g., Bonnie Cramond, John Feldhusen, Jane Pirrto, Jonathan Plucker, Joseph Renzulli, Robert Sternberg, Paul Torrance, and Donald Treffinger). It is natural that the practical ideas and models they have developed based on educational psychology to enhance creativity in gifted education has generality in application for all learners.

How advances in gifted education have contributed to innovation education

To understand the advances made in gifted education with respect to creativity enhancement and development, we need to put them in the context of both the psychological research on creativity and the rise of gifted education and ensuing changes in the American history. On the research side, there are three broad phases of creativity research (Sawyer, 2006b). The first phase roughly started with Guilford's (1950, 1967) presidential speech at the 1950 annual convention of the American Psychological Association (APA), and epitomized by the Utah Conferences held between 1950 and 1963. The focus during this period was on identifying personal characteristics conducive to creativity and nurturing creative potential of individuals by developing these characteristics (e.g., MacKinnon, 1962). The second phase started in the wake of cognitive revolution with a focus on underlying cognitive and motivational processes that lead to creative products (e.g., Getzels & Csikszentmihalyi, 1976; Finke, Ward, & Smith, 1992; Weisberg, 1999), and social conditions that either facilitate or hinder these processes (Amabile, 1983). The third phase started in late 1990s and continues to date, with a focus on synergistic group dynamics and distributed cognitive and social processes leading to creativity (e.g., Dunbar, 1997; Sawyer, 2003). Different zeitgeists during these distinct historical periods have had a strong impact on how creativity enhancement in education is conceptualized and implemented.

On the educational side, the rise of gifted education in the US was prompted by the launch of the first satellite (the Sputnik) in the former Soviet Union in 1957. The focus was on a unique group of gifted children, the "creatively gifted" (Getzels & Jackson, 1962). In its later development, leading scholars in gifted education have paid increasing attention to the dynamic, developmental, and contextual nature of creativity (Renzulli, 1978, 1986, 2005), hence the role of educators in creating optimal conditions for developing students' creative potential.

First phase: from IQ to creativity – a shifting focus

IQ had been the gold standard for gifted identification since Louis Terman (1925), who launched the first ever longitudinal study of over 1,500 children whose IQs were 135 or above. Guilford (1950) suggested that correlations between intelligence measures and creativity measures were low enough to justify treating them as distinct constructs. Guilford's (1967) theory of Structure of Intellect was one of the major theories that provide a model for the differentiation. But it was Getzels and Jackson (1962) who made a strong argument, based on their research and Guilford's theory, that the "creatively gifted" would be overlooked by exclusively using the IQ criterion for identification purposes. Although the "creatively gifted" Getzels and Jackson identified turned out to have a mean IQ of 127, not "average" by any standard, the notion that there is a distinct quality called "creativity" that is different from but as equally important as "intelligence" gained currency. It was cemented by the Marland Report (Marland, 1972), which provided an "official" definition of giftedness and identified "creative giftedness" as a distinct feature of gifted potential (see also Gagné, 2005; Tannenbaum, 1997 for further elaboration). But it was Paul Torrance who truly laid the foundation for something we might call "education for creativity."

Paul Torrance

Torrance was one of the earliest scholars who systematically conducted research and developed ideas on how teachers can foster creativity in their classroom teaching. His 1963 book entitled *Education and the Creative Potential* was a mixture of theoretical exposition and empirical research attempting to map out social and educational conditions that either facilitate or inhibit creative thinking and the long-term development of creativity. Although building on Guilford's (1950) notion of divergent production as a source of creativity, and Getzels and Jackson's (1962) focus on the "creatively gifted," Torrance's approach was much richer in his conception of creativity (e.g., defining creativity as finding and resolving discrepancies), and much more educationally and developmentally oriented; that is, how to create an environment conducive to the flourishing of the creative potential in all children. Torrance contrasted "learning and thinking creatively" with "learning by authority":

> A child learns creatively by questioning, inquiring, searching, manipulating, experimenting, even by aimless play; in short, by always trying to get at the truth. Learning and thinking creatively take place in the process of sensing difficulties, problems, and gaps in information; in making guesses or formulating hypotheses about these deficiencies; in testing these guesses and possibly revising and retesting them; and finally in communicating the results.... We learn by authority when we are told what we should learn, when we accept an idea as true on the word of some authority.
>
> *(Torrance, 1963, p. 47)*

In his 1970 book *Encouraging Creativity in the Classroom*, Torrance further expounded what teachers can do to promote creativity. He identified several pedagogical steps to engage creative learning, such as heightening anticipation for what the learner will be exposed to, sustaining the momentum by creating the unexpected, building creative skills, going beyond textbooks, classrooms, and curricula.

Major contributions

Torrance's work epitomizes early seminal contributions to innovation education in the mid-20th century. First, creativity was seen as originated from the primary ability to produce novel ideas, ideas deviating from norms and familiar ones; this ability is a different kind than intelligence as we know based on IQ tests. Guilford's influence was palpable. Second, some individuals display strong "creative needs," such as intellectual curiosity, the need for meeting challenge, the need to give oneself completely to a task (i.e., task commitment), the need to be honest and search for the truth, and the need for being oneself (Torrance, 1970, Chapter 2). Teaching that responds to these "creative needs" would naturally bring students' creative ability into play. Third, the unruly childhoods of Edison, Franklin, the Wright Brothers were used as prototypes of the youth of the creatively gifted. Fourth, divergent thinking tests were considered a good instrument for measuring individual differences in creative potential (Torrance, 1972). Although his work was clearly influenced by the zeitgeist of his times, being an educator himself, Torrance's legacy for creativity or innovation education is much richer than the above theoretical abstractions. His focus on teaching for creativity went beyond the "creatively gifted" to reach out to all children, and the pedagogy he advocated for "learning and thinking creatively" pre-dates the later movement of inquiry learning (Aulls & Shore, 2008). In effect, his education focus went beyond the then popular psychometric view of creativity in considering a combination of endogenous and exogenous factors, not the least of which is the intrinsic motivation to know (Torrance, 1970).

Second phase: from traits to processes – teaching creative problem solving and encouraging and supporting the development of creativity

Although in his later years, Torrance was exposed to the cognitive revolution and used ideas about creative problem solving developed by cognitive psychologists (e.g., Newell, Shaw, & Simon, 1962), Torrance's times were dominated by the psychometric view of traits and their contributions when creativity is concerned, be it IQ, divergent production, or personality traits of creative individuals (MacKinnon, 1962, 1978). This trend changed after cognitive psychology took hold (e.g., Newell & Simon, 1972); gradually the focus shifted from what characteristics creative people possess to the issue of what people do when they engaged in tasks involving complex problem solving and creativity. Getzels and Csikszentmihalyi (1976) conducted a longitudinal study on art and found problem finding (or representation) and intrinsic motivation to be predictive of creative productivity in arts in later years. In contrast, Finke et al. (1992) developed their "creative cognition" approach based on their lab experimentation; they specified two phases of creative cognition, generate and explore (hence the Geneplore Model). About the same time, Amabile (1983) developed her "social psychology of creativity," focusing on social-contextual influences on performance on creative tasks. Taken together, processes gained primacy over traits, a move that had an impact on gifted education as well.

In gifted education, Guilford and Torrance continued to influence scholar discourse, research, and practice regarding how to define, identify, and nurture creativity in 1970s and 1980s (e.g., Feldhusen & Treffinger, 1986). However, a major shift occurred during that period from treating creativity as an aptitude variable to an outcome variable; that is, learning and thinking creatively became a major educational goal for the gifted (Feldhusen & Treffinger, 1986; Gallagher, 1975; Renzulli, 1977; Renzulli & Callahan, 1973). This change was important because creativity, however defined, was seen not as an enduring characteristic of the person but as developmental in nature, subject to educational interventions. Gifted education researchers developed two distinct approaches to enhancing creativity, each with its own theoretical underpinnings.

The first approach is based on the problem solving model, specifying steps involved in solving complex problems. This approach treats creativity as a set of teachable skills that can be trained and modeled. For example, Treffinger (quoted in Feldhusen & Treffinger, 1986, p. 45) identified six steps of creative problem solving:

1 selecting parameters;
2 listing attributes for each parameter;
3 developing evaluation criteria;
4 examining many combinations;
5 checking up on other resources; and
6 following up on promising ideas.

To scaffold students to be creative problem solvers, Feldhusen and colleagues developed the Purdue Three-Stage Model of creative enrichment (Feldhusen & Kolloff, 1986; Feldhusen & Treffinger, 1986). Stage One activities are teacher-directed, aimed to develop basic, discrete thinking skills necessary for higher-level thinking; Stage Two activities are teacher-guided and designed to develop broader strategies for tackling given, circumscribed problems, but there is more self-direction on the part of students; Stage Three activities are independent projects by which students are given more freedom and self-direction to use skills developed in the previous two stages in tackling more realistic, open-ended problems (see Treffinger & Isaksen, 2005, for a review of recent developments in the Creative Problem Solving model).

Alternatively, the second approach sees creativity as an emergent property when people, through their high task commitment, bring their abilities, skills, and knowledge to bear upon a task that has authentic meaning and potential impact on an audience. The three-ring conception of giftedness (gifted manifestations as the interplay of above average ability, task commitment, and creativity), developed by Renzulli (1978, 1986), represents such an approach. It treats creativity as a contextually emergent capability of making authentic inquiry to its fruition. This capability cannot be dissociated from one's domain-specific and general abilities, task commitment (a distinct motivational component), and environmental support for such a productive activity. Renzulli (1986) also distinguished between two kinds of giftedness: schoolhouse giftedness in terms of good lesson learners and test-takers, and creative-productive giftedness in terms of producing solutions and fashioning products that have a real impact on an audience. In so doing, Renzulli departed from the Guilford tradition, which defines creativity as divergent thinking. In stressing the importance of task commitment, he also differed from a purely "creative cognition" approach. His Enrichment Triad model (Renzulli, 1977) identified three types of enrichment: Type I activities expose students to a variety of real life topics and domains of human endeavor not often featured in school curriculum; strengths and interests can be identified. Type II activities teach students relevant skills necessary to carry out independent inquiries; abilities and commitment can be assessed to determine suitability for Type III activities; Type III activities are independent or group investigative projects meant to produce tangible products of real impact; while a student's "above average ability" may be gauged by cognitive ability tests, task commitment and creativity can only be assessed during the productive process.

Major contributions

Theoretical contributions during this period include a shift in focus from traits to developmental processes leading to creativity, from person accounts of creativity to process accounts of creativity. This shift led to pedagogical innovations such as the Schoolwide Enrichment Model (Renzulli & Reis, 1997), Creative Problem Solving (Treffinger & Isaksen, 2005), initially designed for gifted students but later applied to all students. Because of these changes, there is a shift in assessing creativity. Beyond the objective, psychometric measurements such as the *Torrance Test of Creative Thinking* (TTCT), other techniques were developed, such as Consensual Assessment Technique that incorporates expert judgments (Amabile, 1982), and performance assessment based on authentic inquiry (Renzulli & Reis, 1997). A complete model of person, process, and product necessitates a new view of creativity assessment that involves evidence of changing behaviors, strategies, and performance in a more dynamic, contextualized fashion.

How advances in innovation education can contribute to gifted education

Gifted education has long been a pioneer for teaching and learning for creativity and innovation (Renzulli, 1977; see Tomlinson & Callahan, 1992). However, it also tends to see itself as having a separate identity, apart from the rest of education, in the name of serving "special needs" of gifted students. Conceptualized this way, gifted education has been somewhat insulated from a broader educational perspective. The third phase of research on creativity goes beyond a focus on the differential creative potential of individuals, which characterizes research in mid-20th century (the first phase), or on cognitive and motivational processes involved in creative thinking, which characterizes research in late 20th century (the second phase); the new trend is characterized by exploring new possibilities for creative learning through social interaction, technological support, and personalization of learning and knowledge. This body of research is conducted outside of gifted education, and, as I submit, can potentially make important contributions to gifted education.

The third phase: naturalizing creativity or the inherently creative nature of inquiry activities

The thrust of this new wave of research on teaching and learning for creativity comes from a realization, attained a long time ago, that learning can be truly a creative act or a form of creative cognition in that learning is generative (Bruner, 1960) and that novelty in thinking can be engendered through learning (Torrance, 1963). This new movement is poised to "naturalize" creativity. Naturalizing creativity means that creativity is not some kind of special processes humans deployed for special purposes (making creations) but is the product of natural human quests for meaning, truth, and optimality. For instance, Sawyer (2006b) quoted Heisenberg's remarks that "science is deeply rooted in conversations" (p. 276) as a strong support for collaborative and synergistic creativity (see also Sawyer, 2003); that is, creative activity is fundamentally social and interactive. Naturalizing creativity is also based on the argument that creativity origins with situated actions rather than mere ideation; namely, creativity does not start with novel ideas but with meaningful tasks, actions, and interactions from which creativity emerges over time (Sawyer, 2006b). Consequently, creativity enhancement efforts need to be repositioned seamlessly in daily transactions in natural settings. Several educational innovations take on this tack.

Historically, learning has been defined as an act of absorbing knowledge created by others, with the teacher serving as a medium. There is a correspondence between the teacher's input and students' output. Scardamalia and Bereiter (2006) challenged this notion of learning. In their model of the Knowledge Building community that engages students in what they called "creative knowledge work" (p. 98), the line between learning and creative thinking is blurred. For example, the work of fifth graders on Gregor Mendel's problem of genetics is seen as "continuous with that of Gregor Mendel, addressing the same basic problem" (p. 98). A key concept meant to materialize this transformation is "idea improvement." Pedagogical features that ensure its effectiveness are high levels of student control and collective cognitive responsibility, coupled with its "launching pad": Knowledge Forum, a technological platform for organizing publicly expressed ideas (Zhang, 2012). This way, the role of learners as creative agents is redeemed. Different from static, individualistic conceptions of intelligence and creativity, based on which giftedness is attributed to individuals, this new approach is committed to "relational ontology" of human functioning (Barab & Plucker, 2002; Gresalfi, Barab, & Sommerfeld, 2012). So construed, gifted learners are those who continually engage in an active, critical way of learning through which information is transformed and new insights into the world are achieved (Dai, 2012; Gee, 2007; Perkins, 2009). Learning in this sense is not merely preparation for creativity. Learning is a way of keeping an innovative edge.

One question that has long plagued research on learning and creativity and divided researchers is how to deal with the fact that school-age children are in the process of developing their knowledge and skills and may not have the cognitive infrastructure to build their creative representations and thoughts on important matters (e.g., Kirschner, Sweller, & Clark, 2006). In gifted education, the problem is presented as a gap between what Renzulli (1986) called two kinds of giftedness: "knowing a lot" through textbooks and secondary sources in childhood on the one hand, and creative productivity in adulthood on the other. Schwartz and Bransford (1998; Bransford & Schwartz, 1999) built on Hatano's work and proposed efficiency and innovation as two dimensions of adaptive expertise. Seeking innovation without efficiency, one will end up as a frustrated novice; achieving efficiency without an innovative spirit, one will end up as a routine expert. They argue that there is an "optimal adaptivity corridor" leading to adaptive expertise (Bransford et al., 2006, p. 27). Building a curriculum that balances the acts of building efficiency while developing an innovative edge is a fundamental task for gifted education that can ultimately fill in the gap between two kinds of giftedness (Renzulli, 1986).

Another way of "naturalizing" creativity is to highlight personal creativity as ubiquitous to human beings when they are allowed to freely choose and develop their repertoire of knowledge, skills, and values (Runco, 2010; see also Beghetto & Kaufman, 2010; Collins & Halverson, 2009). An advantage of this conception for school-age populations is that we emphasize demonstration and nurturance of creative potential and agency, rather than mature creative "performance" or "product." Ultimately, little "c" in the form of personal knowledge (Polanyi, 1958) is a primary source of eminent adult creative productivity (big "C"; Csikszentmihalyi, 1996). New advances in technology clearly enhance the opportunity for developing personal creativity, when student learning is less dictated by a fixed school curriculum and more "customized" based on individuals' strengths and interests (Collins & Halverson, 2009). Many digitally engendered or -enhanced resources, platforms, and tools can be used to help children think more creatively and participate in knowledge creation (Craft, 2010). This new vision represents a new direction gifted education can take, from serving "special needs" of advanced students beyond the confines of regular classroom, as it currently stands, to building a personal (and sometimes social) network of ideas, skills, values, and worldviews for advanced students to enhance their personal creativity.

Major contributions

Torrance (1963) and Renzulli (1977) foretold the "naturalization" of creativity. But the recent work has explicated many principles and avenues to realizing creative potential. It takes two forms, to leverage the power of a community in building new understandings and cognitive apparatus, and to develop personal creativity by carving one's own niche. Assessment of creativity is taking an increasingly flexible approach, tracking processes rather than merely gauging products, moving away from parametric assumptions of individual differences to contextualized diagnosis of progress and shortfalls (e.g., Shute & Kim, 2012). Instead of using individuals as a unit of analysis, dialogic interactions and collaborative discourses become an empirical basis for assessing creative dynamics (Sawyer, 2006a). Gifted education apparently needs to reposition itself in the midst of these fundamental changes.

Prospects of interactions between gifted education and innovation education

If the purpose of innovation education is identifying, developing, and transforming child talent into adult innovation (Shavinina, this volume), gifted education should be the central part of this endeavor (see Delcourt & Renzulli, this volume; Fox, this volume; VanTassel-Baska, this volume). As I suggest in the previous sections, explorations in gifted education and innovation education have a history of reciprocation given their overlapping concerns. Gifted education, as conventionally defined, is concerned with those students who demonstrate unusual capability and potential, such that their educational needs are typically not well met within the regular classroom (Marland, 1972; Ross, 1993). Defined as such, gifted education shares the concerns of turning talents into creative use with innovation education, while having the extra tasks of identifying the *most promising* youth for interventions. In the following section, I attempt to identify areas in which the fruitful reciprocation can continue.

A life-span perspective on the development of creativity and innovation

A common trend in gifted education and innovation education, as Shavinina (this volume) defines it, is to take a life-span developmental perspective (Matthews, 2009). This means that we see high-level creativity demonstrated in the production of novel and valuable ideas and tangible artifacts not as a characteristic possessed by few individuals but as the outcome of a prolonged developmental process, involving a unique formation of knowledge, skills, dispositions, and values vis-à-vis a particular line of work. The traditional, psychometric view of intelligence and creativity can be recast in this developmental framework as indicative of the probability with which some may be more inclined than others to engage in novel ways of thinking and doing that prove valuable. For example, conceptions of fluid and crystallized intelligence (Cattell, 1971), of fluid analogizing (Geake, 2008) are still meaningful for understanding individual differences in their creative potential and inclinations. The tradition of understanding creativity as underpinned by particular cognitive structures and processes can also be incorporated into a developmental account. For example, in Finke et al.'s (1992) Geneplore Model of creative cognition, generating new possibilities entails pre-inventive cognitive structures (i.e., unique organization of knowledge); explorations of these possibilities involve evaluation and decision on the most promising avenues. These cognitive processes leading to real life creativity last for months and years, even decades (e.g., Charles Darwin), and are best characterized as developmental in nature in that the cognitive system undergoes changes in itself while acting upon a particular aspect of the world. Finally the tradition of motivation research would find its own niche in contributing to a life-span developmental account by explicating what propels some individuals to pursue their unique visions, take calculated risks, and stay at the edge of chaos (Dai & Renzulli, 2008).

From this life-span perspective, what can educators do to make changes that respond to the need for developing creative talents and an innovative ethos in education? In the following section, I propose four imperatives as an agenda for innovation education.

Identifying and cultivating creative potential: the curriculum imperative

Naturalization of creativity means that creative agency can manifest itself everywhere in every aspect of human life. A life-span perspective on education for innovation would naturally pay attention to those transitions, from childhood to adolescence, and from adolescence to adulthood, where both challenges and opportunities are present for creativity. Individuals' creative potential is enhanced when differential development in terms of various trajectories, pathways, and niches are encouraged and supported. The rationale for differential development is that when individual strengths and interests are identified and supported, the end result is more cognitive diversity in the talent pool, hence the better chance for innovation. In that regard, a one-size-fits-all education coupled with rigid, uniform evaluation standards is detrimental to the development of creativity and ultimately innovations in practical settings.

Does gifted education have a distinct place in innovation education? Are there individual differences in creative potential? The answer is definitively yes. There are prerequisites for pursuing a particular creative career leading to eminent contributions (Ackerman, 2003; Lubinski, Webb, Morelock, & Benbow, 2004). One cannot be a creative engineer without a strong mathematics background, just as one cannot become an eminent composer without a strong foundation of musical knowledge and skills. However, creative potential is not unitary and can take many forms and shades. Even within the same domain, creativity may entail different sets of skills and penchants. Thus classic music composers differ from jazz musicians, and molecular biologists differ from more "naturalistic" biologists. Because of this diversity, generic divergent thinking tests cannot capture a unique aspect of creative potential, just like generic intelligence tests cannot capture specific intellectual propensities. Therefore, educators are better off when identifying high creative potential through student performance in a particular domain, be it a traditional school subject (e.g., mathematics) or non-traditional topic (robotics or computer animation). Selectivity and high-level excellence are what distinguishes gifted education from general education (Dai, 2010). But that does not negate a common vision for all educators: to look for a unique combination of knowledge, skills, dispositions, and values in individual students that eventually leads to distinct "personal knowledge" (Polanyi, 1958), and unique representations or visions of the world (Shavinina, 2009), which is the foundation for innovation.

Playing the whole game: the pedagogical imperative

As alluded earlier, there are two schools of thought on how to nurture creativity; one believes that creative processes are discrete and teachable, and should be explicitly taught. The other believes that creativity is a by-product of inquiry and action, and can only be nurtured in a holistic manner, Feldhusen's (Feldhusen & Kollof, 1986) and Renzulli's (1977) enrichment programs incorporate both features (engaging students in authentic activities, and provide some structural guidance as needed). There is an increasing realization in cognitive psychology that how to think (process) can never be completely separated from what to think about (content) (Anderson, 1987); in other words, it is erroneous to treat "higher-order thinking" as a separate process that can be brought to bear upon whatever content one is dealing with. On the other hand, higher-order thinking can be deliberately engaged through instructional guidance if the learning goal is deep understanding. Scardamalia and Bereiter's (2006) "knowledge building" engages learning and thinking that is by nature generative (i.e., creative). Schwartz and Bransford (1998) provide ways of determining when to give students full freedom to explore and "when to tell" in building adaptive expertise. Lehrer and Schauble (2006) suggest engaging students in model-based reasoning and imagination as a way of building deep understandings. It should be pointed out that cognitive modeling is fundamental for creativity, whether it takes the form of analogical thinking (Holyoak & Thagard, 1995) or embodied mental simulation (Barsalou, 2003). In addition, a certain dose of critical thinking is always present in order to generate ideas and solutions that are novel and valuable (i.e., creativity) (Fairweather & Cramond, 2010; Langer, 2012). Consequently, the best metaphor for a pedagogy of creativity is that of "playing the whole game" (Perkins, 2009). Any distinct domain of human practice is a "game" with a particular design; learning for creativity means learning how the game is designed to achieved its goals, how to play the game (not merely learning about the game), and how the game might be improved by modifying its components or design features (Gee, 2007; cf. Sawyer, 2010). This kind of generative learning involves gaining new perspectives on a game, building instruments for tackling problems the game presents, and reflecting on how the game should be played (Dai, 2012). The process is fundamentally social in that participants who play the same game share information and build on each other's ideas (Gee, 2007; Zhang, 2012).

Would the whole game be played differently for gifted and talented students, given that they typically achieve automaticity faster, develop deeper insights into the issue at hand, and taking more promising avenues with less instructional guidance (Borkowski & Peck, 1986; Kanevsky, 1990; Steiner, 2006)? In other words, does Aptitude-Treatment Interaction (ATI) (Cronbach & Snow, 1977) matter? A strong version of the ATI argument states that students with Aptitude A would be best served when matched with Treatment A, but not Treatments B or C; thus, those who prefer a more creative way of learning would benefit most from a creative style of teaching (Grigorenko & Sternberg, 1997). A weak version of the ATI argument states that while certain pedagogical strategies would work for most students, they carry value-added benefits for gifted and talented students. For example, there is a plenty of evidence that inquiry-based learning can benefit all students (Hmelo-Silver, Duncan, & Chinn, 2007). The Integrated Curriculum Model developed by VanTassel-Baska and her colleagues (VanTassel-Baska, 1986; VanTassel-Baska & Brown, 2007) leverages three cognitive components in enhancing high-end learning: (a) advanced content, (b) high-level process and product work, and (c) intra- and interdisciplinary concept development and understanding. Such a model is not exclusively applicable to gifted students, but gifted students clearly stand to gain more from such an inquiry-based curriculum (Aulls & Shore, 2008; see also Ceci & Papierno, 2005). Thus, there is no "gifted pedagogy," a pedagogy suitable only for gifted and talented learners (Tomlinson, 1996), but different doses of instructional guidance may be needed depending on how much support is needed for the learner to advance to the next level of competence (Vygotsky, 1978). Knowing that gifted and talented students are often autodidactic (i.e., self-taught and self-educated in skill development), what Collins and Halverson (2009) prescribed as the 21st century learning, customization, learner control, and interactivity, may be especially beneficial to gifted learners. On the other hand, because of the more advanced levels which gifted learners characteristically attain, it is also crucial that mentorship experiences with experts in a field be provided (Grassinger, Porath, & Ziegler, 2010) so that the learners will develop deep insights into a domain of human endeavor and social practice.

Developing resources, tools, and support for "junior version" innovations: the capacity building imperative

Although it is not always feasible for school-age students to engage in "real play" in work settings when it comes to exercising their newly acquired knowledge and skills, according to Perkins

(2009), we can always design some "junior versions" of a game so that students will get a feel of what the game is about and how to play it. Lave and Wenger (1991) aptly called it "legitimate peripheral participation." How to develop resources, tools, and support systems to enable students to play the whole game is a practical as well as intellectual challenge. At the conceptual level, what makes a "whole game?" Can we define its boundary, or can one move from one "game" to another? What are the essential elements for such learning, or, more accurately, an apprenticeship game? Gee (2007) drew inspirations from the video gaming community. Sawyer (2003, 2010) got clues from jazz musicians and artists at work. In general, of course, we can roughly equate "playing the whole game" with inquiry and project-based learning (Hmelo-Silver et al., 2007; Krajcik & Blumenfeld, 2006), in which students have to tackle real world problems (at least their "junior" versions) similar to what scientists, artists, and scholars, other professionals deal with in work settings.

Conceptualizing learning for creativity is one thing; practically making it happen is another. It seems essential to build an infrastructure of support for playing the whole game, with resources earmarked, tools available, and coaches on the side. Digital technologies clearly have created new possibilities for this endeavor (e.g., Barab, Gresalfi, & Ingram-Goble, 2010). What practical constraints on the part of students as well as the learning environment are involved in "playing the whole game" effectively needs to be fully understood before we can put to test the proposition that this is a viable approach to enhancing creativity. How can practical innovations in gifted education contribute to the capacity building effort for playing the whole game in general? One example is Renzulli Learning System (RLS), a fully integrated technological system for creating and managing inquiry and creative activities for primary and secondary school students (Renzulli & Reis, 2009). The system begins by providing a computer-generated profile of each student's academic strengths, interests, and learning styles. A search engine then *matches* internet resources to the student's profile by subject area, grade level, state curricular standards, and degree of complexity. Then a management system called the Wizard Project Maker guides students in the *application* of knowledge to teacher- or student-selected assignments, independent research studies, or creative projects that individuals or small groups would like to pursue. Although the system is designed to help teachers differentiate curriculum for gifted students, it can be used to engage all willing students to engage in projects chosen by themselves or suggested by teachers. There is much to learn about the support system for developing creativity and innovation. For instance, each year more than 1,500 high school students nationwide in the US apply for Intel Science Talent Search program (www.intel.com/about/corporateresponsibility/education/sts/index.htm). The basis for selecting finalists and final winners is the products of their scientific projects. It would be highly instructive to know how what kind of resources, tools, and technical and social support systems were behind these undertakings, and what makes some efforts more successful than others.

Developing a taxonomy of creativity-innovation enhancement: the assessment imperative

Finally, in order to have effective modules or curricula of innovation education, we need a system of assessment that is conceptually sound and technically reliable in gauging the effectiveness of educational interventions. In the past assessment of creativity was plagued by the confusion surrounding how to define creativity. Some researchers adhere to product-based criteria (Csikszentmihalyi, 1996) and others advocate use of person-centered criteria (Runco, 2010). Fortunately, the field has come to realize that creativity and its practical incarnation, innovation, can take many forms and shades. Indeed it is meaningful to talk about different magnitude and degrees of creativity (Beghetto & Kaufman, 2010). Furthermore, in the spirit of naturalizing creativity, it is no longer tenable to treat creativity as a personal characteristic; rather, whether social and pedagogical conditions are conducive to creative thinking and expression of creative agency should also be assessed.

In developing such an assessment system for creativity-innovation enhancement, several conceptual issues need to be considered. First, the age-old person-process-product distinction should be honored. The focus on creativity enhancement can be personal qualities (based on dispositions and characteristics; see Runco, 2010), optimal conditions and processes (based on some critical aspects of the creative process; see Treffinger & Isaksen, 2005), and product qualities (based on professional standards; see Amabile, 1982). Second, the competence-performance distinction is also relevant to creativity enhancement. According to Vygotsky (1978), *performance before competence* is a developmental trajectory that needs to be captured in assessment; namely, initial fledgling performance may not show the kind of sophistication desired but may reflect a burgeoning creative talent nevertheless. Third, an assessment system needs to distinguish domain-specific and domain-general characteristics. Personal attributes can be domain-general (e.g., broad interests, polymaths, propensity for risk-taking), but creative products (innovations) are always domain-specific. If the main goal is to improve products, domain-specific criteria are more important. If, on the other hand, the purpose is to arouse the creative spirit and create an ethos of exploring knowledge frontiers (e.g., building a knowledge creation community; Scardamalia & Bereiter, 2006), then an exclusive product focus in assessment may be inappropriate. In general, a trajectory of increasing differentiation (i.e., becoming more domain-specific, more fine-tuned to task requirements and criteria) can be followed, whereby strengths and interests become increasingly focused, and technical proficiency and the value of the novelty in thinking and products more manifest. The logic is the same as the one used by Beghetto and Kaufman (2010), who argued that big-C (i.e., transformative creativity) always evolves from little-c (everyday creativity; Richards, 2007) and mini-c (informal, interpretive creativity), and there are many levels and degrees in-between (e.g., pro-c, standing for professional creativity, which characterize most of scholarly, artistic, and practical innovations). As big-C is rare and often beyond what is "educable," the assessment for creative enhance may target pro-c (professional creativity).

Technically, creativity and innovation is likely multidimensional, rather than psychometrically unitary. The traditional method of indexing creativity or creative potential using a single number might have to give place to a more complex analysis of performance and behavior, similar to cognitive diagnostic assessment (Leighton & Gierl, 2007). Modeling adaptive expertise in problem solving and reasoning provides a good benchmark for such diagnosis (Shute & Kim, 2012). When dynamic problem solving is assessed, non-obtrusive methods need to be used. Shute (2011) uses stealth assessment in computer game play situations, an assessment innovation that is potentially capable of capturing "teachable moments" than otherwise possible (see also Gee & Shaffer, 2010). Measurement of creative potential and even "creative giftedness" may rely on an appropriate assessment of the product, but understanding of how it comes about will rely fundamentally on investigating the underlying process. The primary goal of developing such an assessment system for innovation education is to assess and access creativity in person, process, and product (Feldhusen & Goh, 1995) so that the workforce we produce is capable of self-direction and innovative work in a knowledge economy. For gifted education, such a system would help identify (and sometimes select) high creative potential through their authentic performance for further advancement. The traditional notion of identifying the "creatively gifted" once and for all should be put to rest. There is no litmus test or shortcut for assessment, as creative potential evolves and changes, and indeed even dwindles if not nourished.

Summary and conclusion

In this chapter, I delineate the historical and theoretical connections between gifted education and innovation education. The changing focus since the 1950s, from personal attributes to cognitive and motivational processes to social dynamics, provides rich heuristics as to how human creative potential can be harnessed and nurtured. It is clear that person, process, content, and context are all important elements to reckon with in the equation of innovation education. What is worth noting is an increasing realization that creativity is a result of the natural human tendency to seek truth, optimality, and variation. Scholars and researchers in gifted education clearly have contributed many practical ideas as to how we can cultivate this creative potential through education. Current thinking puts more emphasis on personal creativity (little c) rather than epoch-making creativity (big C). This focus is appropriate for education, whose main charge is to lay a good foundation for talented students so as to increase the likelihood of creative productivity in their adulthood. For that purpose, I suggest four educational imperatives on curriculum, pedagogy, capacity building, and assessment, respectively, which can put creativity and innovation back to the educational landscape as a priority. In an age of accountability, how well a school protects and nurtures students' creative potential should be a main criterion for judging its efficacy, as the vitality of the new generation is at stake.

References

Ackerman, P. L. (2003). Aptitude complexes and trait complexes. *Educational Psychologist, 38*, 85–93.
Amabile, T. M. (1982). Social psychology of creativity: A consensual assessment technique. *Journal of Personality and Social Psychology, 43*, 997–1013.
Amabile, T. M. (1983). *The Social Psychology of Creativity*. New York: Springer-Verlag.
Anderson, J. R. (1987). Skill acquisition: Compilation of weak-method problem situations. *Psychological Review, 94*, 192–210.
Aulls, M. W., & Shore, B. M. (2008). *Inquiry in Education: The Conceptual Foundations for Research as a Curricular Imperative*. New York: Erlbaum.
Barab, S. A., Gresalfi, M., & Ingram-Goble, A. (2010). Transformational play: Using games to position person, content, and context. *Educational Researcher, 39*, 525–536.
Barab, S. A., & Plucker, J. A. (2002). Smart people or smart context? Cognition, ability, and talent development in an age of situated approaches to knowing and learning. *Educational Psychologist, 37*, 165–182.
Barsalou, L. W. (2003). Situated simulation in the human conceptual system. *Language and Cognitive Processes, 18*, 513–562.
Beghetto, R. A., & Kaufman, J. C. (2010). Broadening conceptions of creativity in the classroom. In R. A. Beghetto & J. C. Kaufman (Eds.), *Nurturing Creativity in the Classroom* (pp. 191–205). Cambridge, UK: Cambridge University Press.
Borkowski, J. G., & Peck, V. A. (1986). Causes and consequences of metamemory in gifted children. In R. J. Sternberg & J. E. Davidson (Eds.), *Conceptions of Giftedness* (pp. 182–200). Cambridge, UK: Cambridge University Press.
Bransford, J. D., Barron, B., Pea, R. D., Meltzoff, A., Kuhl, P., Bell, P., et al. (2006). Foundations and opportunities for an interdisciplinary science of learning. In R. K. Sawyer (Ed.), *The Cambridge Handbook of the Learning Sciences* (pp. 19–34). Cambridge, UK: Cambridge University Press.
Bransford, J., & Schwartz, D. L. (1999). Rethinking transfer. A simple proposal with multiple implications. In A. Iran-Nejad & P. D. Pearson (Eds.), *Review of Research in Education*, vol. 24 (pp. 61–100). Washington, DC: American Psychological Association.
Bruner, J. S. (1960). *The Process of Education*. Cambridge, MA: Harvard University Press.
Cattell, R. B. (1971). *Abilities: Their Structure, Growth, and Action*. Boston: Houghton Mifflin.
Ceci, S. J., & Papierno, P. B. (2005). The rhetoric and reality of gap closing: When the "have-nots" gain but the "haves" gain even more. *American Pshchologist, 60*, 149–160.
Collins, A. M., & Halverson, R. (2009). *Rethinking Education in the Age of Technology*. New York: Teachers College Press.
Craft, A. (2010). *Creativity and Education Futures: Learning in a Digital Age*. Sterling, VA: Trentham Books.

Cronbach, L. J., & Snow, R. E. (1977). *Aptitudes and Instructional Methods: A Handbook for Research on Interactions*. New York: Irvington.
Csikszentmihalyi, M. (1996). *Creativity: Flow and the Psychology of Discovery and Invention*. New York: HarperCollins.
Dai, D. Y. (2010). *The Nature and Nurture of Giftedness: A New Framework for Understanding Gifted Education*. New York: Teachers College Press.
Dai, D. Y. (2012). From smart person to smart design: Cultivating intellectual potential and promoting intellectual growth through design research. In D. Y. Dai (Ed.), *Design Research on Learning and Thinking in Educational Settings: Enhancing Intellectual Growth and Functioning* (pp. 3–40). New York: Routledge.
Dai, D. Y., & Renzulli, R. S. (2008). Snowflakes, living systems, and the mystery of giftedness. *Gifted Child Quarterly, 52*, 114–130.
DeHaan, R. G., & Havighurst, R. J. (1957). *Educating the Gifted*. Chicago: University of Chicago Press.
Dunbar, K. (1997). How scientists think: On-line creativity and conceptual change in science. In T. B. Ward, S. M. Smith, & J. Vaid (Eds.), *Creative Thought: An Investigation of Conceptual Structures and Processes* (pp. 461–493). Washington, DC: American Psychological Association.
Estrin, J. (2009). *Closing the Innovation Gap: Reigniting the Spark of Creativity in a Global Economy*. New York: McGraw-Hill.
Fairweather, E., & Cramond, B. (2010). Infusing creative and critical thinking into the curriculum together. In R. A. Beghetto & J. C. Kaufman (Eds.), *Nurturing Creativity in the Classroom* (pp. 113–141). Cambridge, UK: Cambridge University Press.
Feldhusen, J. F., & Goh, B. E. (1995). Assessing and accessing creativity: An integrative review of theory, research, and development. *Creativity Research Journal, 18*, 231–247.
Feldhusen, J. F., & Kolloff, M. B. (1986). The Purdue three-stage model for gifted education. In R. S. Renzulli (Ed.), *Systems and Models for Developing Programs for the Gifted and Talented* (pp. 126–152). Mansfield Center, CT: Creative Learning Press.
Feldhusen, J. F., & Treffinger, D. J. (1986). *Creative Thinking and Problem Solving in Gifted Education*. Dubuque, IA: Kendall/Hunt Publishing Company.
Finke, R. A., Ward, T. B., & Smith, S. M. (1992). *Creative Cognition: Theory, Research, and Applications*. Cambridge, MA: The MIT Press.
Gagné, F. (2005). From gifts to talents: The DMGT as a developmental model. In R. J. Sternberg & J. E. Davidson (Eds.), *Conceptions of Giftedness*, 2nd ed., (pp. 98–119). Cambridge, UK: Cambridge University Press.
Gallagher, J. J. (1975). *Teaching the Gifted Child*, 2nd ed., Boston: Allyn & Bacon.
Geake, J. G. (2008). High abilities at fluid analogizing: A cognitive neuroscience construct of giftedness. *Roeper Review, 30*, 187–195.
Gee, J. P. (2007). *What Video Games have to Teach Us About Learning and Literacy*. New York: Palgrave/Macmillan.
Gee, J. P., & Shaffer, D. W. (2010). Looking where the light is bad: Video games and the future of assessment. *Edge, 6*, 3–19.
Getzels, J. W., & Csikszentmihalyi, M. (1976). *Creative Vision: A Longitudinal Study of Problem Finding in Art*. New York: Wiley.
Getzels, J. W., & Jackson, P. W. (1962). *Creativity and Intelligence: Explorations with Gifted Students*. New York: Wiley.
Grassinger, R., Porath, M., & Ziegler, A. (2010). Mentoring the gifted: A conceptual analysis. *High Ability Studies, 21*, 27–46.
Gresalfi, M., Barab, S. A., & Sommerfeld, A. (2012). Intelligent action as a shared accomplishment. In D. Y. Dai (Ed.), *Design Research on Learning and Thinking in Educational Settings: Enhancing Intellectual Growth and Functioning* (pp. 41–64). New York: Routledge.
Grigorenko, E. L., & Sternberg, R. J. (1997). Styles of thinking, abilities, and academic performance. *Exceptional Children, 63*, 295–312.
Guilford, J. P. (1950). Creativity. *American Psychologist, 5*, 444–454.
Guilford, J. P. (1967). *The Nature of Human Intelligence*. New York: McGraw-Hill.
Halpern, D. F. (2008). Is intelligence critical thinking? Why we need a new definition of intelligence. In P. C. Kyllonen, R. D. Roberts, & L. Stankov (Eds.), *Extending Intelligence: Enhancement and New Constructs* (pp. 293–310). New York: Routledge.
Hmelo-Silver, C., Duncan, R. G., & Chinn, C. A. (2007). Scaffolding and achievement in problem-based and inquiry learning: A response to Kirschner, Sweller, and Clark (2006). *Educational Psychologist, 42*, 99–107.
Holyoak, K. J., & Thagard, P. (1995). *Mental Leaps: Analogy in Creative Thought*. Cambridge: MA: The MIT Press.

Kanevsky, L. (1990). Pursuing qualitative differences in the flexible use of problem-solving strategy by young children. *Journal for the Education of the Gifted, 13*, 115–140.

Kirschner, P. A., Sweller, J., & Clark, R. E. (2006). Why minimal guidance during instruction does not work: An analysis of the failure of constructivist, discovery, problem-based, experiential, and inquiry-based teaching. *Educational Psychologist, 41*, 75–86.

Krajcik, J. S., & Blumenfeld, P. C. (2006). Project-based learning. In R. K. Sawyer (Ed.), *The Cambridge Handbook of the Learning Sciences* (pp. 317–333). Cambridge, UK: Cambridge University Press.

Langer, J. A. (2012). The interplay of creative and critical thinking in instruction. In D. Y. Dai (Ed.), *Design Research on Learning and Thinking in Educational Settings: Enhancing Intellectual Growth and Functioning* (pp. 65–82). New York: Routledge.

Lave, J., & Wenger, E. (1991). *Situated Learning: Legitimate Peripheral Participation*. Cambridge, UK: Cambridge University Press.

Lehrer, R., & Schauble, L. (2006). Cultivating model-based reasoning in science education. In R. K. Sawyer (Ed.), *The Cambridge Handbook of the Learning Sciences* (pp. 371–387). Cambridge, UK: Cambridge University Press.

Leighton, J., & Gierl, M. (Eds.) (2007). *Cognitive Diagnostic Assessment for Education: Theory and Applications*. Cambridge, UK: Cambridge University Press.

Lubinski, D., Webbs, R. M., Morelock, M. J., & Benbow, C. P. (2004). Top 1 in 10,000: A 10-year follow-up of the profoundly gifted. *Journal of Applied Psychology, 86*, 718–729.

MacKinnon, D. (1962). The nature and nurture of creative talent. *American Psychologist, 17*, 484–495.

MacKinnon, D. (1978). *In Search of Human Effectiveness: Identifying and Developing Creativity*. Buffalo, NY: Bearly.

Marland, S. P. (1972). *Education of the Gifted and Talented: Report to the Congress of the United States by the U.S. Commissioner of Education*. Washington, DC: Government Printing Office.

Matthews, D. J. (2009). Developmental transitions in giftedness and talent: Childhood to adolescence. In F. D. Horowitz, R. F. Subotnik, & D. J. Matthews (Eds.), *The Development of Giftedness and Talent Across the Life Span* (pp. 89–108). Washington, DC: American Psychological Association.

Newell, A., Shaw, J. C., & Simon, H. A. (1962). The process of creative thinking. In H. E. Gruber, G. Terrell, & M. Wertheimer (Eds.), *Contemporary Approaches to Creative Thinking* (pp. 65–66). New York: Atherton Press.

Newell, A., & Simon, H. A. (1972). *Human Problem Solving*. Englewood Cliffs, NJ: Prentice-Hall.

Partnership for 21st Century Skills. (2008). 21st century skills education and competitiveness guide. Retrieved online at www.p21.org/documents/21st_ century_skills_education_and_competitiveness_guide.pdf.

Perkins, D. N. (2009). *Making Learning Whole: How Seven Principles of Teaching Can Transform Education*. San Francisco: Jossey-Bass.

Polanyi, M. (1958). *Personal Knowledge: Toward a Post-Critical Philosophy*. Chicago: University of Chicago Press.

Renzulli, J. S. (1977). *The Enrichment Triad Model: A Guide for Developing Defensive Programs for the Gifted and Talented*. Mansfield Center, CT: Creative Learning Press.

Renzulli, J. S. (1978). What makes giftedness? Re-examining a definition. *Phi Delta Kappan, 60*, 180–184, 261.

Renzulli, J. S. (1986). The three-ring conception of giftedness: A developmental model for creative productivity. In R. J. Sternberg & J. E. Davidson (Eds.), *Conceptions of Giftedness* (pp. 53–92). Cambridge, UK: Cambridge University Press.

Renzulli, J. S. (2005). The three-ring conception of giftedness: A developmental model for promoting creative productivity. In R. J. Sternberg & J. E. Davidson (Eds.), *Conceptions of Giftedness*, 2nd ed., (pp. 98–119). Cambridge, UK: Cambridge University Press.

Renzulli, J. S., & Callahan, C. M. (1973). *New Directions in Creativity*. New York: Harper & Row.

Renzulli, J. S., & Dai, D. Y. (2001). Abilities, interests, and styles as aptitude for learning: A person-situation perspective. In R. J. Sternberg & L. F. Zhang (Eds.), *Perspectives on Learning, Thinking, and Cognitive Styles* (pp. 23–46). Mahwah, NJ: Lawrence Erlbaum.

Renzulli, J. S., & De Wet, C. F. (2010). Developing creative productivity in young people through the pursuit of ideal acts of learning. In R. A. Beghetto & J. C. Kaufman (Eds.), *Nurturing Creativity in the Classroom* (pp. 24–72). Cambridge, UK: Cambridge University Press.

Renzulli, J. S., & Reis, S. M. (1997). *Schoolwide Enrichment Model: A How-to Guide for Educational Excellence*. Mansfield Center, CT: Creative Learning Press.

Renzulli, J. S., & Reis, S. M. (2009). A technology-based application of the Schoolwide Enrichment Model and high-end learning theory. In L. Shavinina (Ed.), *International Handbook on Giftedness* (pp. 1203–1223). New York: Springer.

Resnick, L. B. (2010). Nested learning systems for the thinking curriculum. *Educational Researcher, 39*, 183–197.
Richards, R. (Ed.). (2007). *Everyday Creativity and New Views of Human Nature*. Washington, DC: American Psychological Association.
Ross, P. O. (1993). *National Excellence: A Case for Developing America's Talent*. Washington, DC: US Government Printing Office.
Runco, M. (2010). Education based on a parsimonious theory of creativity. In R. A. Beghetto & J. C. Kaufman (Eds.), *Nurturing Creativity in the Classroom* (pp. 235–251). Cambridge, UK: Cambridge University Press.
Sawyer, R. K. (2003). *Improved Dialogues: Emergence and Creativity in Conversation*. Westport, CT: Greenwood.
Sawyer, R. K. (2006a). Analyzing collaborative discourse. In R. K. Sawyer (Ed.), *The Cambridge Handbook of the Learning Sciences* (pp. 187–204). Cambridge, UK: Cambridge University Press.
Sawyer, R. K. (2006b). *Explaining Creativity: The Science of Human Innovation*. Oxford, UK: Oxford University Press.
Sawyer, R. K. (2010). Learning for creativity. In R. A. Beghetto & J. C. Kaufman (Eds.), *Nurturing Creativity in the Classroom* (pp. 172–190). Cambridge, UK: Cambridge University Press.
Scardamalia, M., & Bereiter, C. (2006). Knowledge building: Theory, pedagogy, and technology. In R. K. Sawyer (Ed.), *The Cambridge Handbook of the Learning Sciences* (pp. 97–115). Cambridge, UK: Cambridge University Press.
Schwartz, D. L., & Bransford, J. (1998). A time for telling. *Cognition and Instruction, 16*, 475–522.
Shavinina, L. (2009). A unique type of representation is the essence of giftedness: Toward a cognitive-developmental theory. In L. Shavinina (Ed.), *International Handbook on Giftedness* (pp. 231–257). New York: Springer.
Shute, V. J. (2011). Stealth assessment in computer-based games to support learning. In S. Tobias & J. D. Fletcher (Eds.), *Computer Games and Instruction* (pp. 503–524). Greenwich, CT: Information Age Publishing.
Shute, V. J., & Kim, J. Y. (2012). Does playing the World of Goo facilitate learning? In D. Y. Dai (Ed.), *Design Research on Learning and Thinking in Educational Settings: Enhancing Intellectual Growth and Functioning* (pp. 243–267). New York: Routledge.
Steiner, H. H. (2006). A microgenetic analysis of strategic variability in gifted and average-ability children. *Gifted Child Quarterly, 50*, 62–74.
Tannenbaum, A. J. (1997). The meaning and making of giftedness. In N. Colangelo & G. A. Davis (Eds.), *Handbook of Gifted Education*, 2nd ed., (pp. 27–42). Boston: Allyn & Bacon.
Terman, L. M. (1925). *Genetic Studies of Genius: Vol. 1, Mental and Physical Traits of a Thousand Gifted Children*. Standford, CA: Stanford University Press.
Tomlinson, C. A. (1996). Good teaching for one and all: Does gifted education have an instructional identity? *Journal for the Education of the Gifted, 20*, 155–174.
Tomlinson, C. A., & Callahan, C. M. (1992). Contributions of gifted education to general education in a time of change. *Gifted Child Quarterly, 36*, 183–189.
Torrance, E. P. (1963). *Education and the Creative Potential*. Minneapolis, MN: University of Minnesota Press.
Torrance, E. P. (1970). *Encouraging Creativity in the Classroom*. Dubuque, IA: Wm. C. Brown Company.
Torrance, E. P. (1972). Predictive validity of the Torrance Tests of Creative Thinking. *Journal of Creative Behavior, 6*, 236–252.
Toynbee, A. (1964). Is America neglecting her creative minority? In C. W. Taylor (Ed.), *Widening Horizons in Creativity* (pp. 3–9). New York: Wiley.
Treffinger, D. J., & Isaksen, S. G. (2005). Creative problem solving: The history, development, and implications for gifted education and talent development. *Gifted Child Quarterly, 49*, 342–353.
VanTassel-Baska, J. (1986). Effective curriculum and instruction models for talented students. *Gifted Child Quarterly, 30*, 164–169.
VanTassel-Baska, J., & Brown, E. (2007). Toward best practice: An analysis of the efficacy of curriculum models in gifted education. *Gifted Child Quarterly, 51*, 342–358.
Vygotsky, L. S. (1978). *Mind in Society: The Development of Higher Psychological Processes*. Cambridge, MA: Harvard University Press.
Weisberg, R. W. (1999). Creativity and knowledge: A challenge to theories. In R. J. Sternberg (Ed.), *Handbook of Creativity* (pp. 226–250). Cambridge, UK: Cambridge University Press.
Zhang, J. (2012). Designing adaptive collaboration structures for advancing the community's knowledge. In D. Y. Dai (Ed.), *Design Research on Learning and Thinking in Educational Settings: Enhancing Intellectual Growth and Functioning* (pp. 201–224). New York: Routledge.

5

Innovation education meets conceptual change research

Conceptual analysis and instructional implications[1]

Stella Vosniadou and Panagiotis Kampylis

NATIONAL AND KAPODISTRIAN UNIVERSITY OF ATHENS, GREECE, AND INSTITUTE FOR PROSPECTIVE TECHNICOLOGICAL STUDIES (IPTS), EUROPEAN COMMISSION, SPAIN

Summary: We argue that science education designed to promote conceptual change can play an important role in an overall program of innovation education. Teaching science from a conceptual change point of view requires the development of explicit, reflective and metacognitively guided knowledge construction and the cultivation of learning strategies for the deliberate reorganization of knowledge, such as analogy making and model-based reasoning. We argue that the creation of a classroom environment that fosters conceptual change activities in the process of learning science can help create in students a 'conceptual change know-how schema' which can generalize to other situations supporting knowledge-revision processes and the creation of new ideas and products.

Key words: Conceptual change, innovation education, science education, metaconceptual awareness, instructional analogies, model-based reasoning.

Introduction

Conceptual change research investigates knowledge acquisition processes in situations where the new, to-be-learned, information is very different from learners' prior beliefs and requires the creation of new concepts (Hatano & Inagaki, 2003). Research on conceptual change has emerged in recent years as an important area in educational psychology. Conceptual change instruction attempts to promote metaconceptual awareness and intentional learning and to teach students the cognitive mechanisms that can help them engage in deliberate knowledge restructuring. In this chapter we focus on science education and argue that when science is taught from a conceptual change perspective it has the potential to promote the formation of new concepts and to develop future innovators in science and technology. The pages that follow start with a clarification of what conceptual change research is and does. We continue to argue that similar mental processes are involved in scientific discovery and technological innovation and that they both require conceptual change. In the last section we describe the basic elements of a conceptual change approach to teaching science and explain how this type of science education can play an important role in an overall program of innovation education.

Conceptual change research – definitional issues

Concepts are not static. They change with learning and development in students but also in situations of scientific and technological development through the practices of scientists and inventors. There can be many kinds of conceptual changes, from the small and mundane which may involve the addition of an instance to an existing concept, to the most radical, which may involve changes in theory and the generation of new concepts. We will use the term *conceptual change* to refer to the latter, more radical types of changes. Conceptual change can be the outcome of scientific discovery and technological innovation, but can also be the product of teaching, as, for example, in situations that involve the learning of science.

The problem of conceptual change became first apparent by philosophers and historians of science in their attempts to explain how scientific theories change. According to Kuhn (1970) normal science operates within sets of shared beliefs, assumptions, commitments and practices that constitute 'paradigms'. Discoveries emerge over time that cannot be accommodated within the existing paradigm. When these anomalies accumulate, science enters a period of crisis which is eventually resolved by a revolutionary change in paradigm. Many scientific revolutions, such as the Newtonian theory in physics, the Copernican theory in astronomy and the Darwinian theory in biology can be seen as the products of radical conceptual change. In these cases new theories are generated to explain known and new phenomena and how new concepts are formed (Thagard, 1992).

Ideas about conceptual change from the history and philosophy of science were soon brought to developmental psychology through the work of Susan Carey (1985) and to science education through the work of Michael Posner and his colleagues (Posner, Strike, Hewson, & Gertzog, 1982). By the late 1970s it had become apparent that students bring to the science education task alternative frameworks, preconceptions or misconceptions some of which are rather robust and difficult to extinguish through teaching (e.g., White & Gunstone, 2008). In some cases these alternative frameworks appeared to be similar to earlier theories in the history of science, such as, for example, the impetus theory[2] in mechanics (McCloskey, 1983). Posner et al. (1982) drew an analogy between the concepts of normal science and scientific revolution offered by philosophers of science such as Kuhn (1970) and Piaget's (1970) concepts of assimilation and accommodation, and derived from this analogy an instructional theory to promote 'accommodation' in students' learning of science. According to Posner et al. (1982) students need to undergo a radical conceptual change when it comes to learning scientific concepts like *force*, *heat* and *energy*.

Over the years, a significant body of research emerged to investigate the processes of conceptual change, the learning mechanisms involved in the generation of new concepts and the instructional strategies that can promote it. The theoretical and methodological discussions that have taken place in this process have been some of the most interesting in the field of learning and instruction, raising important questions about the nature of knowledge, its organization and its revision. Although the beginnings of conceptual change research can be traced to scientific discovery in physics and physics education, this research is by no means restricted to physics but makes a larger claim about learning that transcends many domains of knowledge and can apply, for example, to biology (Inagaki & Hatano, 2002), psychology (Wellman, 2002), history (Leinhardt & Ravi, 2008), political science (Voss & Wiley, 2006), medicine (Kaufman, Keselman, & Patel, 2008), environmental learning (Rickinson, Lundholm, & Hopwood, 2009) and mathematics (Vosniadou & Verschaffel, 2004).

Some researchers are not persuaded that there is a need to distinguish 'conceptual change' processes from learning in general. We argue, however, that while conceptual change is undeniably a form of learning, it is important to differentiate it from other types of learning because it requires fundamental changes in the content and organization of existing knowledge and hence the development of new learning mechanisms for deliberate knowledge restructuring and the generation of new concepts. Most learning is implicit and additive involving mainly the enrichment of prior knowledge. Conceptual change cannot however be achieved through the use of implicit, enrichment-types of learning mechanisms alone. In fact, the use of enrichment-types of mechanisms in situations that require conceptual change can often lead to the creation of misconceptions or, what we call, 'synthetic models' (Vosniadou, Vamvakoussi, & Skopeliti, 2008). For example, young children often interpret the information regarding the spherical shape of the Earth to mean that the Earth is circular but flat like a pancake or that the Earth is spherical but people live on flat ground inside it (Vosniadou & Brewer, 1992). These types of misconceptions are synthetic constructions which suggest that students are implicitly assimilating the new information regarding the spherical shape of the Earth into their intuitive model of a flat Earth. Similarly, erroneous strategies used by students in mathematics, such as the common mistake that $1/3 + 1/3 = 2/6$ (National Mathematics Advisory Panel, 2008), reveal the implicit interference of natural number operations in fraction addition (see Vamvakoussi & Vosniadou, 2010, for many more examples).

It is therefore important to develop instruction to help students distinguish the instances of learning that require conceptual change from those that do not and alert them as to the use of the appropriate learning strategies. Even more importantly, instruction needs to develop in students the kind of intentional and metacognitively guided learning that will help them identify situations requiring conceptual change themselves. Finally, it is important to teach students new learning strategies that can be used for the conscious and deliberate reorganization of existing knowledge (Vosniadou, 2008; Vosniadou & Mason, 2012).

What is the relevance of conceptual change for technological innovation?

Innovation is usually defined as the implementation of creative ideas into new products, processes and services (Shavinina, 2003, 2009b). Innovation is closely linked to creativity with which it is often confused (Kahl, Fonseca, & Witte, 2009). For some scholars (e.g., West, 2002), innovation is a two-component, but non-linear process, encompassing the development of creative ideas as well as their application in practice. According to this view, creative ideas are the foundation on which innovation is built. In other words, creative thinking, personal and/or collective, is the starting point for innovation; it is a necessary but not sufficient condition. Moreover, although the term innovation usually characterizes the creation of new technological inventions, it can also apply to non-technological and social innovations in public or private services (e.g., European Commission, 2008).

Technological innovation and scientific discovery both involve the generation of new concepts. Some may argue that inventors are different from scientists in that they create new technologies to accomplish practical goals rather than discover new theoretical ideas. However, as Thagard and Croft (1999) have persuasively argued, the cognitive processes implicated in scientific discovery and technological innovation are fundamentally similar (see also Nersessian, 2008). According to Thagard and Croft (1999, p. 137), both scientific discovery and technological innovation, 'involve the generation of new concepts, although technological concepts usually apply to newly created devices rather than to discovered or postulated entities'. In both cases, the important and distinguishing characteristic of the discovery process is that it involves *conceptual change* (Thagard & Croft, 1999).

In their interesting article, Thagard and Croft (1999) analyze two examples, one of scientific discovery and one of technological innovation to explicate the common conceptual change processes involved. The case of scientific discovery is the bacterial theory of ulcers. The bacterial theory of ulcers was formulated by Warrant and Marschall who discovered a new bacterium in the stomachs of people with gastritis which they named *Helicobacter pylori*. This discovery led them to the hypothesis that ulcers may be caused by a bacterial infection rather than by stress or acidity. This in turn led them to propose that ulcers can be treated with antibiotics. Initially this controversial hypothesis was viewed with a great deal of skepticism. However, it is now widely accepted that antibiotics can be used in the treatment of many ulcers to eradicate the bacteria responsible for them.

The example of technological innovation used by Thagard and Croft (1999) is the construction of Java. Java, one of the most successful new programming languages, was created by James Gosling in 1991 as part of a project at Sun Microsystems. Its success is attributed mostly to the discovery of a new concept, which is called 'applet'. Java applets are programs that can be downloaded from the web. Java's innovation was that applets are platform independent and can be downloaded on to computers with different CPU chips and operating systems.

Based on the above analysis, as well as on more extensive studies that compared 200 cases of scientific discovery and technological innovation, Thagard and his colleagues concluded that they all involve the generation of new concepts, namely *conceptual change*. It follows that research on how to promote conceptual change through education should be relevant to innovation education. Indeed, the innovation research community has recognized the importance of investigating the development of scientific talent and recommendations have been made to fund programs that study these developmental processes as an important aspect of IE (Shavinina, this volume).

Innovation education

Theorists, scholars, researchers, policy-makers and educators deploy a range of claims about innovation education, which emerge from different theoretical and academic traditions, scientific fields, policy contexts and sociocultural perspectives. Research on IE can be separated into three categories: (i) gifted education, (ii) creativity training programs and (iii) case studies.

Gifted education programs

Shavinina (this volume, 2009a) describes eleven interrelated and overlapping components of IE that refer mainly to the identification and development of the gifted and talented. These include new gifted education programs, new programs for the development of entrepreneurial giftedness, programs to develop the metacognitive abilities of the gifted and their abilities to implement things – the so-called executive abilities – new programs to study the scientific talent of Nobel laureates, and so on. According to Shavinina (this volume), the proposed components of IE constitute the *know-what* part of IE and can be implemented through a cluster of ICT – the *know-how* part of IE – which she calls High Intellectual and Creative Educational Multimedia Technologies (HICEMTs). With their general and specific sets of characteristics – which are described in detail in Shavinina (this volume, 2009a) – HICEMTs seem to be a promising way to deliver IE because they provide students with exceptional opportunities for personalized, differentiated, and flexible learning and actualization of their own mental potential and the subsequent development of their intellectual and creative abilities.

Although an important part of IE centers on the development of those individuals who are exceptionally gifted and talented (Shavinina, 2009b), other perspectives emphasize the importance of IE addressed to all students. In fact, current sociocultural approaches to education argue that all humans with normal capacities are able to produce innovative ideas, at least in some domain (Sawyer, 2006) and therefore have the potential to become innovators. According to this argument, schools 'should prepare all students to participate in complex creative systems, in which they will need to work collaboratively, at multiple levels of organization, to build knowledge together' (Sawyer, 2006, p. 43).

Creativity training programs

Creativity training programs involve training in specific heuristics or strategies that help students achieve cognitive flexibility and facilitate the generation of new ideas. The effectiveness of these strategies for performing creative tasks leads to feelings of efficacy and motivates people to engage to creative work. Scott, Leritz, and Mumford (2004a, b) reviewed 156 creativity training programs and categorized them into four clusters. The first cluster is *idea production training*. It covers several forms of creativity training – such as situated idea production or computer-based idea production – that focus mainly on idea generation, divergent thinking, ideation, elaboration and brainstorming. The second cluster is *imagery training*. It is the most common type of training focusing on expressive activities and imaginative exercises. The third cluster, *cognitive training*, incorporates processing activities such as conceptual combination, analogies and metaphors. The fourth cluster is *thinking skills training*; it focuses on the critical aspects of thinking and incorporates processes such as problem finding, idea evaluation, solution monitoring, constraint identification and metacognition.

Case studies

The research on creativity training programs is of great importance and can provide valuable insights for innovation education content and procedures. According to Scott et al. (2004b, p. 383), however, creativity training should not be viewed merely as the implementation of a predetermined set of heuristics or strategies but rather as a flexible and holistic attempt to enhance creative thought and innovative action based on 'real-world' cases. This conclusion agrees with the suggestions of Thorsteinsson and Denton (2003) who presented one of the few implementations of IE in Icelandic schools, where it has become an independent area of curriculum.

The implemented model of IE is based on *problem-based project work* and *ideation*: formulating a problem or need discovered in one's environment, finding a solution and bringing it to a realization by employing integrated knowledge gained not only across the school curriculum but also the real-life experiences. In other words, this model of IE has a cross-curricula character based on innovation processes rather than subject content and it is strongly connected with real-life needs and experiences. Moreover, Thorsteinsson and Denton (2003; see also Thorsteinsson, this volume) report that IE can be a foundation for students' ethical development and provide specific examples that support that viewpoint.

Other studies also report success in teaching innovation related cognitive skills in real world situations. For example, Beyers (2010) reports preliminary results from a pilot study that was conducted with Grade 10 students from South African schools who were exposed to a high-tech rapid-prototyping environment. Students followed a design process consisting of five steps: investigate, design, make, evaluate and communicate. Preliminary results from this study indicate that the participants were able to operate effectively in a post-constructivist environment by applying their collective prior knowledge, skills, attitudes and values in order to produce innovative solutions to the challenge provided by using a range of ICT available to them.

Finally, innovation education research has produced specific recommendations for policy makers (Burke & Grosvernor, 2003; Cachia, Ferrari, Ala-Mutka, & Punie, 2010; Kampylis, Saariluoma, &

Berki, 2011) which involve, amongst others, changes in assessment, initial education and in-service training for teachers, and for integrating and utilizing ICT and digital media in a creative and innovative way. Last, but not least, changes are proposed in the overall educational culture of stakeholders such as policy makers, educational authorities and parents. These recommendations emphasize the need for a holistic strategy for revising all the areas of formal education in order to establish a supporting environment for creativity and innovation.

Teaching science from a conceptual change point of view as part of an overall program of innovation education

As we saw in the previous section, recommendations for innovation education usually aim at the development in students of certain domain general cognitive skills across the whole curriculum. We would like to propose here that, in addition to the above recommendations, innovation education can be greatly enhanced by teaching subject matter content – and specifically teaching science – from a conceptual change point of view.

Taking a conceptual change point of view means teaching science in ways that make explicit to students the need to revise their knowledge and to create new concepts. It aims at helping students become aware that their prior knowledge needs to be substantially restructured both in terms of its content and its structure, and that new concepts need to learned. This requires the development of new learning strategies which can be helpful in the gradual reorganization of knowledge, rather than in its enrichment.

Changing existing beliefs, when they are inconsistent with currently accepted scientific explanations, has proven to be a difficult task to accomplish with instruction. Constructivist theory is based on the premises that new information can be built on what is already known. It does not tell us what to do when new information is incompatible with what is known. This is an important problem for science instruction but it is also a central problem for innovation education. The discovery of new technological artifacts often requires radically new ways of looking at things, ways that break the barriers established by existing lines of thought.

We argue that through the explicit and conscious experiences of conceptual change in science classes, students can gradually become capable of abstracting a 'conceptual change know-how schema' and generalize it to other learning situations. This type of 'how to engage in conceptual change' compact knowledge can help students distinguish situations where radical knowledge revision is required and apply appropriate learning strategies to make it possible.

Instruction for conceptual change requires long term planning, substantial changes in curricula and extensive sociocultural support (Hatano & Inagaki, 2003; Nersessian, 2008; Vosniadou, 2003, 2007); we do not have space here to describe all the aspects of an educational environment that fosters conceptual change in science (see Vosniadou & Mason, 2012). For this reason, we will focus in this chapter on three important factors that have been associated with conceptual change instruction: the first concerns the whole educational environment which should exhibit a focus on the creation of students' metaconceptual awareness and the enhancement of intentional learning; the last two describe two learning mechanisms that can be helpful in knowledge restructuring processes – the use of instructional analogies, and explanatory models.

Creating an educational environment that enhances metaconceptual awareness

Metaconceptual awareness is awareness of the intellectual content of one's beliefs, as compared to metacognitive strategies which have to do with procedural knowledge designed to deal, for example, with comprehension failure. Many students are not aware of their beliefs, particularly when these beliefs involve intuitive knowledge about the physical world which is constantly confirmed by everyday experience. For example, Vosniadou and her colleagues have shown how the implicit belief in a top-down organization of space together with an 'up-down gravity' concept (i.e. that non-supported objects will fall 'down' until they find some kind of support) can seriously constrain young children's representations of the spherical shape of the Earth and create interesting misconceptions (Vosniadou & Brewer, 1992). Similarly, as was mentioned earlier, the implicit belief that force is a property of inanimate objects that have been set in motion can seriously interfere in the understanding of the Newtonian theory in mechanics even in college students.

Instruction for conceptual change needs to make explicit such intuitive and implicit beliefs that constrain the acquisition of new knowledge and seriously limit innovation. Many scientific and technological innovations are based on the elimination of prior beliefs which inhibited new forms of thinking. Teaching science from a conceptual change point of view helps students understand that science ideas are not simple and absolute but complex and continuously evolving. Instruction of this sort can also be instrumental in developing students' epistemic beliefs. There is emerging but strong empirical evidence that supports the indirect and direct role of epistemic beliefs on knowledge revision processes (Stathopoulou & Vosniadou, 2007; Vosniadou & Mason, 2012).

An instructional strategy commonly used to promote metaconceptual awareness and belief change is cognitive conflict. In situations of cognitive conflict students are asked to make predictions or explanations of phenomena and are then presented with some contradictory information, such as the results of empirical experiments, demonstrations or data summaries. Some less direct ways to introduce dissonance is through argumentation in class discussions or in peer collaboration tasks where information is provided that allows students to discover conflicts between their ideas and the normative view.

There is a large and controversial literature on the effects of cognitive conflict (Clement, 2008; Limon, 2001). Some researchers report negative results with the use of cognitive conflict as an instructional strategy (Tillema & Knol, 1997), but many studies report considerable success. Positive are also the results coming from the use of refutations in scientific text. Refutational text explicitly states readers' alternative conceptions about a topic, directly refutes them and introduces scientific conceptions as viable alternatives (Hynd, McWhorter, Phares, & Suttles, 1994). Refutational texts have been found to be more effective than standard expository texts for knowledge revision in various science domains. In physics, for example, they were more supportive for learning meaningfully new concepts about Newtonian mechanics (Kendeou & van den Broek, 2007), energy (Diakidoy, Kendeou, & Ioannides, 2003), and light and vision (Mason, Gava, & Boldrin, 2008).

How does cognitive conflict work? Experiencing psychological conflict has been related to increases in arousal potential that motivate the learner to attempt to resolve the conflict (e.g., Berlyne, 1960). It generates what Hatano and Inagaki (2003) call 'epistemic motivation' to check and revise prior knowledge. Limon (2001) argues that cognitive conflict situations demand from students a higher level of cognitive engagement which may increase their motivation for conceptual change, resulting in deeper processing of information.

Cognitive conflict cannot produce conceptual change by itself but must be used together with knowledge building strategies in the context of a rich learning environment that fosters and sustains conceptual change under substantial sociocultural support (Clement, 2008). Inagaki and Hatano (1977), for example, showed that introducing a controversial experiment in the class and asking students to make a prediction can be helpful in generating fruitful class discussion that can lead to a deeper comprehension of the scientific concept. Dialogical interaction, argumentation, collaboration, classroom discussion and meaningful practices around carefully designed curricula based on students' learning progressions are the means of developing metaconceptual awareness, and prolonged motivation for change. It is through these kinds of socioculturally based practices that students need to be guided by teachers to a new form of explicit, intentional learning that has the potential to make them learn and think like scientists.

Developing knowledge restructuring learning strategies – instructional analogies

Instructional analogies are explicit analogies in which a less familiar concept or explanation is introduced by appealing to its relational similarity to another concept from a different but more familiar domain. For example, the concept of electricity can be made meaningful through an analogy to water flow (Gentner & Gentner, 1983). In this analogy, electricity is mapped to water, wires to pipes, batteries to reservoirs, current to flowing water, voltage to pressure and resistors to narrow constrictions in pipes. Learners can transfer the relevant information from the familiar base to the unfamiliar target and use it in order to better understand how electricity works. Analogical reasoning provides an important, constructivist, mechanism that can facilitate knowledge revision and conceptual change. Instructional analogies allow learners to transfer existing representations from a familiar domain to start the knowledge acquisition process or revise prior knowledge in a different area of thought (Vosniadou, 1989).

Most of the research on the use of analogies in science education shows positive results, but not always. It appears that it is important to teach students how to use analogies for belief revision because they do not always know how to do this on their own (Clement, 2008; Glynn, 2007). Glynn (2007) discussed the problems that follow when analogies are used without sufficient elaboration that ensures that students understand them and use them effectively. He developed the Teaching-With-Analogies Model (TWA) to help teachers lead the students through the steps of comparing the base and the target, identifying the relevant features, mapping similarities and drawing conclusions. Analogies can be combined with dissonance producing strategies for better effects. According to Clement (2008) using dissonance strategies together with analogies can be more powerful for conceptual change than using analogies alone. Vosniadou and Skopeliti (submitted) have come to similar conclusions.

Model-based reasoning

According to Nersessian (2008), a model is an idealized and schematic abstraction that represents the physical system to which it refers by 'having surrogate objects and properties, relations, behaviors, or functions of these that are in correspondence with it' (p. 394). Models can be internal representations in the form of 'mental models', or external representations such as diagrams, graphs, maps, computer models and simulations, or even cultural artifacts like a globe. The focus is here on the construction of external models, and in particular on what Clement (2008) calls explanatory models. An explanatory model provides a description of a hidden, non-observable mechanism that explains how the system works and about the causes behind observable cases. Model-based reasoning involves the construction or retrieval of a model and the derivation of inferences through model manipulation.

There is a growing literature that shows that even preschool children are capable of engaging in model-based activities that help them construct new representations and understand new phenomena in science (e.g., Wiser & Smith, 2008; Clement & Steinberg, 2002). Wiser and Smith (2008) argue how model-based activities allow children to visualize decomposing matter into tiny pieces that continue to exist even if they are not directly observable by the senses, and as a result to slowly create a compositional model of matter according to which any piece of matter, however small, has weight and occupies space.

By constructing explanatory models students translate the verbal, abstract theories and explanations of science into concrete representations that can be explored and examined. The modeling activity becomes the means for understanding underlying mechanisms, correcting faulty representations and making new predictions. In addition, the construction of models is an activity that has the potential to increase students' interest and maintain their reasoning for long periods of time.

To sum up, it has been argued that with extensive sociocultural support science teaching classrooms can be designed that teach students specific learning strategies that help them reorganize prior knowledge in the process of acquiring subject-matter knowledge. This type of science instruction has the potential to create a 'conceptual change know-how schema' which students can learn to apply to other subject-matter domains in the classroom and more generally.

Conclusions

Teaching science from a conceptual change point of view can provide an additional, subject-specific means of enhancing an overall program of innovation education. This type of instruction requires creating in students vivid conceptual change experiences as they learn specific strategies for knowledge reorganization in the process of acquiring subject-matter knowledge with extensive sociocultural support. Teaching students how to gradually reorganize prior knowledge and create new concepts in the process of learning science can create the necessary metaconceptual awareness, belief revision and concept acquisition strategies that can promote creativity and innovation in other domains of thought. The explicit and conscious experiences of conceptual change that students can have in the process of learning science can in time enable them to create a 'conceptual change know-how schema' allowing them to recognize situations that require knowledge restructuring and apply learning strategies that have proven to be fruitful in the past to generate new theoretical concepts as well as new practical, technological devices.

Notes

1 The views expressed in this chapter are the sole responsibility of the authors and do not necessarily reflect the views of the European Commission.
2 In the Middle Ages, an interesting theory was proposed known as the 'impetus theory'. The most articulated view of the theory was that of Buridan's (1300–1388). According to Buridan, when an object is set in motion an 'impetus' (or 'vis' and 'forza' in Latin) is imparted into the object. This 'impetus' keeps the object in motion for some time after it has lost its contact with the agent. As the impetus gradually dissipates, the object slows down, until it finally stops or falls to the ground due to its weight (Franklin, 1978). The impetus theory resembles a common misconception found in children and adults that there is a force within inanimate objects that have been set in motion even when the objects have lost their contact with the original mover. This force gradually dissipates and finally runs out as the object slows down and stops.

References

Berlyne, D. (1960). *Conflict, Arousal, and Curiosity*. New York: McGraw-Hill.
Beyers, R. N. (2010). Nurturing creativity and innovation through FabKids: A case study. *Journal of Science Education and Technology, 19*(5), 447–455.
Burke, C., & Grosvenor, I. (2003). *The School I'd Like: Children and Young People's Reflections on an Education for the 21st Century*. New York: Routledge-Falmer.
Cachia, R., Ferrari, A., Ala-Mutka, K., & Punie, Y. (2010) *Creative Learning and Innovative Teaching: Final Report on the Study on Creativity and Innovation in Education in EU Members States*, Joint Research Centre (JRC), European Commission, Luxembourg (JCR 62370). Retrieved September 6, 2011, from http://ftp.jrc.es/EURdoc/JRC62370.pdf.
Carey, S. (1985). *Conceptual Change in Childhood*. Cambridge, MA: Bradford Books, MIT Press.

Clement, J. (2008). The role of explanatory models in teaching for conceptual change. In S. Vosniadou (Ed.), *International Handbook of Research on Conceptual Change* (pp. 417–452). New York: Routledge.

Clement, J., & Steinberg, M. (2002). Step-wise evolution of models of electric circuits: A 'learning-aloud' case study. *Journal of the Learning Sciences, 11*(4), 389–452.

Diakidoy, I. N., Kendeou, P., & Ioannides, C. (2003). Reading about energy: The effects of text structure in science learning and conceptual change. *Contemporary Educational Psychology, 28*(3), 335–356.

European Commission. (2008). *Lifelong Learning for Creativity and Innovation: A Background Paper*, Slovenian EU Council Presidency, Retrieved September 25, 2011, from www.sac.smm.lt/images/12%20Vertimas%20SAC%20Creativity%20and%20innovation%20-%20SI%20Presidency%20paper%20anglu%20k.pdf.

Franklin, A. (1978). Inertia in the middle ages. *Physics Teacher, 16*(4), 201–207.

Gentner, D., & Gentner, D. R. (1983). Flowing waters or teeming crowds: Mental models of electricity. In D. Gentner & A. L. Stevens (Eds.), *Mental Models* (pp. 99–129). Hillsdale, NJ: Erlbaum.

Glynn, S. M. (2007). Methods and strategies: Teaching with analogies. *Science and Children, 44*(8), 52–55.

Hatano, G., & Inagaki, K. (2003). When is conceptual change intended? A cognitive-sociocultural view. In G. M. Sinatra & P. R. Pintrich (Eds.), *Intentional Conceptual Change* (pp. 407–427). Mahwah, NJ: Lawrence Erlbaum Associates.

Hynd, C., McWhorter, J., Phares, V., & Suttles, C. (1994). The role of instructional variables in conceptual change in high school physics topics. *Journal of Research in Science Teaching, 31*(9), 933–946.

Inagaki, K., & Hatano, G. (1977). Amplification of cognitive motivation and its effects on epistemic observation. *American Educational Research Journal, 14*(4), 485–491.

Inagaki, K., & Hatano, G. (2002). *Young Children's Thinking About the Biological World*. New York: Psychology Press.

Kahl, C. H., Fonseca, L. H. d., & Witte, E. H. (2009). Revisiting creativity research: An investigation of contemporary approaches. *Creativity Research Journal, 21*(1), 1–5.

Kampylis, P., Saariluoma, P., & Berki, E. (2011). Fostering creative thinking: What do primary teachers recommend? *Hellenic Journal of Music, Education, and Culture, 2*(1), 46–64.

Kaufman, D. R., Keselman, A., & Patel, V. L. (2008). Changing conceptions in medicine and health. In S. Vosniadou (Ed.), *International Handbook of Research on Conceptual Change* (pp. 295–327). New York: Routledge.

Kendeou, P., & van den Broek, P. (2007). Interactions between prior knowledge and text structure during comprehension of scientific texts. *Memory and Cognition, 35*(7), 1567–1577.

Kuhn, T. S. (1970). *The Structure of Scientific Revolutions*. Chicago: Chicago University Press.

Leinhardt, G., & Ravi, A. (2008). Changing historical conceptions of history. In S. Vosniadou (Ed.), *International Handbook of Research on Conceptual Change* (pp. 328–341). New York: Routledge.

Limon, M. (2001). On the cognitive conflict as an instructional strategy for conceptual change: A critical appraisal. *Learning and Instruction, 11*(4–5), 357–380.

McCloskey, M. (1983). Naive theories of motion. In D. Gentner & A. Stevens (Eds.), *Mental Models* (pp. 229–324), Hillsdale, NJ: Erlbaum.

Mason, L., Gava, M., & Boldrin, A. (2008). On warm conceptual change: The interplay of text, epistemological beliefs, and topic interest. *Journal of Educational Psychology, 100*(2), 291–309.

National Mathematics Advisory Panel. (2008). *Foundations for Success: The Final Report of the National Mathematics Advisory Panel*. Washington, DC: U.S. Department of Education.

Nersessian, N. J. (2008). Mental modeling in conceptual change. In S. Vosniadou (Ed.), *International Handbook of Research on Conceptual Change* (pp. 391–416). New York: Routledge.

Piaget, J. (1970). *Science of Education and the Psychology of the Child*. London: Longman.

Posner, G. J., Strike, K. A., Hewson, P. W., & Gertzog, W. A. (1982). Accommodation of a scientific conception: Toward a theory of conceptual change. *Science Education, 66*(2), 211–227.

Rickinson, M., Lundholm, C., & Hopwood, N. (2009). *Environmental Learning: Insights from Research into the Student Experience*. Dordrecht: Springer.

Sawyer, R. K. (2006). Educating for innovation. *Thinking Skills and Creativity, 1*(1), 41–48.

Scott, G. S., Leritz, L. E., & Mumford, M. D. (2004a). Types of creativity training: Approaches and their effectiveness. *Journal of Creative Behavior, 38*(3), 149–179.

Scott, G., Leritz, L. E., & Mumford, M. D. (2004b). The effectiveness of creativity training: A quantitative review. *Creativity Research Journal, 16*(4), 361–388.

Shavinina, L. V. (2003). *The International Handbook on Innovation* (1st ed.). Amsterdam; Boston: Elsevier Science.

Shavinina, L. V. (2009a). High intellectual and creative educational multimedia technologies for the gifted. In L. V. Shavinina (Ed.), *International Handbook on Giftedness*, 1st ed., (pp. 1181–1202). New York: Springer.

Shavinina, L. V. (2009b). Innovation education for the gifted: A new direction in gifted education. In L. V. Shavinina (Ed.), *International Handbook on Giftedness*, 1st ed., (pp. 1257–1267). New York: Springer.

Stathopoulou, C., & Vosniadou, S. (2007). Conceptual change in physics and physics related epistemological beliefs: A relationship under scrutiny. In S. Vosniadou, A. Baltas & X. Vamvakoussi (Eds.), *Re-Framing the Conceptual Change Approach in Learning and Instruction* (pp. 145–165). Oxford, UK: Elsevier.

Thagard, P. (1992). *Conceptual Revolutions*. Princeton, NJ: Princeton University Press.

Thagard, P., & Croft, D. (1999). Scientific discovery and technological innovation: Ulcers, dinosaur extinction, and the programming language Java. In L. Magnani, P. Nersessian & P. Thagard (Eds.), *Model-Based Reasoning in Scientific Discovery* (pp. 125–137). New York: Plenum.

Thorsteinsson, G., & Denton, H. (2003). The development of innovation education in Iceland: a pathway to modern pedagogy and potential value in the UK. *Journal of Design and Technology Education, 8*(3), 172–179.

Tillema, H., & Knol, W. (1997). Collaborative planning by teacher educators to promote belief change in their students. *Teachers and Teaching, 3*(1), 29–45.

Vamvakoussi, X., & Vosniadou, S. (2010). How many decimals are there between two fractions? Aspects of secondary school students' reasoning about rational numbers and their notation. *Cognition and Instruction, 28*(2), 181–209.

Vosniadou, S. (1989). Analogical reasoning and knowledge acquisition: A developmental perspective. In S. Vosniadou & A. Ortony (Eds.), *Similarity and Analogical Reasoning* (pp. 413–422). New York: Cambridge University Press.

Vosniadou, S. (2003). Exploring the relationships between conceptual change and intentional learning. In G. M. Sinatra & P. R. Pintrich (Eds.), *Intentional Conceptual Change* (pp. 377–406). Mahwah, NJ: Lawrence Erlbaum Associates.

Vosniadou, S. (2007). Conceptual change and education. *Human Development, 50*(1), 47–54.

Vosniadou, S. (2008). *International Handbook of Research on Conceptual Change*. New York: Routledge.

Vosniadou, S., & Brewer, W. F. (1992). Mental models of the Earth: A study of conceptual change in childhood. *Cognitive Psychology, 24*, 535–585.

Vosniadou, S., & Mason, L. (2012). Conceptual change induced by instruction: A complex interplay of multiple factors. In K. R. Harris, S. Graham & T. Urdan (Eds.), *APA Educational Psychology Handbook: Individual Differences, Cultural, and Contextual Factors in Educational Psychology*, vol. 2 (pp. 221–246). Washington, DC: American Psychological Association.

Vosniadou, S., & Skopeliti, I. (submitted). Instructional analogies in conceptual restructuring processes. *Cognitive Psychology*.

Vosniadou, S., Vamvakoussi, X., & Skopeliti, I. (2008). The Framework Theory Approach to the problem of conceptual change. In S. Vosniadou (Ed.), *International Handbook of Research on Conceptual Change* (pp. 3–34). New York: Routledge.

Vosniadou, S., & Verschaffel, L. (2004). Extending the conceptual change approach to mathematics learning and teaching. In L. Verschaffel & S. Vosniadou (Guest Eds.), *Conceptual Change in Mathematics Learning and Teaching*, special issue of *Learning and Instruction, 14*(5), 445–451.

Voss, J. F., & Wiley, J. (2006). Expertise in history. In K. A. Ericsson, N. Charness, P. Feltovich & R. R. Hoffman (Eds.), *The Cambridge Handbook of Expertise and Expert Performance* (pp. 1746–2424). New York: Cambridge University Press.

Wellman, H. M. (2002). Understanding the psychological world: Developing a theory of mind. In U. Goswami (Ed.), *Handbook of Childhood Cognitive Development* (pp. 167–187). Oxford: Blackwell.

West, M. A. (2002). Sparkling fountains or stagnant ponds: An integrative model of creativity and innovation implementation in work groups. *Applied Psychology: An International Review, 51*(3), 355–424.

White, R. T., & Gunstone, R. F. (2008). The conceptual change approach and the teaching of science. In S. Vosniadou (Ed.), *International Handbook of Research on Conceptual Change* (pp. 619–628). New York: Routledge.

Wiser, M., & Smith, C. L. (2008). Teaching about matter in grades K-8: When should the atomic molecular theory be introduced? In S. Vosniadou (Ed.), *International Handbook of Research on Conceptual Change* (pp. 205–239). New York: Routledge.

6

New brain-imaging studies indicate how prototyping is related to entrepreneurial giftedness and innovation education in children

Larry R. Vandervert and Kimberly J. Vandervert-Weathers

AMERICAN NONLINEAR SYSTEMS, USA, AND SPOKANE PUBLIC SCHOOLS, USA

Summary: In this chapter it is argued that recasting Shavinina's (2008, 2009, Chapter 3, this volume) conception of entrepreneurial giftedness as the processes of prototyping supports her contentions and permits innovation education to be studied in a scientific fashion and to be broken down into manageable components for curriculum design. This argument is strongly supported by (a) new brain-imaging studies that show that the brain creates progressively more efficient mental and physical *prototypes*, and (b) *deliberate practice* effects of the Khan Academy. To illustrate how prototyping occurs in children, Camp Invention, a nationwide innovation education program for children in the United States, is reviewed. In conclusion, it is suggested that although Camp Invention is a state-of-the-art innovation education program, it might focus more on (1) prototyping skills, and (2) how students can increase their own capacities to envision new ideas and new products.

Key words: Brain-imaging, Camp Invention, cerebellum, deliberate practice, innovation, Khan Academy, prototyping.

Introduction

Shavinina (2008, 2009, Chapter 3, this volume) proposed that the most important element of innovation education is the development of *entrepreneurial giftedness*, or simply *entrepreneurial ability*. In her proposal, entrepreneurial giftedness referred to talented individuals who have succeeded in business by creating new ventures with at least a minimal financial reward or who demonstrated an exceptional potential ability to succeed in such ventures. Shavinina (2008, Chapter 3, this volume) identified the following five cardinal characteristics in the development of entrepreneurially gifted individuals, they: (1) constantly generate ideas on how to make money; (2) love to generate and implement real-life projects with at least a minimal financial reward; (3) love doing real business plans with predicted financial outcomes; (4) work passionately and hard on executing their plans; (5) wish to do "real" things that bring money and try to do whatever possible to cut unnecessary steps. Beginning from this five-point perspective on innovation education, how can entrepreneurial giftedness, or entrepreneurial ability, be developed in the future?

The purpose of this chapter is to describe three tightly interrelated bases upon which innovation education might develop in the future. *First*, it will be proposed in this chapter that Shavinina's above five key components of innovation education, can be fully recast within the overall processes that makeup *prototyping*. This recasting does not diminish Shavinina's contentions but rather supports them by tying them directly in with science, technology, engineering and mathematics (STEM) subject matter and hands-on sequences of achievements. Prototyping, it will be argued below, not only includes the various prototyping activities themselves, but (1) the *desire* to generate new ideas and (2) the desire to manipulate them, through stages of increasing efficiency—two desires, which it will be shown below, became inherent in humans through the evolution of the human *urge* and *capacity* to innovate. Recasting the components of entrepreneurial giftedness as features of prototyping permits classic accounts of innovation to be studied in a scientific fashion, to be tied into imaging studies on how the human brain innovates, and to be broken down into manageable components for curriculum design. That is, in the future an understanding of the motivational, cognitive and behavioral components of prototyping that underlie innovation and entrepreneurial giftedness can provide curriculum developers with straightforward, scientifically based strategies that help ensure innovation education actually results in the development of innovators.

In the *second* basis for the future of innovation education, it will be shown how new studies that look inside the brain (brain-imaging) can explain how and why the human brain evolved to naturally produce *iterative successions of ideas, tools and ornaments* (prototyp*ing*) which result in the optimization of their efficiencies, that is, that produce innovations of useful/trading/monetary value. Thus prototyping not only is the underlying process of innovation but, due to the efficiencies it produces, is also the connect-up to the "business" components of entrepreneurial giftedness—more will be presented on this tight connection between prototyping and entrepreneurial giftedness below (see Vandervert, 2006, 2007, 2011a, 2011b).

Third, to apply the above connections between prototyping and entrepreneurial giftedness to an exemplary program in innovation education, prototyping techniques will be suggested that can be emphasized in the ongoing and future curriculum development of Camp Invention, a huge national program of innovation education in the United States. In the future, could programs like Camp Invention help develop the entrepreneurial abilities of not only gifted children, but of all children?

How entrepreneurial giftedness can be recast as curriculum friendly prototyping skills

In this chapter it is proposed that the above five manifestations of entrepreneurial ability identified by Shavinina (Chapter 3, this volume) are the result of a single underlying behavioral/cognitive tendency and acquired ability to engage in *prototyping*. That is, the five manifestations of entrepreneurial ability or giftedness are exactly what take place in prototyping. A *prototype* is an early model or sample envisioned and "built" for the purposes of testing the workability of a concept (including engineering, software and business concepts) or to act as a thing to be replicated in manufacturing or simply from which to learn. Prototyp*ing* consists of a propelled, iterative series of prototypes, each of which provides the testing and learning necessary to (1) improve the workability and efficiency of the originally envisioned goal and/or (2) perhaps modify the envisioned goal toward a more workable direction.

All of Shavinina's five manifestations of entrepreneurial ability or giftedness clearly take place within the processes of prototyping, and *prototyping is the only place where all five manifestations can be seen occurring in an integrated fashion*. The famous entrepreneur and inventor, Thomas Edison, provides the greatest possible case in point. Edison's Menlo Park "invention factory" *was all about (1) constantly generating new ideas to make money, (2) real-life projects, (3) working hard, (4) predicted financial outcomes and (5) wishing to do real things that brought money and cut unnecessary steps.* His overall goal was "to invent useful things every man, woman, and child in the world wants ... at a price they could afford to pay" (Josephson, 1992, p. 314). While we can summarize Edison's achievements as entrepreneurial giftedness, in fact, it is well known that all of Edison's entrepreneurial productivity when broken down into behavioral and cognitive activities was accomplished within a framework of prototyping *and could only have been accomplished within a framework of prototyping*. Prototyping was the skeletal structure of actual innovative events in Edison's lab. The absolute necessity of prototyping to invention and innovation is captured in Edison's most famous quote, "Genius is 1% inspiration and 99% perspiration." For Edison, the 99% perspiration was never a random expenditure of energy, as his prolific notes attest; it was always a careful, arduous process of prototyping. The electric light, for example, was the result of countless prototyping versions (thousands of different filaments were tried) both by others before Edison and in Edison's lab.

Prototypes, either a physical model or a simple sketch, drew Edison and his team at the Menlo Park lab toward higher and higher efficiencies and monetary value. Prototyping is the embodiment of entrepreneurial giftedness. That is, the processes of prototyping inherently carries with it the ideas and goals of Shavinina's (Chapter 3, this volume) five manifestations of entrepreneurial giftedness. Thus, prototyping is not simply a disembodied mechanical process but is actually an evolved brain-based, sequential scheme of thinking. Prototyping is pervasive in human advancement. Prototyping is even the sequential, efficiency-optimizing scheme of thinking that underlies redrafting in writing wherein new ideas and new ways of stating them appear (often to the surprise of the writer) only during the redrafting process—just as Edison's inventions and innovations came into being *only* during a sequential "redrafting" process of prototyping.

With an understanding that entrepreneurial giftedness can be boiled down to prototyping, it is no surprise that as Massachusetts Institute of Technology's Michael Schrage (1999) pointed out forward-looking modeling and prototyping are the engines that drive innovation and performance in the world's best companies:

> Virtually every significant marketplace innovation of this century is the direct result of extensive prototyping and simulation. Consider, for example, the airplane, the animated motion picture, the transistor, the microprocessor, the personal computer, the software spreadsheet, recombinant DNA biotechnology, junk bonds, leveraged buyouts, the Internet and its World Wide Web, financial derivatives, and synthetic securities, and index funds and yield management.
>
> By shifting from physical clay to virtual clay, every major automobile company has radically reengineered its design and production process. Boeing's breakthrough 777 jet [and the 787] was built around the breakthrough digital prototypes. Walt Disney can't produce feature-length animations without storyboards. Microsoft could not enjoy the market share and margins it does without its strategic deployment of beta-version software. Merrill Lynch's [Merrill Lynch now operates as a subsidiary of Bank of America] ability to model synthetic securities in simulated over-the-counter markets makes it far easier to sell its novel financial instruments to Fortune 500 treasuries. Wherever you look for the fundamental dynamics driving innovation, you find *innovators managing models* [italics added].
>
> *(pp. 11–12)*

The upshot of what Schrage is saying is that prototyping is the best way to produce new value (the best way to entrepreneurial success), *because* it comprises in fact the very processes that underlay entrepreneurial productivity. But *why* do prototypes lead to such great innovative value in the first place?

Prototyping begins in the brain

The evolution of the human urge and capacity to innovate: prototypes harness the brain directly

In this section we address why and how prototypes would be so powerfully innovative. In recent work, one of the authors (along with Paul Schimpf, Washington State University, and Hesheng Liu, Martinos Biomedical Imaging Center, Boston, MA) has described adaptive processes in the human brain that produce creativity and innovation (Vandervert et al., 2007).[1] Here is what was found:

1. Based upon a huge volume of new imaging research on the brain it was found that when people work on well-defined problems their brains naturally build new neurological connections related to those problems. These new neurological connections represent decomposed and re-combined *models* (see Schrage above) and configurations of hands-on sequences toward better and quicker ways both of doing everyday things and of achieving goals toward creativity and innovation.
2. Imaging studies indicate that these models are constantly *blended* as people continue to work hard at problems. During focused work on prototyping, this blending process finds new, better ways of doing things with each new iteration of the process, and thus, along the way, the *prototyping* process spontaneously results in new inventions and innovations in ideas, art, science and technology. As long as the prototyping processes are continued the innovative blending and re-blending process never ends. Each new version in the prototyping process is recycled through the blending process making the subsequent versions constantly higher-efficiency versions of the goal.
3. The decomposition, re-combining and blending processes are the brain's sources of new insights that occur suddenly either while people are directly involved in solving a problem, or, since the blending process continues until the problem is solved, when they are relaxed and thinking about something else. These insights often come to mind with the familiar exclamation of "aha!"

In sum, brain imaging studies show that there is a general sequence of steps or "achievements" within the innovation processes—*this natural sequence of the brain is harnessed in the processes of prototyping*. Whenever we repeat a movement or a thought while working on a goal or a prototype, a new connection is established in the brain that creates a slightly improved mental prototype or simulation of that activity. This prototyping process is at work for athletes, scientists, writers, music composers, inventors, innovators and all students alike. In special regard to students, we propose the repetitive "mental prototyping" or mental rehearsal involving the cerebellum and the cerebral cortex explains the success of the self-driven, individualized patterning of *graduated* repetitions initiated by the teaching methods of the innovative Khan Academy (www.khanacademy.org/about). The Khan Academy provides a framework which could easily be adapted to innovation education. This proposal is based upon the findings of Ericsson and his colleagues which demonstrated how *deliberate practice* (practice aimed toward *constantly elevated* levels of performance) produces experts and exceptional performers (Ericsson, 2002, 2003a, 2003b; Ericsson, Nandagopal, & Roring, 2009; Ericsson, Roring, & Nandagopal, 2007). Ericsson (2002, 2003a, 2003b) has strongly emphasized the finding that, "the essence of expert performance is a generalized skill at successfully meeting the demands of *new* [italics added] situations and rapidly adapting to changing conditions" (Ericsson, 2002, p. 41). Vandervert (2007) argued the deliberate, constantly elevated levels of practice takes direct advantage of how the collaboration of working memory and the cerebellum produces *novel* behavior in each of the elevated levels of new and challenging learning. In this way, we believe the formatting of the graduated lessons of the Khan Academy encourages development of innovative problem solving which will carry over to challenging situations outside the structure of the Khan Academy. The prototyping process involving the cerebellum and the cerebral cortex is why also, for example, if athletes *stay on repetitive, graduated training targets*, they will become highly efficient champions who often develop innovative signature moves and plays; if tinkerers, thinkers,and R&D people *stay on targets* carefully managing their prototypes and ideas, as did, for example, Thomas Edison, they will develop that highly efficient, and extremely high-value "electric light" or "phonograph."

How the human brain became a prototyping "machine"

These prototyping-to-innovation processes have been at work in evolving humans since at least the beginning of the first prototypes of stone tools (Vandervert, 2009, 2011a). That is, the first stone tools, beside being innovations that gave evolving humans new ways of living and surviving, were themselves prototypes for ever newer types of stone tools (Ambrose, 2001). Across hundreds of thousands of years these prototypes were tinkered with by countless generations of humans. During this time, the prototyping human brain evolved to become the basic means of survival for humans and thereby became the way humans innovate. The fact that constant prototyping toward innovations evolved as a (perhaps *the*) special key to *survival* among humans is why, as Shavinina (Chapter 3, this volume) points out, new ideas, new projects and the entrepreneurial aspects of innovation are passionately driven.

By 40,000 to 50,000 years ago prototypes of all kinds had proliferated, and the acceleration of culture reached the proportions of a "creative explosion" (Pfeiffer, 1982). Ambrose (2001) described the explosion of artifacts typical of the Upper Paleolithic period (40,000 years ago):

> Of greater significance [than prototypes of advanced stone tools of this period] are ground, polished, drilled, and perforated bone, ivory, antler, shell and stone, shaped into projectiles, harpoons, buttons, awls, needles, and ornaments. Such artifacts are extraordinarily rare in MP/MSA [Middle Paleolithic/Middle Stone Age] sites but are a consistent feature of Upper Paleolithic (UP) and Later Stone Age (LSA) sites after 40 ka [40,000 years ago].... Traces of more perishable materials, including string and woven fibers that may have been made into nets, ropes, bags, and clothing are also well documented. These innovations are among many that signify modern human behavior, including art, ornamentation, symbolism, ritual burial, sophisticated architecture, land use planning, resource exploitation, and strategic social alliances.
>
> *(p. 1752)*

As it could be said at least 40,000 years ago, the ideas and physical products of culture can be looked upon as interacting webs of constantly changing prototypes, the human brain constantly and quite naturally decomposing and recomposing prototypes toward higher states of utility and efficiency. In every pre-historical and every historical period there were individuals who, like Edison, were gifted entrepreneurs who innovated, manufactured and traded constantly newer things of value. They all followed a brain-based sequence of steps that had evolved to provide the uniquely human advantage of *a sequence of innovative achievements*.

A sequence of achievements for innovation education

The brain-based sequence of steps that produces innovation provides a prototyping commonality across ancient stone-tool makers, Edison and all innovators. It also provides a way to understand the sequential relationships among (1) the mental models (concepts) which contain the visions of innovation, (2) STEM subjects which give these visions real-world frameworks, (3) hands-on prototyping which carry innovative visions forward and (4) the entrepreneurial drive which sustains focus during the sequence of achievements. All of these interrelated factors can be placed in the following sketch of a *sequence of achievements* which, because they can be objectively observed, can be used for monitoring development, for evaluation and for curriculum design in future innovation education:

1. The *first* achievement in the sequence of achievements is that a new product or a newly improved product is imagined or envisioned. The inspiration for the new product or idea appears to arise directly from the brain's blending processes described earlier on page 82, and this inspiration is sustained throughout the innovation process by the sheer promise of the newly useful efficiencies that might be produced and by their promise of entrepreneurial value. Thus, the brain's blending processes and their "push" toward entrepreneurial productivity are apparently behind Edison's often quoted "inspiration" phase of genius, invention or innovation. *It is important to recognize that the inspiration containing the new product or idea is actually part of the prototyping sequence*—the new idea is itself a first embryonic prototype produced in the brain. It is part of the same recycling mechanism in the brain that will continue the prototyping process as described in the next achievement of the sequence. How this first achievement toward innovation can be observed in children will be suggested in the next section of this chapter.

2. In the second achievement of the sequence the new idea or product either immediately or slowly becomes placed into some form of an initial hands-on, physical prototype. In Edison's lab this initial prototype was built from materials and parts commonly in supply in the lab or would be custom made by lab technicians. These initial prototypes are constructed with various degrees of precision and workability requiring science, technology, engineering and mathematics (STEM).[2] This absolutely necessary hook-up between imagination and STEM-level, physical-world prototyping is sometimes overlooked, but STEM knowledge always provided the tangible frameworks by which Edison was able to make his initial prototypes or drawings and by which the world's best companies use prototyping models to make the best new products (see Schrage, page 81 above). Even early stone-tool makers placed their new ideas into frameworks of some sort of rational precursors to STEM knowledge. For example, an early stone-tool maker would, based upon the then current knowledge or properties of various types of stones, angles of percussion and so forth, envision a new sequence of stone flaking or a new carefully measured combination of stone, wood and leather components. That is, the properties of stones and so forth appear to have been the earliest forms of rational thought which eventually gave way to the various principles now codified in STEM subject matter (see Ambrose, 2001, for a proposal that language too may have come about in this way). The foregoing rational frameworks of pre-STEM or, in Edison's case, STEM knowledge suggest to the innovator workable and verifiable hands-on prototype-able action sequences. That is, STEM knowledge brings ideas for new things solidly into the form of a *rational* physical world. The *second* observable achievement, then, is the development of an initial real-world prototype which is sharable with others, and which relies on the use of some degree of STEM subject matter.

3   As prototyping is carried out the envisioned new tool or product gains new levels of efficiency and utility. The *third* achievement is the development of at least a second, improved version of the initial prototype.
4   The efficiency and utility of the new tool or product not only has value to the innovator but to others, and therefore it has trading or market value—it has entrepreneurial value, and, therefore, the new product is marketed. The *fourth* achievement in the sequence of achievements toward innovation is the marketing of the new idea or product.

Can such a sequence of achievements for innovation-entrepreneurial ability be taught to children within a curriculum which capitalizes on the natural efficiencies of prototyping?

The case of Camp Invention: is it a framework for the future of innovation education for all children?

What is the Camp Invention program?[3]

The Camp Invention program is a nationally acclaimed week-long summer enrichment experience for children entering in grades 1 through 6 that is free to schools and organizations nationwide. The Camp Invention program is science based and aligns with national education standards. All curricula are written in a transdisciplinary approach that includes subject areas such as engineering, mathematics, social studies, literacy, history and art; as well as 21st century skills including creative and critical thinking, problem solving, collaboration and communication. The Camp Invention experience involves immersive, hands-on learning that features fun, engaging challenges that encourage participants to take risks, use creative problem solving and develop an entrepreneurial mindset in an open, safe environment. Local schools host the program and activities are led by local teachers to ensure a safe learning environment. The staff to child ratio is 1:8! The Camp Invention program enhances a child's ability to learn through teamwork while cultivating a new appreciation for discovery and creativity—it's learning disguised as fun!

The Camp Invention program was created by the National Inventors Hall of Fame Foundation with support from the United States Patent and Trademark Office. Since its inception in 1990, the program has grown to include more than 1100 sites in 49 states. The Camp Invention program provides curricula with background information, step-by-step instructions and guiding questions to local directors and their staff, as well as marketing support, project materials, staff training, staff salaries, customer service and reservations assistance. Parents pay an established fee for their children to attend. Participating parents, educators and children rave about the program—giving it an approval rating of more than 98%.

Because it is a widely embraced, state-of-the-art innovation education program, Camp Invention was selected for this chapter as a model for what might happen in the future of innovation education. One of the authors, Vandervert-Weathers, has had two years of personal involvement, both as a camp teacher and camp organizer, with the nuts and bolts of Camp Invention while at Seattle Hebrew Academy. So that critical parts of Camp Invention can be examined carefully, an entire sketch of the program is now laid out in some detail.

A nationwide week of summer enrichment

The Camp Invention program runs for a period of one week, and each day up to 140 children rotate between thematic modules that build upon what was learned the previous day. A typical program runs Monday through Friday from 9:00 a.m. to 3:30 p.m.

A day at the Camp Invention program is packed with hands-on activities, brainstorming, experimentation and unbelievable action! Although daily activities are widely diverse and vary based on the program being hosted, as well as the number of participants, here is a quick glimpse of what a typical day of fun at the Camp Invention program looks like...

> **9:00 a.m.** Children are signed in by their parents and join their counselor and age-appropriate group at the *Base Camp* area. During this gathering period each morning, they might build a newspaper tower or learn the Camp Invention cheer.
> **9:15 a.m.** Children move to their first module, during which they might be crash-landed astronauts on an alien planet! After assessing their strange new surroundings, children might be challenged to design shelters and spacesuits that are able to withstand the planet's harsh conditions and acid rain.
> **10:20 a.m.** Children enjoy a morning snack.
> **10:30 a.m.** Children move to their second module, during which they might find themselves in safety goggles and using real tools to take apart and investigate small household appliances, transforming them into incredible new machines that may be used to burst balloons.
> **11:40 a.m.** Half of the children gather in the common area for lunch, while the other half indulge in high-energy games and activities in the gymnasium or outdoors.
> **12:10 p.m.** Children who ate lunch earlier now work off that energy by switching with the other group, who now gather in the common area for lunch.
> **1:05 p.m.** Children move to their next module of the day, during which they might explore Newton's first law of motion by conducting an experiment based on a magician's tablecloth trick or participating in a relay race that demonstrates the concept of inertia.
> **2:10 p.m.** Children move to their final module of the day, during which they might clean up a simulated landfill that is leaking toxic chemicals into groundwater.
> **3:15 p.m.** Children gather in the *Base Camp* area with their counselor, where they might create a marshmallow sculpture or create erupting soda pop fountains until their parents arrive.
> **3:30 p.m.** The fun comes to an end, and children are signed out by their parents.

Each year, Invent Now curriculum writers employ rigorous testing sessions to produce new and innovative curricula that align with national and state education standards and foster creative problem solving and teamwork—21st century life skills imperative to success in a competitive global workforce. In each Camp Invention program, children work in diverse teams, engage in investigations, experiments and engineering challenges that combine *science, technology, engineering and math (STEM)* in fun, hands-on activities.

Camp Invention offers new programs each year. A child could attend every summer, and never complete the same experience twice.

The CREATE Program
The INNOVATE Program
The SPARK Program
The DISCOVER Program

Here's the content of the Innovate program.

The INNOVATE program

In a secret lab, a mysterious scientist has been investigating how to power robotic creatures in the *Power'd*™ module. The scientist requests the help of Camp Invention participants to explore new types of energy to help power the robotic creatures. Children spend the week building, powering and exploring the capabilities of their own fantasy creatures by investigating wind, solar and hydro power. Children explore how they can light an LED with a small fan, waterwheels and lemons. They also create creatures made from upcycled materials that wiggle using motors and batteries to create circuits.

The *Hatched*™ module uses the concept of online virtual worlds to relay economics solutions to given problems! The daily activities have children creating, buying and selling to understand the role of economics in the world. Imagination, creative problem solving and economic principles lead children to create a thriving virtual world for their self-created avatars and other inhabitants of the world. During the week, children work in teams to create a water system, as well as items for the *Hatched* world's residents. Using marketplaces and *Hatched* money, their created items are bought and sold between teams.

The *SMArt: Science, Math & Art*™ module illustrates the beauty of patterns in mathematics. Imagination, creative problem solving and mathematical solutions are fostered as children recreate the beautiful math of Pattern Kingdom! This module uses the mathematical fields of tessellations, topology, minimal surfaces, fractals and angles to demonstrate that math is a subject involving some of the basic questions and areas of our lives. During the week, children work to restore the love of mathematics that the Pattern Kingdom's residents have forgotten by introducing them to some of the more little known areas that help to make mathematics a language that speaks to us all.

Children shake up traditional games to create new experiences that will have them thinking and upcycling their way to a brand new kind of fun! In the *Game On: Power Play*™ module, children are challenged to use nontraditional equipment (such as water balloons) to play classic games. Each day features fresh ideas that will have children mentally and physically engaged and completely entertained. The laughter is contagious as children use fun ways to enhance their level of cooperation and coordination in these innovative, team-building exercises. Children's minds and bodies are put to the test as they combine physical activity, creativity and fun!

During the *I Can Invent: Edison's Workshop*™ module children walk in the footsteps of Thomas Edison as they create and market a multi-step machine. Creative problem solving is fostered as children imagine and assemble the unthinkable! Younger children work in teams to create multi-step inventions using pieces and parts of a broken appliance and other upcycled materials. Older children work in teams to build complicated, multi-step machines that solve a challenge. All children further explore the process of invention as they market their inventions. A participant favorite, children of all ages find this module incredibly challenging and exciting! All of the programs end the week with the I Can Invent: Edison's Workshop.

What Camp Invention already accomplishes toward the development of innovators

A careful reading of the activities of Camp Invention indicates that most of the elements of Shavinina's (2008) entrepreneurial giftedness (or as it has been recast in this chapter, "the *desire* to generate new ideas and manipulate them into a marketable form that is embodied in *the processes of prototyping*") are a part of what children learn in the Camp Invention program. The Camp Invention Program immerses children in highly motivated team work on an interesting, open-ended problem, and the activities in the modules teach them to enjoy it—it helps develop a positive attitude toward innovation and a way to express the basic human prototyping urge in innovation. Thus Camp Invention is well on its way toward becoming a model of innovation education.

Further, Camp Invention sensitizes children to the fact that their inventive work (to be innovative) must have a useful (marketable) application. In the Edison's Workshop module, Camp Invention children create and market new products. Camp Invention's hands-on emphasis represents a strong beginning on teaching the underlying way prototyping works. The fact that children are able to develop any true innovation skills at all is dependent upon hands-on activity *within a framework of brainstorming and experimentation*. Through Camp Invention the appreciation *of* and desire *for* hands-on activity are turned into learnable skills. Not only does hands-on activity directly foster innovation because it is the bridge to prototyping, but as Shavinina (Chapter 3, this volume) points out, it connects in highly learnable ways to science, technology, engineering and mathematics (STEM). The basis of the connections between the hands-on working through of problems, innovation and STEM was described above (see pages 84–85 this chapter) in the findings of Vandervert et al. (2007). These findings completely support the STEM Education Coalition's promotion of "A strong emphasis in learning environments on hands-on, experiential, inquiry-based and learner-centered student experiences and activities, including engineering design processes" (www.stemedcoalition.org/about-us/). The connection between hands-on activity, innovation and STEM is a natural and deeply brain-based one.

Two essential points for curriculum design in the future of innovation education: an emphasis on prototyping and idea-generating skills

While Camp Invention fulfills many educational purposes, the main way it delivers education is through activities organized around innovation. This purpose is highlighted in the above opening paragraph which addresses what Camp Invention is: "The Camp Invention program enhances a child's ability to learn through teamwork and subject immersion while cultivating a new appreciation for discovery." Camp Invention certainly qualifies as "innovation education." The activity and curriculum organization which brings Camp Invention solidly into innovation education as defined in this volume is the *I Can Invent: Edison's Workshop*™ module which closes each of the four different camps. As described earlier, Edison's story of invention and innovation is most essentially about entrepreneurial giftedness based on two essential steps: (1) the dreaming-up of marketable ideas, which are then (2) brought to an actually marketable form through prototyping.

What Camp Invention could do in the future to further enhance the development of young innovators

Thomas Edison's generation of new ideas and new products and then the refining of them to marketability constitute the overarching model for Camp Invention and for innovation education efforts around the world. Within the framework of this "Edison model," we propose two recommendations for the future of innovation education.

First, since Edison is perhaps the world's greatest example of entrepreneurial giftedness and its underlying mechanisms of prototyping (that is, producing successive models toward an intended goal through taking notes, making drawings, storyboarding, simulation, hands-on experimentation and so forth), *actual prototyping techniques should be introduced to children in each of the modules, the earlier in the week, the better*. Early inclusion of prototyping in the modules will help initiate the natural, brain-based innovation optimization processes in children. Improvements in marketability should thus be more forthcoming, as were improvements in marketability in the world's best companies mentioned above by Michael Schrage. When children leave Camp Invention or any innovation education program they should have acquired new skills at (1) recognizing the successive models (forms) acquired from their innovative efforts and (2) recording such continuing improvements.

Second, following Edison's greatest personal strength, the generation of innovative ideas, it is recommended that in the future Camp Invention add a module on how children themselves can develop problems and ideas that would be fun to explore. That is, to develop true innovators in the Edison tradition children must have experience in developing their *own* processes of dreaming-up problems and ideas which can then be explored. At the children's level, this can be accomplished by beginning with a surplus of ideas which the children themselves winnow down to constitute the modules they will actually pursue. This experience will help sensitize them to the fact that innovators have to be problem and idea selectors and generators, not just problem solvers—someone must come up with the original problems to be solved or there would have been, for example, no Thomas Edison. When children leave Camp Invention (or any innovation education program) they should have a new skill of recognizing that new ideas and products they themselves dream up have manageable pathways toward further development and toward marketability.

Conclusions and discussion

Shavinina's (Chapter 3, this volume) components of entrepreneurial giftedness can be recast within the processes of prototyping. This recasting supports Shavinina's contentions by permitting classic accounts of innovation (such as those associated with Thomas Edison) to be studied in a scientific fashion, to be tied into imaging studies on how the human brain innovates, and to be broken down into manageable components for curriculum design. Brain imaging research shows that prototyping is the overall process by which the brain has evolved to blend models of ideas and the tools of both technology and the arts to constantly produce new, more efficient versions.

The great span of human innovation began at least one million years ago with the natural selection of stone-tool innovations associated with simple mental models associated with foraging intertwined with the hands-on action of stone transport and percussion. These innovations were part of the evolution of a brain which was driven to constantly hone thoughts and activities toward greater efficiency and utility. This constant honing of the human capacity for survival through innovation reveals itself in the most intriguing milestones which chronicle the ascent of humans: (1) the hundreds of thousands of years of the progressive development of more and more efficient stone tools, (2) the creative explosion of innovations (including the likely beginnings of entrepreneurial giftedness) which occurred approximately 50,000 years ago (Pfeiffer, 1982), (3) the rise of *historical* human culture (the total collection of practiced innovations in ideas, technology and the arts) and (4) the concomitant rise (with historical human culture) of what might be called an "age of STEM-driven innovation," where as Ambrose (2001) pointed out: "A mere 12,000 years separate the first bow and arrow from the international Space Station" (p. 1752).

The overall evolved brain processes behind all of these innovation milestones, dating back to ancient stone-tool makers, can be shown to closely parallel how *prototyping* produces marketable innovation in the world's best companies and how Thomas Edison used prototyping to produce marketable innovations in his Menlo Park laboratory.

From the overall processes of prototyping a sequence of achievements for innovation education is proposed. This sequence of achievements can be used to monitor progress, evaluate programs and to design future innovation education. In simplified form this sequence of achievements (see full descriptions above on pages 84–85) includes the following:

1 imagine a new idea or product;
2 translate the new idea or product into some rational framework (STEM) to give it a physical form, so that real-world prototyping can begin;
3 develop at least a second, more efficient version of the prototype;
4 determine to whom and how the new idea or product would be marketable, and, within the program, market new idea or product.

Camp Invention is a widely embraced, state-of-the-art innovation education program and is therefore a model for what might happen in the future of innovation education. It is suggested that Camp Invention can improve the monitoring of student development, Camp outcomes and future curriculum design by including more on prototyping experiences (at least through a second-version prototype) and more on the development of students' own capacities to envision new ideas and new products.

Notes

1 Vandervert et al. (2007) and Vandervert (2011a, 2011b) based the descriptions of these mental processes on over a decade of brain-imaging experiments conducted by Hiroshi Imamizu and his colleagues (Imamizu, Higuchi, Toda, & Kawato, 2007; Imamizu & Kawato, 2009; Imamizu & Kawato, 2012; Imamizu, Kuroda, Miyauchi, Yoshioka, & Kawato, 2003; Imamizu et al., 2000). Independently, Yomogida et al.'s (2004) brain-imaging studies strongly supported the findings of Imamizu and his colleagues. All of this imaging research is in turn based upon the landmark findings and proposals of Leiner, Leiner and Dow (1986) who traced the evolutionary emergence of these mental processes back at least one million years.
2 Other real-world prototyping frameworks for innovation include traditions of techniques in art and ornament-making, and music, including systems of music notation and, of course, the instruments and materials of the arts themselves.
3 All descriptions of Camp Invention and Camp Invention curricular activities for this chapter were taken directly from the Camp Invention Internet website: www.invent.org/camp/default.aspx.

References

Ambrose, S. (2001, March 2). Paleolithic technology and human evolution. *Science, 291*(5509), 1748–1753.
Ericsson, K. A. (2002). Attaining excellence through deliberate practice: Insights from the study of expert performance. In M. Ferrari (Ed.), *The Pursuit of Excellence Through Education* (pp. 21–55). Mahwah, NJ: Lawrence Erlbaum Associates.
Ericsson, K. A. (2003a). The acquisition of expert performance as problem solving. In J. E. Davidson & R. J. Sternberg (Eds.), *The Psychology of Problem Solving* (pp. 31–83). Cambridge, UK: Cambridge University Press.
Ericsson, K. A. (2003b). The search for general abilities and basic capacities: Theoretical implications from the modifiability and complexity of mechanisms mediating expert performance. In R. J. Sternberg & E. I. Grigorenko (Eds.), *The Psychology of Abilities, Competencies, and Expertise* (pp. 93–125). Cambridge, UK: Cambridge University Press.
Ericsson, K. A., Nandagopal, K., & Roring, R. W. (2009). An expert-performance approach to the study of giftedness. In L. Shavinina (Ed.), *International Handbook of Giftedness* (pp. 129–153). Berlin, Germany: Springer Science + Business Media.
Ericsson, K. A., Roring, R., & Nandagopal, K. (2007) Giftedness and evidence for reproducibly superior performance: An account based on the expert performance framework. *High Ability Studies, 18*(1), 3–56.
Imamizu, H., Higuchi, S., Toda, A., & Kawato, M. (2007). Reorganization of brain activity for multiple internal models after short but intensive training. *Cortex, 43*(3), 338–349.
Imamizu, H., & Kawato, M. (2009). Brain mechanisms for predictive control by switching internal models: Implications for higher-order cognitive functions. *Psychological Research, 73*(4), 527–544.
Imamizu, H., & Kawato, M. (2012). Cerebellar internal models: Implications for dexterous use of tools. *Cerebellum, 11*(2), 325–335.

Imamizu, H., Kuroda, T., Miyauchi, S., Yoshioka, T., & Kawato, M. (2003). Modular organization of internal models of tools in the cerebellum. *Proceedings of the National Academy of Science, 100*(9), 5461–5466.

Imamizu, H., Miyauchi, S., Tamada, T., Sasaki, Y., Takino, R., Pütz, B., Yoshioka, T., & Kawato, M. (2000). Human cerebellar activity reflecting an acquired internal model of a new tool. *Nature, 403*(6766), 192–195.

Josephson, M. (1992). *Edison: A Biography.* New York: John Wiley and Sons, Inc. www.nps.gov/nr/twhp/wwwlps/lessons/25edison/25edison.htm.

Leiner, H., Leiner, A., & Dow, R. (1986). Does the cerebellum contribute to mental skills? *Behavioral Neuroscience, 100*(4), 443–454.

Pfeiffer, J. (1982). *The Creative Explosion: An Inquiry into the Origins of Art and Religion.* New York: Harper and Row.

Schrage, M. (1999). *Serious Play: Simulate to Innovate.* Boston: Harvard Business School Press.

Shavinina, L. (2008). Early signs of entrepreneurial giftedness. *Gifted and Talented International, 23*(2), 3–17.

Shavinina, L. (2009). Entrepreneurial giftedness. In L. Shavinina (Ed.), *International Handbook on Giftedness* (pp. 793–807). Dordrecht: Springer.

Vandervert, L. (2006). Sky-high innovation in the boardroom: Theory-Z for boards. *Innovation Journal: The Public Sector Innovation Journal, 11*(2), article 5, www.innovation.cc/volumes-issues/vandervert+sky=high4.pdf.

Vandervert, L. (2007). Cognitive functions of the cerebellum explain how Ericsson's deliberate practice produces giftedness. *High Ability Studies, 18*(1), 89–92.

Vandervert, L. (2009). The emergence of the child prodigy 10,000 years ago: An evolutionary and developmental explanation. *Journal of Mind and Behavior, 30*(1/2), 15–32.

Vandervert, L. (2011a). The evolution of language: The cerebro-cerebellar blending of visual-spatial working memory with vocalizations. *Journal of Mind and Behavior, 32*(4), 317–332.

Vandervert, L. (2011b). How Thomas Edison used a results focus to produce constant invention and innovation. *Board Leadership, 2011*(117), 1–8.

Vandervert, L., Schimpf, P., & Liu, H. (2007). How working memory and the cognitive functions of the cerebellum collaborate to produce creativity and innovation. *Creativity Research Journal, 19*(1), 1–18.

Yomogida, Y., Sugiura, M., Watanabe, J., Akitsuki, Y., Sassa, Y., Sato, T., Matsue, Y., & Kawashima, R. (2004). Mental visual synthesis is originated in the fronto-temporal network of the left hemisphere. *Cerebral Cortex, 14*(12), 1376–1383.

7

How can scientific innovators–geniuses be developed?

The case of Albert Einstein

Larisa V. Shavinina

UNIVERSITÉ DU QUÉBEC EN OUTAOUAIS, CANADA

Summary: The first years of a child's life are characterized by a number of sensitive periods—periods of heightened and selective responsiveness to everything that is going on around him or her.[1] Sensitive periods—which constitute the developmental foundation of individual innovation[2]—accelerate the child's development through the actualization of his or her intellectual potential and the growth of cognitive resources.[3] This chapter is about the developmental foundation of innovation in the case of scientific talent. It explores the role of sensitive periods in the development of Albert Einstein. It was found that a number of overlapping sensitive periods distinguished a trajectory of Einstein's educational development and this explains the emergence of his scientific genius. The implications for innovation education are discussed.

Key words: Sensitivity, sensitive periods, innovation, innovator, scientific talent, genius, accelerated child's development, Albert Einstein, Nobel caliber.

Introduction

Although young Albert did not say a word until the age of two or three; was a late and quite a slow speaker; was a loner, a poor student, and was expelled from the Munich school (Gardner, 1993; Isaacson, 2007; Kuznetsov, 1979; Neffe & Frisch, 2007; Shavinina, 2006), today everyone knows Einstein as a brilliant scientist, the greatest mind of the 20th century, who was able to came up with radically innovative theories. But what in his childhood or adolescence predicted the later manifestations of scientific genius–great innovator? The goal of this chapter is to ultimately answer this question.

The chapter proceeds as follows. The first section discusses what developmental psychology and high ability studies tell us about sensitive periods. The second section discusses Albert Einstein's sensitive periods in childhood and adolescence and their impact on the development of his scientific genius. The third section—with four sub-sections—explains the emergence of Einstein's scientific talent. The concluding section summarizes the presented findings and describes their implications for innovation education.

The findings reported in this chapter resulted from the research project entitled *A Study of Early Childhood and Adolescent Education of Nobel Laureates and the Implications for Gifted and General Education: Developing Scientific Talent of Nobel Caliber.* The project was sponsored by the Social Sciences and Humanities Research Council (SSHRC) of Canada. A total of 869 pages of documents regarding early childhood and adolescent education of Albert Einstein alone were collected from all the existing printed sources (i.e., mainly (auto) biographical notes and biographies) published in all languages. This chapter is thus based on the analysis of these documents.

Methodological issues

The data sources used in this case study were Albert Einstein's autobiographical notes and biographies. The use of the (auto) biographical accounts and the case study method can be to a certain extent a controversial matter. However, any attempt at a comprehensive understanding of innovative scientists and other distinguishing individuals is not feasible without reliance on such accounts. These accounts and the case-study method are perfectly suited for capturing the special characteristics of innovative individuals. (Auto) biographical literature is essential for the research and description of persons or events distinguished by their rarity as is the case with exceptional genius of Albert Einstein. Using this literature, psychologists can describe the idiosyncratic features of innovators including the characteristics of their educational development.

Usually, the use of (auto) biographical literature for the study of outstanding individuals has some limitations:

a   The possible subjectivity and contradictions of autobiographers in their accounts of their own thinking processes, psychological states, and the surrounding events, which lead up to and follow their innovative achievements in science. The subjectivity is even more important in the case of biographers resulting from their individual interpretations of thoughts, states, and events. These interpretations may be influenced by their personal attitudes toward the person about whom they write, an attitude potentially swayed in part by whether the latter is living or not.
b   The timing of writing the (auto) biography, normally after an individual has already become a famous personality. In this case the writer relies on vague memories of his thinking processes, which may have likely been weakened or altered over time. It definitely raises the question about the validity and reliability of subjective reports as data (Brown, 1978, 1984; Ericsson & Simon, 1980).

However, these concerns are only partially true in the case of Albert Einstein. First, he wrote the autobiographical notes himself. Second, Albert's sister wrote about his childhood. Third, other people, who knew Einstein well, collected stories about his young years based on their conversations with him or on their personal experience (e.g., Helen Dukas, his secretary, or Elsa Einstein, the second wife of Albert, who was also his cousin and personally remembered their childhood years). This is what makes Albert Einstein's (auto) biographical sources quite reliable data sources. All Einstein's biographers relied on these three main sources of information.

What do developmental psychology and high ability studies tell us about sensitivity and sensitive periods?

Human development is not a smooth process. Instead, it has certain stages or periods. Many scientific findings, both in general and developmental psychology, testify to it (Ananiev, 1957; Case,

1984a, 1984b; Fischer & Pipp, 1984; Flavell, 1984; Piaget, 1952; Sternberg, 1990; Vygotsky, 1956, 1972; Wallon, 1945), especially research on giftedness[4] in the framework of the developmental approach, which reveals an uneven development of the gifted (Bamberger, 1982, 1986; Feldman, 1982, 1986a; Gruber, 1982, 1986; Shavinina, 1997; Silverman, 1993, 1994, 1997; Terrassier, 1985, 1992). In this context, the most interesting ideas can be found in the studies on the asynchronous (Silverman, 1993) and dyssynchronous (Terrassier, 1985) development of gifted children. Moreover, the Columbus group views giftedness in general as an "asynchronous development" (Silverman, 1993, p. 634). The interpretation of giftedness as synonymous with asynchronous development indicates the importance of such a development for the understanding of the nature of giftedness. The Columbus group's approach to giftedness is also a strong evidence of the uneven and periodical development of the gifted and prodigies (Shavinina, 1997). However, this specificity of the development of the gifted has only been described, not explained. Shavinina (1997) suggested that a key to the understanding of their uneven, asynchronous, dyssynchronous, and, therefore, unique development should be seen in a child's age sensitivity.

Examining the individual development of prodigies, Shavinina (1999) concluded that a child's sensitivity plays a central role in the emergence of prodigies and the gifted. *Age sensitivity* is defined as a specific, heightened, and very selective responsiveness of an individual to everything that is going on around him or her (Leites, 1996). Such a definition might seem rather general; however, it appears to be appropriate at the current stage of research on sensitivity of the gifted, when the amount of data is restricted. Indications that age sensitivity takes a certain place in the appearance of the gifted can be found in the literature (Feldman, 1986b; Jellen & Verduin, 1986; Leites, 1960, 1971, 1996; Kholodnaya, 1993; Piechowski, 1979, 1986, 1991; Silverman, 1993, 1994, 1997; Sternberg, 1986). Leites (1971) found that the child's sensitivity and sensitive periods are critical phenomena in the development of prodigies. Kholodnaya (1993) viewed prodigy phenomenon as a result of the specific development of a child during the early years. Piechowski (1991) considered sensitivity as an individual's heightened response to selective sensory or intellectual experiences. He pointed out that unusual sensitivity reveals the potential for high levels of development, especially for self-actualization and moral vigor (Piechowski, 1979, 1986; Jackson, Moyle, & Piechowski, 2009). Feldman (1986b) included unusual sensitivity in his theory of prodigy phenomenon, in the "individual psychological qualities" component. Sternberg (1986) considered "sensitivity to external feedback" as one of the metacomponents of his theory of intellectual giftedness. Sensitivity is also one of the main elements in Jellen and Verduin's (1986) conception of giftedness.

Shavinina (1997) distinguishes between cognitive (i.e., sensitivity to any new information), emotional (i.e., sensitivity to one's own inner world and to the inner words of other people), and social kinds of sensitivity, which intersect with one another, forming mixed kinds of sensitivity. Leites (1971) emphasized that each child's age is characterized by one or numerous kinds of sensitivity. Vulnerability, fragility, empathy, social responsiveness, and moral responsibility are the manifestations of sensitivity. Cognitive sensitivity is extremely important in a child's development in general and in the development of the gifted in particular. Thus, the first years of a child's life are characterized by the ease and stability of the formation of many skills and habits (for example, linguistic abilities; Leites, 1996). Probably, because of cognitive sensitivity, children's knowledge acquisition is very quick; it may take place even from the very first experience. This is especially applicable to exceptionally gifted children: prodigies (see Shavinina, 1999, for examples).

A child's sensitivity is not always the same, it changes with age. Special age periods of the child's heightened sensitivity are defined as *sensitive periods*.[5] Exceptionally favorable inner conditions and extraordinary possibilities for cognitive and intellectual development are presented during sensitive periods (Leites, 1996). The early years of language acquisition by children is a widely cited example of sensitive periods (Shavinina, 1997). Sensitive periods are particularly impressive in the case of prodigies (Shavinina, 2009).

Sensitive periods demonstrate the uniqueness of certain stages in a child's development and the tremendous potential of childhood. They provide temporary favorable conditions for accelerated intellectual development. Such periods occur in each child's age, even at the earliest years. For example, Skuse, Pickles, Wolke, and Reilly (1994) found that the first few postnatal months constitute a sensitive period for the relationship between growth and mental development. It seems that childhood periods prepare and temporarily conserve favorable internal possibilities for the development of exceptional early abilities.

The analysis of the gifted at sensitive periods demonstrates that each child is sensitive in his or her own way (Leites, 1971). Based on examples of sensitive periods in prodigies, Leites (1996) concluded that the specificity of a child's mind depends on the age period in which intellectual abilities appear. He found that in childhood years, the specific "temporary states"—sensitive periods—emerge at each age stage, which manifest significant opportunities for advanced mental development. Zaporozhets (1964) pointed out that "each period of a child's development has its own age sensitivity, and because of that learning ... is more successful in the early years, than in the elder ones" (p. 678).

Rosenblatt (1976) also noticed the existence of sensitive periods and emphasized that all behavioral development (both in people and in animals) is divided into sensitive periods, among which there are internal relations and mutual transitions. He found that the rapidity of the appearance, effectiveness, and duration of a sensitive period depends on the specificity of the previous periods. Furthermore, Rosenblatt demonstrated that certain stages of development appear within a sensitive period itself. The change of sensitive periods interrupts and, at the same time, continues the course of the individual's development. Vygotsky (1956) acknowledged the existence of sensitive periods and emphasized that:

> in these periods certain influences have big impact on all course of the individual development by provoking one or other deep changes. In other periods, the same influences can be neutral or even give opposite impact on child development. Sensitive periods coincide fully with ... the optimal terms of learning.
>
> (p. 278)

A child's heightened level of sensitivity is extremely important for the understanding of his or her development. The unusual three-week period of intensive and non-child questioning (about God, life, the universe, death, her own mortality, and similar questions) in 4½-year-old Jennie studied by Morelock and described by Silverman (1993) is a good example of a gifted child's sensitive period. This period of questioning can be explained by a three-week period of heightened sensitivity to everything unknown to her. It should be added that the following stage in Jennie's mental development was also a clearly distinct period, with a change in her external behavior (e.g., she became quiet) and an incredible shift in her reading ability. Probably, it was her second sensitive period that explains Jennie's transformed behavior. The change of sensitive periods was very positive for her intellectual development, since Jennie reached new levels of cognition (Shavinina, 1999). It is not surprising that psychologists use a concept of "cognitive leap" to describe her remarkable intellectual growth (Silverman, 1993).

Such cases of sensitive periods in highly gifted children allowed Leites (1996) to conclude that even age sensitivity itself can be considered a specific kind of giftedness. The changes of age bring, therefore, unrepeatable determinants of the individual development: sensitive periods. Sensitive periods mean a qualitatively new strengthening of the possibilities for mental growth, which appear during the early childhood years. The strengthening of such possibilities leads to the general heightening of a child's cognitive resources, and this is what greatly accelerates his or her individual—and especially intellectual—development.

Sensitive periods in the individual development of Albert Einstein

Based on the available (auto) biographical sources, at least six powerful sensitive periods can be identified during Einstein's childhood and adolescence. The first was about assembling complicated structures with building blocks and erecting tall towers of playing cards as high as fourteen stories (Hoffman & Dukas, 1972; von Boehm, 2005). As his sister Maja later recalled, little Albert would spend hours patiently building those houses of cards, undeterred by any upsets he may have encountered during his play (Calaprice & Lipscombe, 2005). This testifies to his unusual manual dexterity. During this period young Albert also loved mechanical toys, especially finding out how they worked (Highfield & Carter, 1993; Michio, 2004; Milton, 1983). Because of his father and uncle's electrochemical business, there were plenty of electric dynamos, motors, and gadgets lying around their private factory. Such toys naturally nourished his curiosity. Specifically, little Albert could simply look at the gadgets surrounding the factory and intuitively tried to understand electricity and magnetism (Michio, 2004). The first sensitive period thus definitely stimulated Einstein's interest in science and developed his persistence. This period started in early childhood, well before Albert began to talk, and lasted up to the age of six.

The second sensitive period was a magnetic compass that Albert's father presented to him at the age of four or five. He was sick in bed and the father showed him a small compass. It was intended as an amusing toy, a diversion, but it had a strong impression on the young child. It was the first miracle or "wonder" in his life. Many years later Einstein recollected that great impact, which the compass had on his subsequent intellectual development.

> The development of this world of thought is in a certain sense a continuous flight from "wonder." A wonder of this kind I experienced as a child of four or five years when my father showed me a compass. That this needle behaved in such a determined way did not at all fit into the kind of occurrence that could find a place in the unconscious world of concepts (efficacy produced by direct "touch"). I can still remember – or at least believe I can remember – that this experience made a deep and lasting impression upon me. Something deeply hidden had to be behind things. What man sees before him from infancy causes to reaction of this kind; he is not surprised by the falling of bodies, by wind and rain, nor by the moon, nor by the fact that the moon does not fall down, nor by the differences between living and nonliving matter.
>
> *(Einstein, 1949, p. 9)*

The compass was a fascinating mystery for little Albert. He came to realize that space was not empty! There was something in the void that was gripping the iron needle and holding it fixed. Einstein had no idea what a magnetic field was, but he was enchanted by the invisible force that filled what seemed to be empty space. Where did it come from? As his quote above indicates, the wonder and awe of the compass never left him (Highfield & Carter, 1993; Michio, 2004; Milton, 1983). Einstein's interest in the universe and nature—especially space and time—thus originated from the first "wonder": that small magnetic compass. The capacity to wonder that thus appeared early in Albert's life can, therefore, be considered as one of the outcomes of the second sensitive period.

The third sensitive period was about religion and it began at the age of five. Little Albert was attracted by traditional religion and could not accept his father's skepticism with respect to God and religious rituals (Whitrow, 1967). At this time he even refused to eat pork and criticized his family for not being religious. In spite of his intense religious feelings, he did not receive the traditional religious instruction for which, at that time, he longed. So, acting on his own, little Albert wrote and set to music brief songs in praise of God and sang them in his home and on the street. He identified God with nature. He was carried away by a Spinozistic pantheism. Like the young Goethe, he felt nature an all-embracing, all fulfilling power (Reiser, 1930). This religious feeling later found expression in a profound belief in the essential harmony of the laws of nature (Whitrow, 1967). This sensitive period lasted up to the age of twelve.

When he was twelve or thirteen, his attention was drawn to a series of popular books on natural science which he absorbed with a great interest. One striking effect of the science books on Albert was to make him suddenly antireligious. For a while he became not just a nonbeliever but a fanatical skeptic, profoundly suspicious of authority (Hoffman & Dukas, 1972). The reading of philosophical books also made him skeptical about the existence of God, especially *Critic of Practical Reason* by German philosopher Immanuel Kant (see below). Although eventually Einstein rejected religious ritual and decided not to join any religious group, he became convinced of the great ethical value of the Biblical tradition (Whitrow, 1967).

The fourth sensitive period is related to mathematics. It was uncle Jakob, a brother of his father, who introduced little Albert to algebra and geometry, portraying math as a jolly game, hunting for the animal "x" whose name we do not know (Fölsing, 1997; Highfield & Carter, 1993; Lassieur, 2005). Albert found much pleasure in solving simple problems by ways of his own instead of slavishly following a standard method. Einstein's interest in mathematics was thus first aroused at home and not at school.

Before Albert had started to study geometry, uncle Jakob told him of the Pythagorean Theorem. Albert was fascinated. After strenuous effort he found a way to prove the theorem—an extraordinary feat under the circumstances and one that must have given both him and his uncle intense pleasure. Yet, strangely, this pleasure seems to have been negligible compared with the emotion later aroused in him by a small textbook on Euclidian geometry in which he became entirely absorbed (Hoffman & Dukas, 1972).

When Einstein was twelve years old, Max Talmey, a perceptive medical student who for a while was a weekly visitor of the Einsteins, gave him this book of geometry. Geometry fitted nicely the boy's developed visual abilities. The book provoked a feeling of profound wonder and the child spent many hours on solving mathematical problems presented in that book. Geometry amazed and delighted Albert. He would later refer to the book as the "sacred little geometry book." Geometry was a perfect subject for Einstein, because he could think through a problem step by step and come to a concrete, sure answer. Soon Albert was so good in math that he had surpassed Max in knowledge of the subject (Lassieur, 2005).

As Einstein (1949, p. 9) put it himself,

> At the age of twelve I experienced a second wonder of a totally different nature – in a little book dealing with Euclidean plane geometry, which came into my hands at the beginning of a school year. I remember that an uncle told me about the Pythagorean Theorem before the holy geometry booklet had come into my hands. After much effort I succeeded I "proved" this theorem on the basis of the similarity of triangles; in doing so it seemed to me "evident" that the relations of the sides of the right-angled triangles would have to be completely determined by one of the acute angles.

Albert never forgot the enormous impression that this popular textbook on Euclidean geometry left on him. Later on he wrote: "Anyone who was not transported by this book in youth was not born to be a theoretical searcher" (quoted in Vallentin, 1954, pp. 21–22).

Before Einstein was sixteen, he read books on analytical geometry and calculus and was fortunate in hitting on authors who gave vivid expositions of the power of these methods, without bothering too much about mathematical rigor (Growther, 1955).

The fifth sensitive period for young Albert was about reading scientific books. In addition to integral and differential calculus, as well as analytical geometry, which he mastered all by himself, Albert read popular science books with great interest. These books had not come to him by accident. They had been deliberately put into his hands by Max Talmay, who had long discussions with young Albert, guiding him and widening his intellectual horizons at a crucial formative age (Fölsing, 1997; Hoffman & Dukas, 1972; Mih, 2000). Specifically, Max Talmay had introduced Einstein to Aaron Bernstein's *Popular Book on Natural Science*. Albert wrote later that it was a work which he read with "breathless attention" (quoted in Michio, 2004, p. 43). Bernstein asked the reader to take "a fanciful ride inside a telegraph wire, racing alongside an electric signal at fantastic speeds" (Growther, 1955). Because of this included discussion on the mysteries of electricity, this book had a fateful impact on Albert.

"I also remember," said Einstein to Carl Seelig, one of his chief biographers, "that at the age of 13 I read with enthusiasm Ludwig Buchner's *Force and Matter*, a book, which I later found to be rather childish in its ingenuous realism" (Seeling, 1956, p. 12). As a teenager, Albert was more and more turning his mind to the problems that scientists of that day were struggling with. He was trying to understand space and time. He had already acquired some background knowledge to understand why these two ideas were so puzzling (Milton, 1983). The great problems of the universe, whose eternal laws he aspired to decipher, were thus of great interest to Einstein in adolescence (Seeling, 1956).

However, Albert did not simply read popular books on natural science. He could hold an idea about which he read in his mind for months. Einstein-teenager could examine the problem from different angles and make daring guesses as to the answer. Some authors believe (e.g., Milton, 1983) that this gift—an extraordinary concentration for months and later for years—is what separates a genius from man of talent. It is the insight of a Galileo, a Newton, and an Einstein. These men would doggedly stay with a problem until they saw what the answer must be; then they went back and "prettied it up" with mathematics. At 16, Albert already had his gift—as he would prove some 9 years later (Milton, 1983).

The sixth sensitive period was a philosophical one. When Einstein began teaching himself higher mathematics and acquired a great deal of knowledge, Talmey, in self-defence, had to turn their discussions to philosophy, where he could still hold his own. Recalling those days, Talmey wrote, "I recommended to him the reading of Kant. At that time he was still a child, only thirteen years old, yet Kant's works, incomprehensible to ordinary mortals, seemed clear to him" (quoted in Hoffman & Dukas, 1972, p. 24; as well as in Fölsing, 1997, pp. 22–23; and Mih, 2000, p. 5).

At thirteen-years-old, Einstein thus became seriously interested in philosophy. He particularly studied Immanuel Kant's *Critic of Practical Reason*, whose moral theory of freedom of the will appeared to resonate with him (Calaprice & Lipscombe, 2005). Kant had a great impression on Albert, especially those questions that preoccupied him in his youth such as: What should I do? and What should I know? From this time on, the idea of Kant, according to which God probably does not exist, did not leave Einstein.

It looks like his interest in philosophy was partially determined by his personality. For instance, Einstein-teenager refused to believe anything he read until he had proved it for himself. He questioned everything. "How do we know this is true?" and even "What is truth?" In search for the answers, he thus read Kant and other German philosophers who also wanted to know what truth was. What was difficult for most adults was soon clear to the eager young Albert (Milton, 1983).

Therefore, Einstein's individual development in childhood and adolescence was characterized by six sensitive periods: (1) assembling complicated structures with building blocks, building houses of cards, playing with mechanical toys, and finding out how they worked; (2) the magnetic compass that had a great impact on little Albert; (3) a vivid interest in religion; (4) a growing and life-long interest in mathematics; (5) reading scientific books; and (6) interest in philosophy. How do these six sensitive periods determine the development of Einstein as a scientific innovator–genius?

Explaining the emergence of scientific innovator–genius

Was Einstein scientific genius–radical innovator in his childhood? If we interpret his early manifestations described at the beginning of this chapter in light of the traditional understanding of giftedness, the answer is a firm "No." It is clear that he was a typical underachiever. All relatives and biographers testify to it. Later Albert Einstein himself wrote in his autobiography that "my intellectual development was retarded" (Einstein, 1949). In accordance with the theory of individual innovation (see Chapter 12 of this volume *Do not overlook innovators!: discussing the "silent" issues of the assessment of innovative abilities in today's children–tomorrow's innovators*) and the cognitive-developmental theory of giftedness (Shavinina, 2009), the answer is a definitive "Yes." Young Albert was a gifted underachiever–potential innovator characterized by uneven development in childhood that can be explained by periods of heightened sensitivity or sensitive periods. These periods had played an exceptional role in understanding the emergence of a scientific innovator–genius in Einstein.

Sensitive periods as the developmental foundation of scientific innovator–genius

As was shown above, developmental psychology and high ability studies clearly indicate that sensitive periods provide favorable inner opportunities for the development of a child's mind. Specifically, they accelerate an individual's intellectual growth. Nevertheless, they have their own dialectics. The sad thing is that later such favorable possibilities for individual development will weaken at a fast or slow rate (Leites, 1971). In this light a few interesting questions arise: if sensitive periods constitute an inner developmental mechanism of scientific innovator–genius, can a child be named as gifted if he or she had one or a few sensitive periods? Similarly, can sensitive periods experienced by a child be the predictors of his or her intellectually creative productivity and innovation in adulthood? Certainly, sensitive periods indicate that exceptional development can be possible. Nonetheless, it is not enough. The answer to these questions will be "yes" only if two important requirements are fulfilled in the individual development of a child. First, all *developmental* capacities (i.e., new abilities, habits, skills, qualities, traits, and characteristics acquired during sensitive periods; these capacities can be called *developmental* capacities or acquisitions because sensitive periods are a *developmental* phenomenon in the life of a child; a manifestation of a child's development) should be transformed into the stable *individual* acquisitions. Second, these acquired individual capacities should, in turn, be transformed into the unique cognitive experience of a child.

Although all stages of childhood can be distinguished by the heightened sensitivity of a child, sensitive periods have their own "life story." Sensitive periods emerge, exist, and even disappear during a child's development (Leites, 1971). What is important is what remains in the child at the end of sensitive period(s) when these periods are already over and favorable opportunities for mental development are getting weak either suddenly or gradually. It seems paradoxical, but this is the reality: the favorable possibilities opened up by sensitive periods allow a child to advance significantly in his or her intellectual development by acquiring something new and valuable (i.e., knowledge, skills, habits, and so on), but he or she can also lose these acquisitions when a sensitive period is over. This is a real problem of sensitive periods. This is why researchers differentiate between *developmental* and *individual* aspects of sensitive periods (Leites, 1996).

If, at the end of a sensitive period a child loses almost all the exceptional capacities that he or she acquired during the given period, then one can assert that these capacities were mainly a *developmental* phenomenon (i.e., *developmental* capacities). That is, a certain stage in the development of a child is over and all the extraordinary acquisitions accumulated during this stage via sensitive period(s) are lost. It is a key to the explanation of why so many gifted individuals who manifested exceptional abilities in their childhood become ordinary adults who do not express extraordinary talents, outstanding creativity, and innovation. Gifted children lose their unusual abilities and talents in the process of their own individual development.

At the same time, sensitive periods are also a foundation for the powerful *individual* acquisitions. If new extraordinary capacities acquired during a certain sensitive period remain in the developing child after this period, then one can assert that these capacities have been transformed into the *individual* acquisitions. Only in this case one can assume to a great extent that the child is gifted and he or she has the potential to be an intellectually creative adult or an innovator. What can we say about Albert Einstein in this regard? One can see that Einstein's sensitive periods during childhood and adolescence had great impact on his intellectual development.

The impact of Einstein's sensitive periods on his development

If one looks at specific outcomes of each sensitive period that Albert experienced early in life, then one sees the following results. The first sensitive period—assembling with building blocks, building houses of cards, playing with mechanical toys, and finding out how they worked—developed persistence and unusual ability to concentrate in a little boy. As his sister Maja later recalled,

> Even in a large, quite noisy group, he could withdraw to the sofa, take pen and paper in hand, set the inkstand precariously on the armrest, and lose himself completely in the conversation of many voices stimulated rather than disturbed him.
>
> *(quoted in Isaacson, 2007, p. 24)*

Later he needed this exceptional ability to think for months and years on big scientific questions. It was noted above, that this quality is a hallmark of genius. Extraordinary focus and persistence are among the vital characteristics of innovators. For instance, they never give up after the first failed ventures (for examples see Chapter 16 included in this volume, *The trajectory of early development of prominent innovators: entrepreneurial giftedness in childhood*).

Albert's early interest in mechanical toys stimulated his intuitive[6] understanding of electricity and magnetism (Michio, 2004). As an adolescent, Einstein was able to help with mechanical problems. Thus, when his uncle Jacob had great difficulty with calculations for the construction of some machine, Albert helped him quickly.

> You know,—said Jacob many years later to Albert's friend, Otto Neustatter—it is really fabulous with my nephew. After I and my assistant-engineer had been racking our brains for days, that young sprig had got the whole thing in scarcely 15 minutes. You will hear of him yet.
>
> *(quoted in Hoffman & Dukas, 1972, pp. 27–28)*

Therefore, one can see the lasting impact of the first sensitive period on the development of Einstein's scientific talent. His *developmental capacities* (e.g., building houses of cards and an interest in mechanical toys) were transformed into the stable *individual* acquisitions: patience, a high concentration, and knowledge of mechanics.

With respect to the second sensitive period related to the compass, it is obvious that little Albert for the first time had a presentiment of the mysterious web of nature, of its prodigious power, to which all being is subject. The parents were amazed when they saw the child, usually phlegmatic and absent-minded, show a passionate interest for the little finger that vibrated of its own accord as though propelled by a mysterious force. He had probably already realized the relation existing between an external cause and the effect; he knew that things moved because they were touched. But this finger of the compass behaved in a manner that never ceased to surprise Albert and for which he found no explanation. An infinitesimal seed must have been sown at that moment. "The evolution of our world of ideas [Einstein wrote later in his autobiography, recalling this scene of his early childhood] is in a certain sense a constant struggle against the 'miraculous'" (Einstein, 1949, p. 9). The mystery of the compass's action impressed him to a great extent. It seemed that something outside the ordinary course of nature was present. He was overcome with a feeling of awe, and could remember that he trembled with wonder at the phenomenon, and turned cold. As he subsequently recognized, it was his first acquaintance with the electromagnetic field (Growther, 1955). The magnetic compass—given as a present to little Albert when he was four or five—thus altered his entire life and etched forever in his mind, as well as in the history of science (Isaacson, 2007). This sensitive period thus generated a sense of wonder in Einstein, an unusual curiosity about everything around him leading to the development of an ability to pose interesting questions related to the fundamental issues of nature. His *developmental capacities* (i.e., a sense of wonder) were indeed transformed into the stable *individual* acquisitions: an amazing sense of curiosity.

One of the most important outcomes of Albert's strong religious feelings experienced in childhood—his third sensitive period—was the development of his almost religious devotion to the wonders of nature (Reiser, 1930). Also, his dislike for any authority originated from this time. One can see that in this case, again, Einstein's *developmental* capacities (i.e., intense religious feelings) were transformed into powerful *individual* acquisitions (i.e., religious fondness to the miracles of nature and the universe), which influenced the development of his scientific talent.

The mathematical sensitive period had a long-lasting effect on Albert's intellectual growth as well. Thus, as a secondary school boy, he had solved Pythagorean Theorem by his own power of reasoning without a teacher (Seeling, 1956). Then the book on Euclidean geometry, which Albert discovered at the age of twelve, impressed him enormously. In his "Autobiographical Notes" he speaks rapturously of "the holy geometry booklet" and says:

> here were assertions, as for example the intersection of the three altitudes of a triangle in one point, which – though by no means evident – could nevertheless be proved with such certainty that any doubt appeared to be out of the question. This lucidity and certainty made an indescribable impression on me.
>
> *(quoted in Hoffman & Dukas, 1972, p. 22)*

Einstein's interest in math was such a consuming one that eventually between the ages of twelve and sixteen he learned the elements of mathematics, including the principles of differential and integral calculus, himself (Seeling, 1956). Therefore, his *developmental capacities* (i.e., interest in math) were transformed into the strong *individual* acquisitions: profound math knowledge needed for a future scientist wishing to work at the cutting-edge of physics and a powerful ability for self-study.

The sensitive period related to reading a series of popular books on natural science, which began when Albert was about twelve or thirteen, was a significant one for the development of his scientific genius. Such intellectual experiences—as, for instance, reading Ludwig Buchner's *Force and Matter*—transformed Einstein, who as a child was slow, almost pathologically modest, and bode fair to be an introspective dreamer, into an independent scholar. Distrustful of any attempt on the part of authority to interfere with the freedom of adolescence, he soon found himself in opposition to the Prussian system of power (Seeling, 1956). The impact of this sensitive period can thus be seen in Albert's developing directness and readiness to question everything that others took for granted (Highfield & Carter, 1993). This is a sign of great scientists–outstanding innovators.

His even more developed curiosity resulted from this sensitive period as well: "There is such a thing as a passionate desire to understand, just as there is a passionate love for music. This passion is common with children, but usually vanishes as they grow up. Without it, there would be no nature science and no mathematics." Thus Einstein wrote in 1950, on the occasion of the publication of the theory of generalized gravitation (quoted in Vallentin, 1954, p. 30).

Also, reading of scientific books affected his first writings. For example, some biographers see Albert's first essay as a "harbinger of what was to come." However, others think that this is over-interpreting it (e.g., Fölsing, 1997). In the wake of Heinrich Hertz's epoch-making discovery a great many popular accounts of electromagnetic theory appeared in Germany, and Einstein would have read at least some of these. 'In fact, striking parallels have been found between passages in Einstein's first text and a chapter, *The Revolution in Our Concepts of the Nature of Electrical Effects*, in a popular-science monthly" (Fölsing, 1997, pp. 35–36).

In the case with this sensitive period, one can see again that his *developmental capacities* (i.e., interest in reading popular books on natural science) were transformed into the strong *individual* acquisitions: curiosity, readiness to question everything, and a deep interest in science.

Whitehead has said that philosophy begins with wonder, and, as it was mentioned above, Einstein's first recorded experience of his sense of wonder at the marvels of the nature occurred when his father showed him a magnetic compass. The sight of the needle pointing always in the same direction, whenever the compass case was turned about, made a deep and lasting impression on him (Whitrow, 1967). Here one can see a link between the two sensitive periods: the one related to the compass and another related to philosophy. Albert became an antireligious person after reading philosophical books. At the same time he got even more interested in the fundamental issues of the universe. Reading of Kant further developed his willingness to question everything. These are the three main outcomes of the philosophical sensitive period. As in the cases with other periods, *developmental capacities* of this period (i.e., interest in questions that preoccupied philosophers) were transformed into powerful *individual* acquisitions (i.e., interest in the universe and readiness to question everything).

Therefore, each sensitive period experienced by Albert Einstein in his childhood and adolescence had a significant impact on the development of his scientific talent. However, this is not the whole story about his sensitive periods.

Understanding exceptionality of Einstein's development: a chain of sensitive periods

Analyzing child prodigies at sensitive periods, Shavinina (1999) discovered a chain of sensitive periods in their individual development, which allowed her to conclude that the prodigies' sensitivity does not disappear completely. In this light, Silverman's (1993) conclusion concerning emotional sensitivity seems to be correct. She asserted that "extraordinary levels of sensitivity and compassion do not disappear with maturity. A capacity for rich, intense emotions remains in the personality throughout the lifespan" (p. 642). Shavinina (2009) found that it depends on the kind of sensitivity (i.e., cognitive, emotional, or social). Perhaps, emotional sensitivity, more than any other kind of sensitivity, remains in the individual during his or her life, whereas cognitive sensitivity changes periodically (but certainly it does not disappear in prodigies and the gifted). Such characteristics of the gifted as sensitivity to a new experience and an openness of mind—which are the manifestations of cognitive sensitivity—can be regarded as evidence of this tendency of cognitive sensitivity. These are also the distinguishing characteristics of prominent innovators (see Chapter 36 of this volume, *What can innovation education learn from innovators with longstanding records of breakthrough innovations?*). The availability of cognitive sensitivity throughout the lifespan most likely determines the exceptional abilities of an individual. It was the case for Albert Einstein: he was curious and open-minded all his life (Calaprice & Lipscombe, 2005; Hoffman & Dukas, 1972; Michio, 2004; Miller, 1996; Milton, 1983; von Boehm, 2005).

If sensitivity remains in prodigies and the gifted for a long time (Roedell, 1984), then Shavinina (1999, 2009) assumed that it is possible to assert that new capacities acquired during a certain sensitive period will also remain in them for a long time. These capacities are fortified and developed later, and finally they are transformed into really *individual* acquisitions that have a potential to remain in the person throughout their life. As a result, one can predict the transition of a gifted child into an outstanding adult who will be able to produce extraordinary intellectual and creative performance(s) and exceptional innovative achievement(s).

Furthermore, the developmental trajectory of Albert Einstein indicates a *chain* of sensitive periods: the first period—assembling building blocks, building houses of cards, and playing with mechanical toys—that started at the age of two was followed by his interest in the magnetic compass at the age of four or five. Then were the religious and mathematical sensitive periods, which began when little Albert was five or six years old. Finally, the periods of immense interest in reading scientific and philosophical books appeared when he was twelve or thirteen years old. The individual development of Albert Einstein also demonstrates that his *developmental* capacities (i.e., those new capacities acquired during sensitive periods in childhood) were indeed transformed into powerful *individual* abilities that remained throughout his life (e.g., an amazing curiosity, a high concentration, and an ability to ponder fundamental questions of nature at length, just to mention a few). Therefore, all the above written about sensitive periods shows that they were not a factor, condition, characteristic, feature, or trait in Einstein's development. They should be understood as an inner developmental mechanism of his scientific genius.

Overlapping sensitive periods and resulting unique cognitive experience

Shavinina (1997) found that any development leading to the significant expression of an individual's potential (in the forms of individual innovation, scientific genius, giftedness, creativity, exceptional wisdom, or extraordinary intuition) and resulting in any human achievement is influenced by a number of periods of heightened sensitivity. Perhaps the stages or levels of the gifted's development (Feldman, 1982, 1986a; Gruber, 1982, 1986) as well as prodigies' "mid-life" crisis (Bamberger, 1986) and "crystallizing experience" phenomenon (Walters & Gardner, 1986) correspond to certain sensitive periods (Shavinina, 1999).

Moreover, if we ask ourselves, what is behind "asynchrony" and "dyssynchrony," the answer probably is "sensitive periods." It is interesting to note that there are some indications to sensitive periods in the definition of giftedness given by the Columbus group.[7] For example, "advanced" (that means that something might not be advanced) and "heightened" (correspondingly, something might not be heightened). Furthermore, the asynchrony term itself is also connected to the very essence of sensitive periods in the following way: asynchrony assumes the emergence and disappearance (i.e., beginning and end) of certain qualities forming to some extent disproportionately in child development. Sensitive periods have their beginning and end as well.

What is very interesting in the case of the development of the gifted is not only a chain of their sensitive periods, but also the fact that they overlap. This is a key to understanding Einstein's development that led to the emergence of his scientific genius. The detailed analysis of his sensitive periods presented above demonstrates that his previous sensitive periods never disappeared. Rather, they continued to co-exist with the new periods, which appeared later. For instance, Albert's very first sensitive period in early childhood expressed by his interest in mechanical toys, among other things, continued in his youth and probably led to his famous "thought experiments" with elevators and other mechanical devices. The sense of wonder at the marvels of the nature occurred when Einstein's father showed him a magnetic compass, and this sense never left him. The same is true with respect to his interest in math and scientific and philosophical books. Consequently, one can see that his previous sensitive periods did not disappear at all. Instead, they overlapped and continued to co-exist.

Such an overlapping of age sensitivity and sensitive periods significantly strengthened the foundation for Albert's accelerated intellectual growth and thus advanced it. This meant rapid accumulation of his cognitive resources and the construction of those resources into the unique cognitive experience that continued to enrich itself in the process of the further advanced development governed mostly by heightened cognitive sensitivity. This experience (i.e., experience of the cognitive interaction of an individual with the external world) is the cognitive basis of giftedness (Shavinina & Kholodnaya, 1996) and individual innovation (Shavinina & Seeratan, 2003). Einstein's specific cognitive experience manifested itself in his unique type of representation of everything that was going on around him (i.e., any idea, event, problem, or situation). This is why he saw, understood, and interpreted the world around him in a different manner from the rest of people. His unique individual picture of the world, worldview, or vision is the essence of scientific genius and of breakthrough innovations in science. An important aspect of this unique picture of the world is its objectivization of cognition: he was able to "see the world as it was in its objective reality" (Kholodnaya, 1990, p. 128; Shavinina, 1996). Einstein's unique cognitive experience was ultimately responsible for his extraordinary achievements: the general and special theories of relativity. This account thus explains the emergence of Albert Einstein–scientific genius–radical innovator.

Summary and implications for innovation education

Albert Einstein's extraordinary scientific talent can be explained by a very specific individual development during childhood and adolescence. Particularly, he experienced a chain of overlapping sensitive periods, which greatly accelerated and thus advanced his mental development through the actualization of his intellectual potential and the growth of his cognitive resources resulting in the appearance of a unique cognitive experience that eventually led to the emergence of outstanding scientific genius–great innovator.

The presented perspective on the nature of scientific genius–great innovator has certain implications for innovation education. The analysis of Einstein's sensitive periods clearly demonstrates an immense potential of childhood. Educators should certainly keep this in mind: innovation potential originates from childhood. For instance, fast knowledge acquisition, increased mental activity, and creativity are evident during these periods. Educational influences will thus be more productive at sensitive periods.

Correspondingly, the effectiveness of any instruction will diminish at the end of sensitive periods. Teachers and parents should understand that if there are signs of a child's sensitivity to a certain knowledge domain, this is the best time for accelerated knowledge acquisition. The findings about Albert Einstein's sensitive periods presented in this chapter provide additional evidence in the favor of both accelerated and enriched educational options for today's children–tomorrow's innovators.

The identification of a child's sensitive periods should be one of the elements of the comprehensive approach to the assessment of innovative abilities (see Chapter 12 included in this volume, *Do not overlook innovators!: discussing the "silent" issues of the assessment of innovative abilities in today's children–tomorrow's innovators*). The case of Albert Einstein shows that one can predict an exceptional scientific genius–great innovator only when all developmental capacities are transformed into stable individual acquisitions. Educators should know the differences in developmental and individual specificity of sensitive periods when dealing with potential innovators.

Further research on the developmental foundation of individual innovation—that is, on sensitive periods—is definitely necessary. It will bring profound insights on the nature of scientific innovator–genius and hence advance our understanding of how to better develop innovators.

Acknowledgments

This research was supported by the Social Sciences and Humanities Research Council (SSHRC) of Canada. The findings and opinions presented in this chapter do not reflect the positions or policies of the SSHRC. Larry Vandervert was helpful in providing useful comments on the first draft of the chapter.

Notes

1 Innovators are sensitive individuals. They are able to sense hidden consumers' needs, changes in those needs, new market trends, promising technologies, and so on. For more on this subject see Chapter 36, *What can innovation education learn from innovators with longstanding records of breakthrough innovations?* included in this volume.
2 The model of individual innovation is presented in Chapter 12 of this volume, *Do not overlook innovators!: discussing the "silent" issues of the assessment of innovative abilities in today's children–tomorrow's innovators*.
3 This will lead to the appearance of a unique cognitive experience, which manifests itself in a specific type of representations: that is, how a child sees, understands, and interprets surrounding reality. In other words, this is a unique point of view or a unique vision, which is the essence of individual innovation (for a detailed account see Chapters 12 and 36 mentioned above).
4 Innovation is a special type of giftedness. Therefore, everything written in this chapter regarding gifted children is also applicable to innovators. Child prodigies represent the extreme, extraordinary advanced case of giftedness.
5 Certainly, such a definition of sensitive periods is rather general and other definitions of this construct can also exist in psychology. But, the current state of the research on the sensitive periods of the gifted is not very advanced; therefore, this definition seems to be appropriate.
6 Intuition is one of the components of individual innovation (for details see Chapters 3, 12, and 36 of this volume).
7 According to this definition, "giftedness is *asynchronous development* in which advanced cognitive abilities and heightened intensity combine to create inner experiences and awareness that are qualitatively different from the norm. This asynchrony increases with higher intellectual capacity" (Silverman, 1993, p. 634).

References

Ananiev, B. G. (1957). About the system of developmental Psychology. *Voprosi Psichologii, 5*, 112–126.
Bamberger, J. (1982). Growing-up Prodigies: The midlife crisis. In D. H. Feldman (Ed.), *Developmental Approaches to Giftedness* (pp. 61–77). San Francisco: Jossey-Bass.
Bamberger, J. (1986). Cognitive issues in the development of musically gifted children. In R. J. Sternberg & J. E. Davidson (Eds.), *Conceptions of Giftedness* (pp. 388–413). Cambridge, UK: Cambridge University Press.
Brown, A. L. (1978). Knowing when, where, and how to remember: A problem of metacognition. In R. Glaser (Ed.), *Advances in Instructional Psychology: Vol. 1* (pp. 77–165). Hillsdale, NJ: Erlbaum.
Brown, A. L. (1984). Metacognition, executive control, self-regulation, and other even more mysterious mechanisms. In F. E. Weinert & R. H. Kluwe (Eds.), *Metacognition, Motivation, and Learning* (pp. 60–108). West Germany: Kuhlhammer.
Calaprice, A., & Lipscombe T. (2005). *Albert Einstein: A Biography*. London: Greenwood Press.
Case, R. (1984a). *Intellectual Development: A Systematic Reinterpretation*. New York: Academic Press.
Case, R. (1984b). The process of stage transition: A neo-piagetian view. In R. J. Sternberg (Ed.), *Mechanisms of Cognitive Development* (pp. 19–44). New York: W. H. Freeman and Company.
Einstein, A. (1949). Autobiographical notes. In P. A. Schlipp (Ed.), *Albert Einstein: Philosopher and Scientist* (pp. 3–49). New York: Library of Living Philosophers.
Ericsson, K. A., & Simon, H. A. (1980). Verbal reports as data. *Psychological Review, 87*(3), 215–251.

Feldman, D. H. (1982). A developmental framework for research with gifted children. In D. H. Feldman (Ed.), *Developmental Approaches to Giftedness and Creativity* (pp. 31–45). San Francisco: Jossey-Bass.

Feldman, D. H. (1986a). Giftedness as a developmentalist sees it. In R. J. Sternberg & J. E. Davidson (Eds.), *Conceptions of Giftedness* (pp. 285–305). Cambridge, UK: Cambridge University Press.

Feldman, D. H. (1986b). *Nature's Gambit: Child Prodigies and the Development of Human Potential*. New York: Basic Books.

Fischer, K. W., & Pipp, S. L. (1984). Processes of cognitive development: Optimal level and skill acquisition. In R. J. Sternberg (Ed.), *Mechanisms of Cognitive Development* (pp. 45–80). New York: W. H. Freeman and Company.

Flavell, J. H. (1984). Discussion. In R. J. Sternberg (Ed.), *Mechanisms of Cognitive Development* (pp. 187–209). New York: W. H. Freeman and Company.

Fölsing, A. (1997). *Albert Einstein*. New York: Penguin Group.

Gardner, H. (1993). *Creating Minds*. New York: Basic Books.

Growther, J. G. (1955). *Six Great Scientists: Copernicus, Galileo, Newton, Darwin, Marie Curie, Einstein*. London: Hamish Hamilton.

Gruber, H. E. (1982). On the hypothesized relation between giftedness and creativity. In D. H. Feldman (Ed.), *Developmental Approaches to Giftedness and Creativity* (pp. 7–29). San Francisco: Jossey-Bass.

Gruber, H. E. (1986). The self-construction of the extraordinary. In R. J. Sternberg & J. E. Davidson (Eds.), *Conceptions of Giftedness* (pp. 247–263). Cambridge, UK: Cambridge University Press.

Highfield, R., & Carter, P. (1993). *The Private Lives of Albert Einstein*. London: Faber and Faber Ltd.

Hoffman, B., & Dukas, H. (1972). *Albert Einstein Creator and Rebel*. New York: Viking Press.

Isaacson, W. (2007). *Einstein: His Life and Universe*. New York: Simon & Schuster.

Jackson, P. S., Moyle, V. F., & Piechowski, M. M. (2009). Emotional life and psychotherapy of the gifted in light of Dabrowski's theory. In L. V. Shavinina (Ed.), *International Handbook on Giftedness* (pp. 437–466). Dordrecht: Springer Science.

Jellen, H., & Verduin, J. R. (1986). *Handbook for Differential Education of the Gifted: A Taxonomy of 32 Key Concepts*. Carbondale, IL: Southern Illinois University Press.

Kholodnaya, M. A. (1990). Is there intelligence as a psychological reality? *Voprosu Psichologii*, 5, 121–128.

Kholodnaya, M. A. (1993). Psychological mechanisms of intellectual giftedness. *Voprosi Psichologii*, 1, 32–39.

Kuznetsov, B. G. (1979). *A. Einstein: Life, Death, Immortality*. Moscow: Nauka.

Lassieur, A. (2005). *Albert Einstein: Genius of the Twentieth Century*. Markham, Ontario: Scholastic Canada.

Leites, N. S. (1960). *Intellectual Giftedness*. Moscow: APN Press.

Leites, N. S. (1971). *Intellectual Abilities and Age*. Moscow: Pedagogica.

Leites, N. S. (Ed.) (1996). *Psychology of Giftedness of Children and Adolescents*. Moscow: Academia.

Michio, K. (2004). *Einstein's Cosmos: How Albert Einstein's Vision Transformed our Understanding of Space and Time*. New York: W. W. Norton & Company Inc.

Mih, W. (2000). *The Fascinating Life and Theory of Albert Einstein*. New York: Kroshka Books.

Miller, A. (1996). *Insights of Genius: Visual Imagery and Creativity in Science and Art*. New York: Springer Verlag.

Milton, D. (1983). *Albert Einstein*. New York: Franklin Watts.

Neffe, J., & Frisch, S. (2007). *Einstein: A Biography*. New York: Farrar, Straus & Giroux.

Piaget, J. (1952). *The Origins of Intelligence in Children*. New York: International Universities Press.

Piechowski, M. M. (1979). Developmental potential. In N. Colangelo & R. T. Zaffrann (Eds.), *New Voices in Counseling the Gifted* (pp. 25–27). Dubuque, IA: Kendall/Hunt.

Piechowski, M. M. (1986). The concept of developmental potential. *Roeper Review*, 8(3), 190–197.

Piechowski, M. M. (1991). Emotional development and emotional giftedness. In N. Colangelo & G. Davis (Eds.), *Handbook of Gifted Education* (pp. 285–306). Boston: Allyn & Bacon.

Reiser, A. (1930). *Albert Einstein: A Biographical Portrait*. New York: Albert & Charles. Boni, Inc.

Roedell, W. C. (1984). Vulnerabilities of highly gifted children. *Roeper Review*, 6(3), 127–130.

Rosenblatt, J. S. (1976). Sensitive periods in development: A problem of continuity/discontinuity in development. In *Proceedings of XXI International Congress of Psychology*. Paris: University of Sorbonne.

Seeling, C. (1956). *Albert Einstein: A Documentary Biography*. London: Staples Press.

Shavinina, L. V. (1996). The objectivization of cognition and intellectual giftedness. *High Ability Studies*, 7(1), 91–98.

Shavinina, L. V. (1997). Extremely early high abilities, sensitive periods, and the development of giftedness. *High Ability Studies*, 8(2), 245–256.

Shavinina, L. V. (1999). The psychological essence of the child prodigy phenomenon: Sensitive periods and cognitive experiences. *Gifted Child*, 43(1), 25–38.

Shavinina, L. V. (2006). Was Einstein gifted as a child? invited lecture at the Canadian Museum of Nature on the occasion of the opening of a special exhibition on Albert Einstein (November 23, Ottawa, Ontario, Canada).

Shavinina, L. V. (2009). A unique type of representation is the essence of giftedness: Towards a cognitive-developmental theory. In L. V. Shavinina (Ed.), *International Handbook on Giftedness* (pp. 231–257). Dordrecht: Springer Science.

Shavinina, L. V., & Kholodnaya, M. A. (1996). The cognitive experience as a psychological basis of intellectual giftedness. *Journal for the Education of the Gifted, 20*(1), 4–33.

Shavinina, L. V., & Seeratan, K. (2003). On the nature of individual innovation. In L. V. Shavinina (Ed.), *The International Handbook on Innovation* (pp. 31–43). Oxford, UK: Elsevier Science.

Skuse, D., Pickles, A., Wolke, D., & Reilly, S. (1994). Postnatal growth and mental development: Evidence for a "sensitive period." *Journal for Child Psychology and Psychiatry, 35*(3), 521–545.

Silverman, L. K. (1993). Counseling needs and programs for the gifted. In K. A. Heller, F. J. Mönks, & A. H. Passow (Eds.), *International Handbook of Research and Development of Giftedness and Talent* (pp. 631–647). Oxford, UK: Pergamon Press.

Silverman, L. K. (1994). The moral sensitivity of gifted children and the evolution of society. *Roeper Review, 17*(2), 110–116.

Silverman, L. K. (1997). The construct of asynchronous development. *Peabody Journal of Education, 72*(3&4), 36–58.

Sternberg, R. J. (1986). A triarchic theory of intellectual giftedness. In R. J. Sternberg & J. E. Davidson (Eds.), *Conceptions of Giftedness* (pp. 223–243). Cambridge, UK: Cambridge University Press.

Sternberg, R. J. (1990). *Metaphors of Mind: Conceptions of the Nature of Intelligence*. Cambridge, UK: Cambridge University Press.

Terrassier, J.-C. (1985). Dyssynchrony: uneven development. In J. Freeman (Ed.), *The Psychology of Gifted Children* (pp. 265–274). New York: John Wiley.

Terrassier, J.-C. (1992). Gifted children: Research and education in France. In F. J. Mönks, M. W. Katzko, & H. W. Boxtel (Eds.), *Education of the Gifted in Europe: Theoretical and Research Issues* (pp. 212–216). Amsterdam: Swets & Zeitlinger.

Vallentin, A. (1954). *The Drama of Albert Einstein*. New York: Doubleday & Company.

Von Boehm, G. (2005). *Qui était Albert Einstein?* Paris: Assouline.

Vygotsky, L. S. (1956). *Selected Papers*. Moscow: APN Press.

Vygotsky, L. S. (1972). Age periods in child development. *Voprosi Psichologii, 2*, 53–61.

Wallon, H. (1945). *Les Origines de la Pensée Chez l'enfant*. Paris: P.U.F.

Walters, J., & Gardner, H. (1986). The crystallizing experience: Discovering an intellectual gift. In R. J. Sternberg & J. E. Davidson (Eds.), *Conceptions of Giftedness* (pp. 306–331). Cambridge, UK: Cambridge University Press.

Whitrow, G. J. (1967). *Einstein: The Man and his Achievement*. New York: Dover. Publication

Zaporozhets, A. V. (1964). *Child Psychology*. Moscow: Pedagogica.

Part III
Creativity as a foundation of innovation education

8

From creativity education to innovation education

What will it take?

Joyce VanTassel-Baska
COLLEGE OF WILLIAM AND MARY, USA

Summary: This chapter will trace the development of creativity in research on gifted individuals and link it to the more recent work on innovation education. The chapter will highlight key distinctions in the two processes as well as analyze similarities, using examples of work in curriculum, instruction, and assessment to illustrate key points. While creativity has always been associated with new ideas and methodologies that break from the past, innovation represents an attempt to synthesize past efforts and render them pragmatic in a given field of endeavor. Since education is an applied field, major breakthroughs may almost always be seen as innovations at both development and research levels. The purpose of innovation in education is to implement sound practices faithfully, according to research-based models.

Key words: Innovation education, creativity, innovators, creators, giftedness, gifted individuals, Javits programs, gifted education.

Introduction to the concepts

What do we mean by creativity and what by innovation? Are they the same or different processes? Both constructs demand proof in the real world, the final test of acceptance of an idea or a product by peers and ultimately by a broader audience of consumers. Both also require a set of skills that combines critical thinking with creative thinking and problem solving behaviors to be successful. Both also require non-intellective traits such as motivation, perseverance, and autonomy. Both constructs also demand of an individual heightened motivation and desire to design and develop products that work. The passion to create becomes a central driving force on the road to creativity and innovation. Moreover, individuals who create and innovate must also be extremely hard workers, devoting large amounts of time to the projects they are working on. Finally, people who create and innovate must have a deep knowledge base in their field of endeavor in order to be playful and experimental with the content.

Yet innovators have additional requirements. They must be pragmatic and see how the innovation fits into the real world of existing products and ideas. They must see the product as feasible for use in a given field. Thus innovators are concerned about implementation and application of ideas, not just ideas themselves. They also are visionary, seeing the potential for an idea within an existing market or identifying how to move in the right direction within a field, based on subtle environmental cues. It is through the timely application of products or ideas in systematic ways that positive change occurs in a business, a classroom, or even a society. This is the primary job of innovators—bringing a product or idea to the real world and making it work over time and in many places.

Contemporary examples

Sometimes creators and innovators are one and the same as in the case of Edison who invented and marketed his discoveries. More often, they are different people. Companies have design teams who pass on the product to an implementation team of marketers, distributors, and sales staff who make the product appealing, affordable, and necessary for consumers to purchase. Collaborative teams of creators and innovators is another way to think about the two constructs working together. Theorists, researchers, and practitioners working together can create models of research in practice (Dai, 2011) where the theory may drive design and practice but may also work in reverse, with practice causing a theory to be reworked. In the world of gifted curriculum, collaboration between content specialists, curriculum developers, and gifted specialists yielded stronger products than would have been created with only one expertise applied to the problems of design (VanTassel-Baska & Little, 2011).

Steve Jobs and Steve Wozniak both were creative and innovative yet Jobs became and stayed the CEO of Apple because he had superior innovative skills and understood the importance of marketing new products successfully. He also was more driven and motivated to succeed in the world of business than Wozniak. He was an entrepreneur in the best sense of the word, having a vision for change in the multiple industries of computing, electronics, animation, and music.

In the field of gifted education, we have stressed the development of creative producers but not necessarily innovators. Is the preparation the same or different? Is the preparation of innovators counterproductive to traditional schooling models? After all, both Bill Gates and Steve Jobs were college dropouts, both attending very selective institutions where all the students are gifted. To what extent are the habits of mind associated with innovation different from those employed by a creator? By the age of 12, Bill Gates had taught himself the world of computing to a level at that time only known by fewer than 50 people. Steve Jobs taught himself the business of animation and then refined it into a company called Pixar, a move that no one else had thought to do. Clearly schooling was not an impetus for this kind of knowledge acquisition or use (Isaacson, 2011).

The special cases of Charles Darwin and Sir Francis Galton

In the history of eminent individuals, we see the seeds for such innovative activity—the autodidact is common among poets, writers, artists, and even scientists. It would be fair to say that the father of gifted education, Sir Francis Galton, was himself an autodidact, inventing statistical operations, research methods, and a whole line of inquiry on the inheritance of ability. His cousin Charles Darwin also was an avid autodidact, teaching himself on HMS *Beagle* the strategies he needed to observe animal and plant life in the Galapagos. I would argue that Darwin was a creator while Galton was an innovator, living to shepherd and see his ideas to fruition and established in schools and hospitals around England. His pragmatic interest in application of his ideas was never far from his mind as attested to in his letters while Darwin was more consumed with seeing his ideas get published and letting others make the applications (VanTassel-Baska, in preparation).

In the lives of each of these men, raised as part of England's gentried class in the 19th century, the role of education was background to their desire to understand and apply their knowledge in the real world. Driven by a rage to know and a relentless motivation to delay gratification, each came to contribute to the world albeit at different levels of influence and impact. The work of the innovator may be important at a local level but not easy to generalize to all contexts and time periods. For Darwin, his theory became highly generalizable as multiple applications of evolution came to be seen. For Galton, whose work was more atheoretical, although heavily influenced by Darwin's theory, the practical applications were buttressed by an active research agenda.

So is another distinction between the two the capacity to generate theory as well as do research and apply it to practice? Creators come up with paradigm-shifting ideas that are well articulated for others to apply. This is Kuhn's notion of how science progresses—by the big ideas of a few who gather adherents and then have others apply those ideas in the way of normal science to test their validity. He sees science as revolutionary in this respect, with new ideas having the gravitational pull to change a field and its research agenda. Others, of course, see science as more evolutionary, the "standing on the shoulders of giants" image the most commonly held.

What are the features of creators that don't apply to innovators and vice versa? Perhaps Table 8.1 may be instructive.

These distinctions between creators and innovators assumes that they are not the same person. They also assume that creators are rarer than innovators in any field. For example, in education as an applied field, the number of innovators at all levels of the enterprise of schools and universities far exceeds the creatives who develop theories to impact the thinking about a construct. Our premiere theory-builder and creative in the field of gifted education, Bob Sternberg, has taken upon himself the task of trying to do it all—beyond theory to research and development. Yet even he stopped short of engaging in implementation realities. Many others have used application in schools as their playground, spurred by the edicts of the Javits Act which required research, development, and implementation insights to be a part of the grants.

The contention that creators and innovators differ in their assumptions about the nature of knowledge, about the purpose of their activities, and the habits of mind they bring to the enterprise

Table 8.1 A comparison of creators and innovators

Creators	Innovators
Change basic paradigms in a field.	Change the world of practice, using new paradigms.
Create theories as a way to explain ideas in a connected way.	Create products that illustrate theories.
Work alone to articulate ideas for dissemination.	Work collaboratively to bring products to scale.
Prefer working on theory and researchable questions in the problem solving process.	Prefer working on problem solving from the multiple levels of theory, research, development, and implementation.
Possess the habits of mind of curiosity, skepticism, objectivity, and openness.	Possess the habits of mind of pragmatism, systems thinking, and flexibility.
Assume that knowledge is tentative and can be reshaped for deeper understanding of a phenomenon.	Assume that knowledge can be transformed into an endless variety of products that respond to the needs of people and institutions.
Assume that learning is idiosyncratic, based on prior knowledge and relevant skills and motivation.	Assume that learning is collaborative and dynamic, creating its own momentum in *medias res*.
Assume that ideas and learning alone have currency.	Assume that ideas and learning only have currency when applied to useful ends.

is to suggest that the orientation of creators is distinctive from that of innovators. Creators assume that learning itself is enough, that it is idiosyncratic, and that it is tentative. Innovators, on the other hand, see knowledge as instrumental in the service of the greater good of a society or a profit margin, collaborative and dynamic leading to product outcomes, and utilitarian. Dominant in the habits of mind of creators is skepticism, objectivity, and curiosity while innovators practice systems thinking, flexibility, and pragmatism regularly. The outcomes of creative endeavors often are a change in paradigm in a field, the creation of a new theory for viewing a phenomenon, often accomplished by an individual working alone on an idea. The outcomes of an innovator, on the other hand, are new products that change practice in myriad ways.

Implications of the distinctions for school-based learning

If creators and innovators differ in the ways I am suggesting in this chapter, then perhaps schooling models need to be sensitive to the distinction as well in how we approach optimal learning. Gifted education has always recommended acceleration by grade, independent study, and advanced placement in levels of learning (e.g., university classes early) as the most fruitful patterns to pursue for the highly gifted who are the most likely to becomes the creatives of the next generation (see Park, Lubinski, & Benbow, 2006). While such approaches have shown powerful effects on individual creative productivity, they have not necessarily produced innovators who will influence the practical applications of creation. Perhaps we need to consider a schooling model that honors the 21st century skills of collaboration, communication, critical and creative thinking, problem solving, and metacognition to a greater extent than before. I would argue that our research on creativity has always supported such a direction.

Creativity research in gifted education

Over the past decade, studies have continued to suggest the relationship between critical thinking and reasoning to high level creative production within and across domains (Gardner, 2000; Csikszentmihalyi, 2000). In gifted education, becoming a creative producer in the real world is predicated on the acquisition of a combination of creative thinking, problem solving, and critical thinking within a domain (VanTassel-Baska & Little, 2011).

While earlier studies have shown that students show important gains in content-specific higher order skills such as literary analysis and persuasive writing in language arts (VanTassel-Baska, Avery, Hughes, & Little, 2000) or designing experiments in science (VanTassel-Baska, Bass, Reis, Poland, & Avery, 1998), studies have only recently demonstrated that a content-based intervention provided students with enhanced generic critical thinking and reasoning skills at the elementary level (Bracken, Bai, Fithian, Lamprecht, Little, & Quek, 2003; VanTassel-Baska, Bracken, Feng, & Brown, 2009). Other Javits projects, focused on working with low income students, have also promoted the use of higher level thinking within content areas (Gavin et al., 2007; Swanson, 2006) with positive results.

Most K-12 programs for gifted students include some components of critical thinking as a fundamental part of the curriculum (Chandler, 2004). Only recently, however, have we begun to test the efficacy of curriculum in respect to student growth in this area at various stages of development, being satisfied instead to use proxy outcome data like Advanced Placement (AP) and International Baccalaureate (IB) scores, SAT scores, or even state tests to tell us how well these students are performing at higher levels of thought (VanTassel-Baska & Feng, 2003).

The teaching of creativity, however, is not as prevalent in classrooms due to the emphasis on standards and accountability that do not assess or value the development of creative skills. Still, some evidence suggests that educational programs based on appreciation for creative thinking abilities may in fact facilitate the creativity process in learners over time. Two nascent longitudinal studies have attempted to link creatively oriented gifted programs to later adult productivity. Delcourt (1994) studied 18 secondary students who were identified by Renzulli's Three Ring Conception of Giftedness and were provided with Type III enrichment activities three years after completing a creatively oriented gifted program. All of the students were found to be satisfied with the nature and extent of the project work with which they were engaged (see Delcourt & Renzulli, this volume). Moon, Feldhusen, and Dillon (1994) studied 23 students who participated for at least three years in an enrichment program using the Purdue Three-Stage Model of creative development. They found that all of the students planned to attend college and 78% planned to undertake graduate training. The study noted that aspiration levels for girls were tempered by interest in marriage and children.

Other types of study designs have been used in attempts to correlate creative performance in adulthood with creativity test scores in childhood. Cramond (1994), for example, studied the lifetime productivity of individuals identified at elementary ages by the Torrance Tests of Creative Thinking as having creative potential. Results demonstrated that lifetime creative achievement was moderately correlated with the test scores. Two other variables were found also to have important correlational value: an enduring future career image during childhood and a mentor at some time.

Torrance (1993), in a related study, reported on two exceptional cases of "beyonders" who outperformed any prediction of their success in the adult world. He found that these individuals possessed such characteristics as love of work, perseverance with tasks, lack of concern with being in the minority, enjoyment of working alone, and immersion in work-related tasks. It is interesting to note that all of these characteristics are highly related to the ethics of intrinsic motivation, individualism, and work.

Research on innovation and giftedness

Much of the research on innovation and the gifted has emerged from the work of Shavinna (2003, 2009) and her handbooks that demonstrate how the construct of innovation has been applied to gifted education. Her work has emphasized the importance of multiple factors to be nurtured in students, drawing from instructive real world case examples. These factors include the development of entrepenurial abilities, managerial talent, and time management strategies as well as affective characteristics like courage. Root-Bernstein (2003), to cite another example, sees the major task of new fields of science to be innovative in that basic science must be wedded to the practicalities of technology in the real world in order for it to be useful to a society. The science that underlies genetics and its application to medicine is but one of many examples that illustrate his ideas.

Innovation is the clarion call of the new National Science Board report which calls for priming the pipeline for scientists, technology specialists, engineers, and mathematicians (STEM) who can solve the real world problems we face as well as provide a competitive edge. In the past decade, several national reports have called for increased STEM education, including suggestions for earlier intervention, foci on the most able children, and renewed interest in the importance of spatial ability for STEM innovation. In particular, the National Science Board (2010) details the lack of STEM preparation in schools and outlines an agenda for action in their report, *Preparing the next generation of STEM innovators*. The report notes that, while many others have made recommendations focusing on raising overall performance of America's students, few have "focused on raising the ceiling of achievement for our Nation's most talented and motivated students" (p. 4). The National Science Board further outlines key issues, including the importance of early intervention and that spatial ability is rarely measured or developed in children. Cited in the report, the Business Roundtable (2005) suggests that the problems cannot wait to be addressed:

> One of the pillars of American economic prosperity—our scientific and technological superiority—is beginning to atrophy even as other nations are developing their own human capital. If we wait for a dramatic event—a 21st-century version of Sputnik—it will be too late. There may be no attack, no moment of epiphany, no catastrophe that will suddenly demonstrate the threat. Rather, there will be a slow withering, a gradual decline, a widening gap between a complacent America and countries with the drive, commitment and vision to take our place.
>
> (p. 5)

In another national report, *Rising above the gathering storm* (2007), the National Academy of Sciences (NAS), the National Academy of Engineering, and the Institute of Medicine elucidate that point in terms of the future prosperity of the United States:

> This nation must prepare with great urgency to preserve its strategic and economic security. Because other nations have, and probably will continue to have, the competitive advantage of a low wage structure, the United States must compete by optimizing its knowledge-based resources, particularly in science and technology.
>
> (p. 4)

This report notes that STEM, particularly the technological advancements that it encompasses, have driven the U.S. economy for the past several decades. The authors conclude that the highest priority must be to improve K-12 science education. The National Research Council (2007) reflects that, while standards-based reform has been underway for more than 15 years, improvements in U.S. science education have been lackluster, especially in comparison with other countries. They argue that, "At no time in history has improving science education been more important than it is today" (p. 1). The need to improve science education is great, but part of the solution may lie outside the traditional classroom.

The National Academy of Education (NAE) white paper, *World-class science and mathematics* (2009), affirms this, suggesting that STEM education is vital for the security and economy of the United States. Despite this well-known importance, the United States has yet to make a concerted effort in schools to provide quality STEM education in the post-Cold War era. In the book, *Taking Science to School* (2007), the National Research Council (NRC) analyzed the available data and concluded that the United States is seriously behind in science education. This lack of STEM focus is seen in higher education and the job market, which has an ever-increasing need for highly educated people capable of filling the openings (Shea, Lubinski, & Benbow, 2001). While employers expect to hire 2.5 million STEM workers between 2004 and 2014, there is a national shortage of students graduating from institutions of higher education with degrees in many important STEM fields (American Competitiveness Initiative, 2006).

Given the demand for highly educated people in STEM fields coupled with the fact that they earned about 70% more than the U.S. average in 2005 (Terrell, 2007), it may be surprising that too few people choose to pursue STEM fields in higher education. The reason can be found long before higher education begins. Students who do not prepare well during their K-12 education will likely have a tougher time getting into and succeeding in STEM university. Data from international studies (see Fleischman, Hopstock, Pelezer, & Shelley, 2010) continue to show the United States ranked well below other countries, raising the question of how well we are focusing on an innovative agenda in schools. Countries like Singapore continue to overtly pursue the development of entrepreneurs and innovators, even giving awards to the best each year in this new area of emphasis while the United States does little to pursue an active agenda for its most talented students in STEM areas.

So what is the foundation for preparing leaders who can become the entrepreneurial innovators of tomorrow? The agenda for development depends heavily on the systematic use of different modes of thinking and problem solving, grounded in the real world of problems, issues, and themes.

Teaching to higher level skills

To teach the higher order process skills of critical thinking and creativity to gifted learners is to engage them in lifelong learning skills that provide the scaffolding for all worthwhile learning in the future. It is "teaching them to fish," not providing one to be eaten for only a day. This constructivist approach to learning, however, requires similar approaches to be employed by the teacher, requiring long term investment in learning new ways to think as well as teach. Because higher order thought and creativity is not formulaic, it requires being open to the moment, asking the probing question at the right time, engaging the class in the right activity based on when they most need it, and assessing levels of functioning with regularity. Constructive teaching also requires teachers to provide students with useful models in order to have schema on which to hang their ideas. However, even useful models cannot be taught mechanistically; they must be thoughtfully applied and used idiosyncratically by gifted learners so that the greatest benefits accrue. Finally, teachers must help students understand that real thinking is hard work, that it takes effort over time to improve, and that the outcome is frequently uncertain.

Models that promote creativity and innovation in learning

Selecting models that enhance the learning of these higher order process skills is also desirable since their utility has been proven in countless classrooms, and research suggests that a few selected models used over time enhances learning more strongly than eclecticism (Hillocks, 1999). Several models have proven useful to teachers in addressing the higher order skills of creative and critical thinking in the classroom.

One of the most viable creativity models at a theoretical level is Amabile's (Table 8.2), which focuses on the relative importance of three areas—domain-specific knowledge and the ability to apply it to worthy problems, motivation and interest, and creativity-relevant skills that support contributions to a given domain of learning. Major emphases within her model include a focus on developing products judged to be exemplary by those in the domain and the importance of contexts for nurturing creative behavior (Amabile, 1983, 2001).

Another model that is instructive in addressing creativity is that of Csikszentmihalyi (2000) who studied creativity from the vantage point of adult creators who had made significant contributions to a field of study. He found these individuals to possess a high degree of intrinsic motivation, characterized by a state of flow which had the following characteristics:

- challenging but doable tasks;
- time and space to concentrate on those tasks;
- goal-oriented tasks with a feedback mechanism;
- high level of task involvement to the exclusion of everyday concerns (e.g., eating, sleeping);
- loss of self-consciousness replaced by task orientation;
- time passing unnoticed.

Table 8.2 Amabile's view of creativity

Domain-relevant skills	Creativity-relevant skills	Task motivation
Includes: • Knowledge about the domain • Special skills required • Special domain-relevant "talent"	*Includes:* • Appropriate cognitive style • Implicit or explicit knowledge of heuristics for generating novel ideas • Conducive work style	*Includes:* • Attitudes toward the task • Perceptions of own motivation for undertaking the task
Depends on: • Innate cognitive abilities • Innate perceptual and motor skills • Formal and informal education	*Depends on:* • Training • Experience in idea generation • Personality characteristics	*Depends on:* • Initial level of intrinsic motivation toward the task • Presence or absence of salient extrinsic constraints • Individual ability to cognitively minimize extrinsic constraints

Such characteristics speak to the level and type of connection creative people have to their work.

The work of Perkins (1981) is instructive in teaching creativity as well for he has identified key principles for being creative and teaching others to be. These principles represent a pragmatic way of looking at enhancing the creative skills of individuals over the lifespan. They include the following:

1   Creativity involves traits that make a person creative; the act of creativity calls for traits and behaviors that are not intrinsically creative, such as planning and abstracting.
2   Creativity requires four fundamental acts: planning, abstracting, undoing, and making means into ends.
3   The guiding force that creates a product is purpose or intent.
4   Creating is a process of selecting among many possible outcomes by using such approaches as noting opportunities and flaws, directed remembering, reasoning, looking harder, setting work aside, using schemata, and problem finding.
5   Creativity involves a style, values, beliefs, and tactics that specially favor selecting for a creative product.

Studies of insight have also contributed to our understanding of creativity. Sternberg's work (1988, 2001a, 2001b) in this area has suggested that deep immersion in an area coupled with recognizing an apt analogy and reasoning through it can lead to important understanding and discovery of new solutions to difficult problems. He suggests that the most mysterious aspect of creativity may in fact be described and even taught, given the right context.

Ochse (1990) and other researchers who have studied creative individuals in a number of different fields (Simonton, 1994; Torrance, 1993) have all been struck by the sheer work and effort that creative individuals are willing to devote to their area of specialty. Such individuals are clearly in love with the work but also continue to persevere with it over time in the face of criticism, lack of support, and much time being spent alone. The single variable that these researchers focus on, however, is the capacity and actualization of work over time. Thus the ways to instill creativity in young people may not vary considerably from the fundamental values found basic to schooling. The major differences appear to lie with the following issues:

- *Work autonomy.* Students need to feel that they are planning out their own work, making choices about what they do and how they do it.
- *Time allocation.* Students need work time that is in larger chunks in order to be productive with their projects, many times requiring whole days away from school in order to carry out aspects of learning not possible in a school setting.
- *Mentors.* Students need teachers and other adults in their environment who can counsel and guide their work to be at a high level and contributory to a given area of learning.
- *Supportive environment.* Students need a classroom that is conducive to creative production, one that is open, warm, accepting of experimentation, and of taking risks that may bring failure.
- *Use of creative skills.* Students need to employ the specific skills of fluency, flexibility, and elaboration to work-related tasks.

Creative problem solving

Problem recognition and delineation as a critical element of the creative problem solving process was first identified by Getzels and Csikszentmihalyi (1976) in their pioneering study of artists' approaches to the problem of depicting some aspects of human experiences. They found that creative artists who were able to sustain careers in art were more effective at problem finding not problem solving than less successful fellow students. These findings spawned many models that provided a more balanced perspective between the two types of skills.

Problem solving formally may be described as a series of steps. Beyer (2000) set forth such a model in his broader taxonomy of thinking skills:

1. *Recognize a problem.*
2. *Represent the problem.*
3. *Deliver/choose a solution plan.*
4. *Execute the plan.*
5. *Evaluate the solution.*

The formal steps may or may not characterize students' cognitive activity in a real problem situation. In a sense, they represent an ideal. The steps also define a convergent conception in that a single solution is envisioned, although the language of the model is open to alternative solutions from different problem solvers.

Another complex form of problem solving that involves both critical and creative thinking, widely applied in gifted programs and special extracurricular programs like Olympics of the Mind and Future Problem Solving, is creative problem solving (Isaksen, Treffinger, Dorval, & Nollar, 2000). Six steps or processes characterize the model:

1. mess finding
2. data finding
3. problem finding
4. idea finding
5. solution finding
6. acceptance finding.

The main characteristic of "mess finding" is to sort through a problem situation and find direction toward a broad goal or solution. In "data finding," participants sort through all available information about the mess and clarify the steps or direction to a solution. In "problem finding," a specific problem statement is formulated. "Idea finding" is a processing of many ideas for solution to *the* problem or parts of the problem. "Solution finding" is an evaluation or judgmental process of sorting among the ideas produced in the last step and selecting those most likely to produce solutions. Finally, in "acceptance finding," a plan is devised for implementing the good solution. An adaptation of the creative problem solving model is called "Future Problem Solving." It involves the application of the creative problem solving model to studies of the future and to problems that are now emerging as major concerns (Volk, 2004).

Treffinger, Isaksen, and Dorval (2000) extended the creative problem solving model by suggesting that Stage One should include opportunities for participants to identify their own problem within a specific domain of interest or study. They also suggested that the solution finding stage should involve more than selecting best ideas; it should often involve synthesizing the best ideas into a more complex and creative solution.

Another model that promotes higher level problem solving is problem-based learning, a curriculum and instructional model that is highly constructivist in design and execution. First used in the medical profession to socialize doctors better to patient real world concerns, it is now selectively employed in educational settings at elementary and secondary levels with gifted learners (Gallagher & Stepien, 1996; Gallagher, 1998; Boyce, VanTassel-Baska, Burruss, Sher, & Johnson, 1997). The technique involves several important features:

1. Students are in charge of their own learning. By working in small investigatory teams, they grapple with a real world unstructured problem that they have a stake in and must solve within a short period of time. Students become motivated to learn because they are in charge at every stage of the process.
2. The problem statement is ambiguous, incomplete, and yet appealing to students because of its real world quality and the stakeholder role that they assume in it. For example, students may be given roles as scientists, engineers, politicians, or important project-based administrators whose job it is to deal with the problem expeditiously.
3. The role of the teacher is facilitative not directive, aiding students primarily through question-asking and providing additional scaffolding of the problem with new information or resources needed. The teacher becomes a metacognitive coach, urging students through probing questions to deepen their inquiry.
4. The students complete a Need to Know Board early in their investigation that allows them to plan out how they will attack the problem, first by identifying what they already know from the problem statement, what they need to know, and how they will find it out. They then can prioritize what they need to know, make assignments, and set up timelines for the next phase of work. Such an emphasis on constructed metacognitive behavior is central to the learning benefits of the approach.

These features work together then in engaging the learner in important problems that matter in their world. Many times problems are constructed around specific situations involving pollution of water or air, dangerous chemicals, spread of infectious disease, or energy source problems. Students learn that the real world is interdisciplinary in orientation, requiring the use of many different thinking skills and many different kinds of expertise in order to solve problems.

In order to work through a problem-based learning episode, students must be able to analyze, synthesize, evaluate, and create—all higher level thinking tasks according to Anderson and Krathwohl (2000). The following problem and its levels of complex thinking are illustrative of a problem-based learning episode.

Problem: There is a lack of mass transit into and out of a central city. You are an urban planner, given one month to come up with a viable plan. However, your resources have been used on another project, that of city beautification. A new airport is about to be built 20 miles out from the city, but negotiations are stalled. What do you do?

Higher level skills needed to address the problem include:

1. analysis of what the real problem is—mass transit, airport construction, beautification?
2. synthesis of the aspects of the problem—is there a creative synthesis of each facet of the problems noted?
3. evaluation of alternative strategies to be employed—can I shift funds, can I employ a transportation expert, can I deal with the airport deal?
4. creation of the plan of action that will need to be sold to city council.

Critical thinking and innovation

Just as the literature and models on creativity are relevant to enhancing innovation in students, so too is an emphasis on critical thinking in learning. Without the skills of judgment, students may not be capable of developing a real world product of value. For innovation, the test of feasibility and practicality must be met, leading to the use of higher level skills of judgment, typically found in the world of critical thinking, not creative thinking per se.

The real world of innovation in all professional fields requires students to make nuanced judgments and interpretations about data. An effective model to teach students to enhance these skills is the Ennis Model of Critical Thinking which uses judgment and inference as the centerpiece of the critical thinking process (Ennis, 1996). Although the model has been used more extensively at secondary level, it can be applied with gifted students at upper elementary levels with successful results. An important aspect of this model is the 12 dimensions of critical thinking he derived from a study of the literature and his own philosophically trained education. These are:

- grasping the meaning of a statement;
- judging whether there is ambiguity in a line of reasoning;
- judging whether certain statements contradict each other;
- judging whether a conclusion necessarily follows;
- judging whether a statement is specific enough;
- judging whether a statement is actually the application of a certain principle;
- judging whether an observation statement is reliable;
- judging whether an inductive conclusion is warranted;
- judging whether the problem has been identified;
- judging whether something is an assumption;
- judging whether a definition is adequate;
- judging whether a statement made by an alleged authority is acceptable.

The first dimension of his model involves all aspects of interpretation, whether it is derived by inductive or deductive means. A student activity that aids the development of interpretation might be to have students study proverbs or the sayings of great writers and philosophers. Presented with a statement of import, students could be asked the following questions:

What do the significant words mean?
What does each line of the statement mean?

What situations does the statement refer to?
What ideas about life does it share?
What new applications can you make to the idea that relate to your life and to the society as a whole today?

Another model that has proven helpful to many teachers and other educators in the application of critical thinking to real life has been the use of Richard Paul's elements of reasoning (Elder & Paul, 2004). These elements include the following:

- *Purpose, goal, or end view.* We reason to achieve some objective, to satisfy a desire, to fulfill some need. For example, if the car does not start in the morning, the purpose of my reasoning is to figure out a way to get to work. One source of problems in reasoning is traceable to "defects" at the level of purpose or goal. If our goal itself is unrealistic, contradictory to other goals we have, confused or muddled in some way, then the reasoning we use to achieve it is problematic. If we are clear on the purpose for our writing and speaking, it will help focus the message in a coherent direction. The purpose in our reasoning might be to persuade others. When we read and listen, we should be able to determine the author's or speaker's purpose.
- *Question at issue (or problem to be solved).* When we attempt to reason something out, there is at least one question at issue or problem to be solved (if not, there is no reasoning required). If we are not clear about what the question or problem is, it is unlikely that we will find a reasonable answer, or one that will serve our purpose. As part of the reasoning process, we should be able to formulate the question to be answered or the issue to be addressed. For example, why won't the car start? or should libraries censor materials that contain objectionable language?
- *Points of view or frame of reference.* As we take on an issue, we are influenced by our own point of view. For example, parents of young children and librarians might have different points of view on censorship issues. The price of a shirt may seem too low to one person while it seems high to another because of a different frame of reference. Any defect in our point of view or frame of reference is a possible source of problems in our reasoning. Our point of view may be too narrow, may not be precise enough, may be unfairly biased, and so forth. By considering multiple points of view, we may sharpen or broaden our thinking. In writing and speaking, we may strengthen our arguments by acknowledging other points of view. In listening and reading, we need to identify the perspective of the speaker or author and understand how it affects the message delivered.
- *Experiences, data, evidence.* When we reason, we must be able to support our point of view with reasons or evidence. Evidence is important in order to distinguish opinions from reasons or to create a reasoned judgment. Evidence and data should support the author's or speaker's point of view and can strengthen an argument. An example is data from surveys or published studies. In reading and listening, we can evaluate the strength of an argument or the validity of a statement by examining the supporting data or evidence. Experiences can also contribute to the data of our reasoning. For example, previous experiences in trying to get a car to start may contribute to the reasoning process that is necessary to solve the problem.
- *Concepts and ideas.* Reasoning requires the understanding and use of concepts and ideas (including definitional terms, principles, rules, or theories). When we read and listen, we can ask ourselves, "What are the key ideas presented?" When we write and speak, we can examine and organize our thoughts around the substance of concepts and ideas. Some examples of concepts are freedom, friendship, and responsibility.
- *Assumptions.* We need to take some things for granted when we reason. We need to be aware of the assumptions we have made and the assumptions of others, and to acknowledge the

importance of the beliefs that underlie people's point of view. If we make faulty assumptions, this can lead to defects in reasoning. As a writer or speaker we make assumptions about our audience and our message. For example, we might assume that others will share our point of view; or we might assume that the audience is familiar with the First Amendment when we refer to "First Amendment rights." As a reader or listener we should be able to identify the assumptions of the writer or speaker.

- *Inferences.* Reasoning proceeds by steps called inferences. An inference is a small step of the mind, in which a person concludes that something is so because of something else being so or seeming to be so. The tentative conclusions (inferences) we make depend on what we assume as we attempt to make sense of what is going on around us. For example, we see dark clouds and infer that it is going to rain; or we know the movie starts at 7:00; it is now 6:45; it takes 30 minutes to get to the theater; so we cannot get there on time. Many of our inferences are justified and reasonable, but many are not. We need to distinguish between the raw data of our experiences and our interpretations of those experiences (inferences). Also, the inferences we make are influenced by our point of view and assumptions.
- *Implications and consequences.* When we reason in a certain direction, we need to look at the consequences of that direction. When we argue and support a certain point of view, solid reasoning requires that we consider what the implications are of following that path; what are the consequences of taking the course that we support? When we read or listen to an argument, we need to ask ourselves what follows from that way of thinking. We can also consider consequences of actions that characters in stories take. For example, if I don't do my homework, I will have to stay after school to do it; if I water the lawn, it will not wither in the summer heat.

By applying these elements systematically to different situations and events, students come to reason out both personal and real world problems that they encounter. By converting topics to issues, students also learn the value of questioning all sides of an issue. For example, instead of having students study animal habitats from a topical perspective, why not have them debate the issue of "Should animals have rights?" or "Should we protect endangered species?"

Such a transformation of the focus for debate and discussion as well as project work takes an activity to a higher level of thought and reflection. Moreover, it sets up the possibility for a dialectic which pushes the thinking of the group to a higher level as well.

Paul, through his Foundation for Critical Thinking, has developed a series of templates to use in analyzing the logic of different disciplines and applications of the model to dealing with challenging content in those disciplines. The Center for Gifted Education at William and Mary has adapted his model for use in providing challenging thinking in all core subjects for gifted learners specifically.

Combining critical and creative thinking

Teaching a combination of critical and creative thinking skills through relevant models can also do double duty in respect to learning. It can promote strong content-based understanding at a deeper level as well as teaching the skills of creativity and problem solving. It can successfully model the real world decision-making necessary for innovators to employ. Consider the following outcomes of learning as a result of students dealing with the options that Truman faced in ending the war against Japan.

Outcomes for "ending the war against Japan"

After resolving the problem of "Ending the war against Japan," the student will:

History

- understand the range of choices facing President Truman and the Interim Committee related to a strategy for ending the war with Japan;
- develop a recommendation for ending the war with Japan that is defensible given the war goals of the United States in 1945, the military and diplomatic events between 1941 and 1945, and the evolution of the relationship between the United States and Soviet Union up to 1945;
- explain why President Truman decided to use the atomic bomb in preference to other options open to him in concluding the war with Japan.

Ethics

- make an ethically defensible recommendation regarding the use of the atomic bomb to help end the war with Japan that recognizes the conflicting ethical appeals present in the possible options to end the war.

Critical thinking

- argue a point of view on waging war;
- write an essay that outlines the implications and consequences for the United States based on the outcomes of any given war;
- explain different stakeholders' assumptions about war.

Problem solving

- recognize the gap between the "real" and "ideal" as the area in which problem resolution takes place;
- enlarge his database in preparation for forming decision options;
- generate a resolution for the problem of ending war that is defensible within the context provided by real events and that is ethically acceptable;
- refine personal problem solving strategies to make skills more effective, efficient, and humane through self-evaluation;

Creativity

- apply fluency, flexibility, and elaboration skills to their problem solving behaviors;
- generate original solutions to the problem;
- display positive attitudes for a creative climate.

All of these outcomes are simultaneously achievable within a learning episode where students engage directly with a real world problem in which they take charge of the learning pace, style, and organization. Autonomy in learning shares the stage in this example with collaboration and shared responsibility.

Metacognition

The regulation of specific learning behaviors and deliberately using executive processes in order for deeper learning to be achieved (Schunk, 2000) is also a central aspect of promoting innovation since these behaviors are critical for long term project work and research. Metacognition refers to two types of knowledge—self-knowledge in respect to declarative, procedural, and conditional situations (Bereiter, 2000) and self-knowledge in respect to controlling how knowledge is used—the planning, monitoring, and assessing of the process in oneself (Beyer, 2000). Each aspect is a necessary way of conceptualizing the skills needed for gifted learners to become effective in their thinking and problem solving activities over time and into adulthood.

Research suggests that metacognition is developmental, beginning early but continuing well into adulthood (Snyder, Nietfeld, & Linnenbrink-Garcia, 2011). It also appears to be more advanced in adults than children, in gifted students rather than in typical students especially in transferring the skills to new domains of activity. Metacognition is easier to teach to gifted learners as well and they appear to benefit more from being taught the strategies than other learners. Gifted learners work harder at learning the strategies and appear to be more motivated than non-gifted students. Perhaps this is due to a larger information base that they have which supports metacognitive regulation strategies since we know that metacognition improves with more knowledge in a domain (Sternberg, 2001a).

The findings on metacognition from the research literature strongly suggest the value of direct instruction, collaborative learning across age levels, and reflection techniques such as journaling, discussion, and introspection (Schraw & Graham, 1997). Innovation education requires the development of these skills and strategies as well to heighten flexibility in thinking, and reflection in action, a capacity so valued in successful educational innovations (Dai, 2011).

Conclusion

Gifted education, if it is to be seen as relevant in the next decades, must adopt an agenda that presses on the teaching and learning associated with real world innovation and change. It must embrace the use of technological tools that enhance the application of ideas in all fields. It must systematically teach the higher level skills of thinking and problem solving as routine ways to instruct in all disciplines. It must promote the use of collaborative and dynamic ways to learn that stress options and alternatives over linear paths to a given end. It must promote the use of higher level questions both teacher and by the learner to scale up the inquiry process. Finally, it must acknowledge that the goal of 21st century learning for the gifted is innovation, not just creativity. We need people who have the vision to use the tools, the strategies, and the inventions in ways that make the quality of life for all a higher level experience.

References

Amabile, T. M. (1983). *The Social Psychology of Creativity*. New York: Springer-Verlag.
Amabile, T. M. (2001). Beyond talent: John Irving and the passionate craft of creativity. *American Psychologist, 56*(4), 333–336.
American Competitiveness Initiative. (2006). *American Competitive Initiative: Leading the World in Innovation*. Washington DC: Domestic Policy Council Office of Science and Technology. Retrieved from www.innovationtaskforce.org/docs/ACI%20booklet.pdf.
Anderson, L. W., & Krathwohl D. R. (2000). *Taxonomy for Learning, Teaching, and Assessing: A Revision of Bloom's Taxonomy of Educational Objectives*. New York: Longman.
Bereiter, C. (2000). Keeping the brain in mind. *Australian Journal of Education, 44*(3), 226–238. Retrieved January 21, 2005, from ERIC database.

Beyer, B. K. (2000). *Improving Student Thinking: A Comprehensive Approach.* Boston: Allyn & Bacon.

Boyce, L. N., VanTassel-Baska, J., Burruss, J. D., Sher, B. T., & Johnson, D. T. (1997). A problem-based curriculum: Parallel learning opportunities for students and teachers. *Journal for the Education of the Gifted, 20*(4), 363–379.

Bracken, B. A., Bai, W., Fithian, E., Lamprecht, S., Little, C., & Quek, C. (2003). *Test of Critical Thinking.* Williamsburg, VA: Center for Gifted Education, College of William and Mary.

Business Roundtable. (2008). *Tapping America's Potential: The Education for Innovation Initiative.* Washington, DC: Business Roundtable.

Chandler, K. (2004). *A National Study of Curriculum Policies and Practices in Gifted Education.* Unpublished doctoral dissertation, College of William and Mary, Williamsburg, VA.

Cramond, B. (1994). The Torrance Tests of Creative Thinking: From design through establishment of predictive validity. In R. Subotnik & K. Arnold (Eds.), *Beyond Terman: Contemporary Longitudinal Studies of Giftedness and Talent* (pp. 229–254). Norwood, NJ: Ablex.

Csikszentmihalyi, M. (2000). *Beyond Boredom and Anxiety: Experiencing Flow in Work and Play.* San Francisco: Jossey-Bass.

Dai, D. Y. (2011). Essential tensions surrounding the concept of giftedness. In L. Shavinina (Ed.), *International Handbook on Giftedness* (pp. 39–80). New York: Springer.

Delcourt, M. A. B. (1994). Characteristics of high-level creative productivity. In R. Subotnik & K. Arnold (Eds.), *Beyond Terman, Contemporary Longitudinal Studies of Giftedness and Talent* (pp. 401–436). Norwood, NJ: Ablex.

Elder, L., & Paul, R. (2004). *Guide to the Human Mind: How it Learns, How it Misleams.* Dillon Beach, CA: Foundation for Critical Thinking.

Ennis, Robert H. (1996). *Critical Thinking.* Upper Saddle River, NJ: Prentice Hall.

Fleischman, H. L., Hopstock, P. J., Pelezer, M. P., & Shelley, B. E. (2010). *Highlights from PISA 2009: Performance of US 15 year old Students in Reading, Mathematics, and Science Literacy in an International Context.* Washington, DC: IES-NCES.

Gallagher, S. A. (1998). The road to critical thinking: The Perry scheme and meaningful differentiation. *NASSP Bulletin, 82*(595), 12–20.

Gallagher, S. A., & Stepien, W. J. (1996). Content acquisition in problem-based learning: Depth versus breadth in American studies. *Journal for the Education of the Gifted, 19*(3), 257–275.

Gardner, H. (2000). *The Disciplined Mind: Beyond Facts and Standardized Tests, the K-12 Education that Every Child Deserves.* New York: Penguin Putnam.

Gavin, M. K., Casa, T. M., Adelson, J. L., Carroll, S. L., Sheffield, L. J., & Spinelli, A. M. (2007). Project M3: Mentoring mathematical minds—a research-based curriculum for talented elementary students. *Journal for the Education of the Gifted, 18*(4), 566–585.

Getzels, J., & Csikszentmihalyi, M. (1976). *The Creative Vision: A Longitudinal Study of Problem Finding in Art.* New York: Wiley.

Hillocks, G. (1999). *Ways of Thinking, Ways of Teaching.* New York: Teachers College Press.

Isaacson, W. (2011). *Steve Jobs.* New York: Simon & Schuster.

Isaksen, S. G., Treffinger, D. J., Dorval, K. B., & Noller, R. B. (2000). *Creative Approaches to Problem Solving: A Framework for Change*, 2nd ed. Dubuque, IA: Kendall/Hunt.

Moon, S. M., Feldhusen, J. F., & Dillon, D. R. (1994). Long-term effects of an enrichment program based on the Purdue Three-Stage Model. *Gifted Child Quarterly, 38*(1), 38–48.

National Academy of Education. (2009). *World-class Science and Mathematics.* Washington, DC: National Academy of Education. Retrieved from www.naeducation.org/White_Papers_Project_Science_and_Mathematics_Briefing_Sheet.pdf.

National Academy of Sciences. (2005). *Rising Above the Gathering Storm.* Washington, DC: National Academy Press. Retrieved from www.nap.edu/catalog.php?record_id=11463.

National Research Council. (2007). *Taking Science to School: Learning and Teaching Science in Grades K-8.* Committee on Science Learning, Kindergarten Through Eighth Grade. R. A. Duschl, H. A. Schweingruber, & A. W. Shouse (Eds.), Board on Science Education, Center for Education. Division of Behavioral and Social Sciences and Education. Washington, DC: The National Academies Press.

National Science Board. (2010). *Preparing the Next Generation of STEM Innovators: Identifying and Developing our Nation's Human Capital.* Arlington, VA: National Science Foundation.

Ochse, R. (1990). *Before the Gates of Excellence: The Determinants of Creative Genius.* Cambridge, UK: Cambridge University Press.

OECD (2010). *PISA 2009 results: What Students Know and Can Do—Student Performance in Reading, Mathematics and Science (volume 1).* Retrieved from http://dx.doi.org/10.1787/9789264091450-en.

Park, G., Lubinski, D., & Benbow, C. P. (2007). Contrasting intellectual patterns for creativity in the arts and sciences: Tracking intellectually precocious youth Over 25 years. *Psychological Science, 18*(11), 948–952.

Perkins, D. N. (1981). *The Mind's Best Work*. Cambridge, MA: Harvard University Press.

Root-Bernstein, R. (2003). The art of innovation. In L. V. Shavinina (Ed.), *The International Handbook on Innovation* (pp. 267–278). Oxford, UK: Elsevier Science.

Schraw, G., & Graham, T. (1997). Helping gifted students develop metacognitive awareness. *Roeper Review, 20*(1), 4–8.

Schunk, D. H. (2000). *Learning Theories: An Educational Perspective*, 3rd ed., Upper Saddle River, NJ: Merrill.

Shavinina, L. V. (2003). *The International Handbook on Innovation*. Oxford, UK: Elsevier Science.

Shavinina, L. V. (2009). *International Handbook on Giftedness*. New York: Springer.

Shea, D., Lubinski, D., & Benbow, C. P. (2001). Importance of assessing spatial ability in intellectually talented young adolescents: A 20-year longitudinal study. *Journal of Educational Psychology, 93*(3), 604–614.

Simonton, D. K. (1994). *Greatness: Who Makes History and Why*. New York: Guilford.

Snyder, K., Nietfeld, J., & Linnenbrink-Garcia, L. (2011). Giftedness and metacognition: A short-term longitudinal investigation of metacognitive monitoring in the classroom. *Gifted Child Quarterly, 55*, 181–193.

Sternberg, R. J. (Ed.). (1988). *The Nature of Creativity: Contemporary Psychological Perspectives*. New York: Cambridge University Press.

Sternberg, R. J. (2001a). *Complex Cognition: The Psychology of Human Thought*. Oxford, UK: Oxford University Press.

Sternberg, R. J. (2001b). What is the common thread of creativity? Its dialectical relation to intelligence and wisdom. *American Psychologist, 56*(4), 360–362.

Swanson, J. D. (2006). Breaking through assumptions about low income minority students. *Gifted Child Quarterly, 50*, 11–25.

Terrell, N. (2007). STEM Occupations. *Occupational Outlook Quarterly, 51*(1), 26–33.

Torrance, E. P. (1993). The beyonders in a thirty-year longitudinal study of creative achievement. *Roeper Review, 15*(3), 131–139.

Treffinger, D. J., Isaksen, S. G., & Dorval, K. B. (2000). *Creative Problem Solving: An Introduction*. Waco, TX: Prufrock.

VanTassel-Baska, J. (in preparation). Sir Francis Galton, The father of gifted education. In A. Robinson & J. Jolly (Eds.) *Illuminating Minds*. London: Routledge.

VanTassel-Baska, J., Avery, L. D., Hughes, C. E., & Little, C. A. (2000). An evaluation of the implementation of curriculum innovation: The impact of William and Mary units on schools. *Journal for the Education of the Gifted, 23*(2), 244–272.

VanTassel-Baska, J., Bass, G., Ries, R., Poland, D., & Avery, L. D. (1998). National study of science curriculum effectiveness with high ability students. *Gifted Child Quarterly, 42*, 200–211.

VanTassel-Baska, J., Bracken, B., Feng, A., & Brown, E. (2009). A longitudinal study of reading comprehension and reasoning ability of students in elementary Title I schools. *Journal for the Education of the Gifted, 33*(1), 7–37.

VanTassel-Baska, J., & Feng, A. X. (Eds.). (2003). *Designing and Utilizing Evaluation for Gifted Program Improvement*. Waco, TX: Prufrock Press.

VanTassel-Baska, J., & Little, C. (Eds.) (2011). *Content-based Curriculum for the Gifted*. Waco: TX: Prufrock Press.

Volk, V. (2004). *Confidence Building and Problem Solving Skills: An Investigation into the Impact of the Future Problem Solving Program on Secondary School Students' Sense of Self-efficacy in Problem Solving, in Research, in Teamwork, and in Coping with the Future*. University of New South Wales, Sydney, AU.

9

The three-ring conception of innovation and a triad of processes for developing creative productivity in young people

Marcia A. B. Delcourt and Joseph S. Renzulli

WESTERN CONNECTICUT STATE UNIVERSITY, USA, AND THE UNIVERSITY OF CONNECTICUT, USA

Summary: Although innovation is typically viewed as a process that always begins with a creative idea and ends with new or improved products, there are other factors that contribute to designing purposive tasks that can be organized into systematic plans for transforming ideas into tangible outcomes. In this chapter we discuss three interrelated components of creative productivity—above-average ability in a particular domain, creativity, and task commitment. The interaction between and among these components of the innovative process are necessary to provide the strategies for developing products or performances that can become audience- or consumer-valued products. Next, we discuss three types of educational services for promoting innovation in young people. These services consist of exposing students to areas of potential interest and task commitment, providing them with the methodological skills to pursue their interests in a professionally authentic manner, and providing the opportunities, resources, and encouragement to see their ideas through to fruition.

Key words: Creative-productive, creativity, innovation, motivation, above-average, task-commitment.

Introduction: meet the innovators

Max is a sixth grade writer who has produced an extensive book every year since the second grade. His vocabulary is beyond that of many high school students and, he is, of course, an avid reader. His literary tastes are varied and include poetry, fantasy, science fiction, as well as biographies. He has begun to illustrate his works and is presently featuring a child inventor in his new trilogy. His school program is adapted to include time for his writing, resulting in him already having published two short stories for a children's magazine. He is developing a blog to provide tips for young authors.

As a tenth grader, Amber experiments with the mathematical equations used to calculate changes in black holes and then shares this information via the Internet with physicists in other parts of the world. This talent for logical and insightful thinking is recognized and validated for Amber when scientists provide constructive feedback for her ideas, helping her to obtain new knowledge and skills as well as providing her with an outlet for synthesizing information about this topic. Several Internet sites provide opportunities to "ask the experts." Two such sites for science are www.pitsco.com and the *Scientific American* site www.sciam.com. With the help of scientists located through these websites, Amber became interested in specific calculations related to the black hole phenomenon. She entered her mathematical calculations about this topic in a local science fair competition and was awarded an honorable mention at the state level for her precision in and explanation of her calculations. As a result of her project, she also developed a website to teach other students about astronomy and space travel.

At age 17, Chris's interest in engineering led to his acceptance in a competitive summer internship at a university well known for its faculty which conducts scientific research. Paired with a scientist in chemical engineering, Chris was assigned a project to investigate cancer cell growth. During the seven-week internship, he researched the topic, wrote a program in a computer language he had barely used before, solved the mathematical problem given to him about the growth of a particular cancer cell, then crafted and presented a paper entitled "Computer Simulations and Cancer Research: A New Solution to a Complex Problem?"

The ability to develop innovative ideas can begin at an early age and evolve over time. These three students have multiple characteristics in common, namely, the capacity to work intensely on a specific topic, to apply their natural inclinations and ability toward a specific activity, and to create something that they want to share with others, particularly with an audience who appreciates the topic. While Amber is still establishing her knowledge base in the field of black holes, she is able to produce sound ideas and questions which have received pointed responses from international experts who are nurturing her understanding of astronomy and physics. Her science fair submission and her website represent outlets for sharing her work with others. Max and Chris also created new products, which they disseminated to the public. Children and young adults have the ability to be more than consumers of information. They can be creative producers of high quality products (Delcourt, 2008; Renzulli, 1986; Renzulli & Reis, 1985). To help youths achieve this goal, we need to recognize the characteristics that lead to innovative behaviors and provide the type of environment that is a catalyst for these endeavors.

The three-ring conception of giftedness

Research on creative-productive people has consistently shown that although no single criterion can be used to determine giftedness, individuals who have achieved recognition because of their unique accomplishments and creative contributions possess a relatively well-defined set of three interlocking clusters of traits. These clusters consist of: above-average, though not necessarily superior, ability; task commitment; and creativity (see Figure 9.1). It is important to point out that no single cluster "makes giftedness." Rather, it is the interaction among the three clusters that researchers have shown to be the necessary ingredient for creative-productive accomplishment (Renzulli, 1978). This interaction is represented by the shaded portion in the center of Figure 9.1. It is also important to indicate that each cluster plays an important role in contributing to the display of gifted behaviors. This point is emphasized because one of the major errors that educators continue to make in identification procedures is to overemphasize superior abilities at the expense of the other two clusters of traits. The background of this diagram, referred to as a "houndstooth" pattern, represents the multiple internal and external characteristics that influence each individual.

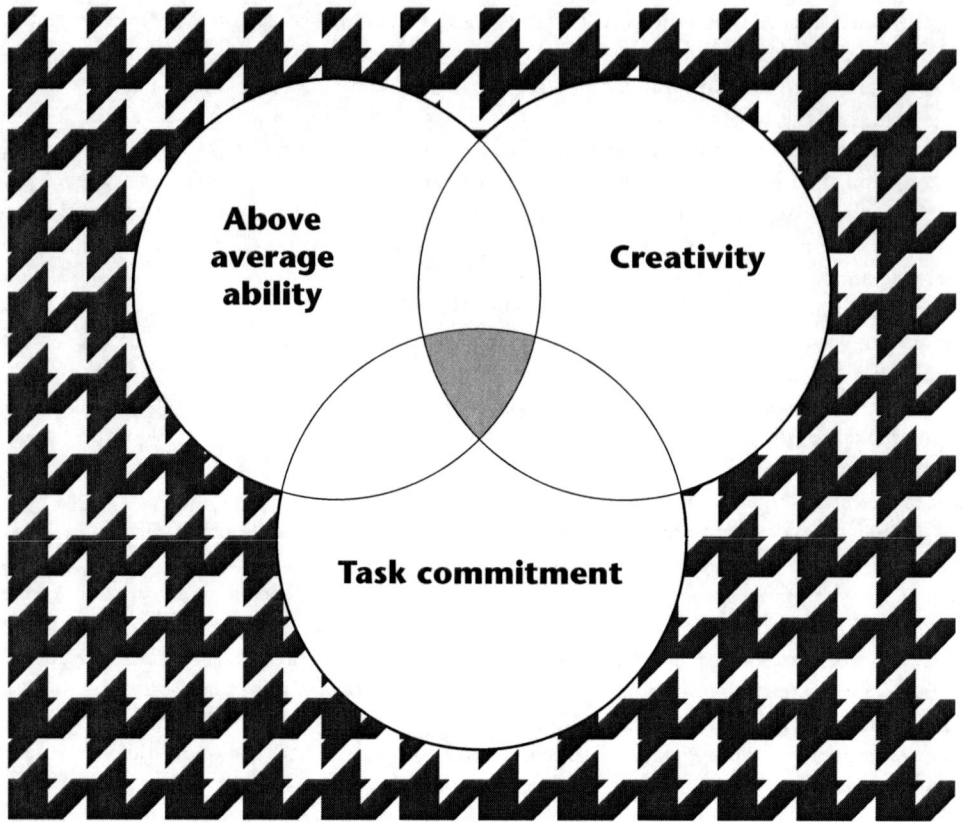

Figure 9.1 The three-ring conception of giftedness.

Above-average ability

Above-average ability can be defined in two ways. One view includes the broad concept of overall capacity to think and perform, while the other perspective targets particular ways in which an individual understands, interprets, and reacts to information in a specialized domain (Csikszentmihalyi, 1990; Renzulli, 2005; Treffinger, 1998).

General ability consists of the capacity to process information, to integrate experiences that result in appropriate and adaptive responses in new situations, and the capacity to engage in abstract thinking. Examples of general ability are verbal and numerical reasoning, spatial relations, memory, and word fluency. These abilities are usually measured by tests of general aptitude or intelligence, and are broadly applicable to a variety of traditional learning situations. The following are examples of general abilities:

- high levels of abstract thinking, verbal and numerical reasoning, spatial relations, memory, and word fluency;
- the adapting to and the shaping of novel situations encountered in the external environment;
- the automatization of information processing; rapid, accurate, and selective retrieval of information.

Specific abilities consist of the capacity to acquire knowledge, skill, or the capacity to perform in one or more activities of a specialized kind and within a restricted range, as indicated in the following examples:

- the application of various combinations of the above general abilities to one or more specialized areas of knowledge or areas of human performance (e.g., the arts, leadership, administration);
- the capacity for acquiring and making appropriate use of advanced amounts of formal knowledge, tacit knowledge, technique, logistics, and strategy in the pursuit of particular problems or the manifestation of specialized areas of performance;
- the capacity to sort out relevant and irrelevant information associated with a particular problem or area of study or performance.

These abilities are defined according to the ways in which human beings express themselves in real-life (i.e., non-test) situations. Examples of specific abilities are chemistry, ballet, mathematics, musical composition, sculpture, and photography. Each specific ability can be further subdivided into even more specific areas (e.g., portrait photography, astrophotography, photo journalism). Specific abilities in certain areas such as mathematics and chemistry have a strong relationship with general ability, and some indication of potential in these areas can therefore be determined from tests of general aptitude and intelligence. They can also be measured by achievement tests and tests of specific aptitude. Many specific abilities, however, cannot be easily measured by tests. It follows that areas such as the arts must be evaluated through one or more performance-based assessment techniques.

Within this model the term *above-average ability* will be used to describe both general and specific abilities. *Above-average* should also be interpreted to mean the upper range of potential within any given area. Although it is difficult to assign numerical values to many specific areas of ability, "well above-average ability" refers to persons who are capable of performance or the potential for performance that is representative of the top 15–20% of any given area of human endeavor.

Task commitment

A second cluster of traits that has consistently been found in creative-productive persons is a refined or focused form of motivation known as task commitment. Examples include the following:

- the capacity for high levels of interest, enthusiasm, fascination, and involvement in a particular problem, area of study, or form of human expression;
- the capacity for perseverance, endurance, determination, hard work, and dedicated practice;
- self-confidence, a strong ego, and a belief in one's ability to carry out important work, freedom from inferiority feelings, drive to achieve;
- the ability to identify significant problems within specialized areas; the ability to tune into major channels of communication and new developments within given fields;
- setting high standards for one's work, maintaining an openness to self and external criticism, developing an aesthetic sense of taste, quality, and excellence about one's own work and the work of others.

Whereas motivation is usually defined in terms of a general energizing process that triggers responses in organisms, task commitment represents energy brought to bear on a particular problem (task) or specific performance area. The terms most frequently used to describe task commitment are perseverance, endurance, hard work, dedicated practice, and self-confidence in one's ability to carry out important work.

Creativity

The third cluster of traits that characterize persons who display gifted behaviors consists of factors usually lumped together under the general heading of "creativity." The following are examples of creative abilities:

- fluency, flexibility, and originality of thought;
- openness to experience; receptiveness to that which is new and different (even irrational) in the thoughts, actions, and products of oneself and others;
- curiosity, speculation, adventurousness, and "mental playfulness";
- willingness to take risks in thought and action, even to the point of being uninhibited;
- sensitivity to detail, aesthetic characteristics of ideas and things;
- willingness to act on and react to external stimulation and one's own ideas and feelings.

As one reviews the literature in this area, it becomes readily apparent that the words *gifted, genius, innovators,* and *eminent creators* or *highly creative persons* are used interchangeably. In many of the research studies designed to understand the characteristics of innovative individuals, those ultimately selected for intensive study were in fact recognized *because* of their creative accomplishments (Bloom & Sosniak, 1981; MacKinnon, 1964; McCurdy, 1960; Nicholls, 1972).

Above-average ability, task commitment, and creativity are the types of characteristics we are looking for in people who are known to be or will be innovators. Children who manifest *or are capable of developing* an interaction among these three clusters require a wide variety of educational opportunities, resources, and encouragement above and beyond those ordinarily provided through regular instructional programs.

Identification of students with the potential for innovation

The Three-Ring Conception of Giftedness was reported to be one of the most prevalent models for identifying potentially gifted or innovative students in the United States (Callahan, Hunsaker, Adams, Moore, & Bland, 1995). School personnel who use this model typically employ one or more instruments that represent each of the three concepts: above-average, though not necessarily superior, ability; task commitment; and creativity. Ideally, multiple instruments and sources of data are used to target students who are capable of creative-productive behavior. Specific processes for identifying these students can be found in supplementary references (Renzulli, 1994). The following sections of this chapter focus on methods used to develop and enhance characteristics of innovative behavior in children and young adults.

The Enrichment Triad Model

Often employed as an educational programming model for gifted and talented students, the Enrichment Triad Model (Renzulli, 1977) is based upon activities and experiences that enhance the regular curriculum to promote creative-productive behaviors leading to innovative ideas and solutions to problems. The model has three types of activities (Type I, Type II, and Type III) that can be presented in a sequence or employed as needed for individuals, small groups, or whole classes of students. Figure 9.2 provides a depiction of the model.

The three-ring conception of innovation

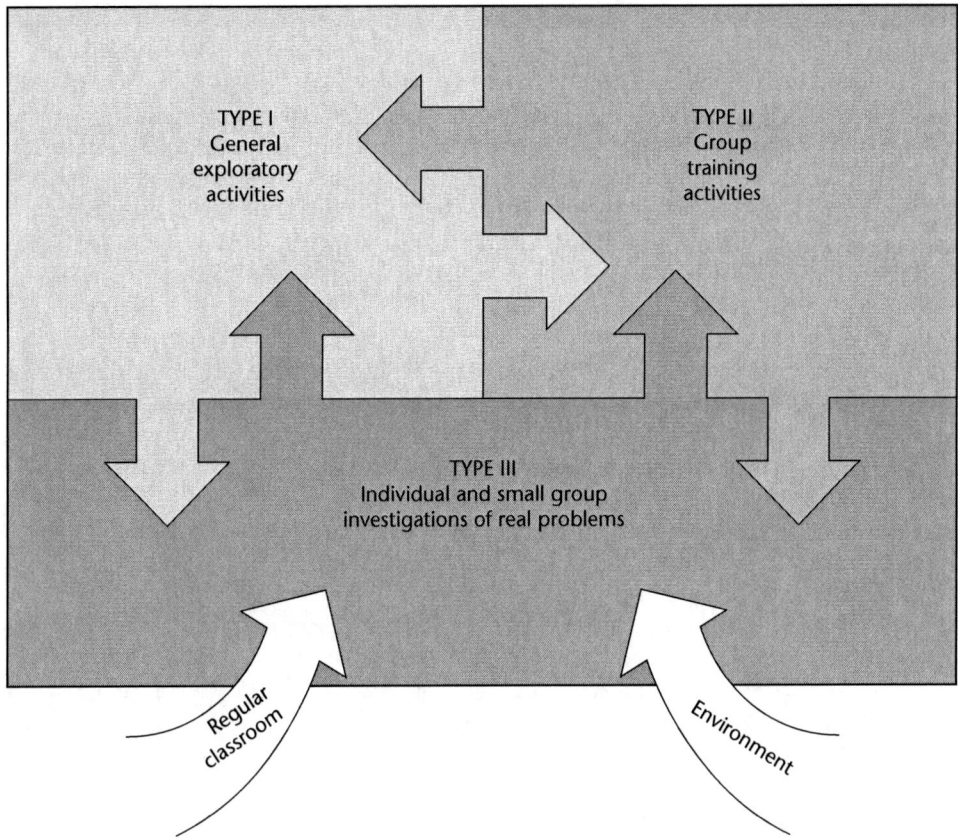

Figure 9.2 Enrichment Triad Model.

Type I enrichment

This aspect of the model consists of general exploratory activities introducing a variety of topics not ordinarily available in the regular curriculum. A classroom can be organized to include a variety of experiences including guest lecturers, demonstrations, interest development centers, book displays, introductory lessons, etc., which allow students exposure to new topics. Examples include the following:

- A local meteorologist presents information about climate change.
- A simulation is presented about animal migration.
- An interest development center is based on the works of the lithographer and painter M. C. Escher, where his works are displayed and activities are planned which allow students to relate Escher's work to that of other artists.
- Books are placed in the classroom library that introduce topics such as entomology, economics, physiology, poetry, etc.
- The teacher introduces the students to topics such as mystery story writing (language arts), "freakonomics" by Levitt and Dubner, 2005 (mathematics, economics, political science), genealogy (social studies), etc.

133

Type I activities are appropriate for all students. Everyone should be given the opportunity to learn and think about new topics. Many children and adolescents lament their lack of interest in anything, mostly because they do not know about a great variety of topics. A survey could be completed by a child to find out where his or her strengths and interests lie. The *Interest-A-Lyzer* for elementary school students (Renzulli, 1997) and the interest inventory for secondary school students (Hébert, Sorensen, & Renzulli, 1997) offer many opportunities for a child to express what he or she would like to do by asking questions such as "You are a famous author about to write your next book, what will it be about?" "Imagine that you can travel to any time in history, where would you go?" "Have you ever made up a new game? Tell about it here." Responses to these and many other questions can help a student identify areas that are already strengths and locate new topics of interest.

An educator could choose an idea from the *Interest-A-Lyzer* and initiate or locate related presentations at a local library or target books that explain how to get involved with the selected topic. By introducing a student to a variety of ideas, it is more likely that he or she will be able to choose an area of interest. Once a student focuses on a particular area such as poetry, chemistry, bird watching, or cars, the next step is to find out the skills needed to pursue this topic. Beginning questions can include: "How do writers get their ideas for a poem?" "What equipment do I need to conduct a chemistry experiment?" "How can I identify the birds in my neighborhood?" "What steps should I take to design a blueprint for a car?"

Students, teachers, and parents can also use a computer-based program called Renzulli Learning (www.RenzulliLearning.com) in order to match student interests, learning styles, and expression styles with a variety of skill development activities and viable project ideas. All three types of enrichment activities can be accessed through this interactive program. After completing the student profiler, hundreds of websites will become available that target a student's interests and skills. The online educational program also promotes academic achievement as indicated by significant growth in reading comprehension, oral reading fluency, and achievement in social studies (Field, 2009).

Type II enrichment

These activities introduce and strengthen "training in thinking and feeling processes, learning how-to-learn skills, research and reference skills, and written, oral, and visual communication skills" (Renzulli, 1994, p. 207). A student could be involved in the following activities:

- measuring rain and snowfall patterns;
- tracking monarch butterfly migration;
- designing tessellations of Escher-like drawings;
- writing poetry;
- developing a family tree.

In addition to accessing online activities through Renzulli Learning, there are excellent resources for developing skills of young researchers (*Chi Square, Pie Charts and Me* by Baum, Gable, & List, 1998; *Think Data* by Renzulli, Heilbronner, & Siegle, 2010; *Looking for Data in All the Right Places* by Starko & Schack, 1992), entrepreneurs (*The New Totally Awesome Business Book for Kids* by Bochner & Bochner, 2007), scientists (*Hands-on Ecology* by Kessler, 2006), writers (*How Writers Work* by Fletcher, 2000), and other professionals. The focus of these materials is to assist in the development of a variety of behaviors similar to those of a practicing professional in a given field of study, allowing students to rehearse and refine the skills necessary to produce high quality, innovative projects.

Type III enrichment

This type of enrichment consists of individual and small group investigations of real problems, based on student interest and commitment. Results are presented to an audience whose members understand and appreciate the topic, such as scientists, editors, or business managers.

Project examples may include the following:

- reviving a wetlands area to attract more wildlife and presenting the project to a local environmental planning board for replication in multiple regions;
- planting shrubbery to feed migrating animals and logging the progress of the migration in a national scientific database;
- organizing an art festival that features Escher-like artwork and presenting the display in a local business;
- creating and publishing a regional poetry journal and submitting it to a national writing competition;
- designing a workshop to assist families in investigating their ancestry and reporting the project to a local historical society.

A mentor, teacher, or parent can be instrumental in helping students throughout the development of a Type III project, specifically by assisting to locate resources, providing background information, or supplying procedural advice. The more projects a student completes, the better he or she becomes at understanding the process and tapping into human and material resources. Students who are experienced in creative-productivity not only generate better projects over time (Delcourt, 2008; Kay, 1994), but demonstrate more acute skills at finding problems worth solving (LaBanca, 2008). Getting ideas or problem-finding is the most crucial stage in creative achievement (Csikszentmihalyi, 1999). In the first phase of self-regulated learning individuals use forethought (Zimmerman, 1998), students sort through potential ideas to identify those worth pursuing. Highly creative-productive students select projects based on the value the topic has to the learner and to a projected audience (Delcourt, 2008). For example, LaBanca (2008) explains that students who are amongst the highest ranked at the Intel International Science and Engineering Fair (ISEF) are problem-solvers who demonstrate high levels of inquiry and are recognized for identifying truly novel problems, rather than for selecting topics that are considered to be merely of technical interest. He classifies science fair projects into four categories: those which provide a review of the literature about a particular topic; technical projects that are used to replicate well-known results, often referred to as "cookbook" projects; technical projects with value that extends the data set and potentially the knowledge base about a known topic; and projects which represent a novel approach to a problem that is of value to the scientific community. Novel approach projects are the goal for students in any realm of study when the outcome is to produce innovative thinking.

Forms of implementation

While types of enrichment can be sequenced according to the numbered activities (I, II, and III), they can also be organized around the kinds of experiences in which a student is prepared to engage. This is represented by the flow of the arrows in Figure 9.2. A student could enter a class with an elementary background about a topic such as botany (Type I), but needs to know how to maintain specific species of plants used for pollination (Type II). Another student might approach a teacher about an idea for a project in photography to be used to illustrate a book (Type III), and needs the time to prepare the book design and layout. Learning should proceed in a natural way from one type of activity to another, as the student's needs change.

The three types of activities can be incorporated into student experiences in a number of ways, which focus on delivering appropriate services to students. All three types of activities can be implemented within the regular classroom structure or take place in another environment, such as a pull-out program, a cluster class, an after-school program, or a mentorship.

For Max, the young writer, the entire library was his Type I resource. He had favorite authors such as Roald Dahl, who not only wrote wonderful fiction stories such as *Charlie and the Chocolate Factory* (Dahl, 1964), but explained his motivation for writing in his autobiographies, *Boy* (Dahl, 1986) and *Going Solo* (Dahl, 1988). Max attended every writing workshop he could fit into his schedule, enrolling in summer camps and Saturday programs. He was heavily influenced by the works of C. S. Lewis, Madeleine L'Engle, and Eoin Colfer. Max's language arts teacher tailored his classroom experiences and homework to his interests, allowing him to submit his draft articles as class assignments and assisting him to locate state writing competitions and national journals as outlets for his manuscripts. Max also began a literary journal at his middle school where he serves as editor-in-chief. His own writing is definitely above the caliber of a typical 11- to 12-year old. His ideas have been judged to be novel and creative by an outside audience of writers who have selected his work for publication.

Amber had difficulty obtaining direct assistance for her interests in physics and astronomy because her local teachers did not have a background in these areas. Her teachers were, however, supportive in making suggestions for resources. She read a great deal on the topic, ultimately turning to works by Stephen Hawking in order to understand the issues related to black holes. She admitted that she was only a novice in this field, but tried to get clarifications for her questions by contacting scientists who were able to supply her with understandable resources and answers to her many questions. While her science fair project might have been considered only a technical project or a technical project with value, based on LaBanca's definitions, she continues to study physics, astronomy, and mathematics. Her potential for innovation is still to be discovered.

Chris is another student who has always been interested in science. When he entered a summer program in engineering, he thought he would be building bridges. He never considered the role of a chemical engineer. His experiences in mathematics and in computer technology were practical skills for his assignment in the summer program because he was able to make a useful computer simulation for the cancer growth model. He jumped right into a Type III activity, brushed up on skills where needed, and succeeded in the task. He also completed other projects at his high school. Chris was a member of a pull-out program where he worked either individually or on a team to complete science-related projects. On one occasion, he was a member of a team that wrote a grant to obtain a laser. When Chris's team was awarded the funding, they built a machine for making holographs and traveled to local and regional schools to demonstrate laser technology.

These are only three examples of how students have engaged in activities related to the Enrichment Triad Model. Students in elementary, middle, and high schools throughout the world participate in programs that use this model to help them develop creative-productive activities.

Catalysts and barriers: why some people become innovators and others do not

Catalysts

Catalysts change over time as students acquire more experience with the entire process of creative-productivity. Haensly and Roberts (1983) described the creative processes of professionals in six different fields. In their study, the subjects reported the following necessary ingredients for successful product development: task commitment, the ability to select an appropriate audience for presenting one's contribution, and energy to overcome obstacles such as a lack of time, money, or cooperation.

While students reported that getting started is the largest hurdle, those who have enjoyed the satisfaction of completing at least one high quality project have had less of a struggle commencing and continuing with subsequent ones (Delcourt, 1993). These students also learned how to use resources effectively, harness their own ability to be motivated, and target appropriate audiences for their work.

Getting started

Choosing something of interest is a key factor in beginning a project. Several researchers have investigated the vital role of problem-finding in the problem-solving process (Csikszentmihalyi, 1999; Getzels, 1987; LaBanca, 2008). A key variable is examining multiple possibilities for prospective projects. Therefore, the best advice is to explore many possible ideas before settling on a topic. Being curious and open minded could lead to a project that is potentially unique and innovative.

Effective use of resources

Resources include equipment, mentors, books, financial support, background information, time, organizational and communication skills, imagination, and appropriate feedback. As individuals work on projects, they learn how to manage their resources and to coordinate their efforts with others to get what they need when they need it. When resources are scarce and students have specific deadlines, their projects might have a limited scope or could be dropped altogether. The inability to access appropriate resources can certainly influence interest in a project. When giving advice about working on a Type III activity, students have indicated that an early assessment of feasibility is important (Delcourt, 1994). Fraenkel and Wallen (2003) provide similar advice for those conducting a thesis. They recommend that a topic be feasible, clear, significant, and ethical, prior to investing too much time in the project.

Seeing a project through to completion

What sustains someone to complete a long-range project? Problem-solving tactics, motivation, and the ability to learn from former projects are three important factors that help students to stay on target. Students who work on Type III investigations become expert problem-solvers. After locating a topic, expert problem-solvers are able to recognize patterns and principles related to their problem, whereas novice problem-solvers only see the more obvious and concrete aspects of an issue (Schoenfeld & Herrmann, 1982). This means that expert problem-solvers are more adept than novices at understanding inferences that lead to innovative solutions. Experts also break their tasks into manageable parts and are able to use these parts flexibly toward completing an end goal (Larkin, Heller, & Greeno, 1980). Dweck (1986) and Good and Dweck (2005) indicated that those who base their motivation on ways to improve their own learning not only sustain themselves through complex situations, but are more likely to achieve their goals. In comparison, individuals who are more concerned about how others perceive them in the completion of their goals are less likely to complete their tasks.

Another major influence on creative-productive activity is past learning from former projects. In a study of creative-productive secondary school students, Delcourt (1993, p. 29) synthesized the following list based on reports of 18 adolescents who were asked what they learned from working on their creative projects in science and the humanities:

1 The project itself resulted in
   a   increased interest and task commitment
   b   improved quality of projects completed later
   c   the ability to get more ideas
   d   better organizational strategies
   e   future selection of more challenging projects
   f   the ability to accept criticism more realistically.

2 Skill acquisition or development occurred in the areas of
   a   research
   b   writing
   c   communication
   d   technical abilities.

3 General personality traits showed improvement in
   a   self-satisfaction
   b   patience
   c   self-assurance
   d   responsibility
   e   attitude toward learning
   f   independence
   g   enjoyment
   h   passion for a topic.

Given the intensity of the activities needed to create the products by Max, Amber, and Chris, it is no surprise that they spend considerable time and energy on their investigations. These students need to know their own work styles and to use their time wisely to meet self-established deadlines as well as those imposed by formal organizations such as science fair committees and publishers.

The role of the audience

It takes someone who is knowledgeable about a topic to provide appropriate feedback. It is also rewarding to present ideas to an appreciative audience. Audience members put an idea into perspective when judging its innovative impact. Certainly, students may need to complete several projects before they understand the process well enough to create products of value, and they can learn a great deal from targeted feedback that can improve their ideas, skills, and final projects.

Barriers

When asked about the criteria related to their least-liked projects, students have referred to barriers related to project completion, such as: a lack of commitment and interest; inadequate time, information, and skills for working on a specific project; and poor selection of human and material resources (Delcourt, 1993). In a recent study of students participating in science fairs, Shore, Delcourt, Syer, and Schaprio (2008) found similar obstacles in the path of project completion such as lack of time, inappropriate resources, compulsory participation, difficulty in selecting a topic, and lack of support. These barriers were so strong and the pressures were so great that 20–25% of the students reported that they cheated in some capacity while completing their science fair submission. Unfortunately, the reported cases of cheating amongst professionals and scientists are also widely documented (Shore et al., 2008). Research into compulsory participation in certain activities when time and resources are limited should be a topic of future investigations.

Tips to foster innovative behaviors

While all projects have stumbling blocks, the biggest one is a lack of encouragement. Students need teachers and mentors to guide them in understanding their strengths and interests, to assist them in developing their skills, and to support their creative productive activities. These actions can lead to the development of innovative behaviors and the following tips can serve as guidelines:

1. Educators have the responsibility to recognize the potential for a child or young adult to be innovative.
2. Students should have the opportunities to identify their strengths and learning styles in order to understand how to develop their potential for innovative behaviors.
3. Problem-finding should be a top priority. Strategies to locate feasible, concise, significant, ethical, and realistic topics should be taught, and enough time should be allocated for this most vital stage of the creative-productive process.
4. Children who have ideas for making positive changes in their environments should be recognized and encouraged.
5. Projects need to be shared with an appropriate audience.
6. When students know that they can make significant contributions through their projects, their self-confidence increases and they are more likely to exhibit these types of behaviors in the future.
7. A child's interests need support and guidance, but children need to understand how to work through their problems and projects on their own.
8. Barriers that are seen as insurmountable can lead students to prematurely end a project or produce an unethical result. Students should have an outlet to explain their obstacles and have a realistic pathway to complete or exit from a project.
9. Students need models of highly effective products in order to see what they can accomplish.
10. The developmental nature of creative productive behavior should be recognized. As students practice these activities they are more likely to become innovative adults.

References

Baum, S. M., Gable, R. K., & List, K. (1998). *Chi Square, Pie Charts and Me*. Unionville, NY: Royal Fireworks.
Bloom, B. S., & Sosniak, L. A. (1981). Talent development vs. schooling. *Educational Leadership, 38*, 86–94.
Bochner, A., & Bochner, R. (2007). *The New Totally Awesome Business Book for Kids*. New York: Newmarket Press.
Callahan, C. M., Hunsaker, S. L., Adams, C. M., Moore, S. D., & Bland, L. C. (1995). *Instruments Used in the Identification of Gifted and Talented Students*. (Research Monograph 95130). Storrs, CT: National Research Center on the Gifted and Talented.
Csikszentmihalyi, M. (1990). The domain of creativity. In M. A. Runco & R. S. Albert (Eds.), *Theories of Creativity* (pp. 190–212). London: Sage Publications.
Csikszentmihalyi, M. (1999). Implications of a systems perspective for the study of creativity. In R. J. Sternberg (Ed.), *Handbook of Creativity* (pp. 313–335). New York: Cambridge University Press.
Dahl, R. (1964). *Charlie and the Chocolate Factory*. New York: Alfred A. Knopf.

Dahl, R. (1986). *Boy*. New York: Puffin Books.
Dahl, R. (1988). *Going Solo*. New York: Puffin Books.
Delcourt, M. A. B. (1993). Creative productivity among secondary school students: Combining energy, interest and imagination. *Gifted Child Quarterly, 37*, 23–31.
Delcourt, M. A. B. (1994). Creative/productive behavior among secondary school students: A longitudinal study of students identified by the Renzulli three-ring conception of giftedness. In R. Subotnik & K. Arnold (Eds.), *Beyond Terman: Longitudinal Studies in Contemporary Gifted Education* (pp. 401–436). Norwood, NJ: Ablex.
Delcourt, M. A. B. (2008). Where students get creative-productive ideas for major projects in the natural and social sciences. In B. M. Shore, M. W. Aulls, & M. A. B. Delcourt (Eds.), *Inquiry in Education Volume II: Overcoming Barriers to Successful Implementation* (pp. 63–92). New York: Routledge.
Dweck, C. (1986). Motivational processes affecting learning. *American Psychologist, 41*(10), 1040–1048.
Field, G. B. (2009). The effects of the use of Renzulli learning on student achievement in reading comprehension, reading fluency, social studies, and science: An investigation of technology and learning in grades 3–8. *International Journal of Emerging Technologies in Learning, 4*(1), 29–39.
Fletcher, R. (2000). *How Writers Work*. Clarion Books. New York: Harper Collins Publishers.
Fraenkel, J. R., & Wallen, N. E. (2003). *How to Design and Evaluate Research in Education* (5th ed.). New York: McGraw-Hill.
Getzels, J. W. (1987). Problem finding and creative achievement. *Gifted Students Institute Quarterly, 7*(4), B1–B4.
Good, C., & Dweck, C. S. (2005). A motivational approach to reasoning, resilience, and responsibility. In R. J. Sternberg & R. F. Subotnik (Eds.), *Optimizing Student Success in School with the Other Three Rs: Reasoning, Resilience, and Responsibility* (pp. 39–56). Charlotte, NC: Information Age Publishing.
Haensly, P. A., & Roberts, N. M. (1983). The professional productive process and its implications for gifted studies. *Gifted Child Quarterly, 27*(1), 9–12.
Hébert, T. P., Sorensen, M. F., & Renzulli, J. S. (1997). *Secondary Interest-a-lyzer*. Mansfield, CT: Creative Learning Press.
Kay, S. (1994). From theory to practice: Promoting problem-finding behavior in children. *Roeper Review, 16*(3), 195–197.
Kessler, C. (2006). *Hands-on Ecology*. Waco, TX: Prufrock Press.
LaBanca, F. (2008). *Impact of Problem Finding on the Quality of Authentic Open Inquiry Science Research Projects*. Unpublished doctoral dissertation. Western Connecticut State University, Danbury, CT.
Larkin, J. H., Heller, J. I., & Greeno, J. G. (1980). Instructional implications of research on problem solving. *New Directions in Teaching and Learning, 1980*(2), 51–65.
Levitt, S. D., & Dubner, S. J. (2005). *Freakonomics: A Rogue Economist Explores the Hidden Side of Everything*. New York: Harper.
McCurdy, H. G. (1960). The childhood pattern of genius. *Horizon, 2*, 33–38.
MacKinnon, D. W. (1964). The creativity of architects. In C. W. Taylor (Ed.), *Widening Horizons in Creativity*. New York: Wiley.
Nicholls, J. C. (1972). Creativity in the person who will never produce anything original and useful: The concept of creativity as a normally distributed trait. *American Psychologist, 27*(8), 717–727.
Renzulli, J. S. (1977). *The Enrichment Triad Model: A Guide for Developing Defensible Programs for the Gifted*. Mansfield Center, CT: Creative Learning Press.
Renzulli, J. S. (1978). What makes giftedness? Re-examining a definition. *Phi Delta Kappan, 60*(3), 180–184.
Renzulli, J. S. (1986). The three-ring conception of giftedness: A developmental model for creative productivity. In R. J. Sternberg & J. E. Davidson (Eds.), *Conceptions of Giftedness* (pp. 53–92). New York: Cambridge University Press.
Renzulli, J. S. (1994). *Schools for Talent Development: A Practical Plan for Total School Improvement*. Mansfield Center, CT: Creative Learning Press.
Renzulli, J. S. (1997). *Interest-a-lyzer: Family of Instruments*. Mansfield, CT: Creative Learning Press.
Renzulli, J. S. (2005). The three-ring conception of giftedness. In R. J. Sternberg & J. E. Davidson (Eds.), *Conceptions of Giftedness* (pp. 246–278). New York, NY: Cambridge University Press.
Renzulli, J. S., & Reis, S. M. (1985). *The Schoolwide Enrichment Model: A Comprehensive Plan for Educational Excellence*. Mansfield Center, CT: Creative Learning Press.
Renzulli, J. S., Heilbronner, N. N., & Siegle, D. (2010). *Think Data*. Mansfield Center, CT: Creative Learning Press.

Schoenfeld, A. H., & Herrmann, D. J. (1982). Problem perception and knowledge structure in expert and novice mathematical problem solvers. *Journal of Experimental Psychology, 8*(5), 484–494.

Shore, B. M., Delcourt, M. A. B., Syer, C. A., & Shapiro, M. (2008). The phantom of the science fair. In B. M. Shore, M. W. Aulls, & M. A. B. Delcourt (Eds.), *Inquiry In Education Volume II: Overcoming Barriers to Successful Implementation* (pp. 93–118). New York: Routledge.

Starko, A. J., & Schack, G. D. (1992). *Looking for Data in all the Right Places*. Mansfield Center, CT: Creative Learning Press.

Treffinger, D. J. (1998). From gifted education to programming for talent development. *Phi Delta Kappan, 79*(10), 752–755.

Zimmerman, B. J. (1998). Developing self-fulfilling cycles of academic regulation: An analysis of exemplary instructional models. In D. H. Schunk & B. J. Zimmerman (Eds.), *Self-regulated Learning: From Reaching to Self-reflective Practice* (pp. 1–19). New York: Guilford Press.

10

New creative education

When creative thinking, entrepreneurial education, and innovative education come together

Fangqi Xu

KINKI UNIVERSITY, JAPAN

Summary: In recent years, innovation became a hot issue around the world. A variety of new words—such as social innovation, open innovation, just to mention a few—were born one after another. But many of them have not a clear concept. Without exception, innovative education is one of them. In this chapter I will first of all explain what innovative education is. Then I will explain its content by comparing it with creative education and entrepreneurial education. Creative education is not a new concept. It has a long history over half a century. However, academic research and practical activities are mainly aimed at the development of creative thinking so far. Nevertheless, it is not enough, because it does not link to entrepreneurship and innovation. So, I proposed a new creative education which including creative thinking, entrepreneurial education, and innovative education, creative personality.

Key words: Creative education, creative thinking, entrepreneurial education, innovative education, venture business, divergent thinking, idea generating.

Introduction: environment around venture and entrepreneurial education

Recently, the allure of venture business has seemingly lost its luster. From America to Asia and all across Europe, investors are growing increasingly reluctant to jump back on the bandwagon and reenter the market. There are several plausible explanations for the slowdown, including the shortage of capital brought on by the dot-com bubble burst, the meltdown and subsequent financial crisis of the Lehman Shock, the continuous rise in the price of raw materials, a national financial crisis in Europe, the tight money policy in place in China, and an inactive economy in Japan. Certainly, each of these contributing factors has had an influence on investors' decisions regarding venture business. However, if we look more closely at the current global economic situation we will come to realize that we need venture business now more than any time in the recent past.

On a positive note, entrepreneurial education for venture business appears to be on the rise, particularly in Japan. And the number of Japanese colleges and universities where courses on venture business and entrepreneurship are offered has been increasing year after year. According to research conducted in 2009 by the Ministry of Economy, Trade and Industry (METI), 1,078 different courses related to venture business were offered at roughly 250 colleges and universities in Japan. That is an increase of over three times the number of classes offered just 10 years ago. It is reassuring to note that the tendency is still increasing and that the number of classes being offered is still on the rise, however the data is not yet complete (see Figure 10.1).

Despite the increase in classes offered on college campuses across Japan, the social environment and outlook around venture and entrepreneurial education is not all sunny. The first part of this chapter will attempt to cast light upon the current backdrop of education with three examples. The second and third parts of this chapter will address traditional and contemporary creative education respectively.

The first example is the change in the number of venture businesses (VB) formed or started in colleges and universities. According to METI, 1,344 venture businesses were formed from 1990 to 2008, but it is clear from the data that future prospects are gloomy and a greater decline can be forecast (see Figure 10.2).

The second example is the change in the number of Initial Public Offerings (IPOs). According to the Venture Enterprise Center (VEC), a state-funded research firm in Japan, the number of IPOs has been decreasing in recent years (see Figure 10.3).

The third illustration is the change in the number of active members associated with the Japan Academic Society for Venture and Entrepreneurs (JASVE). Membership in JASVE has been steadily decreasing for the last several years. Since 2007, the organization has lost about one-third of its members (see Figure 10.4).

Traditional creative education

Creative education started gaining popularity 50 years ago when the Creative Education Foundation (CEF) was established in Buffalo, New York (Xu, 1992). Founded by Alex Osborn, father of the technique popularly known as Brainstorming in 1954, the CEF is the first non-profit organization of its kind to form in the United States. Together with colleagues Sidney Parnes and Ruth

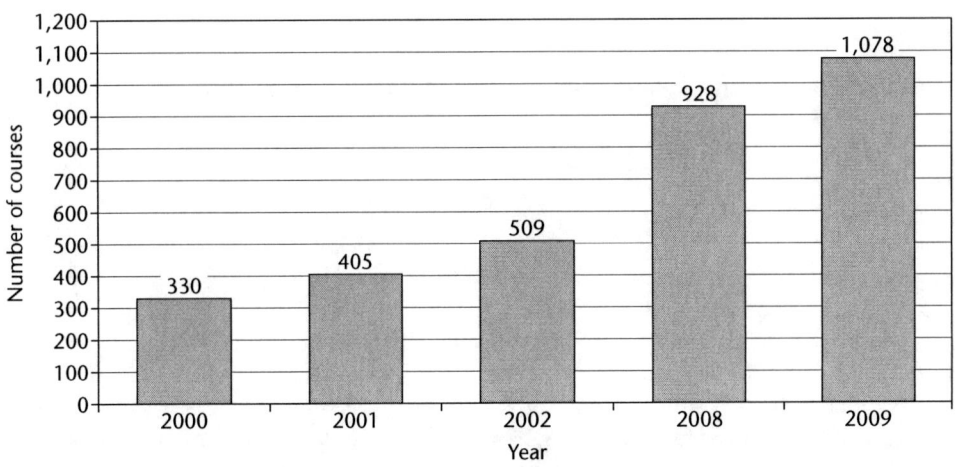

Figure 10.1 Change of the number of courses (source: METI News Release, December 10, 2009).

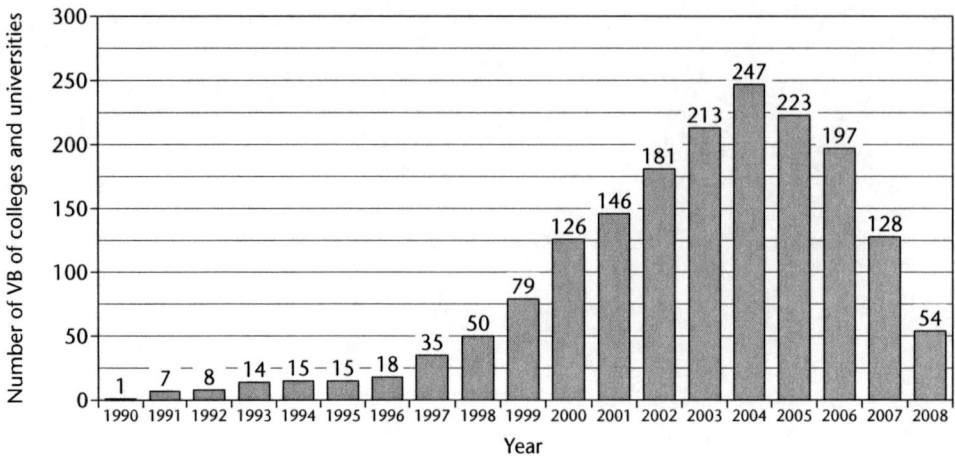

Figure 10.2 Number of VB of colleges and universities (source: METI News Release, December 10, 2009).

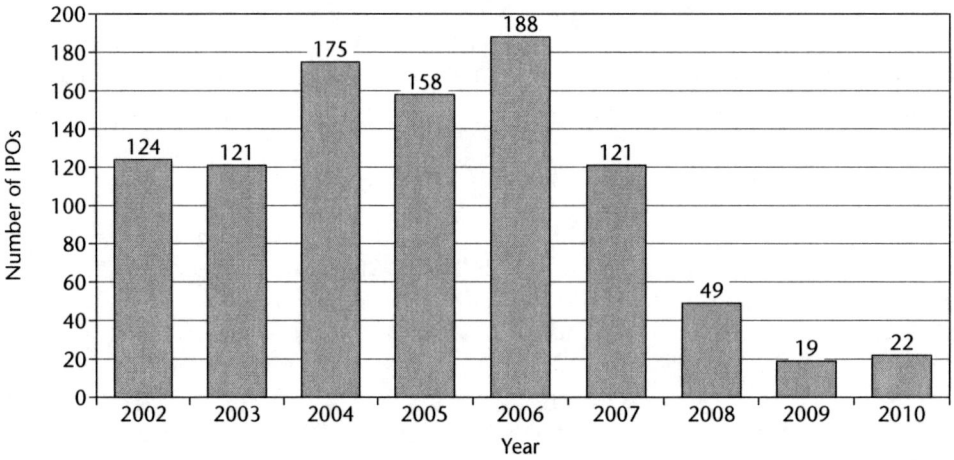

Figure 10.3 Change in the number of IPOs (source: Venture Enterprise Center and Nikkei Daily, October 13, 2011).

Noller, Osborn helped start a mini-movement and the city of Buffalo quickly became the center of creative education in the world. Although the CEF recently moved to Amherst, MA, in 2008, the influence left behind is still alive and thriving at Buffalo State College.

After the CEF held its first Creative Problem Solving Institute (CPSI) in 1955, a paradigm shift in teaching creative thinking began taking place across the globe (Xu, 1992). The movement has since gathered steam and become a part of the main content of creative education in the United States. At front and center is the concept of Brainstorming, a method for creatively solving problems through teamwork and group thinking sessions. When "group think" sessions are productive they can often yield several plausible solutions to a question. Divergent thinking is a characteristic of Brainstorming that encourages participants to come up with as many different possible answers to a question as they can. As a result, divergent thinking has become the mainstream in education

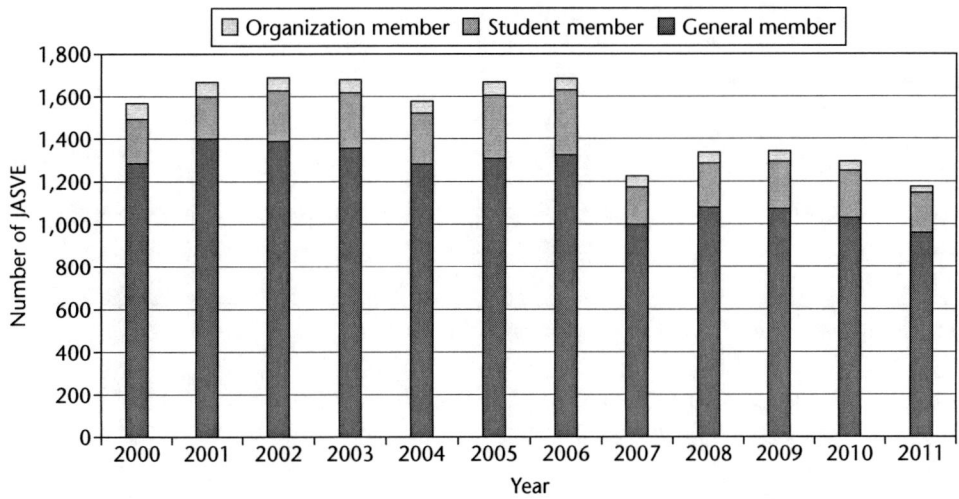

Figure 10.4 Change of the number of JASVE (source: member list of JASVE).

circles in the United States. J.P. Guilford's Structure-of-Intellect (SOI) Model, E.P. Torrance's Torrance Test of Creative Thinking (TTCT) both have their roots firmly grounded in divergent thinking (Sato & Onda, 1978).

Similar shifts have occurred in Europe. Scholars Tudor Rickards in the UK and Horst Geschka in Germany have made tremendous contributions to the field introducing brainstorming techniques and practicing them actively. Rickards' *Electronic Brainstorming* and Geschka's *Brainwriting (635-method)* are two widely used variations on Osborn's initial methodology (Rickards, 1999; Geschka, Moger, & Rickards, 1994). Although the 635-method was developed by Bernd Rohrbach in 1969, Geschka and his colleagues at Battelle group in Frankfurt checked and tested the method in real applications and came up with new variants in 1971 and later. Elsewhere in Europe, Edward de Bono (1969, 1986) pushed the boundaries of creative thinking with two more variations, called Lateral Thinking and The Six Thinking Hats. For the last several decades creative education and the teaching of creative thinking have taken similar forward-moving strides both in Europe and the United States.

New variations on Brainstorming methodology have led to significant breakthroughs in Asia as well, most notably in Japan and China. For example, Card Brainstorming, Mitsubishi Brainstorming, and Card Brainwriting were all pioneered in the 1970s and 1980s in Japan (Takahashi, 2002). Additional developments like the KJ Method, developed by Jiro Kawakita, and the NM Method, developed by Masakazu Nakayama in the 1970s, were in many ways similar to their Western counterparts (Takahashi, 2002). Since then, a boom of research and practice on creative thinking techniques has appeared in nearly every industry circle in Japan. This has trickled down into the application of these techniques into product development, quality improvement action, and employee education.

In October 1991, Xu, with Yuzuru Oshika and Tatsuo Hoashi, sent sets of questionnaires to 500 Japanese enterprises in order to investigate the application of creative thinking techniques. Eighty-nine enterprises returned the sets, reflecting the collection rate of 17.8%. Table 10.1 shows that the percentage of group thinking techniques, including Brainstorming and KJ Method was higher.

Table 10.1 Creative thinking techniques in Japanese enterprises (%)

Brainstorming	86.6
KJ Method	73.0
NM Method	22.4
Others	16.9
Lateral Thinking	13.4
Ichikawa Method	10.1
Synectics	4.4
635 Method	1.1

Source: Xu and Kunifuji (2005).

Influenced by the success of the industry circle, elementary and middle schools have begun applying some of the principles of creative education, which have traditionally focused on training creative thinking.

China, too, has recently begun introducing creative education programs and started teaching creative thinking techniques to students. Many elementary schools and middle schools across the country now offer a course called *Creative Technique* and many colleges and universities have also started teaching *Creative Engineering*. These courses focus on creativity development through the enhancement of people's skill at creative thinking (Xu, 2005, 2009).

So, what then is the problem? Why has traditional creative education ceased being relevant in the last few years? One explanation is that traditional creative education paid too much attention to the methods associated with creative thinking and overlooked its relevance and relationship to entrepreneurial education. At the same time, entrepreneurial education has not met social expectations in part because it has not had support from practitioners/teachers focused on creative education (Xu, 2009).

New creative education

It is my opinion that a new creative education should include three parts: creative thinking, entrepreneurial education, and innovative education. The remainder of this chapter will elucidate upon the three parts (Xu, 2009).

Creative thinking

According to the principle of creativity development, everyone has the capacity for creativity, but that capacity differs from person to person. In traditional settings, creativity development is most often fostered through the early and elementary years of education, but tends to wane as students advance in age and grade.

More time and resources need to be allotted for late-blooming students. So-called creativity development in school education should not be limited to the early years of education, but should continue on as students mature. This will provide opportunities for all students to get attention to the development and enhancement of their ability to think creatively. Obviously, creativity development in advanced students is a challenging prospect, but the rewards are often greater when success is achieved with such students.

Creative thinking is an important part of creativity. It comprises three parts:

a   First, it can help people see things from various viewpoints. For example, when seen from only one pattern or perspective, the choices of problem solving are narrow, and the probability of coming up with a good idea is low. If however, many patterns are used to think about a

New creative education

situation, the choices of problem solving will be far wider and the probability of coming up with a good idea is much greater.

b Second, it can promote creative personality. Creative personality refers to an individual's overall constitution and capacity to be creative. When educators develop and train students' ability to think "outside the box," they are also promoting the development of students' personalities indirectly.

c Third, it stimulates creative desire. A person who is used to creative thinking seeks out creative solutions. Approaching a problem from a creative viewpoint is often rewarded with some form of success.

Figure 10.5 illustrates the relationship among creative thinking, creative personality, creative desire, and creative result.

Entrepreneurial education

Entrepreneurial education at most colleges and universities in Japan is extremely bland. Courses are usually limited in scope and focus primarily on teaching students how to launch a company or how to start a non-profit organization. The curriculum typically consists of creating a business plan, outlining the structure of a business model, the vision, core competence, raising money, using of incubation, employment of key persons, product development, and, finally, a marketing strategy. Lucky students may receive training in other fundamentals like supply chain management, accounts receivable, writing profit and loss statements, balance sheets, cash flows, simple structures of joint-stock company, and IPO rules and regulations.

As noted above, entrepreneurial education in Japan touches primarily on the essential elements of curriculum. Students are mainly taught the basics of entrepreneurial education. This may be precisely what is holding them back. Idea generation is a quintessential skill in entrepreneurial education and yet most students are completely unable to do it. Learning to think creatively is a skill that should be fostered and reinforced throughout a student's education. And yet it is not uncommon for students to graduate classes in entrepreneurship, and be unable to do anything remotely creative; they lack the imagination to come up with ideas, which relate directly or indirectly to business. One way to remedy this is to teach them the techniques of idea generation early on and provide them with ample opportunities for practice at every stage of their education. In addition to teaching them Western techniques, it is critical that students learn Eastern techniques of idea generation as well.

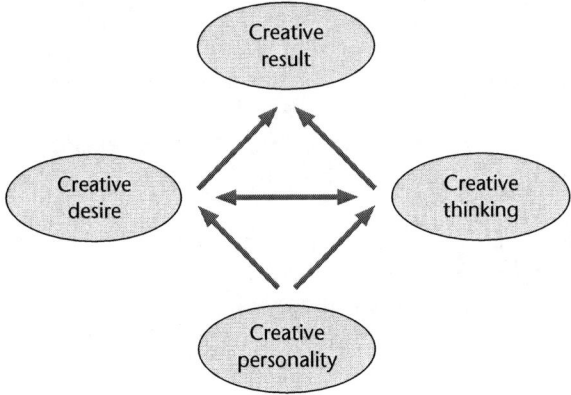

Figure 10.5 The relationship of 4C.

Twenty-six years ago, I taught a class which I called Amateur Invention Seminar in Changzhou City, China. It was a six-month class with two lectures or practices every week. I started the course by teaching the techniques of idea generation including Brainstorming, Brainwriting (635-method), Check List, the KJ Method, and the NM Method. In about two months, the students began to come up with some very interesting ideas. The number of ideas they generated was bigger and the quality of the ideas was higher than before they began learning these methods. Several of the students' ideas were so good that they applied for patents while enrolled in the seminar.

Innovative education

The benefits of entrepreneurial education are clear. Persons with such experience and training appear to have a greater chance for success when launching a venture business than those who have not received such training. That said, entrepreneurial education alone is not enough to succeed in venture business; it is only the first step.

There are several other factors that contribute to success in venture business, but the most important is continuous innovation. According to Ikujiro Nonaka, Professor Emeritus at Hitotsubashi University and the originator of the theory of knowledge management, innovation is a highly individual process of personal and organizational self-renewal (Nonaka & Takeuchi, 1995). Innovation means change, and never being satisfied with the present conditions.

General Electric's (GE) stance on innovative education is well known around the world. It spends upwards of one billion dollars on education every year. GE has a longstanding tradition of training in creativity and innovation for its young engineers. For example, the Creative Engineering Program was the first course in the world on creativity development which was developed in 1937 by A.R. Stevenson (Samstad, 1962; Von Fange, 1963). Since then, GE has launched dozens of similar programs every year. In 2006, it launched a team-based training course called Leadership Innovation and Growth (LIG). For its work, GE has received many honors from the media. One of its most notable honors came in 2008 when *Business Week* awarded it one of the World's Most Innovative Companies that year. No other big company in the world has been able to keep its competitive power as well as GE.

Conclusion

Creative education has a history that spans over half a century. Although it has risen to prominence in recent years, the early days of creative education were slow going. Creative education seemed fixated on creative thinking and had little to offer society or the real world until the 1960s. That all changed when a group of entrepreneurial pioneers broke the mold in Silicon Valley in the middle of the 1970s. Their innovative and forward thinking vision started a paradigm shift still underway today. From then on entrepreneurial education took off like wildfire on the west coast of the United States and eventually extended outward to other countries around the world. After the Lehman Shock, many companies realized the importance of innovation and begun innovative education programs of their own.

To sum-up, this chapter discussed the relationship between creative thinking, entrepreneurial education, and innovative education, and emphasized that any new creative education curriculum should include all of them. The image of new creative education is presented in Figure 10.6.

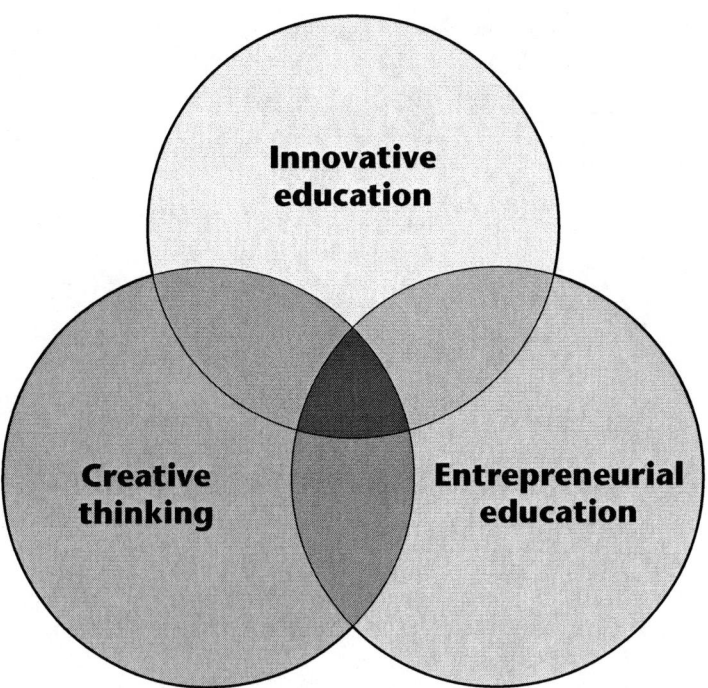

Figure 10.6 The image of new creative education.

Acknowledgments

I want to thank Mr Joshua Cohen who is a native English speaker and a lecturer at Kinki University for his help.

References

De Bono, E. (1969). *New Think: The Use of Lateral Thinking in the Generation of New Ideas* (M. Shirai, Trans.). Tokyo: Kodansha.
De Bono, E. (1986). *Six Thinking Hats* (M. Matsumoto, Trans.). Tokyo: Diamond.
General Electric Company. (2006, 2007, 2008). Annual Reports.
Geschka, H., Moger, S., & Rickards, T. (Eds.) (1994). *Creativity and Innovation: The Power of Synergy*. Darmstadt: Geschka & Partner Unternehmensberatung.
Ministry of Economy, Trade and Industry (2009). Trade and Industry News Release (in Japanese), December 10.
Nikkei Daily. (2011). *News Daily*. October 13.
Nonaka, I., & Takeuchi, H. (1995). *The Knowledge-Creating Company*. New York: Oxford University Press.
Rickards, T. (1999). Brainstorming. In M. A. Runco & S. A. Pritzker (Eds.), *Encyclopedia of Creativity*. San Diego, CA: Academic Press.
Samstad, G. I. (1962). General electric's creative courses. In S. J. Parnes & H. F. Harding (Eds.), *A Source Book for Creative Thinking*. New York: Charles Scribner's Sons.
Sato, S., & Onda, A. (Eds.) (1978). *Creative Talent: Development and Evaluation* (in Japanese). Tokyo: Tokyo Shinri.
Takahashi, M. (Ed.) (2002). *The Bible of Creativity* (in Japanese). Tokyo: Nikkagiren.
Von Fange, E. K. (1963). *Professional Creativity* (Y. Kato & K. Okamura, Trans.). Tokyo: Iwanami Shoten.
Xu, F. (1992). A consideration of creative studies in the United States of America. In Japan creativity society (Ed.), *From Surprise to Flash* (in Japanese). Tokyo: Kyoritsu Shuppan.
Xu, F. (2005). Creative education in China. In K. Yumino (Ed.), *Creativity Education in the World* (in Japanese). Kyoto: Nakanishi.

Xu, F. (2009). The new development of creativity education at colleges and universities. In K. Zhang & Y. Jiang (Eds.), *The Proceedings of 2008 Creativity Education Forum in China* (in Chinese). Beijing: Beijing University of Technology.

Xu, F., & Kunifuji, S. (2005). A comparative research on creativity development between Japanese and Chinese enterprises. In B. Jöstingmeier & H. J. Boeddrich (Eds.), *Cross-Cultural Innovation*. Wiesbaden: Deutscher Universitäts-Verlag.

Part IV
Assessment and identification related issues of innovation education

11

Torrance's innovator meter and the decline of creativity in America

Kyung Hee Kim and Robert A. Pierce

COLLEGE OF WILLIAM AND MARY, USA, AND CHRISTOPHER NEWPORT UNIVERSITY, USA

Summary: IQ tests measure intelligence, and the Torrance Tests of Creative Thinking (TTCT)-Figural is the most widely accepted measure of adaptive creativity, innovative creativity, and creative personality. Creative thinkers who excel in convergent thinking usually exhibit more "adaptive" skills, meaning they readily adapt existing solutions to new scenarios. Those who are stronger in divergent thinking usually exhibit more "innovative" skills: they can create or postulate new solutions. Adaptive creative thinkers create original ideas that are more likely to fit existing paradigms, whereas innovative creative thinkers create original ideas that are more likely to challenge existing paradigms. This chapter describes what the TTCT measures. It also considers implications of research showing that, even as IQ increased annually, TTCT scores continually decreased. Since 1990, adaptive creativity, innovative creativity, and creative personality have significantly diminished in America, with significant implications. Explanations for the decrease in creativity and its implications are discussed.

Key words: Adaptive, creativity, creativity test, innovative, Torrance Tests of Creative Thinking, innovator, innovation.

Introduction: intelligence and creativity

The relationship between creativity and intelligence has been controversial. Some researchers found independence between creativity test scores and IQ (e.g., Gough, 1976; Helson & Crutchfield, 1971; Rossman & Horn, 1972; Torrance, 1977), while others found dependence (Runco & Albert, 1986; Wallach, 1970). One line of reasoning, the *Threshold Theory* (i.e., that IQ and creativity are related up to an IQ of approximately 120), asserts that creativity and intelligence are separate constructs above a minimum IQ of 120 (Barron, 1961; Guilford, 1967; MacKinnon, 1967; Simonton, 1994). Kim's meta-analysis (2005), however, found the threshold theory is not supported. The correlation coefficient between creativity and intelligence is only 0.17, which suggests creativity and intelligence might be separate constructs for the entire spectrum of IQs. Kim found the relationship between creativity and intelligence among young children is even weaker. In a subsequent meta-analysis, Kim (2008b), showing that creativity scores predict creative achievement better than IQs, found that creativity scores better predict creative achievement than IQs. Further, when predicting creative achievement in different domains, Kim (2008b) found that creativity scores better predict achievement in art, science (including mathematics, medicine, and engineering), writing, and social skills than IQ scores. Interestingly, IQs do better predict musical achievement than creativity scores.

Changes in intelligence

Intelligence is increasing. Ceci (1991) and Ceci and Williams (1997) reported education and schooling improve children's IQs. Dickens and Flynn (2001) concluded that intellectual challenges improve IQs. Based on the test norms of the Stanford-Binet and Wechsler tests, Flynn (1984) revealed that IQs have increased in the United States over the decades of the last century, something now referred to as the "Flynn Effect." Flynn (2007) also concluded IQs have also increased worldwide during the past century: IQs on the Wechsler Intelligence Scale for Children (WISC) Arithmetic, Information, and Vocabulary subtests have gained by about three points, and IQs on the Raven's Matrices and on the Similarities subtest of the WISC have gained by about 25 points. Flynn (2007) explained that the increase in IQs might be because of reduced inbreeding, improved nutrition, or increased affluence around the world.

Changes in creativity

If creativity and intelligence are the same thing, creativity should have also increased – just as IQ has increased (Kim, 2005). If intelligence and creativity are independent, then creativity would not necessarily increase with intelligence. Let's now discuss how it can be measured.

Measuring creativity by the Torrance Tests of Creative Thinking (TTCT)

The TTCT was developed in 1966 and re-normed five times. The total sample for all six normative samples included TTCT-Figural scores from 272,599 kindergarten through 12th grade students and adults (see Kim, 2011b for details). The TTCT has two versions, the TTCT-Verbal and the TTCT-Figural, each with two parallel forms, Form A and Form B. It requires 30 minutes to take each part, Verbal and Figural, so speed is important. Artistic quality is not required to receive credit (see Kim, 2006a for details). Because "creative" is not synonymous with "artistic," the TTCT measures creativity in many other ways than artistic ability. The TTCT-Figural has been found to be fair in terms of gender, race, community status, and for individuals with a different language background, socioeconomic status, and culture (Cramond, 1993; Torrance, 1971; Torrance & Torrance, 1972). When predicting creative achievement, Kim (2008b) found scores on the TTCT predict creative achievement better than other measures of creative or divergent thinking. Translated into over 35 languages and used worldwide, the TTCT is utilized extensively in both educational and the corporate world.

The creative thought process involves two types of thinking, as Figure 11.1 shows. Convergent thinking, bringing facts and data together from various sources, uses logic and knowledge to find solutions. Divergent thinking, considering a problem from a variety of different perspectives, discovers and develops original solutions. Creative thinkers stronger in convergent thinking can adapt existing solutions to new scenarios (adaptive skills). Creative thinkers stronger in divergent thinking can create or postulate new solutions (innovative skills). Adaptively creative thinkers are those who try to do things better, whereas innovatively creative thinkers are those who try to do things differently. Adaptively creative thinkers create original ideas within existing paradigms, whereas the original ideas of innovatively creative thinkers challenge the existing paradigm. Kirton (1999), in a similar approach, defines creative thinking as a continuum of styles, ranging from Adaptive preferences for decision-making and problem solving to Innovative preferences. This theory proposes

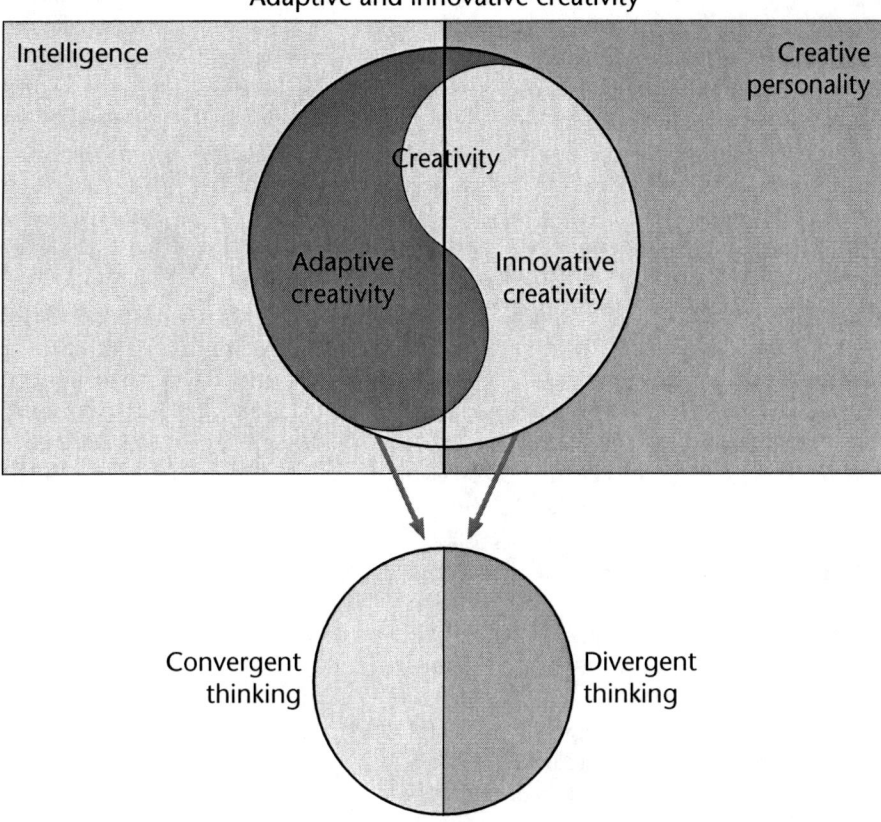

Figure 11.1 Adaptive and innovative creativity in relation to convergent and divergent thinking within intelligence and creative personality.

that individuals tend to have relative preferences for solving problems, independent of their creative ability. Creativity requires both processes and works best when these two thinking processes complement each other. For example, a person engages in divergent thinking to generate many novel ideas, and then engages in convergent thinking to evaluate these ideas and adapt one of these novel ideas to solve a particular problem. Convergent thinking is included in both intelligence (measured by IQ) and creativity (which, because it is not part of intelligence, is not measured by IQ). In addition to convergent and divergent thinking, creative people often possess a "creative personality," perhaps best described as a spark of childlike curiosity or playfulness within them driving them to creative opportunities and results. Risk-taking and a lack of cognitive inhibitions are also generally exhibited in creative personalities. Without a creative personality, creative ability simply will not be achieved. Deriving the TTCT's initial measurements from the divergent-thinking factors referenced in Guilford's (1959) Structure of the Intellect Model (Torrance, 1966, 1974), Torrance later included more measures of creative potential and creative strengths (Kim, 2011a, 2011b). However, other theories of creativity have influenced how to interpret the scores on the TTCT. De Bono (1970) theorized that creative thinking is a synthesis between Lateral thinking and Vertical thinking. A Lateral thinker is idea-generative and can make jumps; a Lateral thinker deliberately chooses to be outside of bounded thought processes; a Lateral thinker prefers looser cognitive situations that allow him or her to break out of the paradigm. A Vertical thinker is idea-selective and is sequential, analytical, careful, and precise; a Vertical thinker takes the data around a problem and analyzes it with defined methodologies to find logical solutions; a Vertical thinker prefers structured situations; a Vertical thinker's focus is on redefining, elaborating, modifying, and improving a paradigm (De Bono, 1970). An easy way to tie the theories together of Bono and Kirton is to analogize Adaptive thinkers to Vertical thinkers and Innovative thinkers to Lateral thinkers.

Sternberg's (1999) propulsion theory of creativity also divides creativity into two categories: contributions that accept and contributions that reject current paradigms. The difference between Kirton's and Sternberg's theories is in perspective. Sternberg examines the products of creativity, whereas Kirton examines the creator's style. Both of these two theories indicate that individuals who score highly on Innovative thinking are more likely to produce creative contributions that attempt to establish new paradigms, and those who score highly on Adaptive thinking are more likely to produce contributions that elaborate on existing paradigms. Kim (2006b) found Innovative thinking factor is associated with Fluency and Originality, whereas Adaptive thinking factor is associated with Elaboration and Abstractness of Titles on the scores of the TTCT. Kim (2010) added Resistance to Premature Closure and Strengths as the third factor, Creative Personality factor.

Other factors are connected to creativity. Kim (Cheng, Kim, & Hull, 2010; Kim, 2009; Kim & Lee, 2007; Lee & Kim, 2010) found relationships between TTCT Adaptive thinking scores and personality type, Confucianism, bilingualism, culture, age, and gender. However, no relationships were found between TTCT Innovative thinking scores and those factors (Kim, 2010). Kim (2005) found among various creativity or divergent thinking tests that the ones that measure only Innovative thinking have the least relationship with intelligence test scores. Cho, Te Nijenhuis, Van Vianen, Kim, and Lee (2010) also found intelligence test scores relate only to Adaptive thinking, and not Innovative thinking, which may indicate Adaptive thinking is more influenced by language, society, and culture, gender, intelligence, and age, for examples, than is Innovative thinking. The current study's results are consistent with these prior findings. Kim (2008a, 2010, 2011a, 2011b) concluded the TTCT-Figural is more than a divergent-thinking test because divergent thinking is only related to Innovative thinking but not related to Adaptive thinking or Creative Personality. This supports Torrance's earlier conclusions (1987, 1988, 1994, 2000, 2002, 2008) that the TTCT-Figural is more than a divergent-thinking test and is a measure of creative thinking since the revision of the TTCT scoring system in 1984. Other researchers (e.g., Cramond, 1993; Johnson & Fishkin, 1999) confirmed that the revised scoring system addresses essential constructs of creative behaviors that are reflective of Torrance's definition of creativity.

Unlike IQ tests, which give a single measure of intelligence, the TTCT gives a *profile* of test results on several subscales. The TTCT-Figural has remained unchanged but was re-normed with each edition of the TTCT manual. The only significant change to the test was scoring procedures, which were revised in 1984. The current form of the TTCT-Figural includes scores for Fluency, Originality, Elaboration, Abstractness of Titles, Resistance to Premature Closure, and 13 creative personality traits that comprise the Creative Strengths subscale, as Table 11.1 shows.

The above subscales are usually grouped together into three main factors of creative potential:

1   For the Lateral/Innovative thinking factor (Fluency and Originality). Fluency measures an ability to produce a number of the relevant ideas. Originality measures an ability to produce a number of statistically infrequent ideas and shows how unique and unusual the ideas are. The scoring procedure counts the most common responses as "0" and all other legitimate responses as "1."

Table 11.1 Descriptions of the subscales of the TTCT-Figural

Factor	Subscale	Description
Innovative/Lateral Thinking	Fluency	Ability to produce a number of relevant ideas
	Originality	Ability to produce a number of statistically infrequent ideas; how unique and unusual the ideas are
Adaptive/Vertical Thinking	Elaboration	Ability to develop and elaborate upon ideas; detailed and reflective thinking; motivation to be creative
	Titles	Ability to perform the thinking processes of synthesis and organization that good titles require, to capture the essence of the information involved, and to distinguish what is important
Personality Traits	Closure	Intellectual curiosity; open-mindedness
	Strengths Checklist	Description
	Emotional expressiveness	Being emotionally expressive
	Movement or action	Being energetic
	Storytelling articulateness	Being talkative
	Expressiveness of titles	Being verbally expressive
	Humor	Being humorous
	Fantasy	Being imaginative
	Extending or breaking boundaries	Being unconventional
	Richness of imagery	Being lively or passionate
	Colorfulness of imagery	Being perceptive
	Synthesis of incomplete figures	Connecting seemingly irrelevant things together
	Synthesis of lines or circles	Synthesizing
	Internal visualization	Seeing things from a different angle
	Unusual visualization	

Note
Titles = Abstractness of Titles; Closure = Resistance to Premature Closure; Strengths = Creative Strengths.

2   For the Vertical Thinking factor (Elaboration and Abstractness of Titles). Elaboration measures an ability to develop and elaborate upon ideas and detailed and reflective thinking, but it also indicates motivation to be creative (Torrance, 2000). Abstractness of Titles measures an ability to produce the thinking processes of synthesis and organization that good titles involve, and, based on the idea that creativity requires an abstraction of thought, measures an ability to capture the essence of the information involved and to know what is important. Abstractness of Titles is related to verbal intelligence (Torrance & Safter, 1990).
3   For the Creative Personality factor (Resistance to Premature Closure and 13 Creative Strengths). Resistance to Premature Closure measures intellectual curiosity as well as open-mindedness. Open-mindedness, which predicts both IQ (DeYoung, Peterson, & Higgins, 2005) and creativity (Feist, 1998), also is the most influential factor on intelligence (Furnham & Thomas, 2004; Furnham & Chamorro-Premuzic, 2006). Finally, the Creativity Personality factor assesses 13 Creative Strengths.

The results of the creativity crisis study

Decrease in Fluency since 1990

Fluency scores decreased by 4.68% from 1990 to 1998 and by 7.00% from 1990 to 2008. The biggest decrease in Fluency scores was for Kindergartners through third graders, and the second biggest decrease was for fourth through sixth graders, which indicate that younger children's ability to produce many ideas significantly decreased since 1990.

Decrease in Originality since 1990

Originality scores increased until 1990, but decreased by 3.74% from 1990 to 1998, and remained static from 1998 to 2008. Originality is the only TTCT subscale that is reflective of different cultures and time so that Torrance instructed to develop and update originality lists culture- and time-specifically. Thus, Kim (2006a) questioned the credibility of Originality scores of the TTCT based on the Originality Lists that Torrance developed in 1984. The continued use of 1984 Originality Lists leads to an expectation that the Originality scores should go up artificially the longer the Originality Lists are not updated. However, the results indicated that the Originality scores decreased from 1990 to 1998 and remained static from 1998 to 2008. Therefore, the results might suggest that Originality scores have actually significantly decreased, but the decrease may have been deflated through the use of outdated scoring lists.

Examining each age group separately revealed that the biggest decrease in Originality scores from 1990 to 2008 was for Kindergartners through third graders. It can be concluded that younger children's ability to produce statistically infrequent, unique, and unusual ideas has significantly decreased since 1990.

Decrease in Creative Strengths since 1990

Strengths scores decreased by 3.16% from 1990 to 1998 and by 5.75% from 1990 to 2008. The decrease of Strengths scores after 1990 might indicate that over the last 20 years, children are becoming less emotionally expressive, less energetic, less talkative or verbally expressive, less humorous, less imaginative, less unconventional, less lively or passionate, less perceptive, less connecting of seemingly irrelevant things, less synthesizing, or less seeing of things in a different angle.

Decrease in Elaboration since 1984

Elaboration scores decreased by 19.41% from 1984 to 1990, by 24.62% from 1984 to 1998, and by 36.80% from 1984 to 2008. The decrease in Elaboration scores since 1984 might indicate that people of all ages are losing their ability to elaborate upon ideas and detailed and reflective thinking over the last 30 years, that they are becoming less motivated to be creative, and that the home, school, or society overall value creativity less.

Decrease in Abstractness of Titles since 1998

Titles scores decreased by 7.41% from 1998 to 2008, a little later than the decreases of other TTCT subscales, which started in 1984 (Elaboration) or in 1990 (Fluency, Originality, and Strengths). Because Titles scores have a positive relationship with verbal intelligence scores, and also because intelligence scores have increased, Titles scores are expected to increase. However, the decrease indicates that the scores might have actually decreased earlier than 1998. This result may indicate that younger children are becoming less capable of the critical thinking processes of synthesis and organization and less capable of capturing the essence of the information to know what is important to a certain problem.

Decrease in Resistance to Premature Closure since 1998

Closure scores decreased by 1.84% from 1998 to 2008. Because Closure scores have a strong positive relationship with intelligence, and also because intelligence has increased, Closure scores are expected to increase. However, the decrease indicates that the scores might have actually decreased earlier than 1998. This result may indicate that younger children are becoming less intellectually curious and less open to new experiences and more narrow-minded.

In conclusion, the results of the creativity crisis study (Kim, 2011b) showed that creativity scores significantly decreased since 1990. Adaptive thinking factor (Elaboration by 17.39% and Titles by 7.41%) decreased more than Innovative thinking factor (Fluency by 7.00% and Originality by 3.74%) or Creative Personality factor (Strengths by 5.75% and Closure by 1.84%). These findings are consistent with previous findings (Cheng et al., 2010; Kim, 2009, 2010; Kim & Lee, 2007; Kim, Lee, Chae, Andersen, & Lawrence, 2011; Lee & Kim, 2010) in that Adaptive thinking factor is more influenced by society and culture than Innovative thinking factor. Individuals may be losing their ability to elaborate upon ideas, less capable of detailed and reflective thinking, and becoming less motivated to be creative. These developments may be because of the home, school, and society overall encourage creativity less. The result may also indicate that younger children are becoming less capable of the critical thinking processes of synthesis and organization and less capable of capturing the essence of the information to know what is important to a certain problem. Creativity scores for kindergartners through third graders decreased the most, and those from the fourth through sixth grades decreased by the next largest amount.

The possible reason for the creativity crisis

The results indicate that children in the United States, especially kindergarten through third grade, are becoming less creative as measured by TTCT. An important question is why. The decrease of creativity for these younger children may arise from their home environment, rather than their schools because kindergarteners and first graders tend to be influenced more by home than school. More children are spending the majority of their time in front of televisions, computers, and videogames, and less time engaging in creative activities.

The decrease may originate from children's earlier experiences or lack of experiences, and so as circumstances and maturity permit, efforts to preserve creativity should begin earlier. Childhood fantasies and play should be discouraged less than they are, as creative imagination develops from children's play (Vygotsky, 1994). Play in the first and second grade predicts divergent thinking in the fifth and sixth grade (Russ, Robins, & Christiano, 1999). Pretended play in young children is considered as improvization, which is crucial in adult creativity. Home, school, or both may unwittingly cause the slump by prematurely restricting childhood play and fantasy.

Children are spending more of their time interfacing with machines instead of people and paper. With cell phones, handheld Internet, DVD players in cars, talking GPS, and hundreds of channels available on ever-increasing numbers of televisions, children are wired into a wealth of information. Regardless of the apparent benefits, the time they spend wired is hurting the children. We speculate that all of this electronic "company" is distracting children, so they are unable to particularly focus and consider such valuable information as they may receive. Much of that information is advertisers broadcasting marketing messages and reinforcing brand imaging, and much of the rest is of no real value or consequence. Research shows that even educational television injures children by serving as a distraction, rather than an enhancement, of a child's creative development.

Video games constitute an ever-increasing part of a child's day. Though video games differ, generally they are set in a fantasy environment – even a realistic one – and hone hand-eye coordination. They may also offer some limited problem solving, such as figuring out how to get through a door or across a chasm. However, whereas programming or designing these games may sometimes be creative, playing them generally limits creativity. Because video games and computer games are programmed, for players, even when they have an enormous number of potential responses or actions in a game, are limited to the playing field. Creative problem solving is not generally a part of these games, because ultimately the number of opportunities for engagement and proactive response are limited.

Over-diagnosis and over-prescribing of Attention-Deficit/Hyperactivity Disorder (AD/HD) medications for children, in attempts to control undesirable behaviors of creative children, may have resulted in decreased creativity (Hartnett, Nelson, & Rinn, 2004; Healey & Rucklidge, 2006; Krautkramer, 2005), as creative children and children with AD/HD share common characteristics: rebelliousness, emotionally expressiveness, spontaneity and impulsiveness, high energy, excitability, and a tendency to become bored quickly. In a misguided attempt to control these (perceived) undesirable and inconvenient behaviors in some, we are unwittingly doping our creative children. Fewer children will survive childhood with their creative abilities intact. Those who do will have less overall creativity. And, as long as this trend continues, outliers, potentially a future Steven Jobs, will be ever more suspect.

The very qualities that cause creative individuals to have problems are the same ones that may facilitate their creative accomplishments. Creative people have many characteristics that, as opposite sides of the same coin, can be viewed as positive or negative. For example: Original/imaginative or bizarre? Independent/persistent or stubborn? Curious or annoying? High energy or hyperactive? Talkative or verbally expressive? Spontaneous or impulsive? Emotionally sensitive or emotionally unstable? Robert Frost daydreamed (was imaginative or bizarre?) so much that he was put out of school. Frank Lloyd Wright went into such trancelike dreams (was original or bizarre?) that one had to shout at him to bring him back. Samuel Taylor Coleridge demonstrated restlessness and verbal diatribes, and Virginia Woolf also demonstrated a tendency to talk on and on (was talkative or verbally expressive?). Was Van Gogh emotionally sensitive or emotionally unstable? Marie Curie was stubborn and rebellious (was independent/persistent or stubborn?). Thomas Edison experienced school problems, in part because of his high energy (had high energy or was hyperactive?). Nikola Tesla's tendency to act without thinking (was spontaneous or impulsive?) caused him to have several scrapes with death and near-tragedies such as: plunging to the earth from the roof of a barn, clutching an umbrella; being chased by a flock of crows or angry hogs; nearly drowning in a vat of hot milk; jumping from a church balcony on to the train of a lady's dress to the sound of a loud rip, and the like.

The history of Torrance's journey to study creativity is instructive. A counselor and high school teacher at a boarding school, Georgia Military College, Torrance read "Square Pegs in Square Holes" (Broadley, 1943). Broadley wrote that Creative imagination is, unless used and directed into the right channels, like a *wild colt* roaming the prairies. When well directed and developed, however, the aptitude can lead to outstanding creative work. While working as a counselor, Torrance noticed that many of the boys at Georgia Military College were there because of discipline problems. Yet, many of these troubled students seemed to display a special quality that, he later realized, was creativity. Torrance stated that these boys were like wild colts, and that they needed to learn to direct their creative energy in positive ways.

Harnessing creativity has significant social implications. Military service in the U.S. Army interrupted Torrance's career in education. Appointed to head a task force to study factors in fighter interceptor effectiveness in Korea, with particular emphasis on the jet aces, he found that the outstanding aces were full of ideas and, just like many of his students at Georgia Military College, were like *wild colts*. When they learned discipline and how to adapt successfully in the Air Force, they not only survived (literally), they also thrived, advancing in the most prestigious career path in the Air Force, serving as fighter pilots. Torrance also had other important findings. In his 40-year longitudinal study, begun in 1958 and finished in 1998 (following the participants for 40 years), and in other studies, Torrance found that individuals who were creatively successful had at least one significant mentor in their lives who recognized, understood, and supported their creative potential. Torrance also found that environments, including home, peer, school, society, and others, influence (encourage or discourage) an individual's creativity.

A possible explanation for upper grade elementary school children might be the lack of purposeful creativity development or the stifling of children's creativity in schools. Since *A Nation at Risk* (National Commission on Excellence in Education, 1983), higher academic standards have been emphasized. The Goals 2000: Educate America Act of 1994 re-emphasized a standards-based educational system where standards and expectations are the same for all students, and the success of reform efforts is evaluated based on students' achievement levels (Duffy, Giordano, Farrell, Paneque, & Crump, 2008). Finally, the No Child Left Behind (NCLB) Act of 2001 holds schools accountable for ensuring that all students achieve proficiency by 2013–2024. The NCLB has received enough strong political and public support to move the entire public school system into a mandatory state assessment mode (Crocker, 2003). The NCLB requires that all states administer annual assessments in reading/language arts and mathematics in Grades 3 through 8.

These assessments may come at a price. Wang, Beckett, and Brown (2006) indicated that while state-mandated testing movement appeared to yield some positive score changes in student learning; however, the increased emphasis on standardized testing may have shifted the emphasis in schools toward more drill exercises and rote learning instead of critical, creative thinking. Gandal and McGiffert (2003) reported that state-mandated standardized tests tend to sample lower level standards and measure lower order thinking skills (Hillocks, 2002). Kohn (1993, 2000) argued that the state-mandated standardized testing system decreases students' natural curiosity and the joy for learning in its own right, which decreases students' intrinsic motivation and increases the need for extrinsic motivation.

Flake et al. (2006) expressed concern that the NCLB forces education to focus on tests, and that education designed to teach children to pass tests does not develop students' appreciation for knowledge or learning. Further, Flake et al. (2006) argued that NCLB stifles teachers' creativity, because the high pressure to cover the content required to produce passing test scores overrides the desire to stimulate children's imagination and curiosity. The education agenda forced upon the education system by NCLB does not value teachers' skills that could encourage the creative application of classroom learning to real life situations. Teaching professionals are reduced to teaching technicians with less ability to make curricular choices or develop creative approaches to engage students because they are required to follow a prescriptive curriculum (Duffy et al., 2008). The high-stakes testing environment created by NCLB has led to the elimination of content areas and activities including electives, the arts, enrichment, gifted programs, foreign language, elementary science, and elementary recess, which leaves little room for imagination, critical or creative thinking, and problem solving (Gentry, 2006). This may eliminate the opportunities for creative students to release their creative energy in school. When their creative needs are not met, they often become underachievers (Kim, 2008c, 2010; Kim & VanTassel-Baska, 2010). Underachievement leads to lower levels of educational attainment. Further, using the data sets from the National Educational Longitudinal Study (National Center for Education Statistics, 1988) and Educational Longitudinal Study (National Center for Education Statistics, 2002), Kim and Hull (2008, 2012) found that high school students who are creative are more likely to drop out than other students.

Creativity and educational standards are not necessarily incompatible. Creativity necessitates a rich knowledge structure. Therefore, ensuring that children acquire necessary facts, skills, and knowledge can promote creativity and the creative process. Preparation is the first stage of the creative process and is the process of accumulating the knowledge necessary to solve a problem (Amabile, 1996; Finke, Ward, & Smith, 1992). The creative process may cement that knowledge into a person's psyche so that the knowledge is not just memorized but is internalized. Once knowledge is established in a personal schema, it can be manipulated in creative ways. For example, Picasso was a skilled and trained realist before he transformed his style into abstract art by incorporating cubism.

Creativity assessments should supplement high-stakes testing (Kaufman & Agars, 2009), and it might contribute to reversing the trend of decreasing creativity in children. Creative pursuits could give children intrinsic motivation to acquire a diverse knowledge base. Therefore, creativity-based learning could increase children's scores on standardized tests. Elevating the status of creativity to be equal to that of content knowledge can provide motivation for schools to help children develop creative skills while simultaneously fostering an appreciation for learning in children.

The clarion call cannot be sounded loudly enough: the pressure against creativity in education comes not merely from NCLB; it also comes from the false premise that in schools creativity is and should principally be the domain of the arts. This myth, for one, pits creativity against traditional and dominant notions of masculinity, which has long valued practicality and toughness. On this (unspoken) basis, funding for the fine and performing arts, so often perceived as feminine and superfluous, is routinely cut during times of fiscal difficulty – typically in the name of "necessity" (Imagine, in the name of necessity, the reaction to a proposal to cut the football team!). Yet, the premise is false: creativity plays crucial roles in almost all fields and can be fostered in almost all academic disciplines. Moreover, creativity is eminently practical and will almost certainly be more so in the century ahead. For Westerners, as the world becomes "flat" (Friedman, 2005) and the "great divergence" ends (Huntington, 1996), creativity may become *the* discriminator between the successful and the unsuccessful, between winners and losers. For Americans, for whom being creative may be the core of American *exceptionalism*, the decline in creativity augurs the decline in the distinctive quality that has historically set the United States apart. Not shackled to a past, immigrants to America were free to adopt, adapt, and innovate.

> What then is the American, this new man?... He is an American, who, leaving behind him all his ancient prejudices and manners, receives new ones from the new mode of life he has embraced, the new government he obeys, and the new rank he holds. He has become an American by being received in the broad lap of our great Alma Mater. Here individuals of all races are melted into a new race of man, whose labors and posterity will one day cause great changes in the world.
>
> *(Crèvecoeur, 1782, p. 54)*

Based upon modern research, Crèvecoeur's (1782) clairvoyance should not be surprising: multicultural environments such as the United States are fertile environments for creativity. Whereas a long tradition of historical research has noted and explained why the West has been particularly creative (Rosenberg & Birdzell, 1986), the United States, especially, has been considered a source of innovation. This distinctiveness was apparent from the earliest days of the Republic. Writing in the early 1780s, Crèvecoeur spoke with immediate prescience. Within a few years, the Constitutional Convention approved the new, revolutionary constitution, which many around the world viewed with wonder and fear (Palmer, 1999; Adelman, 2008). Several decades later, the United States ushered in mass democracy.

The spirit of innovation did not stop there. Some of the greatest contributions in modernity can be attributed to American creativity. The steam-powered ships, blues music, the National Parks, Ford's Model-A, the polio vaccine, light bulbs, computers, the Internet, the micro-computer, and an endless and still growing list of American innovations have shaped the world in which we live. However, perhaps above all else Americans have been creative in the economic and industrial arenas. To be sure, England still deserves pride of place in the early decades of industrialization, but, by the late 19th century, Germany and the United States were its rivals. In the 20th century, the United States unquestionably led the world in economic and industrial innovation and in prosperity (Hall & Preston, 1988). That innovation bred innovation; the 20th century was the American century.

Many indicators, however, not just the decline in creativity scores on the Torrance tests, suggest that that innovative spirit is on the wane. Banking and finance, rather than industrial innovation, now occupy an ever-greater segment of the American economy (Meyerson, 2011). While such a shift to finance may make some Americans significantly richer, it means that less talent and fewer resources are being committed to the kind of innovation that fostered the prosperity of the United States in the first place. Some commentators have noticed an increasing risk-aversion among Americans (Samuelson, 2011), risk being an essential element of creativity. At the same time, other countries are investing heavily in preserving and developing the creative abilities of their students. Does the world really want to know what the world of the future could look like if this American *ingenuity* and innovation diminish anymore?

The finding that Americans are less motivated to be creative is possibly the most disturbing result of the creativity crisis study. The probable cause of this decrease is that creativity is continually less encouraged by home, school, and society overall in the United States. As the American economy loses its totally commanding place in the world economy, many Americans are retreating to safety and security. It stands to reason that fear of risk and aversion to creativity will compound, as the United States keeps producing citizens who tend to be even less tolerant of creative people and of creative expression. If society stifles interests in developing individual differences, creative and innovative thinking, or individual potential, and eliminates the opportunities for creative students to release their creative energy in school, this may cause problems in the future. Those who, despite the odds, preserve and develop their creative abilities will be particularly adversely affected. When children's creative needs are not met, they often become underachievers and show behavioral problems. Underachievement leads to lower levels of educational attainment and later life goal attainment. Further, research shows that high school students who exhibit creative personalities are more likely to drop out of school than other students. On the other hand, imagine if the opposite were true, if within the United States creativity was being fostered: creativity can elevate giftedness (which is measured mainly by IQ) into eminence and greatness.

Time will tell whether the United States is, in terms of creativity and its economic, political, and social implications, at a point of inflection, those rare moments in the history of a country or society or organization when its long-term fate hangs in the balance. One such moment was the 1430s, when the Chinese emperor ordered Chinese ocean-going vessels home, in effect forcing the Chinese to look inward, not to innovate. More analogous to the situation in the United States, perhaps, than the situation in China, where the emperor was powerful enough to *order* a turning point, was Venice in the 16th century. There, merchants held vast accumulated financial capital, intellectual capital, and extended trading networks. However, in responding to the threats of new ocean-going routes to the East and to the discovery of America, these merchants made an interesting and illustrative choice. Rather than creatively leveraging their collective strengths for new opportunities, Venetian merchants defensively retreated to the safety and security of real estate investments on the mainland. However much this may have protected individual merchants and their families in the short term, in the long run their decisions collectively were costly (Cozzi, 1984). In less than 150 years, Venice went from being the premier commercial center in Europe to a fashionable but backwater resort town. Could the same thing happen to the United States? The case of General Motors Corporation (GM) could be telling. At its peak, GM was the largest industrial concern in the world. Rather than creatively leverage its market position and immense capital resources into market dominance in the next wave of automobiles, however, GM continued to invest mostly in the development of its large cars and trucks, all the while more innovative firms such as Toyota slowly chipped away at GM's formerly hegemonic power. On June 1, 2009, GM filed for bankruptcy. Though now reconstituted, GM is no longer the company it was, its new workers no longer enjoying the prosperity that GM workers traditionally did. These examples should cause alarm.

If children's creative potential is diminishing, then their creative production in adulthood will probably decrease as well. Thus, the decrease in creative production will influence the future world economy and standard of living, both in the United States and beyond. American society, industry, and the economy will pay an unknown price in the form of unachieved potential and opportunities lost to foreign innovators. More importantly, on a human scale, a less creative world will have less tolerance for creative people. Thus, individuals born with more creativity will suffer psychologically because realization of full creative potential in individuals enhances their mental health.

Conclusion

Though intelligence is increasing, since 1990 measurable creativity is in decline and the implications are potentially grave. Several factors can probably explain these decreases. Children spend an increasing amount of time interfacing with electronic devices, which have a limited range of possible outcomes. Parents and teachers, too, in their responses to certain characteristics such as rebelliousness, emotional expressiveness, and impulsiveness, may bear some of the blame. Rather than rewarding children for these qualities so crucial to creativity, adults are ever more likely to define them – and label the children possessing them – as problems. As a fix, health care providers are increasingly prescribing drugs such as Ritalin. Also playing a role in the decline in creativity could be the standards movement in education and NCLB. Despite the good intentions of NCLB to provide focus in schools, teachers' responses to the pressures to achieve "adequate yearly progress" may actually be decreasing teacher creativity in the classroom and, thus, opportunities for spontaneity and awakening of student interest. This decline in creativity and the presence of significant forces working against fostering it in young people augur poorly for the United States. From the beginning of its history, America has been a laboratory for creativity, and creativity has been an essential element of the American national character and a source of American prosperity. The declines in creativity since 1990 speak to an alarming trend in American life, one that deserves the collective attention of parents, teachers, and politicians so that this essential quality to a joyful and prosperous life is not diminished beyond repair.

References

Adelman, J. (2008). The age of the democratic revolution: A political history of Europe and America, 1760–1800. *American Historical Review, 113*, 319–340.
Amabile, T. M. (1996). *Creativity in Context*. Boulder, CO: Westview Press.
Barron, F. (1961). Creative vision and expression in writing and painting. In D. W. MacKinnon (Ed.), *The Creative Person* (pp. 237–251). Berkeley, CA: Institute of personality assessment research, University of California.
Broadley, M. E. (1943). *Square Pegs in Square Holes*. Garden City, NY: Doubleday, Doran & Co.
Ceci, S. J. (1991). How much does schooling influence general intelligence and its cognitive components? *Developmental Psychology, 27*, 703.
Ceci, S. J., & Williams, W. M. (1997). Schooling, intelligence, and income. *American Psychologist, 52*, 1051.
Cheng, Y.-L., Kim, K. H., & Hull, M. F. (2010). Comparisons of creative styles and personality types between American and Taiwanese college student and the relationship between creative potential and personality types. *Psychology of Aesthetics, Creativity, and the Arts, 4*, 103–112.
Cho, S. H., Te Nijenhuis, J., Van Vianen, A. E. M., Kim, H. B., & Lee, K. H. (2010). The relationship between diverse components of intelligence and creativity. *Journal of Creative Behavior, 44*, 125–137.
Cozzi, G. (1984). *Il doge Nicolò Contarini: Ricerche sul patriziato veneziano agli inizi del seoicento*. Venice: Istituto per la Collaborazione Culturale.
Cramond, B. (1993). The Torrance Tests of Creative Thinking: From design through establishment of predictive validity. In R. F. Subotnik & K. D. Arnold (Eds.), *Beyond Terman: Contemporary Longitudinal Studies of Giftedness and Talent* (pp. 229–254). Norwood, NJ: Ablex.
Crèvecoeur, H. S. J. (1782). *Letters From an American Farmer*. Reprinted from the original ed., with a prefatory note by W. P. Trent and an introduction by Ludwig Lewisohn. New York, Fox, Duffield, 1904. Retrieved from http://xroads.virginia.edu/.
Crocker, L. (2003). Teaching for the test: Validity, fairness, and moral action. *Educational Measurement: Issues and Practice, 22*, 5–11.
De Bono, E. (1970). *Lateral Thinking: Creativity Step by Step*. New York: Harper & Row.
DeYoung, C. G., Peterson, J. B., & Higgins, D. M. (2005). Sources of openness/intellect: neuropsychological correlates of the fifth factor of personality. *Journal of Personality, 73*, 825–858.
Dickens, W. T., & Flynn, J. R. (2001). Heritability estimates versus large environmental effects: The IQ paradox resolved. *Psychological Review, 108*, 346.
Duffy, M., Giordano, V. A., Farrell, J. B., Paneque, O. M., & Crump, G. B. (2008). No child left behind: Values and research issues in high-stakes assessments. *Counseling and Values, 53*, 53–66.
Feist, G. J. (1998). A Meta-analysis of personality in scientific and artistic creativity. *Personality and Social Psychology Review, 2*, 290–309.
Finke, R. A., Ward, T. B., & Smith, S. M. (1992). *Creative Cognition: Theory, Research, and Applications*. Cambridge, MA: Massachusetts Institute of Technology.
Flake, M. A., Benefield, T. C., Schwarts, S. E., Bassett, R., Archer, B., Etter, F., et al. (2006). A Firsthand Look at NCLB. *Educational Leasdership, 64*, 48–52.
Flynn, J. R. (1984). The mean IQ of Americans: Massive gains 1932 to 1978. *Psychological Bulletin, 95*, 29–51.
Flynn, J. R. (2007). *What is Intelligence? Beyond the Flynn Effect*. New York: Cambridge University Press.
Friedman, T. L. (2005). *The World is Flat: A Brief History of the Twenty-first Century*. New York: Farrar, Straus & Giroux.
Furnham, A., & Chamorro-Premuzic, T. (2006). Personality, intelligence and general knowledge. *Learning and Individual Differences, 16*, 79–90.
Furnham, A., & Thomas, C. (2004). Parents' gender and personality and estimates of their own and their children's intelligence. *Personality and Individual Differences, 37*, 887–903.
Gentry, M. (2006). No child left behind: Neglecting excellence. *Roeper Review, 29*, 24–27.
Gough, H. G. (1976). Studying creativity by means of word association tests. *Journal of Applied Psychology, 61*, 348–353.
Gandal, M., & McGiffert, L. (2003). The power of testing. *Educational Leadership, 60*, 39–42.
Guilford, J. P. (1959). *Personality*. New York: McGraw-Hill.
Guilford, J. P. (1967). *The Nature of Human Intelligence*. New York: McGraw-Hill.
Hall, G. H., & Preston, P. (1988). *The Carrier Wave: New Information Technology and the Geography of Innovation, 1846–2003*. London; Boston: Unwin Hyman.
Hartnett, D. N., Nelson, J. M., & Rinn, A. N. (2004). Gifted or ADHD? The possibilities of misdiagnosis. *Roeper Review, 26*, 73–76.

Healey, D., & Rucklidge, J. J. (2006). An investigation into the psychosocial functioning of creative children: The impact of ADHD symptomatology. *Journal of Creative Behavior, 40*, 243–264.

Helson, R., & Crutchfield, R. S. (1971). Women mathematicians and the creative personality. *Creative Achievements, 32*, 210–220.

Hillocks, G., Jr. (2002). *The Testing Trap: How State Writing Assessments Control Learning*. New York: Teachers College, Columbia University.

Huntington, S. P. (1996). *The Clash of Civilizations and the Remaking of World Order*. New York: Simon & Schuster.

Johnson, A. S., & Fishkin, A. S. (1999). Assessment of cognitive and affective behaviors related to creativity. In A. S. Fishkin, B. Cramond, & P. Olszewski-Kubilius (Eds.), *Investigating Creativity in Youth: Research and Methods* (pp. 265–306). Cresskill, NJ: Hampton Press, Inc.

Kaufman, J. C., & Agars, M. D. (2009). Being creative with the predictors and criteria for success. *American Psychologist, 64*, 280–281.

Kim, K. H. (2005). Can only intelligent people be creative? A meta-analysis. *Journal of Secondary Gifted Education, 16*, 57–66.

Kim, K. H. (2006a). Can we trust creativity tests? A review of the Torrance Tests of Creative Thinking (TTCT). *Creativity Research Journal, 18*, 3–14.

Kim, K. H. (2006b). Is creativity unidimensional or multidimensional? Analyses of the Torrance Tests of Creative Thinking. *Creativity Research Journal, 18*, 251–260.

Kim, K. H. (2008a). Commentary: The Torrance Tests of Creative Thinking already overcome many of the perceived weaknesses that Silvia et al.'s (2008) Methods are Intended to Correct. *Psychology of Aesthetics, Creativity, and the Arts, 2*, 97–99.

Kim, K. H. (2008b). Meta-analyses of the relationship of creative achievement to both IQ and divergent thinking test scores. *Journal of Creative Behavior, 42*, 106–130.

Kim, K. H. (2008c). Underachievement and creativity: Are gifted underachievers highly creative? *Creativity Research Journal, 20*, 234–242.

Kim, K. H. (2009). Cultural influence on creativity: The relationship between Asian culture (Confucianism) and Creativity Among Korean Educators. *Journal of Creative Behavior, 43*, 73–93.

Kim, K. H. (2010). Measurements, causes, and effects of creativity. *Psychology of Aesthetics, Creativity, and the Arts, 4*, 131–135.

Kim, K. H. (2011a). Proven reliability and validity of the Torrance Tests of Creative Thinking (TTCT). *Psychology of Aesthetics, Creativity, and the Arts*.

Kim, K. H. (2011b). The APA 2009 Division 10 debate: Are the Torrance tests still relevant in the 21st Century? Importance of Torrance Tests of Creative Thinking (TTCT) will be continued. *Psychology of Aesthetics, Creativity, and the Arts*.

Kim, K. H. (2011c). The Creativity Crisis: The decrease in creative thinking scores on the Torrance Tests of Creative Thinking. *Creativity Research Journal, 23*, 1–11.

Kim, K. H., & Hull, M. (2008, November). *An Examination of Creative Personality and Anti/pro-creative School Environment as Predictive Factors in High School Underachieving and Dropouts: Using Data From ELS, NELS, and Local High School Students*. Paper presented at the 55th Annual Convention of the National Association for Gifted Children in Tampa, FL, October 29 to November 2, 2008.

Kim, K. H., & Hull, M. (2012). Creative personality and anti-creative environment for high school dropouts: Using data from ELS, NELS, and high school students in MI. *Creativity Research Journal, 24*, 169–176.

Kim, K. H., & Lee, H. E. (2007, August). *Cultural Influence on Creativity: A Comparative Study of Creativity and Creativity Types Between Easterners and Westerners*. Paper presented at the 115th Convention of the American Psychological Association in San Francisco, CA, August 17–20, 2007.

Kim, K. H., Lee, H., Chae, K., Andersen, L., & Lawrence, C. (2011). Creativity and confucianism among American and Korean educators. *Creativity Research Journal, 23*, 357–371.

Kim, K. H., & VanTassel-Baska, J. (2010). The relationship between creativity and behavior problems among underachievers. *Creativity Research Journal, 22*, 185–193.

Kirton, M. J. (1999). *Kirton Adaption-innovation Inventory Manual*, 3rd ed. Berkhamsted, UK: Occupational Research Center.

Kohn, A. (1993). *Punished by Rewards: The Trouble with Gold Stars, Incentive Plans, A's, Raises, and Other Bribes*. Boston, MA: Houghton Mifflin.

Kohn, A. (2000). *The Case Against Standardized Testing*. Portsmouth, NH: Hinemann.

Krautkramer, C. J. (2005). Beyond creativity: ADHD drug therapy as a moral damper on a child's future success. *American Journal of Bioethics, 5*, 52–53.

Lee, H., & Kim, K. H. (2010). Relationships between bilingualism and adaptive creative style, innovative creative style, and creative strengths among Korean American students. *Creativity Research Journal, 22,* 402–407.

MacKinnon, D. W. (1967). Educating for creativity: A modern myth? In P. Heist (Ed.), *Education for Creativity* (pp. 1–20). Berkeley, CA: Center for Research and Development in Higher Education.

Meyerson, H. (2011, October 4). Rescuing America from Wall Street. *Washington Post.*

National Center for Education Statistics. (1988). *National Education Longitudinal Study of 1988 (NELS: 88).* Retrieved from http://nces.ed.gov/surveys/nels88/.

National Center for Education Statistics. (2002). *National Education Study of 2002 (ELS: 2002).* Retrieved from http://nces.ed.gov/surveys/els2002/.

National Commission on Excellence in Education. (1983). *A Nation at Risk: A Report to the Nation and Secretary of Education.* Washington, DC: U.S. Department of Education.

Palmer, R. R. (1999). *The Age of the Democratic Revolution: A Political History of Europe and America, 1760–1800.* Princeton, NJ: Princeton University Press.

Rosenberg, N., & Birdzell, L. E. (1986). *How the West Grew Rich: The Economic Transformation of the Industrial World.* New York: Basic Books.

Rossman, B. B., & Horn, J. L. (1972). Cognitive, motivational and temperamental indicants of creativity and intelligence. *Journal of Educational Measurement, 9,* 265–286.

Runco, M. A., & Albert, R. S. (1986). The threshold theory regarding creativity and intelligence: an Empirical test with gifted and nongifted children. *Creative Child and Adult Quarterly, 11,* 212–218.

Russ, S., Robins, D., & Christiano, B. (1999). Pretend play: Longitudinal prediction of creativity and affect in fantasy in children. *Creativity Research Journal, 12,* 129–139.

Samuelson, R. (2011, October 2). Risk-averse America. *Washington Post.*

Simonton, D. K. (1994). *Greatness: Who Makes History and Why.* New York: Guilford.

Sternberg, R. J. (1999). A propulsion model of types of creative contributions. *Review of General Psychology, 3,* 83–100.

Torrance, E. P. (1966). *The Torrance Tests of Creative Thinking: Norms-technical Manual Research Edition. Verbal Tests, Forms A and B. Figural Tests, Forms A and B.* Princeton, NJ: Personnel Press.

Torrance, E. P. (1967). *Understanding the Fourth Grade Slump in Creative Thinking: Final Report.* Athens, GA: University of Georgia.

Torrance, E. P. (1971). Are the *Torrance* Tests of Creative Thinking biased against or in favor of "disadvantaged" groups? *Gifted Child Quarterly, 15,* 75–80.

Torrance, E. P. (1974). *The Torrance Tests of Creative Thinking: Norms-technical Manual Research Edition. Verbal Tests, Forms A and B: Figural Tests, Forms A and B.* Princeton, NJ: Personnel Press.

Torrance, E. P. (1977). *Creativity in the Classroom.* Washington, DC: National Education Association.

Torrance, E. P. (1987). *Guidelines for Administration and Scoring/Comments on Using the Torrance Tests of Creative Thinking.* Bensenville, IL: Scholastic Testing Services, Inc.

Torrance, E. P. (1988). The nature of creativity as manifest in its testing. In R. J Sternberg (Ed.), *The Nature of Creativity* (pp. 43–73). New York: Cambridge University Press.

Torrance, E. P. (1994). *Creativity: Just Wanting to Know.* Pretoria, South Africa: Benedic Books.

Torrance, E. P. (2000). *Research Review for the Torrance Tests of Creative Thinking Figural and Verbal Forms A and B.* Bensenville, IL: Scholastic Testing services, Inc.

Torrance, E. P. (2002). *The Manifesto: A Guide to Developing a Creative Career.* West Westport, CT: Ablex Publishing.

Torrance, E. P. (2008). *The Torrance Tests of Creative Thinking Norms-technical Manual Figural (streamlined) Forms A & B.* Bensenville, IL: Scholastic Testing Services, Inc.

Torrance, E. P., & Safter, H. T. (1990). *The Incubation Model of Teaching: Getting Beyond the Aba!* Buffalo, NY: Bearly.

Torrance, E. P., & Torrance, P. (1972). Combining creative problem-solving with creative expressive activities in the education of disadvantaged young people. *Journal of Creative Behavior, 6,* 1–10.

Vygotsky, L. S. (1994). Imagination and creativity of the adolescent. In R. Van Der Veer & J. Valsiner (Eds.), *The Vygotsky Reader* (pp. 266–288). Cambridge: Blackwell Publishers (R. Van Der Veer & J. Valsiner, Trans.) (Original work written in 1931).

Wallach, M. A. (1970). Creativity. In P. H. Mussen (Ed.), *Carmichael's Manual of Child Psychology,* vol. 1 (pp. 1273–1365). New York: Wiley.

Wang, L., Beckett, G. H., & Brown, L. (2006). Controversies of standardized assessment in school accountability reform: A critical synthesis of multidisciplinary research evidence. *Applied Measurement in Education, 19,* 305–328.

12

Do not overlook innovators!

Discussing the "silent" issues of the assessment of innovative abilities in today's children–tomorrow's innovators

Larisa V. Shavinina

UNIVERSITÉ DU QUÉBEC EN OUTAOUAIS, CANADA

Summary: This chapter discusses the silent issues related to the identification of innovative abilities such as the theoretical basis of innovation, why the normal notion of assessment that applies to intelligence testing does not necessarily apply to innovation talent, and how the best measurements can be developed. The chapter also discusses what should be included in the comprehensive approach to the psychological assessment of innovative abilities. It describes one of the parts of this approach. The nine methodological and procedural principles, which form this approach, are considered along with the examples of tests. Innovative abilities can be identified in the form of what might be called "trajectory analysis" or "trajectory assessment." Cognitive styles, metacognitive, and extracognitive abilities should also be assessed. The information about a child's sensitive periods—the developmental foundation of individual innovation—and his or her strong interests should be gathered from parents and teachers.

Key words: Innovative abilities, innovation testing, "trajectory analysis," "trajectory assessment," psychological mental context, retrospective and prospective assessment.

Introduction

David Dai and other contributors to this volume emphasized a need to develop assessment tools aimed at the identification of innovative abilities. In spite of the importance of innovation in any area of human endeavor today, it is a disturbing reality that there are no reliable and exact measurements, which would allow us to identify (and thus not overlook!) the hidden innovative talents of children and adolescents. Chances are then pretty high that many potential innovators are overlooked these days and this is an alarming thought for the innovation education community. When one considers the great impact of innovators on the world in general (Shavinina, 2012) and their unique innovative talents in particular, then it is clear that the measurement of innovative abilities is an exceptionally important scientific topic and the task of developing comprehensive and ideal identification methods is a challenging job for innovation scholars in the near future.

Definitely, creativity testing should be an integral part of the comprehensive assessment of innovative abilities (Kim & Pierce, this volume). However, this is only one of the parts. Others should measure entrepreneurial giftedness, wisdom related skills, intuition, excellence, and an ability to meet deadlines. In other words, all the components of the phenomenon of individual innovation (described in Chapter 36, *What can innovation education learn from innovators with longstanding records of breakthrough innovation?* included in this volume) and all the elements of innovation education (discussed in Chapter 3 of this volume, *The fundamentals of innovation education*) should be incorporated in the all-inclusive approach to the assessment of innovative abilities. The characteristics of entrepreneurial giftedness (identified in Chapter 16 of this volume, *The trajectory of early development of prominent innovators: entrepreneurial giftedness in childhood*) offer immense potential for getting at identification of innovation talent. As Larry Vandervert (2012) suggested, today innovative abilities can only be identified in the form of what might be called "trajectory analysis" or "trajectory assessment." For example, a distilled form of trajectory analysis could be performed as children interact with phases of a challenging, open-ended computer game, thus measuring their innovation potential.

This chapter presents one of the promising parts of the comprehensive approach to the assessment of innovative abilities that is based on a theory of individual innovation (Shavinina & Seeratan, 2003). The chapter is organized as follows. The first section briefly describes the theory and emphasizes a specific type of representation or a unique vision as the psychological basis of individual innovation. The second section describes the proposed assessment that is based on nine relatively independent and at the same time interrelated principles. These methodological and procedural principles are presented along with examples of tests, or, to be more precise, subtests.

The theory of individual innovation

Since innovation is a special type of giftedness, Kholodnaya's (1993) and Shavinina and Kholodnaya's (1996) theory of intellectual giftedness may serve as a theoretical foundation for the development of an important part of the comprehensive approach to the psychological assessment of innovative abilities. It has been modified by Shavinina and Seeratan (2003) into the theory of individual innovation.

It has been demonstrated that the main difficulty in understanding the nature of intellectual giftedness and innovation is that their various traits, characteristics, properties, and qualities (i.e., their external *manifestations* in any real activity) have been the subject of research; but the psychological basis (or psychological carrier) of these manifestations has not been investigated (Kholodnaya, 1993; Shavinina & Kholodnaya, 1996; Shavinina & Seeratan, 2003). Attempts to understand the nature of any scientific phenomenon solely on the basis of listening and describing its own characteristics, traits, features, qualities, and properties are unsatisfactory. For instance, contradictions and crises in psychology testify to this (Vekker, 1981). The external manifestations of psychological phenomena can be studied endlessly; however, the real, deep understanding of the phenomena will not be attained.

In this light the only promising solution is to consider individual innovation as the sum of its two important aspects: the external manifestations of individual innovation (i.e., its features, traits, characteristics, properties, and qualities) and its psychological basis (i.e., the psychological carrier of these manifestations). A need for a new research direction was, hence, emphasized. Specifically, Shavinina and Seeratan (2003) argued that there is an urgent need to re-examine researchers' approach to understanding the nature of individual innovation as a psychological phenomenon. It means that scholars should not answer the question "What is innovation?" by listing its characteristics and traits (i.e., its external manifestations). Rather, they should answer the following question:

"What is the carrier (a basis) of the characteristics and traits associated with individual innovation?"

In light of this fundamentally changed point of view, researchers should study an individual's cognitive experience and, more precisely, the specificity of its structural organization. Cognitive experience is a system of the available psychological mechanisms, which forms a basis for an individual's cognitive attitude to the world around and predetermines the specificity of his or her mind (Shavinina & Kholodnaya, 1996). This experience is the psychological basis of individual innovation. In other words, the cognitive experience serves as the psychological carrier of numerous manifestations of innovation.

The structural organization of individual innovation is presented at six levels: the three basic levels (i.e., neuropsychological, developmental, and cognitive foundations of innovation) and the three levels of its manifestations (i.e., its numerous characteristics and traits). See Figure 12.1.

The first level is the neuropsychological foundation of individual innovation, mostly connected to the exceptional neural plasticity of the brain of innovators (for a thorough account of the neuropsychological basis of an innovator's mind see the chapter by Vandervert & Vandervert-Weathers, this volume). The second level is the developmental foundation of innovation, mainly formed by sensitive periods, which significantly accelerate an innovator's development during childhood through the actualization of his or her creative and intellectual potential and the growth of the individual's cognitive resources resulting in the appearance of a unique cognitive experience (see the chapter on Albert Einstein for a detailed analysis of sensitive periods as a developmental basis of innovation, this volume). This experience is the cognitive basis of innovation, the third level in its structural organization.

The cognitive experience of innovators expresses itself in their specific type of representation of everything that is going on around them (i.e., any event, idea, problem, and so on). In other words,

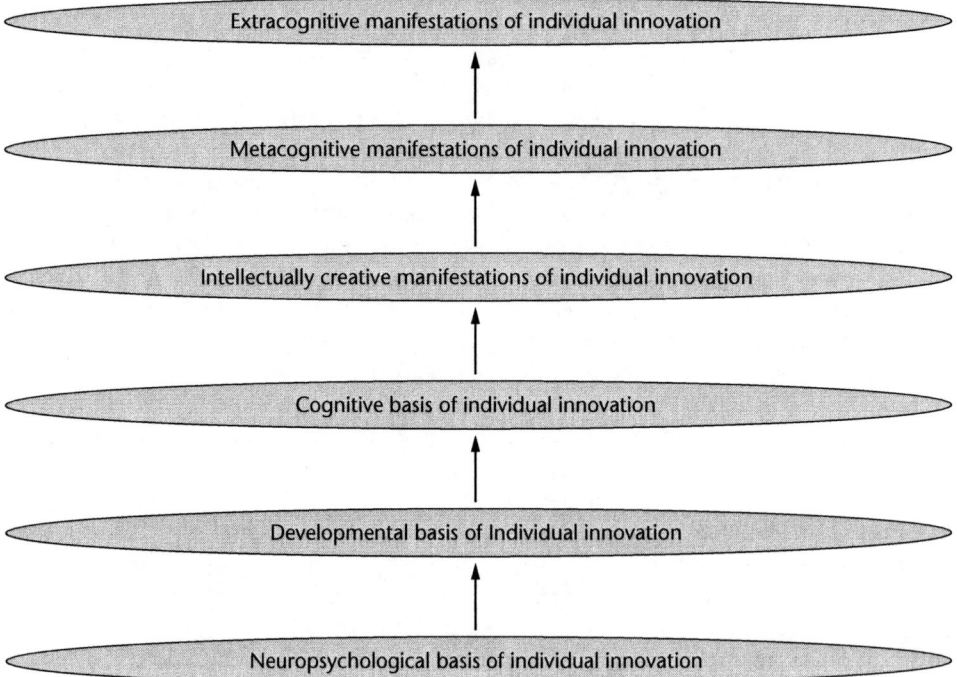

Figure 12.1 The model of the internal structure of individual innovation.

Assessment of innovative abilities

innovators see, understand, and interpret the world around them in a different manner from the rest of people. This unique picture of the world or an individual's unique point of view or a unique vision is the essence of individual innovation. An important aspect of this uniqueness is their objectivization of cognition. It means that innovators are able to see the world "as it was, as it is, and as it will be in its objective reality" (Kholodnaya, 1990, p. 128; Shavinina, 1996; examples of innovators' unique vision can be found in Chapter 36 of this volume, *What can innovation education learn from innovators with longstanding records of breakthrough innovations?*).

The innovators' unique cognitive experience provides a basis for three levels of the manifestations of individual innovation (i.e., its various characteristics and traits). That is, all multiple manifestations of innovation can be categorized in three main groups. These three groups—creative-intellectual, metacognitive, and extracognitive manifestations—represent the fourth, fifth, and sixth levels in the structural organization of innovation, respectively (see Figure 12.1). Innovation is closely related to creativity in that innovation always begins with creative ideas and creativity is thus the first step in innovation process. Many contributors to this volume emphasized this point. Innovators' highly developed creative abilities, as well as metacognition and intuition (i.e., extracognition), are considered in Chapter 36. With respect to intellectual manifestations of individual innovation it should be noted that innovators demonstrate an obvious practical intelligence.

Taken together, these six levels in Figure 12.1 describe the nature of individual innovation. An approach to the psychological assessment of innovative abilities, which will be presented below, should, therefore, be based on the theory of individual innovation briefly discussed above. This theory brings a fresh perspective on the psychological understanding of the essence of individual innovation and, consequently, calls for special methods for the adequate measurement of an individual's innovative potential. A particular emphasis must be on the assessment of an individual's specific type of representations or a unique point of view or a unique vision.

A new approach to the psychological assessment of innovative abilities

It seems reasonable to base the development of the assessment methods for the identification of innovative abilities on the following principles. Innovation tests should:

- examine the *psychological mental context* generated by an individual;
- have an "*open character*";
- test the *basis of innovation* (i.e., *the psychological carrier of the numerous manifestations of innovation*);
- not evaluate *psychological functions* (short- and long-term memory, attention span, and so on); instead, they should assess innovative abilities. Any psychological function, or even their combination, is not individual innovation;
- *avoid the emphasis on speed or speed responses*;
- also measure *innovative potential or hidden abilities*, not only *actual innovative abilities*. It means that innovation testing should not be exclusively *retrospective* in its measurement of an individual's innovative resources; it should be *prospective* as well;
- analyze *cognitive styles*;
- examine an individual's *metacognitive* and *extracognitive abilities*; and
- not be very long, and, as a consequence, time consuming.

These principles concern various facets of individual innovation, both methodological and procedural ones (i.e., *what* should be included in innovation testing and *how* it should be measured). Very roughly, the above-presented principles can be categorized as the methodological and procedural principles.

L.V. Shavinina

Psychological mental context

The first and fundamental methodological principle of a new approach to the development of innovation tests is the requirement to examine the *psychological mental context* that is generated by an individual (Shavinina & Kholodnaya, 1996). Oatley (1978) first mentioned this idea in his research on human representations. The need to examine the psychological mental context generated by people themselves implies that innovation tests do not contain tasks or problems already prepared by test developers.

The ability to solve real problems is one of the important skills of innovators. Nonetheless, it is not a reason to include items on problem solving in innovation tests. First of all, individual innovation has a lot of its internal and external manifestations and it is not possible to assess all of them by any innovation test. Second, and the most important, as it follows from the above presented theory of innovation, innovation is not problem solving. Therefore, innovation tests should not examine human ability to *solve* problems. Or, on the other hand, tests that estimate problem solving abilities should be called appropriately: tests of problem solving abilities, but not innovation tests.

Performing on innovation tests, examinees should not find any correct or quick solutions. Instead, they are asked to generate something that has never existed before; that is, the individual mental context. Examinees are free in their choice of the direction of the generation of their mental contexts. They are also not limited in the content of their mental contexts. Because of that each mental context is really the *individual* mental context. Basing innovation testing on the examination of people's mental context allows psychologists to avoid many misunderstandings, which occur, for example, in standardized intelligence tests.

For instance, criticizing the conventional intelligence tests, Olson (1986) reasonably noted that there is a difference between the meaning of a written word for test developer(s) and how it will be understood by examinee(s). Unfortunately, the absolute coincidence of the interpretations of test developer(s) and examinee(s) is not a case of traditional intelligence tests. This "dissonance of understandings" was also underlined by Sternberg (1985), when he concluded that very low scores on IQ tests "often result from an examinee not quite understanding what is required by the tasks at hand" (p. 300).

Another example of misunderstanding in IQ tests is a situation when "in figural reasoning items, examinees sometimes find that none of the answer options appears to be correct; such an outcome often results from representing the geometric structure of the item in a way different from that intended by the item writer" (Sternberg, 1985, p. 300). It looks like the best way to overcome these problems of misunderstanding is to provide an opportunity for examinees to generate their individual mental contexts by themselves.

The possible examples of innovation intelligence tests aim to measure an examinee's mental context are such tests as "Ideal Computer" and "Conceptual Synthesis."

The "Ideal Computer" test (Shavinina & Kholodnaya, 1996) is used to examine how today's children—tomorrow's innovators see, understand, and interpret the world around as a whole; that is, to assess their individual representations in general. To measure this type of representations, examinees are instructed in the following way:

> Just imagine that there is an ideal computer of the latest generation that knows absolutely everything about everything and can answer any of your questions. You can communicate with it for ten minutes. Please, write down any questions you would like answered.

Scores are based on the following parameters:

a   general number of questions;
b   the number of objective questions, which are directed to the understanding of problems of the

external world and which are connected to the actualization of certain elements of objective knowledge about the world (e.g., "Is there an end of the Universe?" "Will new inexhaustible sources of energy be discovered?" "How can military conflicts be prevented?" "Are there any inhabited worlds in the Universe?");

c   the number of subjective questions, which are related to the actualization of personal problems and which focus on personally important situations (e.g., "When will I get married?" "What will my faith be?" "How many children will I have?" "How can I overcome the negative features of my character?");
d   the number of categorical questions, which are characterized by the most general consideration of all aspects of the surrounding reality and the world around us (e.g., "What will happen to mankind in the future?" "What are the rules of people's interactions?" "What are the natural laws of the structure of the Universe?" "Is there the most general principle in the organization of nature?");
e   the number of concrete questions, which are related to a single concrete fact (e.g., "How many stars are there in the sky?" "When will I die?" "How much does the most modern stereo-system cost?" "What will I get on the math exam?").

"Conceptual Synthesis" is used to examine the conceptual representations of potential innovators. This test is Shavinina and Kholodnaya's (1996) modification of Abracham's test of cognitive synthesis (Arina & Koloskova, 1989). The essence of the test is that examinees are given three words that are not connected to each other by meaning. They belong to three different, remote semantic categories (for example, "sandstorm–computer–safety pin," "lightning–ruler–wheel," "chain–fire–watch," and others). According to the instruction, examinees should establish any possible kinds of connections between these words that would have meaning and write the connections down. The test is proposed five times (three new words are given every time). The scores are for:

a   the complexity of established connections (i.e., total number of all connections in quantitative marks); and
b   the number of connections, which are evaluated by maximum mark.

The estimation criteria (on the example of "chain–fire–watch") are: zero marks, if only two words are connected to each other (for example, "A chain can be changed by fire," "A wrist-watch can be on the chain," and so on); one mark, if connection is established only on the basis of simple enumeration of objects and phenomena or their formal contradiction (for example, "A wrist-watch can be changed in the fire, but a chain cannot," "Fire can be blown out by chain and a wrist-watch can be broken by a chain"); two marks, all three words included in a certain concrete situation (for example, "I see such a picture: Prometheus stands near Olympus with a golden watch on the hand; he is shackled in the chains and his death from fire is coming soon"); three marks, all three words connected with a help of some general categorical concept or with a help of some reason-consequences connections (for example, "Fire is a chain of the subsequent processes of oxidation; a watch is a chain of the subsequent positions of pendulum; and a chain consists of a few connected subsequent links").

One can see that performing these tests examinees do really generate some individual mental context, instead of task or problem solving that constitutes, for example, the essence of conventional intelligence testing. Traditional tests are closed with respect to people's possibility to generate subjective mental context.

"Open character"

The next principle is closely related to the first one. It states that innovation *tests should have an "open character."* It means that their instructions should not indicate or predetermine the possible "space" or direction for the generating of individual mental context. Any such indications can lead to the limited "framework" within which an examinee will attempt to produce his or her mental context. Thus, according to the instruction of the "Ideal Computer" test, examinees can write down any questions they wish. It is not to say, for instance, that examinees should write questions regarding some concrete topic. Likewise, according to the instruction of the "Conceptual Synthesis" test, examinees can establish any possible kinds of connections between three words, as many as they want. Certainly, the more restrictions will be in the instruction, the less complete and rich mental contexts will be produced by examinees.

This principle fits well into the numerous observations of psychologists (Csikszentmihalyi, 1992; Dewey, 1919, 1938; Getzels & Csikszentmihalyi, 1976; Wertheimer, 1959) and great scientists (Einstein & Infeld, 1938; Vernadsky, 1988) about the specificity of their exceptional mental functioning and the importance of certain kinds of mental activity for extraordinary achievements. Specifically, they noticed that an individual's abilities to problem finding and posing the right questions are more important in prominent scientific discoveries than reaching the right solutions to the problems. For example, Albert Einstein (Einstein & Infeld, 1938, p. 92) wrote that

> the formulation of a problem is often more essential than its solution, which may be merely a matter of mathematical or experimental skill. To raise new questions, new problems, to regard old problems from a new angle, require creative imagination and mark real advances in science.

If one looks for a while at conventional intelligence tests, one can see that they mainly aim to solve the already formulated—by others—problems and/or tasks. Traditional IQ tests do not have an open character: they do not provide a maximal "space" for examinees' mental initiative that can express itself in problem finding, new ideas generation, raising new questions, and so on, beyond which is the individual mental context. Examinees should be free to produce their mental context without any restrictions. The existing IQ tests are not suitable to measure innovative abilities. In this light it seems appropriate to give an example of another test, the "Formulation of Problems" test, which is used to measure conceptual representations of innovators (Shavinina & Kholodnaya, 1996).

According to the instruction of the "Formulation of Problems" test, (Kholodnaya, 1983), the examinee is a scientist and the word, given by tester, is the subject of his or her research. The examinee formulates problems, which occur in his or her mind in connection with a given subject. Two words, for example, "illness" and "soil," can be the subjects. Examinees are instructed as follows: "Imagine, you are a scientist. 'Illness' is the subject of your research. Please, write down the problems, which you would like to study in connection with this subject." The parameters considered are:

a   the complexity of all formulated problems in quantitative marks; and
b   the number of problems, which are evaluated by the maximum mark.

The scoring for word "illness" is: zero marks, the problem is being formed on the basis of the examinee's evaluation and subjective impression (e.g., "Where and how is it better to train doctors?" "What is the influence of part of the day on a sick man?"); one mark, the problem is being formed by grouping some specific characteristics or properties or features of the proposed object (i.e., of the term "illness." For example, "What are the reasons for illness?" "What are the methods of illness prevention?"); two marks, the problem is being formed through the conjunction of the term "illness" to another semantic field, which is quite remote from the considered term (for example, "What is the connection between 'illness' and the way of life and profession?" "What is going on with illness on different stages of the development of human being?"). These are usually the most creative and innovative problems. Examinees are given as much time as they need to perform the "Conceptual Synthesis" and "Formulation of Problems" tests.

Basis of individual innovation

The third principle states that innovation tests should first and foremost be directed to the versatile assessment of the various aspects of the basis of individual innovation, but not its numerous manifestations. Innovation has so many internal and external manifestations (Shavinina, 2003) that it is impossible to take all of them into account in any test. In this case the above-considered distinction between the "manifestations" of innovation and its "basis" (i.e., the psychological carrier of these manifestations) is appropriate. In accordance with the theory of individual innovation a need to examine the basis of innovation implies a need to examine an individual's cognitive experience that expresses itself in unique representations.

All of the above-presented tests are directed to the measurement of the fundamental components of an individual's cognitive experience that forms the basis of innovation. For example, the "Ideal Computer" test primarily aims to evaluate a global type of representations (i.e., how an individual sees, understands, and interprets the world around as a whole) and to estimate the degree of objectivization–subjectivization of cognition. The "Formulation of Problems" and "Conceptual Synthesis" tests examine conceptual representations.

The "Ecological Forecast of Future Development of the Earth" test (Shavinina, 1993) is used to study representations of future events. Polak (1973) demonstrated the importance of the concrete and rich images of the future for the advanced development of human civilization. Torrance (1979, 1980) found that creative adolescents are characterized by an amazing ability to forecast future events (see also Crammond et al., this volume). The following instruction is given to examinees: "Imagine that you are a science fiction writer who is able to predict future events. Please, describe your predictions about the ecology of the Earth 50 years from now." Scores are for:

a   the differentiation of forecast (i.e., the number of those concrete aspects of the future that have been predicted); and
b   the optimistic, pessimistic, or neutral forecast (i.e., positive, negative, or neutral vision of the future).

Do not test psychological functions

According to the fourth principle, innovation tests should *not examine the development of psychological functions* (for example, attention or memory). Any psychological function or process or even their combination (i.e., all possible processes of human cognitive system) do not reflect the essence of individual innovation and will not help develop innovators. As one can see from the above examples of innovation tests, they do not measure an individual's concentration of attention or short- and long-term memory or any other psychological function or process. For instance, the "Digit Span" subtest of the WISC-III, which is one of the popular IQ tests, is the example of such a subtest devoted to assess children's mnemonic functions. However, the results of child's short- and long-term memory test tell us very little, if anything, about his or her intelligence or giftedness and definitely nothing about his or her hidden innovative potential.

Conventional intelligence tests can be called "tests of psychological functions," because they mainly examine the extent to which any given psychological function is developed. This is why traditional intelligence tests evaluate those manifestations of intelligence that are strongly related to academic success. This is also why IQ tests measure learning ability and academic achievement, but not intelligence and certainly not innovation.

No emphasis on mental speed

The fifth principle asserts that innovation tests should *avoid the emphasis on speed or speed responses* that characterize many psychometric and even information-processing assessment methods (Silverman, 2009). Although good speed responses can have great value (e.g., for air traffic controllers), in reality any important achievement, including innovation, cannot be obtained during seconds, minutes, hours, or even days (Davidson, 1986; Kholodnaya, 1997; Shavinina & Kholodnaya, 1996; Sternberg, 1985, 1988). Even our daily decisions need and take much more time than is allowed by many items on, say, conventional intelligence tests.

Sternberg (1985) pointed out that one of the critical shortcomings of psychometric tests is their reliance upon the incorrect view that the speed of mental functioning is a crucial aspect of mind. It is often claimed that

> the strict timing of such tests merely mirrors the requirements of our highly pressured and productive society. But for most of us, these seem to be few significant problems encountered in work or professional life that permit no more than the 5 to 50 seconds allowed for a typical test problem on a standardized test.
>
> *(Sternberg, 1985, p. 302)*

Moreover, "the assumption that more intelligent people are rapid information processors also underlies the overwhelming majority of tests used in identification of the gifted, including creativity as well as intelligence tests" (Sternberg, 1985, p. 301).

Rabbitt's (1996) studies have showed that speed is not a global performance characteristic of the cognitive system: in other words, intelligence is not just mental speed. Nevertheless, even much more early, the classic research on reflectivity–impulsivity cognitive style demonstrated that a reflective cognitive style is connected to greater intelligence in comparison with an impulsive cognitive style (Baron, 1982; Kagan, 1966; Kagan, Rosman, Day, Albert, & Phillips, 1964). Recent investigations on reflectivity–impulsivity cognitive style and its relationship to giftedness and, probably, to innovation support this finding (Kholodnaya, 1990; Shavinina & Kholodnaya, 1996).

Any orientation of test developers on speed parameters will lead to the substitution of the real nature of individual innovation by people's rapid responses. In innovation tests described above examinees have either (1) as much time as they need (e.g., "Formulation of Problems" and "Conceptual Synthesis" tests), or (2) a certain time—for example, ten minutes—to accomplish some tests (e.g., the "Ecological Forecast of Future Development of the Earth" and "Ideal Computer" tests). However, time restrains are very flexible in the latter case; the tester may allow additional one to three minutes. After the frequent administration of these tests, it was concluded that ten minutes is enough, because about 97% of all examinees finish them during this time period. For the most part, it is not necessary to say before the administration of these tests that examinees have ten minutes. After ten minutes, the tester can simply say that they should already finish. Moreover, the experience demonstrates that there is nothing wrong in interrupting an examinee's work on these tests after ten minutes, because they have already generated a certain mental context and its basic characteristics are obvious. It means that an individual can present a real "picture" of his or her mental context, generating something essential even within a limited period of time.

Both retrospective and prospective assessment

Conventional tests measure an individual's *actual* level of intelligence, aptitude, learning ability, and so on. That is, they examine *actual* (i.e., existing at the time of measurement) intelligence, aptitude, or learning ability of a person. However, it is not enough in the case of innovative abilities. Innovation tests should also assess *innovative potential*, not only *actual innovative abilities*. Therefore, according to the sixth principle of a given approach, innovation testing should not be exclusively *retrospective* in its measurement of an individual's innovative resources; it should also be *prospective*. In other words, testers should try to predict the further development of an individual's abilities, explaining the conditions under which this development will be possible.

This principle, as well as the trajectory of the actualization and development of hidden innovative abilities, is closely related to Lev Vygotsky's notion of the "zone of potential (or proximal) development." He defined it as "the distance between the actual developmental level and the level of potential development" (1978, p. 86). He insisted that psychologists should be prospective as well as retrospective in the understanding and assessment of an individual's abilities.

> The zone of proximal development defines those functions that have not yet matured but are in the process of maturation, functions that will mature tomorrow but are currently in an embryonic state. These functions could be termed the "buds" or "flowers" of development rather than the "fruits" of development. The actual developmental level characterizes mental development retrospectively, while the zone of proximal development characterizes mental development prospectively.
>
> *(Vygotsky, 1978, pp. 86–87)*

Certainly, it is not easy to develop such tests that could assess the innovative *potential* or hidden abilities. Brown and her colleagues have made a well-known attempt in this direction trying to measure the "zone of proximal development" (Brown, 1987). This is one of the possible directions for the development of innovation tests. Other ways also exist. As it follows from the theoretical foundations that underline this approach to the assessment of innovative abilities, one of its fundamental ideas is a need to examine an individual's representations (i.e., both verbal and visual representations, representations of the future, representations of the world as a whole, and so on). A comprehensive "picture" of the individual pattern of representations—that is, his or her unique point of view or a unique vision—should be one of the results of the psychological measurement of innovation.

As the type of representations is a proto-phenomenon of an individual's mind (Shavinina & Kholodnaya, 1996), this "picture" sheds light on his or her innovative *potential* or hidden abilities. Therefore, one can predict the further development of an individual's innovation on the basis of the information about his or her potential innovative abilities. Objective representations are associated with innovation (see Chapter 36 for examples, this volume). Consequently, if an examinee's pattern of individual representations is highly objective (for example, on the "Ideal Computer" test), then this is one of the good predictors of his or her developing innovation. Surely, in order to be more precise about the specificity of the subsequent innovative growth of a person, psychologists should have as many of such predictors as possible. A given approach to the assessment of innovative abilities allows developers of innovation tests to develop a wide range of such predictors.

L.V. Shavinina

Cognitive styles as an important part of innovation testing

Although cognitive styles belong to the manifestations of individual innovation in accordance with the theory of individual innovation, it is necessary to include the measurement of cognitive styles in innovation tests. Cognitive styles refer to individual differences in human information processing. They provide valuable information about the fundamental cognitive mechanisms, which influence the specificity of an individual's cognitive experience (Shavinina & Kholodnaya, 1996). The seventh principle, therefore, states that innovation tests should examine people's *cognitive styles.*

For instance, introduced by Kagan et al. (1964), reflectivity–impulsivity cognitive style displays individual differences in the speed and accuracy with which people propose and formulate hypotheses and make decisions under conditions of uncertainty. Innovators often work under conditions of high uncertainty, especially when they try to bring new products or services to the marketplace. Usually, there are no markets for breakthrough innovations. It is innovators who create them. The well-known examples include Steve Jobs and iPhone, Akio Morita and the *Sony* Walkman, Fred Smith and *FedEx*, and others.

Experimental studies demonstrate that gifted individuals are characterized by reflective cognitive style (Shavinina & Kholodnaya, 1996). They make fewer errors in the situation of multiple choices according to Kagan's Matching Familiar Figures (MFF) test. From the point of view of basic cognitive mechanisms it testifies to the gifted's accurate analysis of visual space up to the moment of making decisions. Probably, they are more careful in evaluating alternatives, hence making few errors; whereas average people presumably hurry their evaluation and thereby make numerous mistakes. This explains in part why innovators, who are also gifted individuals, are able to make wise decisions. The active character of visual scanning by the gifted indicates, in particular, a capacity to delay or inhibit a solution in MFF test performance containing response uncertainty, and also a capacity to differentiate unimportant and essential features of the external stimulus. Because of that Kagan's MFF test can be included in innovation tests. Other cognitive styles (e.g., cognitive complexity–simplicity) are also important for the understanding of an individual's cognitive experience and, consequently, his or her innovative abilities.

Metacognitive and extracognitive abilities as an integral part of innovation testing

Metacognition provides a lot of important information about an individual's mind. Knowledge about one's own abilities and the whole cognitive set-up, evaluating their efficiency, advantages and limitations, as well as planning, monitoring, and executive control are among important metacognitive abilities (Barfurth, Ritchie, Irving, & Shore, 2009; Brown, 1978, 1987; Flavell, 1976; Pressley, Borkowski, & Schneider, 1987; Shavinina & Kholodnaya, 1996; Shore & Dover, 1987; Shore & Kanevsky, 1993; Sternberg, 1985; see also Chapter 3 of this volume, *The fundamentals of innovation education*).

Furthermore, it is important to know not only what we know, but also what we do not know and how to compensate for what we do not know (for more on compensatory mechanisms see Chapter 36, *What can innovation education learn from innovators with longstanding records of breakthrough innovations?* included in this volume). Innovation tests should, therefore, provide important information about an individual's compensatory mechanisms. Because of this, the eighth principle asserts that the *assessment of innovative abilities should be an integral part of innovation testing*. Kagan's MFF test can be used to measure some aspects of metacognitive abilities in children.

Extracognitive abilities (i.e., specific feelings, intentions, beliefs, and intuition) constitute the highest level of the manifestations of individual innovation (Shavinina & Seeratan, 2003). Chapters

3 and 36 of this volume show that great innovators are distinguished by highly developed intuition and other extracognitive abilities. Innovation tests should, therefore, assess various facets of these abilities (Shavinina & Ferrari, 2004; Shavinina & Seeratan, 2004).

Tests should not be very long

According to the ninth principle of a given approach to the psychological assessment of innovative abilities, innovation tests and their subtests *should not be very long* or time-consuming. Long tests can be increasingly boring for examinees. The absence of the elementary interest in tests implies the emergence of such shortcomings as non-actualization of innovative potential, incomplete display of available abilities, and various signs of psychological defenses. With respect to the innovation tests described above (or their subtests), the performance time for each of them should not exceed ten or 20 minutes.

Some tests do not require any time prescription at all, because they are performed extremely rapidly due to their essence. An example of such type of tests is the "Grouping Dots" test, Shavinina and Kholodnaya's (1996) modification of Frumkina's (1984) test. This test is used to measure visual representations. As human cognitive experience expresses itself not only in the verbal and oral modes, visual representations should also be examined. It was found that gifted individuals see, understand, and interpret the surrounding reality and the world around uniquely in the different cognitive modes: in verbal, oral, visual, and spatial ones (Shavinina & Kholodnaya, 1996). These multiple modes or channels explain in part why innovators, who are gifted individuals, have a unique vision of everything around them. Therefore, it is not possible to understand the nature of an individual's cognitive experience and, consequently, his or her innovation without examining the visual type of representations.

"Grouping Dots" test is directed to assessing perceptual structuring under conditions of perceptual uncertainty. As noticed above, innovators often act under conditions of uncertainty. Examinees are given a piece of paper (15×10 cm) with some multitude of dots on it. The dots are arranged in the following way. Some dots are located close to each other forming gestalts; other dots are randomly placed throughout the piece of paper. According to the instruction, each examinee should group these dots by any convenient and natural way according to his or her point of view. The parameter considered is: the complexity of the accomplished structural reorganizations (in quantitative marks). The scoring criteria are: one mark, global grouping (i.e., any attempts of structuring of a given multitude of dots absent); two marks, the unclear grouping of three basic gestalts; three marks, the clear structuring on the basis of grouping of three basic gestalts including the organization of other dots in certain structures (forms); four marks, the multidimensional structuring (i.e., the grouping of three gestalts, overlapping of multitudes of grouped dots, combining the same dots, which have already been attributed, to different groups).

Conclusions

Based on the theory of individual innovation (Shavinina & Seeratan, 2003), the present chapter discussed the development of innovation testing and described one of the important parts of the comprehensive approach to the psychological assessment of innovative abilities in today's children–tomorrow's innovators. This part of the whole approach consists of nine fundamental principles. The principles underlie the development of innovation tests and subtests, some of which were considered above. According to these principles, the tests of innovative abilities should first of all examine the psychological mental context generated by an individual himself or herself. They should have an "open character," evaluate the basis of innovation (not its numerous manifestations), and allow both retrospective and prospective assessment. These tests should not evaluate psychological functions (e.g., memory or attention) and mental speed, and they should not be long or time-consuming.

Definitely, other parts of the approach should include the assessment tools to measure entrepreneurial giftedness, wisdom, excellence, an ability to meet deadlines, as well as all the components of individual innovation (described in Chapter 36, *What can innovation education learn from innovators with longstanding records of breakthrough innovation?* included in this volume) and all the elements of innovation education discussed in Chapter 3 of this volume, *The fundamentals of innovation education*.

In addition, much work has to be done on the standardization, reliability, and validity of these tests. It is important to point out that due to the fundamentally changed approach to the assessment (for example, in these tests examinees themselves should generate some psychological mental context), traditional procedures—reasoning tests or IQ tests—cannot serve as an external validity criterion. For instance, in the experimental studies of gifted adolescents in science their real intellectual achievements in mathematics and physics were used as external validity criteria. In this case the tests described in this chapter have discriminant validity because they allow us to differentiate between ordinary students and students who are the winners or participants in the International Mathematics Olympiads and/or International Physics Olympiads (Shavinina & Kholodnaya, 1996). Certainly, research on external and discriminant validity must be done. In the case of young innovators the signs of early entrepreneurial giftedness can be viewed as real life achievements (see Chapter 3 and Chapter 16, *The trajectory of early development of prominent innovators: entrepreneurial giftedness in childhood* included in this volume). This chapter allowed Vandervert (2012) to conclude that innovation talent can only be identified in the form of "trajectory analysis" or "trajectory assessment."

In accordance with the theory of individual innovation three levels of the manifestations of innovation should also be assessed (e.g., extracognitive abilities). The information about a child's sensitive periods and his or her strong interests should be gathered from parents and other caregivers. These findings are important for understanding the developmental basis of each child's innovation (for more on sensitive periods as the developmental foundation of individual innovation see Chapter 7 on Albert Einstein, this volume).

To conclude, this chapter outlined one of the parts of the whole approach to the identification of innovation talent in children via the measurement of their intellectual abilities. This approach opens one of the possible directions in the development of innovation testing. Of course, the chapter does not account for all possible facets of the measurement of innovative abilities, and sometimes it is both vague and speculative in some of its formulations. However, it provides a useful attempt to understand how the psychological assessment of innovative abilities can be developed in the future.

Acknowledgments

The work reported herein was supported under the Support for Innovative Projects program (Grant AN-143064) of the *Fonds québécois de la recherche sur la société et la culture* (FQRSC). The findings and opinions expressed in this chapter do not reflect the positions or policies of the FQRSC. I would like to thank Larry Vandervert for his detailed feedback on an early draft of this chapter. It was very helpful.

References

Arina, G. A., & Koloskova, M. V. (1989). The cognitive synthesis test. *Voprosu Psichologii, 3*, 171–172.
Barfurth, M. A., Ritchie, K. C., Irving, J. A., & Shore, B. M. (2009). A metacognitive portrait of gifted learners. In L. V. Shavinina (Ed.), *International Handbook on Giftedness* (pp. 397–420). Dordrecht: Springer Science.
Baron, J. (1982). Personality and intelligence. In R. J. Sternberg (Ed.), *Handbook of Human Intelligence* (pp. 308–351). Cambridge: Cambridge University Press.
Brown, A. L. (1978). Knowing when, where, and how to remember: A problem of metacognition. In R. Glaser (Ed.), *Advances in Instructional Psychology: Vol. 1* (pp. 77–165). Hillsdale, NJ: Erlbaum.
Brown, A. L. (1987). Metacognition, executive control, self-regulation, and other even more mysterious mechanisms. In F. E. Weinert & R. H. Kluwe (Eds.), *Metacognition, Motivation, and Understanding* (pp. 64–116). Hillsdale, NJ: Erlbaum.
Csikszentmihalyi, M. C. (1992). Motivation and creativity. In R. S. Albert (Ed.), *Genius and Eminence* (pp. 19–33). Oxford: Pergamon Press.
Davidson, J. E. (1986). The role of insight in giftedness. In R. J. Sternberg & J. E. Davidson (Eds.), *Conceptions of Giftedness* (pp. 201–222). Cambridge: Cambridge University Press.
Dewey, J. (1919). *How We Think*. Boston: Heath.
Dewey, J. (1938). *Logic: The Structure of Inquiry*. New York: Putnam.
Einstein, A., & Infeld, L. (1938). *The Evolution of Physics*. New York: Simon & Schuster.
Flavell, J. H. (1976). Metacognitive aspects of problem solving. In L. B. Resnick (Ed.), *The Nature of Intelligence* (pp. 231–235). Hillside, NJ: Erlbaum.
Frumkina, R. M. (1984). *Color, Meaning, and Similarity: Aspects of Psycholinguistic Analysis*. Moscow: Nauka.
Getzels, J. W., & Csikszentmihalyi, M. (1976). *The Creative Vision: A Longitudinal Study of Problem-finding in Art*. New York: Wiley.
Kagan, J. (1966). Reflection-impulsivity of conceptual tempo. *Journal of Abnormal Psychology, 71*(1), 17–24.
Kagan, J., Rosman, B., Day, D., Albert, J., & Phillips, W. (1964). Information processing in the child: significance of analytic and reflective attitudes. *Psychological Monographs, 78*, 1 (Whole No. 578).
Kholodnaya, M. A. (1983). *The Integrated Structures of Conceptual Thinking*. Tomsk: Tomsk University Press.
Kholodnaya, M. A. (1990). Is there intelligence as a psychical reality? *Voprosu Psichologii, 5*, 121–128.
Kholodnaya, M. A. (1993). Psychological mechanisms of intellectual giftedness. *Voprosu Psichologii, 1*, 32–39.
Kholodnaya, M. A. (1997). *The Psychology of Intelligence*. Moscow: IPRAN Press.
Oatley, K. (1978). *Perceptions and Representations*. Cambridge: Cambridge University Press.
Olson, D. R. (1986). Intelligence and literacy: The relationships between intelligence and the technologies of representation and communication. In R. J. Sternberg & R. K. Wagner (Eds.), *Practical Intelligence* (pp. 338–360). Cambridge: Cambridge University Press.
Polak, F. L. (1973). *The Image of the Future*. New York: Elsevier.
Pressley, M., Borkowski, J. G., & Schneider, W. (1987). Cognitive strategies: Good strategy users coordinate metacognition and knowledge. In R. Vasta (Ed.), *Annals of Child Development: Vol. 4* (pp. 89–129). Greenwich, CT: JAI Press.
Rabbitt, P. (1996). Intelligence is not just mental speed. *Journal of Biosocial Science, 28*(4), 425–449.
Shavinina, L. V. (1993). Cognitive experience of intellectually gifted individuals. Unpublished doctoral dissertation. Kiev, Ukraine: Institute of Psychology.
Shavinina, L. V. (1996). The objectivization of cognition and intellectual giftedness. *High Ability Studies, 7*(1), 91–98.
Shavinina, L. V. (2003). *The International Handbook on Innovation*. Oxford: Elsevier Science.
Shavinina, L. V. (2012). The emergence of a new research direction at the intersection of talent and economy: The influence of the gifted on economy. *Talent Development and Excellence, 4*(1), 65–88.
Shavinina, L. V., & Ferrari, M. (Eds.) (2004). *Beyond Knowledge: Extracognitive Aspects of Developing High Ability*. Mahwah, NJ: Erlbaum Publishers.
Shavinina, L. V., & Kholodnaya, M. A. (1996). The Cognitive experience as a psychological basis of intellectual giftedness. *Journal for the Education of the Gifted, 20*(1), 3–35.
Shavinina, L. V., & Seeratan, K. (2003). On the nature of individual innovation. In L. V. Shavinina (Ed.), *The International Handbook on Innovation* (pp. 31–43). Oxford: Elsevier Science.
Shavinina, L. V., & Seeratan, K. (2004). Extracognitive phenomena in the intellectual functioning of creative and talented individuals. In L. V. Shavinina & M. Ferrari (Eds.), *Beyond Knowledge: Extracognitive Aspects of Developing High Ability*. Mahwah, NJ: Erlbaum Publishers.

Shore, B. M., & Dover, A. C. (1987). Metacognition, intelligence and giftedness. *Gifted Child Quarterly, 31*, 37–39.
Shore, B. M., & Kanevsky, L. S. (1993). Thinking processes: being and becoming gifted. In K. A. Heller, F. J. Mönks, & A. H. Passow (Eds.), *International Handbook of Research and Development of Giftedness and Talent* (pp. 133–147). Oxford: Pergamon Press.
Silverman, L. (2009). The measurement of giftedness. In L. V. Shavinina (Ed.), *International Handbook on Giftedness* (pp. 947–970). Dordrecht: Springer Science.
Sternberg, R. J. (1985). *Beyond IQ: A Triarchic Theory of Human Intelligence*. Cambridge: Cambridge University Press.
Sternberg, R. J. (1988). *The Triarchic Mind: A New Theory of Human Intelligence*. New York: Viking.
Torrance, E. P. (1979). Gifted students want to be on the latest frontiers. *Gifted Child Quarterly*, May/June, 10–11.
Torrance, E. P. (1980). Creativity and futurism in education: Retooling. *Education, 100*(4), 298–311.
Vandervert, L. R. (2012). Personal e-mail Communication. March 30.
Vekker, L. M. (1981). *Psychological Processes, Vol. 3*. Leningrad: Leningrad University Press.
Vernadsky, V. I. (1988). *Diaries and Letters*. Moscow: Molodaya Gwardiya.
Vygotsky, L. S. (1978). *Mind in Society: The Development of Higher Psychological Processes*. Cambridge, MA: Harvard University Press.
Wertheimer, M. (1959). *Productive Thinking*. Westport: CT: Greenwood Press.

Part V
From advances in giftedness and gifted education to innovation education

13

Innovation education
Perspectives from research and practice in gifted education

Lynn H. Fox
AMERICAN UNIVERSITY, USA

Summary: This chapter reviews major findings, successful strategies, and conclusions from over 50 years of gifted and talented education research and programming. Research findings from study of mathematically precocious students and STEM educators' work, which fosters interest and excellence in science, technology, engineering and mathematics education, are particularly relevant for the emerging Innovation Education discipline. STEM and gifted education discoveries are important for Innovation Education's conceptualization as well as program implementation and practice. Insights and efficacious action plans can be gleamed from the history of gifted education curriculum and instruction as well as by reviewing gifted and talented debates over different approaches and goals. As educators seek to develop program models and strategies for Innovation Education, particularly with respect to the issue of diversity, studying gifted and talented program successes and failures will help in the formulation of operational definitions and successful programs.

Key words: Gifted education, mathematically gifted, SMPY, gender differences, STEM education, problem-based learning, diversity and equity, curriculum differentiation.

Introduction

Regardless of one's view of the relative roles of aptitude, or nurture in the development of "innovators," efforts to create educational programs to foster their development can be informed by the efforts of nearly 50 years of gifted education research. Indeed models or strategies for education for innovation can be derived from the study of successful gifted programs, particularly in the areas of creativity and giftedness in science, technology, engineering and mathematics (STEM).

This chapter will first consider two different approaches to gifted education for and how they might be of interest as models for education for innovators. The first set of programs or interventions are those which target a relatively small portion of the population. These are particularly interesting when resources are limited. In this case the target is a subset of students, ideally those who might have the most potential or are thought to be the most likely to benefit from programs or interventions. For the purpose of this chapter this will be called the "Identification and Facilitation" model of giftedness. This model includes specific programs and some more generic approaches such as STEM schools or summer programs.

The second more inclusive model might be called "Systemic Reform" or "All Children Have Potential." This model focuses on school system-wide educational reform efforts for all age/grade and administrative levels. This approach within the gifted child education movement, especially in the areas of creativity and curriculum differentiation, may be of interest to those who advocate systemic reform to foster a more innovative society.

The chapter will conclude with a discussion of equity and access issues, particularly issues of gender, culture, and ethnicity, that have plagued the gifted education movement and is likely to find a parallel in the Education for Innovation arena.

Programs that focused on identification and facilitation of talent

Much of the past 100 years of research in giftedness has focused on issues surrounding conceptual and operational definitions of giftedness. Some have focused on the psychology of giftedness in terms of both cognitive and affective characteristics from intensive study of gifted adults, while other work has involved the longitudinal tracking of children thought to have potential in childhood and/or those who have participated in special programs.

Within the field of gifted education and research there has been a general shift away from the notion of "general intelligence" and identification of children as gifted based on single measures of global intelligence. Instead the focus has been on either looking for achievement or potential within specific content domains, or upon constructing multi-dimensional views of intelligence and giftedness.

Although there may be a generalizable quality that will produce "innovators," identification of potential innovators within specific domains may be the most productive approach. This debate is likely to echo ones from the gifted movement. From the perspective of efficiency and effectiveness it is likely to be far easier to focus on specific talent domains. It will be easier in terms of developing conceptual definitions, operational definitions, and easier in terms of conceptualizing intervention or program specifics.

The Study of Mathematically Precocious Youth (SMPY)

One program in gifted education that has a long history of research and a variety of successful intervention strategies is the Study of Mathematically Precocious Youth (SMPY). SMPY was begun in the fall of 1971 by Julian Stanley and his associates, Daniel Keating and Lynn Fox, at Johns Hopkins University (Stanley, Keating, & Fox, 1974; Stanley, 1996). Originally its intent was to locate mathematically and scientifically gifted children, but it gravitated to a focus on mathematics as it became clear that mathematical precocity was more readily identified than talent in science. SMPY was the pioneer in creating a talent search approach to gifted identification, an advocate for both radical acceleration and diagnostic-prescriptive instruction, and committed to a program of longitudinal research.

Identification

SMPY sought to find students who reasoned well in mathematics and by virtue of this ability had become precocious in their understanding of mathematics. This identified youth who were not only knowledgeable about mathematics beyond their grade placement but those who were likely to learn mathematics at a pace much faster than the norm. Although widely known for introducing the notion of using "difficult" tests or tests with more "ceiling" for identification when it began to use the Scholastic Achievement Test (SAT), intended for college admission screening, with 7th graders, early work by the SMPY team also included extensive testing of high scoring students on a battery of measures including measures of nonverbal reasoning, spatial ability, interests, and values. The use of the college-admission screening test, SAT, for talent searches among 7th graders was a radical shift from the use of Intelligence tests so often used in gifted education. The Center for Talented Youth (CTY) at Johns Hopkins University continues to conduct the talent search and to offer classes for students and has developed an international presence as well. This talent search approach using the SATs has been embraced by several other universities throughout the United States.

Findings about the characteristics of those with superior mathematical reasoning were:

- Students selected as mathematically precocious based on their performance on tests can succeed in college courses (Keating, Wiegand, & Fox, 1974) and can succeed in accelerated mathematics classes where they might learn an entire Algebra I course in nine weeks meeting just two hours a week (Fox, 1974; George & Denham, 1976).
- The model first piloted for accelerated mathematics classes also works for physics and chemistry classes (Stanley & Stanley, 1986).
- Self-selection, risk-taking, parental support and willingness to work are important factors in determining who will succeed in challenging accelerated programs (Fox, 1974; Fox, Brody, & Tobin, 1984–85).
- Measures of values, interests, and personality indicated that mathematically gifted boys and girls were often similar to other samples of gifted students. Gifted girls were often more like gifted boys than non-gifted cohorts. Indeed there were, as early as grade 7, some similarities in personality profiles of these students and adult samples of creative mathematicians (Fox, 1976a, 1976b; Fox & Denham, 1974; Fox, Pasternak, & Peiser, 1976; Weiss, Haier, & Keating, 1974).

A variety of interventions

Stanley's initial interest was in counseling youths, particularly the extremely precocious students and their families, and when appropriate he encouraged radical acceleration. Indeed SMPY encouraged what was termed a "smorgasbord" of educational opportunities, most of which did involve some acceleration of the rate or level of learning (Benbow & Lubinski, 1997; Stanley et al., 1974). Although not the first to promote early college admission, SMPY was the first to champion radical efforts such as admitting 14 year olds to college straight from junior high school. Through the years Stanley moved toward less dramatic acceleration and endorsed less radical transitions and encouraged efforts to provide special schools for the gifted. In later years Stanley began a project, the Study of Exceptional Talent (SET) to search for even younger prodigies, those who scored 700 or higher on either the mathematics or verbal portions of the SAT before the age of 13 (Stanley, 1988; Brody, 2005).

The original focus of SMPY, which continued with SET, was to find the ideal combination of formal and out of school experiences, largely accelerative ones, for each student based on their needs and abilities. In the early years Stanley had a few students attempt to earn college credit by correspondence courses as well as by taking regular college courses while still fulltime enrolled in middle school. The CTY spinoff talent search portion of SMPY now offers its own online distance programs embracing the new technology in ways that were not available in the early years of SMPY.

Another SMPY innovation was the concept of diagnostic–prescriptive instruction. Students could master a full course of mathematics in a few weeks in the summer or in a series of Saturdays during the school year (Fox, 1974; George & Denham, 1976; Stanley, 1991, 1996). These telescoped classes while highly effective were only rarely implemented by school systems who were concerned about what to do with those who fell to the wayside and did not transition easily back into the regular "lock-step" curriculum. Although SMPY paid relatively little attention to creativity per se, it is likely that most of the students who chose to participate, especially those who attempted to circumvent the lock-step of age-graded dominated education, were those who did indeed possess some of the characteristics associated with creativity such as risk-taking and divergent thinking.

Research

Some of the work in the early years of SMPY included extensive testing of students on a variety of interest and personality measures. Longitudinal follow-up of students is now at a stage where it has produced a great deal of interesting insights into the nature of mathematically precocious youth and the impact of various program opportunities upon them. Findings that may be of interest for innovation education are as follows:

- The SAT scores of 7th graders were excellent predictors of college majors and achievements and career interest after college. The talent search cohorts followed through college were far more likely than the general college population to major in mathematics, computer science or a physical science (Benbow & Lubinski, 1994; Benbow, Lubinski, Shea, & Eftekhari-Sanjani, 2000; Lubinski & Benbow, 1992).
- Those students who took advantage of some of the accelerative options suggested to them based on their scores in the talent search had higher GPAs in college, attended more selective colleges and won more honors than students with similar scores who did not avail themselves of accelerative opportunities (Brody & Benbow, 1987).
- Those students who entered Johns Hopkins University early were more likely than regular students to win honors, earn Master's degrees simultaneously with their Bachelor's degree, and be elected to honor societies such as Phi Beta Kappa (Brody, Assouline, & Stanley, 1990).

The enrichment triad and the school-wide enrichment approach

Elsewhere in this volume there are two chapters which describe the work of Joseph Renzulli at the University of Connecticut. Like SMPY the Renzulli Enrichment models have had a huge impact within gifted education. This approach differs from the SMPY program focus. While SMPY, CTY and SET focus on the top 1% or less of the general population and Renzulli has usually designated his model as appropriate for the top 15 to 20% of a school population. The Renzulli program tends to look at enrichment within specific content areas and has addressed areas other than mathematics and science. It has always included creativity as both a component for identification of students and as one of the curricular and instructional components of the model.

Although the SMPY did create a model of accelerated mathematics classes that was adopted by a few schools or school systems, the major thrust of SMPY's efforts was on suggesting a wide range of possible ways to navigate through and sometimes around the school system. For example, SMPY might encourage students to graduate from high school a year or two early or consider dual enrollment in college during their last years of high school. Renzulli by contrast has designed a program for schools that enriches without necessitating acceleration. One advantage some school systems see with this approach is that students may flow in and out of the program without any disruption to their progress through the curriculum or special administrative adjustments.

Special schools

There are at least three interesting variations of the special schools approach that should be considered. There are Schools for the Gifted, Schools with a specific focus such as STEM schools, and special tracks or programs within schools. The distinctions between and among these are sometimes blurred. A few public schools for the gifted exist like Pine View in Sarasota Florida that encompasses all academically gifted from 2nd to 12th grade using tests of intelligence and achievement to evaluate for admission. Although some accommodations might be made for individuals, it basically assumes it has an advanced program for all students in every subject area, most high school offerings are Advanced Placement courses. A STEM school may allow for more self-selection and not use intelligence measures for admission (Thomas & Williams, 2010). Emphasis is on providing a richer array of courses in mathematics, technology and the sciences than is typically offered in high schools. The International Baccalaureate program would be an example of a school with a special program or track that has its own curriculum and screening criteria. Graduates of this track are often accepted with advanced standing in many colleges and universities.

Extra-curricular programs

Many gifted education opportunities have been presented to students as after-school, weekend, or summer experiences that are independent of their K-12 curriculum. Although a few may allow for subject matter acceleration, most focus on experiences thought to enrich, expand or supplement the regular school curriculum. Some of these are linked to competitions such as the Math Olympiads or Odyssey of the Mind whereas others might be offered by colleges and universities, science centers, visual and performing arts centers, or other organizations. Some like those tied to the Talent Searches may have rigorous screening procedures. Others may be more open to self-selection or less formal forms of screening such as a teacher nomination. Some have rather specific curricular focus such as ones offered along the lines of the NASA space camps, or creative writing programs. Typically these programs do not document outcomes other than student satisfaction with the program. A probable major benefit of participation in extra-curricular programs is the opportunity to interact with other like-minded and like-talented peers. Anecdotal accounts from participants indicate that they develop long-term friendships and the sense of finally finding a peer group that makes them feel like they belong.

Synthesis across the identification and facilitation models

Approaches to gifted education, which emerged from different emphases on enrichment or acceleration, have elements that could be adapted to deliver interventions aimed at promoting the development of innovators. Indeed, such efforts might want to join forces with one or more of the existing programs. For example, SMPY and the other talent search programs already identify potential innovators, especially in the areas of science. Creating summer, weekend or distance-learning opportunities to target these youth with programs that enhance creative problem-solving would be relatively easy. Some might argue that their current offerings already provide this. Special schools or schools within schools with an Innovation focus might be developed. Indeed existing STEM schools should be encouraged to include more emphasis on curriculum and instruction that will promote the development of "innovators." Indeed some of the ideas from the Renzulli models, such as the notion of the "revolving door," could be incorporated into STEM schools to create a flow of students in and out of experiences designed to foster innovation.

The systemic reform approach

If one assumes all children have potential to become innovators then systemic reform is the answer. Among educators and researchers in gifted there have been those who believed "all children are gifted" or a slight variation such that what is advocated for the gifted learner is overall good pedagogical practice. Many would say gifted programs can be easily adapted for use with all learners. There are two broad categories of such programs. One set is focused on how material is taught and the importance of creative thinking or problem-solving skills, the other group is more closely tied to curriculum development and specifically the differentiation of curriculum to meet the needs of all students.

Problem-based learning

Problem-based learning is an instructional approach based on constructivist views of learning. It involves authentic but often unstructured problems posed by teachers to students that lead to explorations in one or more curricular area. Ideally problems are ones for which there are multiple solutions or paths to solutions. It is typically done in ways that encourage collaboration and cooperative learning (Gallagher, 1997). This approach has been tried in elementary, secondary and higher education. More teacher-scaffolding is needed at the elementary school level. At all levels there is an initial period of grappling with the problem to define it, data collection and possibly brain-storming and finally trial and error testing of possible solutions.

Creative problem solving

One very specific program, using a problem-based approach from the field of creativity, is Creative Problem-Solving (CPS). This program originated in the early 1950s with the publication of two books by Alex Osborn: *How to become more creative* (1952) and *Applied imagination* (1953). This is thought to be the work that launched interest in the notion of group problem solving that has come to be known as "brainstorming." It has gone through several iterations since then and for a while was referred to as the Osborn–Parnes model (Treffinger & Isaksen, 2005). Treffinger and associates have launched what is considered the 6.1 version of the model which is the thirteenth iteration since 1952 (Isaksen, Dorval, & Treffinger, 2011). The newest version is said to improve upon previous versions in several ways. First, it moves away from a series of five steps that are followed in sequence to a more dynamic process where steps need not follow a specific order or all be used. Second, the newer version puts more emphasis on the onset where the problem needs to be examined and analyzed to more clearly define it. Third, there is an effort to balance the importance of both divergent and convergent thinking in the process of problem exploration, data gathering, solution generation, and solution evaluation. Finally, there is some effort to de-emphasize the role of teacher/leader and to consider developing more self-awareness of meta-cognition and executive function (Treffinger, Isaksen, & Dorval, 2006).

Project-based learning

Although sometimes treated as interchangeable with problem-based learning, project-learning is sometimes used to describe methods of teaching that are not as inquiry based as most problem-based instruction. For example, in recent years some high schools have implemented Senior Thesis Projects patterned after college-level Senior Thesis Projects. Students may work in different content domains. Projects may be more exploratory and not necessarily framed around finding a solution to a specific problem.

Curriculum differentiation

There are numerous models of curriculum differentiation. Even those specifically designed for the gifted often note that their model can be adapted for all learners, primarily by adjusting the level or depth of study of the content area (VanTassel-Baska & Stambaugh, 2008). What is of interest for education for innovation is how these programs can be implemented within regular schools and classrooms. They provide challenging learning opportunities for all students while controlling or adapting the instruction for a wide range of individual differences in terms of students' prior knowledge and experience with content. In some cases they offer multiple paths to learner outcomes. A brief description of several often used models follows.

The Integrated Curriculum Model (ICM)

The ICM developed by Joyce VanTassel-Baska (1986) was initially developed for high-ability students. It has shown potential to be effective with all learners given certain teaching modifications (VanTassel-Baska & Stambaugh, 2008). The model is built around three dimensions. First is the advanced content focus in basic subjects. Second is work in critical thinking and problem solving. Third, it seeks to promote both intra- and interdisciplinary concept development. It has been implemented in hundreds of schools in several different content areas and in both elementary and secondary programs.

There is a growing body of evaluative research including experimental and quasi-experimental studies that demonstrate the effectiveness in terms of a variety of different outcomes. It has been shown to be effective with problem-based science curriculum, in language arts, reading, social studies and on measures of critical thinking (VanTassel-Baska & Brown, 2007; VanTassel-Baska, Feng, & de Brux, 2007; VanTassel-Baska, Feng, & Evans, 2007; VanTassel-Baska & Stambaugh, 2008).

The Parallel Curriculum Model (PCM)

Tomlinson (1995, 1996) has proposed numerous ways for teachers to differentiate curriculum and instruction in the classroom by adjusting levels of difficulty of material around the same broad content area and utilizing individualized and cooperative group processes to deliver instruction. In collaboration with others Tomlinson (Tomlinson & McTighe, 2006; Tomlinson et al., 2002) has proposed one specific model that is particularly useful for adapting for the needs of the advanced learner within the regular classroom. PCM stresses the importance of linking core curriculum to professional standards at the national, state or local level for each content domain. Second, it supports an interdisciplinary approach to support connectedness among and between domains of knowledge. It promotes an emphasis on the application of knowledge. It emphasizes the need for connections between individual learners' lives and the curriculum. Although her work has been widely used in professional development workshops and adopted by many teachers, there is as yet, a paucity of research to address the impact of this approach.

Successful intelligence

Robert Sternberg has written extensively on the constructs of intelligence, giftedness and creativity. In his early writings he focused on the Triarchic model of giftedness as a combination of analytical ability, creative thinking and practical intelligence (Sternberg, 1986). Some of his more recent writings have focused on his notion of *Successful Intelligence* that depends upon an integrated set of abilities, the ability to understand one's own strength and weakness and the metacognitive awareness of how to compensate for weakness or maximize strengths; it requires the ability to either adapt to, select or shape the environment; and it requires the balance of the analytical, creative and practical aspects of intelligence (Sternberg, 1997, 2008). A related model for giftedness in leadership, Wisdom, Intelligence, and Creativity Synthesized (WICS) expands upon successful intelligence. What may be of particular interest for innovation are his concept of types of creative leadership: conceptual replication, redefinition, forward incrementation (i.e., moving a group project forward when it encounters problems), advance forward incrementation, redirection (i.e., moving a project forward but in a new direction), reconstruction/redirecton, reinitiation (i.e., when a project must transform itself or die), and synthesis. Sternberg also suggests ten skill sets, or mind sets, related to creative leadership and has given examples of how these might be taught or developed (Sternberg, 2000a, 2000b, 2005). Some of these seem to echo the types of components discussed by others in terms of creative problem solving or creative personality such as the ability to redefine problems or questions, tolerance for ambiguity, ego-strength or willingness to take risks. Other ideas for teaching are more unique such as teaching "how to sell your ideas or creation," or "how to surmount obstacles." Some projects that have evolved from this work have also dealt directly with the issues of diversity and inclusion or access and will be discussed in the next section.

Synthesis across the systemic models

Many of the goals espoused for education for innovation could readily be met through the use of curriculum and program models that have been developed over the past 50 years. Teaching with differentiated curriculum, creative problem solving and the notion of successful intelligence and WICS all seem to have direct applications for programs aimed at increasing flexible thinking, divergent thinking, and problem-solving strategies and skills.

Issues of diversity

Issues of diversity have been a major source of controversy both within and outside the gifted education community. It is sometimes the reason school systems abolish or abandon gifted and talented programs (Robinson, 2008). Topics for Innovation Education are research on gifted women and girls including barriers to success, especially in the STEM areas, and research on low-income or ethnically diverse students, sometimes referred to as the culturally and linguistically different (CLD).

Women and girls

Efforts to discover and nurture potential innovators can be informed by research on gender in both gifted and STEM education programs. Although women and girls comprise slightly more than half the general population, they were often under-represented in programs for the gifted. Unfortunately their lack of achievements were for many years treated as normal. Despite efforts to increase the participation of women and girls in all educational arenas as well as in the worlds of business, science, applied sciences, technology, the arts, social sciences and sports, discrepancies continue in favor of males in terms of career outcomes and recognition for eminence (Arnold, 1993, 1995; Kirschenbaum & Reis, 1997; Piirto, 1991; Reis, 2002, 2003; Subotnik & Arnold, 1996).

Differences in aptitude and achievement, especially as measured in early and middle childhood, are small but do tend to increase in adolescence (Hyde & Lindberg, 2007). This has led to research on the importance of social and cultural factors in shaping gender outcomes. In fact, many achievement differences such as scores on tests, course-taking in high school, college majors and career choices have changed over the past 40 years. For example, the numbers of girls and boys enrolled in Advance Placement (AP) courses and their scores on the (AP) tests have become more similar over time. A few differences remain, notably in Physics, Computer Science and the higher level of examination for Calculus (BC). Woman now outnumber men in bachelor degree programs in almost every field except engineering, computer science and physics. The gap does remain, however, in all STEM areas except Biology at the PhD levels.

Studies of the mathematically precocious continue to find rather large differences at the highest levels of performance. On the SAT the ratio of males to females is about 2:1 for the tenacious 7th graders who enter a talent search and score 500 or above on the Mathematics portion of the SAT; this becomes 4:1 at the cutoff score of 600 and about 13:1 for scores at or above 700. Yet among these high scorers there are few if any differences on verbal measures or measures of abstract problem solving (Fox & Cohn, 1980; Lubinski & Benbow, 1992; Lubinski, Benbow, & Morelock, 2000; Stanley, Benbow, Brody, Dauber, & Lupkowski, 1992). There are, however, huge differences on spatial ability or mechanical comprehension measures. It is possible that these latter measures capture differences in early learning that is socially driven such as play with dolls versus play with building materials or trucks. Longitudinal follow-up of students from SMPY may provide some insight into how spatial abilities contribute to career choices and to achievement levels within those fields.

Studies of gifted students typically find gender in the affective domain which may relate to achievement outcomes (Fox, 1977; Freeman, 2004). Certainly in the research on gender in STEM areas several hypotheses have been explored. In the following sections the barriers to achievement will be discussed as well as some interventions that have been tried to improve the educational opportunities and experiences for gifted girls.

Gender-related barriers to success

Freeman (2004) argues that socialization accounts for most if not all of the outcomes differences as evidenced by the fact that gender gaps in outcomes vary from country to country in size and directions. Research has point to a variety of factors that create barriers for gifted girls and women (Fox & Soller, 1999, 2007; Kerr & Foley-Nicpon, 2002; Reis, 2002, 2003),

Self-confidence

Numerous studies have focused on self-confidence or lack thereof among gifted women and girls, especially in the STEM areas. Despite public recognition for achievement in high school, female valedictorians showed a decline in self-confidence during the college years (Arnold, 1993, 1995). Lack of confidence is not just in the STEM areas. Piirto (1991) found that although there were few gender differences among gifted art students, females lacked self-confidence and often choose to teach rather than face the challenges of being a professional artist.

Family and career conflict

Many researchers have argued that the failure of gifted women to achieve is a result of the conflicting demands of career and family responsibilities. Piirto (1991) suggested that childbearing occurs during peak production years and this accounts for differences in productivity between creative male and female artists. The necessity to achieve early and to continue producing in creative endeavors is a problem for women who choose to have families and thus are not able to be as single-minded in the pursuit of their career.

Interviews with hundreds of gifted women conducted by Reis (2002) suggest how family and career demands impact creative achievement and the attainment of eminence. While creative women must balance work and family, men tend to define an end goal and move persistently to complete the goal. Arnold (1993, 1995) found that two-thirds of the women high school valedictorians planned at some point to reduce or interrupt their careers due to child rearing responsibilities.

Interests and values

Interests and values differ among the gifted. One study of the personalities of mathematically gifted youth found more similarities than differences between gifted girls and boys (Benbow & Lubinski, 1994), but notable differences were found on measures of career interests and related values. Even though both males and females scored high on measures of interest in investigative careers and theoretical values, gifted girls far more than boys had equally strong, competing interests and values, particularly social and artistic ones (Fox, 1976a, 1976b, 1977; Fox & Denham, 1974; Fox et al., 1976). Yet these differences are sometimes not as great between the sexes in samples of mathematically gifted adolescents as they are in the general population.

Stereotypes

Although most scientists may believe in social more than biological explanations of gender differences in adult achievements, the general population still harbors some sex appropriate notions of behavior with underlying assumptions of "biology as destiny." Both parents and teachers sometimes discourage girls' interest in science, engineering and technology while discouraging boys from developing their aesthetic interests and talents. Stereotype images of the role and duties of a "scientist" may also lead to well-intentioned but misguided advice about course-taking and careers from educators and family. The notion of stereotypic threat and its implications for minorities and women has been widely discussed (Steele, 1997; Steele & Aaronson, 1995). In STEM areas this phenomenon is reinforced by the lack of role models and mentors for women and girls.

Bias

Stereotypes can lead to discrimination. Teachers may reward intellectual curiosity and academic achievement for boys more than girls, especially in STEM areas (Bianco, Harris, Garrison-Wade, & Leech, 2011; Fox, Sadker, & Engle, 1999). It has been suggested that giftedness goes unrecognized among females and some minority group males and this leads to under-representation in gifted programs (Crombie, Gouffard-Bouchard, & Schneider, 1992; Daignault, Edwards, Pohlman, & McCabe, 1990). A recent study found that when teachers were given student profiles to read and make recommendations for admittance to gifted programs, they were more likely to recommend a student who was designated as male than female, even though the profiles were otherwise identical (Bianco et al., 2011).

Interventions

A number of program models have been used with girls, particularly ones targeting above average ability and gifted ones in the areas of mathematics and science. Some efforts have started as early as the elementary school years and others have focused on middle school, high school and college. Accelerative mathematics programs have been successful with some girls but the many of the most successful interventions have focused on counseling about career opportunities with mentors and role models (Fox, 1977; Fox & Brody, 1980; Fox et al., 1984–85; Fox & Tobin, 1988; Fox & Soller, 2007; Kerr & Kurpius, 2004).

Research on CLD students

CLD students are generally under-represented in programs for the gifted and talented because they tend to score below cutoff scores on intelligence tests and many achievement measures. Although approximately 60% of the elementary and secondary school population is white, about 73% of those identified as gifted are white (Ford & Whitney, 2009). While Asian/Pacific Islanders make up less than 5% of the population, they account for almost 8% of the gifted. The remaining 19 to 20% of the gifted are Black, Hispanic/Latino or Native American but they are approximately 36% of the total school population.

Efforts to include a more diverse group of students in gifted programs have included using non-verbal tests, checklists and teacher and parent nominations. Unfortunately only a limited amount of research documents the success in programs of students identified by such alternative methods. Two approaches that have shown promise and should be of interest for identification of potential innovators use performance-based assessments. One project was developed in South Carolina in elementary and secondary schools. The second approach has two variations based on Robert Sternberg's notion of Successful Intelligence and his WICS model of creative leadership and have been tried in several settings, largely with young adults.

South Carolina

In 1998, the State Department of Education in South Carolina commissioned the Center for Gifted at the College of William and Mary to develop assessment procedures that would improve upon their ability to identify African-American and low socio-economic status students. The Center created a series of open-ended problems that are analyzed in terms of the process skills the students use to attack the problems (VanTassel-Baska, Bracken, Feng, & Brown, 2009).

A six-year follow-up study found an increase from 18 to 23% in the participation of low-income students and an increase from 11 to 14% for African Americans. Although the students identified by the alternative tasks did not scores as well as those identified by traditional tests on post-achievement measures in mathematics and language arts in the follow-up the difference in performance was small. Several positive outcomes were reported such as improved self-esteem and motivation for those identified by the alternative method (VanTassel-Baska, Feng, & de Brux, 2007; VanTassel-Baska, Feng, & Evans, 2007; VanTassel-Baska et al., 2009).

Successful Intelligence and WICS

Based on the model of Successful Intelligence, Sternberg developed a test that assessed analytical, creative and practical skills. In an effort called Project Rainbow, the test was administered to about 1000 students in two high schools and 13 colleges. The test proved to be almost twice as effective as the SATs in predicting freshman year in college GPAs and substantially reduced difference in scores among students of different ethnic and cultural groups (Sternberg, 1997, 2007). In this instance the assessment was administered as a proctored test.

A few years later in a project called Kaleidoscope the questions were incorporated into the college application form at Tufts University. Tasks include writing a short story to go with a provocative title that is provided, or take a historical event and discuss the possible impact had the outcome been different, and invent a new product or create an advertising plan for a new product. Using the WICS model the answers were scored on rubrics that separately assessed the dimensions of wisdom, practical intelligence and creativity. The results were that 30% more African-American students were admitted and 15% more Hispanic/Latino students than in the previous year (Sternberg, 2007, 2008). Predictions that using this method would result in lower SAT scores for the freshman class than in previous years were not supported, indeed the reverse was true.

Lesson to be learned about diversity

Clearly the types of performance-based assessments used by VanTassel-Baska and colleagues in South Carolina and by Sternberg and colleagues in multiple settings could readily be adapted to screen for students with potential for innovation. Indeed the notion of scoring for dimensions of analytical, creative and practical intelligence would seem to have value for innovation education. Tasks such as design a new product or a plan to market a new product seem very appropriate. These tasks might be used in conjunction with more traditional measures of science and mathematics achievement to facilitate placement for the right level of challenge in programs. If use of these non-traditional assessments leads to greater diversity, especially in terms of socio-economic and cultural diversity, they would be of great value.

The problem with gender is somewhat different. Although there may be some issues in finding equal numbers of girls and boys if the SMPY approach is used, more in grade level measures of achievement are not likely to show this discrepancy. If self-selection is part of the process for participating in the program, then gender could become an issue depending on how the program is described and advertised. Careful thought needs to be given to recruitment strategies to ensure female participation. Retention in programs may be enhanced for girls if attention is paid to providing career counseling with role models and mentors to help gifted girls and women understand their choices and opportunities. Teachers need to be aware of the issues of gender and be sure they are encouraging participation equally for males and females as well as incorporating examples of the work of both male and female scientists and mathematicians in their materials (Fox, Engle, & Soller, 1999; Fox, Sadker, & Engle, 1999; Fox & Soller, 1999; Fox & Tobin, 1988; Fox & Zimmerman, 1985).

Conclusion

If one is looking to create a program that targets and nurtures those with great potential, then the ideal model for innovation education would be a hybrid of several gifted programs summarized in this chapter. For example, using SMPY's identification of talent supplemented with either the VanTassel-Baska or Sternberg problem-based assessments should yield a group of students who are intellectually able, creative, motivated and diverse. Delivery of the program might use a combination of the STEM school with the diagnostic–prescriptive approach to allow for the appropriate pace or level of instruction in the content while using Renzulli's approach of having students flow in and out of experiences. Certainly some of the existing gifted curriculum and creative problem-solving materials already in use in some schools could be incorporated into programs for innovation.

For those more interested in systemic reform and inclusion, a focus on problem solving for innovation across the curriculum is possible. Models for teacher training based on the notion of differentiated instruction are already developed and in use. Materials and curriculum guides such as those already developed and discussed herein are available that focus on problem finding and problem solving. Although the issues of diversity would seem less problematic in a systemic reform effort, care needs to be given to inclusiveness in subtle ways. Attention should be paid to providing diverse role models and ways to motivate or empower girls and CLD students.

The goal of increasing productivity and creative problem-solving capacity within the social spheres of life as well as the sciences is important. Innovation Education can learn from its STEM and gifted education forerunners. Careful reading and application of gifted education research, particularly research on the mathematically talented, provides a guide for both planning and implementation.

References

Arnold, K. (1993). Undergraduate aspirations and career outcomes of academically talented women. *Roeper Review, 15*(3), 169–174.

Arnold, K. (1995). *Lives of Promise*. San Francisco: Jossey-Bass Publishers.

Benbow, C. P., & Lubinski, D. (1994). Individual differences amongst the mathematically gifted: Their educational and vocational implications. In N. Colangelo, S. G. Assouline, & D. L. Ambrose (Eds.), *Talent Development: Proceedings from the 1993 Henry B. and Jocelyn Wallace National Research Symposium on Talent Development* (pp. 83–100). Dayton, OH: Ohio Psychology Press.

Benbow, C. P., & Lubinski, D. (1997). Intellectually talented children: How can we best meet their needs? In N. Colangelo & G. A. Davis (Eds.), *Handbook of Gifted Education*, 2nd ed. Boston: Allyn & Bacon.

Benbow, C. P., Lubinski, D., Shea, D. L., & Eftekhari-Sanjani, H. (2000). Sex-differences in mathematical reasoning ability at age 13: Their status 20 years later. *Psychological Science, 11*(6), 474–480.

Bianco, B. H., Harris, B., Garrison-Wade, D., & Leech, N. (2011). Gifted girls: Gender bias in gifted referrals. *Roeper Review, 33*(3), 170–181.

Brody, L. E. (2005). The study of exceptional talent. *High Ability Studies, 16*(1), 87–96.

Brody, L. E., Assouline, S. G., & Stanley, J. C. (1990). Five years of early entrants: Predicting successful achievement in College. *Gifted Child Quarterly, 34*(4), 138–142.

Brody, L. E., & Benbow, C. P. (1987). Accelerative strategies: How effective are they for the gifted? *Gifted Child Quarterly, 31*(3), 105–110.

Crombie, G., Bouffard-Bouchard, T., & Schneider, B. H. (1992). Gifted programs: Gender differences in referral and enrollment. *Gifted Child Quarterly, 36*(4), 213–214.

Daignault, M., Edwards, A. L., Pohlman, C., & McCabe, A. (1990). Selection for giftedness programs: Why the gender imbalance? In J. L. Ellis & J. M. Willinsky (Eds.), *Girls, Women and Giftedness* (pp. 61–64). New York: Trillium Press.

Ford, D. Y., & Whitney, G. W. (2009). Recruiting and retaining underrepresented gifted students. In S. I. Pfeiffer (Ed.), *Handbook of Giftedness in Children: Psychoeducational Theory, Research and Best Practices* (pp. 293–308). New York: Springer.

Fox, L. H. (1974). A mathematics program for fostering precocious achievement. In J. C. Stanley, P. Keating, & L. H. Fox (Eds.), *Mathematical Talent: Discovery, Description, and Development* (pp. 101–125). Baltimore, MD: Johns Hopkins University Press.

Fox, L. H. (1976a). Sex differences in mathematical talent: Bridging the gap. In D. P. Keating (Ed.), *Intellectual Talent: Research and Development* (pp. 183–214). Baltimore, MD: Johns Hopkins University Press.

Fox, L. H. (1976b). The values of gifted youth. In D. P. Keating (Ed.), *Intellectual Talent: Research and Development* (pp. 273–284). Baltimore, MD: Johns Hopkins University Press.

Fox, L. H. (1977). Sex differences: Implications for program planning for the academically gifted. In J. C. Stanley, W. C. George, & C. H. Solano (Eds.), *The Gifted and the Creative: A Fifty-year Perspective* (pp. 113–138). Baltimore, MD: Johns Hopkins University Press.

Fox, L. H., & Brody, L. (1980). An accelerative intervention program for mathematically gifted girls. In L. H. Fox, L. Brody, & D. Tobin (Eds.), *Women and the Mathematical Mystique* (pp. 64–78). Baltimore, MD: Johns Hopkins University Press.

Fox, L. H., Brody, L., & Tobin, D. (1984–85). The impact of intervention programs upon course taking and attitudes in high school. In S. F. Chipman, L. R. Brush, & D. M. Wilson (Eds.), *Women and Mathematics: Balancing the Equation* (pp. 249–274). Hillsdale, NJ: Lawrence Erlbaum Associates, Inc.

Fox L. H., & Cohn, S. (1980). Sex differences in the development of precocious mathematical talent. In L. H. Fox, L. Brody, & D. Tobin (Eds.), *Women and the Mathematical Mystique* (pp. 94–111). Baltimore, MD: Johns Hopkins University Press.

Fox, L. H., & Denham, S. A. (1974). Values and career interests of mathematically and scientifically precocious youth. In J. C. Stanley, D. P. Keating, & L. H. Fox (Eds.), *Mathematical Talent: Discovery, Description, and Development* (pp. 140–175). Baltimore, MD: Johns Hopkins University Press.

Fox, L. H. Engle, J., & Soller, J. (1999). Gifted girls and the math/science mystique. *Understanding Our Gifted*, 11(2), 3–7.

Fox, L. H., Pasternak, S. R., & Peiser, N. L. (1976). Career-related interests of adolescent boys and girls. In D. P. Keating (Ed.), *Intellectual Talent: Research and Development* (pp. 273–284). Baltimore, MD: Johns Hopkins University Press.

Fox, L. H., Sadker, D. L., & Engle, J. L. (1999). Sexism in the schools: Implications for the education of gifted girls. *Gifted and Talented International*, 14(2), 66–79.

Fox, L. H., & Soller, J. (1999). Mathematically gifted girls: Bridging the gender gap. *Gifted Education Press Quarterly*, 13(1), 2–7.

Fox, L. H., & Soller, J. (2007). Issues in gender equity for gifted education. In S. Klein, C. Kramare, B. Richardson, L. Fox, & D. Pollard (Eds.), *Handbook for Achieving Gender Equity through Education* (pp. 573–582). Mahwah, NJ: Lawrence Erlbaum Associates, Inc.

Fox, L. H., & Tobin, D. (1988). Broadening career horizons of gifted girls. *Gifted Child Today*, 11(1), 9–13.

Fox, L. H., & Zimmerman, W. (1985). Gifted women. In J. Freeman (Ed.), *The Psychology of Gifted Children* (pp. 219–243). London: Wiley & Sons.

Freeman, J. (2004). Cultural influences on gifted gender achievement. *High Ability Studies*, 15(1), 7–23.

Gallagher, S. A. (1997). Problem-based learning: Where did it come from, what does it do, and where is it going? *Journal for the Education of the Gifted*, 20(4), 332–362.

George, W. C., & Denham, S. (1976). Curriculum experimentation for the mathematically talented. In D. P. Keating (Ed.), *Intellectual Talent: Research and Development* (pp. 183–214). Baltimore, MD: Johns Hopkins University Press.

Hyde, J. S., & Lindberg, S. M. (2007). Facts and assumptions about the nature of gender differences and the implications for gender equity. In S. Klein, C. Kramare, B. Richardson, L. Fox, & D. Pollard (Eds.), *Handbook for Achieving Gender Equity through Education* (pp. 19–32). Mahwah, NJ: Lawrence Erlbaum Associates, Inc.

Isaksen, S. G., Dorval, K. B., & Treffinger, D. J. (2011). *Creative Approaches to Problem Solving: A Framework for Change*, 3rd ed. Thousand Oaks, CA: Sage Publications.

Keating, D. P., Wiegand, S. J., & Fox, L. H. (1974). Behavior of mathematically precocious boys in a college classroom. In J. C. Stanley, D. P. Keating, & L. H. Fox (Eds.), *Mathematical Talent: Discovery, Description, and Development* (pp. 176–185). Baltimore, MD: Johns Hopkins University Press.

Kerr, B., & Foley Nicpon, M. (2002). Gender and giftedness. In N. Colangelo & G. Davis (Eds.), *Handbook of Gifted Education* (pp. 493–505). New York: Allyn & Bacon.

Kerr, B., & Kurpius, S. E. R. (2004). Encouraging talented girls in math and science: Effects of a guidance intervention. *High Ability Studies*, 15(1), 85–102.

Kirschenbaurm, R. J., & Reis, S. M. (1997). Conflicts in creativity: Talented female artists. *Creativity Research Journal*, 10(2&3), 251–263.

Lubinski, D., & Benbow, C. P. (1992). Gender differences in abilities and preferences among the gifted: Implications for the math-science pipeline. *Current Directions in Psychological Science*, 1(2), 61–66.

Lubinski, D., Benbow, C. P., & Morelock, M. J. (2000). Gender differences in engineering and physical sciences among the gifted: An inorganic–organic distinction. In K. A. Heller, F. J. Monks, R. J. Sternberg, & R. F. Subotnik (Eds.), *International Handbook of Giftedness and Talent*, 2nd ed. (pp. 633–648). Oxford: Elsevier Science Ltd.

Osborn, A. F. (1952). *How to Become More Creative: 101 Rewarding Ways to Develop Potential Talent*. New York: Scribners.

Osborn, A. F. (1953). *Applied Imagination: Principles and Procedures of Creative Problem Solving*. New York: Scribners.

Piirto, J. (1991). Why are there so few? (Creative women: Visual artists, mathematicians, musicians). *Roeper Review*, 13(3), 142–147.

Reis, S. M. (2002). Toward a theory of creativity in diverse creative women. *Creativity Research Journal, 14*(3&4), 305–316.

Reis, S. M. (2003). Gifted girls, twenty-five years later: Hope realized and new challenges found. *Roeper Review, 25*(4), 154–157.

Robinson, N. (2008). Two wrongs and two rights: Reason and responsibility. *Journal for the Education of the Gifted, 26*(4), 321–328.

Stanley, J. C. (1988). Some characteristics of SMPY's "700–800 on SAT-M before age 13 group": Youths who reason extremely well mathematically. *Gifted Child Quarterly, 32*(1), 205–209.

Stanley, J. C. (1991). An academic model for educating the mathematically talented. *Gifted Child Quarterly, 35*(1), 36–42.

Stanley, J. C. (1996). In the beginning: The study of mathematically precocious youth (SMPY). In C. P. Benbow & D. Lubinski (Eds.), *Intellectual Talent: Psychometric and Social Issues* (pp. 225–235). Baltimore, MD: Johns Hopkins University Press.

Stanley, J. C., Benbow, C. P., Brody, L. E., Dauber, S., & Lupkowski, A. (1992). Gender differences on eighty-six nationally standardized achievement and aptitude tests. In N. Colangelo, S. G. Assouline, & D. L. Ambrose (Eds.), *Talent Development: Proceedings from the 1991 Henry B and Joclyn Wallace National Research Symposium* (pp. 42–61). Unionville, NY: Trillium Press.

Stanley, J. C., Keating, D. P., & Fox, L. H. (Eds.) (1974). *Mathematical Talent: Discovery, Description, and Development*. Baltimore, MD: Johns Hopkins University Press.

Stanley, J. C., & Stanley, B. S. K. (1986). High-school biology, chemistry or physics learned well in three weeks. *Journal of Research in Science Teaching, 23*(3), 237–250.

Steele, C. M. (1997). A threat in the air: How stereotypes shape intellectual identify and performance. *American Psychologist, 52*(6), 613–629.

Steele, C. M., & Aronson, J. (1995). Stereotype threat and the intellectual test performance of African Americans. *Journal of Personality and Social Psychology, 69*(5), 797–811.

Sternberg, R. J. (1986). A triarchic theory of intellectual giftedness. In R. Sternberg & J. Davidson (Eds.), *Conceptions of Giftedness* (pp. 223–243). Cambridge: Cambridge University Press.

Sternberg, R. J. (1997). *Successful Intelligence*. New York: Plume.

Sternberg, R. J. (2000a). Giftedness as developing expertise. In K. A. Heller, F. J. Monks, R. J. Sternberg, & R. F. Subotnik (Eds.), *International Handbook of Giftedness and Talent*, 2nd ed. (pp. 55–66). Oxford: Elsevier Science Ltd.

Sternberg, R. J. (2000b). Creativity and giftedness: Identifying and developing creative giftedness. *Roeper Review, 23*(2), 60–64.

Sternberg, R. J. (2005). Broader conceptions of leadership. *Roeper Review, 28*(1), 37–44.

Sternberg, R. J. (2007). Cultural dimensions of giftedness and talent. *Roeper Review, 29*(3), 160–165.

Sternberg, R. J. (2008). Applying psychological theories to educational practice. *American Educational Research Journal, 45*(1), 150–165.

Subotnik, R. F., & Arnold, K. D. (1996). Success and sacrifice: The costs of talent fulfillment for women in science. In K. D. Arnold, K. D. Noble, & R. F. Subtonik (Eds.), *Remarkable Women: Perspectives on Female Talent and Development* (pp. 263–280). Cresskill, NJ: Hampton Press, Inc.

Thomas, J., & Williams, C. (2010). The history of specialized STEM schools and the formation and role of the NCSSSMST. *Roeper Review, 32*(1), 17–24.

Tomlinson, C. (1995). *How to Differentiate Instruction in Mixed Ability Classrooms*. Alexandria, VA: Association for Supervision and Curriculum Development.

Tomlinson, C. A. (1996). Good teaching for one and all: Does gifted education have an instructional identity. *Journal for the Education of the Gifted, 20*(2), 155–174.

Tomlinson, C. A., Kaplan, S. N., Renzulli, J. S., Purcell, J., Leppien, J., & Burns, D. (2002). *The Parallel Curriculum: A Design to Develop High Potential and Challenge High-ability Learners*. Washington, DC: National Association for Gifted Children.

Tomlinson, C. A., & McTighe, J. (2006). *Integrating Differentiated Instruction and Understanding by Design*. Alexandria, VA: Association for Supervision and Curriculum Development.

Treffinger, D. J., & Isaken, S. G. (2005). Creative problem-solving: History, development and implications for gifted education and talent development. *Gifted Child Quarterly, 49*(4), 342–353.

Treffinger, D. J., Isaksen, S. G., & Dorval, K. B. (2006). *Creative Problem Solving: An Introduction*, 4th ed. Waco, TX: Prufrock Press.

VanTassel-Baska, J. (1986). Effective curriculum and instruction models for talented students. *Gifted Child Quarterly, 30*(4), 164–169.

VanTassel-Baska, J., Bracken, B., Feng, A., & Brown, E. (2009). A longitudinal study of enhancing critical thinking and reading comprehension in Title I classrooms. *Journal for the Education of the Gifted, 33*(1), 7–37.

VanTassel-Baska, J., & Brown, E. F. (2007). Toward best practice: An analysis of the efficacy of curriculum models in gifted education. *Gifted Child Quarterly, 51*(4), 342–358.

VanTassel-Baska, J., Feng, A. X., & de Brux, E. (2007). A study of identification and achievement profiles of performance task-identified gifted students. *Journal for the Education of the Gifted, 31*(1), 7–34.

VanTassel-Baska, J., Feng, A. X., & Evans, B. L. (2007). Patterns of identification and performance among gifted students identified through performance tasks: A three-year analysis. *Gifted Child Quarterly, 51*(3), 218–231.

VanTassel-Baska, J., & Stambaugh, T. (2008). Curriculum and instructional considerations in programs for the gifted. In S. I. Pfeiffer (Ed.), *Handbook of Giftedness in Children: Psychoeducational Theory, Research and Best Practices* (pp. 347–366). New York: Springer.

Weiss, D. S., Haier, R. J., & Keating, D. P. (1974). Personality characteristics of mathematically precocious boys. In J. C. Stanley, D. P. Keating, & L. H. Fox (Eds.), *Mathematical Talent: Discovery, Description, and Development* (pp. 126–139). Baltimore, MD: Johns Hopkins University Press.

14

An application of the schoolwide enrichment model and high-end learning theory to innovation education

Ruth E. Lyons and Sally M. Reis

UNIVERSITY OF CONNECTICUT, USA

Summary: This chapter summarizes the ways in which the implementation of enrichment pedagogy has created opportunities for student successes at a small, urban Academy of high potential youth, known as the Renzulli Academy. The use of the Schoolwide Enrichment Model has resulted in creative and innovative learning challenges for students who attend the academy. In this chapter, we discuss the philosophy of the Renzulli Academy, the curriculum used to challenge these advanced learners, our preliminary identification procedures, and the ways in which we focus on innovation and creativity to foster a culture of achievement and innovation.

Key words: Renzulli Academy, enrichment pedagogy, high potential, urban, schoolwide enrichment, creativity, self-regulation, motivation, gifted.

Introduction

The North End of Hartford, Connecticut, is historically known for being a place to avoid; as both violence and criminal activity has given this neighborhood a reputation that is hard to live down. However, every day students from across the city of Hartford climb aboard multiple busses to attend a school where they are academically engaged and challenged, as well as accepted for being smart. The name of the school is the Renzulli Academy and it was founded in 2009, when Hartford Public School administrators recognized the need for a gifted and talented program to challenge students who are above average achievers and want to learn.

In September of 2009, the Renzulli Academy opened its doors to 60 gifted and highly creative Hartford public students in grades 4, 5, and 6. The Academy was originally organized as a school within a high poverty school that housed another 350 students. During the fall of 2011, in the third year of implementation, the Renzulli Academy had grown to include 180 students in grades 4–8. The future plan includes continued growth into a full K-12 Academy for 500 students. The teachers in the Academy have a specialization in gifted education, a commitment to work with urban students of poverty, strong content knowledge, and an interest in developing students' creativity. Academy students are identified using the Talent Pool Model, a flexible identification approach, designed to correct the underrepresentation that exists nationally in gifted and talented programs to be combated (Renzulli & Reis, 1997). Using data from the 2006 Elementary and Secondary School Civil Rights Survey, Ford, Grantham, and Whiting (2008) found that African American students are underrepresented by about 51% and Hispanic students by about 42% in gifted programs, relative to their proportion in the nation's schools. A majority of Renzulli Academy students are from high poverty families and the vast majority are culturally and linguistically diverse, also known as high potential/low income students (HP/LI). HP/LI students represent approximately three million young people from low-income families that score in the top quartile on their earliest achievement tests. Research studies have shown, however, that by the time these students reach the fifth grade, nearly half of the HP/LI students formerly in the top tier in reading have made little or no progress, as their math and reading scores are low, graduation rates hover around 50%, college attendance rates are stagnant, and 41% who enter college do not finish (College Board, 2010; Hemphill & Vanneman, 2011; Nord et al., 2011; Wyner, Bridgeland, & DiIulio, 2007). The diminished achievement of these students has a profound effect on their future college and career opportunities and a major goal of the Renzulli Academy is reach these students and create an atmosphere in which they can achieve at high levels, and be admitted to competitive high schools and colleges. In this chapter, we describe the philosophy of the Renzulli Academy, the curriculum used to challenge these advanced learners, our preliminary identification procedures, and the ways in which we implement a focus on innovation and creativity to foster a culture of achievement and innovation.

Philosophy

The philosophy of the Renzulli Academy is based on the Schoolwide Enrichment Triad Model (SEM) (Renzulli & Reis, 1985, 1997), based on over three decades of research and field-testing (Renzulli & Reis, 1994; Reis & Renzulli, 2003), that combines the previously developed Enrichment Triad and Revolving Door Identification Models. The SEM has been implemented in school districts worldwide, and extensive evaluations and research studies indicate the effectiveness of the model which VanTassel-Baska and Brown (2007) called one of the mega-models in the field (Renzulli & Reis, 1994; Reis & Renzulli, 2003). Previous research has demonstrated that the model is effective at serving high-ability students in a variety of educational settings and works well in schools that serve diverse ethnic and socioeconomic populations, exactly the population targeted by this school (Renzulli & Reis, 1994; Reis & Renzulli, 2003).

The Schoolwide Enrichment Triad Model

The Schoolwide Enrichment Triad Model (SEM) has been implemented in thousands of schools across the country as well as internationally and research on the SEM has been conducted in these widely differing schools with consistently positive results relating to increased student achievement, attitudes toward learning, engagement, and creative productivity (Renzulli & Reis, 1994; Reis & Renzulli, 2003). In the SEM, students receive several kinds of services, all core features that are guaranteed at the Renzulli Academy. First, interest, learning styles, and product style assessments are conducted with talent pool students using the program Renzulli Learning (www.renzullilearning.com). Each student creates a profile that identifies their unique strengths and talents and teachers can identify patterns of student's interests, products, and learning styles across the three classes. These methods are being used to both identify and create students' interests and to encourage students to develop and pursue these interests in various ways. Learning style preferences assessed include projects, independent study, teaching games, simulations, peer teaching, programmed instruction, lecture, drill and recitation, and discussion. Product Style preferences include the kinds of products students like to do, such as those that are written, oral, hands-on, artistic, displays, dramatization, service, and multimedia.

The Enrichment Triad Model assists teachers and students to encourage creative productivity (Renzulli & Reis, 1997). This model operates on the idea of exposing students to various topics, developing the ability to apply knowledge content, and, finally, instilling the hope for students to take self-selected interest areas and develop them into projects and work that will hold social capital. There are three types of enrichment in this model; Type I, Type II, and Type III. Type I Enrichment focuses on exposing students to content that is ordinarily not covered in the regular curriculum (Renzulli & Reis, 1997). At the Academy teachers alternate organizing Type I experiences such as webinars, speakers, documentaries, or focused activities. Type II Enrichment consists of promoting the development of thinking and feeling processes. These activities often train students to learn how-to skills and advanced level communication skills. Most of this enrichment is integrated into lessons at the Academy, as teachers plan lessons using this model. Type III Enrichment activities provide opportunities for students to apply knowledge, creative ideas, and task commitment to a self-selected interest area (Renzulli & Reis, 1997). Students are encouraged to develop authentic products that will implement and impact change on the audience of their work. All students in the Academy have time to create an independent study/Type III innovative opportunity block that is also incorporated into the philosophy of the school.

All of the Academy teachers have attended Confratute, a summer institute on enrichment-based differentiated teaching and gifted pedagogy, as well as taken coursework in the gifted and talented field (www.gifted.ucon.edu/confratute). The knowledge about gifted and high creative students helps to make the Academy unique. This specialization that teachers have attained in enrichment-based differentiated teaching and pedagogy enables them to address the unique needs of these students with both understanding and sensitivity. In many ways, the teachers' knowledge and understanding of talented and highly creative students is what makes the Academy special, for many of the new students have come from schools where they have been misunderstood. Students have consistently stated that attending the Academy has enabled them to have teachers who understand and care about them for the first time. This caring environment is created with the students' needs in mind; support, challenge, rigor, respect, honesty, and attention paid to creativity and innovation.

The role of the parent is also crucial to the success of the students and the Academy itself. Parents are expected to attend meetings that introduce the key elements of SEM; having the parents understand the uniqueness and operational philosophy of the Academy is very important. The goal is for parents to feel as if their child is an integral part of the Academy; that their involvement is what makes their child and the Academy successful. Parent involvement in a gifted child's life is crucial for the exposure to the field or area of interest. Many successful creative producers have supportive parents to thank for enrolling them in lessons, driving them to museums, workshops, or camps, and often also funding these experiences. It is the parent that often needs to facilitate the opportunities and provides the access to unique experiences that enrich the lives of the students. At the Academy we work to ensure that all students have the opportunity to develop their gifts and talents, not just the students who may have access to additional resources provided by some parents. In offering evening opportunities and allowing students time and access we hope to be able to create the atmosphere where students grow and parents support.

Creativity and innovation are key elements at the Academy where we offer students a different way to approach education; one that is academically challenging and also one that pushes them to think outside the box and be inventive in the ways they approach both content and products. We deeply believe in allowing students to have choice when it comes to demonstrating mastery. In having choice students are able to use any creative element they feel necessary to show what they have learned. In removing the prescribed way of presenting mastery, students are pushed to cultivate their own ideas that often far exceed the expected. A multitude of teaching strategies and project options enabled us to develop these attributes in our students. For example, students at the Academy are continually in the midst of project work, all of which has an element of student choice. Within each content area, students are asked to develop the ability to think about the best possible ways to promote their own learning and are challenged to demonstrate that in a creative way.

All students attend a SEM orientation, which includes the key concepts and foundational aspects of the curriculum used in our school. Ensuring that the students understand why and what approach is being used is important in helping students learn to advocate for their educational needs. Being able to be an advocate for their own needs gives them the ownership and knowledge that it is acceptable for students to approach a problem from a different perspective. We want students to have the freedom to be innovative with how they solve problems and look for connections across contents. It is a goal of SEM, that students learn how they are most successful in a classroom setting.

The ability to create Type III projects and meaningful, creative work has had to be taught and demonstrated. For example, each year students at the Academy participate in a state-wide competition entitled, The Connecticut Invention Convention. This is a competition that challenges students to create a solution to everyday problems by inventing that solution. Students are required to create a visual display that shows key elements of their invention as well as a replication of the invention. The first year that the students participated in this competition, creating such a display was not a common experience and so we had to arrange for students to view presentations from other students as well as professionals who were highly creative and innovative. For many students, just being given the freedom and opportunity to produce a creative project was motivation enough whereas other students needed more scaffolding and exposure to possibilities. In this case, students were shown the work of other students and in some cases the teacher would actually walk the students through the process from start to finish; although this was a somewhat timely endeavor, we found it to be of critical importance, as many students did not have the support to encourage an unknown process.

Having a collaborative approach was a by-product of having common language that leads to effective communication between parents and teachers (Renzulli & Reis, 1997). Parents understood the use of a profile to identify students' interests, learning, and product styles, as well as the use of various forms of enrichment and differentiated learning activities. Using SEM as the foundation enabled us to achieve an easier planning and organization of activities, as well as a more effective and efficient allocation of resources for enrichment. Additional information on the Schoolwide Enrichment Model is found in Delcourt and Renzulli's chapter in this volume.

Identification

Identification of potential Renzulli Academy students occurs with a review of all eligible grade-level student records that score at the mastery level on the Connecticut Mastery Test (CMT), the state assessment test. Students who placed at the highest level on academic achievement tests administered by the state as well as school records were identified as a preliminary talent pool, and follow-up meetings were held with parents. This initial identification method is referred to as the Renzulli Talent Pool Identification Model. Using the results from the CMT as the initial screening allowed applications to be sent out to students who scored in the top 85th percentile in the state of Connecticut. Parents and teachers are also asked to submit nominations of students who display the elements of the three-ring conception of giftedness, above average abilities, creativity, and task commitment (Renzulli, 1978). In addition to the student application, teachers also complete a Scale for Rating the Behavioral Characteristics of Superior Students. Students are asked to complete essays that are included in the application and these are also included in the decision to offer admission to students who demonstrate above average ability and creativity that might lead to innovative work. Finally, families attend an informational meeting prior to the start of school in which they commit to volunteer for a minimum block of time to help and support Academy Programs.

In the SEM, a talent pool of 15–20% of above-average ability/high-potential students is identified through a variety of measures, including achievement tests, teacher nominations, assessment of potential for creativity, and task commitment, as well as alternative pathways of entrance (self-nomination, parent nomination, etc.), and these measures were used in the Academy, as the highest levels of achievement tests were used as the primary screening and the first step for the identification in this school.

As the Academy continues to grow, so will the need for further identification methods and models. When the expansion to a lower elementary school is in place the identification will need to match the programming. Identification for this age level will include observations in both kindergarten and first grade, the use of early childhood checklists, performance-based assessment of various academic assignments and tasks, and tests such as the Cognitive Abilities Test™ (CogAT®), which has been designed to evaluate the level and pattern of cognitive development of students in Kindergarten through 12th grade. The CogAT includes subtests from verbal, quantitative, and nonverbal batteries. This will be the first student aptitude exam process used in the city of Hartford and the identification system has been approved by the Superintendent, Assistant Superintendent, and Board of Education.

Since the Academy's definition of giftedness is the Renzulli Three Ring Conception of Giftedness, it is very important that our identification measures incorporate those essential qualities (Renzulli, 1978). We believe that gifted learners demonstrate or have the potential to demonstrate above average ability, task commitment, and creativity. In our application process there are elements dedicated to each attribute. In using achievement scores as an initial indicator of student readiness, we try to identify students who have shown mastery in reading and have the ability to perform at high levels in math and writing. Within our writing samples, we look for students who show a spark of creativity and understanding of the written language. The third attribute, task commitment, is a more elusive one to measure, as some of our students have not attended classes in which they had an opportunity or a desire to demonstrate either their creativity or task commitment. However, we attempt to capture this through the use of the Scale for Rating the Behavioral Characteristics of Superior Students (SRBCSS), a teacher recommendation developed by Dr. Renzulli.

The use of teacher recommendation has been one successful measure, for students accepted using teacher recommendations account for a portion of the student body who may otherwise have not been identified. Since gifted students from low-income families may achieve at lower levels than students from higher income families, and because they may need assistance with their learning skills, teachers may not see these students as being gifted or as having potential (VanTassel-Baska et al., 2004). Additional teacher development about the attributes of gifted learners may be helpful in a more holistic approach. The Academy is known as a place in which the focus is on talent development, where all students who have all met the baseline criteria for admission receive the SEM intervention, which includes a continuum of services and opportunities for engagement and innovation.

Curriculum

Creating an atmosphere of challenge and high expectations is a crucial element of the Academy's focus within curriculum. The curriculum and instructional program adopted for the Academy combined the philosophy and work of Renzulli and Reis (1997) and the work of Sandra Kaplan, that seeks to add depth and complexity for gifted and high potential students (2009). The SEM structure has been infused across all content areas, as both enrichment and opportunities for independent and small group study are used to enrich and extend the regular content curriculum across the content areas. Each content area curriculum combined the depth and complexity advocated by Kaplan (2009) with the ideas included in the Multiple Menu Model (Renzulli, Leppien, & Hayes, 2000).

This curriculum is derived from large-scale national research studies carried out by researchers at the National Research Center at the University of Connecticut. Using nationally validated curriculum helps to prepare students at the Academy to be competitive scholars nationally, instead of focusing only on local or state-wide tests and standards. Thus, we set out to achieve our vision of the school, to enable academically talented and high potential youth in Hartford to achieve at the highest levels of performance, graduate with advanced knowledge base possible, and prepare them for admission to competitive colleges and universities in the world so they will contribute as ethical, productive members of a global society whose talents will make a positive difference in our world.

An important characteristic of the Academy's curricular approach based on the SEM is on accessing students' strengths, interests, and talents. When students complete the Renzulli Learning personality profile, teachers access information about the type of learner each student is, as well as the interest areas of each student, and their expression style. Learning styles and strengths are talent indicators and manifest themselves in several ways (Renzulli & Smith, 1978; Sternberg, 1988). The curriculum options at the Academy demonstrated that when educators provide high potential students with choice and teaching to their interests, as well as with using advanced content and hands-on activities, students' motivation to learn increases (Reis, Gentry, & Maxfield, 1998).

Mathematics

Students at the Renzulli Academy participate in an advanced mathematics curriculum called Project M^3, Mentoring Mathematical Minds (Gavin et al., 2007; Gavin, Casa, Adelson, Carrol, & Sheffield, 2009). This program emerged during a collaborative research effort coordinated by Dr. Katherine Gavin and a team of national experts in the fields of mathematics, mathematics education, and gifted education. Using a project-based approach, Project M^3 offers depth and complexity of math concepts taught across grade levels to high-ability students focusing on curriculum units with advanced mathematics. The program includes advanced math curriculum with projects and investigations to foster creativity, critical thinking, and problem-solving skills that lead to higher math and problem scores than comparison group students (Gavin et al., 2009). For example, in place value, students move beyond using tens, hundreds, and thousands and take part in a simulated archaeological dig and discover unusual calculations carved into rock. Using creative problem-solving skills, students are asked to determine which place value system was used by these people. Hartford Schools math standards have been integrated into classroom preparation time each day and compacted for students as part of this process.

Reading/language arts/writing

The Schoolwide Enrichment Model in Reading (SEM-R) (Reis et al., 2007; Reis, Eckert, McCoach, Jacobs, & Coyne, 2008) has also been integrated into the Renzulli Academy as the core reading/language arts program. This approach, developed by Dr. Sally Reis and a team of reading and gifted education specialists, focuses on reading acceleration and enrichment for talented readers through engagement in challenging, self-selected reading, accompanied by instruction in high-level thinking and reading strategy skills. A second core focus of the SEM-R is differentiation of reading content and strategies, and the integration of more challenging reading experiences and advanced opportunities for metacognition and self-regulated reading. In other words, the SEM-R program challenges and prepares students who are talented in reading to begin reading increasingly challenging books in school and to continue this reading at home.

The goals of the SEM-R approach are to encourage children to both enjoy the reading process by giving them access to high-interest, self-selected books that they can read for periods of time at school and at home and to develop independence and self-regulation in reading through the selection of these books as well as the opportunity to have individualized reading instruction. Our teachers expect that all students will improve in reading fluency and comprehension through practice with reading comprehension strategies in challenging, self-directed reading and every student is required to have at least one book like this that they carry with them throughout the day and read when they have free time.

The SEM-R includes three phases. During the Phase 1 "exposure" phase, teachers read short selections from high-quality, engaging literature to introduce students to a wide variety of titles, genres, authors, and topics. With these read-alouds, teachers provide instruction through modeling and discussion, demonstrating reading strategies and self-regulation skills and using higher-order questions to guide discussion in these book hooks. Early in the SEM-R at the Renzulli Academy, these Phase 1 activities lasted about 20 minutes per day but this phase decreased in length over the course of the year when students could spend more time on Phase 2. Currently, all students read for about 50–60 minutes each day of challenging, self-selected books.

Phase 2 of the SEM-R emphasizes the development of students' ability to engage in supported independent reading (SIR) of self-selected, appropriately challenging books, with differentiated instructional support with teacher conferences. During Phase 2, students select books that are at least 1 to 1.5 grade levels above their current reading levels. Students access strategies for recognizing appropriately challenging books, and they were guided and encouraged to select challenging books in areas of personal interest to promote engagement. Over the course of the intervention, students initially read for 5–15 minutes a day during Phase 2; over time they extended SIR to 20–25 minutes, and finally to 45 minutes each day. During this in-class reading time, students participate in individualized reading conferences for five to seven minutes with their teachers; on average, each student participated in one to two conferences per week. In these conferences, teachers assessed reading fluency and comprehension and provided individualized instruction in strategy use, including predicting, using inferences, and making connections. For the most advanced readers, conferences focused less on specific reading strategies and more on higher-order questions and critical concepts.

In Phase 3, students move from teacher-directed opportunities to self-choice activities over the course of SEM-R. Activities include opportunities to explore new technology, discussion groups, practice with advanced questioning and thinking skills, creativity training in language arts, learning centers, interest-based projects and book chats. These experiences provide time for students to pursue areas of personal interest using the Internet to learn to read critically and to locate other reading materials, especially high-quality, challenging literature. Options for innovative and creative independent study using Renzulli Learning are also made available for students during Phase 3. The length of Phase 3 varies, with more or less time devoted to Phase 3 on particular days based on progress in independent reading and needs for time to be devoted to independent projects and activities.

All teachers use high interest books focusing on creativity in reading to support their SEM-R implementation and the teachers augmented their collections as needed, choosing literature based on students' interests and experiences. Each bookmark includes about three to five questions addressing a particular literary element, theme, genre, or other area of study and is tied to advanced reading strategies as well as state standards. Creativity is also incorporated into these bookmarks to enable teachers to integrate creative questioning into Phase 1 discussions and Phase 2 conferences to promote higher-order thinking. Based on a decade of research, the SEM-R has been demonstrated to be effective at increasing achievement in reading and encouraging talented readers to read more challenging material for longer periods of time (Reis et al., 2008; Reis & Morales-Taylor, 2010). Results of randomized studies suggest SEM-R is particularly effective for urban talented students (Reis et al., 2008; Reis & Morales-Taylor, 2010) and for students who speak English as a second language (Reis & Housand, 2009).

Science

The science curriculum is based upon challenging standards and big ideas, applied to units of study across the grade levels with options for innovative student projects. Using the Multiple Menu approach (Renzulli et al., 2000) and Kaplan's (2008, 2009) work, a curriculum map has been created with essential questions and big ideas across content area units such as habitats and the water cycle. Science units also introduce project-based work that employs the scientific method with applications for innovative student projects and products. Students study key concepts and principles in science based on grade-level standards and then depth and complexity will be added to enable students to work actively on innovative science projects by forming a hypothesis and applying the scientific method to project-based learning and inquiry experiences in science. Enrichment is scaffolded across these units with Type I, II, and III opportunities in science. The goal each year is to expose students to key concepts in science and for each student to complete an advanced science project in an area of interest using data collection methods and the scientific method.

Social Studies

Social Studies is taught by adding depth and complexity (Kaplan, 2009) to the grade-level standards, as well as by infusing enrichment into the content area using the Enrichment Triad Model, and also by requiring a project based on advanced content acquisition, primary sources, and interests each year. A curriculum map has been developed for each grade level to enable these academically talented students to demonstrate and/or acquire knowledge of the grade-level social studies curriculum, as well as to engage in authentic historical research. Units of study on the curriculum map include explorations about topics such as Native Americans, Connecticut history, geography and map skills, and government. A focus on big content ideas is integrated into these units that also introduce students to the critical thinking and problem-solving skills that can be applied to innovative work. A social studies project is required each year during the second semester of each year, culminating with a National History Day project for all students in sixth grade and above. In fourth grade, for example, students are required to complete a research project about a significant person or place in Connecticut history incorporating the use of primary sources and at least one big content idea. The products may be expressed in students' areas of strength and choice, such as dramatic, written, display, technological, auditory, or in any combination of student preferences and will be completed during the last marking period of fourth grade. In sixth grade, advanced themes from the National History Day Competition are integrated with standards-based instruction and all students are required to complete an historical project of sufficient quality that it can be submitted to the regional competition.

Enrichment clusters

Enrichment clusters, another component of the Schoolwide Enrichment Model, is targeted for groups of students who share common interests, and who are grouped together during specially designated time blocks to work with an adult who shares their interests and who has some degree of advanced knowledge and expertise in the area. A series of clusters was planned and implemented for all students in the Academy. Students completed an interest inventory developed to assess their interests, and an enrichment coordinator, one of the Academy teachers, tallied all of the major families of interests and then recruited teachers and other professionals in the school to facilitate enrichment clusters based on these interests, with innovative choices such as drama, history, creative writing, drawing, music, archeology, and other areas. Students select their top choices for the clusters and scheduling is completed to place all children into their first, or in some cases, second choice. Research has also suggested that the use of enrichment clusters results in higher use of advanced thinking and research skills in gifted and in other students as well (Reis et al., 1998).

Academic competitions

All students at the Academy are also encouraged to participate in a wide spectrum of academic competitions. We believe that enabling students to expand on classroom content and facilitate real world connections and context for instruction to occur makes learning more meaningful. A goal of the Academy is individualized instruction that adds depth and complexity; supplementing the school's curriculum with such opportunities as academic competitions provides one element of this challenge. Such competitions produce an innovative approach to learning as the students are in charge of their own learning; they are the directors of what they learn and what they gain from the experience.

National History Day (NHD) is a national history competition that makes history come alive for America's youth by engaging them in the discovery of the historic, cultural, and social experiences of the past. Students are given the opportunity to express their learning through hands-on experiences and presentations. Each year the theme for the NHD competition changes which allows students the freedom to study a given subject within the realm of that year's given theme. The choice of topic allows for students to become invested in project-based curriculum and instruction provided by NHD. Teachers use NHD to provide students the structure to approach project-based learning from a research perspective. Ensuring that the students have this experience gives them the confidence to approach other projects in the same way. Having significant positive changes allows for real innovation to take hold and become a characteristic within the students. We have witnessed students who have developed the ability to approach tasks and assignments with more creativity and task commitment due to the nature of the individualized programs and influences the academic competitions have upon that.

Students at the Academy participate alongside their suburban, high socioeconomic peers, who often have much more access to materials and resources. Students at the Academy have placed in the top placements in regional competitions. Participation within academic competitions helps to foster talent, for many students the engagement in such competitions is the first time that they realize the true talent they have within a domain or field. For many, participation in these competitions is a crystallizing experience.

Our goal at the Academy is to help students realize their potential and develop young leaders within their areas of interest. Competing in academic competitions is an important aspect for our students to gain a competitive edge that is a very important attribute in the workplace today. We want our students to be collaborative, creative, and driven; and we believe that enrichment and academic competitions foster this development. Students at the Academy participate in science fairs, state-wide invention conventions, writing competitions, and also participate in Odyssey of the Mind.

Odyssey of the Mind is another option for students to develop creative problem solving, as teams of students select a problem, create a solution, and then present their solution in a competition against other teams (Micklus & Micklus, 2011). Participation in Odyssey of the Mind helps us to offer creative problem-solving opportunities and to foster original and divergent thinking. The teachers involved in running the OM teams promote creativity by challenging teams to solve a problem that has more than one solution. Students develop a sense of self-respect and respect for others during the team's preparatory process. It is our hope that through this opportunity that students will become more comfortable with creative problem solving and confident risk takers.

Despite our mission as an academically focused program, our students have also demonstrated a strong interest and talent in the arts. We have worked diligently to focus our enrichment clusters around student interests and have offered a dance cluster, improvization cluster, and visual arts cluster. Incorporating student interest is at the heart of what we do and is the foundation of the SEM (Renzulli & Reis, 1997), which includes trying to create opportunities for our students to explore and develop their own creativity. During one of the initial cluster meetings, the students wanted to organize a talent show. In enrichment clusters the participants seek out authentic measures of success and products; a show was a great way to showcase the talent we were seeing in our students. The students planned, practiced, and procured all of the necessary elements to make the show happen. Teachers worked tirelessly to ensure that each student's creative endeavor was properly performed and practiced. The spring show that was held at the end of our second year was the event that attracted more parents than any previous gathering. These types of events are an important outlet for students' creativity but also for parents and families to see and appreciate their children's talents. The arts provide excellent training for talent development as students learn about opportunities to practice and then develop talents. This spring show is a celebration of students' creativity that has become a tradition at the Renzulli Academy.

The transition to high expectations: a case study of Abigail

When students initially enter the Academy they are often disillusioned with education. Many of the students have equated school with minimal effort and a boring, uninspiring curriculum. Too many American schools have been forced to focus on test improvement for those at the bottom of the achievement continuum that has translated into classrooms focusing on remediation, not rigor. Too many gifted students spend their days relearning material that they have already mastered (Reis, Westberg, Kulikowich, & Purcell, 1998; Subotnik, Olszewski-Kubilius, & Worrell, 2011). Our experiences and knowledge of Renzulli students have led us to understand that many come from classrooms where they have been in classes with students whose ability level has spanned multiple grade levels and little opportunity has existed for creativity and innovation.

The first month of school is quite difficult for new students as they enter a new world in which the bar has been raised and expectations have increased dramatically. Students are asked to increase the production and quality of their work in a very short time. For many students, this is the time when they start to understand the need for explicit instruction on self-regulation and study skills. However, as we also start to see students becoming accountable for their education, we watch as they become excited about learning. This is, by far, one of the most magical parts of the Academy's academic life.

Abigail, for example, started in the Academy during its pilot year, as a 6th grader and at the beginning of the year. It soon became apparent that Abigail had tremendous potential that had not been tapped. Assignments often lacked any sign of effort and effective presentation of her work was lacking. However, the creativity and intensity in her work was clear, as was the fact that Abigail had never been asked to redo an assignment or asked to add higher-level thinking, defend her answers, or been challenged to show her learning in a unique way. She had mastered the idea of "school" and was quiet, helpful, and organized. All of these are laudable characteristics but we wanted more, as we urged Abigail to demonstrate higher-level thinking and creativity in her work.

Abigail has flourished while at the Academy and much of her success is due to the curriculum options and approaches we offer. One of these approaches, based on the Enrichment Triad Model, enables us to introduce content using three types of enrichment experiences: Type I, II, and III. Type I is a general exposure to content that is not usually covered in the regular curriculum. At the Academy we dedicate a portion of Friday to Type I presentations have included holocaust survivors, to Entertainment and Sports Programming Network (ESPN) journalists, to the senator of the district our school is located within, to documentaries on global warming, to webinars, virtual fieldtrips, and skyping with experts in the field. This weekly opportunity enables students to try and find that spark and interest. It was, for example, through a Type I of various clips from the Discovery Channel's Shark Week in 2009 that Abigail discovered a passion for sharks. Her mother explained that sharks had been an interest of Abigail's but one that she had not really had the opportunities to pursue and her mother was happy that the Academy was able to foster her interest and pursue an advanced topic. Two years later, Abigail became a Type I presenter on sharks, which has turned into a passion. After arranging for other students to watch a webinar on sharks, she shared her expertise and knowledge with fellow Academy students and answered advanced questions on this topic, like a practicing professional. It was very rewarding to watch her share her newly found passion with other students.

Abigail appears to have a creative nature and we are convinced that having the opportunity to express that and use her talents on a daily basis has enabled her to win several awards. She was, for example, a finalist at the Connecticut Invention Convention and also was awarded first place at the regional National History Day competition.

Abigail's invention convention project was entitled, the Tech Pet Feeder, a contraption that allowed a pet owner to text their pet's food bowl and upon receipt of the text message, release food for the pet. From the beginning of the process Abigail knew that she wanted to create an invention that would assist an animal, and she played around with several ideas, ultimately deciding to create a digitalized pet feeder as she had witnessed her own dog become sick from overeating. Having students solve real world problems is a goal of the Invention Convention as well as the Academy and a major part of the SEM. The Academy is founded on the principle that students should explore their interests and develop their learning around these areas of personal interest. Abigail's participation in National History Day provides evidence of the importance of personalization of interest, as the theme was Debate and Diplomacy in History: Successes, Failures, and Consequences. The approach that NHD encourages is for students to decide on a topic, conduct their own research, and use primary and secondary sources to determine the consequences that played out in history within their selected topic. The students then have five options of how to tell their story, through a display, documentary, live performance, written report, or website. Students are eligible to compete at regional, state, and national contests. At her request, Abigail worked with two other girls to research the Salem Witch Trials, finding websites that included the court transcripts and documents, which helped them to write an original script that played out the court proceedings of Ann Putnam. There were three roles; Judge Hawthorne, Ann Putnam, and an afflicted victim. The girls made connections to witch hunts that continue to occur today such as the Witch camps in Ghana, in which women had been forced to flee their homes and seek refuge in these camps based solely on accusations. The resulting performance was both captivating and replete with historical relevance. The girls performed their skit many times for different visitors to the Academy, demonstrating a poignant display of our creative approach to education and our belief that all of our students should have the chance to achieve high levels of creativity and success.

Watching Abigail, now a confident 8th grader and part of the pioneer graduating class at the Renzulli Academy, we reflect on how she has grown into a confident and academically successful young woman who envisions the mission of the Renzulli Academy. It is in through this type of success, that we continue to have profound confidence in our SEM approach.

Innovation and creativity within the Academy walls

Allowing students the space and time to become creative producers is the integral aspect of the SEM approach on which our curriculum of the Academy is based. Developing creative producers is one of the most important goals of the Academy and the core curriculum of the school. A creative producer generates new knowledge or art forms and significantly alters a field with his or her work; he or she also advances new paradigms or revolutionize existing ones (Simonton, 1996). If we are to promote these types of behaviors, it is critical that that we encourage Renzulli Academy students to be risk takers and feel safe in doing so. Many factors contribute to enabling creativity to flow, including the environment, the space, and the access. The environment must support student risk taking and creative endeavors, which are celebrated. Students must have adequate space and time, as well as opportunity to develop their ideas and to create innovative products. Students must have the access to supplies, materials, and personnel and they must feel supported enough to take the gamble of venturing into open-ended choices about projects and innovative opportunities such as National History Day, science fair projects, and inventions. This is a difficult thing to develop especially with a population of students who are extremely aware of the challenges they face in their surroundings.

The environment in which some Renzulli students live and grow up can be characterized as often chaotic. It has been suggested that these environments facilitate creative producers by engendering characteristics that help individuals meet the demands of creative careers or jobs that involve tackling ill-defined, unstructured, and complex patterns (Albert 1978; Goertzel & Goertzel, 2004; Subtonik et al., 2011). That, too, is both a function and goal of the Renzulli Academy.

Conclusion: other notable accomplishments

The Renzulli Academy has demonstrated the benefits of programs and schools like this for high-potential and low-income students. With support from the Hartford Public Schools Administration and Board of Education, the Hartford Public School System has become a place where high potential students can flourish and become the creative producers of the future.

By the end of the first year of the Academy, as a result of enrichment and challenging content, several notable achievements occurred. First, an expected academic achievement increase occurred, as during the second year of its existence, for example, students at the Academy achieved the highest CMT scores in Hartford Public Schools. In 2010, 89% of the students body scored at goal or mastery level; and in 2011, 95% of the student body scored either at goal or mastery level. An increase was also realized in student creativity, innovation, motivation, and self-regulation with the majority of the students. For example, all students in specific grade levels participated in

Connecticut History Day, the Connecticut Invention Convention, Connecticut Science Fair, and smaller samples of the most academically advanced students participated in other academic competitions.

The future benefits that are anticipated for students at the Academy are bold and broad. As our first class enters high school, we believe that Renzulli Academy students will apply and gain acceptance at competitive colleges and universities. And, above all, the Renzulli Academy students have learned to love learning as well as understand the rigors and joys of creative and productive work that we hope and believe will be an integrated part of their future academic and career paths.

References

Albert, R. S. (1978). Observation and suggestions regarding giftedness, familial influence and the achievement of eminence. *Gifted Child Quarterly, 28*(2), 201–211.

College Board (2010). *2010 College Bound Seniors: Total Group Profile Report.* New York: College Board.

Ford, D. Y., Grantham, T. C., & Whiting, G. W. (2008). Another look at the achievement gap: Learning from the experiences of gifted black students. *Urban Education, 43*(2), 216–238.

Gavin, M. K., Casa, T. M., Adelson, J. L., Carroll, S. R., & Sheffield, L. J. (2009). The impact of advanced curriculum on the achievement of mathematically promising elementary students. *Gifted Child Quarterly, 53*(3), 188–202.

Gavin, M. K., Casa, T. M., Adelson, J. L., Carroll, S. R., Sheffield, L. J., & Spinelli, A. M. (2007). Project M3: Mentoring mathematical minds: Challenging curriculum for talented elementary students. *Journal of Advanced Academics, 18*(4), 566–585.

Goertzel, V., & Goertzel, M. G. (2004). *Cradles of Eminence,* 2nd ed. Scottsdale, AZ: Great Potential Press.

Hemphill, F. C., & Vanneman, A. (2011). *Achievement Gaps: How Hispanic and White Students in Public Schools Perform in Mathematics and Reading on the National Assessment of Educational Progress* (NCES 2011–459). Washington, DC: National Center for Education Statistics, Institute of Education Sciences, U.S. Department of Education.

Kaplan, S. N. (2008). Curriculum consequences: If you learn this, then. *Gifted Child Today, 31*(1), 41–42.

Kaplan, S. N. (2009). The grid: A model to construct differentiated curriculum for the gifted. In J. S. Renzulli, E. J. Gubbins, K. S. McMillen, R. D. Eckert, & C. A. Little (Eds.), *Systems and Models for Developing Programs for the Gifted and Talented* (pp. 235–252). Mansfield Center, CT: Creative Learning Press.

Micklus, S., & Micklus, C. (2011). *Odyssey of the Mind: 2011–2012 Program Guide.* Sewell, NJ: Creative Competitions, Inc.

Nord, C., Roey, S., Perkins, R., Lyons, M., Lemanski, N., Brown, J., & Schuknecht, J. (2011). *The Nation's Report Card: America's High School Graduates* (NCES 2011 462). U.S. Department of Education, National Center for Education Statistics. Washington, DC: U.S. Government Printing Office.

Reis, S. M., Eckert, R. D., McCoach, D. B., Jacobs, J. K., & Coyne, M. (2008). Using enrichment reading practices to increase reading fluency, comprehension, and attitudes. *Journal of Educational Research, 101*(5), 299–314.

Reis, S. M., Gentry, M., & Maxfield, L. R. (1998). The application of enrichment clusters to teachers' classroom practices. *Journal for Education of the Gifted, 21*(3), 310–324.

Reis, S. M., & Housand, A. (2009). The impact of gifted education pedagogy and enriched reading practices on reading achievement for urban students in bilingual and English-speaking classes. *Journal of Urban Education, 6*(1), 72–86.

Reis, S. M., McCoach, D. B., Coyne, M., Schreiber, F. J., Eckert, R. D., & Gubbins, E. J. (2007). Using planned enrichment strategies with direct instruction to improve reading fluency, comprehension, and attitude toward reading: an evidence based study. *Elementary School Journal, 108*(1), 3–24.

Reis, S. M., & Morales-Taylor, M. (2010). From high potential to gifted performance: Encouraging academically talented urban students. *Gifted Child Today, 33*(4), 28–38.

Reis, S. M., & Renzulli, J. S. (2003). Research related to the schoolwide enrichment triad model. *Gifted Education International, 18*(1), 15–40.

Reis, S. M., Westberg, K. L., Kulikowich, J. M., & Purcell, J. H. (1998). Curriculum compacting and achievement test scores: What does the research say? *Gifted Child Quarterly, 42*(2), 123–129.

Renzulli, J. S. (1978). What makes giftedness? Reexamining a definition. *Phi Delta Kappan, 60*(3), 180–184, 261.

Renzulli, J. S., Leppien, J., & Hayes, T. (2000). *The Multiple Menu Model*. Mansfield Center, CT: Creative Learning Press.

Renzulli, J. S., & Reis, S. M. (1985). *The Schoolwide Enrichment Model: A Comprehensive Plan for Educational Excellence*. Mansfield Center, CT: Creative Learning Press.

Renzulli, J. S., & Reis, S. M. (1994). Research related to the schoolwide enrichment triad model. *Gifted Child Quarterly, 38*(1), 7–20.

Renzulli, J. S., & Reis, S. M. (1997). *The Schoolwide Enrichment Model: A How-to Guide for Educational Excellence*, 2nd ed. Mansfield Center, CT: Creative Learning Press, Inc.

Renzulli, J. S., & Smith, L. H. (1978). *Learning Styles Inventory: A Measure of Student Preference for Instructional Techniques*. Mansfield Center, CT: Creative Learning Press.

Simonton, D. K. (1996). Creative expertise: A life-span developmental perspective. In A. K. Anders (Ed.), *The Road to Excellence* (pp. 227–253). Mahwah, NJ: Erlbaum.

Sternberg, R. J. (1988). A three-faceted model of creativity. In R. J. Sternberg (Ed.), *The Nature of Creativity* (pp. 125–147). New York: Cambridge University Press.

Subotnik, R. F., Olszewski-Kubilius, P., & Worrell, F. C. (2011). Rethinking giftedness and gifted education: A proposed direction forward based on psychological science. *Psychological Science in the Public Interest, 12*(1), 3–54.

VanTassel-Baska, J., & Brown, E. F. (2007). Toward best practice: An analysis of the efficacy of curriculum models in gifted education. *Gifted Child Quarterly, 51*(4), 342–358.

VanTassel-Baska, J., Brown, E., Worley, B., & Stambaugh. T (2004). *An Analysis of State Policies in Key Program Development Components*. Williamsburg, VA: Center for Gifted Education, College of William and Mary.

Wyner, J. S., Bridgeland, J. M., & DiIulio, J. J. (2007). *Achievement Trap: How America is Failing Millions of High-achieving Students from Lower-income Families*. Landsdowne, VA: Jack Kent Cooke Foundation.

Zimmerman, B. J., Bandura, A., & Martinez-Pons, M. (1992). Self-motivation or academic attainment: The role of self-efficacy beliefs and personal goal-setting. *American Educational Research Journal, 29*(3), 663–676.

15
Future problem solving as education for innovation

Bonnie L. Cramond and Elizabeth C. Fairweather

UNIVERSITY OF GEORGIA, USA

Summary: With economic prosperity being linked to a nation's creativity, more pressure has been put on educators to nurture creativity and innovation in students. However, educators often feel overwhelmed by the prospect of fitting one more thing into their already standards-crowded curriculum. How can they teach academic content, basic skills, attend to the social and emotional needs of the students, and nurture creativity, too? Future Problem Solving is offered as a program that can be infused into the existing content to support the learning of academic content, introduce and reinforce critical skills, address social and emotional needs, and promote the tendency toward innovative thinking, too.

Key words: Creative, innovative, thinking skills, dispositions, social, emotional, Future Problem Solving, Creative Problem Solving, Common Core Standards.

Introduction

Recent reports from economists (Economic Commission for Africa, 2009; Florida, 2002 & 2005; Friedman, 2005) have emphasized the need for innovation in any society that hopes to retain an economic advantage in the coming years. Responding, reports on education have likewise emphasized the need for infusing skills to develop creativity and innovation into the curriculum in order to prepare a workforce for the future (National Center on Education and the Economy, 2007; National Advisory Committee on Creative and Cultural Education, 1999).

Educators, caught in the push for basic skills and high stakes testing, lament that they cannot teach to the standards and teach students to be innovative, too. What schools need is a method of teaching standards in the various subject areas that will also nurture creative thinking and innovative dispositions. The Future Problem Solving Program International (FPSPI) is a program that can meet both needs.

What is education for innovation?

For years, our educational institutions have emphasized the transmission of knowledge and skills to students, and we have measured students' success by the degree and accuracy by which they could retain and reproduce this information. When problems were presented, they had set answers and methods to solve them. Original methods of solution and unconventional answers were discouraged and their proponents often punished (De Simone, 1968). Is it any wonder that many inventors and innovators have been those without much formal education?

In fact, Baumol (2005, p. 34) noted that

> a review of the biographies of the most celebrated of these innovators shows, in a surprising share of these cases, a most remarkable absence of rigorous technical training and, in many cases, little education at all. The obvious names of yore—Watt, Whitney, Fulton, Morse, Edison and the Wright brothers—illustrate the point.

Other more contemporary inventors such as Walt Disney, Buckminster Fuller, Bill Gates, Steve Wozniak, Steve Jobs, and Michael Dell famously dropped out of college.

Lidwell (2005) interviewed another college dropout, Dean Kamen, inventor of the Segue as well as hundreds of other inventions, asking if education stifles creativity and innovation. Kamen responded:

> I think that most of the people who succeed in some extraordinary way, and most of the people who fail in some extraordinary way, tend to be people who did really well or really poorly in school. I think that school systems are really good at telling people how to do "okay" in the world. That is what their curriculum is about. That is what their institutional capability is about. That is what the people who run them are about. This is not to say that there isn't a world full of hugely talented teachers working hard every day to make a difference and change kids' lives. It is simply to say that the bureaucracy of the educational system limits the ability of educators to address the fringes. So if somebody was a good "B" student in school, you can be pretty sure that he or she is a good, average person. If a person gets A pluses in everything, or F minuses in everything, you can be pretty sure that he or she is an unusual person. Unusual people wrap around the ends of the bell curve. The system does not deal well with them. And I am not sure that the A+ person and the F− person are particularly different. It does not surprise me that when you look at people later in life, the people who got A pluses and F minuses end up doing substantially differently than the average people who are doing well in a system designed to accommodate the center of the bell curve.

Similarly, Baumol (2005) argued that the type of education that provides technical competency and mastery to future innovators is not the same as, and may even be antithetical to, the type of education that would promote creativity. He contended that by indoctrinating students into the paradigms and methods of the field, unorthodox thought is stymied. Such an observation led De Simone (1968) to ask, "Must inventiveness be sacrificed to education?" (p. 83).

Simonton (1997) reanalyzed the data that Cox (1926) had collected on 302 eminent people who spanned four centuries and six nations to determine the factors that affected their achievement. Through the use of modern statistical methods and by applying new concepts to create variables from the raw data, Simonton was able to look at the preludes to success in a different way. Pertinent to this discussion, he found that education had an inverted-U association with the ranked imminence of the creative individuals studied. In other words, some education helped, but not too much. In a later discussion of this study, he concluded that the ideal amount of education is somewhere in the last half of undergraduate school (Simonton, 2003, p. 365).

Yet, the individuals studied by Cox and Simonton were from another time. Even the contemporary innovators mentioned grew up in a different world. It is evident that some degree of knowledge and technical skill is needed for innovation. Perhaps Simonton (1997) is correct that there is an optimal level of education that is enough to know the field, but not so much as to indoctrinate one into it. But, this optimal level may be changing, especially in more technical fields. Baumol (2005) has predicted that the amount of necessary knowledge to innovate will grow in this increasingly complex world. "No carpenter such as John Harrison (who solved the longitude problem) no mere bicycle repairperson such as the Wright brothers, can any longer hope to contribute, for example, today's mind-boggling medical breakthroughs" (p. 50).

So the question may not be whether to educate for competence or innovation, but how to educate for both. In what Kuhn (1959) termed "The Essential Tension," he argued that both tradition and innovation are important in groundbreaking scientific research. "Very often the successful scientist must simultaneously display the characteristics of the traditionalist and of the iconoclast" (p. 23).

One way to teach content and encourage creativity is to involve students in using real methods and data toward the solution of authentic problems that they care about (Shaffer & Gee, 2007). There are several methods that might fit these criteria. For example, Tribus (1971) argued that innovation in engineering can be encouraged by having students creatively solve realistic problems early in their college careers. But what of teachers who are trying to promote creative thinking in the elementary and high school years? How might they find suitable problems in a variety of fields and offer opportunities for research and engagement in a number of areas? The Future Problem Solving Program International offers a systematic approach that provides the research bases as well as a method for addressing problems in a wide range of fields. It provides educators with a resource-rich program for assisting their students in finding, researching, and offering solutions to problems in the world or in their own community.

What is the Future Problem Solving Program International?

The Future Problem Solving Program International (FPSPI) is a program designed to stimulate student thinking about the future in the context of solving problems that have many possible solutions. Students can participate in either the competition components of the program or in a noncompetitive way. They work in teams or individually to understand and solve problems related to world topics provided by FPSPI or issues generated by the students themselves. In working to solve the problems students are guided to "develop the ability ... to design and promote positive futures using critical and creative thinking" (FPSPI, 2011, Welcome page).

The competitive components of FPSPI comprise an annual competition with two main categories involving problem solving (FPSPI, 2011). The first is the Global Issues Problem Solving (GIPS) category, which comprises practice and competitive problem solving experiences during the academic year so that students can become very familiar with the problem solving process. For each problem, students in grades equivalent to U.S. grades 4 through 12 study a different global issue provided by FPSPI. Examples of global issues include advantages and disadvantages of genetic testing and planning for natural and man-made emergencies in the 21st century. With the first two problems of the academic year, students work with their facilitator, without time limits, as they learn the steps of the problem solving process. Then, the third problem is a qualifying problem; if a team scores high enough on the qualifying problem, it moves on to the affiliate competition and another problem, and, if successful there, to the international competition.

The second competitive problem solving category of FPSPI is the Community Problem Solving (CmPS) component, open to the same grade levels as GIPS (FPSPI, 2011). In this category, students focus on issues relevant to their own communities. They select and define a problem in the community (local, state, or nation) that they would like to address and use the problem solving process to find real solutions that they actually apply to ameliorate the problem. Students become active problem solvers rather than just theoretical ones in this case. Teams with projects that receive the highest ratings are invited to the international conference.

A third competitive component is Scenario Writing. This is a short story writing competition in which individuals address the same issues as in the Global Problem Solving competition, but they do so through writing about the problem, possible solutions, and ramifications. Winners at the affiliate level can compete at the International Bowl in On-Site Scenario Writing.

The noncompetitive aspects of FPSPI include Action-based Problem Solving (AbPS) (FPSPI, 2012) and Curriculum-based Future Problem Solving (FPSPI, 2011). AbPS is very similar to CmPS, though it is open to students in all grades and is noncompetitive. Through a more simplified problem solving process, teacher-facilitators direct students in selecting a community issue to which they can actually apply their solutions.

Curriculum-based Future Problem Solving is a means for infusing the process into the curriculum without a competitive aspect. Teachers who understand the Future Problem Solving process can employ it in their classrooms to address problems in many areas of study. They may also purchase instructional materials from FPSPI (2011) to use in specific subjects with specific grade levels.

FPSPI teaches a problem solving method based on the Creative Problem Solving (CPS) model developed by Osborn and Parnes and adapted by Torrance (Cramond, 2009). CPS provides a method for finding solutions to open-ended problems through the following three main steps (which, in turn, comprise smaller steps):

1. *Understanding the challenge.* This includes the processes of gaining background knowledge about the topic or situation, understanding the situation, sifting through the information to determine the challenges in the situation, and paring these down or synthesizing them to select an underlying problem.
2. *Generating ideas.* During this step, students come up with possible solutions to the underlying problem.
3. *Preparing for action.* Through this final step, students learn to establish criteria by which they can make judgments about the viability of their solutions and use the criteria to select the best solution. Also, students learn to formulate a plan to implement the solution.

How does FPSPI develop skills necessary for innovation?

Throughout the academic year, students have several opportunities to experience the problem solving process in an in-depth and meaningful way. In addition to learning the process as a whole, students learn individual creativity skills. Specific creativity strategies are important for helping students accomplish the individual components of the problem solving process (Barlow, 2011). For example, students can be introduced to various divergent thinking strategies as they try to generate solutions to the underlying problem. In fact, FPSPI has produced a publication that offers a curriculum for teaching the future problem solving process and contains many suggestions for developing creativity skills in each step of the process (Barlow, 2011). This document, *The Problem Solving Experience: Classroom Curriculum Design to Promote Problem Solving in the 21st Century* (Barlow, 2011), outlines a plan for preparing students for problem solving and implementing problem solving experiences that are included in the document. It also contains a brief section on how some of the problem experiences can be integrated into the curriculum.

Despite the boosts to creative thinking that result from participating in FPSPI, teachers might feel that they cannot spend valuable classroom time participating in future problem solving because they need to devote all of their instructional time to teaching standards mandated by the district or national curriculum. On the contrary, integrating FPSPI may allow teachers to more effectively teach basic skills while engaging students in meaningful learning. In other words, teachers may also be building better readers and writers while they are developing innovators.

Various components of the problem solving process provide opportunities to teach fundamental skills that are emphasized in today's standards-based classrooms. First, research conducted to understand the problem, specifically, to gain background knowledge, advances students' non-fiction reading abilities. As students read text material about the topic under investigation, they experience practice with analyzing text structure, grasping main idea, and synthesizing information from multiple texts. The best part of using FPSPI to cover these skills is that students have an authentic platform for engaging in the skills. Also, as different issues are encountered throughout the academic year, different aspects of each skill can be highlighted. For instance, one form of note-taking can be taught during step one on the first problem, another type during step one on the second problem, and so on, so that students learn several methods of taking notes. Or, teachers can scaffold one particular skill throughout the progress of the problems during the year.

Below are specific skills, related to standards enunciated in the Common Core Standards for English Language Arts and Literacy in History/Social Studies, Science and Technical Subjects (Common Core Standards Initiative, 2011) for the United States that are involved in the future problem solving process:

1. determine main idea, analyze its development throughout the text, and summarize the text;
2. provide details from the text to support inferences drawn from the text;
3. consult multiple texts to gather information for problem solving; determine similarities and differences of multiple texts covering the same topic;
4. explain an author's organizational text structure and how it supports the author's purpose or argument;
5. determine an author's perspective, point of view, and purpose;
6. distinguish between fact and opinion;
7. use elements such as table of contents, headings, glossaries, etc.;
8. examine the correlation between primary and secondary sources related to a specific topic;
9. take notes from texts and organize gathered information;
10. evaluate an argument proposed in a text.

As students select an underlying problem and then develop a plan of action, they gain practice in still more language arts skills including conducting research and synthesizing and applying text information. The following specific skills, related to language arts standards enunciated in the Common Core Standards (Common Core Standards Initiative, 2011), can be practiced during the future problem solving process:

1. make predictions based on background knowledge;
2. synthesize information from multiple sources to write about a topic;
3. obtain information from relevant electronic sources through use of effective search terms;
4. express ideas in information gathered from various texts;
5. carry out research using multiple resources to develop knowledge in specific areas.

Covering these language arts skills alone would be enough to justify integrating future problem solving into the classroom from a curricular standpoint. But, FPSPI is structured so that it can easily cover content area curriculum standards as well. Each year, the GIPS topics range widely and can be connected with the school curriculum at several points. Although the specific topics differ each year, they are always drawn from three major strands: business and economics, science and technology, and social and political issues (FPSPI, 2012). For example, the GIPS topics for the 2011–2012 academic year included the following (FPSPI, 2011): effects of technology, globalization, and changing attitudes toward work on the workforce; effects of vanishing coral reefs and the benefits/costs of preserving them; the ethics of going to war over human rights and the balance between human rights and prevention of terrorism; free trade vs. fair trade; and what types of medicines should be produced and at what cost.

It is easy to see how these fit into the curriculum with two social studies examples. The free trade topic fits perfectly with economics standards espoused by the National (U.S.) Council for Social Studies (Council for Economic Education, 2011) which recommend that students in grade 8, for instance, be able to identify the benefits and costs of trade barriers and free trade as well as examine historical examples of trade barriers and what effects these trade barriers had during the time periods and situations in which they were enacted. The human rights topic also touches on social studies standards, this time in the area of civics and government. Human rights are typically covered at several points in U.S. curriculums with respect to individual versus government rights. Also, though, the civics curriculum generally addresses how nations interact with each other, and this may be addressed with this human rights problem.

It seems evident that the topics on technology, ecology, medicine, and aspects of terrorism also lend themselves to instruction in science. Similar standards to those in language arts, science standards require students to be able to: read appropriate text; cite evidence; analyze a writer's purpose; integrate data; distinguish among facts, reasoned judgment, and speculation; and compare findings from different sources (Common Core Standards Initiative, 2011). All of these skills, and more, are required by the FPSPI process.

With the CPS or AbPS it is even easier to make connections to the curriculum since the problem situations are not set by FPSPI but are selected instead by individual teams themselves. As a result, a teacher might select an area of the curriculum on which he/she would like to focus and present a problem situation in that area to students. For example, if state and local government is covered in the curriculum, students could look at local issues that may be addressed in some way through governmental means. These might include zoning issues in an area of rapid growth, lighting restrictions in a beach area, or providing low cost housing. In addition to content standards, during the first part of this component, researching the problem and devising a solution plan, students practice the important reading skills delineated above. In addition, though, as they implement the plan, as called for in these two components, the students engage in still more important language arts skills that involve writing.

When teachers use Curriculum-based Problem Solving, they can choose the problems to fit the content they are teaching and still use the processes of FPSPI to integrate future problem solving into the classroom. For example, a teacher charged with instructing a class of fifth grade students about space exploration, including having students describe technologies used for space exploration, explain the purpose of space exploration, and identify benefits, spin-offs, and by-products of space exploration, used the process as a cumulative activity. Students engaged in a future problem solving activity looking at problems that might be related to space exploration. The students were presented the following topic, or problem situation: space exploration provides many benefits for humans. However, it is costly, and space is beginning to get crowded with materials used for exploration. Also, countries are competing for dominance in space exploration, which could result in conflicts on earth or in space. What issues need to be resolved to minimize the negative aspects of space exploration?

Students worked through the future problem solving process in groups of four. First they conducted searches on the Internet, particularly the National Aeronautics and Space Administration website, as well as read periodicals and books on the topic of space exploration. They took notes on the texts they read and then settled on the following as problems to solve: what to do with space garbage, territorial disputes over space for satellites, and how to create an international organization devoted to space exploration.

Within their groups, students then used various creative strategies to generate solutions. Mainly the students brainstormed possible solutions, but some students utilized a modified form of listing and synthesizing problem components. Other students adapted solutions to similar problems. Final solutions included using the moon as a garbage "space" fill, sending all space garbage to a black hole, creating an international space organization through the United Nations, dividing the earth's atmosphere into lanes and assigning lanes to different countries for satellite use. Groups wrote up their plans and presented them to the class for feedback. Then they wrote up final proposals and published them in a notebook.

How does FPSPI develop innovative dispositions and meet affective needs?

If the FPSPI achieved the cognitive goals as stated, it could be considered an effective curricular tool. However, over the years it has become clear that part of learning to think critically and creatively comes from people's dispositions toward such activities (cf. Ennis, 1987). One definition of a thinking disposition is "a reasoned motivation for a certain thinking pattern, a thinking quality (open-mindedness, depth, and systematic thinking, etc.) imbued with reasoned motivation" (Harpaz, 2007, p. 1847). Perkins, Jay, and Tishman (1993) have described dispositions as being composed of abilities, sensitivities, and inclinations. In other words, true innovation comes from a combination of thinking skills and the motivation to be innovative, the awareness of when to seek novel solutions, and the desire to find and ask the right questions.

Harpaz (2007) discussed three ways to teach thinking: the skills approach, the dispositions approach, and the understanding approach. The first two approaches are characterized by an emphasis on either skills or dispositions, and the third is presented as a method of teaching that focuses on big ideas, knowledge in context, and investigative learning. It emphasizes the individual's construction of knowledge, and, therefore, incorporates content, as in traditional education, as well as thinking skills, and dispositions. If one accepts this classification, it should be apparent that FPSPI more clearly fits in the third category. There is certainly an emphasis on content as each problem is situated in a field and requires acquisition of knowledge in order to address the situation. There is an emphasis on using thinking skills, a problem solving process, and thinking dispositions are modeled and reinforced.

So, what are the dispositions that are vital to innovation, and how are they nurtured?

In this regard, we can take some guidance from research on creative individuals. Although there have been a number of studies on the characteristics of creative individuals (c.f. Tardiff & Sternberg, 1988), some characteristics are specific to certain fields. As Simonton (2009) indicated, the traits and dispositions of individuals who make breakthroughs in science, for example, tend to be different from those of individuals who exhibit creativity in the arts. However, there are some common dispositions that are generally related to good thinking across fields (Tishman, Jay, & Perkins, 1993, p. 148):

1 To be open-minded—explore alternative views and generate multiple ideas
2 To engage in sustained intellectual curiosity—wonder, find problems, observe and formulate questions
3 To seek understanding—find connections and explanations, recognize inconsistency, and seek clarity
4 To be planful and strategic—set goals and follow plans
5 To be intellectually careful—seek precision, thoroughness
6 To seek and evaluate reasons—question the given, seek evidence, weigh and assess information
7 To be metacognitive—awareness and control of one's mental processes, reflective.

The FPSPI promotes these dispositions by providing opportunities for students to find problems, define them, research them, and pursue them to completion. In addition, the coaches and winning teams model the dispositions for participants.

Yet, the greatest rewards that the participants immediately recognize are the affective ones. The opportunities to work with a team to investigate a problem, devise a solution, and, in some cases, implement the solution, provide many affective benefits to participants. For those who get to travel to state, regional, or international competitions, the benefits are increased. The positive benefits to young people's motivation, self-confidence, ability to deal with competition, and its sometimes subjective judgments, are valuable components of good academic competitions (Ozturk & Debelak, 2008). FPSPI is such a competition.

Treffinger (2011) noted that FPSPI provides opportunities for students to engage with others of high ability and similar interests, through which they receive affirmation that being smart and creative are admirable. Also, they are motivated and challenged by seeing what other students, like themselves, can accomplish.

FPSPI and current trends in education

More support for integrating the FPSPI program into the classroom is found in the fact that it fits so well with current trends and recommended best practices in education. One such trend, problem based learning (PBL), emphasizes student oriented development and resolution of problems that have multiple possible solutions (Barrell, 2007). The impetus behind PBL is that it involves students with real-life situations they are more likely to be interested in thinking about and investigating. Many of the ideas presented in PBL programs are part of the future problem solving process as laid out in CmPS competitions and AbPS materials. Consistent with the PBL concept, students engaging with either of these FPSPI components can direct their own inquiry into problem situations by examining issues at their school and local community or, if appropriate for the grade and ability level of the students, even the global community. Like FPSPI, problem based learning calls for group investigation of a problem situation and then problem posing and solving. To conclude the PBL experience, the groups put in place their final solutions. A difference between FPSPI and PBL is that most PBL situations contain all of the information needed to solve the problem, and there is often a real solution to the problem.

Another trend in education is service learning. Service learning can be defined as a strategy "that integrates meaningful community service with instruction and reflection to enrich the learning experience, teach civic responsibility, and strengthen communities" (National Service Learning Clearinghouse, 2012). Promoted because it teaches higher level thinking skills along with civic responsibility, service learning is also motivating for young people because it is authentic and empowering. Several evaluation studies have shown that service learning can positively impact content understanding, thinking skills, communication skills, creativity, and cognitive moral development (c.f. Hurd, 2006; Melchior & Bailis, 2002). The Community-based Problem Solving and the Action-based Problem Solving components of FPSPI are good examples of service learning.

A third major trend is the globalization of education. As the world becomes smaller, it is increasingly important for students to learn how our actions are interconnected in the larger world. The Global Issues Problem Solving component of FPSPI is specifically designed to address the larger issues that affect the world. Also, as the FPSPI program grows internationally, students have opportunities to interact with other students in participating countries that are addressing the same problems. At the time of this writing, 2012, the Future Problem Solving Program International involves thousands of students from Australia, Canada, Great Britain, Hong Kong, India, Japan, Korea,

Malaysia, New Zealand, Portugal, Russia, Singapore, Turkey, and the United States (FPSPI, 2012). Thus, students involved in the program are able to gain insight into the views of students from other very different cultures.

Is FPSPI effective?

It wouldn't matter how well the program met standards and fit with educational trends if it were not effective. In an international evaluation (Treffinger, Selby, & Crumel, 2011a), a team of evaluators surveyed key stakeholders in the program about the three competitive components of FPSPI: (a) Global Issues Problem Solving (GIPS), (b) Community Problem Solving (CmPS), and (c) Scenario Writing (SW). The three basic questions that they sought answers to were:

1. To what extent does FPSPI meet its stated goals?
2. What are the strengths of the program and what areas warrant improvement?
3. What is the impact of the program on its participants?

They designed surveys for each stakeholder group—affiliate directors, coaches, students, parents, and alumni—and collected responses via a web-based survey site. Affiliate directors, coaches, and the International Office invited all participants to respond, and a number of reminders were sent out. The team

> received responses from participants in the United States and several international affiliates; specifically, responses came from 34 Affiliate Directors, 48 program Alumni, 220 Coaches in 33 Affiliates, 633 students from 27 Affiliates, and 195 parents representing 23 Affiliates. There were responses from the eight largest Affiliates, 14 of the largest 15 (≥100 teams) and 17 of the 27 smallest Affiliates (<100 teams).
>
> *(Treffinger, Selby, & Crumel, 2011b, p. 1)*

From these data they concluded that (Treffinger et al., 2011b, pp. 1, 3):

1. All three of the program components were rated above average or higher for meeting the following 12 stated program goals:
    - developing teamwork and collaboration (working together, cooperating with each other);
    - developing leadership skills;
    - enhancing the skills of preparing and delivering materials and/or presentations that communicate ideas effectively;
    - showing evidence that team members are able to apply FPS skills in other situations;
    - developing the skills needed to manage time effectively;
    - fostering creative thinking (the ability to generate many, varied, and unusual options);
    - fostering critical thinking (the ability to sort and sift information or to focus one's thinking);
    - developing research and inquiry skills (the ability to gather information from many and varied sources);
    - using a deliberate process for Creative Problem Solving methods and tools;
    - developing skills in listening and following directions;
    - learning about complex issues that will shape the future; and,
    - developing an active interest in the future.

2. Respondents indicated strong overall satisfaction with the program. Common strengths mentioned included:

- easy and fair goals, rules, and procedures
- feedback and evaluation
- value in traveling and competing
- overall organization of events
- fun.

The most often listed areas for improvement include:

- ongoing growth of the program through marketing
- recruitment and retention of volunteers and participants
- stress from time management and multiple demands
- need for more training and support for coaches
- funding
- efficiency of communication
- technology-related concerns, for example, about the website
- location and logistics of the International Competition
- evaluation and feedback.

It is interesting to note that the areas listed for improvement are primarily institutionally related rather than about the program itself, with the exception of evaluation and feedback, which was also listed as a strength.

3. All respondent groups provided evidence of the positive impact that FPSPI had in a number of different ways and for each of the different stakeholder groups. Respondents wrote anecdotal accounts of the impact the program had on the cognitive, psychological, and social development of the students. They expressed appreciation for the cross-cultural experiences, opportunities to learn a process for problem solving, and encouragement to become involved in global and local challenges and issues.

The evaluators concluded that from the evidence offered by the respondent groups, that the FPSPI program went beyond its stated goals in helping students learn a variety of other life skills such as time and project management, leadership, autonomy, and social skills. Aren't these skills the tools of innovators? Further, they concluded, "that the respondents provided evidence (albeit informal, anecdotal evidence) that participation in FPSPI has had positive impact on young people—in personal relationships, in subsequent academic experiences, and in their work or career experiences" (Treffinger et al., 2011b, p. 4).

Conclusion

Will the teaching of FPSPI ensure that we create innovators? It is too much to expect any educational program to ensure us of this; there are too many intangibles. Education attempts to prepare students for future careers, but cannot ensure that individuals will use the training appropriately.

A metaphor may be useful here. We have a friend who enjoys the racetrack. Being mathematically minded, he has enjoyed creating prediction formulas, along with his chemist cousin, to foretell which horses would win which races. They would collect years of racing forms and test their formulas against the race results. Each year, they would further refine their formula by adding another variable: track condition, length of race, breeding, jockey, etc. They were constantly quantifying factors and adding them in. In spite of these comprehensive and increasingly complex formulas, they didn't win enough to make them rich. The formulas were not faulty; there are just too many variables, many of which are hard to quantify, but above all else, horses are sentient beings. All conditions can be right, but if the horse doesn't feel like running, for whatever reason, he won't.

So, too, educators can equip students with the tools and model the dispositions, but if students choose not to use them, they won't. Also, the very nature of innovation prescribes its rarity. There are so many factors that impact whether a particular innovative idea will break through the gatekeepers to fruition. But, if we want to increase the likelihood that individuals will be innovative, shouldn't we provide them the skills, dispositions, and practice to do so?

References

Barrell, (2007). *Problem Based Learning: An Inquiry Approach*. Thousand Oaks, CA: Corwin Press.

Baumol, W. J. (2005). Education for innovation: Entrepreneurial breakthroughs versus corporate incremental improvements. In A. B. Jaffe, J. Lerner, & S. Stern (Eds.), *NBER Book Series Innovation Policy and the Economy*, vol. 5 (pp. 33–56). Cambridge, MA: National Bureau of Economic Research/ MIT Press. Retrieved from www.nber.org/papers/w10578.

Common Core Standards Initiative. (2011). *Common Core Standards for English Language Arts and Literacy in History/Social Studies, Science, and Technical Subjects*. Retrieved from www.corestandards.org/assets/CCSSI_ELA%20Standards.pdf.

Council for Economic Education. (2011). *Voluntary National Content Standards in Economics, 2nd edition*. New York: Council for Economic Education.

Cox, C. (1926). *The Early Mental Traits of Three Hundred Geniuses*. Stanford, CA: Stanford University.

Cramond, B. (2009). Future problem solving in gifted education. In L. Shavinna (Ed.), *Handbook on Giftedness* Part 2, (pp. 1143–1156). New York: Springer.

De Simone, D. V., U.S. Department of Commerce (1968). Education for innovation. *Spectrum, IEEE, 5*(1), 83–89.

Ennis, R. (1987). A taxonomy of critical thinking dispositions and abilities. In J. B. Baron & R. J. Sternberg (Eds.), *Teaching Thinking Skills: Theory and Practice* (pp. 9–26). New York: W. H. Freeman and Co.

Barlow, M. (2011). *The Problem Solving Experience: Classroom Curriculum Designed to Promote Problem Solving in the 21st Century*. Melbourne, FL: Future Problem Solving Program International. Retrieved from http://fpspimart.org/index.php?main_page=product_info&cPath=7&products_id=241.

Economic Commission for Africa. (2009). *Economic Report on Africa 2009: Developing African Agriculture Through Regional Value Chains*. Addis Ababa, Ethiopia. Retrieved from http://new.uneca.org/Portals/era/2009/ERA2009_ENG_Full.pdf.

Florida, R. (2002). *The Rise of the Creative Class: And How it's Transforming Work, Leisure, Community and Everyday Life*. New York: Basic Books.

Florida, R. (2005). *The Flight of the Creative Class: The New Global Competition for Talent*. New York: Basic Books.

Friedman, T. (2005). *The World is Flat*. New York: Farrar, Straus, & Giroux.

Future Problem Solving Program International. (2012). Retrieved from www.fpspi.org/index.html.

Harpaz, Y. (2007). Approaches to teaching thinking: Toward a conceptual mapping of the field. *Teachers College Record, 109*(8), 1845–1874.

Hurd, C. A. (2006). *Is Service-learning Effective? A Look at Current Research*. Retrieved from https://docs.google.com/viewer?a=v&q=cache:rJjA1Wsak4oJ:tilt.colostate.edu/sl/faculty/Is_Service-Learning_Effective.pdf+effectiveness+service-learning+Hurd&hl=en&gl=us&pid=bl&srcid=ADGEESgxXjBwdvrYuD5BOLeDaLnugeVzDknXzXzvHdglHlllWur6hF3YaL6V0nTGwVwqjU4lDV4QU-U68DhhA8UuaMeGdGy-IdPBtlZq5tUcCW2qfjgZSGsEti68LIwcmW4AI3_fhq5rR&sig=AHIEtbRlAjEp3k_ystKiDzIYXJt2joOuYw.

Kuhn, T. (1959). The essential tension: Tradition and innovation in scientific research. In C. Taylor (Ed.), *The Third University of Utah Research Conference on the Identification of Scientific Talent* (pp. 21–30). Salt Lake City, UT: University of Utah Press.

Lidwell, W. (2005, October 13) More from Dean Kamen. *Make, 04*. Retrieved from http://makezine.com/extras/29.html.

Melchior, A., & Bailis, L.N. (2002). Impact of service-learning on civic attitudes and behaviors of middle and high school youth: Findings from three national evaluations. In A. Furco & S. H. Billig (Eds.), *Service-Learning: The Essence of the Pedagogy* (pp. 201–222). Greenwich, CT: Information Age Pub.

National Advisory Committee on Creative and Cultural Education. (1999). *All Our Futures: Creativity, Culture and Education*. London: DFEE. ED440037.

National Center on Education and the Economy. (2007). *Tough Choices or Tough Times: The Report of the New Commission on the Skills of the American Workforce*. Washington, DC: Jossey-Bass. Retrieved from http://www.josseybass.com/WileyCDA/WileyTitle/productCd-0787995983,descCd-description.html.

National Service Learning Clearinghouse (2012) *What is Service-Learning?* Retrieved from www.servicelearning.org/what-service-learning.

Ozturk, M. A., & Debelak, C. (2008). Affective benefits from academic competitions for middle school gifted students. *Gifted Child Today, 31*(2), 48–53.

Perkins, D. N., Jay, E., & Tishman, S. (1993). Beyond abilities: A dispositional theory of thinking. *Merrill-Palmer Quarterly, 39*(1), 1–21.

Shaffer, D. W., & Gee, J. P. (2007). Epistemic games as education for innovation. *British Journal of Educational Psychology Monograph Series II, Number 5—Learning Through Digital Technologies, 1*(1), 71–82.

Simonton, D. K. (1997). Biographical determinants of achieved eminence: A multivariate approach to the Cox data. In D. K. Simonton (Ed.), *Genius and Creativity: Selected Papers* (pp. 79–94). Greenwich, CT: Ablex.

Simonton, D. K. (2003). When does giftedness become genius? And, when not? In N. Colangelo & G. A. Davis (Eds.), *Handbook of Gifted Education*, 3rd ed. (pp. 358–370). Boston: Allyn & Bacon.

Simonton, D. K. (2009). Varieties of (scientific) creativity: A hierarchical model of domain-specific disposition, development, and achievement. *Perspectives on Psychological Science, 4*(5), 441–452.

Tardiff, T. Z., & Sternberg, R. J. (1988). What do we know about creativity? In R. J. Sternberg (Ed.), *The Nature of Creativity: Contemporary Psychological Perspectives* (pp. 429–440). New York: Cambridge University Press.

Tishman, S., Jay, E., & Perkins, D. N. (1993). Teaching thinking dispositions: From transmission to enculturation. *Theory into Practice, 32*(3), 147–153.

Treffinger, D. J. (2011). *Future Problem Solving Program International: Catalyst for Talent Recognition and Development*. Retrieved from www.fpspi.org/PDF/FPSPI/Talent Development.pdf.

Treffinger, D. J., Selby, E. C., & Crumel, J. H. (2011a). *Evaluation of the Future Problem Solving Program International*. Sarasota, FL: Center for Creative Learning.

Treffinger, D. J., Selby, E. C., & Crumel, J. H. (2011b). *Future Problem Solving Program International (FPSPI) Program Evaluation Report: Executive Summary*. Retrieved from www.mnfpsp.org/FILES/2012/FPSPI-Eval-ExecutiveSummary.pdf.

Tribus, M. (1971). Education for innovation. *Engineering Education, 61*(5), 421–423.

16

The trajectory of early development of prominent innovators

Entrepreneurial giftedness in childhood

Larisa V. Shavinina

UNIVERSITÉ DU QUÉBEC EN OUTAOUAIS, CANADA

Summary: This chapter focuses on early development of gifted entrepreneurs–innovators. The cases of Richard Branson, Warren Buffett, Steven Case, Michael Dell, Bill Gates, and Sam Walton are presented, which demonstrate that the trajectories of their entrepreneurial giftedness originated from childhood. The chapter considers these cases showing that an early developmental path is common for these gifted entrepreneurs and discusses the impact of early manifestations of entrepreneurial giftedness on the subsequent development of their entrepreneurial talent.

Key words: Entrepreneurial giftedness, early development, innovation talent, Branson, Buffett, Case, Dell, Gates, Walton.

Introduction

As was emphasized in the chapter on the fundamentals of innovation education (Shavinina, this volume), the essence of innovation coincides with entrepreneurial giftedness to a great extent and the development of entrepreneurial giftedness in today's children should be an important element of innovation education. *Entrepreneurial giftedness* refers to talented individuals who have succeeded in business by creating new ventures with at least a minimal financial reward (*fulfilled* entrepreneurial giftedness) or who demonstrated an exceptional potential ability to succeed (*prospective* entrepreneurial giftedness; Shavinina, 2009). Because research on entrepreneurial giftedness is in its initial stage and scholars do not know a lot about this phenomenon, it seems reasonable to use the concepts "entrepreneurial giftedness," "entrepreneurial talent," and "entrepreneurial ability" interchangeably.

This chapter is about early development of entrepreneurial giftedness in the cases of such well-known entrepreneurs as Richard Branson, Warren Buffett, Steven Case, Michael Dell, Bill Gates, and Sam Walton.[1] Case studies are the data source about various aspects of entrepreneurial giftedness of these innovative businessmen. The case study method is ideally suited to the investigation of individuals characterized by their rarity, which is the case of distinguished entrepreneurs. Case studies can describe their particular characteristics and experiences more thoroughly than is possible with any other research methodology. Case studies provide a holistic view of the subject (Foster, 1986; Frey, 1978) and allow researchers to develop and validate theories grounded in direct observation of specific individuals (Merriam, 1988). This study thus adopts a very different sampling procedure, *significant samples* (Simonton, 1999), because the developmental story of each prominent entrepreneur is itself a significant sample. This means that rather than sample randomly from the entire population of entrepreneurs, even highly accomplished ones, the present research focuses on the eminent exemplars of entrepreneurial giftedness relying on the existing (auto) biographical sources and published interviews.

From the methodological point of view, the use of (auto) biographical literature for the study of outstanding individuals incurs the possibility of the following limitations:

- The possible subjectivity and contradictions of autobiographers in their accounts of their own thinking processes, psychological states, and the surrounding events, which lead up to and follow their achievements.
- The timing of writing the (auto) biography, normally after an individual has already become a famous personality. In this case the writer possibly relies on vague memories of his or her thinking processes, which may have likely been weakened or altered over time. It definitely raises the question about the validity and reliability of subjective reports as data (Brown, A., 1978, 1987; Ericsson & Simon, 1980).

However, these concerns are minimized in the cases of Richard Branson, Warren Buffett, Steven Case, Michael Dell, Bill Gates, and Sam Walton in the following ways. First, they either wrote their (auto) biographies themselves (as Sam Walton did) and some even at a relatively young age (e.g., being age 43 as was Richard Branson or even younger as was Michael Dell) when details were still quite fresh in their minds, or they read the manuscripts of their biographies written independently by others and then approved them as, for instance, did Warren Buffett. Second, because parents and other relatives of Richard Branson and Michael Dell were alive at the time of writing their books, Richard and Michael consulted with them regarding the accuracy of details. In the case of Warren Buffett and Bill Gates their sisters were alive and biographers checked details with them, as well as with parents of Bill Gates. Third, some of them (e.g., Richard Branson) have always documented their lives, and the (auto) biographies are, therefore, based on numerous notebooks, which they "fill in every day" (Branson, 2002, p. 511). This is what makes, for example, Branson's (2002) autobiography an especially reliable data source.

The chapter proceeds as follows. The first section briefly discusses the existing literature on entrepreneurial giftedness. The second section presents the case studies of early development of entrepreneurial giftedness in Richard Branson, Warren Buffett, Steven Case, Michael Dell, Bill Gates, and Sam Walton. This section does not discuss the implications of these early developments. Its goal is only to present real examples of business ventures undertaken by these entrepreneurs in childhood. (They also undertook many business ventures during adolescence; however, this is the topic for another chapter.) The third section discusses the implications of the findings of these case studies; it discusses especially the impact of early entrepreneurial giftedness on the subsequent development of their entrepreneurial talent. A broader discussion of this impact appears in Chapter 36 of this volume, *What can innovation education learn from innovators with longstanding records of breakthrough innovations?* The final section summarizes the presented findings and highlights their importance for the fields of innovation education and giftedness, as well as practical implications aiming to develop entrepreneurial ability of today's children–tomorrow's innovators.

What do we know about entrepreneurial giftedness?

The Nobel Prize winner, Herbert Simon stated in 2002 that serious systematic research on entrepreneurship is just beginning (Simon, 2002). As research on entrepreneurship in general was analyzed in Shavinina (2009), it will not be discussed here. In the context of the given chapter it is important to note that the study of entrepreneurial giftedness in particular is an uncharted research territory. A few years ago Shavinina (2006) pioneered research on the subject by exploring the micro-social[2] factors in the development of entrepreneurial giftedness. Specifically, she analyzed the family milieu, "significant others," and great contemporaries in the case of Richard Branson, the most successful entrepreneur in the UK. She found that parents (i.e., nuclear family) and other relatives (i.e., extended family), "significant others," and great contemporaries are among the micro-social factors conducive to the appearance of exceptional entrepreneurial ability of Richard Branson. Shavinina (2006) identified *general supportive factors* such as talented parents with a wide range of interests, who loved each other and their children a great deal. Parents trusted in the children, had open, sincere relationships with them, treated them as equals, supported them in all possible ways, and set up high ethical standards for them. These factors are necessary but not sufficient for the development of entrepreneurial giftedness. Something else must be in play, particularly aimed at the actualization and growth of entrepreneurial ability.

In this regard, Shavinina (2006) also identified *specific factors*, which were uniquely aimed at the development of Richard's entrepreneurial talent. These factors include: mother as a role-model of entrepreneur, early exposure to challenges, rule-breaking and "change the world" attitude, parental love of adventure, hard working, and sense of teamwork, as well as independence in thoughts and actions. The two lines of evidence (which overwhelmingly support) that the micro-social factors identified were indeed important and predetermined the development of Richard's entrepreneurial giftedness, were: (a) the personality and creativity traits of Richard-adult, and (b) the long-lasting effect of those factors on both Richard-person and his entrepreneurial career. The unique combination of the identified micro-social factors greatly contributed to the appearance of Richard Branson—eminent, highly accomplished entrepreneur.

The second direction of research on entrepreneurial giftedness is related to the study of its early signs. As it was considered in the chapter on the fundamentals of innovation education (Shavinina, this volume), it will not be described here. The proposed research on the trajectories of early development of entrepreneurial giftedness will complement and extend the existing studies by further exploring the nature of this interesting scientific phenomenon.

Early development of entrepreneurial giftedness: the six case studies

The focus of this research described in this chapter is on the cases of early entrepreneurial giftedness of Richard Branson, Warren Buffett, Steven Case, Michael Dell, Bill Gates, and Sam Walton. Why were these six individuals chosen for analysis? First of all, they all are eminent businessmen distinguished by a long-standing record of entrepreneurship (i.e., *fulfilled* entrepreneurial giftedness). In this context it does not matter whether they founded and successfully developed one company (as Michael Dell or Bill Gates did) or a group of companies (as Richard Branson did).

Second, nearly all of them are exceptional in other ways as well. For instance, Richard Branson is the most successful entrepreneur in the UK.[3] Born in 1950, he is the founder and owner of the *Virgin* group of companies. By 2010, the *Virgin* group included 250 companies, which successfully function in various areas of business. *Virgin Airways*, *Virgin Direct*, *Virgin Money*, *Virgin Books*, just to mention a few. Besides that, Branson accomplished world records in crossing the Atlantic by boat and in hot-air balloons (Branson, 2002). Certainly, an individual with such a remarkable record of achievements in entrepreneurial business is of great interest to giftedness researchers. However, this is not the whole story. The significance of the study of Richard Branson for the fields of innovation education and giftedness consists in his twice exceptionality during childhood and adolescence. Being dyslexic and underachiever in school, yet demonstrating practical intelligence and creative abilities, Branson was able to achieve an unbelievable success in business. Another important aspect is that Richard started his first entrepreneurial venture—*Student* magazine—from scratch: he did not have money. The same was true for *Virgin Shops* and *Virgin Records*, his second and third ventures, respectively. Also, Richard's entrepreneurial motivation deserves the attention of innovation education specialists and high ability scholars, because it turns out that creativity is the driving force of all his ventures. Business can be "a creative enterprise in itself. If you publish a magazine, you are trying to create something that is original, that stands out from the crowd. ... Above all, you want to create something you are proud of" (p. 57).

Warren Buffett is the world's greatest investor and the second richest person on the planet. However, high ability researchers should study Warren Buffett not only because of his entrepreneurial giftedness, but also because he was a mathematical prodigy from early childhood. He liked numbers very much. Friends and acquaintances have said that young Warren was a mathematical prodigy and was able to add large columns of numbers in his head—a talent he still occasionally shows off to business associates these days (Lowenstein, 1996; Schroeder, 2009). Therefore, we are dealing with a unique type of dual giftedness in the case of Warren Buffett.

The existing (auto) biographical sources on Steven Case and Michael Dell demonstrate that they both were highly intelligent individuals and their brainpower was mainly focused on creating new, diversified business ventures from early childhood (see below for examples; Ashby, 2002; Dell, 1999). For instance, Jacques Nasser calls Michael Dell "a gifted new leader" and Fred Smith, the founder of FedEx "a genius in the computer world," just to mention two (Dell, 1999).

Bill Gates epitomizes a unique case of the coincidence of intellectual, entrepreneurial, and leadership types of giftedness. Not surprisingly, the pertinent literature contains a lot of information about Gates' highly developed intelligence and precociousness (Dickinson, 1997; Manes & Andrews, 1994; Sherman, 2000; Wallace & Erickson, 1992). As Shavinina (2012a) analyzed this issue in detail elsewhere, it will not be discussed here. For the purposes of the given chapter, only examples of early development of Bill's entrepreneurial giftedness will be presented below.

Likewise, Sam Walton embodies an extraordinary case of the coincidence of entrepreneurial, intellectual, and leadership types of giftedness and should also be included in the study of early development of entrepreneurial giftedness (Walton, 1992; Teutsch, 1991).

Taking all of the foregoing into account, it seems, therefore, justified to include the cases of Richard Branson, Warren Buffett, Steven Case, Michael Dell, Bill Gates, and Sam Walton for research on early development of entrepreneurial giftedness.

Entrepreneurial giftedness originates from childhood: Richard Branson

The following two examples of early entrepreneurial giftedness are the most compelling in the case of Richard Branson. The first took place at the age of 12 when he decided to follow his "mother's example[4] and make some money" by growing Christmas trees (Branson, 2002, p. 37).

> Undeterred by the school's lack of faith in my ability with numbers, I saw an opportunity to grow Christmas trees... I went round to talk Nik[5] into the plan. He was also on holiday from his school....We would plant 400 Christmas trees in the field.... By the Christmas after next, they would have grown to at least four feet and we would be able to sell them. Nik and I agreed to do the work together, and share the profits equally.

> That Easter we furrowed the ground and planted the 400 seeds. … We worked out that, if they all grew to six feet, we would make £2 a tree, creating a grand total of £800, compared with our initial investment of just £5 for the seeds. In the following summer holiday, we went to investigate the trees. There were one or two tiny springs above ground, but the rest had been eaten by rabbits. We exacted dire revenge and shot and skinned a lot of rabbits. We sold them to the local butcher for a shilling each, but it wasn't quite £800 we had planned.
>
> *(Branson, 2002, p. 37)*

Nevertheless, Richard did not give up after his very first venture failed. Soon he saw another excellent business opportunity and focused entirely on it.

The following Christmas Nik's brother was given a budgerigar as a present. This gave Richard the idea for

> another great business opportunity: breeding budgies! For a start, I reasoned, I could sell them all year round rather than just during the fortnight before Christmas. I worked out the prices and made some calculations about how fast they could breed and how cheap their food was, and persuaded my father to build a huge aviary.
>
> *(Branson, 2002, pp. 37–38)*

He wrote to Dad from school and explained the financial implications:

> So few days now until the holidays. Have you ordered any material we might want for our giant budgerigar cage? I thought our best bet to get the budgerigars at reduced rate would be from Julian Carlyon. I feel that if the shops sold them for 30sh., we would get say 17sh. And we could buy them off him for 18 or 19sh. Which would give him a profit and save us the odd 10sh. per bird. How about it?
>
> *(Branson, 2002, p. 38)*

His father

> reluctantly built the aviary and the birds bred rapidly. However, I had overestimated the local demand for budgies. Even after everyone in Shamley Green had bought at least two, we were still left with an aviary full of them. One day at school I got a letter from my mother breaking the bad news that the aviary had been invaded by rats which had eaten the budgies. It was only many years that she confessed she had been fed up with cleaning out the aviary so on day had left the cage door open and they all escaped. She didn't try too hard to recapture them.
>
> *(Branson, 2002, p. 38)*

These childhood business ventures demonstrate that Richard was indeed an entrepreneurially gifted child from a young age (Brown, M., 1994). This early experience had a profound impact on the subsequent development of his entrepreneurial talent (to be discussed below).

The developmental trajectory of early entrepreneurial giftedness: Warren Buffett

Examples of early development of entrepreneurial giftedness are abundant in the case of Warren Buffett. Thus, he made the first few cents from selling packs of chewing gum at six years of age (Schroeder, 2009). Around the same time his family went for vacations to Lake Okoboji, and Warren managed to buy a six-pack of Cokes for twenty-five cents; then he waddled around... selling the sodas at five cents each, for a nickel profit. Back in Omaha, he bought soda pop from his grandfather's grocery and sold it door-to-door on summer nights while other children played in the street.

(Lowenstein, 1996, p. 10)

As mentioned above, Warren was a mathematical prodigy and this influenced his entrepreneurial behavior. One episode from his childhood is particularly amazing in this respect. When Warren was seven, he was hospitalized with a mysterious fever. Doctors removed his appendix, but he remained so ill that the doctors feared he would die. Even when his father fetched his favorite noodle soup, Warren refused to eat. But left alone, he took a pencil and filled a page with numbers. These, he told his nurse, represented his future capital. "I don't have much money now," Warren said cheerfully, "but someday I will and I'll have my picture in the paper" (Lowenstein, 1996, p. 11). Purportedly in his death throes, Warren sought succor not in dad's soup but in dreams of money.

He started to do money early in his life. For instance, by the time he was nine or ten, Warren was selling used golf balls at a golf course with his friend—until someone reported them and they got kicked by police. Being ten-years old, he got a job selling peanuts and popcorn at the University of Omaha football games (Schroeder, 2009).

At age ten, Warren's father took him to New York. The boy knew what he wished to see: "I told my dad I wanted to see three things," one of which was New York Stock Exchange (Schroeder, 2009, p. 58). Meeting with the senior partner of the investment bank Goldman Sachs, Sidney Weinberg, who was the most famous man on Wall Street, and a member of the Stock Exchange, At Mol made a huge impression on Warren. A vision of his future was planted in New York: he wanted to make money.

> It could make me independent. Then I could do what I wanted to do with my life. And the biggest thing I wanted to do was work for myself. I didn't want other people directing me. The idea of doing what I wanted to do every day was important to me.
>
> *(Schroeder, 2009, p. 59)*

By age 11 Warren was thus fascinated by stocks as other boys were by model aircraft. His childhood bedroom was full of copies of the *Wall Street Journal*, annual reports, and stock analyses. It was his favourite reading (Ross & Holland, 2006). He made his first investment in the stock market when he was just 11 years old. Specifically, he bought three shares of Cities Service for himself and three for his sister Doris, at $38 a share. "I knew then he knew what he was doing," Doris would recall. He "lived and breathed numbers" (Lowenstein, 1996, p. 12). But Cities Service plunged to 27. They sweated it out, and the stock recovered to 40, whereupon Warren sold, netting, after commission, his first $5 of profit in the market.

By age 13, Buffett was running his own businesses as a paperboy and selling his own horseracing tip sheet. He was an enthusiastic paperboy for the *Washington Post*, covering five paper routes and delivering 500 newspapers before school. His gross earnings were that of a full-time adult worker (Ross & Holland, 2006).

This list of Warren's early entrepreneurial ventures is impressive and clearly testifies to his outstanding entrepreneurial ability in childhood.

Entrepreneurial flair at a young age: Steven Case

Steve was an entrepreneur from a young age. "I think I was an entrepreneur before I knew what an entrepreneur was," said Case many years later (Gendron, 2004, p. 309). "From a relatively early age I got interested in business" (Academy of Achievement, 2004, p. 1). At six he started a juice business with his brother Dan. They collected limes from the backyard for free and sold limeade for 2 cents a cup. At this age Steve had already set up a perfect business model, which consisted in zero cost in exchange for high profits. Steve and Dan formed *Case Enterprises* before their teenage years. They started calling their bedrooms "offices," and typed out advertising circulars late into the evening. Steve would often wake up his brother in the middle of the night with a new idea. He loved to come up with new ideas and see if they would get implemented. Steve thus demonstrated a unique and rare talent for being a creator and innovator at the same time (Shavinina, 2003, 2007). From the beginning, Steve had the ideas, and Dan had the money. "Even if it was five bucks to start some business, I never had any money," Steve says. "So Dan would give me five bucks and suddenly own fifty percent of my idea" (Ashby, 2002, p. 10). He brainstormed with Dan, searching for ways to make money. Making the most out of the daily contact he had with people on his paper route, "he sold seeds, Christmas cards and watches, among other things" (R. Brown, 2000, p. 4).

Steve and Dan did not make much money, but it did not really matter to them. "We made a fortune, tens of dollars," Dan joked later. Steve and Dan didn't need extra cash, because their parents were well off, but the challenge of marketing and sales excited the young boys. "It was clear I'd be an entrepreneur," Steve said later in an interview (Ashby, 2002, p. 11).

Early path of entrepreneurial giftedness: Michael Dell

The developmental trajectory of Michael Dell during his childhood provides a strong evidence of his early entrepreneurial giftedness. The most impressive example of Michael's early entrepreneurial ventures is his own stamp auction, which he organized at the age of 12. The story is as follows. The father of his best friend in Houston was a pretty avid stamp collector, so naturally Michael and his friend wanted to get into stamp collecting, too.

> To fund my interest in stamps, I got a job as a water boy in a Chinese restaurant two blocks from my house. I started reading stamp journals just for fun, and soon began noticing that prices were rising. Before long, my interest in stamps began to shift from the joy of collecting to the idea that there was…"a commercial opportunity."
>
> *(Dell, 1999, p. 3)*

> It was obvious to me from what I'd read and heard that the value of stamps was increasing, and being a fairly resourceful kid, I saw this as an opportunity. My friend and I had already bought stamps at an auction, and since I knew even then that people rarely did something for nothing, I assumed that the auctioneers were making a decent fee. Rather than pay them to buy the stamps, I thought it would be fun to create my own auction.
>
> *(Dell, 1999, p. 4)*

He, therefore, decided to create his own auction where he "could learn even more about stamps *and* collect a commission in the process" (Dell, 1999, p. 4). Michael got neighbors to consign their stamps to him, and then advertised "Dell's Stamps" in *Linn's Stamp Journal*. Finally, he typed, with one finger, a 12-page catalogue (he did not yet know how to type, nor had a computer) and mailed it out. Michael made $2,000 on his very first business venture (Friedman, 2009). He was only 12 years old.

It is interesting to point out that even earlier in life Michael had been fascinated with the idea of eliminating unnecessary steps. Thus, when he was in third grade, he sent away for a high school diploma. An eight-year-old Michael had seen the advertisement in the back of a magazine: "Earn your high school diploma by passing one simple test," it said. He liked third grade, but trading nine years of school for "one simple test" seemed like a pretty good idea to him (Dell, 1999, p. xv). When a woman from the testing company came to visit, both she and Michael's parents decided that he applied to take the test as a joke. But he was quite serious. ... It is not, therefore, surprising that he later started a company based on eliminating the middleman (i.e., bypassing the dominant method of computer distribution). *Dell Computer Inc.* sells computers directly to customers, deals directly with its suppliers, etc., all without the unnecessary and inefficient presence of intermediaries (Holzner, 2006). Early entrepreneurial giftedness of Michael Dell, as well as its impact on his future success in business, is thus quite evident.

The manifestations of early entrepreneurial giftedness of Bill Gates

Bill Gates demonstrated obvious signs of entrepreneurial giftedness in childhood. One example of this is a contract he designed, at age 11, to give him access to his sister's baseball glove. The contract states: "When Trey[6] wants the mitt, he gets it" (Boyd, 1995, p. 15). He paid the grand total of $5 on acceptance of terms, signing in imitation of his father's signature with yet another variant of his name "William H. Gates Jr." Apparently, Bill was already showing interest in legal formalities: at the bottom of the page was a blank line with the legend "witness sign here" (Manes & Andrews, 1994, p. 21). Bill's father was a lawyer, and he learned most of his legalese from dad. Some biographers believe that the fact that young Bill had a contract, and the one-sided terms of the arrangement, are a good indicator of the type of entrepreneur he would become (Boyd, 1995).

While in elementary school, Bill distributed newspapers. He designed a scam to make more money. As a promotion, a drug-and-discount store would include crude keys and some of them would open a prize box at the establishment. Bill carefully scrutinized the keys he had to deliver, determined which ones were valuable, and took them to the store to get his prize (Manes & Andrews, 1994).

At the age of 12, Bill was part of the Contemporary Club, which was formed by the Laurelhurst elementary schoolchildren, which got together to discuss several topics and play games. One of the games that Bill played was "Risk," which involved taking over the world and establishing world domination (Manes & Andrews, 1994).

In sixth grade Bill prepared a report entitled "Invest with Gatesway Incorporated" for his economics class. To help out, Bill's grandma helped him with the cover letter and his father got him an interview with one of the company's principals where he worked. In this report, Bill described himself as an inventor and concluded that if his idea is good and he is able to hire good people and raise enough money, he will be successful. He got an A1 in this report (Manes & Andrews, 1994).

Few months after Lakeside, the private school that Bill attended, bought a computer; Computer Center Corporation (CCC) offered computer time on better machine at a lower cost. Bill and his friends became so experienced that they were able to temporarily stop the computer from working. They also changed some files on the computer to make it look as though they were not spending much time there. Someone at CCC noticed these changes and the boys were banned from using the computer for several weeks (Sherman, 2000).

At Lakeside, Bill and some other friends joined to form the Lakeside Programmers Group (LPG) with the goal of making money by using computers. Bill read a lot of business magazines to prepare himself for this opportunity (Dickinson, 1997). "The average grade school student did not spend time reading about business, but Bill was far from average" (Dickinson, 1997, p. 18). At one point in time Bill was thrown out from the LPG, but he came back after his friends realized that they needed his business knowledge, and Bill thus became the president of the group (Wallace & Erickson, 1992). It is clear that Bill was highly competitive in different areas including business.

CCC hired the LPG to find bugs by crashing the computer system and then fix those bugs. It was the first job of 12-years-old Bill. In exchange for the work, CCC gave the boys unlimited computer time (Sherman, 2000). Bill became very skilled at this crashing and fixing and thereby obsessed with programming. He skipped many gym classes to be with computers. After some time CCC closed down, but the experience provided Bill a very valuable experience in the business world of computers.

At the age of 13, Bill Gates had a vision of what the future of computers would be and often discussed their unlimited possibilities with his friend Paul Allen (Dickinson, 1997). For instance, he asked Paul: "Don't you think that someday everybody will have one of these things? And if they don't, couldn't you deliver magazines and newspapers and stuff through them? I mean, I wonder if you could make money doing something like that?" (Schlender, 1995, p. 46). Bill Gates was just 13-years-old, but he already had an objective vision of the future and knew that computers would change the world.

Bill's interest in software thus originated from childhood and his subsequent entrepreneurial ventures were related to it.

> I wrote my first software program when I was thirteen years old. It was for playing tic-tac-toe. The computer I was using was huge cumbersome and slow and absolutely compelling. I realized later part of the appeal was that here was an enormous, expensive, grown-up machine and we, the kids, could control it. We were too young to drive or to do any of the other fun-seeming adult activities, but we could give this big machine orders and it would always obey. Computers are great because when you're working with them you get immediate results that let you know if your program works. It's feedback you don't get from many other things. That was the beginning of my fascination with software.
>
> *(Myhrvold & Rinearson, 1995, pp. 1–2)*

Bill Gates, therefore, acquired a useful experience in both computer programming and entrepreneurship well before he co-founded *Microsoft*. His early entrepreneurial giftedness clearly indicated the area of his specialization in business: software development.

The origins of Sam Walton's entrepreneurial giftedness

Sam Walton—the man who created the largest retail empire in the world, *Wal-Mart*—proved to be industrious and hard-working starting from the age of ten. He milked cows and sold the milk to the local store owner (Walton, 1992). Then while he was in town, he delivered newspapers. He thought that he was already awake, and there was plenty of time before school to get the newspaper route done. Sam thus wanted to kill two birds with one stone. After school he tried to find other jobs to earn money. While his brother and other kids played in the street, Sammy was making money by cleaning porches, raking leaves, or shelling peas. Whatever anyone wished done, Sam was ready to do it. This is the origin of his leadership. Leadership comes from within an individual, who is not afraid of taking on a task and finding a way to get the job done. Sam was exactly that person from childhood years. It is very important to mention his leadership because "from the start, Sam was the leader, the genius" (Teutsch, 1991, p. 12). Like Bill Gates, Sam Walton represents a rare case of the coincidence of entrepreneurial, leadership, and intellectual types of giftedness.

While his leadership and intellectual giftedness is considered elsewhere (Shavinina, 2012c), a few above-mentioned examples of Sam's entrepreneurial undertakings undoubtedly testify to his developing entrepreneurial giftedness in childhood.

Discussion

The above-described cases of Richard Branson, Warren Buffett, Steven Case, Michael Dell, Bill Gates, and Sam Walton demonstrate one of the possible paths of the development of entrepreneurial talent, namely: the trajectory of *early* development of entrepreneurial giftedness. The end point of this trajectory is well known: all these individuals achieved unbelievable successes in business by creating new, highly profitable companies with Bill Gates at the top as the richest man in the world. What in particular was the impact of their early business ventures undertaken in childhood on the subsequent development of their entrepreneurial giftedness?

As to Richard Branson, while neither of his childhood ventures had the effect of making money as he wanted, they did teach him "something about maths."

> I found that it was only when I was using real numbers to solve real problems that maths made any sense to me. If I was calculating how much a Christmas tree would grow, or how many budgies would breed, the numbers then became real and I enjoyed using them. Inside the classroom I was still a complete dunce at maths. I once did an IQ test in which the questions just seemed absurd. I couldn't focus on any of the mathematical problems, and I think that I scored about zero. I worry about all the people who have been classified as stupid by these kinds of tests. Little do they know that often these IQ tests have been dreamt up by academics who are absolutely useless at dealing with the practicalities of the outside world. I loved doing real business plans—even if the rabbits did get the better of me.
>
> *(Branson, 2002, pp. 38–39)*

Also, Richard's involvement in early business ventures determined the development of certain personality traits, which distinguish gifted entrepreneurs. These traits include, but are not limited to: love of challenges and adventures, creativity, initiative, independence, rule-breaking attitude, and strive for excellence (Shavinina, 2008). It is interesting to see a striking similarity between Branson-entrepreneur in childhood and Branson-entrepreneur in adulthood (Alcraft, 1999; Shavinina, 2006).

Warren Buffett learned useful lessons from his entrepreneurial ventures in childhood. As mentioned above, he purchased three shares of Cities Service at $38 a share for himself and his older sister. Although the stock fell to just over $27, he held his shares until they rebounded to $40, unfortunately selling them before they climbed to $200. The experience taught him one of the basic lessons of investing, which he has successfully been applying in adulthood: patience is a virtue. However, the first and foremost outcome of Warren's early development of entrepreneurial giftedness was a clear understanding that finance and investment are *his* areas in business. This is probably in great part due to Buffett's amazing mathematical ability briefly noted earlier.

As it follows from the previous section, creating new businesses was a hallmark of Steve Case's childhood.

> I'm not sure I knew what an entrepreneur was when I was ten, but I knew that starting little businesses and trying to sell greeting cards or newspapers door-to-door or just vending machine kind of thing is—there's just something very intriguing to me about that.
>
> *(Academy of Achievement, 2004, p. 1)*

Steve's entrepreneurial giftedness originated in childhood and predetermined the subsequent development of his entrepreneurial talent in adolescence and adulthood.

> I was not an outstanding student. I did a reasonable amount of work. I got generally good—pretty good grades, but I was not that passionate about getting straight A's. I was more passionate about starting businesses... I enjoyed high school and college, and I think I learned a lot, but that was not really my focus. My focus was on trying to figure out what businesses to start.
>
> *(Academy of Achievement, 2004, p. 1)*

Steve's early entrepreneurial experience allowed him to better understand himself. For instance, he found that "managing a mature business is not my thing" (Goodell, 1996, p. 2).

The very first business venture of Michael Dell had a lasting effect on the subsequent development of his entrepreneurial talent. Thus, the roots of the famous "direct model" of *Dell Computer Inc.* lie in his stamp auction, when Michael first experienced the power and the rewards of being direct (i.e., eliminating the middleman). He also learned an important lesson that "if you've got a good idea, it pays to do something about it" (Dell, 1999, p. 4). This is why Michael Dell founded a company with a cool idea in mind: to "eliminate the middleman," that is, to bypass the dominant way of computer distribution. The above-presented examples of Michael Dell's entrepreneurial giftedness demonstrate that early business ventures during his childhood lie at the heart of the corporate strategies of *Dell Computer Inc.*, which revolutionized the computer industry. This is mainly his famous direct approach that is considered a great business innovation: the company sells PCs directly to customers, deals directly with its suppliers, and so on, all without the unnecessary and inefficient intermediaries.

Likewise, if one looks at early business experience of Bill Gates, one can conclude that his entrepreneurial giftedness in childhood obviously determined his success in adulthood. For instance, his very first entrepreneurial venture related to software business and *Microsoft* is about the same industry. It is interesting to note that Bill Gates was not only an entrepreneur in childhood; he was also a subject matter expert, that is, he wrote computer programs himself. It became one of his distinguishing characteristics as an entrepreneur and businessman in adulthood. In other words, he is not just a co-founder and the chairman of the leading software company in the world; he is a gifted computer programmer as well who can write codes and check programs written by his talented employees. This is why he is highly respected by computer industry experts. There is, therefore, an amazing similarity between Bill-child and Bill-adult, a famous entrepreneur and businessmen.

Sam Walton's early entrepreneurial ventures are evidence of his entrepreneurial giftedness in childhood. The examples of his entrepreneurship reveal an individual, who would not afraid any job that could raise money. This is also true for other gifted entrepreneurs considered in this chapter. In the case of Sam Walton his leadership giftedness influenced the development of his entrepreneurial talent.

Taking together, the cases of Richard Branson, Warren Buffett, Steven Case, Michael Dell, Bill Gates, and Sam Walton demonstrate that entrepreneurial giftedness is a real scientific phenomenon, where scientific study should be intensified. Specifically, these six cases are about early development of entrepreneurial giftedness. The trajectory of such development indicates that entrepreneurial giftedness indeed originates from childhood and has a long-standing impact on the subsequent development of entrepreneurial talent of these famous entrepreneurs that determined their eventual success in business. How does entrepreneurial giftedness come to such fruition?

First of all, each of them learned the basics of business. Thus, Richard Branson gained knowledge in doing real business plans for real projects. Warren Buffett learned about stocks, Bill Gates about software, and so on. Michael Dell discovered that direct business model is a great way of making money. The powerful educational element was thus embedded in the above-considered early entrepreneurial projects of these great entrepreneurs. It helped them develop their entrepreneurial giftedness to its fullest extent.

Second, "crystallizing experience" (Walters & Gardner, 1986) had a place during early development of their entrepreneurial giftedness. "Crystallizing experience" refers to remarkable and memorable unusual encounters between a developing person and a particular field of endeavour (p. 307). This phenomenon manifests itself in changes in "an individual's concept of the domain, his performance in it, and his view of himself" (Walters & Gardner, 1986, p. 309). This reorganization of the individual experience later becomes a foundation for great achievements. Thus, when Warren Buffett at age ten met Sidney Weinberg, the most famous investor in New York, this event definitely crystallized his experience. Especially what Weinberg did at the end of the visit made a huge impression on Warren. "As I went out, he put his arm around me and he said, 'What stock do you like, Warren?' He'd forgotten it all the next day, but I remembered it forever" (cited in Schroeder, 2009, p. 58). Warren would never forget that Sidney Weinberg, a big shot on Wall Street, had paid such attention to him and seemed to care about his opinion. Similar events had a place in life of almost all gifted entrepreneurs.

Third, all these six individuals realized in late childhood that their professional future will be related to entrepreneurship and business; and not, say, to science or arts. Steve Case expressed it well in summarizing his entrepreneurial experience in childhood, "It was clear I'd be an entrepreneur" (Gendron, 2004, p. 2). Similarly, on his way back from New York where a ten-year-old Warren attended the New York Stock Exchange and met two prominent investors, he decided that investment and finance were exactly what he wanted to do in life (Schroeder, 2009).

Fourth, the nature of entrepreneurial experience in childhood predetermined to a great extent either the choice of an *area* of specialization in business (e.g., investment for Buffett or computer software for Gates) or a *way* of doing business (e.g., Dell founded a company on the fundamental idea of eliminating intermediaries that is at the center of the famous direct business model rooted in his very first entrepreneurial venture: the stamps auction). It also looks like those gifted entrepreneurs, whose childhood business ventures were more or less diversified (e.g., Richard Branson), will become a sort of versatile businessman, who is able to found and lead a group of companies. For instance, Richard's *Virgin* group includes 250 companies, which he founded. However, to be more precise, it seems like almost every great entrepreneur in childhood tried to go in a variety of directions and do multiple businesses aimed at making money. Steve Case is an excellent example. As the previous section indicates, he did a lot of entrepreneurial things as a child. Nevertheless, he became well known as the founder of the *America Online Company* (aol.com). Probably, the willingness and ability of children-entrepreneurs to undertake many different entrepreneurial ventures and, hence, acquire a diversified business experience is one of the distinguishing characteristics of early entrepreneurial giftedness.

Fifth, entrepreneurial undertakings of Richard Branson, Warren Buffett, Steven Case, Michael Dell, Bill Gates, and Sam Walton in childhood helped to "build" their personalities of entrepreneurially gifted individuals. This means that their persistence, determination, motivation to achieve, and discipline were further developed. These are the distinguished characteristics of all the gifted (Shavinina, 1995) including the entrepreneurially gifted. For instance, the six entrepreneurs had to be very well organized and highly disciplined for their early entrepreneurial projects to succeed. Warren Buffett and Sam Walton are good examples, who had to wake up early in the morning in order to deliver newspapers before going to school. The childhood ability of these six gifted entrepreneurs not to stop after the first failed project and go ahead with the implementation of other ideas contributed to the development of their persistence and determination to succeed.

Finally, the micro-social factors (i.e., family, friends, and school) obviously contributed to the development of Richard Branson, Warren Buffett, Steven Case, Michael Dell, Bill Gates, and Sam

Walton's entrepreneurial giftedness in childhood. It was either the mother and the aunt who acted as the role models of entrepreneur for young Richard, or the father who took the ten-year-old Warren to the New York Stock Exchange that had a profound effect on the boy, or the oldest brother of Steven Case who was the partner in Steve's entrepreneurial ventures by constantly supporting his ideas and "investing" in them, or the friend's father who fueled Michael's interest in stamp collecting and stamp auctions, or the grandma who helped Bill with a report for his economics class, or the father-entrepreneur of Sam who was able to make profit at a time when many Americans did not have anything to eat. These examples are encouraging and obviously demonstrate what parents and teachers of the gifted can do today in order to develop entrepreneurial giftedness in their children. Such cases should be incorporated in innovation education.

Summing-up

This chapter is about the early developmental path of entrepreneurial giftedness by way of the examples of such eminent entrepreneurs–innovators as Richard Branson, Warren Buffett, Steven Case, Michael Dell, Bill Gates, and Sam Walton. Focusing on instances of entrepreneurship in childhood, it was found that they all started "to do business" early in life by trying to undertake ventures with money-making potential. They all wanted to make profit since a very young age. The trajectory of early development of entrepreneurial giftedness was common for these gifted entrepreneurs. It was also found that such a developmental path influenced the subsequent development of their entrepreneurial talent in a number of ways discussed above. A tireless wish to start new businesses is an important characteristic of early entrepreneurial giftedness (Shavinina, 2012b).

The research presented is significant for the advancement of the fields of giftedness and innovation education in many ways. First, it fills an obvious niche in high ability studies by exploring a special type of giftedness: entrepreneurial giftedness. Second, it examines the nature of entrepreneurial giftedness by studying its early developmental trajectory. This can be considered a new direction in research on entrepreneurial giftedness, which has not yet been investigated. Third, the initial finding of this new direction is that entrepreneurial giftedness originates from childhood and has a profound impact on the subsequent development of an individual's entrepreneurial talent and the eventual success in business. Fourth, parents and teachers concerned with the versatile development of children's abilities get useful ideas for practical applications. The case studies of early entrepreneurial giftedness of Richard Branson, Warren Buffett, Steven Case, Michael Dell, Bill Gates, and Sam Walton give educators a clear insight on how this type of giftedness may look in childhood and how it can be developed. This should be done as a part of innovation education. Certainly, future research is needed for a comprehensive understanding of the multidimensional nature of entrepreneurial giftedness.

Acknowledgments

The research presented herein was supported under the Support for Innovative Projects program (Grants AN-129135 and AN-143064) of the *Fonds québécois de la recherche sur la société et la culture* (FQRSC). The findings and opinions expressed in this chapter do not reflect the positions or policies of the FQRSC. Special thanks to Larry Vandervert for his encouraging comments on the first draft of this chapter.

Notes

1. The names of entrepreneurs are mentioned in an alphabetical order in this chapter.
2. There is the differentiation between micro-social and macro-social factors. The micro-social factors refer to the influence of such social institutions as family, school, university, and proximal social surrounding (e.g., childhood friends). The macro-social factors refer to those societal, cultural, and historical contexts in which individuals live (i.e., the contemporary Zeitgeist). Macro-social factors often operate through micro-social ones.
3. It should be emphasized that in this chapter I do not analyze the impact of the talented individual on his or her micro- and macro-social environment. Also, the negative influence of the social context on forming high abilities has been considered elsewhere and it is not analyzed here (Csikszentmihalyi, 1996; Howe, 1990).
4. The mother was a role model of an entrepreneur for Richard and that was her major impact on her son's developing entrepreneurial giftedness. As the influence of micro- and macro-social factors on the development of his entrepreneurial talent was analyzed in details elsewhere (Shavinina, 2006), this issue will not be discussed in this chapter.
5. Nik Powell was his best friend.
6. Trey was a nickname of Bill Gates.

References

Academy of Achievement. (2004). Steve Case interview: Co-founder, America Online (pp. 1–8). www.achievement.org/autodoc/page/cas1int-1. Downloaded on May 3, 2010.

Alcraft, R. (1999). *Heinemann Profiles: Richard Branson*. Jordan Hill, Oxford: Reed Educational and Professional Publishing.

Ashby, R. (2002). *Steve Case: America Online Pioneer*. Brookfield, CT: Twenty-First Century Books.

Boyd, A. (1995). *Smart Money: The Story of Bill Gates*. Greensboro, NC: Morgan Reynolds Incorporated.

Branson, R. (2002). *Losing My Virginity: The Autobiography*. London: Virgin Books.

Brown, A. L. (1978). Knowing when, where, and how to remember: A problem of metacognition. In R. Glaser (Ed.), *Advances in Instructional Psychology*, vol. 1 (pp. 77–165). Hillsdale, NJ: Erlbaum.

Brown, A. L. (1987). Metacognition, executive control, self-regulation, and other even more mysterious mechanisms. In F. E. Weinert & R. H. Kluwe (Eds.), *Metacognition, Motivation, and Understanding* (pp. 64–116). Hillsdale, NJ: Erlbaum.

Brown, M. (1994). *Richard Branson: The Inside Story*. London: Headline.

Brown, R. (2000). Meet the new boss. *CED, 26*(2), 4.

Csikszentmihalyi, M. (1996). *Creativity*. New York: HarperCollins Publishers.

Dell, M. (1999). *Direct from Dell*. New York: HarperBusiness.

Dickinson, J. D. (1997). *Bill Gates: Billionaire Computer Genius*. Springfield, NJ: Enslow Publishers.

Ericsson, K. A., & Simon, H. A. (1980). Verbal reports as data. *Psychological Review, 87*(3), 215–251.

Foster, W. (1986). The application of single subject research methods to the study of exceptional ability and extraordinary achievement. *Gifted Child Quarterly, 30*(1), 333–337.

Frey, D. (1978). Science and the single case in counselling research. *Personnel and Guidance Journal, 56*(5), 263–268.

Friedman, L. S. (2009). *Business Leaders: Michael Dell*. Greensboro, NC: Morgan Reynolds Publishing.

Gendron, G. (2004). Practitioners' perspectives on entrepreneurship education: An interview with Steve Case, Matt Goldman, Tom Golisano, Geraldine Laybourne, Jeff Taylor, & Alan Webber. *Academy of Management Learning and Education, 3*(3), 302–314.

Goodell, J. (1996). The fevered rise of America Online. *Rolling Stone, 744*, 60–66.

Holzner, S. (2006). *How Dell Does It: Using Speed and Innovation to Achieve Extraordinary Results*. New York: McGraw-Hill.

Howe, M. J. A. (1990). *The Origin of Exceptional Abilities*. Cambridge, MA: Blackwell.

Lowenstein, R. (1996). *Buffett: The Making of an American Capitalist*. New York: Main Street Books.

Manes, S., & Andrews, P. (1994). *Gates: How Microsoft's Mogul Reinvented an Industry and Made Himself the Richest Man in America*. New York: Simon & Schuster.

Merriam, S. B. (1988). *Case Study Research*. San Francisco: Jossey Bass.

Myhrvold, N., & Rinearson, P. (1995). *The Road Ahead: Bill Gates*. New York: Viking.

Ross, E., & Holland, A. (2006). *100 Great Business and the Minds Behind Them*. Naperville, IL: Sourcebooks, Inc.

Schlender, B. (1995). What Bill Gates really wants. *Fortune Magazine.* January 16, p. 46.

Schroeder, A. (2009). *The Snowball: Warren Buffet and the Business of Life.* New York: Bantam Books.

Shavinina, L. V. (1995). The personality trait approach in the psychology of giftedness. *European Journal for High Ability, 6*(1), 27–37.

Shavinina, L. V. (Ed.) (2003). *The International Handbook on Innovation.* Oxford: Elsevier.

Shavinina, L. V. (2006). Micro-social factors in the development of entrepreneurial giftedness. *High Ability Studies, 17*(2), 225–235.

Shavinina, L. V. (2007). Comment l'innovation peut-elle accroitre la performance organisationnelle? In L. Chaput (Ed.), *Modèles Contemporains en Gestion: Un Nouveau Paradigme, La Performance* (pp. 167–197). Le Delta I, Québec, Canada: Presses de l'Université du Québec.

Shavinina, L. V. (2008). Early signs of entrepreneurial giftedness. *Gifted and Talented International, 23*(2), 3–17.

Shavinina, L. V. (2009). On entrepreneurial giftedness: Where did all great entrepreneurs come from? In L. V. Shavinina (Ed.), *International Handbook on Giftedness* (pp. 793–807). Dordrecht, Netherlands: Springer Science.

Shavinina, L. V. (2012a). The coincidence of intellectual, leadership, and entrepreneurial giftedness: What insights does Bill Gates reveal on these types of giftedness? (In preparation).

Shavinina, L. V. (2012b). When intellectual and entrepreneurial giftedness come together: A rare type of double giftedness. (In preparation).

Shavinina, L. V. (2012c). What does Sam Walton tell us about giftedness? On the coincidence of intellectual, leadership and entrepreneurial giftedness. (In preparation).

Sherman, J. (2000). *Bill Gates: Computer King.* Brookfield, CT: Millbrook Press.

Simon, H. A. (2002). Achieving excellence in institutions. In M. Ferrari (Ed.), *The Pursuit of Excellence Through Education* (pp. 181–194). Mahwah, NJ: Erlbaum.

Simonton, D. K. (1999). Significant samples: The psychological study of eminent individuals. *Psychological Methods, 4*(4), 425–451.

Teutsch, A. (1991). *The Sam Walton Story: An Inside Look at the Man and His Empire.* New York: Berkley Books.

Wallace, J., & Erickson, J. (1992). *Hard Drive: Bill Gates and the Making of the Microsoft Empire.* New York: HarperCollins Publishers.

Walters, J., & Gardner, H. (1986). The crystallizing experience: Discovering an intellectual gift. In R. J. Sternberg & J. E. Davidson (Eds.), *Conceptions of Giftedness* (pp. 306–331). Cambridge: Cambridge University Press.

Walton, S. (1992). *Made in America.* New York: Doubleday.

Part VI
The role of teachers, parents, and schools in the development of innovators

17

Educating wizards
Developing talent through innovation education

Sarah J. Noonan
UNIVERSITY OF ST. THOMAS, USA

Summary: The educational conditions favoring the realization and use of a capacity or aptitude for innovation may be described using the results of research related to creativity, effective learning and teaching, and motivation. These include (1) interactions with *expert teachers* in a variety of settings, (2) access to a *challenging curriculum*, (3) opportunities to receive *skilled coaching* to develop talents, and (4) *opportunities to satisfy individual needs and achieve developmental goals* in innovator-friendly environments. To realize a talent, individuals must use their gifts to innovate; gifts remain underdeveloped, ignored, or perhaps even unknown to individuals without opportunities to know a gift exists as well as see how it may be realized in the world (Gagné, 2010). The author illustrates these points through a case study of Harry Potter, an adolescent with unrealized special powers, as portrayed in *Harry Potter and the Sorcerer's Stone* (Rowling, 1997).

Key words: Innovation education, expert teachers, wizards, Harry Potter, challenging curriculum, achievement motivation.

Introduction

J. K. Rowling, author of the internationally famous Harry Potter series, delivered a memorable commencement address at Harvard University on the topic of imagination and failure (Rowling, 2008). Several important points about innovation were evident from her description of becoming a writer: (1) *education and experience* (including failure) were important sources of her creativity, and (2) *a love and passion for doing something personally meaningful* led to her notable achievement as an innovative writer.

Rowling studied the "classics" in college to pursue her fascination with English literature and Greek mythology, despite its apparent lack of currency in the real world, and against the advice of her parents (Rowling, 2008). She agreed to study German to satisfy her parents' desire for a marketable career, but soon changed her mind: "Hardly had my parents' car rounded the corner at the end of the road than I ditched German and scuttled off down the Classics corridor" (p. 10).

After college, Rowling worked for Amnesty International while pursuing her career as a writer, learning viscerally about the human capacity for evil and good through stories of unspeakable violence told by its victims. Rowling's view of her personal circumstances as well as her capacity for moral imagination changed. She learned an important lesson about imagination from the victims of atrocities.

> Imagination is not only the uniquely human capacity to envision that which is not, and therefore the fount of all invention and innovation. In its arguably most transformative and revelatory capacity, it is the power that enables us to empathise with humans whose experiences we have never shared.
>
> (Rowling, 2008, p. 24)

Perhaps the most notable line in Rowling's speech with regard to innovation education occurred when she declared her personal passion: "I was convinced that the only thing I wanted to do, ever, was to write novels" (Rowling, 2008, p. 9). Rowling knew her future vocation at a young age and pursued it at every opportunity, using her education and life experience as raw material for her books. Like Harry Potter, the protagonist in her famous book series, Rowling benefitted from her education when she pursued her interests with single-minded purposefulness (Csikszentmihalyi, 1996) and made a decision to be creative (Sternberg, 2004). Much can be learned from studies of creativity and creative people to identify important elements regarding innovation education.

Creativity and innovation

Rowling's personal story about her creativity closely aligns with findings from Csikszentmihalyi's (1996) 30-year study of creativity: "Creativity results from the interaction of a system composed of three elements: a culture that contains symbolic rules, a person who brings novelty into the symbolic domain, and a field of experts who recognize and validate the innovation" (p. 6). Creative people express their talents in a domain, using knowledge of its structure in new ways to innovate. The culture or "domain" works in partnership with creative individuals to produce innovation.

> To say that Thomas Edison invented electricity or that Albert Einstein discovered relativity is a convenient simplification.... Edison's or Einstein's discoveries would be inconceivable without the prior knowledge, without the intellectual or social network that stimulated their thinking, and without the social mechanisms that recognized and spread their innovations.
>
> (p. 7)

The third element, expert validation, occurs when an innovation results in an *"effective surprise…, producing a shock of recognition following which there is no longer astonishment"* (Bruner, 1962/1979, p. 18). Bruner suggested criteria for expert validation, describing three types of effectiveness: (1) predictive, the innovation adds significant value to the field, (2) formal, new relationships and "ways of putting things together not before within reach" become visible (p. 19), and (3) metaphoric, "connecting domains of experience that were before apart, but with a form of connectedness that has the discipline of art" (pp. 19–20).

Innovators break structures and rules, recombine ideas in new and here-to-before unknown ways, apply new metaphors to gain insight, and liberally draw from a variety of disciplines and experiences to develop new ideas. Deep knowledge of a domain, combined with individual imagination and creative effort, produce innovations of proven value (Csikszentmihalyi, 1996). Ultimately external validation recognizes the contribution made to the domain, even if not acknowledged or appreciated during the inventor's lifetime (Sternberg, 2003).

Individuals use certain innate or nurtured capacities (gifts) within a domain, applying creativity to innovate (Gagné, 2010). When others see an excellent performance or innovation, the application of a gift in a domain serves as evidence of talent (Gagné, 2010). The distinction between a gift and a talent raises critical awareness regarding the role of innovation education: to realize a talent, individuals must use their gifts to innovate; gifts remain underdeveloped, ignored, or perhaps even unknown to individuals without opportunities to know a gift exists as well as see how it may be realized in the world. Even "gifted" individuals need an education to develop their talents to innovate.

While many assume students with natural abilities or aptitudes need little or no support to achieve their dreams, too many talented students predictably languish and experience talent loss due to a lack of recognition of their talents and support to nurture their creativity (Kim, 2008). A key point regarding creativity and innovation education here: education provides access to knowledge needed to *realize a passion* within a particular domain, while at other times, education *introduces passions* to unsuspecting creative students largely through engaging their personal curiosities and interests (Csikszentmihalyi, 1996).

The educational conditions favoring the realization and use of gifts bestowed or earned to innovate may be identified using the results of research related to creativity, effective learning and teaching, and motivation. Combining research from a variety of disciplines, I address the following question with regard to innovation education in this chapter: "What kind of innovation education nurtures creativity and its expression?"

The short answer to the above question with regard to student learning and education involves at least some of the following components: (1) interactions with *expert teachers* in a variety of settings, (2) access to a *challenging curriculum*, (3) opportunities to receive *skilled coaching* to develop talents, and (4) *opportunities to satisfy individual needs and achieve developmental goals* in innovator-friendly environments (at least long enough for something magical to happen). I refer to the above combination of these elements as "engaging pedagogy".

Edgerton used the term, "pedagogies of engagement", to describe the teacher's "capacity to engage students actively with learning in new ways" (as cited in Shulman, 2002, p. 38). I use the term engaging pedagogy to refer to active and appealing learning events responsive to the spectrum of interests and needs found in individuals and within a cohort of learners likely to foster achievement, talent development, and innovation in all students.

In the remainder of this chapter I use the first book in the Harry Potter series, *Harry Potter and the Sorcerer's Stone* (1997), to illustrate some important "facts" about innovation education. I selected Harry Potter's story to illustrate the principles of innovation education for several reasons: (1) his story captured the imagination of an entire generation, now graduating from college and entering the workforce, (2) the challenges faced in developing his talent resemble "typical" challenges experienced by creative people during adolescence and young adulthood, (3) his educational program offered opportunities and disappointments worthy of discussion and comparison, and (4) readers may enjoy, and later recall concepts related to innovation education due to the power of story in learning and leadership (Noonan, 2007).

If you have not yet read any of the Harry Potter books, do not worry. This short synopsis of *Harry Potter and the Sorcerer's Stone* (1997) provides enough background to get started:

> Harry Potter has never played a sport while flying on a broomstick. He's never worn a cloak of invisibility, befriended a giant, or helped hatch a dragon. All Harry knows is his miserable life with the Dursleys, his horrible aunt and uncle, and their abominable son, Dudley. Harry's room is a tiny closet at the foot of the stairs, and he hasn't had a birthday party in eleven years. But all that is about to change when a mysterious letter arrives by owl messenger: a letter with

an invitation to a wonderful place he never dreamed existed. There he finds not only friends, aerial sports, and magic around every corner, but a great destiny that's been waiting for him… if Harry can survive the encounter.

(Rowling, 1997, back cover)

Educating wizards at Hogwarts School and elsewhere

Harry Potter, *a capable student in need of a good education*, received his invitation to attend the Hogwarts School for Witchcraft and Wizardry on his eleventh birthday (Rowling, 1997). Despite Harry's impressive family background as a child of deceased and heroic wizards, Harry's natural "gifts" needed development to realize his talent for innovation. Assuming innovators (past and present) at one time possessed a "gift" in need of development, research regarding talent development may foster the capacity and desire of *all students* to innovate in a talent domain (Gagné, 2010). Harry needed contact with expert teachers and skilled coaches (Gentry, Steenbergen-Hu, & Choi, 2011) and a challenging curriculum with opportunities to experiment with technique and format to release his creative potential (Reis & Renzulli, 2010).

Expert teachers

At the Hogwarts School for Witchcraft and Wizardry, Harry soon found there was a lot to learn about magic from teachers with differing levels of expertise and appeal. Descriptions of his teachers and educational program as seen through the eyes of a budding adolescent, bring the cast of characters to our attention. "A dumpy little witch called Professor Sprout" taught students about the night sky at midnight on Wednesdays and on the care of plants used in magic (Rowling, 1997, p. 133). Professor McGonagall imposed firm discipline at the first class meeting and then impressed students by turning her desk into a pig and then back into a desk again (Rowling, 1997). Viewed as a capable teacher, students took "complicated notes" from her lecture and then attempted to turn a match into a needle during a practice exercise. Only one student could do it.

While Harry found some value in every class, he clearly distinguished experts from novice teachers:

> The class everyone was looking forward to was Defense Against the Dark Arts, but Quirrell's lessons turned out to be a bit of a joke.… His turban, he told them, had been given to him by an African prince as a thank-you for getting rid of a troublesome zombie, but they they weren't sure they believed this story.
>
> *(Rowling, 1997, p. 134)*

Harry's colorful description of his teachers raises an important point: if you want to learn about the most effective teachers and teaching, just ask the students. Gentry et al. (2011) conducted a study of teacher effectiveness, asking talented students to nominate the best teachers using surveys measuring "constructs of appeal, challenge, choice, enjoyment, interest, meaningfulness, and self-efficacy" (p. 113). An appealing curriculum and student-friendly classroom attract student interest and provide enjoyment in learning. A challenging curriculum offers "rigor, depth, and complexity" (p. 113). Students experience learning as meaningful when the content proves valuable to their present and future lives. Making choices and gaining competence round off the list identified by Gentry et al. (2011).

When interviews with exemplary teachers of gifted students were analyzed, four themes regarding teacher effectiveness emerged. Teachers (1) knew their students well and were personally interested in them, (2) established high expectations for their teaching and student work, (3) planned meaningful learning activities, connecting the curriculum to student lives and also providing choices, and (4) exhibited passion "for their students, teaching, and for their content" (Gentry et al., 2011, p. 116). When engaging teachers possess a passionate interest in their subject *and* they pass on this love of their discipline to their students, predictable student creative achievement occurs (Torrance in Kim, 2008).

Challenging curriculum

The "general education program" for wizards at Hogwarts School consisted of content and methods from a variety of disciplines, such as history, anthropology, horticulture, and science (Rowling, 1997). Theory was combined with "hands-on" learning experiences to help students increase their magical powers. In some cases the classes and curriculum were highly regarded, while in other cases more substance was clearly needed.

For example, Harry complained about the History of Magic class because Professor Binns "droned on and on while they [the students] scribbled down names and dates, and got Emeric the Evil and Erick the Oddball mixed up" (Rowling, 1997, p. 133). What might happen if students learned history in ways that allowed them (1) to understand events in history as recurring themes and challenges in human life, (2) to trace the roots of present day issues and experience to the past and learn from it, (3) to use knowledge to examine the issues in their lives now and in the future, and, finally, (4) to awaken a desire to know more?

Professor Binns would need a "curriculum upgrade" (Jacobs, 2010), to offer more rigorous and meaningful content. For example, Binns might introduce a dilemma involving the wise and ethical use of magic by asking this question at the beginning of his History of Magic class: "If you had to choose between (1) killing one person to save the lives of five others and (2) doing nothing even though you knew that five people would die right before your eyes if you did nothing—what would you do?" (Justice with Michael Sandel, n.d., p. 1)? The challenging question would intrigue Harry and his classmates, causing them to examine how the "special powers" granted to them should be wisely used (a good lesson for all to consider).

Adolescents question adult values and begin to recognize shades of gray in moral issues; however, they lack coping skills for managing difficult issues (Caskey & Anfara, 2007). Professor Binns and other teachers like him should locate topics and investigations to increase critical thinking and student interest, providing a more engaging pedagogy and, potentially, more reasons to study history. Realistically, teachers do not teach an entire discipline; they select topics and develop a curriculum, aimed at meeting the demands of various stakeholders (including the discipline) in their assigned classes. The most important stakeholder, the student, should exert some influence on the curriculum based on developmental goals and interests.

Jarvis (2009, p. 238) developed an exercise to test the fitness of curriculum offered to talented individuals. Teachers were first asked to develop a profile of a famous person (past or present) in the "famous five exercise", and then evaluate the effectiveness of the curriculum using the following questions:

a Would the curriculum provide sufficient *challenge* to allow the individual to progress in his or her areas of evident advanced ability?
b Would the curriculum address the individual's areas of *strength* (based on his or her later accomplishments) that might not be evident at this time?
c Does the curriculum provide this individual with the opportunity to work in an area of *interest* that is likely to ignite a passion for one or more aspects of the field?

d   What *supports* should be provided for this student to allow him or her to perform at the highest possible level?
e   What *modifications* to the curriculum are indicated by the responses to (a)–(d)?

The criteria used for this exercise easily applies to innovation education.

Because the innovator and the partner (the talent domain) work together, exposure to the domain requires teachers to engage students in the study of recurring themes, processes, and problems encountered in a field. The "core curriculum" must be worthy of students and the discipline, fostering a deep understanding of the knowledge domain as well as attracting students to become scholars and inventors in their chosen domain. In addition to a challenging core curriculum, innovation education curriculum should help students discover their interests and experience the work of the discipline.

The Schoolwide Enrichment Model (SEM) serves as a proven model to raise student achievement and foster talent development (Ries & Renzulli, 2010). The Triad Model, part of SEM, involves a three-stage process: first, students explore topics of interest; next, students participate in group training activities to master content and methods, such as problem solving, investigation, and invention in a domain; and, finally, students experience opportunities to conduct individual and group investigations, applying their creativity to a domain (Ries & Renzulli, 2010; see also Delcourt & Renzulli; Lyons & Reis, this volume). The model

> encourage[s] creative productivity on the part of students by exposing them to various topics, areas of interest, and fields of study… further train[ing] them to *apply* advanced content, process-training skills, and methodology training to self-selected areas of interest.
>
> *(Ries & Renzulli, 2010, p. 44)*

Harry and his friends at Hogwarts School investigated topics of interest and discovered their magical powers, even though ridiculous and uncaring teachers, bureaucratic and dim-witted tyrants, nonsensical regulations, too many useless tests, and even a lack of access to the academic library served as initial barriers to learning (Rowling, 1997). Harry thrived in his new environment because of the opportunities to pursue his chosen profession, access to plentiful resources, and a certain degree of freedom to pursue his curiosities. He explored the school and its mysterious forests, sought ways to acquire magical powers, and, finally, met his destiny on the fateful day when he learned to fly.

Skilled coaching and talent development

Harry did not have a lot of good to say about his teachers, classes, or school until he learned to fly. Then something magical happened.

> He mounted the broom and kicked hard against the ground and up, up he soared; air rushed through his hair and his robes whipped out behind him – and in a rush of fierce joy he realized he'd found something he could do without being taught – this was easy, this was *wonderful*.
>
> *(Rowling, 1997, p. 148)*

Professor McGonagall observed Harry's first flight, quickly recognized his natural talent, and fostered his talent for flying through mentorship, skilled coaching, acceleration, and opportunities to sharpen his gift in an extracurricular sport, playing Quidditch (a competitive team flying game resembling soccer; Rowling, 1997).

While Harry learned he could fly, he still needed guidance and practice to fully develop his technique. Professor McGonagall recruited him to play on a team, added three days of flying practice to his class schedule, and observed him preparing for the first big game of Quidditch. Skilled coaching provided by expert teachers and mentors allows individuals to practice and develop their talents, receive individual advising, and participate in making choices about learning.

Favorable educational conditions, such as interactions with expert teachers, a challenging curriculum, and skilled coaching would be incomplete without an explanation of the motivational pathways and pedagogy attracting students to learning. Students may have access to a good educational program but fail to engage in learning, leaving their opportunities and potential success at the school door due to competing motivational forces. A brief review of pathways inviting students to learn and sustain their effort follows.

Motivational pathways and pedagogy

Imagine Harry Potter standing with his peers at the front door of Hogwarts School, deciding on his greatest need and most predictable route to fulfilling it. Should Harry pursue his love of learning and study topics of greatest interest to him? Would it be wise to follow his peers down the path leading to friendships and acceptance, taking the path marked belonging? Should the underconfident and poorly educated Harry sign up for literacy classes, designed to increase his academic prowess and future success as a student by taking the achievement route? Perhaps Harry can't resist the pull of the path allowing him to be master of his universe, where choices and freedom attract him down Autonomy Road?

Harry Potter, like other humans, possesses "innate psychological needs for competence, autonomy and relatedness" and expends energy to meet these goals (Deci & Moller, 2005, p. 579). Learning activities and experiences available in the classroom environment serve as potential ways to meet human needs based on the desire for learning and achievement (competence), belonging and autonomy. Motivational pathways (my term) attract students to learning, providing choices to meet basic human needs.

Distinguishing between different motivational pathways serves an important purpose for innovation education: the human needs attracting students toward a particular pathway offer ways to attract and sustain engagement in learning. Effective teachers plan accordingly, fostering the development of potential wizards in their classrooms. A brief description of four motivational pathways to learning follows.

For the love of learning

Appealing, interesting, novel, experiential, challenging, and personally meaningful episodes of learning attract and sustain students' intrinsic motivation to learn long enough for learning to occur (Frick, 1992). Issued as invitations to students, careful scripting of an engaging lesson draws in even the most reluctant learner, overcoming typical barriers to learning because of the inherently interesting topic or activity involved.

Personal interest reliably attracts some students with a history of interest in the subject matter; *situational interest* temporarily attracts all students (Frick, 1992). Certain appealing factors in learning attract student interest. This includes a match with the students' developmental stage and life experiences (Arnett, 2000), the desire to experience something new and complex (stimulates impulse to explore; Silvia, 2008), becoming involved in intrinsically interesting topics (Frick, 1992), encountering themes of universal significance (Hidi, 1990), and/or experiencing challenges to beliefs and perspectives (Frick, 1992).

Because interest may be categorized as an emotion, the study of "interesting" topics elicits cognitive and emotional responses (Silvia, 2008), making more of a long lasting impression in the learner's mind. Interesting topics or activities stimulate a love of learning, fostered when students experience something new and complex (Silvia, 2008), value the activity (Wigfield & Wagner, 2006), encounter meaningful and personally relevant goals and tasks (Urdan & Turner, 2006), experience humor (Dohn, Madsen, & Malte, 2009; Robertson, Blain, & Cowan, 2005), and/or seek and experience challenge (Fredricks, Blumenfeld, & Paris, 2004).

Lessons designed to lure students to learning inspire imagination and involve intriguing topics, thrilling activities (like flying a broom), or descriptions of a harrowing experience, whet students' appetite for learning. Harry Potter's interest in learning was based on his desire to acquire magical skills, including flying a bloom, making himself invisible, and exploring forbidden areas of the school in search of dark and powerful secrets. A powerful secret known by effective teachers: students can be attracted to learning, at least temporarily, when the tasks presented to them are inherently interesting (Frick, 1992). This might be thought of as a recovery route for disengagement.

Achievement

Achievement provides a feeling of accomplishment, influencing the students' view of their capacities as well as actual demonstrations of learning (Elliot, 2005). Pedagogy related to achievement-oriented competence emphasizes making progress on students' ability to be successful with academic tasks and tests. Students follow a path toward achievement when they hold favorable estimations of their success (Wigfield & Wagner, 2006), experience learning tasks presented at the right level of challenge (not too difficult or easy; Schultheiss & Brunstein, 2005), feel in control of the task, and possess an internal versus external locus of control regarding the work (Schultheiss & Brunstein, 2005).

Students must have a reasonable "hope of success greater than their fear of failure" (Schultheiss & Brunstein, 2005), evaluate something as comprehensible (Silvia, Henson, & Templin, 2009), use "generative learning strategies" (such as highlighting texts, providing examples to remember principles; Lee, Lim, & Grabowski, 2010), and/or use metacognition (thinking about thinking) and strategy instruction to support their understanding and application of content (Urdan & Turner, 2006).

Many intervention programs designed to raise student achievement on standardized tests focus on achievement-oriented competence. Lesson plans and interventions emphasize procedures and strategies employed by successful learners to master the overarching concepts and processes of the discipline. Effective teachers unlock the secrets of strategies used *by the most capable learners* to achieve challenging objectives. Achievement-oriented competence may be associated with learning-to-learning skills (the how of learning) and also include strategies to monitor goals and obtain predictable results.

Successful achievement-oriented teachers demystify the skills needed to analyze a case study, reduce a complicated problem to a sizable problem, use a concept map to organize information, or keep track of an argument and evidence in debate by teaching students procedures and patterns of thinking. They provide practice and timely feedback, helping students learn strategies and ways of thinking to increase their awareness regarding how to monitor their performance (Waters & Schneider, 2010).

Harry Potter needed the patient guidance of teachers to fill in the gaps of a limited education while living with his aunt and uncle (Rowling, 1997). His experience in the world was limited to school learning and a single trip to the zoo. He had a lot of catching up to do at Hogwarts School. He often felt embarrassed when called on by a professor who found his background limited. This caused him considerable anxiety.

> "Potter!" said Snape suddenly. "What would I get if I added powdered root of asphodel to an infusion of wormwood? *Powdered root of what to an infusion of what?* Harry glanced at Ron, who looked at stumped as he was; Hermione's hand had shot into the air. "I don't know sir," said Harry. Snape's lips curled into a sneer. "Tut, tut – fame clearly isn't everything."
>
> (Rowling, 1997, p. 137)

Harry valued his access to the library and literally consumed books to locate answers to his question, making rapid progress when his love of learning was combined with skilled coaching and practice.

Essentially, achievement-oriented pedagogy supports students when they acquire the general habits of capable students through direct teaching, tutoring, and apprenticeship. Harry experienced more enjoyment once the "basic" classes were mastered. During the end-of-the year examinations, students took written exams and also practical tests of their accomplishments, demonstrating mastery of the curriculum.

> It was sweltering hot, especially in the large classroom where they did their written papers. They had been given special, new quills for the exams, which had been bewitched with an Anti-Cheating spell. They had practical exams as well. Professor Flitwick called them one by one into his class to see if they could make a pineapple tapdance across the desk. Professor McGonagall watched them turn a mouse into a snuffbox – points were given for how pretty the snuffbox was but taken away if it had whiskers.
>
> (Rowling, 1997, p. 202)

While interest attracts students to love learning and achievement provides students with a set of stable skills to learn throughout their lives, the experience of belonging allows individuals to feel at home in a safe and supportive environment.

Belonging

Perhaps one of the most important aspects of Harry's experience at Hogwarts has yet to be described. Because Harry was an orphan, he was left in the care of his aunt and uncle (Rowling, 1997). They failed to love Harry and instead treated him poorly. Harry nearly starved for food and attention. To make matters worse, a mean-spirited and spoiled cousin posed a constant threat to his existence at home and school.

Unfortunately, students bullied Harry at his neighborhood school, making him feel like an outsider (Rowling, 1997). He wore ill-fitting and worn clothing and lacked the basic essentials and experiences for his development. He rarely left home except to attend school and was denied material possessions. Suffering from "stereotype threat" (Aronson & Steele, 2005), Harry felt threatened by real and imagined attacks from bullies due to his family background, ethnic identity, and socio-economic status. Harry's talents were also apparent to his peers and they taunted and attacked him for being different. Harry protected himself by being vigilant against threats, taking a few blows as needed to assert his autonomy, and adopting protective strategies to guard against the tyranny of peers and foster parents.

Harry belonged to a special group of people needing protective support due to his status as a wizard with magical powers and impoverished circumstances (Rowling, 1997). No one recognized his talents or supported his development until "Hagrid, Keeper of the Keys and Ground at Hogwarts" arrived with a birthday cake and a letter inviting him to attend the Hogwarts School for Wizards. You see, Mr. Dursley, his foster father, "didn't approve of imagination" (p. 5) or wizards and kept his family background from him. The letter offered Harry a chance to escape his circumstances, claim his identity, and engage in an education matching his developmental goals and talents.

Lacking a feeling of belonging and safety in his home life, Harry suffered from psychological and physical abuse (Rowling, 1997). He formed a new family of allies at school, beginning with two best friends and his first mentor from the other world, Hagrid. More would follow; however, Harry made remarkable gains with the support of just a few caring individuals. Based on Maslow's hierarchy of needs, feelings of belonging and safety dominate and must be satisfied before higher order needs, such as self-actualizaion, earn our attention (as cited in Johnson & Johnson, 2009).

While the need to belong cannot be ignored or minimized, the need to experience autonomy may be particularly important because of the human desire to make choices and exercise control over one's life (Deci & Moller, 2005).

Autonomy

Student regulation of effort with regard to accomplishing learning tasks results in increased achievement (Fredricks, Blumenfeld, & Paris, 2004). When students experience challenges to their beliefs and perspectives (Frick, 1992) and tackle activities and goals related to their developmental stage and goals (Arnett, 2000; Kegan, 1994), they meet their autonomy needs. Even presenting a small amount of choice has proven benefits (Arnett, 2000; Kegan, 1994).

Harry Potter used his invisible cloak to explore the grounds at Hogwarts School (Rowling, 1998). He discovered a mirror allowing him to see his deceased parents and returned to it regularly, hungry for parental love and also fulfilling the adolescent goal of forming a personal identity. Kegan (1994) described life projects as those developmental tasks requiring mastery at any particular stage of life. To the degree possible, the "life projects" of students earn consideration in the selection of content and methods.

For example, adolescents engage in a search for an authentic identity, closely examining the socialized influences of family and culture (Kegan, 1994). Reading biographies of young people and adults attracts students at this life stage because the stories inform their pursuit of identity. Learning focused on identity development and self-representation increases student engagement (Renninger, 2009). Autonomy-oriented pedagogy incorporates knowledge of the developmental challenges into the design of learning and teaching plans (Arnett, 2000). Offering choices regarding topics, methods of investigation, and tools, or innovative ways to present the results of research increase student engagement.

While the metaphor of a pathway implies separate routes, effective teachers use engaging pedagogy to help students get more than one need met at a time. An episode of learning might arouse interest due to novelty, support skill acquisition, offer opportunities for peer collaboration and friendship, and provide choice in roles, topics, and projects, incorporating several attractors to learning simultaneously.

Student engagement in learning

Once students make an initial decision to "pay attention" or engage, they do not sustain their effort continuously *no matter what happens*. Students make decisions to engage in learning daily, hourly, or even within the same class period based on a complex set of motivational factors (Tsai, Kunter, Lüdtke, Trautwein, & Ryan, 2008). Like the rest of us, students assess their progress and estimate their chances of success, monitor their enjoyment and achievement, manage competing choices, and decide whether to apply greater, equal, or lesser amounts of effort.

Effective teachers plan appealing pathways and guard against a disappointing experience. Establishing motivating conditions likely to help students engage in learning and sustain their effort to achieve and innovate serves as the most fundamental challenge of teaching. If teachers planned their lessons, ignored these factors, and failed to diagnose the likely motivational pathways attracting students to learning, some students might disengage from learning early in the process or, worse yet, never even attempt it.

While all pathways attract students to learn at various times, some will have greater appeal than others based on *individual needs, interests, and development* and also the *social context* where learning takes place. Active engagement in learning should be viewed as a series of individually controlled student decisions to apply the right level of attention and effort needed to achieve goals. Favorable conditions support talent development; however, individuals must decide to get in the game. Gagné (2010) described motivation as decision making related to establishing goals based on individual needs and interests and volition as "an action control process" to implement a goal (p. 83). Innovation education serves as an environmental catalyst (Gagné, 2010) in helping individuals identify their interests, grow in knowledge of a domain, and receive the support needed to progress in their innovative work. Students have the most important role to play in their success.

Summary

Rowling's (1997) story about Harry Potter, an adolescent with unrealized special powers in *Harry Potter and the Sorcerer's Stone*, may prove to be more fact than fiction with regard to educating imaginative minds. The opportunities and roadblocks experienced by Harry Potter at Hogwarts School illustrate how certain educational conditions foster talent development and innovation, using educational research and effective practice to support these claims.

As a "coming of age" narrative, Harry Potter's story brings attention to the potential within us to develop our talents, when given opportunities and support to realize our potential. Teachers share the responsibility with others to create the most favorable conditions possible to foster innovation for all students. These conditions include contact with expert teachers, a challenging curriculum, skilled coaching, and the opportunity to meet basic needs in a safe, nurturing, and creative environment. Motivation science (Pintrich, 2003) allows innovation educators to attract and sustain student interest and engagement in learning, a necessary requirement to acquire domain knowledge for innovation. Applying liberal doses of *engaging pedagogy* facilitate increased student achievement and talent development.

Innovation educators should follow Horton's "two-eyed theory of teaching" as a model for fostering talent development: keep "one eye on where people are, and one eye on where they can be" (1998, p. xx). As it turns out, expert teachers use their education and experience to creatively challenge their students, *motivated by their love of and for their students*. The same care in fostering talented students should be applied to the development of teachers in their profession. Can you imagine the changes ahead?

References

Arnett, J. J. (2000). Emerging adulthood: A theory of development from the late teens through the twenties. *American Psychologist, 55*(5), 469–480.

Aronson, J., & Steele, C. M. (2007). Stereotypes and the fragility of academic competence, motivation and self-concept. In A. J. Elliot & C. S. Dweck (Eds.), *Handbook of Competence and Motivation* (pp. 436–456). New York: Guilford Press.

Bruner, J. S. (1962/1979). *On Knowing: Essays for the Left Hand.* Boston: Belknap Press of Harvard University Press.

Caskey, M. M., & Anfara, V. A., Jr. (2007). Research summary: Young adolescents' developmental characteristics. Retrieved September 22, 2010, from www.nmsa.org/Research/ResearchSummaries/DevelopmentalCharacteristics/tabid/1414/Default.aspx.

Csikszentmihalyi, M. (1996). *Creativity: Flow and the Psychology of Discovery and Invention.* New York: HarperCollins Publishers.

Deci, E. L., & Moller, A. C. (2005). The concept of competence: A starting place for understanding intrinsic motivation and self-determined extrinsic motivation. In A. J. Elliot & C. S. Dweck (Eds.), *Handbook of Competence and Motivation* (pp. 579–597). New York: Guilford Press.

Dohn, N., Madsen, P., & Malte, H. (2009). The situational interest of undergraduate students in zoophysiology. *Advances in Physiology Education, 33*(3), 196–201.

Elliot, A. J. (2005). A conceptual history of achievement goal construct. In A. J. Elliot & C. S. Dweck (Eds.), *Handbook of Competence and Motivation* (pp. 52–72). New York: Guilford Press.

Fredricks, J., Blumenfeld, P., & Paris, A. (2004). School engagement: Potential of the concept, state of the evidence. *Review of Educational Research, 74*(1), 59–109.

Frick, R. (1992). Interestingness. *British Journal of Psychology, 83*(1), 113. Retrieved from Academic Search Premier database.

Gagné, F. (2010). Motivation within the DMGT 2.0 framework. *High Ability Studies, 21*(2), 81–99.

Gentry, M., Steenbergen-Hu, S., & Choi, B. (2011). Student-identified exemplary teachers: Insights from talented teachers. *Gifted Child Quarterly, 55*(2), 111–125.

Hidi, S. (1990). Interest and its contribution as a mental resource for learning. *Review of Educational Research, 60*(4), 549–571.

Hidi, S., & Renninger, K. (2006). The four-phase model of interest development. *Educational Psychologist, 41*(2), 111–127.

Kim, K. (2008). Underachievement and creativity: Are gifted underachievers highly creative? *Creativity Research Journal, 20*(2), 234–242.

Jacobs, H. H. (2010). *Curriculum 21: Essential Education for a Changing World.* Alexandria, VA: ASCD.

Jarvis, J. M. (2009). Planning to unmask potential through responsive curriculum: The "Famous Five" exercise. *Roeper Review, 31*(4), 234–241.

Johnson, D., & Johnson, F. (2009). *Joining Together: Group Theory and Group Skills,* 10th ed. Boston: Allyn & Bacon.

Justice with Michael Sandel, Episode 01 (n.d.). Retrieved July 5, 2011, from www.justiceharvard.org/2011/03/episode-01/#watch.

Kegan, R. (1994). *In Over Our Heads: The Mental Demands of Modern Life.* Cambridge, MA: Harvard University Press.

Lee, H., Lim, K., & Grabowski, B. (2010). Improving self-regulation, learning strategy use, and achievement with metacognitive feedback. *Educational Technology Research and Development, 58*(6), 629–648.

Noonan, S. J. (with Fish, T. L.). (2007). *Leadership Through Story: Diverse Voices in Dialogue.* Lanham, MD: Rowman and Littlefield Education.

Pintrich, P. R. (2003). A motivational science perspective on the role of student motivation in learning and teaching contexts. *Journal of Educational Psychology, 95*(4), 667–686.

Reis, S. M., & Renzulli, J. S. (2010). Opportunity gaps lead to achievement gaps: Encouragement for talent development and schoolwide enrichment in urban schools. *Journal of Education, 190*(1/2), 43–49. Retrieved from EBSCO*host*.

Renninger, K. (2009). Interest and identity development in instruction: An inductive model. *Educational Psychologist, 44*(2), 105–118.

Robertson, J., Blain, N., & Cowan, P. (2005). The influence of friends and family vs the Simpsons: Scottish adolescents' media choices. *Learning, Media, and Technology, 30*(1), 63–79.

Rowling, J. K. (2008). "Failure and Imagination." Harvard University, Boston, MA. June 5.

Rowling, J. K. (1997). *Harry Potter and the Sorcerer's Stone.* New York: Scholastic.

Schultheiss, O. C., & Brunstein, J. C. (2005). An implicit motive perspective on competence. In A. J. Elliot & C. S. Dweck (Eds.), *Handbook of Competence and Motivation* (pp. 31–51). New York: Guilford Press.

Shulman, L. (2002). Making differences. *Change, 34*(6), 36–44. Retrieved from Academic Search Premier database.

Silvia, P. J. (2006). *Exploring the Psychology of Interest.* New York: Oxford University Press.

Silvia, P. (2008). Interest: The curious emotion. *Current Directions in Psychological Science, 17*(1), 57–60.

Silvia, P., Henson, R., & Templin, J. (2009). Are the sources of interest the same for everyone? Using multilevel mixture models to explore individual differences in appraisal structures. *Cognition and Emotion, 23*(7), 1389–1406.

Sternberg, R. J. (2003). *Wisdom, Intelligence, and Creativity Synthesized*. Cambridge, UK: Press Syndicate of the University of Cambridge.

Sternberg, R. J. (2004). Teaching college students that creativity is a decision. *Guidance and Counseling, 19*(4), 196–200. Retrieved from EBSCO*host*.

Tsai, Y., Kunter, M., Lüdtke, O., Trautwein, U., & Ryan, R. M. (2008). What makes lessons interesting: The role of situational and individual factors in three school subjects. *Journal of Educational Psychology, 100*(2), 460–472.

Urdan, T., & Turner, J. (2006). Competence motivation in the classroom. In A. J. Elliot & C. S. Dweck (Eds.), *Handbook of Competence and Motivation* (pp. 297–317). New York: Guilford Press.

Waters, S. H., & Schneider, W. (2010). Common themes and future challenges. In S. H. Waters & W. Schneider (Eds.), *Metacognition and Strategy Use, and Instruction*. New York: Guilford Press.

Wigfield, A., & Wagner, A. L. (2006). Competence, motivation and identity development during adolescence. In A. J. Elliot & C. S. Dweck (Eds.), *Handbook of Competence and Motivation* (pp. 222–239). New York: Guilford Press.

18

Where did all great innovators come from?

Lessons from early childhood and adolescent education of Nobel laureates in science

Larisa V. Shavinina

UNIVERSITÉ DU QUÉBEC EN OUTAOUAIS, CANADA

Summary: This chapter presents some of the main results from the project about early childhood and adolescent education of Nobel laureates in science: the exceptional roles of parents and of teachers in developing innovators–geniuses. Winning a Nobel Prize represents the pinnacle of accomplishment possible in one's field of expertise. Despite the ever-increasing role of science in society and the importance of Nobel laureates in contemporary science, it should be acknowledged that their childhood and adolescent education has never been studied. The discovery of the principles involved in the educational development of Nobel laureates will allow educators to accordingly improve, develop, modify, and transcend areas in the current curriculum in an attempt to cultivate scientific talent, of Nobel caliber, in future generations of potential innovators.

Key words: Innovator, scientific innovation, Nobel laureates in science, early childhood and adolescent education.

Introduction

> I consider early childhood events as most essential to a man's scientific and philosophical development.
>
> *(Konrad Lorenz)*[1]

Nobel laureates during their childhood encompassed a wide range of abilities, including the gifted, gifted underachievers, and children without any special talents. Their divergent trajectories of talent development ultimately led to the same result: amazing scientific innovations–great discoveries, which testified to the outstanding minds of those who made them. Eventually, all the trajectories led to the same point: zenith in science. A question is how and why this happened, and what lessons can be derived for the education of today's children–tomorrow's innovators.

The findings reported in this chapter resulted from the research project entitled *A Study of Early Childhood and Adolescent Education of Nobel Laureates and the Implications for Gifted and General Education: Developing Scientific Talent of Nobel Caliber*. The project was sponsored by the Social Sciences and Humanities Research Council (SSHRC) of Canada. More than 12,000 pages of documents regarding early childhood and adolescent education of Nobel laureates in science were found from all the existing printed sources (i.e., mainly (auto) biographies and book chapters) published in all languages. Nobody could predict such a wealth of data at the beginning of the project. This chapter is thus based on the analysis of these documents about their early childhood and adolescent education.

Although the statistical analysis is underway, it is, however, clear that the four main findings of the project are related to the important roles of parents, teachers, special events (e.g., the Great depression in the case of Nobel laureates in economy or war in the case of early Nobel laureates), and proximity-related issues (e.g., proximity of nature in the case of Nobel laureates in physiology or medicine, laboratories in the case of Nobel Prize winners in chemistry, and public libraries and museums in the case of almost all laureates) in developing scientific innovators of Nobel caliber. Specifically, this chapter will focus on the role of parents and of teachers in cultivating geniuses–innovators. They are described in the second and third sections, respectively.[2] The first section presents a brief literature review of the role of the micro-social environment in developing children's talents. The concluding section summarizes the findings and discusses their implications for innovation education.

What does literature tell us about the role of micro-social factors in the development of high ability?

Research demonstrates that micro-social factors significantly contribute to the development of exceptional talents. Scholars found that many children showing signs of giftedness in their childhood do not manifest an outstanding level of achievement and performance in school and adult life (Howe, 1990, 1993; Tannenbaum, 1986). One explanation for this discrepancy between promise and fulfillment can be the family milieu and educational opportunities (Lewis & Michalson, 1985). Families provide opportunities for the mental growth of their children, and thus play an essential role in talent development. All human achievements are considered to be interaction between inner potential and the resources and opportunities provided by the surrounding environment (Gardner, 1993; Howe, 1990, 1993; McCurdy, 1992; Sternberg & Lubart, 1995).

Specifically, researchers found that the family milieu plays a crucial role in providing good learning opportunities (Feldman, 1986, 1993). Parents of gifted children are almost always highly educated; mothers' educational level is very important. Often parents of talented children are their first teachers (Howe, 1990, 1993; McCurdy, 1992; Storfer, 1990). Parents maintain the optimal balance of freedom and pressure that most favorably influences a child's motivation to learn (Lewis & Michalson, 1985). Gifted children get a lot of encouragement, guidance, and support from parents (Feldman, 1986, 1993; Gardner, 1993).

Family *values* and family *climate* also facilitate the development of high ability. For example, prodigies are generally born into families that recognize and value the talent when it emerges (Feldman, 1986; Robinson & Clinkenbeard, 1998).

The socio-economic status of the family is another significant variable, because it enables a wide diversity of educational and cultural experiences. Many gifted children come from the middle and upper socio-economic classes (Howe, 1990; Gardner, 1993; Lewis & Michalson, 1985).

Contact with *role models* and/or *significant others* is of great importance for the development of giftedness: these people may be parents, relatives, teachers, and the like. They manifest the abilities and skills in some areas of talent, the values and attitudes associated with it, and the self-image required for successful performance (Cropley & Dehn, 1996; Simonton, 1978). The opportunity of personal interaction with *great contemporaries* or eminent models is very important for the fulfillment of an individual's gifted potential.

Although school influences the development of high abilities in many ways, I will not analyze the existing literature in detail here. Rather, I will only mention its major influence, namely *formal education*. Research shows that a key role of school consists in traditional knowledge acquisition (i.e., formal education; Cropley & Dehn, 1996; Simonton, 1978; Sternberg & Lubart, 1995), and an individual's knowledge base plays a critical role in the development of giftedness (Chi & Greeno, 1987; Kholodnaya, 1997; Schneider, 1993; Shavinina, 1997, 1999; Shavinina & Kholodnaya, 1996; Rabinowitz & Glaser, 1985).

Overall, research demonstrates that gifted children get the greatest amount of parental investment that is usually manifested in a high degree of attention focused upon the child and in abundant love from parents (Feldman, 1986; Howe, 1990, 1993; McCurdy, 1992). The educational level of parents and socio-economic status of the family also matter. Often the gifted have a few "significant others" or role models in their lives who influence mainly via formal education. School may stimulate or inhibit developing talents.

What is the role of parents in developing scientific innovators?

One of the main findings[3] of the project on early childhood and adolescent education of Nobel laureates is that family played the most important role in the development of their scientific and innovative talents. It mainly includes (1) encouraged, supporting parents, which valued education and loved to read, and/or (2) their professional occupations related to science, and/or (3) homes full of books and scientific toys/kits that allowed scientific experimentation at home. Early childhood and adolescent education of each Nobel Prize winner was characterized by at least one of these three factors.

Supporting and encouraging parents, who valued education and loved to read

Parents supported children's interests in science and other subjects, as well as encouraged them to pursue those interests. As Wolfgang Ketterle, a winner of the 2001 Nobel Prize in physics, put it,

> My parents supported all our[4] interests in music, sports and sciences. As they hadn't been exposed to many of these activities themselves, they did not steer us in certain directions, but rather observed our interests and then reinforced and supported them.
>
> *(Ketterle, 2012)*

Russell A. Hulse, who won a 1993 Nobel Prize in physics, praised his parents in the same way by saying that "My parents fostered and supported this[5] interest, and I thank them very much for being my first and, by far, most uncritically supportive funding agency" (Hulse, 2012).

Providing a generally happy childhood, parents supported children in their other endeavors as well and demonstrated a great deal of interest in their pursuits. Christiane Nüsslein-Volhard, a winner of the 1995 Nobel Prize in physiology or medicine, convincingly expressed it in her autobiography:

> I had a happy childhood, with many stimulations and support from my parents who, in postwar times, when it was difficult to buy things, made children's books and toys for us. We had much freedom and were encouraged by our parents to do interesting things. I remember that my father showed much interest in what we did, and thereby had a great influence in

our performances, without being particularly ambitious (although good grades at school were more or less a matter of course). I tried to explain to him what we did in mathematics, and we discussed Goethe's scientific papers.

(Nüsslein-Volhard, 2012)

Likewise, Arvid Carlsson, who won the 2000 Nobel Prize in physiology or medicine, described his family milieu as "a stable environment with loving and supportive parents" that characterized his happy life during childhood and youth (Carlsson, 2012).

Parents valued education, regardless of whether they were educated themselves or not. For instance, Rosalyn Yalow, a winner of a 1977 Nobel Prize in physiology or medicine, wrote about her parents: "Neither had the advantage of a high school education but there was never a doubt that their two children would make it through college" (Yalow, 2012). Likewise, Stanley Cohen, who won a 1986 Nobel Prize in physiology or medicine, praised his parents for developing his educational motivation: "My father was a tailor and my mother, a housewife. Though of limited education themselves, they instilled in me the values of intellectual achievement and the use of whatever talents I possessed" (Cohen, 2012).

Because parents valued education, they encouraged children to do their best in school. Herbert Kroemer's (a 2000 Nobel Prize in physics) parents examplified it perfectly:

> Both came from simple skilled-craftsmen families. Neither had a high-school education, but there was never any doubt that they wanted to have their children obtain the best education they could afford. My mother, in particular, pushed relentlessly for top performance in school: simply doing well was not enough.... Despite their insistence on excellence, my parents never pushed me in any particular academic direction; I was completely free to follow my inclinations, which ran towards math, physics, and chemistry. When I finally told my parents that I wanted to study physics, my father merely wondered what that is, and whether I could make a living with it.
>
> *(Kroemer, 2012)*

Likewise, Horst L. Stormer, a winner of the 1998 Nobel Prize in physics, pointed out in his autobiography:

> There was never a doubt in my parents' mind that their sons would receive the best possible education. Although none of my forefathers graduated from high school, my parents regarded highly the merits of a good education as a tool for social advancement. In their value system knowledge always ranked above wealth.... To enter "Gymnasium", at ten, required the passing of a test. I was accepted and from then on commuted for eight years, five km each way, to the "Goethe Gymnasium" in the neighboring town.
>
> *(Stormer, 2012)*

In a similar way Claude Cohen-Tannoudji, who won a 1997 Nobel Prize in physics, highly acknowledged the parental impact on his intellectual growth:

> My parents lived a modest life and their main concern was the education of their children. My father was a self-taught man but had a great intellectual curiosity, not only for biblical and talmudic texts, but also for philosophy, psychoanalysis and history. He passed on to me his taste for studies, for discussion, for debate, and he taught me what I regard as being the fundamental features of the Jewish tradition—studying, learning and sharing knowledge with others.
>
> *(Cohen-Tannoudji, 2012)*

In the case of Steven Chu's (a 1997 Nobel Prize in physics) family education, it

> was not merely emphasized, it was our raison d'être. Virtually all of our aunts and uncles had Ph.D.'s in science or engineering, and it was taken for granted that the next generation of Chu's were to follow the family tradition.
>
> *(Chu, 2012)*

Similarly, William D. Phillips, a winner of the 1997 Nobel Prize in physics, stressed an exceptional role of parents in his educational development:

> I clearly remember the value my parents placed on reading and education. My parents read to us and encouraged us to read. As soon as I could read for myself, walking across town to the library became a regular activity.... Although they had no particular knowledge or special interest in science, they supported mine.
>
> *(Phillips, 2012)*

The family milieu of George H. Hitchings (a 1988 Nobel Prize in physiology or medicine) had an equally great impact in him:

> I enjoyed a warm and loving home environment. A high standard of ethics prevailed in our family, together with a thirst for knowledge and an urge to teach. In their schooling, my mother and father were limited, but they were avid readers, especially my father. It is clear to me in retrospect that he would have been a scientist had opportunities been more easily attainable.
>
> *(Hitchings, 2012)*

Often it was mothers, who instilled a passion for reading in their kids from a early age. For example, Richard J. Roberts (the 1993 Nobel Prize in physiology or medicine) is "a passionate reader, having been tutored very early by my mother. I avidly devoured all books on chemistry that I could find" (Roberts, 2012).

Even in the case of low-income families, parents strongly encouraged children to read and did whatever possible in order to find books for their kids. The case in point is Carl E. Wieman, who won a 2001 Nobel Prize in physics. As he wrote in his autobiography,

> Most of my childhood was spent in the woods of Oregon where lumber was the sole industry.... Much of my youth was spent wandering around in the forests of towering Douglas fir trees. I also spent much of my time reading and picking fruit and fir cones to earn spending money. Every Saturday my family would make a long expedition to the nearest town to do the week's worth of shopping. A stop at the public library was always part of these trips. Although I was unaware of it at the time, my parents must have made special arrangements for their children to use the library since we lived far outside the region it was supposed to serve. The librarians would also overlook the normal five-book limit and allow me to check out a large pile of books each week that I would then eagerly devour. That experience has left me with a profound appreciation for the value of public libraries. At the time I was quite envious that my friends had televisions while we did not, but in retrospect I am very grateful that I spent this time reading instead of watching TV.[6]
>
> *(Wieman, 2012)*

A public library played an equally important role in childhood education of Isidor Rabi, who was from a poor family. The children's books in the Carnegie Library, the local branch of the Brooklyn

Public Library in New York were fun to read, and Rabi read them all. Books of science were organized by subject and he started at the beginning, with A for astronomy. Decades later, Rabi could say, "That was what determined my later life more than anything else—reading a little book on astronomy" (quoted in Rigden, 1987, p. 22).

Likewise, a public library was a crucial facet of Rosalyn Yalow's childhood:

> I was an early reader, reading even before kindergarten, and since we did not have books in my home, my older brother, Alexander, was responsible for our trip every week to the Public Library to exchange books already read for new ones to be read.
>
> *(Yalow, 2012)*

In the case of the families, which were economically well, homes were full of books. As a result, children liked to read a lot. "I was... interested in reading everything that came my way, hiding a book behind the desk while the other students learned from class," noticed John C. Mather (the 2006 Nobel Prize in physics) about his elementary school years. He also pointed out that his "parents also enjoyed reading aloud from various books, including biographies of Darwin and Galileo" (Mather, 2012). In addition,

> We did have a Bookmobile, a traveling library from the County that visited the farms every couple of weeks, and I borrowed as much as I could. I started reading about optics, and I saved my allowance and ordered some lenses from Edmund Scientific and assembled small refractor telescopes.
>
> *(Mather, 2012)*

This is why he had a lot of opportunity to learn science, even in a very rural setting.

Professional occupations of parents influenced Nobel laureates-to-be

Parents also influenced the development of scientific innovators of Nobel caliber in children via their professions. Particularly, professional occupations of many parents were related to science. Thus, the fathers of Niels Bohr (1922),[7] Kenneth G. Wilson (1982), Ernst Ruska (1986), Wolfgang Paul (1989), Steven Chu (1997), and Eric A. Cornell (2001), just to mention a few Nobel laureates in physics, were university professors. Fathers of the following Nobel Prize winners in physiology or medicine were also professors: Thomas H. Weller (1954), Daniel Bovet (1957), Edward Tatum (1958), Feodor Lynen (1964), Haldan K. Hartline (1967), Hamilton O. Smith (1978), Alfred G. Gilman (1994), and Arvid Carlsson (2000), just to mention a few.

George E. Palade, a winner of the 1974 Nobel Prize in physiology or medicine, excellently summarized the main impact of such a home milieu: "My father... was professor of philosophy and my mother... was a teacher. The family environment explains why I acquired early in life great respect for books, scholars and education" (Palade, 2012). Hamilton O. Smith echoed this description by saying:

> At home, an atmosphere of intense intellectualism was maintained. My father was perpetually working and writing. At the same time, my mother struggled to establish herself as a writer.... She, in particular, imbued us with a respect and desire for the creative life.
>
> *(Smith, 2012)*

Parents often challenged their children with scientific problems that sparked kids' interest in science. For instance, Eric A. Cornell recalled that

> Some nights, especially in the early summer when the late evening light kept my west-facing bedroom from getting very dark, I had trouble falling asleep at my appointed bedtime.... My father would come in and suggest to me a "problem" to think about. Stewing over these problems was supposed to help me go to sleep. It never did that, but it did get me in the lifelong habit of thinking about technical issues at all sorts of random moments in my daily life, and not only (or even primarily) during scheduled "thinking time." Some of my father's bedtime problems I now recognize as classic physics brainteasers.
>
> *(Cornell, 2012)*

The fathers of John L. Hall (a 2005 Nobel Prize in physics), Tadeus Reichstein (a 1950 Nobel Prize in physiology or medicine),[8] Allan M. Cormack (1979), David H. Hubel (1981), Rita Levi-Montalcini (1986), Susumu Tonegawa (1987), Sir James W. Black (1988), Linda B. Buck (2004) were engineers. Fathers of a great majority of Nobel laureates in medicine and physiology were medical doctors: Hans Krebs (1953), Hugo Theorell (1955), André F. Cournand (1956), Maurice Wilkins (1962), Gelald M. Edelman (1972), Jean Dausset (1980), Torsten N. Wiesel (1981), Gertrude B. Elion (1988), and Harold E. Varmus (1989), just to mention a few. Similarly, many parents—both mothers and fathers or just one of them—were teachers.

Professional occupations of parents thus determined an early and deep interest of their children in science, books, and other intellectual pursuits.

Scientific experimentation originates from home

Many Nobel Prize winners came from homes full of scientific kits/toys that allowed scientific experimentation at home. Legos with its building blocks, electricity kits, chemistry sets, electronic kits, TV sets, and other similar toys sparked an unusual curiosity and interest in science and technology. Thus, recollecting his childhood years, Wolfgang Ketterle wrote in the autobiography:

> My explorations of the technical world started with Legos, with which I was quite creative in constructing moving objects with the basic building blocks that were then available. (Legos have become much more fancy since then!) I remember playing with electricity kits, doing repairs of household appliances, and using my father's power tools for woodworking projects. Explorations into chemistry were done in our basement, sometimes with friends, and my parents must have had quite a bit of confidence in my abilities when they allowed me to experiment with explosive mixtures. (I was quite impressed when such a mixture was able to melt metal.) Other projects included taking old radios and a TV set apart and combining a portable radio and a vacuum tube audio amplifier to create stereo sound. I was interested in learning more about electronics, but I was disappointed that the electronic kits explained only how to put the parts together, not how they really worked.
>
> *(Ketterle, 2012)*

In the same way William Phillips described how his home life helped develop his early interest in science:

> Almost as far back as I can remember, I was interested in science. I assembled a collection of bottles of household substances as my "chemistry set" and examined almost anything I could find with the microscope my parents gave me.... Science was only one of the passions of my childhood, along with fishing, baseball, bike riding and tree climbing. But as time went on,

Erector sets, microscopes, and chemistry sets captured more of my attention than baseball bats, fishing rods, and football helmets.

(Phillips, 2012)

Many parents entertained children by using various scientific sets. The case in point is Hamilton O. Smith, who praised parents for fostering his intellectual pursuits: "My mother and father... entertained us with arithmetic problems and a small Gilbert chemistry set" (Smith, 2012).

Eric A. Cornell used to build "model rockets." According to him,

> it was fun to watch the rocket blast into the air, suspenseful to wonder if the parachute would open to bring the rocket safely back. I didn't really enjoy the assembling the model kits very much, and usually I couldn't be bothered to paint the thing, or even to stick on the decals. A more vivid memory for me was designing a model of my own. Besides the store-bought kits, the Estes Model Rocketry company in those days also sold by mail various sizes of cardboard tubing, balsa-wood sheets, nosecones, and gun-powder rocket engines. Estes also published a terrific little booklet full of quantitative design tips. A key issue in rocket design is to make sure that the center of mass is well forward from the fins, lest the rocket be aerodynamically unstable. My father showed me how (after a candidate design was laid out on graph paper) to calculate the center of mass of the assembly based on the masses and distribution of the component parts. I designed an over-sized, under-powered, clunky sort of rocket. I didn't care how high it would go—I wanted it to rise slowly enough that I could watch to see if its orientation wobbled during the flight. On its maiden flight it lifted off the ground with all the ponderousness of a Saturn V, rising steady and true but rolling slightly about its long axis (had I glued the fins on crooked?) as it gained altitude. The engine burn completed, and then the parachute popped and my creation drifted with the wind to land on the roof of a schoolhouse.

(Cornell, 2012)

Russell A. Hulse's childhood home was also full of various sets and scientific toys. As he wrote many years later,

> science was a defining part of my approach to life for as far back as I can remember... I ran through a seemingly endless series of interests involving chemistry sets, mechanical engineering construction sets, biology dissection kits, butterfly collecting, photography, telescopes, electronics and many other things over the years.

Such sets and kits developed children's interest in science. John C. Mather well expressed it:

> By the time I was in fourth grade (age = grade + 5 years) I was already pretty sure I liked scientific and engineering things, including electronics. For Christmas I got a one-tube radio kit, and then I saved my allowance for a 5-tube shortwave Heathkit radio that I put together so I could listen to exotic languages and broadcasts from far-away places.

(Mather, 2012)

Many Nobel laureates had home laboratories to do their experiments.[9] "My brother and I spent many hours in our basement laboratory stocked with supplies purchased from our paper route earnings", recollected Hamilton O. Smith (2012). Similar pursuits occupied a young John R. Vane (a 1982 Nobel Prize in physiology or medicine):

> At the age of 12, my parents gave me a chemistry set for Christmas and experimentation soon became a consuming passion in my life. At first, I was able to use a Bunsen burner attached to my mother's gas stove, but the use of the kitchen as a laboratory came to an abrupt end when a minor explosion involving hydrogen sulphide spattered the newly painted decor and changed the colour from blue to dirty green!
>
> Shortly afterwards, my father, who ran a small company making portable buildings, erected a wooden shed for me in the garden, fitted with bench, gas and water. This became my first real laboratory, and my chemical experimentation rapidly expanded into new fields.
>
> *(Vane, 2012)*

Richard J. Roberts had a similar childhood experience that crystallized his desire to become a scientist:

> I received a chemistry set as a present. I soon exhausted the experiments that came with the set and started reading about less mundane ones. More interesting apparatus like Bunsen burners, retorts, flasks and beakers were purchased. My father, ever supportive of my endeavors, arranged for the construction of a large chemistry cabinet complete with a formica top, drawers, cupboards and shelves. This was to be my pride and joy for many years. Through my father, I met a local pharmacist who became a source of chemicals that were not in the toy stores. I soon discovered fireworks and other concoctions. Luckily, I survived those years with no serious injuries or burns. I knew I had to be a chemist.

Living on a farm provided exceptionally good opportunities for experimentation. Godfrey N. Hounsfield's (a 1979 Nobel Prize in physiology or medicine) childhood experience epitomizes it rather well:

> At a very early age I became intrigued by all the mechanical and electrical gadgets which even then could be found on a farm; the threshing machines, the binders, the generators. But the period between my eleventh and eighteenth years remains the most vivid in my memory because this was the time of my first attempts at experimentation, which might never have been made had I lived in a city. In a village there are few distractions and no pressures to join in at a ball game or go to the cinema, and I was free to follow the trail of any interesting idea that came my way. I constructed electrical recording machines; I made hazardous investigations of the principles of flight, launching myself from the tops of haystacks with a home-made glider; I almost blew myself up during exciting experiments using water-filled tar barrels and acetylene to see how high they could be waterjet propelled. It may now be a trick of the memory but I am sure that on one occasion I managed to get one to an altitude of 1000 feet!
>
> *(Hounsfield, 2012)*

Therefore, family played an exceptionally important role in the development of scientific innovators of Nobel caliber in three main ways, namely:

- Encouraging and supporting parents, for whom education and love to read were of paramount importance (even if they had little formal education themselves). They expected children to excel.
- Their professional occupations were related to science. And/or:
- Homes were full of books and scientific toys/kits that allowed scientific experimentation at home.

The families thus greatly influenced early childhood and adolescent education of Nobel laureates in at least one of these ways. These findings support studies, which demonstrated a crucial role of family milieu in developing children's talents (Feldman, 1986, 1993; Gardner, 1993; Howe, 1990, 1993; McCurdy, 1992; Storfer, 1990).

That special teacher

The findings demonstrate that almost each Nobel Prize winner had at least one exceptional teacher in elementary or secondary school. These teachers are highly credited for sparking a potential of innovator–genius in the Nobel laureates via actualizing children's curiosity and inspiring them to learn more about science and other subjects of interest. As Eric A. Cornell (2012) put it:

> Some of my classes in high school were pretty interesting and I benefited from having several very intelligent and inspiring teachers. Among these were John Samp, a physics teacher, and JoAnn Walther, an English teacher. After the Nobel Prize announcement, I got back in touch with them and was delighted to learn that they are still (as of 2001) teaching at my old high school.

Such teachears did a perfect job of developing students' interest in science. "A great chemistry teacher… Mr. Mondzak, excited my interest in chemistry", wrote Rosalyn Yalow in her autobiography (Yalow, 2012).

What was so exceptional about teachers of Nobel laureates? There are a few distinguishing characteristics of those teachers. First of all, they were *teachers with love and curiosity for the subjects they were teaching*. This love for the subjects was contagious and sparked imagination and interest of future Nobel laureates. Zhores I. Alferov, a winner of the 2000 Nobel Prize in physics, very well expressed this:

> In the post-war particular situation I attended an only boy's school in the destroyed Minsk-city, and was lucky in having an excellent physics teacher there Yakov Borisovich Meltserson. He delivered lectures on physics for us, rather naughty boys, and we were sitting quiet and listened attentively. The teacher loved physics devotedly and had a gift of making our imagination work. His explanation of the cathode oscilloscope operation and talk on radar systems greatly impressed me. When finishing the school I took his advice which institution to choose for education and that was a celebrated Ul'yanov Electrotechnical Institute in Leningrad.
>
> *(abbreviated to LETI; Alferov, 2012)*

Likewise, Horst L. Stormer highly appreciated his special teacher:

> One of my teachers stood out, Mr. Nick. He taught math and physics. A new teacher, basically straight out of college, young, open, articulate, fun, he represented what teachers could be like. His love and curiosity for the subjects he was teaching was contagious. As 15 or 16 year-olds, we read sections of Feynman's Lecture Notes in Physics in a voluntary afternoon course he offered.
>
> *(Stormer, 2012)*

Second, teachers of Nobel laureates were *enthusiastic, inspiring, and challenging teachers with a playful spirit*. "They might not have been the best teachers pedagogically, but their intellects and their visions inspired us", told Daniel C. Tsui (a 1998 Nobel Prize in physics) about his teachers (Tsui, 2012). Similarly, Wolfgang Ketterle praised his great teacher: "There was one mathematics teacher,

A. Strobel, who was inspirational. He challenged me with special problems, and tried to teach the class to approach mathematical problems in a playful rather than formal spirit" (Ketterle, 2012). Likewise, Carl E. Wieman wrote about his outstanding teacher: "Mr. Tobias did a great deal to kiddle my interest in science with his enthusiasm and knowledge. I still remember his explanations (far better than any of the material from my college courses!) of the structure of atoms" (Wieman, 2012). The impact of this remarkable teacher was so great that now Carl E. Wieman is leading a $12-million, five-year initiative at the University of British Columbia, Canada, to look for the best ways to teach science: his research interests have switched from the nature of matter to what is the matter with science and engineering at universities that causes so many students to drop them (Anderssen & McIlroy, 2009).[10]

Third, teachers of Nobel laureates were *gifted, excellent teachers*, especially when they introduced new subjects. As Steven Chu (2012) put it,

> My physics teacher, Thomas Miner was particularly gifted. To this day, I remember how he introduced the subject of physics. He told us we were going to learn how to deal with very simple questions such as how a body falls due to the acceleration of gravity.

It was important for future Nobel laureates that very first introduction to new subjects were made by unusual teachers.

Christiane Nüsslein-Volhard highly appreciated excellence of her teachers as well:

> I enjoyed high school where I learned a lot from excellent teachers.... School education was good and interesting, particularly German literature, mathematics and biology. We had very engaged teachers, mostly women. In the final class our biology teacher discussed many modern topics with us such as genetics, evolution, and animal behavior. I remember that I tried to develop a new theory about evolution, when we discussed Darwin at school.
>
> *(Nüsslein-Volhard, 2012)*

Fourth, teachers of Nobel laureates *taught differently*. They deviated in their teaching from the accepted norm of the day. The case in point is Steven Chu again, who wrote about his math teacher:

> Geometry was the first exciting course I remember. Instead of memorizing facts, we were asked to think in clear, logical steps. Beginning from a few intuitive postulates, far reaching consequences could be derived, and I took immediately to the sport of proving theorems.
>
> *(Chu, 2012)*

In the same way Tim Hunt (a 2001 Nobel Prize in physiology or medicine) praised his school education because of great teachers:

> At the age of 14, I moved... to Magdalen College School, Oxford, where science played a much larger role in the curriculum. I loved Chemistry in particular, largely because the teacher, Colonel Simmons *was much more concerned with principles than facts*, although a thoroughly practical man himself. *We were allowed considerable freedom,* and on more than one occasion started fires from distilling volatile flammable solvents. One became adept at avoiding injury. Later on, biology... came to the fore when a young teacher called Terence Doherty took just three of us for Zoology. We dissected my brother's pet rabbit when it died, which was a treat after all the formalin-fixed dogfish.
>
> *(Hunt, 2012; italics added)*

Fifth, teachers provided *advanced, enriched, and accelerated instruction*. They went beyond the scope of prescribed curriculum: being initiative and creative, they taught what they thought was the best. William D. Phillips pointed this out well: "Dedicated and concerned teachers taught us things that were not part of the ordinary elementary school curriculum, like French and advanced mathematics. … Interested teachers continued to provide me with advanced instruction" (Phillips, 2012).

Teachers tolerated accelerated education and had a deep impact on future Nobel laureates. For example, this was the case of Hamilton O. Smith:

> I completed high school in three years largely due to a wonderful science teacher, Wilbur E. Harnish, who allowed me to complete chemistry and physics during the two summers preceding ninth grade. Two other teachers… influenced my development profoundly: Vynce Hines, who taught me the beauties and rigor of plane geometry and Miles C. Hartley, who gave me a sound foundation in algebra.
>
> *(Smith, 2012)*

Sixth, teachers manifested *interest in students and encouraged them to succeed*. Teachers were attentive to the interests of future Nobel laureates and strongly encouraged them to learn in more depth subjects of their interest. As Frederick Reines, who won the 1995 Nobel Prize in physics, recalled many years later: "I was strongly encouraged by a science teacher who took an interest in me and presented me with a key to the laboratory to allow me to work whenever I wanted" (Reines, 2012).

Sir Paul Nurse, a winner of the 2001 Nobel Prize in physiology or medicine, had encouraging teachers as well:

> I enjoyed my time at primary school because my teachers made the world seem such an interesting place and encouraged my innate curiosity.… At age 11 in 1960, I moved to an academic state secondary school… I had an excellent Biology teacher, Keith Neal, who encouraged his pupils to study natural history and to do real experiments. I had a great time investigating the pigments of different mutant fruit flies by following experimental protocols published in *Scientific American*.
>
> *(Nurse, 2012)*

Finally, teachers of Nobel laureates *went beyond the classroom practice*. They did many extra things for their students. For instance, one of these extra things changed the life of Martinus J. G. Veltman (the 1999 Nobel Prize in physics): "My physics teacher came to my home and suggested parents to send me to the University.… Since then I have found out that many physicists owe their career to a good teacher" (Veltman, 2012).

The school teacher of Richard J. Roberts also did his "extra" and thus greatly influenced the development of the scientific talent of the future Nobel laureate:

> At St. Stephen's junior school I encountered my first real mentor, the headmaster Mr. Broakes. He must have spotted something unusual in me for *he spent lots of time* encouraging my interest in mathematics. He would produce problems and puzzles for me to solve and I still enjoy the challenge of crossword and logical puzzles. Most importantly, I learned that logic and mathematics are fun!
>
> *(Roberts, 2012; italics added)*

Therefore, at least one school teacher had an exceptional impact on the development of Nobel laureates. Almost without exception the Nobel Prize winners credited the actions of a key teacher in their lives. Those teachers all went the extra mile—they did something exceptional. These findings thus support research demonstrating that highly able children during their school years had gifted teachers. This is a teacher who is gifted in his or her ability to inspire and support truly meaningful learning (Porath, 2009).

Conclusions

This chapter was mainly focused on the exceptional role of parents and teachers in cultivating geniuses–innovators of Nobel caliber. It was found that family played the most important role in the development of their scientific and innovative talents. Specifically, Nobel laureates had encouraging and supporting parents, who valued education and loved to read to children. Professional occupations of many parents were related to science. Usually, Nobel laureates lived in homes full of books and scientific toys/kits that allowed scientific experimentation at home. Early childhood and adolescent education of each Nobel Prize winner was characterized by at least one of these three factors, which greatly accelerated the development of the future laureates' abilities.

Each Nobel laureate also had at least one unique teacher during the school years. This uniqueness consisted in the fact that those teachers passionately loved the subjects they were teaching and taught in a manner that was different, in that, it provided advanced, enriched, and accelerated instruction. These were gifted and excellent teachers with a playful spirit. Teachers manifested interest in students and encouraged them to succeed by inspiring and challenging them. They went beyond the classroom practice and did many extra things for their students, which had great impact on their lives. This is what made them *special* teachers.

Taken together, a good family milieu and special teachers greatly influenced developing scientific and innovative talents of future Nobel laureates. These findings have important educational implications for today's children. This chapter clearly reveals what parents and teachers should and can do if they are really concerned with nurturing kids' innovative abilities.

It is interesting to note that these findings match the conditions that led to Ericsson's findings on deliberate practice (Ericsson, Nandagopal, & Roring, 2009). Specifically, the parental encouragement and influence of a great teacher promote the increased, long-term deliberate practice, which result in high accomplishments in any field of human endeavor.

Acknowledgments

This research was supported by the Social Sciences and Humanities Research Council (SSHRC) of Canada and the Templeton Foundation/the Institute for Research and Policy on Acceleration of the Belin-Blank International Center for Gifted Education and Talent Development of the University of Iowa. The findings and opinions presented in this chapter do not reflect the positions or policies of the granting agencies. I wish to thank Larry Vandervert for his review of this chapter and suggestions for improvement.

Notes

1 Konrad Lorenz is a winner of the 1973 Nobel Prize in physiology or medicine.
2 Due to space constraints, examples are taken only from the Nobel laureates in physics and physiology or medicine. Because of the same reason supporting quotes (with one exception) are taken from their autobiographies available on the Nobel Foundation at www.nobelprize.org/nobel_prizes.
3 The chapter presents the two main groups of findings from the project along with some quotes from Nobel laureates, which support and exemplify those findings.
4 The word "our" refers to Wolfgang Ketterle himself, his older brother, and a younger sister.
5 In this quote the word "this" refers to Russell A. Hulse's interest in science.

6 The chapter by Lyons and Reis (this volume) mentions how today's parents try to prevent children from excessive watching TV and focus them on reading books.
7 In this paragraph the year mentioned in parentheses means the year of the awarding the Nobel Prize in physics.
8 Starting with Tadeus Reichstein, the year mentioned in parentheses in this paragraph refers to the year of the awarding the Nobel Prize in physiology or medicine.
9 This is expecially true in the case of Nobel laureates in chemistry, which are not considered in this chapter.
10 Carl Wieman has found that most introductory science courses actually reduce interest in science. "If it is presented as memorizing a bunch of facts, and removed from the real world, then you do not see physics as a tremendously exciting thing to spend your life on." On the other hand, giving students an invigorating challenge, such as figuring out how to generate and distribute wind power in a community, could be a good way to teach them the fundamentals about electricity: "Instead of memorizing basic laws and equations, you would start with a problem you want to solve." While Carl Wieman's work is concentrated on university students, high school and middle school are probably even more important, he says: "If you have a bad high-school teacher, a bad high-school experience, and do not know people who support (science) ... you will decide on something else" (Anderssen & McIlroy, 2009, p. F1).

References

Alferov, Z. I. (2012). *Autobiography*. Downloaded from www.nobelprize.org/nobel_prizes/physics/laureates/2000/alferov.html on January 10.
Anderssen, E., & McIlroy, A. (2009). Is Canada losing the lab-rat race? How the country can spark that next, crucial generation of potential Nobel contenders. *Globe and Mail* (p. F1).
Carlsson, A. (2012). *Autobiography*. Downloaded from www.nobelprize.org/nobel_prizes/medicine/laureates/2000/carlsson.html on January 10.
Chi, M. T. H., & Greeno, J. G. (1987). Cognitive research relevant to education. In J. A. Sechzer & S. M. Pfafflin (Eds.), *Psychology and Educational Policy* (pp. 39–57). New York: New York Academy of Sciences.
Chu, S. (2012). *Autobiography*. Downloaded from www.nobelprize.org/nobel_prizes/physics/laureates/1997/chu.html on January 10.
Cohen, S. (2012). *Autobiography*. Downloaded from www.nobelprize.org/nobel_prizes/medicine/laureates/1986/cohen.html on January 10.
Cohen-Tannoudji, C. (2012). *Autobiography*. Downloaded from www.nobelprize.org/nobel_prizes/physics/laureates/1997/cohen-tannoudji.html on January 10.
Cornell, E. (2012). *Autobiography*. Downloaded from www.nobelprize.org/nobel_prizes/physics/laureates/2001/cornell.html on January 13.
Cropley, A. J., & Dehn, D. (1996). A broader view of giftedness. In A. J. Cropley & D. Dehn (Eds.), *Fostering the Growth of High Ability* (pp. 3–20). Norwood, NJ: Ablex.
Ericsson, K. A., Nandagopal, K., & Roring, R. W. (2009). An expert performance approach to the study of giftedness. In L. V. Shavinina (Ed.), *International Handbook on Giftedness* (pp. 129–154). Dordrecht: Springer Science.
Feldman, D. H. (1986). *Nature's Gambit: Child Prodigies and the Development of Human Potential*. New York: Basic Books.
Feldman, D. H. (1993). Cultural organisms in the development of great potential. In R. H. Wozniak & K. W. Fischer (Eds.), *Development in Context* (pp. 225–251). Hillsdale, NJ: Erlbaum.
Gardner, H. (1993). *Creating Minds*. New York: Basic Books.
Hitchings, G. H. (2012). *Autobiography*. Downloaded from www.nobelprize.org/nobel_prizes/medicine/laureates/1988/hitchings.html on January 10.
Hounsfield, G. N. (2012). *Autobiography*. Downloaded from www.nobelprize.org/nobel_prizes/medicine/laureates/1979/hounsfield.html on January 10.
Howe, M. J. A. (1990). *The Origin of Exceptional Abilities*. Cambridge, MA: Blackwell.
Howe, M. J. A. (1993). The early lives of child prodigies. In G. R. Bock & K. Ackrill (Eds.), *The Origins and Development of High Ability* (pp. 85–105). Chichester, UK: Wiley.
Hulse, R. A. (2012). Downloaded from www.nobelprize.org/nobel_prizes/physics/laureates/1993/hulse.html on January 10.
Hunt, T. (2012). *Autobiography*. Downloaded from www.nobelprize.org/nobel_prizes/medicine/laureates/2001/hunt.html on January 10.
Lewis, M., & Michalson, L. (1985). The gifted infant. In J. Freeman (Ed.), *The Psychology of Gifted Children* (pp. 35–57). Chichester, UK: Wiley.

Ketterle, W. (2012). *Autobiography*. Downloaded from www.nobelprize.org/nobel_prizes/physics/laureates/2001/ketterle.html on January 10.

Kholodnaya, M. A. (1997). *The Psychology of Intelligence*. Moscow: APN Press.

Kroemer, H. (2012). *Autobiography*. Downloaded from www.nobelprize.org/nobel_prizes/physics/laureates/2000/kroemer.html on January 10.

McCurdy, H. G. (1992). The childhood pattern of genius. In R. S. Albert (Ed.), *Genius and Eminence* (pp. 155–169). Oxford, UK: Pergamon Press.

Mather, J. C. (2012). *Autobiography*. Downloaded from www.nobelprize.org/nobel_prizes/physics/laureates/2006/mather.html on January 10.

Nurse, P. (2012). *Autobiography*. Downloaded from www.nobelprize.org/nobel_prizes/medicine/laureates/2001/nurse.htm on January 13.

Nüsslein-Volhard, C. (2012). *Autobiography*. Downloaded from www.nobelprize.org/nobel_prizes/medicine/laureates/1995/nusslein-volhard.html on January 10.

Palade, G. E. (2012). *Autobiography*. Downloaded from www.nobelprize.org/nobel_prizes/medicine/laureates/1974/palade.html on January 11.

Phillips, W. D. (2012). *Autobiography*. Downloaded from www.nobelprize.org/nobel_prizes/physics/laureates/1997/phillips.html on January 11.

Porath, M. (2009). What makes a gifted educator? A design for development. In L. V. Shavinina (Ed.), *International Handbook on Giftedness* (pp. 825–838). Dordrecht: Springer Science.

Rabinowitz, M., & Glaser, R. (1985). Cognitive structure and process in highly competent performance. In F. D. Horowitz & M. O'Brien (Eds.), *The Gifted and Talented* (pp. 75–97). Washington, DC: APA.

Radford, J. (1990). The problem of the prodigy. In M. J. A. Howe (Ed.), *Encouraging the Development of Exceptional Skills and Talents* (pp. 32–48). Leicester, UK: British Psychological Society.

Reines, F. (2012). *Autobiography*. Downloaded from www.nobelprize.org/nobel_prizes/physics/laureates/1995/reines.html on January 11.

Rigden, J. S. (1987). *Rabi: Scientist and Citizen*. New York: Basic Books.

Roberts, R. J. (2012). *Autobiography*. Downloaded from www.nobelprize.org/nobel_prizes/medicine/laureates/1993/roberts.html on January 11.

Robinson, A., & Clinkenbeard, P. R. (1998). Giftedness: An exceptionality examined. *Annual Review of Psychology, 49*, 117–139.

Schneider, W. (1993). Domain-specific knowledge and memory performance in children. *Educational Psychology Review, 5*(3), 257–273.

Shavinina, L. V. (1997). Extremely early high abilities, sensitive periods, and the development of giftedness. *High Ability Studies, 8*(2), 245–256.

Shavinina, L. V. (1999). The psychological essence of the child prodigy phenomenon: Sensitive periods and cognitive experience. *Gifted Child Quarterly, 43*(1), 25–38.

Shavinina, L. V., & Kholodnaya, M. A. (1996). The cognitive experience as a psychological basis of intellectual giftedness. *Journal for the Education of the Gifted, 20*(1), 4–33.

Simonton, D. K. (1978). The eminent genius in history. *Gifted Child Quarterly, 22*(2), 187–195.

Smith, H. O. (2012). *Autobiography*. Downloaded from www.nobelprize.org/nobel_prizes/medicine/laureates/1978/smith.html on January 11.

Sternberg, R. J., & Lubart, T. (1995). *Defying the Crowd*. New York: Free Press.

Storfer, M. D. (1990). *Intelligence and Giftedness*. San Francisco: Jossey-Bass.

Stormer, H. L. (2012). *Autobiography*. Downloaded from www.nobelprize.org/nobel_prizes/physics/laureates/1998/stormer.html on January 11.

Tannenbaum, A. J. (1986). Giftedness: A psychosocial approach. In R. J. Sternberg & J. E. Davidson (Eds.), *Conceptions of Giftedness* (pp. 21–52). Cambridge, UK: Cambridge University Press.

Tsui, D. C. (2012). *Autobiography*. Downloaded from www.nobelprize.org/nobel_prizes/physics/laureates/1998/tsui.html on January 11.

Vane, J. R. (2012). *Autobiography*. Downloaded from www.nobelprize.org/nobel_prizes/medicine/laureates/1982/vane.html on January 11.

Veltman, M. J. G. (2012). *Autobiography*. Downloaded from www.nobelprize.org/nobel_prizes/physics/laureates/1999/veltman.html on January 11.

Wieman, C. E. (2012). *Autobiography*. Downloaded from www.nobelprize.org/nobel_prizes/physics/laureates/2001/wieman.html on January 11.

Yalow, R. (2012). *Autobiography*. Downloaded from www.nobelprize.org/nobel_prizes/medicine/laureates/1977/yalow.html on January 11.

19

Settings and pedagogy in innovation education

Svanborg R. Jónsdóttir and Allyson Macdonald

UNIVERSITY OF ICELAND, ICELAND

Summary: This chapter introduces findings from research on innovation and entrepreneurial education (IEE) in Icelandic compulsory schools. IEE is an emerging curriculum area in Iceland and a key feature concerns young people developing creativity and using knowledge to meet needs or solve problems that they have identified. The interaction between IEE and the settings into which it is introduced are considered, as well as the characteristics of settings, which support or hinder the development of IEE and the types of pedagogy it requires. Four different settings for IEE were identified: *dormant, enclosed, developing* and *feasible*. These vary according to level of support from colleagues and leadership for learning, showing an interaction between the characteristics of the school setting and pedagogic practice in the classroom. An ability to create a supportive but flexible learning environment conducive to creativity is critical for innovation education.

Key words: Innovation education, creativity, curriculum change, social ecology, pedagogy, settings.

Introduction

> The issue of the limits to human enterprise has been left out of our thoughts and practices for so long that by now it has become all but incomprehensible and indeed ineffable.
>
> *(Bauman, 2007, p. 77)*

As innovation discourse emerged in national policy and the employment sector in Iceland it presented a view of the modern citizen as capable of action, creative and innovative, technologically and environmentally literate and entrepreneurial. Such a citizen is responsible, independent and collaborative, able to think and act as a complete person using skills and knowledge from life and school interactively. Innovation education is a curriculum area where learners can influence their own learning and have the resources to develop their own projects. Innovation and the practical use of knowledge was defined in the national curriculum of 1999 as follows:

> Innovation and the utilization of knowledge is about activating student ideas on how to utilize the knowledge and skills in each subject to solve problems, meet needs, or create other goods that make a difference. The purpose is to strengthen students' ethical sense and initiative through creative work where the student is trained in systematic methods to develop his or her ideas

from the first glimpse of an idea to the finished product. In this context, product means various kinds of goods, services, leisure activities, knowledge and other things that have market value.

(Ministry of Education, 1999, p. 31)

Innovation education began developing at grassroots level in Icelandic compulsory schools in the early 1990s and formed part of the official curriculum in 1999 (Gunnarsdóttir, this volume). Although its official name is 'innovation and the practical use of knowledge' in practice the term 'innovation education' (this is *nýsköpunarmennt* in Icelandic) is more commonly used in compulsory schools and 'entrepreneurship education' or 'entrepreneurial education' (this is *frumkvöðlamennt* in Icelandic) at upper secondary level. In this chapter we use the term 'innovation education' or the abbreviation IEE when talking about 'innovation and entrepreneurial education' regardless of school level.

Through our research in this area we have realized the interconnection of 'innovation' and 'entrepreneurship' (enterprise) in IEE (Jónsdóttir, 2008). The latter is more business oriented than typical school practice, thus the Scottish approach called 'enterprise education' is also relevant (Deuchar, 2008; Paterson, 2009). The early foundation of IEE in schools lies in creative work and honing creative skills (Figure 19.1), and with time the emphasis moves towards actualizing ideas, being enterprising and developing entrepreneurship skills and knowledge. The balance between creativity and enterprise may alter with the age of the learner but both are present in varying degrees.

Jónsdóttir's previous research on the early years of IEE showed that under 10% of Icelandic compulsory schools offered innovation education as a special subject in the school year 2003–2004 (Jónsdóttir, Thorsteinsson, & Page, 2008). Several factors appeared to influence the development of IEE in schools, such as the special approach needed for working with IEE, attitudes of school administrators and school ethos, clear messages in the curriculum, assessment methods and access to

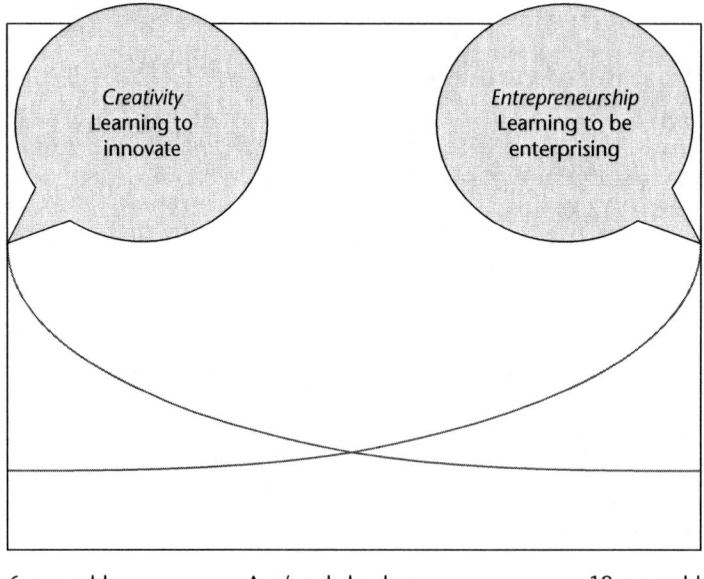

Figure 19.1 Two foundations of innovation and entrepreneurial education (created by Macdonald & Jónsdóttir in 2007).

information and teaching materials in IEE. The purpose of the follow-up research was to find out what IEE looks like when offered to learners.

The particular purpose of this chapter is to consider the interaction between an emerging curriculum area (ECA) and the settings into which such a curriculum is introduced, and identify characteristics of settings, which support or hinder the emergence of the new area. In this chapter we introduce literature on curriculum and pedagogy relevant to change at school and classroom level. We discuss the methodology of the research and introduce and discuss some of the findings most of which can be traced to three case studies of IEE in schools carried out by the first author. We begin however with a short overview of creativity and innovation in education.

Creativity and innovation in education

Research has shown that educating for creativity and innovation is possible (Colangelo, Assouline, Croft, Baldus, & Ihrig, 2003; Gunnarsdóttir, 2001; Reis & Renzulli, 2003; Smith, 2003; Vandervert, 2003).

Much of the research concerns new challenges for teachers and learners in educational settings. Learners in IEE experience more freedom than in traditional classes and have more control of tasks than in other subjects (Jónsdóttir, 2006). Jeffrey and Woods (1997, 2003) show that creativity demands innovation, relevance, ownership and control. Learners in IEE are active and creative and the teacher supports them in developing their ideas and avoids giving them the correct solutions (Jónsdóttir, 2006). To successfully allow learner agency and creativity, teachers need to know when to 'step back' and organize an enabling context that comprises flexible time and space in order to nurture the development of possibility thinking (Cremin, Burnard, & Craft, 2006). The innovation education teacher is the 'flexible teacher' that supports the learners and is willing to accept that the learner is sometimes the specialist regarding his or her idea (Jónsdóttir, 2005). The role of the innovation teacher is social constructivist, an educator who facilitates active learning as opposed to the teacher who 'feeds' the learners with finite knowledge. The teacher's role is crucial for the success of the ideation process of the learners and the development of the idea into solution and product (Gunnarsdóttir, 2001). The learners are key players in innovation education since the power structure of the classroom changes from a traditional arrangement to more autonomy and freedom for learners (ibid.). Teachers should become conscious of the educational philosophies which guide them and what effects these have on their teaching (Bjarnadóttir, 1993; Engelsen, 1993). A creative educator who fosters creativity in learners is willing to reflect critically on practice and be adaptable (Craft, 1997, 2011). For teachers to adapt to a role where the learners are creators and decision makers is a process that has proved difficult to sustain in traditional forms of schooling (Tyack & Cuban, 2001). In accepting a position as the 'flexible teacher' teachers can start by reflecting on their professional philosophies and whether they fit this role.

Changing education

In this age of globalization there are several ECAs such as sustainability education, environmental education or peace education. These ECAs generally do not fit into the shapes and forms of traditional education as they require a more flexible timetable, and a creative cross-curricular or interdisciplinary approach (Craft, 2011), 'happy classrooms' (Noddings, 2003, p. 240), learning activities that reach beyond the classroom and into society and as yet unformed ideas of what might be appropriate assessment (Macdonald, 2011). Burke and Grosvenor (2003) say: 'the organisation of learning has never before been so rigidly organised and ... subject boundaries have never been more strictly observed'.

School level change

Several factors influence the development and implementation of ECA and other school innovations (Fullan, 1982; MacDonald & Walker, 1976; Tyack & Cuban, 2001; Yueh, Cowie, Barker, & Jones, 2010). The understanding that education and change must be seen as a complex whole has been developing in educational research for some time. Goodlad (1975) advocated seeing and understanding the school as an ecological system rather than a mechanical one, with connecting parts that make a whole but each changing in unforeseen ways.

Bronfenbrenner (1979) considered human development as being conditioned by social systems that interact and influence the potential and direction of a person's development. We adapt the argument and state that the feasibility of developing an IEE pedagogy and finding supportive school settings is enhanced or limited by the social ecology of the interacting systems. We also look at the development of IEE in schools as a gradual process as feasible innovation suggests. Rogan and Grayson (2003) argued that a new curriculum cannot be implemented in one big leap but needs to be a series of smaller steps. Building on the 'zone of tolerance', theories of school development and Vygotsky's (1987) 'zone of proximal development' they developed the notion of a *zone of feasible innovation* (Rogan, 2007; Rogan & Grayson, 2003) based on three constructs:

1 curriculum change in the classroom,
2 the capacity of the school to support change and
3 support for change from outside agencies.

Schools are different from one another and therefore the process of change is context specific and varies across schools. Teachers' views towards changes differ, but like-minded teachers can form a learning community to chart new ideas and practices for their schools and thus make the changes more likely to be deep and lasting. Rogan and Grayson (2003) emphasize the need to acknowledge the diversity of schools and advise caution in categorizing schools, noting that any categorization scheme is at best a broad generalization. It can be helpful for policy-makers and researchers if they take care not to use it to label schools but to understand and serve their needs. Rogan and Grayson also advise against endorsing 'deficit models' when approaching curriculum change. Instead of listing school weaknesses, it is more useful to identify its strengths or practices that are in line with the proposed changes. The capacity to support change within a school and beyond it goes hand in hand with the development of teaching and learning practices in the classroom.

The three constructs allow for the construction of three rubrics, with several factors within each construct. A school can be placed at one of four developmental levels for each factor on its path from basic to ideal implementation of the proposed change. Change can occur in manageable steps, moving from one level to the next for each factor and is most likely to succeed when it proceeds just ahead of existing practice (ibid.). The proposed *zone of feasible innovation* takes into account all constructs, including teacher and learner factors, resources and school ecology and management as well as outside support. The zones are formed by the characteristics of neighbouring levels, above and to either side. Thus, for example, if resources are scantily developed they limit the zone of development of school management (Table 19.1).

In this research we combined Bronfenbrenner's systems approach and the developmental ideas from Rogan and Grayson to build a 'social ecology' for IEE and understand change in schools as occurring in small steps in interacting systems (Table 19.1).

In the main research, the first author spent considerable time in classrooms trying to understand whether key characteristics of IEE were actually evident in learning situations being presented as IEE. The capacity at the school to introduce and work with the demands for curriculum change

Table 19.1 Prototype rubric for understanding teacher and school development

4	Ideal					
3						
2						
1	Basic					
	Levels	Personal factors	Microsystem	Mesosystem	Exosystem	Macrosystem

Source: built on Bronfenbrenner (1979) and Rogan and Grayson (2003).

was also assessed, as was the support from the local community and at national level. Of great interest to us was how or whether change occurred in classrooms. This requires an analysis of personal factors and the micro- and mesosystem (Bronfenbrenner, 1979).

Change in classrooms

To develop our understanding of IEE in classrooms we used the theories of the sociologist Basil Bernstein to identify power (classification of roles) and control (framing of communication) in the IEE classroom, eliciting different modes of pedagogy (Jónsdóttir & Macdonald, 2011).

Bernstein introduced two concepts, *classification* and *framing*, that explain the translation of power and power relations and the form it takes in the control of relationships. *Classification* is used to categorize the construction of a social space such as school subjects or roles such as teachers vs learners, home and school (Bernstein, 2000). Power is embedded within a category, which can be strongly or weakly classified. Clear boundaries indicate strong classification and fuzzy boundaries indicate weak classification or weak power. IEE has several unclear boundaries, with regard to role and to subject matter, and thus is not in a strong position within a school with regard to prevailing classification of the teacher as an expert and of subjects of a disciplinary nature.

Control establishes legitimate forms of communication between categories. Control carries the boundary relations of power and socializes individuals into the appropriate relationships, the legitimate communications. *Framing* refers to where control is located. In strong framing the transmitter has explicit control but in weak framing the acquirer has more apparent control (Bernstein, 2000; Bolton, 2008). Strong framing indicates that control is located in a category that has power, for example, a teacher or a school subject, and weak framing indicates control shared between categories, for example, by a teacher and a learner or among several subjects (Macdonald & Jóhannsdóttir, 2006). Framing regulates relations within a context; it refers to the relationship between transmitters and acquirers (Bernstein, 2000, p. 12).

For curriculum change to occur in a school, it must undergo adaptation to the school's traditions and may itself provoke change within the school. It must also be understood and practised by the teacher such that the learner undergoes changed experiences. In this study the school and the classroom were under study. We needed to look for examples of IEE in schools, indications of change, the sort of settings that supported change and the nature of IEE lessons.

Methodology

We felt that an ethnographic approach was appropriate to find answers to our questions as it is ideal for the study of culture (Bogdan & Biklen, 2003). The ethnographic tradition considers it possible to integrate a series of ethnographic observations by relating them to each other and to a cultural whole. A 'combinative ethnography' aims to bring into public space hidden elements of the pragmatic condition of individuals by relating the part to the whole (Baszanger & Dodier, 2004). Pettinger (2005) explains combinative ethnography as a research strategy that entails looking at several sites rather than one and is intended to take account of the complex and multi-faceted nature of a field using data from different sources.

In particular, three case studies by the first author were used in which we could record process, structure and different contexts. By entering the field and hearing how practitioners experience and interpret working with IEE we could draw on different voices, extract values and attitudes and contextualize their experiences. Case study as empirical inquiry is especially relevant when the boundaries between the phenomenon and context are not clearly evident (Yin, 2009). IEE is highly contextual in nature and can therefore be expected to emerge differently in different settings. Therefore three cases of IEE, in different settings, were chosen to identify variations and consider how IEE and their contexts influence each other. Observations of lessons were undertaken, interviews with teachers, principals and learners were conducted and school curricula and other texts analysed. Data was gathered from autumn 2006 till spring 2009.

The three schools selected for the research were chosen from the limited number of schools that offered IEE as a formal programme and because they were willing and ready to take part in the research. The schools are City School, located in Reykjavík the capital of Iceland, and two rural schools, Country School and Trio School (all pseudonyms). Four teachers were the main informants in City School, one in Country School and one in Trio School. All personal names in the research are pseudonyms except the names of two pioneers of IEE, Rósa and Kolbrún, used with their consent.

To gather more data on the pedagogy of IEE, interviews were also taken with seven teachers from seven schools, other than the case study schools. These included design and crafts teachers and teachers trying new teaching materials for innovation education. Two of the teachers were pioneers in working with IEE in Icelandic compulsory schools.

The data was initially analysed for themes and then re-analysed using Bernstein's concepts. We constructed a 'language of description' for the relevant activities and elements in IEE consulting other researchers that had used Bernstein's theories in research in schools and classrooms (Bolton, 2008; Hoadley, 2006; Morais, 2002; Neves & Morais, 2001, Nsubuga, 2009). The language for description consisted of a set of criteria for elements in IEE lessons with indicators of framing and classification of several aspects of school and classroom practice. These indicators were then used in a further analysis of the data and findings to get a deeper understanding of the social relations (Table 19.2).

Findings

Innovation education in the classroom

Analysis of data from interviews with 13 IEE teachers using Bernstein's concepts of classification and framing showed that the IEE in the classroom is characterized by different strengths of classification and framing (Jónsdóttir & Macdonald, 2011). To analyse this further we developed a 2-D model of the interaction of classification of roles and framing of communication, using the four indicators developed for each (Figure 19.2, Table 19.1) (Jónsdóttir & Macdonald, 2011). On the horizontal axis, the continuum is from strong classification of roles of teachers and learners where power lies with the teacher, to weaker boundaries and power shared between teachers and learners. On the vertical axis, we move from strong framing of communication between teachers and learners, which means that teachers control interaction, towards weaker framing, where teachers relinquish some control (Figure 19.2). Four modes of pedagogy emerge from this model: *transmissive, controlled, progressive* and *emancipatory*.

Table 19.2 Classification and framing in IEE lessons

Classification – strength of boundaries			
C++	C+	C–	C– –
Learners have very limited agency and are receivers. The teacher is the specialist and sets criteria of roles. The control in lessons is with the teacher.	Teacher controls most aspects of lessons and is the specialist. Learners have agency within certain well defined areas.	Learners have agency in defined areas and are aspiring innovators. Learner and teacher communication is often on equal footing though the teacher has the power to decree.	Learners have ample agency and are innovative, i.e., creative and active. Learner and teacher roles are often interchanged; learners are experts and teachers learners.

Framing – nature of interaction between teacher and learners			
F++	F+	F–	F– –
Teacher takes/makes decisions in developing solutions.	Teacher suggests choices in development of ideas or influences learner choice.	Learner with teacher's support develops his or her idea and learner makes final choices.	Learner controls the development of his or her ideas and teacher supports.

Source: Jónsdóttir & Macdonald (2011)

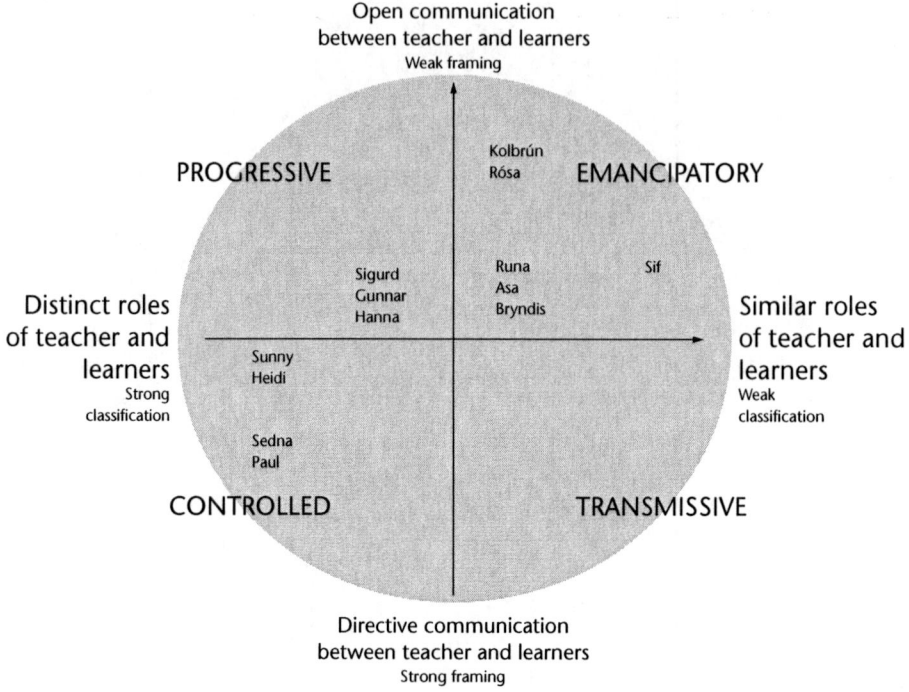

Figure 19.2 A model for analysing IEE pedagogy (Jónsdóttir & Macdonald, 2011).

Although the key idea in the IEE curriculum and the main advice in teaching materials and IEE courses (Jónsdóttir, 2011) is for the teachers to be sensitive to the agency of the learner in developing ideas and that the teacher should often 'step back' it could be seen that teachers did this in different ways, in different degrees and for different elements. The transmissive mode was not found in this research but the other three modes of pedagogy were identified (Figure 19.2) (Jónsdóttir & Macdonald, 2011).

In the *controlled* mode, the control of learner behaviour is distinctly in the hands of the teacher (Figure 19.2). The teacher has authority over students in controlled lessons and uses strong framing in the selection of content and approach, choosing the content, tasks, needs to address, methods and materials to use. Some freedom for creativity and agency may be given to students in the development of ideas and some in pacing. Teachers prescribe and control opportunities for learners to be creative. The macro and micro elements of learning are more or less designed and controlled by the teacher. Examples of a controlled pedagogy in IEE are found in the practice of Sedna and Paul and Sunny and Heidi.

In the *progressive mode* learners have considerable freedom and agency within the lesson (Figure 19.2), though the teacher is undoubtedly the designer of the learning opportunities and leads the lesson. The overall frame of time and content is controlled by the teacher, but learner agency is supported within lessons and learners can decide and control different tasks and elements especially in the development of ideas. Where learner agency is allowed, teachers are supporters rather than experts. Learners are aspiring innovators; they are creators of knowledge as inventors and can sometimes be explorers and experimenters. Examples of the progressive mode are Hanna's, Sigurd's and Gunnar's teaching.

Pedagogy in innovation education

Observing *emancipatory* pedagogy (Figure 19.2) is like moving into a workshop or a place of work with a democratic and creative atmosphere. Learners experience these lessons as a lifting of restrictions and get opportunities to influence their environments. Learners select the location of their work, and learners and teacher freely communicate and take on each other's roles; learners talk together, help each other and teachers learn about student ideas. Learners are explorers and creators of knowledge as inventors and they work autonomously and responsibly alone and with others.

In Trio School a workshop setting was identified in IEE lessons where learners could move freely and work on their ideas independently or with their mates. The descriptions of Rósa and Kolbrún also indicated learner choice of locations, experiments and research as well as a flow of creativity (see in more detail in Jónsdóttir, 2009).

Both Kolbrún and Rósa cross the boundaries of classroom work easily, indoors/outdoors, between teachers, boundaries of subjects and time. Both included other teachers in the IEE work, class teachers, science and crafts teachers. Kolbrún would often take the students to the seaside or to local factories and Rósa would allow students to work in the library or in the classroom or she would allow them to go to the beach near the school to investigate its ecology. Both emphasize the use of real problems and materials and topics that the students are interested in and find important.

Innovation education in school settings

In the case studies and interviews we came to realize the different ways in which school leaders could facilitate IEE, by their approach to school organization. Four categories of school organization were identified: from *rigid* or *firm* on the left to *flexible* or *receptive* on the right (Figure 19.3). Support from the school leaders and other staff for the practice of IEE ranged from *neutral* through *accepting* to *encouraging*, then *enabling* to *participatory* (Figure 19.3).

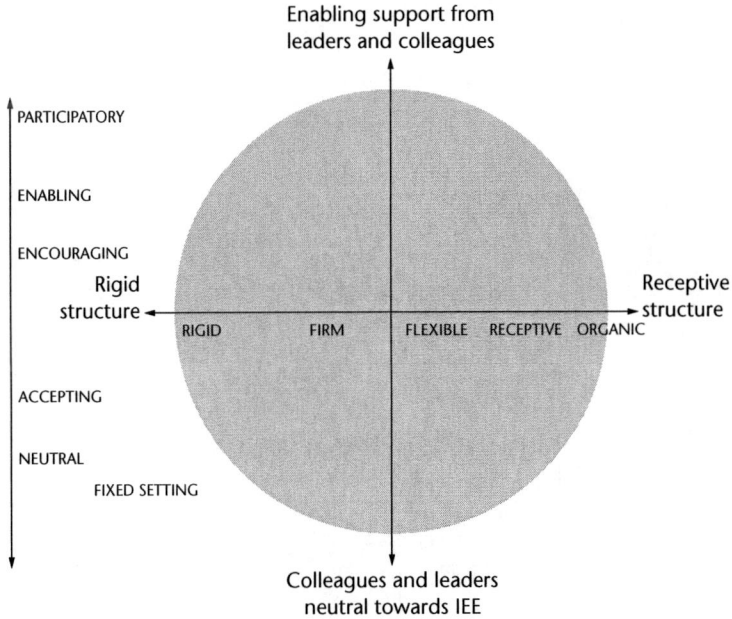

Figure 19.3 School settings – different conditions for IEE.

When these two characteristics of school settings, organization and support, are mapped against each other, four main types of settings emerged. These settings were labelled *dormant*, *enclosed*, *developing* and *feasible*.

Six of the schools provided an *enclosed setting* (Figure 19.4) building on a *rigid* or *firm* organization and *neutral* or *accepting* views of IEE. The basic order of the school is intact (Thomson, 2010) and IEE has no impact on infrastructure or staff other than the IEE teachers. This was the case in Sedna's and Paul's school where the day-to-day organization of the school seemed to be unchanged although IEE was offered on a limited basis. In Sigurd's, Hanna's and Country School there was support from administrators for IEE as a part of the curriculum, including crossing boundaries of school and work life, integrating knowledge areas and organizing suitable time slots, but no lasting changes were made to the basic order.

The *developing setting* is one that even though the structure is *firm*, administrators and colleagues *accept* teachers working with IEE, are willing to make adjustments to the day-to-day operations, permit collaboration for developing IEE in the school and external support is mediated. The developing setting could support agents of IEE within the school that may leave lasting changes in expertise and skills that will not disappear even if one IEE teacher leaves the school. This could be seen in the developing setting of Rósa's and Trio School where administrators had organized longer time slots for IEE than traditionally, collaboration of teachers on IEE existed and there was flexibility to work outside the school when needed as well as support for the input of specialists.

The *feasible setting* (Figure 19.4) is one where IEE as intended by the curriculum becomes possible, a kind of exemplary setting. The infrastructure is *flexible* with regard to planning, time and space, and there is some measure of *encouragement* from other teachers, even collaboration, and support from leaders. This setting offers realistic and feasible arrangements, which facilitate working with IEE. It is a setting where administrators know IEE and what it needs, the school has a group of teachers working on developing IEE and other members of staff know what it is about. This was

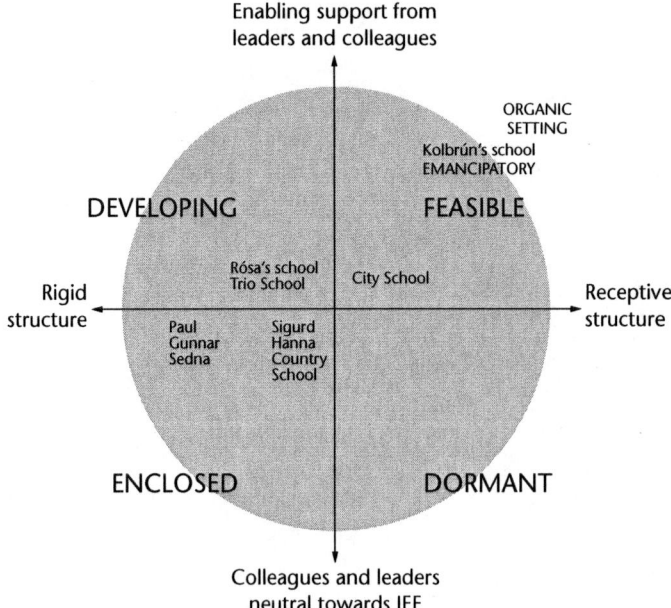

Figure 19.4 Distribution of schools working with IEE across four settings.

the case in City School that had a deliberate plan for opening up longer time slots than traditionally and an emphasis on blurring boundaries of subjects by integrating arts and crafts into lessons and learning.

A promising version of the feasible setting is what we call an *emancipatory setting* (Figure 19.4) where organization is at its most *receptive* and *enabling* views are in place; this interaction creates conditions that emphasize learner agency and creativity by offering IEE in different forms. Leaders acknowledge the importance of IEE, highlighting it in school curricula, taking part in introducing it to society and supporting autonomous learning related to IEE inside and outside the school. Colleagues take active part in IEE. The *emancipatory* setting is receptive to learner needs and society and can vary time and location from day to day. This was seen in Kolbrún's case where she was both the leading IEE teacher and the principal of a small rural school with possibilities for easy contact with the natural and social environment. The curriculum in her school emphasized creativity and innovation and connections with natural environment and local society (Jónsdóttir, 2009).

A tantalizing setting was sometimes glimpsed in Kolbrún's perhaps unique setting, the *organic setting* (Figure 19.4). Kolbrún was the principal of a small rural school and main IEE teacher so she had a perfect position to pursue her holistic view of education such that the school sometimes functioned as a living organism in organic conditions that could respond in a timely and appropriate way to the interest and needs of the learners, allowing them to be creative and innovative. This school community reminded us of the settings described in Summerhill (Neill, 1960) and in the English primary school Prestolee as it operated in 1918–1953 (Burke, 2005).

Interaction between pedagogy and settings

Teachers use a pedagogy based on their professional attributes and personal inclinations. Some of the teachers adapted to the IEE pedagogy like 'a fish to water' such as Rósa, as it fitted her personality and her professional approach in teaching. On the one hand she was inventive and enthusiastic in trying exploratory approaches with learners and on the other she was supported by the principal within the structure of the otherwise traditional setting. In a different example we see that Heidi did not allow as much autonomy and freedom for learners as the other three teachers working with IEE in her school even though she worked within the same flexible and supportive setting of City School. Her personal approach made less use of the IEE components of learner agency and freedom and the potential of the flexible setting than the other teachers. In Sedna's school there was some interest from administrators to expand IEE in the school but they had not yet joined forces to collaborate on this.

Discussion

School structure influences the range of learning opportunities in the curriculum and the timetable and if these are too rigidly organized IEE will find it hard to thrive. The views of leaders and colleagues can soften the edges of a setting, allowing more flexibility and supporting a version of IEE practice in which some curriculum aims are met. Where the timetables were *rigid* they could be mellowed towards flexibility with positive support from leaders who create leeway when needed, moving the setting to a *firm* but not rigid organization.

There are also examples of the provision of school support but strong framing and classification in the classroom, where the teacher rather than the learner drives the IEE process, not opening the lesson up to student-initiated ideas, solutions and products. In such cases opportunities for creative work are lost even though students might have taken part in the design process.

Deliberate interaction of those teachers, having knowledge and experience of working with IEE with the administrators in the schools, supports the change process. They could work collaboratively and expand the scope of IEE and its influence on learners. One way to help schools bring about change is to provide specialist input to help teachers towards a new understanding of teaching and learning and to support them in practice.

Compulsory schools are managed through strong classification of subjects and roles and strong framing of communications and thus are almost always at odds with an emerging curriculum area, which calls for other arrangements and values. Some of the schools with traditional arrangements were willing to be more flexible and change parts of the organization. The role of the principal is vital as well as the understanding and support of other colleagues in accommodating the need to cross boundaries. Schools that deliberately work to weaken classification, are able to develop a culture that enables conditions for weak framing of communication. Such a culture is essential for *progressive* and *emancipatory* modes of IEE pedagogy.

Although the school settings are of vital importance it was also clear that the personal and professional attitudes of the teachers and skills affect implementation of IEE. The modes of IEE pedagogy can be seen as different levels of classroom implementation where the *transmissive* mode is the least developed and most unlikely. Some examples of *controlled* pedagogy do allow for aspects of IEE to develop into the *progressive* mode with weaker framing of communication between teachers and learners. To move from the progressive to the *emancipatory* mode requires a further change, towards even weaker insulation of roles, where teachers can also be learners and vice versa. Such development needs interaction and support from the social ecology of the school with the principal as the prime leader providing vision and direction as well as orchestrating the development of the school culture (Larsen & Samdal, 2008; Thomson & Sanders, 2010)

In a similar way as research in business has shown that the culture of organizations determines whether innovation and creativity is welcomed and supported (Amabile, 1998), this research shows that the social ecology of the school, through the culture of the school settings and teacher action, influences the development of IEE. The creative and flexible process that IEE needs was available to some extent within City School as their organization was built around longer slots of time, and supported integration of knowledge areas and visits to places outside school. Although the organization in Rósa's and Trio School was more rigid the administrators catered for flexibility in lesson time for IEE, organized collaboration of teachers and supported cooperation with and learning from the local environment and society. Such settings and interactions are conducive for meaningful and creative learning and thus support the agency of learners and their creative work.

Differences were detected in school ethos, policy, construction of the school curriculum and timetabling decisions. The *interdisciplinary* and *integrative* nature of IEE is somewhat of a challenge to the segmented organization of schools in general. City School had a policy of crossing boundaries of subjects and age groups and was therefore more accommodating to IEE than the other two case schools. Rosa's, City School and Trio School had a team of innovation education teachers that made up the 'communal cluster' (Shulman & Shulman, 2004) necessary for dealing with ECA but the cluster members still needed to expand and distribute their acquired expertise within the school. While necessary, dedicated and enthusiastic individual teachers are not sufficient to balance the increased workload of reacting to change demands without risking exhaustion (Ballet & Kelchtermans, 2008). The support of collective teacher efficacy in IEE may enhance and sustain IEE development (Goddard, How, & Hoy, 2004).

With more general knowledge about IEE within the schools and direct support from leaders and colleagues, all of the schools within the *enclosed setting* have the potential to become a *developing setting*. Some teachers working with IEE wanted to be the 'specialists in IEE' in their schools, as was the case with Sedna, who worked on IEE in her role as the crafts teacher. This approach limits the collective understanding and flexibility with which IEE thrives. The next formal step for the schools in the developing setting is to create some flexibility in the timetable and for IEE knowledge to be valued as part of the curriculum. The specialist support provided by the first researcher in her role as adviser to the case study schools proved to be an important element in developing and strengthening their capacity to work with IEE. Securing and arranging for specialist support are practical measures that administrators can take to help implement IEE and other ECA in their schools.

Concluding remarks

Teachers with a creative frame of mind and skills suitable for working with IEE enhance the capacity for innovation and creativity. Such teachers work towards balancing freedom and structure in their pedagogy.

The professionalism of the teacher working with IEE lies in using the 'pedagogical palette' (Bernstein, 2000, p. 56) to support and enhance creativity by consciously adjusting framing according to the aims and goals of IEE. If teachers are to be able to tune a suitable strength of framing and classification in their pedagogy it is important that teacher education includes the development of an understanding of the impact of classification and framing and how these affect creative, emancipatory learning. The new curricular area IEE takes into account the interests and knowledge of learners and integrates different knowledge areas. An individual teacher can use progressive pedagogy but it will not develop further or be sustained without a flexible supportive school setting. The social ecology creates or constrains conditions for change. The transformative nature of IEE where learners control the innovation process, brings with it the chance to change thinking and action within the institutions, as administrators and colleagues begin to understand and appreciate the empowerment IEE can deliver. Enhancing IEE needs to be an iterative process of strengthening teacher capacity to work with IEE and developing the ethos of the school with the support of the school leaders.

References

Amabile, T. (1998). How to kill creativity. *Harvard Business Review*, 76(5), 76–87.
Ballet, K., & Kelchtermans, G. (2008). Workload and willingness to change: disentangling the experience of intensification. *Journal of Curriculum Studies*, 40(1), 47–67.
Baszanger, I., & Dodier, N. (2004). Ethnography: relating the part to the whole. In D. Silverman (Ed.), *Qualitative Research: Theory, Method and Practice* (pp. 9–34). London: Sage Publications.
Bauman, Z. (2007). *Liquid Fear*. Cambridge, UK: Polity.
Bernstein, B. (2000). *Pedagogy, Symbolic Control and Identity*, 2nd ed. Lanham, MD: Rowman & Littlefield Publishers.
Bjarnadóttir, R. (1993). *Leiðsögn: liður í starfsmenntun kennara* [*Guidance: a Part of Teachers' Professional Education*]. Reykjavík: Rannsóknarstofnun Kennaraháskóla Íslands.
Bogdan, R. C., & Biklen, S. K. (2003). *Qualitative Research for Education: An Introduction to Theory and Methods*, 4th ed. Boston: Allyn & Bacon.
Bolton, H. (2008). *Comparing pedagogy linked to success in art and science: usefulness of Bernstein's theory of pedagogy, and a question (a draft)*. Paper presented at the the Fifth International Basil Bernstein Symposium. Retrieved from www.google.com/search?client=safari&rls=en&q=Comparing+pedagogy+linked+to+success+in+art+and+science:&ie=UTF-8&oe=UTF-8.
Bronfenbrenner, U. (1979). *The Ecology of Human Development*. Cambridge, MA: Harvard University Press.
Burke, C. (2005). 'The school without tears': E. F. O'Neill of Prestolee. *History of Education: Journal of the History of Education Society*, 34(3), 263–275.
Burke, C., & Grosvenor, I. (2003). *The School I'd Like: Children and Young People's Reflections on an Education for the 21st Century*. London: RoutledgeFalmer.
Colangelo, N., Assouline, S., Croft, L., Baldus, C., & Ihrig, D. (2003). Young inventors. In L. V. Shavinina (Ed.), *The International Handbook on Innovation* (pp. 281–292). Oxford, UK: Elsevier Science Ltd.
Craft, A. (1997). Identity and creativity: educating teachers for postmodernism? *Teacher Development*, 1(1), 83–96.

Craft, A. (2011). *Creativity and Education Futures: Learning in a Digital Age*. Stoke on Trent, UK: Trentham Books.

Cremin, T., Burnard, P., & Craft, A. (2006). Pedagogy and possibility thinking in the early years. *Thinking Skills and Creativity, 1*(2), 108–119.

Deuchar, R. (2008). 'All you need is an idea!': the impact of values-based participation on pupils' attitudes towards social activism and enterprise. *Improving Schools, 11*(1), 11–19.

Engelsen, B. U. (1993). *Når fagplan møter lærer*. Oslo: Ad Notam Gyldendal.

Fullan, M. (1982). *The Meaning of Educational Change*. Toronto: OISE, Ontario Institute for Studies in Education.

Goddard, R. D., How, W. K., & Hoy, A. W. (2004). Collective efficacy beliefs: theoretical developments, empirical evidence, and future directions. *Educational Researcher, 33*(3), 3–13.

Goodlad, J. I. (1975). *The Dynamics of Educational Change: Towards Responsive Schools*. New York: McGraw-Hill Book Company.

Gunnarsdóttir, R. (2001). *Innovation education: defining the phenomenon*. Unpublished doctoral thesis, University of Leeds, Leeds, UK.

Hoadley, U. (2006). Analysing pedagogy: The problem of framing. *Journal of Education, 40*, 15–34.

Jeffrey, B., & Woods, P. (1997). The relevance of creative teaching: pupils' views. In A. Pollard, A. Filer & D. Thiessen (Eds.), *Children and their Curriculum: The Perspectives of Primary and Elementary School Children* (pp. 58–68). London: Routledge.

Jeffrey, B., & Woods, P. (2003). *The Creative School: A Framework for Success, Quality and Effectiveness*. New York: RoutledgeFalmer.

Jónsdóttir, S. R. (2005). *Ný námsgrein verður til: Nýsköpunarmennt í grunnskóla* [The Emergence of a New School Subject: Innovation Education in Compulsory Schools]. Unpublished master's thesis, University of Iceland, Reykjavík [in Icelandic].

Jónsdóttir, S. R. (2006). *The role of the teacher in innovation education*. Paper presented at the Scottish Educational Research Association Conference, November 2005, Perth, Scotland.

Jónsdóttir, S. R. (2008). Two sides of the same coin: Innovation education and entrepreneurial education in Iceland. *Bulletin of Institute of Vocational and Technical Education, 5*, 109–118.

Jónsdóttir, S. R. (2009). Using knowledge creatively. *Netla*. Retrieved from http://netla.khi.is/greinar/2009/003/index.htm.

Jónsdóttir, S. R. (2011). *The location of innovation education in Icelandic compulsory schools*. Unpublished doctoral thesis, University of Iceland, Reykjavík. Retrieved from http://hdl.handle.net/1946/10748.

Jónsdóttir, S. R., & Macdonald, A. (2011). Looking at the pedagogy of innovation and entrepreneurial education with Bernstein. *Sérrit Netlu – Menntakvika 2011*. Retrieved from http://netla.hi.is.

Jónsdóttir, S. R., Thorsteinsson, G., & Page, T. (2008). The ideology of innovation education and its emergence as a new subject in compulsory schools. *Journal on School Educational Technology, 3*(4), 75–84.

Larsen, T., & Samdal, O. (2008). Facilitating the implementation and sustainability of Second Step. *Scandinavian Journal of Educational Research, 52*(2), 187–204.

Macdonald, A. (2011). *How and why does education for a sustainable future call for new views of assessment*. Paper presented at the European Distance E-learning Network Conference, Dublin, June 2011.

Macdonald, A., & Jóhannsdóttir, T. (2006). *Fractured pedagogic discourse: teachers' responses to educational interventions*. Paper presented at the European Conference on Educational Research, September 2006, Geneva. Retrieved from www.leeds.ac.uk/educol/documents/159994.htm.

MacDonald, B., & Walker, R. (1976). *Changing the Curriculum*. London: Open Books.

Ministry of Education (1999). *National Curriculum for Compulsory Schools: Information and Technology Education*. Reykjavík: Ministry of Education [translation into English].

Morais, A. M. (2002). Basil Bernstein at the micro level of the classroom. *British Journal of Sociology of Education, 23*(4), 559–659.

Neill, A. S. (1960). *Summerhill: A Radical Approach to Child Rearing*. New York: Hart.

Neves, I., & Morais, A. (2001). Texts and contexts in educational systems: studies of recontextualising spaces. In A. Morais, N. Isabel, B. Davies & H. Daniels (Eds.), *Towards a Sociology of Pedagogy: The Contribution of Basil Bernstein to Research* (pp. 223–249). New York: Peter Lang.

Noddings, N. (2003). *Happiness and Education*. Cambridge, UK: Cambridge University Press.

Nsubuga, Y. N. (2009). *The Integration of Natural Resource Management into the Curriculum of Rural Under-resourced Schools: A Bernsteinian Analysis*. Unpublished doctoral thesis, Rhodes University, Grahamstown, South Africa.

Paterson, M. (2009). Are Scottish primary schools becoming more enterprising? *Scottish Educational Review, 41*(1), 36–50.

Pettinger, L. (2005). Representing shop work: a dual ethnography. *Qualitative Research, 5*(3), 347–364.

Reis, S. M., & Renzulli, J. (2003). Developing high potentials for innovation in young pepople through the Schoolwide Enrichment Model. In L. V. Shavinina (Ed.), *The International Handbook on Innovation* (pp. 333–346). Oxford, UK: Elsevier Science Ltd.

Rogan, J. (2007). An uncertain harvest: a case study of implementation of innovation. *Journal of Curriculum Studies, 39*(1), 97–121.

Rogan, J., & Grayson, D. (2003). Towards a theory of curriculum implementation with particular reference to science education in developing countries. *International Journal of Science Education, 25*(10), 1171–1204.

Shulman, L. S., & Shulman, J. H. (2004). How and what teachers learn: a shifting perspective. *Journal of Curriculum Studies, 36*(2), 257–271.

Smith, G. F. (2003). Towards a logic of innovation. In L. V. Shavinina (Ed.), *The International Handbook on Innovation* (pp. 17–30). Oxford, UK: Elsevier Science Ltd.

Thomson, P. (2010). *Whole School Change: A Literature Review*, 2nd ed. Newcastle upon Tyne, UK: Creativity, Culture and Education.

Thomson, P., & Sanders, E. (2010). Creativity and whole school change: an investigation of English headteacher practices. *Journal of Educational Change, 11*(1), 63–83.

Tyack, D., & Cuban, L. (2001). *Tinkering Toward Utopia: A Century of Public School Reform*. Cambridge, MA/London: Harvard University Press.

Vandervert, L. R. (2003). The neurophysiological basis of innovation. In L. V. Shavinina (Ed.), *The International Handbook on Innovation* (pp. 17–30). Oxford, UK: Elsevier Science Ltd.

Vygotsky, L. S. (1987). *Thought and Language*. Cambridge, London: MIT Press.

Yueh, M.-C., Cowie, B., Barker, M., & Jones, A. (2010). What influences the emergence of a new subject in schools? The case of environmental education. *International Journal of Environmental and Science Education, 5*(3), 265–285.

Yin, R. K. (2009). *Case Study Research: Design and Methods*, 4th ed. Thousand Oaks, CA: Sage.

20
Exploring innovative schools with preservice teachers

Michael Kamen and Deborah Erickson Shepherd

SOUTHWESTERN UNIVERSITY, USA, AND MERIDIAN SCHOOL, USA

Summary: An elective course is described in which preservice teachers have an opportunity to study and visit innovative schools. Students examine the philosophical, theoretical, and pragmatic aspects of these schools. They explore issues surrounding the implementation and maintenance of an innovative school. These include clarity of mission, leadership, governance, and creating a learning community. Some of the schools have a social justice mission with underserved student populations, while others differ from typical schools in their radically different philosophies and pedagogical approaches. The class visits a number of very different schools, discussing the philosophy and pedagogy of each school with teachers, administrators, and, often, students. Some anecdotal reflections from former students, including classroom teachers who took this course as preservice teachers, are presented along with examples of schools visited. The first-hand experiences visiting schools help students clarify and expand their philosophies and enable creative and innovative practice as they enter their careers.

Key words: Innovative schools, learning community, preservice teachers, sustainability, teacher preparation, innovative teachers, school visits, school mission, social justice.

Introduction

With the growth of high-stakes standardized testing, many educators believe that schools have become narrower in focus (Berliner, 2011; Ritchhart, 2004), and we have a climate in which innovation is often difficult to initiate and maintain in public schools. Giles and Hargreaves (2006) propose that external accountability is one of the factors that inhibits the sustainability of innovative schools. It is also argued that teachers' own school experiences as students (Blanton, 2003) and the cultural nature of teaching (Stigler & Hiebert, 1999) explain considerably more of the variance of enacted teaching philosophy, pedagogy, and "success," than do preservice education classes. Stigler and Hiebert (1999) report on their conclusions about the cultural nature of teaching from the *Trends in Mathematics and Science Study* video study of classroom practice.

> We contend, as do other educational researchers, that although teachers learn some things about teaching from their formal training, mostly they learn from simple cultural participation. After all, teachers spend at least thirteen years in classrooms, as students, before they even enter a teacher-preparation program.

(p. 83)

As teacher educators, we hope to prepare future teachers who are innovators in their classrooms, rather than future teachers who work diligently at doing an excellent job maintaining the status quo. This chapter describes an elective course in which preservice teachers have an opportunity to study innovative schools. Students examine the pragmatic aspects of these schools with a focus on the theoretical and philosophical assumptions that create the school's ethos. The class visits a number of very different schools, discussing the philosophy and pedagogy of each school with teachers, administrators, and, when possible, students.

Innovation education and creative teaching

The goal of preparing innovative teachers connects to the emerging field of innovation education. Initiatives to promote creativity and innovative adults have grown in popularity in recent years. A prominent example comes from Iceland. The Icelandic Ministry of Education added Innovation Education (IE) as a new subject in their national curriculum (Thorsteinsson, 2002). Thorsteinsson states that the field of innovation education assumes that everyone is creative and that "the theory of innovation work emphasizes that the individual use their [sic] powers of creation to mould [sic] their environment. Innovation work is intended to encourage this aspect of a child's character and thereby strengthen the stability of future societies" (p. 179).

Jónsdóttir, Page, Thorsteinsson, and Nicolescu (2008) describe IE as a pedagogical concept as implemented in Icelandic schools.

> The IE process is iterative with an overlying direction leading from "finding needs" to "presentation of solutions." Innovation has to do with the usefulness of ideas and/or how they can be implemented as solutions to problems encountered in daily life ... the emphasis is on enhancing creative activities of students through direct connections to everyday life.
>
> *(p. 455)*

Ritchhart (2004) makes a parallel argument that creative teaching should be understood in a broader sense than the common images from popular media of "charismatic and eccentric teachers." Ritchhart asserts that "creative teaching practices are effective and innovative in promoting the acquisition of skills, knowledge, and understanding" (p. 3). He also argues that in a "creative classroom" students' creativity is also being supported through open-ended assignments and more opportunities for independent thinking and expression.

Combining these ideas, we argue that supporting the development of innovative teachers assumes a natural creative ability, and one can empower the actualization of innovative teaching through helping future teachers understand a wider range of pedagogical and theoretical approaches and applying those approaches to solving educational design problems. We also argue that encouraging education students to go through a process parallel to the IE process of recognizing needs, developing a solution, and presenting the "product" in the context of education may have a powerful impact on their thinking about teaching and the nature and structure of schools. Jónsdóttir et al. (2008) discuss that the context of innovation education may be either

- an invention that may be considered completely new;
- an improvement of an existing product or system; or
- a diffusion of an existing innovation into a new application.

(p. 454)

It is considered best practice in preservice teacher education to place students in schools for in-depth field experiences. Teacher preparation programs often select school placements carefully for observations, field requirements, internships, and student teaching. When possible, criteria may include an excellent cooperating teacher, a diverse school population, supportive staff, and pedagogy consistent with what is taught in methods classes. At its best, these experiences help students learn best practices from dedicated and skilled teachers in authentic and diverse contexts, but they typically occur within the mainstream public education community.

The wisdom of this approach seems obvious, and we do not argue to change this tradition. However, we do argue that the focus on helping young teachers being mentored into skilled and dedicated teachers is a necessary but not sufficient ingredient for innovation. In addition, Shantz (1995) reports that these experiences often present traditional pedagogies stating, "Many faculties of education design curricula that espouse new and innovative methodology and then place students in field experience situations that are traditional in nature" (p. 339). Drawing from the above list by Jónsdóttir et al., the goal is for future teachers to create, improve, or apply innovations. We assert that the probability of such innovative activities taking place will be enhanced from a much wider repertoire of schools, missions, philosophies, and pedagogical approaches.

Our experience is that university students enrolled in *Innovative Schools* have difficulty fully conceptualizing schools that are very different from what they experienced in their own elementary and high school education. As might be expected, their initial analyses of schools we describe, read about, or study online are framed in a traditional school paradigm. They have difficulty critiquing a school in terms of its own goals, rather than through their assumptions and prevailing culture of schools.

Through site visits and readings about schools that meet a variety of needs (from learning differences to social economic status to dissatisfaction with or failure in traditional schools), students broaden their views of the purpose of schools, who they serve, theory, pedagogy, and school-wide structures. Students apply these ideas in class by identifying specific needs and developing new approaches, improvements, or applications as part of their final school design project. It is hoped this will support these future teachers' creativity and innovation in their own classrooms and schools or other professional activities.

Course description

Innovative Schools is a course that provides preservice teachers (and other interested students) with an opportunity to explore, investigate, and describe innovation in education. It is in its third year as a full-semester course at Southwestern University in Central Texas. The *Innovative Schools* class has its roots in a graduate course Michael developed at a large state university in the southeast. He had been assigned a course entitled *The Elementary School Program*. Students in this course were certified teachers with a wide range of experience. With the desire, as Shantz (1995) states, to "teach and encourage students to think beyond the present and be innovative" rather than "perpetuate the current system" (p. 339), he redesigned the course to focus on innovative schools rather than on typical public programs, schools, and administrative structures. In class he was reporting on some schools he had visited in New York City while working on his master's degree. From this discussion the idea emerged to plan a trip to NYC to visit these schools. Several students from that class accompanied him to visit schools in New York over spring break, and he continued to offer this opportunity for several years. This was a powerful experience for the students, and some described it as transformational for their educational philosophy and practice.

Michael reconfigured this experience for students at his current university as an undergraduate May term/spring break course focusing on the school visits. More recently, he offered the course as a full-semester course using a mix of on-campus classroom instruction and site visits to innovative schools in the area. He invited Deborah, a local teacher with both traditional public school and innovative charter school experience, to participate in the class to offer educational insights and perspective.

Learning outcomes

It is our hope that students in the course will bring more innovation into their professional lives from participating in the course. While this applies to a wide range of majors including, for example, art, political science, psychology, and sociology, we have an understandably strong interest in promoting innovation among future teachers in the class. Ritchhart (2004) argues that it is important to take a broad view of innovative teaching that moves well beyond creative teaching as a "personality trait" to a "broad and accessible form of teaching practice that can be understood on three levels: curricular, instructional, and student" (p. 2).

The student learning outcomes for the course as stated in the syllabus (Fall 2011 version) are:

- Students will demonstrate in their reflective journals an increasing ability to recognize and articulate unique features of a school and connections to explicit and implicit theoretical and philosophical underpinnings.
- Students will show an understanding of issues related to sustainability of institutional innovation. These include pressures that make it difficult for a school to maintain an innovation. Students will describe how schools have been successful and unsuccessful in maintaining innovation.

Assignments

According to Klein and Riordan (2011) who investigated a professional development program to prepare teachers to teach in innovative ways, reflection is a key component in an effective program. Each student in the *Innovative Schools* class keeps a reflective journal to document observations and reflections, as well as make connections to their own educational experiences after each visit. The reflections are posted on an online discussion forum on the university course-management system. Students respond to the postings of their classmates with questions and dialogue about each others' reflections.

The experiences of students in the *Innovative Schools* class are further broadened by class readings about innovative educators. Throughout the years, we have struggled to find readings that students could embrace that support the outcomes of the class. Every semester classes have read *The Power of Their Ideas: Lessons to America from a Small School in Harlem* (Meier, 2002). In this book, Deborah Meier writes about her experiences and philosophies in founding a highly successful school in East Harlem. Among other readings tried throughout the years was *Summerhill School: A New View of Childhood* (Neill, 1992) written by the founder of the progressive Summerhill School in England. *Schools for Growth: Radical Alternatives to Current Educational Models* (Holtzman, 1997) provided another opportunity to read about innovative education, including chapters on the Golden Key School and the Sudbury School. John Dewey's *Pedagogic Creed* adds a historical perspective. Students found that much of Dewey's philosophy about educating children is still considered innovative today.

Students choose a book on an innovative school or related topic to read and share with the class. Highlights of each book are reported and students lead a discussion about important ideas in the book. Students submit a paper outlining the salient issues from the book and making connections to innovative education.

Students also research an innovative school of their choice and become an expert on it. They present their school to the class including an on-line tour of the school using school web pages. Students create handouts that describe the school, highlighting what is unique about the school, its mission, and its theoretical frame. The students design and lead an activity that represents an innovative aspect of their school.

As a culminating project, students design their own school, drawing from experiences, resources, and information gathered throughout the course, demonstrating an understanding of the multiple elements that contribute to a school's success in implementing its mission and achieving its goals. These include curriculum, physical facility design, schedule, governance, professional development, leadership, clarity of purpose, etc. Creative, traditional, and innovative ideas are incorporated into school designs.

Site visits

Numerous studies show that field trips have long-term positive effects for students' learning (Anderson, Kisiel, & Storksdieck, 2006; Anderson, Picitelli, Weir, Everett, & Tayler, 2002; Anderson, Storksdieck, & Spock, 2007). We make the argument that innovative school site visits have an outcome for our students similar to that of well-conceived field trips have for K-12 students. The positive impact of site visits is supported by student evaluations of the *Innovative Schools* course, in which 83% of students who wrote comments stated that the most valuable aspect of the class was the school visits. One student wrote, "The school visits were extremely helpful and couldn't have been replaced by any amount of classroom lecture or discussion. Being in the schools was the cornerstone of this course."

Innovative Schools has been taught in several formats (short courses with school visits in NYC, with school visits optional, and in its current iteration as a semester-long course with "local" school visits integrated into the course syllabus and schedule). No two classes visited the same set of schools, had the same guest speakers, or even had the same experience when visiting the same school as a previous class. However, school visits always comprise the foundation of the *Innovative Schools* class. Schools are selected to help students see outside the realm of their experiences and to provoke student insights and creative thinking about what it means to be a school. When possible, students are given an opportunity to provide input into which schools are chosen to visit each semester, but a variety of types of schools and philosophies are always included.

The structure of school visits vary according to each school but usually follow a similar schedule. Visits begin by meeting with an administrator who gives an overview of the school, including history, philosophy, and description of curriculum. The class then tours the school, visiting classes and speaking with students and teachers. At the end of the visit, the class reconvenes for a question and answer session with the school administrator. The students value the follow-up discussions and tend to think of numerous questions after seeing classrooms.

Several themes frame the innovations students see during site visits and learn about during their research. These themes describe the way students' assumptions and thinking about schools are challenged and include social justice, pedagogical approaches, student democracy, and school structure. The following sections describe what these themes mean to students in the course, and examples are provided of how selected schools implement their versions of the theme. The intention is to illustrate a sample of the variety of contexts that may push students to think differently about schools, teachers, children, and the multiple purposes of formal (or not so formal) education in society.

Themes

Social justice

Social justice emerged as a theme throughout the numerous sections and versions of this course. The students gained understanding about schools that serve the needs of students from marginalized groups (income, culture, gender, race, and disability) and related issues. In some schools social justice and activism are integrated into the curriculum. Student response to schools serving low-income students has been very positive throughout all of the iterations of this course. The students are particularly enthusiastic about the opportunities these schools create for children and express respect and appreciation for teachers and administrators who work in the schools with this mission. Many expressed a desire to work in such a school. The class has visited a wide variety of schools and contexts with this theme as part of their mission. A brief description of several of these schools and who they serve follows.

Several are charter schools (state-funded schools that operate independently of school districts). One such school is Rapoport Academy (often the first school visited by the class). It was founded specifically to serve low-income students and their families by specifying the intake boundaries of the school to encompass an area of town that contains lower socio-economic housing and families. The founder and superintendent (recently retired) of the school is a charismatic woman who has given a great deal of time to *Innovative Schools* class visits. She talks about how she was motivated by low test scores in the low-income African American neighborhoods. She was horrified to learn that the overall test scores in these neighborhoods started out low and *decreased* as the students progressed through the local public school district. She clearly states that the mission of the Rapoport Academy is to give the students who attend an equal opportunity. The school has eliminated the achievement gap for their students, and graduating seniors leave ready for higher education with many college credits under their belt. The students in *Innovative Schools* hear about and observe thoughtful and evolving innovations, including partnerships with a local technical college, local businesses, the public library, a fitness center, a college-preparatory focus, and school-wide activities such as recycling and gardening. The students see an energetic and successful model of how the mission is implemented through these initiatives and through an ongoing evaluation and improvement focus.

A very different school, Manhattan Country School, has been visited during every NYC version of the course. Manhattan Country School describes itself as "a private school with a public mission." Started in response to the assassination of Dr. Martin Luther King, the school was designed from the start to cross racial and economic boundaries. In the years that followed, the school has developed a multi-cultural/social justice curriculum in which the mission of the school can be seen on many levels. An innovative tuition plan was developed without scholarships in which families pay a "fair share" based on their incomes, and no grade or class has a racial or ethnic majority. Students in *Innovative Schools* hear about the kindergarten children visiting their classmates' homes, including both apartments in housing projects and penthouse apartments. They learn about the family history museum made by one class and the importance of school-wide jobs in building community. One class organizes the school's participation in New York City's Martin Luther King, Jr. celebration, while another class is given the job to review children's books and select a book that will receive an award for promoting the values advocated by Martin Luther King, Jr. A science teacher might explain to the *Innovative Schools* class how they look at the biology of race. As with Rapoport Academy, the students are impressed with the staff's passion and the school's commitment to the causes of equity and social justice. It is our hope that the students come to understand how the successful enactment of the social justice agenda in schools requires an intentional, systemic, and detailed approach to the curricular and administrative decisions and planning, based on the mission.

Innovative Schools classes have also visited KIPP (Knowledge is Power Program) Austin Public School, which also focuses on providing college preparation environments for low-income, underserved students. KIPP uses advertising to target and attract low-income families. KIPP students attend school 67% more than their counterparts in traditional schools, due to extended days, Saturday school, and shorter summers. Their teachers are available to students by phone nightly for homework support or questions.

The students noted similarities and admired the mission and dedicated staff at both Rapoport Academy and KIPP. However, the philosophies differed in some interesting ways. KIPP uses a metaphor of a sports team in their response to student behavior and academic effort issues. *Innovative Schools* students expressed concerns when they were told about a punitive application of this metaphor, in which students who do not follow the rules are "benched" and thus limited in their interactions with other students. They reacted much more favorably to Rapoport's focus on the individual's responsibility to the group and the expectations that students accept responsibility when transgressions occur by discussing their behavior with their classmates and finding a way to make up for it.

Overarching themes among these schools are high expectations, responsibility, and preparation for college. The students in *Innovative Schools* relate easily to the college-preparatory focus. As students at a selective liberal arts university, it is natural for them to share this value with the schools they visited. Their understanding of what it means to perform well and succeed in college facilitated their recognition and appreciation of the innovations related to college prep. These include extensive college counseling programs, postgraduate support for tutoring and mentoring during college, college visits starting in upper elementary school, and the creation of post-secondary educational plans.

Pedagogical approaches

The school visits offered opportunities for students to observe and discuss a variety of pedagogical approaches. Some clearly present themselves to students because they are so different from what the students experienced in school themselves. Other pedagogical activities were more subtle to the students and needed discussion to elicit a deeper analysis.

Several schools featured problem-solving models for math, featuring open-ended problems and an emphasis on student-invented strategies. Students were also interested in the schools that had an inquiry focus. In these schools, children are taught types of questioning to help frame student inquiry. The questions they generate require investigation and response to discover answers. In one school, the students were able to see student inquiries and their results for all classes featured on bulletin boards in hallways. Many of the activities and tasks seemed familiar to the students and required some processing to fully appreciate how different these schools are from the typical school in this world of high-stakes testing. The reflective activities and discussions helped the students to understand the rationale of these approaches.

The *Innovative Schools* class also visits a local Waldorf school, which offers theory and a set of educational practices based on the work of Rudolf Steiner. The *Innovative Schools* students were uncomfortable when they learned that Waldorf families sign an agreement to avoid technology, including television, movies, and computers, at school and at home in the elementary and middle-school years, due to a concern that it hampers a child's imagination and creativity. The classes are always impressed with the examples of student-created and illustrated textbooks based on the lectures of their teachers. The story-based lessons and student-created textbooks clearly were viewed as innovative by the students. However, they questioned the effectiveness of the overall Waldorf education and wondered if it was too sheltered of an environment. The class had the opportunity to speak with a current Southwestern University student and Waldorf alum, who described her educational experiences at Waldorf and her transition to university. This had a positive effect on the opinions of the class about the Waldorf philosophy. Meeting a "regular" person who is an alum eased their concerns about how sheltered students might be.

Most schools incorporate art, music, and physical education into their students' schedules as separate "special" classes. Waldorf educators think of the arts as having the same importance as traditional academic subjects and so the arts, including a physical movement class called "Eurythmy" which is the art of movement, as well as music, art, and drama, are considered essential. These are incorporated into the curriculum as well as handiwork such as knitting, weaving, and woodworking. At Meridian (a new charter school in the process of implementing an International Baccalaureate curriculum) students attend these "specials," but Meridian's creative arts teachers also plan with classroom teachers in order to integrate the arts into the curriculum as well. Another charter school, the Austin Discovery School, has partnered with a Drama for Schools program to encourage creativity in both teaching and in students' responses. Teachers integrate drama into their teaching of core subjects instead of using a traditional lecture format.

A child-centered philosophy permeated the environment of many schools visited. Several referred to the education of the "whole child," believing that students learn better in an environment that takes their emotional well-being into account. Almost all schools visited had an emphasis on character development. Most had the character traits they promoted on display school-wide as well as in classrooms.

Students were impressed by schools that differentiated the curriculum to meet individual needs of students. They found that some emphasized the use of formative assessment to inform teaching. Students observed teaching methodologies that enable students to perform at their ability levels, including open-ended assignments, inquiry projects, individual conferencing, and small group work. The students' understanding of pedagogical approaches and the theoretical and philosophical underpinnings, grow from the numerous observations, discussions, presentations, and readings.

Student democracy

The course begins with a video produced by the Sudbury Valley School in Massachusetts describing the benefits of freedom and self governance for children and teens. Sudbury is an extreme example of a democratic school. Allowing children the freedom to control their own education is a radical educational philosophy that created significant dissonance with the students in *Innovative Schools*. Students had strong and often conflicting reactions. They appreciated the freedom and choices but worried about children getting enough guidance.

After viewing the Sudbury video, students were eager to visit the local Sudbury School that had recently opened with a few students. They were then confronted with the reality of the Sudbury democratic philosophy when they learned that our visit must be approved by the students and teachers in the school at their weekly school meeting. These school meetings also establish and ratify school rules to insure that student freedoms and rights are preserved. A judicial committee works with students as needed to settle disagreements. When we visited, the primary activity observed was Sudbury students playing video games, although there was one student studying for the SAT. *Innovative Schools* students were very concerned about what they saw, referring to it as "expensive day care", and wondered how the students would get a real education. The instructors tried to help the students frame their critique in terms of the school's values and stated philosophy, rather than their understanding of the purpose of school. It is interesting that even though all the students were uncomfortable with the level of freedom at the Sudbury School, most designed a final project school with more student choice opportunities than they had in their own school experience. Other schools had some student choice opportunities, albeit much more limited or guided. Three of the charter schools we visited encouraged children to set their own academic goals, prepare a portfolio of their accomplishments, and lead parent/teacher conferences.

School structure

Over the course of the semester, students see a variety of non-traditional groupings in the schools visited. At another local charter school, NYOS (Not Your Ordinary School), teachers loop with their classes, teaching each class for two years. Others have multi-age classrooms consisting of a combination of two grades in each class. Waldorf core teachers follow their classes from first grade to eighth grade. Students also had the opportunity to learn about a Golden Key School in Russia that Michael had recently visited. These schools are structured to support Lev Vygotsky's theories about learning and development. For example, an application of the Zone of Proximal Development is supported at the Golden Key schools by multi-aged family groups with children from ages three to ten in one class.

The governing structure of the schools visited allowed students to consider the impact of governance on innovation. Pedagogical decisions for the Waldorf school are made by a group of Waldorf teachers who serve for three years on a board called the "College of Teachers." NYOS is training its staff to participate in building a professional learning community. Giles and Hargreaves (2006) argue that a school structured as a learning community for the staff contributes to the sustainability of innovative schools stating, "the learning organization and professional learning community model may provide a more robust resistance to conventional processes of the attrition of change and of surrounding forces of change" (p. 124). NYOS also includes teachers on its governing board. Sudbury's governing philosophy allows all school members (students and staff) to be involved in the running of the school by putting school decisions up for debate and vote during school meetings. In the students' final school design products, they tend to think through and write about class-level structure more than a school-wide or governance structure. On reflection, we plan to be more intentional in encouraging the *Innovative School* students to investigate specific school-wide innovations including governance and organizational structures.

Supporting reflections and connections

After each visit, a structured reflection time during class encouraged students to discuss each school's innovative characteristics. Anderson et al. (2007) report that "sharing experiences with others and discussing those experiences also reinforce memories of visits" (p. 202), thus increasing the possibility that these memories will influence future decisions with regard to innovative education. The following example illustrates how students gain insight about the schools through the online conversations.

Jennifer initially reported that the "School didn't strike me as being particularly innovative. I didn't dislike the school, but overall it seemed to me like a high quality, traditional college-prep public school."

Other students' reflections showed a range of viewpoints about the degree of innovation, including Rebecca reporting that the "School was definitely innovative ... I really admire the way that the school formulated their own curriculum and had weekly meetings where all the teachers got together to create the kids' lesson plans."

Another student, Sarah wrote that the

> School is innovative, but in an incredibly structured way. The inquiry-based method they use engages all the students, doing exactly what innovative education is said to do. It finds a way to

reach all students so that they might achieve their highest potential. This innovation combined with the standardized innovation of the IB program defines why this was my favorite of all the innovative schools we have visited.

After reading these reflections, Jennifer added another comment,

> After thinking about the school a little longer, I realized that I didn't give it enough credit for its innovation. Even though teachers in normal public schools can and do find ways around the curricula they are prescribed by school district administrators, the school actually has a framework in which this deviation from cookie-cutter curricula is expected and encouraged of all teachers, rather than just the few teachers who are willing to deal with the difficulties associated with going against the norm.

Design-a-school project

The culminating project, to design an innovative school, is consistent with the pedagogical values of Innovation Education (IE) described by Jónsdóttir et al. (2008), which involve "searching for needs and problems in the student's environment and finding appropriate solutions or applying and developing known solutions" (p. 454). Jónsdóttir et al. list the steps in the process of innovation, which include identifying needs, brainstorming, choosing solutions, ideation drawing or modeling of a solution, making a description, and, finally, a presentation of the solution (p. 454). The design proposal for the project includes the philosophy, theoretical underpinnings, curriculum, governance, and physical layout. Students integrate ideas from schools visited as well as schools researched and shared. The possibilities are endless, and students have consistently planned creative, innovative schools. Examples of elements that students combine and apply include preparing political leaders, student choice, connection to the environment, community internships, civic engagement, mini courses, thematic courses, project-based learning, gardening, partnerships with colleges, wilderness experiences, integration of the arts, self-assessment, supporting a learning community for the staff, and shared leadership.

Student reflections and comments

Several former students from the course were contacted and asked to discuss the course or send a reflection. This is not presented as a comprehensive evaluation, but rather as anecdotal reports from some students who felt the course had a lasting influence on their perspective and practice as educators.

Katie, a student who participated in one of the New York City courses from Southwestern University is now teaching at a local charter school. She had a field experience at this school and requested it for one of her student teaching placement. We have used this school as a site visit for the local *Innovative Schools* class. Katie describes the class:

> I will be the first to admit we, as educators, have much work to do; however, despite the failings of some schools, there are countless schools in our country doing extraordinary things for their students. I was fortunate enough to be a witness to this by visiting several schools in New York City and surrounding areas. I traveled with students from my college over Spring Break in 2007 for a class called *Innovative Schools*. On our week long trip, we toured nine schools where we got to meet and talk to teachers and students.

From Katie's perspective, understanding the common attributes of these successful yet very different schools contributed to her development as a teacher.

> The schools were all very different, yet they shared a few qualities. First and foremost, the schools had strong leadership and a solid mission or purpose. Each school had faculty and staff that were on the same page and believed in what the school was trying to do. Along with leadership, the schools seemed to have a strong sense of community, especially in the case of the administration and teachers really trying to get to know each and every one of their students. The schools made a huge effort to find the strengths and interests of each individual child. They really knew their students as human beings, not just as kids they needed to get through the system. This seems to be one of the biggest differences I see from "innovative" schooling when compared to "traditional" schooling. The focus is on the students, not the teacher or the funding or the test scores. I saw a recurring presence of "student-centeredness" in all of the schools we visited. I still remember the students from The East Harlem School at Exodus House who served as our tour guides. Not only were they well-spoken, polite and energetic, they really knew so much about the school, its mission and why their education was so important. You could tell they felt supported at their school and wanted to tell others about their positive experience.

Her choice of where to student teach and accept a job was influenced by her experience and her image of herself as, in the words of Clandinin and Connelly (1992), a "curriculum maker" rather than the "conduit" to implement curriculum. Katie continues,

> I will never forget this trip, not only because of the amazing schools I got to see, but also because it inspired me to student teach at an innovative school in Austin, NYOS Charter School. I currently teach first grade at that same school. Like these other schools, my school places a huge emphasis on knowing each and every student's individual needs. Instead of following a curriculum mandated by a district, we get to create our own curriculum based on the needs of our students. It's hard work, but very rewarding! I would recommend this class to any future educator because it is an inspiration to see the uniqueness of these schools, and learn new, innovative ideas. There is no "one perfect way" to teach our students, but we should never stop trying to improve our schools.

Another student participating in a May term *Innovative School* class in New York reported an ongoing influence on her practice.

> I took many courses in the education department during my years at Southwestern, and they all prepared me exceptionally well to be a successful and innovative classroom teacher, but I truly feel that Dr. Kamen's *Innovative Schools* class provided me the initiative and experiences necessary to be an adaptive and responsive educator. In my four years in the classroom, I often referred to my experiences from our trip to New York. The classrooms I observed and the teachers and administrators that we met with were truly inspirational and were able to reach so many children with very little resources. I was able to take things that I saw in each of the schools and implement small pieces of their structure, philosophies, and academic strategies into my own classroom. *Innovative Schools*, in my opinion, is a class that all students who plan to go into education or plan to work with children in any situation should take. The research required and the experiences and lessons learned while going into different schools in New York City have been invaluable to me as an educator.

Two students still enrolled in the certification program (Lisa and Mary) were interviewed about the on-campus semester-long version of the course they took as freshmen. They reported the field trips as being of great value. From their perspective, the experiences of seeing a variety of schools first-hand helped them develop their own educational philosophy, gave meaning to theories discussed in other courses in their program, and helped them to see value in incorporating field trips into their own unit planning. From the school visits and discussions, they came away with a number of concepts and themes that they believe could be applied in a variety of school environments. These include:

- the power of choice
- valuing students
- high expectations/standards
- passionate teachers
- positive classroom atmosphere
- integration of art
- different children require different approaches.

Course evaluations from the first two years that it was offered as a full semester on-campus course (with local field trips) were reviewed. While there was a typical range of overall feelings about specific aspects of the academic part of the course (e.g., differing opinions about the reading selections), most students rated it as very good or excellent. A consistent theme throughout the evaluations was the value of the field trips. Most respondents mentioned the field trips as something they liked about the course.

Discussion and conclusion

The familiarity students have with schools is both an opportunity and challenge in helping them to develop an innovative and creative approach to teaching. As the class visits schools, the differences in what students have experienced can be striking. The more radical departures from the mainstream can seem irresponsible or even fantastic to some students. These always produce rich reflections and passionate critiques. A school such as Waldorf with eight-year looping, amazing art development and integration, the banning of television, and a "strange" movement program certainly help students think outside of their realm of experience. At the same time, schools that seem innovative to the instructors may have innovations that are far more subtle to the students. When students are sitting at tables doing work, reading, or writing, it is not always evident that the activities were developed by teachers (rather than from a textbook teachers' guide), connected to art class, or the result of a set of inquiry activities. The reflective journals, discussions in class, school research, readings, and final design project, we argue, make the innovations more explicit and meaningful. It is our hope that the first-hand experiences visiting the wide variety of schools with connection to theory will help students clarify and expand their philosophy and enable creative and innovative practice as they enter their careers. We view the class as not only about innovative schools but also as an innovation education experience.

> Freedom, the power of decision, initiative, enhancement of thinking and creative thinking, an attention to the environment switched on, a connection with life outside school, learning the methods of the inventor and the ethos of a creative workplace, were all things that the students experienced in the IE classes.
>
> *(Jónsdóttir et al., 2008, p. 466)*

We hope that our students' "connections with life outside" (of traditional schools) will help them confront the forces that narrow the curriculum and allow them to embrace creative approaches to teaching.

References

Anderson, D., Kisiel, J., & Storksdieck, M. (2006). Understanding teacher's perceptions on field trips: discovering common ground in three countries. *Curator, 40*(3), 365–386.

Anderson, D., Picitelli, B., Weir, K., Everett, M., & Tayler, C. (2002). Children's museum experiences: identifying powerful mediators of learning. *Curator, 45*(3), 213–231.

Anderson, D., Storksdieck, M., & Spock, M. (2007). Understanding the long-term impact of museum experiences. In Falk, J. H., Dierking, L. D., & Foutz, S. (Eds.), *In Principle, In Practice: Museums as Learning Institutions* (pp. 197–215). Lanham, MD: AltaMira Press.

Berliner, D. (2011). Rational responses to high stakes testing: the case of curriculum narrowing and the harm that follows. *Cambridge Journal of Education, 41*(3), 287–302.

Blanton, P. (2003). Constructing knowledge. *Physics Teacher, 41*(2), 125–126.

Clandinin, D. J., & Connelly, F. M. (1992). Teacher as curriculum maker. In Jackson, P. (Ed.), *Handbook of Curriculum Research* (pp. 363–401). New York: Macmillan.

Dewey, J. (1897). *My Pedagogic Creed*. New York: E.L. Kellogg and Co.

Giles, C., & Hargreaves, A. (2006). The sustainability of innovative schools as learning organizations and professional learning communities during standardized reform. *Educational Administration Quarterly, 42*(1), 124–156.

Holzman, L. (1997). *Schools for Growth: Radical Alternatives to Current Educational Models*. Mahwah, NJ: Lawrence Erlbaum Associates, Inc.

Jónsdóttir, S., Page, T., Thorsteinsson, G., & Nicolescu, A. (2008). An investigation into the development of innovation education as a new subject in secondary school education. *Cognition, Brain, Behavior: An Interdisciplinary Journal, 7*(4), 453–468.

Klein, E. J., & Riordan, M. (2011). Wearing the "student hat:" Experiential professional development in expeditionary learning schools. *Journal of Experiential Education, 34*(1), 35–54.

Meier, D. (2002). *The Power of their Ideas: Lessons for America from a Small School in Harlem*. Boston: Beacon Press.

Neill, A. S. (1992). *Summerhill School: A New View of Childhood*. New York: St. Martin's Press.

Ritchhart, R. (2004). Creative teaching in the shadow of the standards. *Independent School, 63*(2), 32–41.

Shantz, D. (1995). Teacher education: Teaching innovation or providing an apprenticeship? *Education, 24*(3), 339–334.

Stigler, J. W., & Hiebert, J. (1999). *The Teaching Gap: Best Ideas from the World's Teachers for Improving Education in the Classroom*. New York: Free Press.

Part VII
Research on mathematical talent and innovations in math education for developing innovators

21

The dynamic curriculum
A fresh view of teaching mathematics for inspiring innovation

Mark Saul

NEW YORK UNIVERSITY, USA

Summary: In many contexts, standard models of curriculum do not serve well the goal of motivating gifted students to think innovatively. This chapter offers an alternative model, in which curriculum is seen as dynamic, and the teacher as responding interactively to the unfolding needs of the students. Examples from working classrooms, using mathematical content, are discussed.

Key words: Innovation, curriculum, mathematics, gifted, innovation education, dynamic curriculum.

Introduction

In some ways, the very notion of 'curriculum' acts to discourage innovative thought. Curriculum is sometimes thought of as a body of knowledge to be delivered ('scope and sequence'), or a set of activities to be enacted with students. Tied to specific objectives and concrete measures, curriculum in this view is an enactment of a syllabus. We will call this view of curriculum a 'static' view.

This view of curriculum, essentially an industrial or productivity model, works well with students in those situations where our goal is to supply them with tools of thought. A gifted student ready for algebra in the fifth grade is served well by a ninth grade algebra course, taught from a fixed syllabus, within a predetermined time line, and resulting in a standardized test or assessment. A student who is ready to explore the axioms of Euclidean geometry does well with the traditional course in the subject. We can increase the depth of the course with new problems. We can supply such a course earlier than usual for the gifted student. We can increase the pace of the course so that the students master the material in a matter of weeks or days rather than months. But we still have a goal in mind: a specific set of theorems, structured by a predetermined set of techniques.

Curriculum thus delivered sets the scene for innovation, but does not necessarily stimulate it. To develop innovative students, we must ask also: what is the next step? How do we provide educational situations for students to put their cognitive tools to work? That question is not easily addressed within the framework of traditional static curriculum. A new vision of what the term 'curriculum' means is required, which we will call 'dynamic curriculum'.

Well, not entirely new. A number of writers have commented on the possibility of this form of curriculum. Smith (1996) lists four ways to look at curriculum:

1   curriculum as a body of knowledge to be transmitted;
2   curriculum as an attempt to achieve certain ends in students (product);
3   curriculum as process;
4   curriculum as praxis.

The straw man we've been talking about so far speaks from Smith's first and second points of view. The dynamic approach to curriculum is closer to Smith's conceptions of curriculum as process or as praxis. Stenhouse (1975, p. 142) characterizes this point of view as follows:

> The idea is that of an educational science in which each classroom is a laboratory, each teacher a member of the scientific community.... The crucial point is that the proposal [proposed curriculum item] is not to be regarded as an unqualified recommendation but rather as a provisional specification claiming no more than to be worth putting to the test of practice. Such proposals claim to be intelligent rather than correct.

In the classroom

Let us look into the classroom-as-laboratory to see how this view of curriculum plays out. What follows is a strand of dynamic curriculum, with comments about the learning it catalyzed in several classrooms.

Some of these are 'gifted' third grade classrooms, whose students have been identified in various ways as being above average in ability. Others are average high school classrooms, where some students sometimes show sparks of intuition. Still others are after-school or lunchtime 'math circles' for gifted fifth grade students with a particular interest in mathematics. (For more information on math circles, see www.mathcircles.org.)

The mathematical topic is magic squares, a classic and well-worked out area in recreational mathematics (see, for example, Andrews, 1917). The initial task is to arrange the numbers 1 through 9 in a 3×3 array so that each row, column, and diagonal has the same sum (the 'magic sum').

One way to do this is as follows:

2	9	4
7	5	3
6	1	8

The literature is rich in this topic: there are sophisticated algorithms for creating magic squares of any dimension. A mastery of such a 'canned' algorithm might be a good topic for a curriculum with a static design. But for a more dynamic approach to the subject, it is better to have students construct the magic squares themselves.

In this case, students were given ordinary playing cards with the numbers 1 through 9 on them. They were asked to deal them out in a 3×3 array. It is of course highly unlikely that the array they deal out will be 'magic'. And, in fact, if the teacher gives the cards out in just that order, students will end up with the initial array

1	2	3
4	5	6
7	8	9

The first objective given to the students is to make just the rows have equal sums. Most students are able to accomplish this by 'evening out' the rows: trading high cards for low so the sums get closer, and converging on a solution. Some will needed a hint (simply to describe the 'trading' strategy); others will lose focus and need more guidance. But in general, this is a good initial approach for most students.

Even at this stage, we can introduce notions of logic and proof. Every time we balance the rows, their sum is 15. Sometimes students deduce this logically, but more often they come to the conclusion empirically, by noticing that the process of balancing made the rows converge on a sum of 15. The point can be clarified by classroom discussion. The sum of all the numbers is 45, so each of three equal rows (the 'magic constant') must be 15.

Students are then asked to make the column sums equal. They can do this by rearranging numbers within each row. This leaves the row-sums intact. This step introduces the notion of an *invariant* with respect to a transformation, a notion basic to much of mathematics. But scaffolding is often required. Students typically see the logic behind the process, but only a few generate the idea themselves.

> Nancy, a high school student, was confused. Having gotten the rows to sum to 15, she was working on the column sums. She got the middle column to sum to 15, but could not get the other two columns to balance out by 'trading around'. The hint she needed was that the middle column was the wrong set of three numbers: she had to destroy the sum of this column by trading some of its number out. That is, she needed to 'back out' of the situation she was in, in order to get all three columns to sum to 15. Even very high ability students sometimes need help with this subtle point.

Finally, students are asked to balance the diagonals as well. Most students need a hint for this step. The easiest hint to give them is that they can 'sort' or 'shuffle' rows and columns without changing their sums, to get diagonals with equal sums. Students rarely discover this technique for themselves, and even understanding it constitutes an insight. It turns out that if they've balanced rows and columns, then it takes fewer than three of these 'shuffles' to balance the diagonals as well.

Having constructed a magic square, students can probe its properties. In a classroom situation, the solutions will generally look different. New insights can be gained by asking what the 'different' solutions have in common.

For example, students quickly notice that the number 5 is always in the center. Must this always happen? Can they give a 'proof' that it must? These are somewhat more difficult problems than deriving the magic constant.

> Michele, a good eighth grade student, expressed her argument using the algebra she already knew: two rows and two diagonals pass through the center square. Since the sum of each is 15, the sum of all these squares (counting the center square multiple times) is 60. This counts the center square four times, and each of the other squares exactly once. So $15 + 3C = 60$, and $C = 45$. Wayne, a gifted third grade student, struggled a bit with the same question but was able to construct essentially the same argument, expressed without algebraic variables. And, in fact, only the very last step of Michele's analysis involved algebra.

The magic squares that result from student constructions will all have something else in common: the even numbers will always be in the corners. This fact is harder to derive, but it is also possible to make an argument for it without using algebra.

Magic squares are intrinsically intriguing to students. As we have seen, students with very little background can engage in meaningful mathematical reasoning about them. But we can go further: we can use these intriguing objects to illustrate some basic yet profound mathematical ideas.

Extension I: group theory and symmetry

This extension probes important basic notions of geometry, in ways which are rarely addressed in the standard curriculum.

Students following the 'curriculum' given above will get a variety of solutions to this initial magic square problem. There are eight possible arrangements of the numbers 1 through 9 into a magic square, corresponding to the eight symmetries of a square. That is, from any one solution, you can get to any other by rotation (four rotations, including a rotation through zero degrees) or line reflection (in the two diagonals and in a horizontal and vertical midline of the square).

The teacher can put several different solutions to the magic square problem on the blackboard and asks students to compare them. Students can visualize the geometric transformations—rotations and line reflections—that take each into the other. They can work out all eight positions, and practice describing the transformation that takes any one into any other.

In the course of describing these transformations, students frequently notice that certain squares transform into certain others either using one single transformation, or a combination of two transformations. For example, we can get from position A to position C (below) either with a single rotation, or a pair of line reflections:

A $\longrightarrow$ C

2	9	4
7	5	3
6	1	8

(rotate through 90 degrees)

6	7	2
1	5	9
8	3	4

or

A $\longrightarrow$ B $\longrightarrow$ C

2	9	4
7	5	3
6	1	8

(reflect horizontally)

6	1	8
7	5	3
2	9	4

(reflect in diagonal)

6	7	2
1	5	9
8	3	4

This circumstance offers an introduction to the notion of composition of transformations, and opens the way to group theory, an important area of mathematics. Briefly, every object has certain symmetries, and the these symmetries can be combined ('composed'). The set of symmetries of an object, together with the operation of composition, form a mathematical structure known as a *group* (see, for example, Scott (1964) or Gaglione (2012)): the composition of two elements does not generate an element outside the group, and every element has an 'inverse' in the group that puts the square back in its original position. Here, students are actually studying the dihedral group D_4, the group of eight symmetries of a square. The construction of magic squares leads naturally to the construction of this group.

Several important aspects of group theory arise very quickly out of this simple situation. One of the most interesting, and poorly understood, aspects is that the group is not *commutative*: if we compose two symmetries in a different order, the result will be different. So, in the example above, we have:

A		B		C

2	9	4	(reflect	6	1	8	(reflect in	6	7	2
7	5	3	horizontally)	7	5	3	diagonal)	1	5	9
6	1	8		2	9	4		8	3	4

but:

A		B′		C′

2	9	4	(reflect	2	7	6	(reflect in	4	3	8
7	5	3	horizontally)	9	5	1	diagonal)	9	5	1
6	1	8		4	3	8		2	7	6

a different result. The first order results in a rotation by 90 degrees clockwise, the second by 90 degrees counterclockwise. The difference here is in the direction of rotation, but that is not consistently the case. In general, it is not easy to 'predict' the outcome if we change around the order in which the symmetries are combined.

A complexity that arises immediately is that some symmetries do 'commute' with each other: the composition of two rotation is, in fact, commutative. It is only when you combine a rotation with a line reflection that you disturb the commutativity. There are two insights that we can elicit from this situation. The first is that of a *quantifier*. It is not enough to find that sometimes composition is commutative, and sometimes not. For a group to be commutative, *any* two elements of the group must commute.

The second insight is a glimpse of the notion of a *subgroup*. We can study the rotational symmetries of a square without looking at the line reflections, because the composition of two rotations is again a rotation. The rotations form a subgroup within the larger group, and this subgroup is in fact commutative.

To clinch the idea, students can be asked how many symmetries an equilateral triangle has. There are six symmetries: three rotations and three line reflections. But they are harder to spot than the symmetries of a square: the reflections are in oblique lines and students find it hard to distinguish them from rotations. It turns out the group of symmetries of an equilateral triangle is the smallest possible group which is not commutative.

In general, a regular polygon of n sides has 2n symmetries, which form a group with 2n elements. Students who get this far are well on their way to studying an important mathematical subject.

Extension II: linear algebra

The magic square activity can be extended in another direction. Having constructed a magic square with the numbers 1 through 9, students can be asked to construct one with the numbers 2 through 10. They often do this in just the same way, using 'trading' algorithms. But some of them will have the insight that if they simply add 1 to each entry in their original magic square, they will solve the new problem. Of course, students who have used the trading method can see, with hindsight, how their work might have been done more simply.

Students can then be asked why the magic constant turns out to be 18. (They can see that it is ⅓ the sum of the new numbers, or that they have added 1 to each number in the original square, so added 3 to the magic constant.)

Generally, we can add the same constant to each term of a magic square, and its magic persists. Students can get to this generalization, for example, by being asked to form a magic square of the numbers 101, 102, 103, ... 109. Because of the form of the numbers, it will be easy for them to see that they can just add 100 to each entry of the original square. An easy generalization allows them to create a magic square for any nine consecutive integers.

Next, we can ask students to create a magic square from the numbers 3, 6, 9, 12, 15, 18, 21, 24, 27. Some may see quickly that they need only multiply each entry of the original magic square by 3. And if they want to use the numbers 4, 7, 10, 13, 16, 19, 22, 25, 28, they can multiply by 3, then add 1. In fact, this method allows them to form a magic square with any nine numbers that form an arithmetic progression: if the general term is $aN + b$, they can multiply the original square by a, then add b.

This generalization can be phrased more abstractly: if we drop the requirement that the numbers in a magic square be distinct, then we can consider an array of nine 1s (or of any constant) as a magic square. And it is not hard to see that the sum of two magic squares is also magic. So we effect the addition of a constant to each term of a magic square by adding two magic squares. It is also not hard to see that multiplying each entry in a magic square preserves its magic.

That is, 3×3 magic squares, including the 'constant' squares, can be added and scaled, and they remain magic. They thus form the mathematical object called a *vector space*, or *linear space*. The field of linear algebra is devoted to the study of this structure. There are canonical ways to think about it, which give further insights into magic squares. For example: what is the dimension of the space? This translates into asking which entries of a magic square must be given to recover the entire square. Certainly if one entry is removed, we can figure out what it must be. By pursuing this idea, students can find that the dimension is three: certain triples of entries will determine the rest of the square. This also means that linear combinations (sums and scalings) of three particular squares will generate the entire set: they form a *basis* for the linear space.

And there's more. A linear space of three dimensions can be represented by ordered triples of numbers; that is, by points in three-dimensional space. Students can explore the correspondence between the various magic squares and the points that represent them. In particular, they can look at the geometric relationships among the points representing different 'versions' of the same square: squares derived from a given square by rotation or line reflection. This exploration leads to the notion of *orbifolds*, another idea which becomes important in topology.

Extension III: game theory and isomorphism.

Students who have studied magic squares stand to reap a bonus if they are introduced to the following game, which we call the 'Fifteen Game'. The numbers 1 through 9 are written on cards, and students take turns choosing a card, which then becomes unavailable to the other player. The first player who can collect three cards adding up to 15 (exactly three cards, and exactly the sum 15) is the winner.

> A class of seventh graders is playing the 'Fifteen Game'. Two students are playing at the front of the room, and the rest of the class is calling out suggestions for possible moves. A chorus develops: 'Take the 5! Take the 5!' The student in front doesn't quite understand, but follows the crowd.
>
> His opponent sees deeper into the game, and takes the 2. She has figured out that even numbers are better than odd numbers. The first player then must block, and the game is almost determined. One student shouts out, 'Block her! It's like tic-tac-toe!'

For most students, the notion of a strategy is itself a concept which arises only from experience in playing. Before the concept is fully formed, students typically talk about 'the proper response', or 'I'll probably win if ...', or 'every time I choose 6 I win ...', without a fully developed concept of a strategy: a plan for playing which always results in a certain outcome. The vignette above is typical about how the strategy emerges. Students usually have two important insights. The first is that it is a game of 'blocking'. One must be thinking of what pairs of numbers one's opponent has, and must choose a third number so that the opponent does not get a sum of 15. A second insight is to note that if two opponents understand the game equally well, the game will be a draw.

The comment that the game 'feels' like tic-tac-toe is a common one. And in fact, the game is exactly tic-tac-toe, in a new guise. Students can see this by playing a game and keeping track of the players' choice of numbers by putting an 'x' or an 'o' over each entry in a magic square. They will find themselves playing tic-tac-toe, and all their 'feelings' about how to play will be justified by their experiences with that game.

Suddenly, the game of 15 is familiar. There is a one-to-one mapping between positions in tic-tac-toe and positions in 15, and the strategies for moving from one position to another are identical. That is, there is an *isomorphism* between the two games. From the point of view of strategy, they are identical. The notion of isomorphism pervades mathematics, and is used exactly as it is here: to make an unusual situation into one which is familiar.

Discussion

We take Smith's (1996) ideas about the nature of curriculum as product as a springboard for discussion. Following Taba (1962), Smith includes the following steps in formulating what we are calling a static curriculum:

Step 1: diagnosis of need
Step 2: formulation of objectives
Step 3: selection of content
Step 4: organization of content.

The magic squares activity demonstrates that a dynamically conceived curriculum does not align well with these steps.

Step 1: the need in dynamic curriculum is always very general—to make the students think. More to the point is the note in Step 2.

Step 2: the discord here arises from the fact that there may be different objectives for different students, even within the same classroom. For example, at the stage of construction of magic squares in which students have the rows equal and are getting columns to be equal, some students will see only a particular strategy for solving this specific problem. Others will understand that they are finding an operation under which the row-sum is invariant. Still others will quickly see that the number 5 must be in the center, even numbers at the corners, and so on—insights which facilitate the construction, but are not essential to it.

Step 3: certainly the selection of content is important. But the content is selected, not in order to attain a specific goal, but to be able to reach toward a hierarchy or range of goals.

There are several more points illustrated by the magic squares activity. Perhaps most obviously, it can be undertaken as early as second or third grade, and can lead to insights for undergraduate majors in mathematics. This is one important characteristic of dynamic curriculum, and one reason why such activities are as rare as they are valuable.

Step 4: organization of any content is critical to the smooth running of a classroom. And a teacher must—as we have above—decide what should happen first, what should come next, and so on. But the teacher does not decide at this point what the goal or outcome of each stage of the activity should be. Indeed (Step 2) the outcome may be different for different students.

Center for mathematical talent: eliciting innovation in unexpected places

One of the unexpected results of adopting a dynamic view of curriculum is that it allows us access to innovative ideas for students often not thought of as having these. If we adopt a static view of curriculum, we will first be selecting goals for students, then help them meet these goals. But perhaps the goals are too low, or the students have their own ideas of what is important to understand. A dynamic view of curriculum changes the nature of goals and expectations for students.

The Center for Mathematical Talent, at the Courant Institute of Mathematical Sciences, New York University, seeks to extend our views of who our innovators will be, by running programs in and after the public schools. Two vignettes from the writer's own experience will illustrate how a dynamic approach to curriculum can work in these settings.

> Lawrence is a gifted student in a third-grade gifted classroom. He loves mathematics and throws himself intensively into every activity. In working on the magic squares, for instance, he quickly constructed the simplest square (with the integers from 1 through 9). Because he was then ahead of his peers, I asked him to construct a magic square with the integers from 2 through 10. This he did, retracing his thinking of the original construction. But when I showed him that he could have simply added one to each entry of the original square, he immediately exclaimed, 'We can make a million of them!'

What Lawrence meant was that he saw the generalization immediately. He saw how he could construct a magic square containing any nine consecutive integers. And when I asked him to construct a magic square of the even integers 2 through 18, he saw this generalization quickly as well. His peers never got that far. But because the curriculum was open, goals were fluid, and expectations high, we were able to elicit advanced ideas from Lawrence, ideas that for him were innovative.

Yet when Lawrence was working with transformations of magic squares, he came to a barrier. His mind did not work as well spatially as it did algebraically, and his ideas were not nearly as innovative. Psychologists tell us (see Wai, Lubinski, & Benbow, 2009) that spatial ability is not well correlated to quantitative ability. A dynamic curriculum can accommodate this fluctuation. Lawrence was held down neither by a curriculum with a low ceiling nor his own lower spatial skills.

> Jose is a fourth-grade student in a remedial class. He has been held back by language (he is not a native speaker of English) and by diagnosed learning disabilities. Yet there is room

in a dynamic curriculum for Jose to shine. We did exercises (preparatory to a study of multiplication) in counting two- and three-dimensional arrays of blocks. Jose was adept at this, visualizing parts of a three-dimensional diagram that were not visible. Most of his peers had much more trouble interpreting these diagrams. Yet Jose could not make the transition easily from direct counting to repeated addition to simple multiplication.

It was because we invented new situations for students to work with and gave them the opportunity to apply their own insights that we were able to tease out this ability in Jose. In later lessons, I was able to use this ability to help him approach multiplication. A curriculum with dynamic elements solves many teaching problems. At virtually every level, it can be differentiated, without breaking up the classroom structure.

This vignette of work with Jose also brings up an important issue in education for innovation: the 'innovation' brought by a student in a classroom need not be innovative in a larger context. 'Innovation' is a term which relates the insight to its audience.

Competitions: a 'locomotive' for innovation

Competition in mathematics is a very old tradition (Kenderov, 2006). Stokes' Theorem was first posed as a problem in the Cambridge 'Tripos' competition (Katz, 1979, p. 149). The algebraists of the Italian renaissance had competitions in solving polynomial equations (see O'Connor & Robertson, 2005). Archimedes' 'Cattle Problem' (Dörrie, 1965) was posed as a challenge to another mathematician.

Preparation for competitions resists formalization. Competition problems are difficult partly because they are given in a context-free environment: the solver does not have clues to the solution which are sometimes given by considering the context in which the problem arises. And so a dynamic view of 'curriculum' is particularly suitable to math team preparation.

A vignette, drawn from the writer's own experience, will make the point more clearly, and illustrate another aspect of this view of curriculum. For many years I was among a group of teachers coaching the New York City mathematics team. This team was a high-achieving group of high school students who competed successfully in on-site national competitions such as the New York State Mathematics League (www.nysml.com) and the American Regions Mathematics League (www.arml.com). Part of the work consisted in training them for 'power questions', Olympiad-style questions that 15 students worked on together for an hour. The training consisted simply in giving these problems to the team, along with 'farm teams' of younger or less experienced students. This circumstance allowed me a rare opportunity to listen to their thinking and observe the group dynamic.

The experience was always the same: a quiet period of ingestion of the problem, followed by conversation. The phase after that was of group organization, which was the most important part of the training. Students would find ways to split the problem into sub-problems and work on them in small groups, with certain designated 'scribes' or 'pushers' keeping the work on track and making sure that it was recorded coherently. Then there was a phase of individual and small group work, and finally a synthesis phase, whose importance was again organizational.

Team coaches also noted a social development of the team, which followed remarkably the descriptions given for instance by Tuckman (1965). This social development took place over several practice sessions. The intellectual stages outlined above took place more quickly, sometimes within 20 minutes, during the high-focused solution of a specific mathematical problem.

It is the phases of conversation and exploration that are germane here. The important observation is that all groups, of whatever ability, went through the same process: initial ingestion, chaotic brainstorming, reorganization, more brainstorming in small groups, then a synthesis. The only difference in the groups was the length of time each phase took (and therefore the level of achievement within the competition's time frame). The more experienced, more 'gifted' students invariably spent less time in ingestion and more in brainstorming, less in organization (they had had experience) and more in exploration. The path they followed was the same, but the rate at which they trod it was different.

The insights, too, usually followed similar paths: the creative process could be described in the same way for almost all groups. The first step was the accrual of specific examples, on the level of 'observations'. Then generalizations were made, and formulated as hypotheses, which were tested against more experience, much as investigations in natural science proceed. The mathematical step—proof—usually came last, as a synthesis of the examples generated. The important observation here is again that the outline of the process was typically the same in all groups, except for the rate at which the different groups went through the process.

'Punctuation' of this process sometimes occurred. Sometimes a student recognized a situation he or she had dealt with before. Sharing an argument already known could accelerate the group activity. Sometimes a student had a deep insight or clever argument that was his own, rather than the result of group activity. Sometimes a student had an 'Aha' experience (Liljedahl, 2008), a sudden flash of intuition, that accelerated the group process. These sporadic exceptions occurred more frequently in the highest achieving groups, and contributed significantly to their success. Perhaps they are manifestations of a still higher level of innovative thinking, setting certain individuals apart from their high-achieving but less imaginative peers. But we can also read these incidents as exceptions to the more usual processes of thought, processes which seem to be the same on each level of achievement.

The central insight of this experience is that there is a level at which human minds are very similar. While they work at different rates, and bring different experiences to the situation, the competition situation (and probably others) catalyzes a synergy which is made possible by this basic similarity. Synergy is one mechanism through which innovative thought emerges from past experience.

Quantum: an interactive journal

Among the challenges of constructing a dynamic curriculum is the need for actual materials: sources of problems or activities that will elicit innovative ideas in the students. The tension between traditional notions of curriculum and a more dynamic view of learning is particularly strong here. The written word is a conservative medium, and anything committed to paper tends to stay there.

One solution to this problem can be found in a Russian tradition of popular scientific journalism. The Russian journal *Kvant*, started in 1960 (see kvant.mccme.ru) by outstanding mathematicians and scientists, strives to bring cutting-edge research, or topics related to the ongoing work of mathematicians and physicists, to pre-college audiences. For 12 years (1989–2001), an English-language version was published, called *Quantum*, by the National Science Teachers Association (see www.nsta.org/quantum/), supported by the National Science Foundation.

The writing style of *Quantum*, influenced by *Kvant*, was very different from the more typical science journalism of mainstream media. It included straightforward exposition on an elementary level, but also problems based on the exposition for the reader to solve. The journal thus became interactive: readers could put down the article and solve a problem, then continue reading. Solutions to the problems were typically printed in a later issue of the journal.

This writing style elicited innovative thinking in several ways. Most clearly documented were new solutions to problems by readers, which were regularly printed by the editors. In the Russian context, often one problem or article would beget another, in a slow version of the process of synergy described above in connection with mathematical contests.

Difficulties and unsolved problems associated with dynamic curriculum

The vision outlined above of curriculum for innovation education is one that will take some work to achieve. At the simplest level, mathematical content that addresses different levels of understanding simultaneously is difficult to construct. The example given above is one of several that exist, but was pulled together out of decades of clinical experience. This sort of curriculum cannot be written at the desk or in the armchair. One must try it out in a variety of venues, with students of different background and teachers of varying experience.

Too, it is important to note that this sort of curriculum is heavily dependent on content. We who work in mathematics are particularly fortunate that in our field it is possible to use very simple material to ask very difficult questions. In the physical sciences or the humanities, these sorts of experiences may prove much more difficult to organize. In our example, consistency, or unity of purpose, is almost guaranteed by the structure of mathematics: it is logic-driven on every level.

Another aspect of dynamic curriculum, related to content, is that it makes great demands on teachers' understanding of content. The student who is ready to make a group operation table of symmetries of a square in grade five must have a teacher who understands this concept and can bring it to the student. That teacher must recognize the student's readiness despite the non-standard language in which his/her thoughts may be couched. Equally important, the same teacher must know the limits of another student, who may not be ready to deal with symmetries as transformations, objects in themselves, distinct from the objects they are applied to.

Finally, it is difficult to find appropriate methodologies for evaluating the learning resulting from a dynamic curriculum. An evaluation tool must be personalized to the student being evaluated. And it must address more than specific skills or results: it must strive to assess the ability of the student to use the result, synthesize new results, or apply them in unusual situations. Indeed, one aspect of this sort of curriculum that makes the term 'dynamic' appropriate is that it eludes the usual forms of evaluation.

Conclusion

The notion of a dynamic curriculum is one that teachers of gifted students are presented with naturally. Unless they are constrained, gifted students will lead the teacher down unexpected paths, paths that the teacher is required to follow, with energies that the teacher is required to exploit. Indeed, creative minds do this constantly, and it behooves us to encourage and not confine this energy in school work. It is vital that we find ways to construct curriculum which achieves this goal.

References

American Regions Mathematics League (2012). www.arml.com, accessed on March 23, 2012.
Andrews, W. S. (1960 [1917]). *Magic Squares and Cubes*. New York: Dover.
Dörrie, H. (1965). Archimedes' *Problema Bovinum*. In H. Dörrie (Ed.), *100 Great Problems of Elementary Mathematics* (pp. 3–7). New York: Dover.
Gaglione, T. (2012). An introduction to group theory. Downloaded from www.usna.edu/Users/math/wdj/tonybook/gpthry/node1.html, on March 24, 2012.
Katz, V. J. (1979). A history of Stokes' theorem. *Mathematics Magazine, 52*(9), 146–156.
Kenderov, P. S. (2006). *Competitions in Mathematics Education. Proceedings of the International Congress of Mathematicians*, Madrid, Spain. Also available on www.icm2006.org/proceedings/Vol_III/.../ICM_Vol_3_76.pdf, downloaded March 24, 2012.

Liljedahl, P. (2008). *The AHA! Experience: Mathematical Contexts, Pedagogical Implications.* Saarbrücken, Germany: VDM Verlag.
National Association of Math Circles (2012). www.mathcircles.org, accessed on March 23, 2012.
New York State Mathematics League (2012). www.nysml.com, accessed on March 23, 2012.
O'Connor, J. J., & Robertson, E. F. (2005). "Niccolò Fontana Tartaglia." MacTutor History of Mathematics Archive, University of St Andrews. Downloaded on March 24, 2012.
Scott, W. R. (1987 [1964]). *Group Theory.* New York: Dover.
Smith, M. K. (1996). Curriculum Theory and Practice. Downloaded from *The Encyclopaedia of Informal Education,* www.infed.org/biblio/b-curric.htm, on February 23, 2012.
Stenhouse, L. (1975). *An Introduction to Curriculum Research and Development.* London: Heinemann.
Taba, H. (1962). *Curriculum Development: Theory and Practice.* New York: Harcourt Brace and World.
Tuckman, B. (1965). Developmental Sequence in Small Groups. *Psychological Bulletin, 63*(6), 384–399.
Wai, J., Lubinski, D., & Benbow, C. P. (2009). Spatial ability for STEM domains: aligning over 50 years of cumulative psychological knowledge solidifies its importance. *Journal of Educational Psychology, 101*(4), 817–835.

22

School textbooks as a medium for the intellectual development of children during the mathematics teaching process

Marina A. Kholodnaya and Emanuila G. Gelfman

INSTITUTE OF PSYCHOLOGY, RUSSIAN ACADEMY OF SCIENCES, RUSSIA, AND
TOMSK STATE PEDAGOGICAL UNIVERSITY, RUSSIA

Summary: One way of developing students' intelligence is through the design of new school textbooks and educational material of a type that meets the requirements of a psycho-didactic approach. As part of the "Mathematics, Psychology, Intelligence" education project (MPI), mathematics textbooks and educational material (study books, practical work, workbooks for independent study, computer software) for middle school pupils (Years 5 to 9) were developed within the framework of an "enrichment" teaching model. The basic purpose of this model is the intellectual development of students through their mathematical education using specially constructed educational texts. The specific nature of these educational texts lies in the fact that, at the same time as conveying structures of formal mathematical knowledge, they also (1) support development of the basic components of students' mental experience (including cognitive, conceptual, metacognitive and intentional experience) and (2) create conditions for students to employ their own individual cognitive styles. Psycho-didactic specifications for educational texts were formulated and a psycho-didactic typology of educational texts was developed.

Key words: Mathematics education, psycho-didactics, students' intellectual development, educational texts.

Introduction

As a result of contemporary social challenges, people's intellectual abilities are beginning to be considered a key factor in the progressive development of society. Therefore, the task of shaping the intellectual resources of the rising generation within a comprehensive school education framework relates to a number of national priorities. A high level of intellectual resources in school graduates lays the foundation for willingness to engage in innovative activity in subsequent stages of their education and professional careers.

Tackling the issue of schoolchildren's intellectual development means taking account of two basic conditions: we need first to review criteria used to evaluate the effectiveness of the educational process, and second to develop new teaching technologies. In other words, developing willingness to engage in innovative activity presupposes the introduction of innovative forms and methods of teaching.

Criteria to assess the effectiveness of the teaching process are, in our view, the following *intellectual qualities* of the schoolchild. We consider these to be preconditions for willingness to engage in innovative activity (CICSU):

> **C** – Competence. Intellectual competence means those particular features of knowledge organisation which ensure the ability to take effective decisions in specific spheres of activity, including diversity, structuredness, flexibility, efficiency, trans-situationality, categorical character, unity of declarative and procedural knowledge, understanding of one's own knowledge.
> **I** – Initiative. Intellectual initiative is the desire to discover new information, to advance ideas and to master different spheres of activity independently and of one's own accord.
> **C** – Creativity. Intellectual creativity is the process of creating what is subjectively or objectively new, based on the ability to generate original ideas, to use non-standard methods and show tolerance towards unusual and "impossible" situations.
> **S** – Self-regulation. Intellectual self-regulation is the ability to control voluntarily one's own intellectual activity and purposively construct a process of self-teaching.
> **U** – Uniqueness of mind-set. Uniqueness of mind-set is the individually distinct mode of cognitive relation to immediate outside occurrences, including the expression of individual cognitive styles and the extent to which individual cognitive preferences are formed.

It is possible to shape these intellectual qualities in an individual within the framework of innovative teaching technologies developed on the basis of the *psycho-didactic approach*.

The psycho-didactic approach as an alternative to the subject-centred approach in school education

In the traditional system of teaching, constructed on the basis of the *subject-centred* approach, criteria for judging the effectiveness of students' education are mainly associated with the level of knowledge awareness, abilities and skills obtained from each school subject. Within innovative teaching technologies developed on the basis of the *psycho-didactic* approach, the emphasis fundamentally shifts: the criteria for evaluating the effectiveness of the educational process become those changes in the student's intellectual and personal realm which characterise his or her development as a productive, self-sufficient and active individual.

Psycho-didactics is the area of pedagogy in which content, forms and methods of teaching are designed based on the integration of psychological, didactic, methodological and thematic (corresponding to different school subjects) knowledge, with priority given to using patterns of psychic personality development as a basis for organising the teaching process and general educational environment (Davydov, 1966; Panov, 2007; Gelfman & Kholodnaya, 2006; Gelfman et al., 1997; Kidron et al., 2010; Brousseau, 1997; Malara & Navarra, 2003; and others).

The result of psycho-didactic work is a qualitatively new pedagogical product combining psychological, didactic, methodological and thematic knowledge, in the form of a new type of school environment, innovative educational technologies, a developmental method of education, a new generation of school textbooks and so on. At the base of the psycho-didactic approach in its direct sense lies pedagogical engineering; that is, the process of designing, constructing and exploiting pedagogical products, inherently oriented towards developing the mental resources of each schoolchild. The basic purpose of psycho-didactics is to create the conditions for students' psychological growth on the basis of improving teaching effectiveness in a given school subject.

There are various routes to implementing the psycho-didactic approach in school education: using "didactic situations" which build up students' knowledge during the learning process, including the use of metaphor and emotional context (Broussau, 1997); an orientation towards comprehension of the learning material and formation of ideas by selecting mathematical tasks and hypotheses on the basis of their influence on the process of education ("Hypothetical Learning Trajectory" – HLT) (Simon, 1995; Simon & Tzur, 2004); the development of basic cognitive operations, such as identification, combination and construction, as the basis of conceptual teaching (the RBC model) supported by the personal experience of the student (Hershkowitz, Schwarz, & Dreyfus, 2001; Bikner-Ahsbahs, 2004); the development of students' creative thought skills (Burke & Williams, 2008) and others. In our view, the key direction in psycho-didactics that guarantees an innovational teaching regime is the *psychologically grounded design of the content of school education*.

Accordingly, the question of the requirements for school textbooks is particularly relevant. Within the traditional subject-centred approach the purpose of the mathematics textbook is reduced to a strict, sequential account of established mathematical knowledge adapted to the age of the students. The form these traditional textbooks take is of a reference book and a workbook of mathematical problems. The implication is that the teaching of mathematics should be carried out by means of mathematics itself, since mathematical knowledge in and of itself possesses an essential developmental effect.

From the point of view of the psycho-didactic approach, the new generation school textbook is a *polyfunctional psychological system*, realising the goals of an array of new functions (Gelfman & Kholodnaya, 2006).

1. The informational function (informing students about the various areas of academic knowledge, taking into account the accessibility of textbook information; a specific form of structuring information, in the sense of the relationship between reference, explanatory and problem-based texts; an orientation towards comprehension of educational material; a balance of extending and condensing in the text; a unity of declarative and procedural knowledge).
2. The directive function (the availability of instructive information and organisation of "enriching" repetition, where past material is repeated at the same time as assimilating new; the creation of conditions for students' investigative activity; their stimulation towards independent work; inclusion of means for routine diagnostics).
3. The developmental function (the creation of conditions for concept formation; the development of general intellectual skills, including the ability to reason, justify, prove, criticise, take rational decisions and so on; the building up of a reflexive position in the schoolchild; motivation towards educational activity; the development of students' creative facilities).
4. The communicative function (problematisation of an educational text, including using various different kinds of question; its interactive character; the expressive style of its presentation).
5. The educational function (the presence of knowledge relating to a cultural worldview, of methodological and historical-scientific knowledge; an initiation through the text of an evaluating attitude in the students towards educational material; personal relevance of an educational text, taking into account the student's personal experience; an orientation towards the formation of his or her personal qualities).
6. The function of differentiation and individualisation of learning (taking into account through the educational text the students' individual tempos of learning, their individual cognitive styles, their different inclinations and levels of interest in the subject and their opportunities to choose a regime of educational activity).

The question naturally arises: can a textbook fulfil all of these functions? Using a traditional textbook it undoubtedly cannot be done. New generation textbooks are necessary, constructed on a fundamentally different basis. First of all, the development of the textbook's content, structure and form should take into account the *requirements of the psycho-didactic approach*; that is, each element of the textbook (ways of presenting educational information, the sequence and arrangement of the material, the style of presentation and so on) should have a specific psychological target and ensure a definite developmental psychological effect. Second, the textbook should be part of a system of supplementary educational material (study books for pupils, practical work on various types of task, workbooks, computer software and so on) forming an educational and methodical set (EMS), providing a varied and enriched educational space for each school subject.

The next question for discussion then, is how, within the framework of the psycho-didactic approach, is the development of students' intelligence effected?

Enrichment of mental (intellectual) experience as the psychological basis for developing students' intelligence

In our view, the psychological basis of intellectual development is the process of enriching students' mental (intellectual) experience during the education process. Figure 22.1 shows a structural model of intelligence, illustrating its particular organisational features from the point of view of the composition and construction of mental experience (Kholodnaya, 2002).

In accordance with the structural model of intelligence put forward, three levels (or layers) of mental experience can be identified, each of which has its own function:

1. *Cognitive experience*. The mental structures allowing storage, regulation and transformation of available and incoming information, facilitating its reproduction in the psyche of the cognisant individual as a stable, ordered aspect of his or her surroundings. Their basic purpose is operative processing of a flow of information about immediate impressions at various levels of cognitive activity, including the level of conceptual experience.
2. *Metacognitive experience*. The mental structures allowing accomplishment of involuntary and voluntary regulation of information processing, and also conscious management of the work of one's own intelligence. Their basic purpose is to control the state of individual intellectual resources and the course of individual intellectual activity.
3. *Intentional (emotional-evaluative) experience*. The mental structures which lie at the base of individual intellectual inclinations. Their basic role consists in the fact that they predetermine subjective criteria of choice for specific knowledge domains, the direction of decision making, choice of sources of information, means for its representation and so on.

In turn, the particular organisational features of cognitive, conceptual, metacognitive and intentional experience determine the characteristics of the individual's intelligence at the level of productivity of intellectual activity (personal intellectual abilities – convergent and divergent abilities, educability, and stylistic characteristics of intelligence, and also integral intellectual abilities – competence, talent and wisdom as phenomena of intellectual giftedness), and on the level of individual distinctiveness of mind-set (personal cognitive style in the form of a combination of preferences for modes of intellectual activity) (Kholodnaya, 2002, 2004).

In identifying intelligence with the particular organisation of individual mental experience, we can say that any pupil is "full" of his or her own mental experience, which will predefine the character of his or her intellectual activity in a given situation. The content and structure of this experience is different in each pupil, therefore children certainly differ in their intellectual resources.

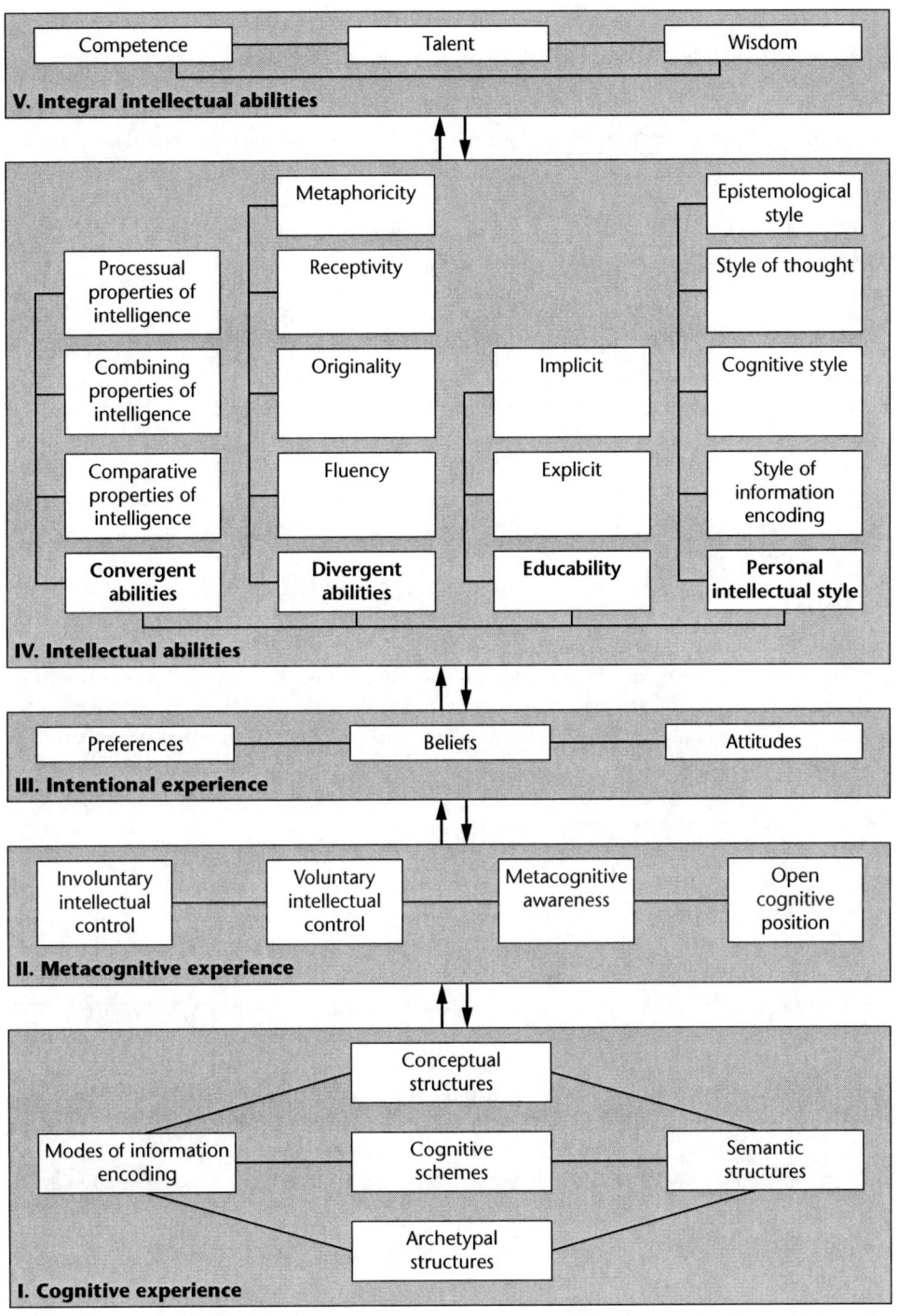

Figure 22.1 A structural model of intelligence illustrating its particular organisational features from the point of view of the composition and construction of mental experience (Kholodnaya, 2002).

However, they all need suitable conditions to be created for their intellectual development through the maximum possible enrichment of their individual mental experiences.

"Enrichment" means, first, that the basic components of every student's mental experience are developed (including cognitive, conceptual, metacognitive and intentional experience) as a basis for developing his or her intellectual abilities, including his or her integral intellectual abilities; and, second, that conditions are created to allow students to display their individual cognitive styles and individual cognitive inclinations.

In short, therefore, the essence of the problem of the student's intellectual development can be represented by the following set of propositions:

- Each student is the bearer of mental experience, and consequently possesses certain initial intellectual resources; by virtue of this individual distinctiveness in content and structure of mental experience, each student is "clever in his or her own way".
- The target of pedagogical influence in school education is a distinct set of components of individual mental experience (including its cognitive, conceptual, metacognitive and intentional components).
- The mechanisms for an individual's intellectual growth are connected to processes occurring in the area of individual mental experience and the enrichment and deepening in complexity of all of its components.
- Each student has his or her own range of potential capacity for intellectual strengths, and the task of the teacher lies in providing him or her with the necessary aid by means of individualisation of school and extra-curricular activity.
- Criteria for evaluating the effectiveness of the educational process, alongside knowledge, ability and skill (KAS), are connected with the extent to which the individual's intellectual qualities, such as competence, initiative, creativity, self-regulation and uniqueness of mind-set (CICSU) are formed.

Subsequently, solving the problem of developing students' individual intellectual resources presupposes:

1. Creating conditions for the actualisation of each student's available mental experience (taking into account preferred modes of information encoding, available cognitive schemes, the particularities of the bases of present knowledge, the level of common sense and formed academic understanding, the particularities of intellectual self-regulation, individual cognitive preferences, the individual tempo of learning and so on).
2. The creation of conditions for the increase, enrichment and deepening in complexity of that student's individual mental experience, within its maximum possible limits (conditioning the mind to work with various different modes of information encoding, widening the repertoire of cognitive schemes, differentiation and integration of verbal and non-verbal semantic structures, enrichment of conceptual frameworks, developing the ability to exercise involuntary and voluntary control of one's own intellectual activity, forming open cognitive positions and a high level of metacognitive awareness, creating conditions for mastery of a broad range of different intellectual styles).

Educational texts as a means for students' intellectual development

The key factor influencing the development of students' intelligence is the *content of school education*. In turn, the unit of subject content in schools is the *educational text*.

The text is the most valuable element of culture and the most important constituent of the educational process. Many experts have discussed the special role of the text in the intellectual development of the individual, examining the text as a "thinking structure" (V. V. Ivanov), a "model for thought adventures" (L. E. Gendenshtein) and a "conversation-partner" (M. M. Bakhtin).

Texts (academic, historical-cultural, fictional, educational) are not linear. An educational text should be built as a multidimensional semantic space, within which the pupil-reader can imaginatively roam in different directions. The structure of the educational text, in addition to its "nucleus" (the specific subject information) includes:

- *Context*. Students should have the opportunity to move "horizontally" across the text, expanding their acquaintance with the study material through the use of various forms of presentation of material (verbal, visual, practical, emotional-metaphorical), texts which are heterogeneous in content (stating, explanatory, problem-discursive, multiple-choice), inclusion of off-subject texts (in the form of narrative elements, elements of game situations, applied physics, ecology and psychology material).
- *Subtext*. Students should have the opportunity to move "vertically" up and down the text, which presupposes the segmentation of the text by different stages of complexity, both in content and in the type of activity (the use of texts and tasks of varying levels of difficulty; the inclusion of normative texts with demonstration models of activity and plain texts; education through performance, research, project work or creative activity and so on).
- *Implied text*. Students should have the opportunity to move down "deep" into the text, that is, to draw out the deeper sense of the text: that which is not expressed verbally, through the connotative meaning of words; the content of their own personal experience; associative links and their own imaginations.

In the area of school education interest in texts is connected to an understanding of their role in effective teaching, particularly in the context of *reader-oriented theory*, in which the reader actively constructs meaning (concepts) in the process of reading a text. This includes where a student works with mathematical textbooks (Weinberg & Wiesner, 2011).

Within the "Mathematics, Psychology, Intelligence" educational project (MPI), mathematics textbooks and educational material were developed for students at middle school (Years 5–9) on the basis of an "enrichment" model of education. The model's basic purpose is students' intellectual development through the medium of mathematical education by specially constructed educational texts (Gelfman et al., 2002; Gelfman & Kholodnaya, 2006). Mathematics teaching is carried out using the educational and methodical set (EMS), in keeping with modern pedagogical ideas about the organisation of educational spaces of learning activity.

Included in the MPI project's educational and methodical set are: for Years 5 and 6, textbooks, study books including narrative texts and practical study workbooks, activity books for independent work, electronic educational resources (practice exercises, tests, a library of cartoons, a collection of mathematical games and an electronic reference guide), teachers' books; for Years 7–9, textbooks, books of mathematical problems, teachers' books.

Educational texts for all the elements of the EMS have been developed using the "enrichment" model of mathematics teaching, taking into account the basic positions of activity-based, personality-oriented and competency-based approaches to organising content for modern mathematical education in schools, which represent a concretisation of the psycho-didactic approach.

The *activity-based approach* is implemented in the "enrichment" model whereby the students take an active role in the process of their education in mathematical knowledge and methods for solving mathematical problems:

- The textbook is supplied with navigational tools (a navigation bar with customised icons), which allow the learner to use all of the individual components of the EMS in accordance with his/her educational needs; the navigational icons manage the student's interaction with textbook content and links to other study material (study books, workbooks and electronic resources), involving him or her in various types of educational activity (performance, research, project work, creative work).
- When working with the textbook and educational materials the student is offered the opportunity to exercise independence at various stages of the study of mathematics (for example, in the textbooks work with texts is organised in ways such as an "appeal to the reader", aimed at the initiation of independent activity in the student).
- Conditions are created for the student to develop effective methods of educational-cognitive activity, namely: mastering algorithms, forming an ability to solve text-based problems, developing willingness to choose rational methods of problem-solving and using various types of analysis for the same educational problem and so on.
- In textbooks and study books, alongside statements of "ready-made" mathematical knowledge, the process by which it was derived is shown (new knowledge is introduced gradually, including the stage of motivation to learn a new mathematical concept, the discussion stage, the generalisation stage and conclusions).
- The textbooks and educational material contain applied material, developing an interest in the practical applications of mathematics and demonstrating the role of mathematical knowledge in real situations.

The *personality-oriented approach* to the organisation of educational material is implemented in the "enrichment" model in the following way:

- Educational information is presented in various forms (verbal, visual, practical, emotional-metaphorical), allowing students with varying cognitive styles to grasp the material successfully.
- The active use of students' personal (including day-to-day) experience (both in the stage of mastering the theoretical sections of the textbook and when developing problem solving skills).
- The textbooks and educational materials have a dialogic character that cultivates in the student a readiness to express his or her opinion and to justify and defend his or her point of view.
- Individualised teaching is achieved by means of the textbook and educational material (components of the EMS allow consideration to be given to students' individual cognitive needs and inclinations and allow the selection of an individual trajectory of self-education).

The *competence-based approach* in the "enrichment" model is covered by the following aspects:

- It uses a thematic principle to organise textbooks and educational materials, which simultaneously allows both deepening and broadening of the students' knowledge, and also builds content appropriate to the topic using various types of systematisation of information.
- Simultaneous formation of both declarative knowledge (knowing *that*), and also procedural knowledge (knowing *how*) is planned for.
- The content of textbooks and educational materials (the sequence of study for each topic, selection of questions and educational tasks) is constructed in such a way as to help form a reflexive position (a conscious, volitional attitude of the student towards the process of education).
- The educational material trains schoolchildren in the right way to respond to contradictions.

- Skills of planning, goal-setting, self-monitoring, prediction, evaluation, substantiating and generalisation are developed through the text as the basis of competency level for mastering educational knowledge.
- Conditions are created for students to be able to employ the theoretical knowledge they have acquired to various practical situations (including through the use of project work).

In our view, therefore, a promising route to the intellectual development of schoolchildren is through their work with specially constructed *developmental educational texts* which fulfil psycho-didactic requirements.

Psycho-didactic requirements for educational texts

We formulated *psycho-didactic requirements* for mathematical educational texts, fulfilled by developing corresponding textbooks and educational material as part of the MPI project.

1. A *thematic organisation* of mathematics course content (each textbook and study book is written on a specific topic from the curriculum of Years 5–9). For example, the textbook for Year 5 includes the topics "Natural numbers and decimals" and "Positive and negative numbers"; the textbook for Year 6 the themes "Solving equations", "The divisibility of numbers", "Rational numbers" and "Coordinates, graphs and symmetry".

    The thematic principle takes into account the possibility of a sequential development of topics, but also allows the implementation of an "immersion" teaching technique by deepening and broadening relevant material, thanks to the use of educational texts of various types (instructive, explanatory, narrative, historical-cultural and psychological).

2. *The multilevel structure of the educational text*, using various forms of presentation of educational information (verbal, visual, practical, emotional-metaphorical), various routines to master mathematical ideas (logical rationalisation, the analysis of real-life practical situations, using "unreal" aspects of mathematical knowledge), various sorts of educational activity (performance, research, project work, creative work) and various forms of self-monitoring.

3. *The interactive character of educational texts*. Educational texts are constructed as dialogues with the pupil-reader: they include various different forms of question on hypothetical problem situations, oriented towards discussion and debate of alternative points of view.

4. *An orientation towards comprehension* of mathematical facts and ideas. Educational texts are constructed taking into account the patterns of concept-forming processes (the development of various modes of information encoding, formation of cognitive schemes of mathematical concepts and methods of mathematical activity, work with the semantics of mathematical language, work with concept indicators, the establishment of various interconnections between concepts, the different stages in the formation of concepts, including the stage of motivation in the introduction of new mathematical concepts).

    Educational texts help the development of general intellectual skills (the skill to make an argument, evaluate, justify, plan, predict, react to contradictions, research and so on). The text of articles in the textbook is constructed according to the following model: first of all it ensures assimilation of the theoretical material, and only after this does the student move on to solving tasks in the practical workbook.

    Presentation of the material is sequential and "slow", with detailed discussion of various aspects of the mathematical object introduced. This allows the student to evaluate the gaps in his or her knowledge and the reasons behind it, to perceive the opportunity to solve the same task by a number of different means, and so on.

5   *Students' independent activity* in the process of acquiring new mathematical knowledge (the text "sets free" the student to go forward, allowing him or her independently to master a given question by him- or herself; a gradual transfer of the goal-setting function of educational activity to the student him- or herself and the stimulation of the student towards independent generation of educational text).

6   *Running diagnosis of the dynamics of students' educational-cognitive activity.* As the means of running testing in the educational texts various diagnostic material is used: multi-level educational tasks with different levels of difficulty (levels I and II); three options for controlled work relating to the students' preferred form of monitoring (of the type "Calculate …"; "Prove that …"; "Write an account of … and give examples"); "Test yourself" sections for self-testing of own knowledge and so on.

7   *Differentiation and individualisation of instruction* for students with various levels of previous education and various cognitive styles through diverse forms of presentation of the material, also taking into account their individual cognitive inclinations and preferences. Through the educational text the student is given the opportunity to choose how to master the material (through game-playing, performance, research, project work or creative activity), the degree of difficulty and various types of self-monitoring.

8   *Using the student's personal experience* by taking into account his or her everyday impressions, common-sense knowledge, willingness to trust his or her own intuitive judgements in the analysis of educational information.

9   *The creation of a psychologically comfortable regime for intellectual work.* What we mean by a psychologically comfortable regime for intellectual work is that type of education which evokes feelings of satisfaction and interest, leading to each student having a sense of success in his or her learning activity. In particular, narrative texts in study books (stories in which well-known figures from children's literature appear) help give emotional support to students with educational and personality problems (above all, those children who were unsuccessful in mathematics at primary school).

It is important to emphasise that the implementation of the psycho-didactic approach is necessary not just to improve the quality of learning of the school subject and create conditions for the intellectual growth of the student, but also to encourage a positive attitude towards different school subjects. This is especially urgent in the study of mathematics, since data show that children experience fear of mathematics, which is transformed into a negative attitude towards the mathematics teacher (Picker & Berry, 2001).

We will try to illustrate the construction of educational texts on the example topic of "Quadratic equations", in particular the theory of the relation between the roots of a quadratic equation and its coefficients – Vieta's theorem. In order to see the difference in the way the educational material treats this topic it will suffice to analyse a small number of school textbooks.

One of the educational texts is a formal one, rich in scientific facts and information, and has the character of a reference book. The scientific fact (normative knowledge) is communicated at once: "For a reduced quadratic equation the correct theorem …" Thereafter examples of the application of the theorem are given.

Another educational text, which begins with a detailed review of a single example, is a typical example of an explanatory-illustrative text:

> The reduced quadratic equation $x^2-7x+10=0$ has the roots 2 and 5. The sum of the roots is equal to 7, and the product is equal to 10. We see that the sum of the roots is equal to the second coefficient, taken with the reverse sign, but the product is equal to the absolute term.

Hereafter the authors direct the students' attention to the fact that this property is inherent to any quadratic equation and then formulate and prove Vieta's theorem.

A third educational text on this theme, constructed according to psycho-didactic requirements, appears as follows:

Studying quadratic equations, you will probably have noticed already that information about their roots is hidden in the coefficients. One or two 'secrets' have already become clear to us.

The availability or absence of roots in a quadratic equation depends on the discriminant which is given by the coefficients of the equation. The roots of the equation can be found by a formula using the coefficients of the quadratic equation.

How else are the roots and the coefficients of a quadratic equation connected? In order to uncover these connections it is useful to observe the coefficients and roots of various quadratic equations.

Exercise 1. Solve the equations:

$$x^2 + 5x + 6 = 0; \qquad (22.1)$$

$$x^2 - 5x + 6 = 0. \qquad (22.2)$$

Compare the coefficients of these equations, and then the roots. What connections between the roots and the coefficients do you notice in the equations?

Do these equations support your findings:

$$x^2 - 7x + 6 = 0; \qquad (22.3)$$

$$x^2 + 7x + 6 = 0; \qquad (22.4)$$

$$x^2 + 8x + 6 = 0; \qquad (22.5)$$

$$x^2 - x - 6 = 0? \qquad (22.6)$$

Try to formulate your findings and write them algebraically.

When searching for patterns researchers often record their observations in tables (see Table 22.1). This helps to reveal these patterns. We suggest you also complete one of these tables.

	Equation	p	q	x_1	x_2	x_1+x_2	x_1-x_2
1	$x^2+5x+6=0$						
2	$x^2-5x+6=0$						
3	$x^2-7x+6=0$						
4	$x^2+7x+6=0$						
5	$x^2+8x+6=0$						
6	$x^2-x-6=0$						

Has this table helped you to discover new connections between the roots and coefficients of quadratic equations?

Does it make sense to include the standard form reduced equation $x^2 + px + q = 0$ in the table?

Compare your findings about the connections between the roots and coefficients of reduced quadratic equations with the findings shown in the following theorem.
[Thereafter follows the formulation of Vieta's theorem.]

Clearly the study of each of these educational texts acquaints students with a well-known scientific fact. However, students are brought to this normative knowledge by different routes; its acquisition is accompanied by different forms of intellectual activity. In our view, only the third educational text can be attributed to the category of developmental educational texts, insofar as it is organised as a dialogue with the student-reader, is oriented towards the comprehension of mathematical facts, promotes development of the ability to reason (analyse, compare, generalise, draw conclusions) and stimulates investigative activity.

A psycho-didactic typology of educational texts

Based on the structural model of intelligence (Figure 22.1) we have prepared developmental educational texts of various types for a school mathematics course (Years 5–9): each type of educational mathematical text conforms to a specific component in the structure of mental experience and is oriented towards its development (Gelfman *et al.*, 1996; Gelfman & Kholodnaya, 1997, 1999, 2006).

In particular, we have presented in educational texts paths to enrichment of *components of cognitive experience* (the formation of various modes of information encoding, cognitive schemes, semantic structures), *components of conceptual experience* (taking into account patterns of concept acquisition), *components of metacognitive experience* (the development of involuntary and voluntary intellectual control, metacognitive awareness, an open cognitive position) and *components of intentional (emotional-evaluative) experience* (the creation of conditions for the actualisation of cognitive preferences, convictions, attitudes of the students).

Table 22.1 shows a *psycho-didactic typology of educational texts* (using the example of mathematics educational texts).

It should be emphasised that each type of text (and here these are "microtexts") appears as an element of the whole text of the textbook or educational and methodical set, which corresponds to the above-formulated pysco-didactic requirements for educational texts.

Conclusion

The trial use in schools of a new generation of mathematical textbooks and educational material, developed by using the psycho-didactic approach, has an important consequence: what comes to the fore is the task of shaping the individual intellectual resources of school graduates; the level of development of these provides a base for their willingness to engage in innovative activity in future life, both in their professional work and in life as a whole.

Table 22.1 Psycho-didactic typology of educational texts

Forms of mental experience	Components in the structure of mental experience	Characteristic features of educational-cognitive activity	Types of educational texts
Cognitive experience	Modes of information encoding	Verbal-symbolic mode of information encoding	• text – mastering mathematical notation • text – looking for formulae • text – reaching formulation
		Visual mode of information encoding	• text – developing normative image • text – motivation for image • text – development of image • text – classification of image • text – activation of individual visual experience
		Practical mode of information encoding	• text – laboratory work • text – practical situation
		Sensory-emotional mode of information encoding	• text – emotional impression • text – metaphor • text – game
	Cognitive schemes	Cognitive schemes of mathematical concepts and methods of mathematical activity	• text – introduction of focus-example • text – framework • text – procedure • text – summary
	Semantic structures	Semantics of mathematical language	• text – the meaning of term • text – systematisation of the meaning of terms • text – translation from the language of mathematical symbols into the native language • text – microcomposition
	Conceptual structures	Patterns of mathematical concept formation	• text – identification of concept indicators • text – choice of concept indicators • text – establishing connections between concepts • text – motivation for concept • text – categorisation of concept • text – enrichment of the content of the concept • text – transferring the concept to a new situation • text – consolidation of content of concept

Table 22.1 Continued

Forms of mental experience	Components in the structure of mental experience	Characteristic features of educational-cognitive activity	Types of educational texts
Metacognitive experience	Involuntary and voluntary intellectual control	Planning	• text – programme • text – goal selection • text – constructing a plan • text – problematisation
		Prediction	• text – developing hypotheses • text – prediction in uncertain situations • text – predicting the result of a function
		Self-control	• text – methods of self-control • text – selection of methods of self-control • text – looking for errors
	Metacognitive awareness	Awareness of methods of mathematical activity, developing representations about one's intellectual resources	• text – reflection on solving methods • text – self-appraisal of one's knowledge and abilities • text – educational self-testing • text – independent composition of text • text – psychological commentary
	Open cognitive position	Readiness to work with contradictory information	• text – contradiction • text – alternative • text – conflict of opinions • text – impossible situation
Intentional (emotional-evaluative) experience	Preferences	Choice of methods of study	• text – individual cognitive style • text – selection of methods of action • text – selection of cognitive position
	Convictions	Actualisation of personal experience	• text – conjecture • text – creative work
	Attitudes	Value attitudes towards educational material	• text – history of mathematics • text – mathematics in the wider world • text – major lines in the development of mathematics

Acknowledgments

We are very grateful to Mary Bailes for her excellent translation of our chapter from Russian into English. She made the extra efforts during her wonderful assignment and eventually the final chapter in English looks even better than in its original Russian.

References

Bikner-Ahsbahs, A. (2004). Towards the Emergence of Constructing Mathematical Meanings. *Proceedings of the 28th Conference of the International Group for the Psychology of Mathematics Education* (pp. 119–126).

Brousseau, G. (1997). *Theory of Didactical Situations in Mathematics.* Dordrecht: Klumer.

Burke, L. A., & Williams, J. M. (2008). Developing young thinkers: An intervention aimed to enhance children's thinking skills. *Thinking Skills and Creativity, 3*, 104–124.

Davydov, V. V. (1996). *The Theory of Developmental Education.* Moscow: INTOR.

Gelfman, E., Demidova, L., Kholodnaya, M., Lobanenko, N., & Wolfengaut, J. (1996). Concept formation process and an individual child's intelligence. In H. M. Mansfield, N. A. Pateman & N. Descamps-Bernarz (Eds.), *Mathematics for Tomorrow's Young Children* (pp. 151–163). Dordrecht: Kluwer Academic Publishers.

Gelfman, E., Kholodnaya, M., & Cherkassov, R. (1997). From didactics of mathematics to psycho-didactics. In N. A. Malara (Ed.), *International View on Didactics of Mathematics as a Scientific Discipline* (pp. 102–107). Proceedings WG25, ICME–8. Modena, Italy: University of Modena.

Gelfman, E., & Kholodnaya, M. (1997). On development of metacognitive experience of students. *Proceedings of the European Research Conference on Math Education* (pp. 57–62). Czech Republic: Charles University.

Gelfman, E., & Kholodnaya, M. (1999). The role of ways of information coding in students' intellectual development. In I. Schwank (Ed.), *European Research in Mathematics Education: Proceedings of the First Conference of the European Society for Research in Mathematics Education*, vol. 2 (pp. 38–48). Osnabrück, Germany: Forschungsinstitut für Mathematikdidaktik.

Gelfman, E. G., Demidova, L. N., Gilina, E. I., Lobanenko, N. B., & Malova, I. E. (2002). *The Enriching Model of Education in the MPI Project: Problems, Reflections, Solutions. Methodological Instructions for Teachers*, vol. 1. Tomsk, Russia: Tomsk University Press.

Gelfman, E. G., & Kholodnaya, M. A. (2006). *Psycho-Didactics of School Textbooks: The Intellectual Nurture of Students.* St Petersburg, Russia: Piter.

Hershkowitz, R., Schwarz, B., & Dreyfus, T. (2001). Abstraction in context: Epistemic actions. *Journal for Research in Mathematics Education, 32*, 195–222.

Kholodnaya, M. A. (2002). *The Psychology of Intelligence: Paradoxes of Research.* St Petersburg, Russia: Piter.

Kholodnaya, M. A. (2004). *Cognitive Styles: On the Nature of the Individual Mind.* St Petersburg, Russia: Piter.

Kidron, I., Bikner-Ahsbahs, A., Cramer, J., Dreyfus, T., & Gilboa, N. (2010). Construction of knowledge: need and interest. In M. M. F. Pinto & T. F. Kawasaki (Eds.), *Proceedings of the 34th Conference of the International Group for the Psychology of Mathematics Education*, vol. 3 (pp. 169–176). Belo Horizonte, Brazil: PME.

Malara, N., & Navarra, G. (2003). *ArAl Project: Arithmetic Pathways towards Favouring Pre-Algebraic Thinking.* Bologna, Italy: Pitagora Editrice.

Panov, V. I. (2007). *Psycho-Didactics of Educational Systems: The Theory and Practice.* St Petersburg, Russia: Piter.

Picker, S. H., & Berry, J. S. (2001). Investigating pupils' images of mathematicians. *Proceedings of the 25th Conference of the International group for the Psychology of Mathematics Education*, vol. 4 (pp. 49–56). Utrecht, Netherlands: Utrecht University.

Simon, M. (1995). Reconstructing mathematics pedagogy from a constructivist perspective. *Journal for Research in Mathematics Education, 26*, 114–145.

Simon, M., & Tzur, R. (2004). Explicating the role of mathematical tasks in conceptual learning: An elaboration of the Hypothetical Learning Theory. *Mathematical Thinking and Learning, 6*, 91–104.

Weinberg, A., & Wiesner, E. (2011). Understanding mathematical textbooks through reader-oriented theory. *Educational Studies in Mathematics, 76*, 49–63.

23

The interfaces of innovation in mathematics and the arts

Bharath Sriraman and Kristina Juter

UNIVERSITY OF MONTANA, USA, AND KRISTIANSTAD UNIVERSITY, SWEDEN

Summary: The chapter outlines human innovation in architecture and art with an emphasis on mathematical creativity and innovation, e.g., the work of Buckminster Fuller who was inspired by popular psychology and human consciousness in his creations. Architectural creation in society is tightly connected to geometry, topology and other parts of mathematics. Buildings and art are results of human minds linking abstract mathematical representations and concrete physical structures. For such links to occur, inventors need to be able to work in interdisciplinary settings. We present some findings from the literature in the light of fostering creative innovators in mathematics related to the arts.

Key words: Architecture, Buckminster Fuller, design education, Euclidean and non-Euclidean geometry, interdisciplinarity, Mandelbrot.

Introduction

Emmer (1993) claimed creativity to be a bridge between art and mathematics exemplified (www.olats.org/colloque/textes/texte7.shtml) in his traveling exhibition *The Eye of Horus: Art and Mathematics* (Emmer, 1990). A shared language is needed to make bridges between science, art and metaphysics (Thiessen, 1998). Scientific phenomena and concepts need to be possible to isolate for better understanding nature which also is represented in artistic, philosophical and theological settings. Thiessen (1998) boiled the crucial factors of interdisciplinary creativity down to three issues: "a common purpose, a common language, and a shared model" (p. 47). Root-Bernstein (2003) suggests that "innovation is a process of survival of the fittest in which multiple variations of ideas are selected by social, economic, cultural and other factors" (p. 267), and argues that while artists think of possibilities and possible worlds, scientists are often hemmed in by domain limitations and have to work within this world. Mathematics as a discipline is often compared to art especially in its aesthetic component (Brinkmann & Sriraman, 2009), and the intermarriage of the world of art and mathematics is often manifested in the world of architecture and more recently in non-linear art installations involving digital media (Brosz, Carpendale, Samavati, Wang, & Dunning, 2009). The question of how to foster and develop innovators in mathematics and science in general, is by and large unanswered, even though there is an increasing body of developmental literature analyzing eminent samples such as Nobel laureates (Shavinina, 2003). In this chapter we make a case for cultivating visualization, geometric modes of thinking, which are common traits of polymaths (Sriraman, 2009a) with the implication that it can be cultivated in schooling.

Visualization and geometry

Visualization has been, and still is, essential for the development and understanding of mathematics. The historical development was often documented and justified visually and the border between art and mathematics is not clear-cut. One example was the 15th century artist and mathematician Piero della Francesca who used mathematics to get perspectives and bodies perfect (Kline, 1953). Bill (1993) states further that every work of art is based on geometrical relations in the plane or in space which is evident from, for example, Robinson's sculptural work in bronze and steel inspired by mathematical knots (Brown, 2005) and Collins' (2005) work where he used Gaussian curvature and various three-dimensional bodies to create wooden figures. The wholeness of the figures gives the beholder an opportunity to regard convexity from one side as concavity from the other.

Emmer (1993) discussed the mutual enrichment of the two disciplines from each other where artists use visual approach based technology and mathematicians apply visual representations. Bill (1993) described the need for added visual aid in the development of mathematical reasoning: "Though mankind's power of reasoning had not reached the end of its tether, it was clearly beginning to require the assistance of some visualizing agency" (p. 8).

Emmer (2005b) compared the pleasures that may come from experiencing an artistic expression, architecture in this case, with that of a beautiful mathematical proof. Art and mathematics has an important distinction in that art often has an emotional side to it whereas mathematics does not (Brown, 2005), a statement Klein (1953) did not fully agree with. Both aversion and joy can be tightly linked to mathematics if a person has studied enough mathematics to be able to work with it. According to Klein, mathematics consists of compositions rather than discoveries in the same way art does, but taking pleasure in art does not require any pre-knowledge. Brown (2005) mentions two other similarities of art and mathematics, one being the notion of structure and the other form. The mathematician can represent complex forms in more ways than can be depicted by a sculpture with the aid of logic deduction but also imagination. Brown states mathematics to be an art of the mind, independent in its nature but with possible real outcomes. One outcome can be seen in architecture, e.g., the Guggenheim museum which is an example of the results of non-Euclidean geometry that changed how humans can interpret space (Emmer, 2005a). Euclidean geometry allows certain interpretations of the world whereas other geometries, e.g., fractal geometry, hyperbolic geometry and the development of computer graphics, open doors for architects to make different interpretations adjusted to certain locations and expressions. The ability to imagine complex three-dimensional images, such as buildings or bridges, rests on spatial intuition (Catastini & Ghione, 2003). Perspective techniques and topographical measurements as well as properties of sight are theorized by Euclid in his work on optics. Proficiency in these fields, logically and intuitively, links the abstractness of mathematics and the concreteness of physics.

Global projections

For centuries man has tried to create a projection from the Earth (a near-sphere) on to a two-dimensional map that preserves distance, angle and relative size. Arguably the closest anyone has come is projecting the globe on to a regular icosahedron. This was first done by Buckminster Fuller in 1954. But polyhedrons were first discovered by the ancient Greeks.

Euclid was able to construct a regular icosahedron using merely straightedge and compass. There are several different ways to do so, as noted by Sir Thomas Heath in the footnotes of Euclid's Book XIII, Proposition 16 construction. Heath offers an alternative proof to Euclid's construction, and Heath shows the alternate constructions and proofs done by both Pappus and H. M. Taylor.

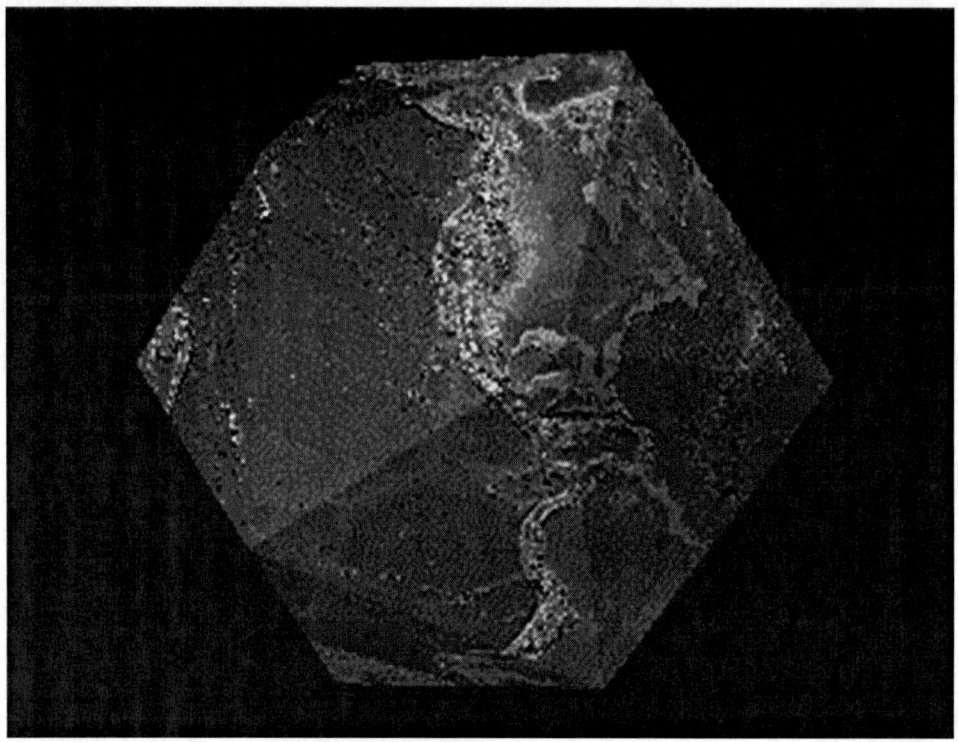
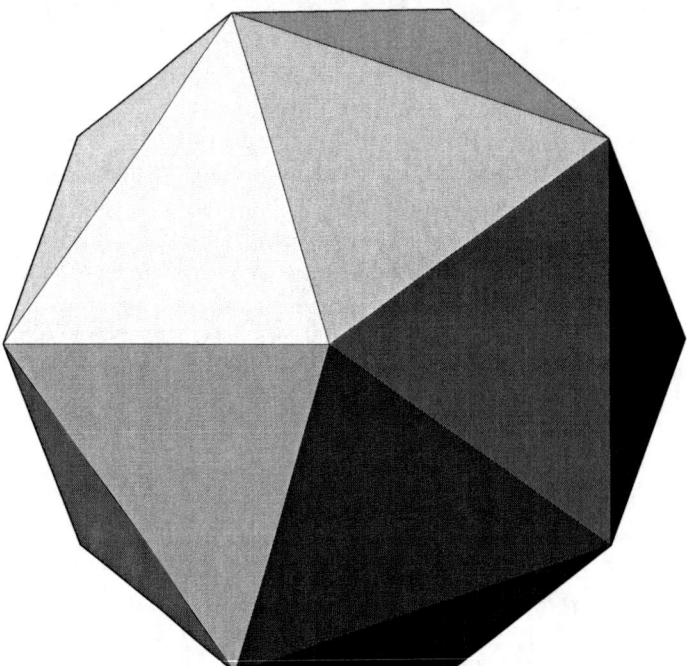

Figure 23.1 Projecting the globe onto a regular icosahedron.

The interfaces of innovation

A regular icosahedron is a 20-sided polyhedron consisting of 20 equilateral triangles. Although Euclid's construction differs from those of Pappus and Taylor, all three (as well as Heath in his proof) utilize the construction of two regular pentagons that are polar opposites. The center of each of these pentagons is elevated off the plane of the pentagon. Hence each triangle that is created by connecting the vertices of the pentagon to the center is in a separate plane. For if any two or more of these triangles were in the same plane, the icosahedron would fail to have 20 surfaces. In between these two pentagons (which together form ten of the 20 triangles of the icosahedron) are ten more triangles which each share exactly one vertex and one side with each of the top and bottom pentagons, as well as two sides with the triangles on either side. The icosahedron has 30 edges and 12 vertices, with five equilateral triangles meeting at each vertex (*Icosahedron*, 2010), and can be represented by its vertex figure, which is the figure created when a corner—i.e., one of the pentagons—is sliced off (see Figure 23.2).

The icosahedron is one of the five platonic solids, which are each regular polyhedrons: the tetrahedron (consisting of four equilateral triangles), cube (six squares), octahedron (eight equilateral triangles), dodecahedron (12 regular pentagons), and the icosahedron (20 equilateral triangles). An interesting way to picture an icosahedron is through its Cartesian Coordinates. As can be extrapolated from the figure below, the Cartesian Coordinates for the 12 vertices of an icosahedron centered at (0,0,0) with side lengths of 2 are: (0, ±1, ±φ), (±1, ±φ, 0) and (0, ±φ, ±1). The origin is the intersection point of the three planes, and

$$\varphi = \frac{-1+\sqrt{5}}{2},$$

the golden ratio. Hence these vertices form five sets of three concentric mutually orthogonal golden rectangles (*Icosahedron*, 2010).

The Greeks discovered these three-dimensional shapes, and Euclid clearly understood the intricacies and great detail required to construct one with simply a compass and straightedge. However, it took about 2,000 years for man to project the globe on to one.

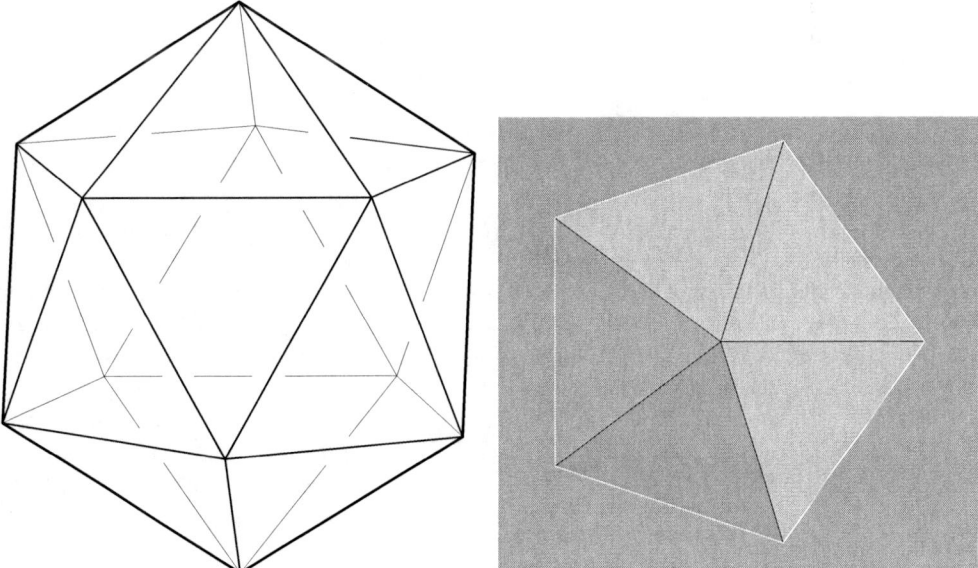

Figure 23.2 Regular icosahedron.

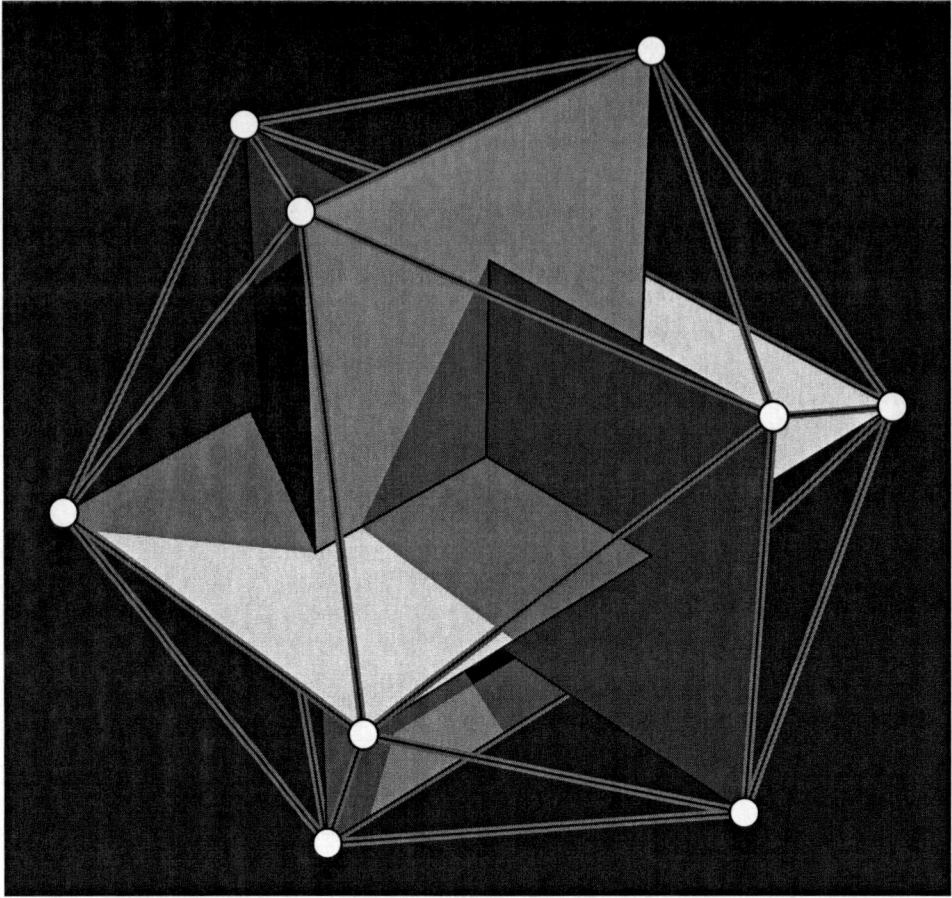

Figure 23.3 Dymaxion maps.

Buckminster "Bucky" Fuller was the first man to project a map of the Earth on to polyhedrons. Fuller called them Dymaxion Maps. He began with a non-regular polyhedron, a cuboctahedron, which consists of six squares and eight triangles. Fuller patented his projection in 1946, and later in 1954 he developed the projection on to an icosahedron (Marks, 1973). This projection is unique and has remained the only one of its kind that is both conformal as well as re-foldable back into an icosahedron. Maps for navigational purposes need to preserve conformality, i.e., angles of constant bearing crucial for plotting correct navigational courses on charts. In other words a line of constant bearing on a Mercator map is a rhumb line on the sphere. Conformality as achieved by Gerardus Mercator (1512–1594) came at the price of the distortion that occurred when projecting the sphere on to a flat piece of paper. Mercator was limited by the Calculus available at that time period, namely the difficulty of integrating the secant function. Hence Mercator's map, although conformal distorted landmasses as one moves further and further away from the equator (e.g., Greenland appearing as big as Africa!). Fuller's map on the other hand does not distort land masses and preserves conformality.

An interesting property of Fuller's Dymaxion Map on the icosahedron is it contains no top or bottom. Fuller claimed (until his death in 1983) that the Earth actually has no up or down, no top or bottom, or North or South. Instead, he claimed, there are merely "in" and "out" gravitational forces; the planets and stars simply create "in" forces toward the gravitational center, or "out" forces away from the gravitational center. Fuller attributed the notions of North and South to cultural bias (Lorance, 2009). We believe Fuller was mistaken in his complete attribution to cultural bias, for the Earth not only revolves around the sun, but rotates about its axis, creating a magnetic field. This field gives us the feeling that there is an up or down to the Earth as experienced when one holds a compass in hand—hence our notions of North and South.

Regardless of his cultural opinions, Bucky Fuller left an indelible mark on mapmaking (as well as in other fields). There are over 43,000 "nets" for the icosahedron. A net of a polyhedron is one of the ways in which it can be folded or unfolded along its edges to form the three-dimensional figure or lay it out flat in two dimensions. Fuller intended for different nets of his Dymaxion Maps on the icosahedron to represent different views and possibilities for the Earth (Marks, 1973). For instance, if it is unfolded one way the continents of Earth can be grouped together, forming a net something similar to Pangea, surrounded by ocean. In contrast, the triangles can be unfolded in a different pattern to show the continents forming a border around ocean in the middle. This brings up an interesting notion: could these different nets bring about insight into tectonic plate patterns, and possible previous or future locations of significant land masses on our planet?

Of course the pertinent question is: can we make a better projection of the globe on to a two-dimensional surface than an icosahedron provides? If a 20-sided polyhedron gives us such a good projection, why wouldn't a 30-sided one give us a better one? We hypothesized about these possibilities and here is what we came up with.

When dealing with projections of a sphere (such as the Mercator projection), chances are that area, distance and/or angle accuracy are sacrificed. Yet Fuller's projection of the globe on to an icosahedron diminishes these inaccuracies greatly. It would be interesting and valuable to find a polyhedron that could maintain accuracy of angles, distance and area, and come even closer to a sphere than an icosahedron.

This leads us to our conjecture: would increasing the number of sides of an icosahedron maintain the accuracy of the world projection on to the new polyhedron? We could increase the number of sides by partitioning each equilateral triangle into four equilaterals. This would be done by having the midpoints of each side of the equilaterals define an inscribed equilateral. Thus the newly created sections outside the inscribed equilateral will also themselves be equilaterals. The center equilateral's endpoints will be on the sphere that previously surrounded the icosahedron.

We could also increase the number of sides by partitioning each equilateral into three sections such that the center of the circle plus two endpoints of the triangle form a section/isosceles triangle. This will have the center point raised from the plane to the surface of the sphere creating a tetrahedron.

Well let's suppose that by partitioning each equilateral triangle into four equilaterals maintains both accuracy of the area and angles or even improves the accuracy. Consequently our new object contains 80 equilateral triangles that are one-quarter the size of the former triangles. Thus by dividing the current equilaterals up again, the second iteration would either maintain the accuracy of the former object or improve upon the accuracy. Thus we could repeat these iterations over and over again. We claim this will lead to a fractal effect and the object would eventually converge into a sphere. But continuing this process until the object becomes a sphere leads to loss in accuracy of conformality—hence the iterations work best when stopped at an icosahedron. The same argument could be stated for the tetrahedron approach. Therefore, we assert that any approach that increases the sides of an icosahedron while keeping the initial shapes the same will result in an ultimate loss of area, distance and/or angular accuracy.

It's interesting to note that a dodecahedron takes up more area in a surrounding sphere (≈66%) than an icosahedron does (≈60%) (*Icosahedron*, 2010). Hence adding more surfaces, vertices or even surface area may have a paradoxical effect. However, though we may assume coming closer to the sphere in area would create a better projection of the globe on to a two-dimensional surface, this is clearly not the case. For if it were, the dodecahedron would produce a better projection than the icosahedron. Clearly there are other pertinent factors at hand here. Perhaps if those factors are pinpointed and surmounted, we can make a better projection of the earth on to a planar map. In the next section, a different and more natural geometry is outlined because of its intricate connection to the world of art and architecture.

Fractal art and geometry

Benoit Mandelbrot (1924–2010), a very recognizable name in 20th century mathematics because of his seminal contributions to the development of fractal geometry repeatedly emphasized the need to reorient our perspectives to better understand the world around us. He often very humbly characterized himself as an "accidental" mathematician. In spite of his early interest and precocity in the study of geometry he was "encouraged" by the French university establishment to embrace formalization which led him to leave the *École Normale Supériere*. He wrote

> I spent several years doing all kinds of things and became, in a certain sense, a specialist of odd and isolated phenomena... I did not know or care in which field I was playing. I wanted to find a place, a new field, where I could be the first person to introduce mathematics. Formalization had gone too far for my taste, in the mathematics favored by the establishment.
>
> *(Mandelbrot, 2001, p. 192)*

Mandelbrot made his astonishing mathematical discovery when working on an economics problem accidentally handed to him by a friend. Economists had long attempted to make sense of (and predict) stock market fluctuations and had proposed theories based on existing data which did not hold up when tested with primitive computers. Mandelbrot viewed fluctuations from the perspective of changing scales. That is the time scale can be in days or months or years. He suggested that the interchangeable nature of the time scales was the key to understanding the fluctuations. This was an astonishing act of innovation based on a simple rule of nature.

> I cooked up the simplest mathematical formula I thought could explain this phenomenon... [making] no assumptions about people, markets or anything in the real world. It was based on a "principle of invariance", the hypothesis that, somehow economics is a world in which things are the same in the small as they are in the large except, of course for a suitable change of scale.
>
> *(Mandelbrot, 2001)*

In most, perhaps all of nature, we encounter a kind of deterministic chaos in a world described by "fractal geometries" which have "become a way of measuring qualities that otherwise have no clear definition: the degree of roughness or brokenness or irregularity in an object" (Gleick, 1987, p. 98). Just as the rules of perspective were developed by the artists in the Italian Renaissance before becoming formalized as projective geometry (Sriraman, 2009b), the world of architecture and art has long known the existence of self-similarity as a simple rule. In the latter part of the 20th century, it became possible for machines to produce geometric art based on very simple formulae which resemble Old Masters paintings, Cubist paintings or Beaux Arts architecture. Cubism itself followed Paul Cezanne's famous proclamation that "Everything in nature takes its form from the sphere, the cone, and the cylinder" and these shapes are used to depict the object of the painting.

The discovery that self-similarity was an inherent property of nature as mathematically conceptualized by Mandelbrot (1981, 1989) was long written about and expressed by poets, satirists, writers, philosophers and numerous religious traditions. For instance, in Southern India, Kolam is an art form used by women to decorate the entrance to homes and courtyards. These art forms go back over 6,000 years and consist of self-similar patterns repeated in different scales in very sophisticated fashion.

Fostering for mathematical innovation

Mathematical knowledge, as well as many other types of knowledge, includes symbolic representation which must be intentionally learned as opposed to knowledge from inherited chemical brain impulses prompted by outer stimuli (Csikszentmihalyi, 1996). The symbolic systems of say calculus or a certain type of music permits us to see patterns developed through *cultural evolution* (p. 37). Csikszentmihalyi found that creativity cannot be studied only on a personal basis. Individuals, e.g., teachers, editors and peers, surrounding a person respond to his or her creative innovations through assessments and implementations of the innovations and ideas. A system of three interacting instances describes creative development; a culture with symbolic rules, an individual introducing a new idea using the symbols of the culture, and a set of experts to recognize and validate the new idea. Thiessen's (1998) three components aforementioned can be combined with Csikszentmihalyi's three instances; a common purpose stated by the culture, a common language expressed in symbols to state new ideas, and a model shared by experts and novices through participation and validation.

School is one instance of possible expertise where children can develop creativity through ideas tested and assessed by teachers and peers. As the children develop and specialize their skills, the specialists may be provided outside the school, e.g., judges in competitions or journal editors. Teaching mathematically gifted students means to take cognitive, social and affective perspectives, and mathematical content issues into account (Leikin, Berman, & Koichu, 2009). Teachers hence need stable knowledge in all these domains to provide fruitful learning environments for the students. The teacher is the main inspiration to students learning mathematics (Berman & Leder, 2009). Carefully selected problems, discussion groups and competitions stimulate gifted students to develop solution strategies and to generalize results. They become part of the culture of mathematicians through researchers' and teachers' validation of their work.

Combining areas of study can also be used in school to increase levels of creativity. Dalke, Grobstein and McCormack (2006) studied interdisciplinary work in biology, physics and literature and found that the combination of perspectives enables studies of intersections where otherwise unveiled discoveries would remain. The joint focus of the task at hand is thus seen through a combination of lenses, metaphorically speaking. This synergy effect was used in a study by Luftig (2000) where 615 students in grades 2, 4 and 5 were taking part. The students took part in a multi-disciplinary program for arts in education. The students met artists (experts) and had one hour of arts (e.g., art, music, drama or dance) added to their curricula each week. Arts were used in the teaching of other subjects (e.g., mathematics and science) and arts were studied with the aid of other subjects. Pre- and post-tests were conducted to measure students' creativity, fluency, abstractness of titles (for drawings), elaboration, resistance to closure (keeping a task going to enable original ideas to take form), academic achievement (reading and mathematics), self-esteem and appreciation of the arts. The results clearly point to an enhanced level of creative thinking from the integrated elements of arts among the students. The students' mathematical comprehension abilities were also better in the integrated arts group than in the control groups. The results on mathematics concepts were however not gender symmetrical as the boys in the integrated arts group had a considerably higher score than all other groups, and the females in the same group by far scored lower than all other groups. The students in the integrated arts group, nevertheless, scored better in mathematics seen in total than the students in the control groups did.

Beswick (2008) suggests a model of learning situations as complex systems. Teachers can use the model to create situations for their students where creativity, individual, as well as the whole group, develop. These systems consist of, possibly independent, actors interacting in a dynamical environment upheld by its participants. The complexity of the systems brings unpredictability as well as a lack of standard strategies for the teacher to take into account when teaching.

Interdisciplinarity as a means of innovation in mathematics and architecture

Soygenis and Erktin (2010) give examples from the Byzantine and Islamic cultures in Turkey to make the case for the presence of prototypical design structures in a student's day to day surroundings. They argue for the strong relationship between architectural design and geometry and report on a program (Archimath) to build curriculum on topics that integrate mathematics and architecture at the elementary school level. Other such initiatives that create architectural awareness at the elementary school level are found in UNICEF sponsored projects in Sweden (www.childfriendly-cities.org/en/search-view?ProductID=742). The claim of the Archimath program is change in attitudes toward the physical environment, with previous indifference being replaced by more curiosity on the place of mathematics in real life as found in architectural designs.

Mastandrea, Bartoli and Carrus (2011) suggest that prototypical objects from the world of art and architecture evoke a preferential aesthetic response. These researchers hypothesized that there was a pre-existent aesthetic preference for certain types of architecture (based on familiarity from the surroundings) that produced an automatic preferential response. This was confirmed through an Implicit Association Test (IAT) in which the reaction times for associating positive words to figurative art and classical architecture and the reaction times for associating positive words to abstract art and contemporary architecture were measured. The former reaction times were shorter than the latter which the authors explain with familiarity and proto-typicality, but also the fact that figurative art and classical architecture is less complicated in its structure than abstract art and contemporary architecture. Innovation includes seeing new developmental possibilities and proficiency in assessing art and architecture in terms of structures, and aesthetics is an aspect of that. A complex artistic composition implies a higher number of elements to structure and relate to one another than a less complex composition and a person trained to handle complexity is hence better equipped for that kind of work. Even though the positive association could be attributable to the presence of more symmetry and less complexity in classical design forms in comparison to contemporary forms, the point we are trying to make here is that there seems to be automaticity, acceptance and an acknowledgment of architectural designs that can be capitalized on in the curriculum.

Interdisciplinary courses may be used to allow for Thiessen's (1998) three factors, namely a common purpose, language and model and to avoid domain limitations (Root-Bernstein, 2003). In a round table discussion (Moran et al., 2002) on the role of mathematics in the architecture curriculum at the university level Krüger (2002) described a curriculum change tried out at the faculty of architecture at the University of Coimbra. The initial curriculum included nine courses from other departments. The new curriculum included only four from other departments with the result of students learning mathematics relevant for architecture within themes such as space syntax and architectonics. The interdisciplinary approach allowed for a combined development which boosted rather than hindered by the domain specificity of the topics. In the same roundtable Glass (2002), an architect teaching mathematics to liberal arts students said:

disciplined habits of thought, of problem solving through the desire to achieve an elegant rational solution to a problem is perhaps the greatest aspect of what mathematics can contribute to the design education, and that elegance may be the link between mathematics and art that we have been talking about.

(p. 95)

We contend that engineering oriented math courses which tend to be Calculus heavy are not the kinds of design education courses for those interested in architecture. The latter would need more geometry through interdisciplinary courses that situate the mathematics contextually in the architectural curriculum.

Conclusion

We have argued that architectural creation is closely linked to mathematics where abstract mathematical representations result in physical constructions. Interdisciplinary work with problems, e.g., within construction or aesthetics, may evoke solutions from the combination of various disciplines that would have been overseen within just one discipline. We conclude that students studying to work within architecture, or other areas requiring artistic proficiency, would benefit from learning mathematics in an interdisciplinary context relevant to their coming profession to endorse artistic and creative development of our society.

Acknowledgments

The section on Global Projections was developed by University of Montana undergraduate scholars, Carlos Morales and Tyler Cherry in Math 439: Euclidean and Non Euclidean Geometries, under the supervision of Professor Bharath Sriraman.

References

Berman, A., & Leder, G. (2009). The pleasure of teaching the gifted and the honor of learning from them. In R. Leikin, A. Berman & B. Koichu (Eds.), *Creativity in Mathematics and the Education of Gifted Students* (pp. 3–10). Rotterdam, Netherlands: Sense Publishers.

Beswick, K. (2008). Fostering creativity by establishing the conditions for complex emergence. Proceedings of the Discussion Group 9: Promoting Creativity for all Students in Mathematics Education *Proceedings of the 11th International Congress on Mathematical Education, Monterrey, Mexico, July 6–13, 2008* (pp. 127–132).

Bill, M. (1993). The mathematical way of thinking in the visual art of our time. In M. Emmer (Ed.), *The Visual Mind: Art and Mathematics* (pp. 5–10). Cambridge, MA: MIT Press.

Brinkmann, A., & Sriraman, B. (2009). Aesthetics and creativity: An exploration of the relationship between the constructs. In B. Sriraman & S. Goodchild (Eds.), *Festschrift Celebrating Paul Ernest's 65th Birthday* (pp. 57–80). Charlotte, NC: Information Age Publishing.

Brosz, J., Carpendale, S., Samavati, F., Wang, H., & Dunning, A. (2009). Art and nonlinear projection. In C. Kaplan & R. Sarhangi (Eds.), *Proceedings of Renaissance Banff II – Bridges 2009: Mathematics, Music, Art, Architecture, Culture* (pp. 105–114).

Brown, R. (2005). Johan Robinson's symbolic sculptures: Knots and mathematics. In M. Emmer (Ed.), *The Visual Mind II* (pp. 125–139). Cambridge, MA: MIT.

Catastini, L., & Ghione, F. (2003). The geometry of sight: From Euclid's optics to the renaissance perspective. In M. Emmer & M. Manaresi (Eds.), *Mathematics, Art, Technology and Cinema* (pp. 53–66). Berlin: Springer-Verlag.

Collins, B. (2005). Geometries of curvature and their aesthetics. In M. Emmer (Ed.), *The Visual Mind II* (pp. 141–158). Cambridge, MA: MIT.

Csikszentmihalyi, M. (1996). *Creativity: Flow and the Psychology of Discovery and Invention*. New York: HarperCollins Publishers.

Dalke, A., Grobstein, P., & McCormack, E. (2006). Exploring interdisciplinarity: The significance of metaphoric and metonymic exchange. *Journal of Research Practice, 2*(2), Article M3.

Emmer, M. (Ed.) (1990). *L'occhio di Horus: Arte e Matematica*, Rome: Istituto Enciclopedia Italiana.

Emmer, M. (Ed.) (1993). *The Visual Mind: Art and Mathematics*. Cambridge, MA: MIT Press.

Emmer, M. (2005a). Mathland: From topology to virtual architecture. In M. Emmer (Ed.), *Mathematics and Culture II* (pp. 65–80). Berlin: Springer-Verlag.

Emmer, M. (2005b). Visual mathematics: Mathematics and art. In M. Emmer (Ed.), *The Visual Mind II* (pp. 59–90). Cambridge, MA: MIT.

Glass, C. (2002). In Moran, J. F., Brangé, J., Wassell, S., Vianello, M., Martins, A., & Krüger, M. (2002). Nexus 2002 round table discussion: Mathematics in the architecture curriculum. *Nexus Network Journal, 4*(3).

Gleick, J. (1987). *Chaos: Making a New Science*. New York: Penguin Books.

Icosahedron. (2010). Mathworld. Retrieved April 20, 2011 from http://en.wikipedia.org/wiki/Icosahedron.

Klein, M. (1953). *Mathematics in Western Culture*. Harmondsworth, UK: Penguin.

Krüger, M. (2002). In Moran, J. F., Brangé, J., Wassell, S., Vianello, M., Martins, A., & Krüger, M. (2002). Nexus 2002 round table discussion: Mathematics in the architecture curriculum. *Nexus Network Journal, 4*(3).

Leikin, R., Berman, A., & Koichu, B. (2009). *Creativity in Mathematics and the Education of Gifted Students*. Rotterdam, Netherlands: Sense Publishers.

Lorance, L. (2009). *Becoming Bucky Fuller*. Cambridge, MA: MIT Press.

Luftig, R. L. (2000). An investigation of an arts infusion program on creative thinking, academic achievement, affective functioning, and arts appreciation of children at three grade levels. *Studies in Art Education, A Journal of Issues and Research, 41*(3), 208–227.

Mandelbrot, B. (1981). Scalebound or scaling shapes: A useful distinction in the visual arts and in the natural sciences. *Leonardo, 14*(1), 45–47.

Mandelbrot, B. (1989). Fractals and an art for the sake of science. *Leonardo, 2*(Supplemental Issue), 21–24.

Mandelbrot, B. (2001). The fractal universe. In K. H. Pfenninger & V. R. Shubnik (Eds.), *The Origins of Creativity* (pp. 191–212). Oxford, UK: Oxford University Press.

Marks, R. (1973). *The Dymaxion World of Buckminster Fuller*. Garden City, NY: Anchor Books.

Mastandrea, S., Bartoli, G., & Carrus, G. (2011). The automatic aesthetic evaluation of different art and architectural styles. *Psychology of Aesthetics, Creativity, and the Arts, 5*(2), 126–134.

Moran, J. F., Brangé, J., Wassell, S., Vianello, M., Martins, A., & Krüger, M. (2002). Nexus 2002 round table discussion: Mathematics in the architecture curriculum. *Nexus Network Journal, 4*(2), 81–100.

Root-Bernstein, R. (2003). The art of innovation: Polymaths and the universality of the creative process. In L. Shavinina (Ed.), *The International Handbook on Innovation* (pp. 267–278). Oxford, UK: Elsevier.

Shavinina, L. V. (2003). Understanding scientific innovation: The case of Nobel laureates. In L. Shavinina (Ed.), *The International Handbook on Innovation* (pp. 445–457). Oxford, UK: Elsevier.

Soygenis, S., & Erktin, E. (2010). Juxtaposition of architecture and mathematics for elementary school students. *International Journal of Technology and Design Education, 20*(4), 403–415.

Sriraman, B. (2009a). Paradoxes as pathways into polymathy. *ZDM—The International Journal on Mathematics Education, 41*(1&2), 29–38.

Sriraman, B. (2009b). A historic overview of the interplay of theology and philosophy in the arts, mathematics and sciences. *ZDM—The International Journal on Mathematics Education, 41*(1&2), 75–86.

Thiessen, B. (1998). Shedding the stagnant slough syndrome: Interdisciplinary integration. *Creativity Research Journal, 11*(1), 47–53.

24
NASA press releases and mission statements
Exploring the mathematics behind the science

Sten Odenwald

NATIONAL INSTITUTE OF AEROSPACE/NASA, USA

Summary: NASA press releases contain a wealth of quantitative information and understated mathematics, which can be used to stimulate student interest in mathematics and science. Students are often curious about space themes, such as the search for extraterrestrial life, black holes, or space exploration. For this reason, press releases about discoveries in space make mathematics relevant and interesting to students beyond the mundane application problems so common in modern-day mathematics textbooks.

The overarching goal of the *SpaceMath@NASA* program is to assist students in seeing the mathematics behind the scientific discovery. We will describe this program and how we 'reverse-engineer' press releases to uncover the often simple mathematics suitable for on-grade-level learners in grades 3–12.

Key words: Mathematics problems, space, astronomy, press releases, newspapers, applied mathematics, education, STEM curriculum.

Bridging the science and math gap

In 2000, the National Commission on Mathematics and Science Teaching for the 21st Century published a provocative report 'Before it's Too Late', which called the preparation of US students in math and science unacceptable, and made the points that (1) if the United States is to remain competitive in an integrated global economy, it must improve its math and science performance, and (2) the most direct route to improving student performance is through better teaching. This led, in 2001, to the No Child Left Behind Act (NCLB), which was intended to apply rigorous achievement goals to the nation's schools. Schools that failed to meet the targeted test results (called Adequate Yearly Progress) in math, reading, and science were subjected to increasing penalties. By 2014, all students in the United States must attain specific, challenging, math and science goals to be determined by each state. Each student will be subjected to frequent assessments, and teachers will be required to systematically improve classroom performance on these assessments while the performance goals continue to increase each year to meet the 2014 national deadline. But this test-based approach to excellence had its unfortunate, but perhaps inevitable, consequence. According to Cramond and Fairweather (this volume) 'Educators caught in the push for basic skills and high-stakes testing lament that they cannot teach to the standards and [at the same time] teach students to be innovative too'.

As this environment of 'teaching to the test' became increasingly adopted, some educators and scientists felt that it was actually adversely impacting student performance and understanding of science. Reading and math goals were comparatively easy to measure, but science is far more than just the memorization of facts. According to Tugel (2004, p. 1), NCLB has emphasized the performance in reading and math with the 'unfortunate consequence of further marginalizing science in some districts, particularly at the elementary level'. According to Hazelwood (2005, p. 9) the direct consequence of this trend is that 'Many schools are now looking for ways to bring science back into the spotlight and the core curriculum.' Many school systems have actually reduced their science content, and are now struggling to find ways to reintroduce science into the curriculum, but at the required levels mandated by NCLB (Lord, 2006).

One of the suggested ways is by connecting science with language arts to emphasize how science and literacy are interdependent (Tugel, 2004). A variety of popular classroom approaches to science education were analyzed by Fensham (2006), who identified a number of common elements. Students responded very well when science was presented as a 'story' involving people, situations, and action. They particularly enjoyed the real-world situations that formed the core of the science or technology, and the science was clearly presented, and had a larger context. In 2006, the Programme for International Student Assessment (PISA) looked into the international context of student interest in science and found, not surprisingly, that the enjoyment of science in the early grades played a central role in their long-term career goals (Ainley & Ainley, 2011). National surveys of parents and students are almost unanimous in their belief that demonstrating real-life applications in science can help make science education more interesting for US students (Teaching with Contests, 2010), and that the Internet should be used more extensively to make interesting science education materials available to teachers, and to help parents engage in their child's education. The challenge remains to find an inexpensive program that captures the positive elements of stimulating student interest, while complying with post-NCLB education standards that favor a cross-curricular approach to presenting subject matter.

The sound of one hand clapping

Most scientists do not start out as mathematicians, but begin their formative years passionately pursuing some topic that interests them. These interests can include dinosaurs, rocks, insects, human anatomy, astronomy, space travel, science fiction, and, yes, even flying saucers! Yet, during the grade-school experience of most students and young scientists, mathematics is formally presented as a separate train traveling on different tracks from any developing curiosity about science topics. This dearth of mathematics in science-based topic areas changes rather dramatically when students suddenly encounter mathematics in high school chemistry and physics courses.

Despite the fact that students may have been operating at a very high level of sophistication in mathematics during their high school years, they had not been equally inspired to engage in the art of problem solving at a commensurable level. Even the problems they were assigned to puzzle over with their mathematical competencies were not even problems of their own choosing. They were problems assigned by a teacher or professor, and often contrived to lead to a specific answer in order to exercise a particular well-defined mathematics skill. As 'artists', they had acquired the finest oil paints, brushes and canvases, but never fully understood how mathematical problem solving techniques applied to the messy real world. They were certainly not encouraged to explore the art of problem solving on their own, so without the help of assigned application problems, students kept staring at blank canvases not really knowing what to do with their palette of mathematical tools.

Students are increasingly taught what might be considered 'toolsmanship' in mathematics, but are virtually left on their own to discover how to creatively solve real-world problems. Teachers call this 'compartmentalization' and remark, 'Why is it that my students can read a graph or "solve for X" in Math Class, but one hour later in their science class they act as though they have never seen a graph or an equation before?'

In fact, according to the very first sentence of the paper by Abbott and Nantz (1994, p. 22), 'In spite of much rhetoric and many adaptations of core curricula, one of the old problems still plagues us: students compartmentalize knowledge and fail to make lasting connections between subjects.' Clearly, compartmentalization of students' science and mathematics knowledge and skills is not surprising when viewed in this larger context. The challenge we face as educators is to find stimulating approaches that work in the classroom to break down these artificial walls.

The absence of math problems with obvious couplings to students' interest in astronomy (or science in general) can be a hindrance to further development as a scientist. Students can fall into the trap of seeing mathematics as a rigid system of knowledge that does not easily map into the very messy and changing world around them. This fear of using mathematics playfully can make it difficult to understand how to launch a question and see a pathway to arriving at a mathematical answer, even in retrospect when the approach is very clear. There is nothing wrong with fitting data with an empirically based regression formula, without any knowledge of the underlying physical reasons for that particular functional form.

For example, I recall in high school wanting to predict the position of a satellite of Jupiter on a given day from tabulated data provided in *Sky and Telescope* magazine. Even though a sinusoidal graph was provided in the magazine, I did not see how to extract the functional form of this sinusoid from the graph. My advanced math teacher told me it would require Fourier analysis; a topic that was understandably far beyond my skills at that time. Today I can look at that same graph and by simple visual inspection determine the amplitude, frequency and phase of the required sine function that adequately fits the data perfectly. I would have been thrilled to have had this insight available to me as a high school student. My teacher's advice was technically correct for an exact solution, but hopeless as a working first approximation, which was all I needed. Dispirited in the apparent complexity of the problem, I put the project aside and never pursued creating my own ephemeredes for periodic astronomical events in our solar system. An important positive experience in the growth of an astronomer had been excised from my life.

How can you possibly teach this kind of a skill to children immersed in the modern teaching-to-the-test environment?

Following a long and happy career in astronomy, I had the luxury to return to these elementary issues as I began my second career at the National Aeronautics and Space Administration (NASA) as an 'Education and Public Outreach Lead'. Doing more of the same science popularization at NASA seemed, quite frankly, a waste of my talents, especially at the elementary levels of discourse required to satisfy national education benchmarks in science. My unique skill as an astronomer was not that I could 'explain' stellar evolution or space weather to a middle school student. This could be done by reasonably articulate NASA educators without the benefit of a PhD in astrophysics. No, my unique skill was that I could easily see the mathematics behind the science content. It was this singular realization that completely redirected my education efforts at NASA!

In 1997 we embarked on a program of creating math enrichment resources for students and teachers. The ones already available at NASA seemed to focus on below-grade-level learners, especially in middle and in high school. There seemed to be plenty of NASA education resources useful for math remediation, but no products for on-grade-level students, and certainly not at the high school level. We began to see how the modern situation in math and science education was strikingly parallel to what I had experienced first hand as a K-12 student over 40 years earlier. We were still teaching science shorn of its mathematical underpinnings in logic and data analysis, and, conversely, mathematics shorn of the richness of its applications to science comprehension!

Through its many images, videos and online products, NASA has a huge resource base that can be called upon to stimulate student learning. What is astonishing is just how vast NASA's implicit mathematics resources are, and how very little of this has made its way into the hands of teachers. NASA primarily produces education resources to support science content education. Ironically for an institution such as NASA, where mathematics is so vital, the science content we produce for educators has been largely cleansed of its mathematical elements. This is often at the request of the teachers themselves. Ultimately, NASA tends to produce and distribute the products that teachers request. This continues to reinforce the idea that science and math are compartmentalizable subjects, which makes problem solving that crosses the science–math threshold even more difficult to stimulate. Because problem-solving skills are so vital to student maturation as young scientists and engineers in STEM careers, a more aggressive approach may be called for. One of the best ways to get someone to learn a new skill is to 'model' that skill yourself!

In 1977, Stanford University psychologist Albert Bandura (1977) offered an interesting 'Social Learning' approach to learning and skill retention. In order for students to learn a new behavior, they must see that behavior modeled.

> Learning would be exceedingly laborious, not to mention hazardous, if people had to rely solely on the effects of their own actions to inform them what to do. Fortunately, most human behavior is learned observationally through modeling: from observing others one forms an idea of how new behaviors are performed, and on later occasions this coded information serves as a guide for action.
>
> *(Cherry, 2011)*

The impeccable logic of this observation appealed to me instantly.

Watching others carefully is, after all, how we learn to speak, dance, and master thousands of other skills. Consciously or otherwise, we see how other people perform a skill, and then we mimic the behavior. A similar truism also applies to math education and problem solving. We have to actively *show* students the art of problem solving with mathematics before they 'get it', and make no mistake about it, problem solving is an art form. It requires creativity, insight, and inspiration to look into some aspect of the real world and parse what you see and know about it into a succinct, defining paragraph. It is also an art to turn a descriptive paragraph into a symbolic statement, and then to come up with the steps needed for solving the problem. Finally, problem solving requires perseverance and constant checking, just as an artist spends hours getting a particular rose on a canvas to look exactly right. It is exactly in mastering this artistic skill of problem solving that NASA can help the most!

Although children often profess an interest in space science, especially black holes and the search for 'aliens', ironically, mathematics textbooks only offer about 4% of their application problems in space-related subject areas. For example, the book *Pre-Algebra* by Glencoe Publishing has about 420 application problems of which only 12 are space related, which is a space-related share of about 3%. Similar problem counts and percentages persist among algebra (4%) and calculus (4%) textbooks. The vast majority of the applications problems found in common US textbooks are in economics or other consumer-based topic areas. Since more than two-thirds of the US economy is based on consumerism and not on scientific or space research, this poor showing by space science in applications problems is perhaps not surprising. It does, however, reinforce the notion among students that scientists do not really need to know mathematics. This could directly support a misconception that claims made by scientists about, for example, global climate warming, evolution theory, or even big bang cosmogenesis are possibly not based upon a solid mathematical and logical foundations. All topics in science are, therefore, subject to linguistic debate as though science were an extension of a comparative literature class. Some students, and adults, may even subscribe to the extreme idea that science is 'just another' belief system. This stealthy misconception, that science and math are different enterprises, should be a real concern among science educators. Despite a passion for science as a highly motivated child, a student can travel a long way in the curriculum and come into contact with virtually no examples of mathematics applied to science is a believable way. A similar 'desertification' of mathematics in space science persists today in the 21st century!

A brief history

In 1997, the NASA Imager for Magnetosphere to Auroral Global Exploration (IMAGE) satellite education program decided to create math enrichment resources for students and teachers rather than the traditional posters, bookmarks, and other science content products. Initially, they began by developing math guides for middle school students covering topics in space weather; the area of research covered by the IMAGE mission. Between 1998 and 2003 the program produced *Solar Storms and You, Northern Lights and Solar Sprites*, and several other guides combining space weather topics with elementary math and geometry. The popularity of these guides soon led to the idea of posting a new math problem each week covering a different area of solar and space research. This online resource, *Space Science Problem of the Week*, featured unusual problems for middle school students in solar science, auroras, magnetic storms, and occasionally astrophysics and planetary science topics too. The problems in a one-page format were eventually collected together into annual math guides beginning with *Exploring Space Science Mathematics*, followed by *Space Math I, Space Math II, Space Math III*, and so on through *Space Math VII* in 2011. Teachers and NASA mission educators soon began to request collections of problems on specific topics to support their particular mission, so math guides such as *Solar Math, Earth Math, Black Hole Math*, and other topics quickly followed.

Growing teacher interest and support for mathematics-related, real-world problems, has recently collided with the evolution of more test-oriented classroom environments even in middle school classrooms. In response to this trend, the IMAGE mission's *Space Science Problem of the Week* was redesigned after the loss of the IMAGE satellite in 2005 to include more targeted math resources. These resources involved 'one-page' math problems spanning the entire topic space of pre-algebra and high school on-grade-level instruction, while offering an interesting science topic as the problem's theme.

The education program of the US/Japanese Hinode solar observatory stepped in to continue this popular program. So, beginning in 2005, the Weekly Space Math program was officially renamed *SpaceMath@NASA*, with an entirely new website and domain name (http://spacemath.gsfc.nasa.gov). With the ending of the Hinode education efforts in 2009, *SpaceMath@NASA* became solely funded by NASA's Science Mission Directorate (SMD) through a series of three-year education grants in 2008 and 2010.

By working in partnership with other missions, and with content areas in other NASA Directorates beyond SMD, we have succeeded in developing a broader spectrum of mathematics subject areas than what SMD missions normally consider. Currently, *SpaceMath@NASA* has incorporated images and data from SMD missions such as the Solar Dynamics Observatory, Mars Global Surveyor, Spitzer Space Telescope, Hubble Space Telescope, in addition to data on various aspects

Figure 24.1 The home page of SpaceMath@NASA showing some of its many features (source: http://spacemath.gsfc.nasa.gov). Teachers may register at the website to gain access to over 25 space-related problem books, and over 400 individual math problems, all provided with answer keys to facilitate immediate classroom use.

of radiation effects to humans (Exploration Systems Mission Directorate) and technology (SMD, Space Systems Mission Directorate, and the Aerospace Systems Mission Directorate).

SpaceMath@NASA does much more than merely create ad hoc mathematics problems to articulate a specific science principle. Instead, we work closely with one NASA resource that is updated daily, and that has an enormous public distribution – its press releases! Thankfully, NASA press releases such as the one shown in Figure 24.2 are not stripped of their mathematical or quantitative content unlike the many other press releases and announcements that you might find in the news media. In addition to provocative and inspiring quotes from the researchers themselves, NASA press releases often contain numerical information, occasional graphs in many forms, and of course the dazzling imagery returned by its many satellites, spacecraft, and astronauts.

All that one needs to do to couple the excitement of science discovery and space exploration with mathematics is to essentially reverse-engineer NASA press releases! There is good reason to think that this approach might be viable in bridging the gap between math and science education.

Mathematics behind the Science

Figure 24.2 Front page of NASA (www.nasa.gov) dramatically showing a news announcement about the discovery of a planet orbiting two stars (September 16, 2011). SpaceMath@NASA published a series of math problem about this discovery three days later.

Can newspapers be a magic bullet?

One common way in which students and parents come into contact with science is through the newspaper, or its electronic equivalent. The use of newspapers in the classroom is not, however, a new idea. Students who participate in these 'Newspapers in Education' (NIE) programs show a marked increase in reading competency, the ability to discriminate facts from opinions, and a heightened interest in current events (Rhoades & Rhoades, 1985). A variety of Internet sites now provide extensive suggestions about how to use newspapers in the classroom in topics spanning the entire curriculum.

Teachers increasingly find that newspapers help bridge the gap between the classroom and the 'real' world. They are extremely flexible and adaptable to all curriculum areas and grade levels, and they make learning fun. According to Vockell and Cusick (1995, p. 359), 'Newspapers give students the opportunity to apply skills used in the classroom and to be exposed to more up-to-date information than that found in textbooks.'

DeRoche (1991) has also found that by 1990 approximately three million students and 16,000 schools were participating in the Newspapers in Education program, which is supported by nearly half of all domestic newspapers in the US. Moreover, students have a measurably improved performance in reading, vocabulary development, and content awareness as a result of frequent access to newspapers. 'On average, students who use newspapers in school scored 10% better on standardized reading tests, and a 29% increase in test scores was found among low-income, non-English speaking or minority students' (Bruder, 2006, p. 1).

A few of the suggested math activities recommended by Newspapers in Education include having students locate ten items in newspapers and calculate the difference between the regular and sales price, have them scale up recipes for larger serving sizes, and follow stock activity by graphing daily share prices (NIE, 2011).

The *SpaceMath@NASA* program was expanded in 2009 to include mathematics problem supplements to many NASA press releases, based on previous successes with disseminating 'one-page' math challenges since 2004. By building on existing programs, and scientific data from across NASA's Science Mission Directorate, we have also succeeded in creating a new generation of space-related math problems and resource guides that support and enhance mathematics instruction. More importantly, *SpaceMath@NASA* dramatically shows students that mathematics is the key to a universe of new opportunities for thinking about the world around them.

Extracting mathematics from press releases

Although originally designed to provide enrichment math problems in space science and heliophysics (our IMAGE legacy), NASA's involvement in space, and students' interests, are much broader in scope. *SpaceMath@NASA* has fully evolved to provide math problem content that covers the full spectrum of NASA's involvement in space research.

As described by Odenwald (2011, 2012), problems are derived from a variety of 'breaking' topic areas and are often coupled directly to press releases posted on the NASA front page (www.nasa.gov) as shown in Figure 24.2. Once a press release appears, it is carefully read for content and a math problem is developed based on information or science content provided in the press release. These problems can include verifying quantitative information in the press release. No two press releases offer the same mathematical topics, so one has to be creative in reading these documents and deciding what the math element might be. We also take advantage of NASA videography, often posted on YouTube (www.youtube.com) to create problems related to speed and displacement, such as the one shown in Figure 24.3 involving the highly publicized launch of the Space Shuttle Atlantis.

For other press releases and discoveries, it helps to be an astronomer to sort through the bewildering field of science and math concepts. There can be many different math principles camouflaged in the same news story, which is a boon to math and science teachers looking for 'extension' activities for more advanced students. It can also benefit students to see how one topic progressively leads to another! Here are a few elementary examples:

Problem 425: NASA Dawn spacecraft returns close-up image of asteroid Vesta
July 18, 2011: 'After traveling nearly four years and 1.7 billion miles (2.8 billion kilometers), Dawn also accomplished the largest propulsive acceleration of any spacecraft'. Analysis: this is a simple speed = distance/time problem that students can solve to get an average speed for the *Dawn* spacecraft of 700 million km/year. This can be further converted into kilometers/day (1.9 million km/d) or kilometers/hour (80,000 km/hr) and compared to the Space Shuttle orbital speed of 28,000 km/hr.

Problem 402: NASA finds Earth-sized planet candidates in the Habitable Zone
February 2, 2011:

> The findings increase the number of planet candidates identified by Kepler to-date to 1,235. Of these, 68 are approximately Earth-size; 288 are super-Earth-size; 662 are Neptune-size; 165

Space Shuttle Atlantis (STS-135) Launch Speed   8

This sequence of images shows the historic launch of the Space Shuttle Atlantis (STS-135) on July 8, 2011 at 11:29 a.m. EDT, from launch pad 39A at the NASA Cape Canaveral Space Center.

From bottom to top, the image times are 11:29:15.0, 11:29:16.0, 11:29:17.0, 11:29:18.0, and 11:29:19.0. The length of the space shuttle orbiter (not the red fuel tank) is 37 meters.

The launch sequence can be seen in the video located at:

Problem 1 - Using a millimeter ruler, what is the scale of an individual image in meters/mm?

Problem 2 - Measure the height in meters between the tip of the red shuttle fuel tank and a fixed location near the bottom of each frame.

Problem 3 - Graph the height of the fuel tank versus elapsed time beginning at T=0 in the bottom (first) image.

Problem 4 - About what was the average speed of the Shuttle in the top image in A) meters/sec? B) miles per hour?

Space Math   http://spacemath.gsfc.nasa.gov

Figure 24.3 An example of a math problem (Problem 431) from SpaceMath@NASA showing the simple science content and how appropriate math problems are derived. The answer key for this problem is provided as a second-page to the PDF file, which may be accessed at the website.

are the size of Jupiter and 19 are larger than Jupiter. Of the 54 new planet candidates found in the habitable zone, five are near Earth-sized. A total of 156,435 stars were surveyed.

Analysis: students can bar-graph the data by planetary size, then determine the percentage of Earth-sized planets in the full candidate sample (1,235) that were in the Habitable Zones of their stars (five) to get 0.4%. As an extension, if it is assumed that the Milky Way contains 40 billion stars similar to the ones in the Kepler survey, students can estimate from $(40\text{ billion}/156435) \times 0.004$ that there are one million Earth-sized planets in the Milky Way that are in their Habitable Zones.

Problem 397: NASA research finds 2010 tied for warmest year on record
January 12, 2011: 'To measure climate change, scientists look at long-term trends. The temperature trend, including data from 2010, shows the climate has warmed by approximately +0.36F per decade since the late 1970s'. Analysis: from the graph that accompanies the press release, students can determine the slope of the linear regression line through the data from 1970 to 2010 using a ruler, or the two-point formula. They can then determine the units for this slope in degrees Celsius per decade, then convert this to degrees Fahrenheit per decade using $F = 9/5C + 32$ to confirm the value in the press release of +0.36F/decade. As an extension, they can extrapolate this rate to the year 2050.

SpaceMath@NASA also develops problems outside the NASA press release pipeline when the particular story has considerable public interest. The British Petroleum Gulf oil spill of 2010 led to several problems for estimating the amount of oil generated by using the published live video stream from the oil well. These timely problems were extensively used by teachers to show how simple math can be used to make important estimates several weeks in advance of the official government-sanctioned numbers. In the aftermath of the Japan Tsunami of 2011, a number of problems were developed related to radiation issues including dosimetry, half-life, and the hazards of interplanetary travel.

Although individual problems are published to the *SpaceMath@NASA* website through out the academic year (September to June), individual special-topic books are also designed concurrently. These books include end-of-the-year compilations of posted problems, as well as an increasing variety of special-topic books on topics as diverse as black holes, remote sensing, and astrobiology. In many instances, these books have been requested by specific NASA education programs.

For example, the math resource book *Astrobiology Math* shown in Figure 24.4, was requested by the NASA Astrobiology Institute and will now be reprinted by them to serve the needs of their teacher workshop participants and interested students. *Transit Math* was requested by the Sun-Earth Day program to support their Transit of Venus theme in 2012. *Space Weather Math* was requested by the Challenger Learning Center to support their teachers and students in their continuing simulation of space operations experiences with technical accuracy in space weather forecasting.

Video

Recognizing that students are immersed in a multimedia world, in 2010, *SpaceMath@NASA* partnered with *NASA eClips* to create a series of video programs supplemented by math content. The *eClips* program has nearly 200 short-segment video segments lasting three to four minutes, and available online (http://spacemath.gsfc.nasa.gov/media.html) covering all aspects of NASA science and technology development and innovation. Thirty episodes have been featured in a series of modules that identify math extensions to the video content in each program. Students in grades 6, 7 and 8 are exposed to a topic presented with conventional fast-paced and visually interesting content, and are

Mathematics behind the science

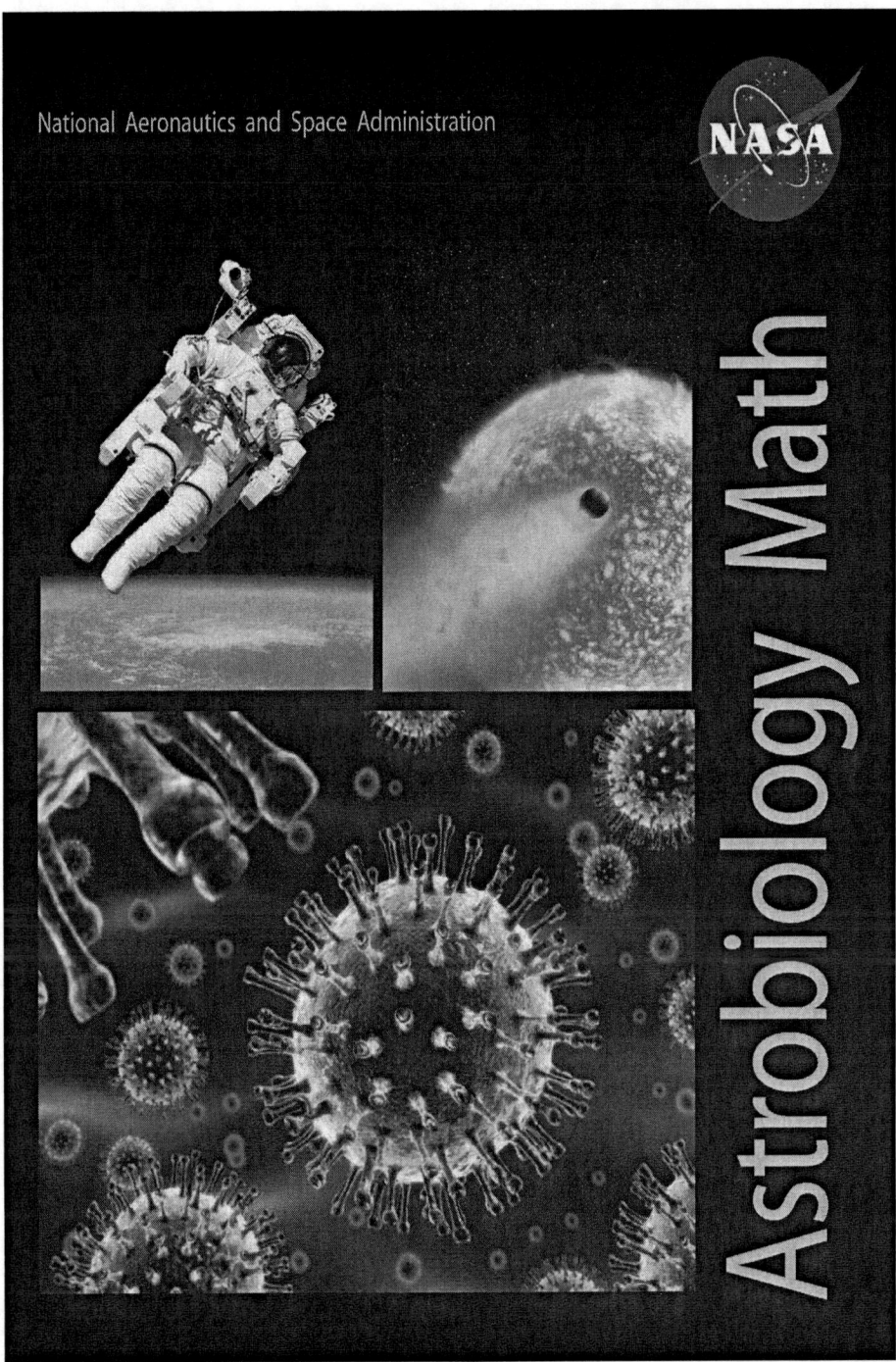

Figure 24.4 The cover of the problem book Astrobiology Math published in 2011. This book was requested by the NASA Astrobiology Institute and contains 50 problems covering the search for Earth-like planets, basic biology and chemistry, and Drake's Equation.

S. Odenwald

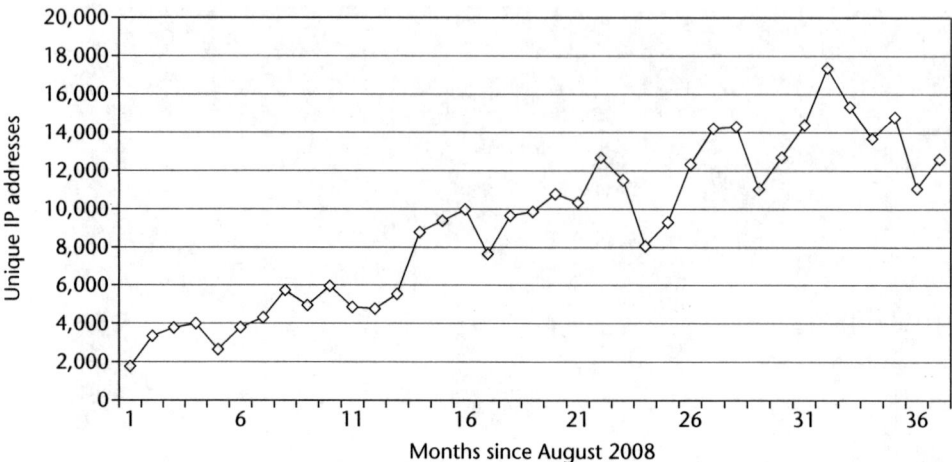

Figure 24.5 Graph of monthly changes in unique IP addresses between August 2008 and August 2011. The periodic dips between July and September are correlated with summer vacation when formal educators and students are predictably less active users.

introduced to math problems that build upon one or two aspects of the program content. For example, the program *Kepler: Earth-like Worlds* describes the excitement of the search for Earth-like planets, while the associated problems cover the statistics and elementary physics behind the exoplanet discoveries.

Impact

The popularity of *SpaceMath@NASA* has grown dramatically since 2008 as the figures and table, below, illustrate. We employ two-monthly statistics to gauge the activity of the *SpaceMath@NASA*

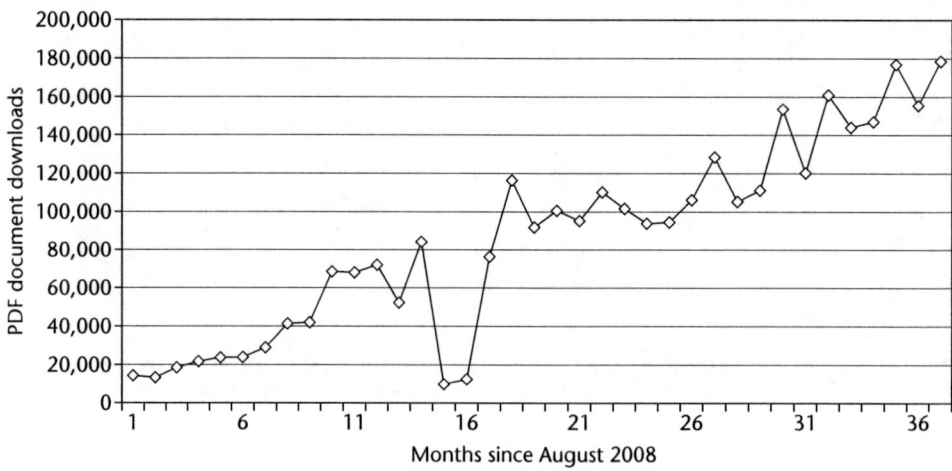

Figure 24.6 Graph of the number of monthly problem downloads between August 2008 and August 2011.

website: the number of unique Internet Provider addresses (called IP addresses) and the number of PDF documents downloaded. Unique IPs, presented in Figure 24.5, is a measure of distinct identifiable visitors using separate computers, each with their own web address and is very helpful in assessing how large the population of visitors is. However, different people using the same computer are counted as one visitor, which underestimates the number of students visiting the website if only one or two computers are available in the classroom. The number of PDF files served is a direct measure of the activity of the website because the vast majority of the website consists of 434+ PDF files and over 30 PDF-formatted books. Figure 24.6 shows the number of monthly PDF files downloaded since 2008.

The trends clearly show that the number of users and downloaded documents increases by about 50% each year. In terms of cumulative problem requests since 2004, in February 2011, the three-millionth PDF problem file was downloaded from this website. This was followed in September, 2011 by the four-millionth.

We can also gauge visitor interest by assembling a list of the 'Top-20' items in each category. Table 24.1 lists the most popular individual problems in terms of the cumulative number of downloads between 2008 and 2011. The total number of downloads represented by these top-ranked problems is 215,619. This is about 9% of all one-page, individual problem PDF files downloaded from at *SpaceMath@NASA* during the same time period.

Table 24.2 lists the most popular book-length files ranked by their cumulative downloads between 2008 and 2011. The total number of downloads represented by these top-ranked books is 467,494. This is about 84% of all book-length PDF files offered at *SpaceMath@NASA*. Since 2010, book-length resources have overtaken and surpassed the downloads of the top individual problems (compare Tables 24.1 and 24.2).

Table 24.1 Top-20 ranked individual math problems according to title

Rank	Downloads	Title of problem
1	17,795	Unit conversion exercises – I
2	14,973	The relative sizes of the sun and stars
3	12,421	How fast does the sun spin?
4	11,453	When is a planet not a planet?
5	11,367	Space Shuttle launch trajectory
6	11,001	Scientific notation – an astronomical perspective.
7	10,807	Time zone mathematics.
8	10,782	Unit conversion exercises – II
9	10,780	Exploring distant galaxies
10	10,478	Radon gas in the basement
11	10,266	Time zone math
12	10,225	Scientific notation – I
13	9,826	The International Space Station: follow that graph!-
14	9,721	The solar tsunami!
15	9,490	CME kinetic energy and mass
16	9,435	Solar eclipses and satellite power
17	9,358	Angular size and velocity
18	9,276	Are the Van Allen belts really deadly?
19	8,405	An introduction to space radiation
20	7,790	Super-fast solar flares!!

Table 24.2 Top-20 ranked math resource books according to title

Rank	Downloads	Book title
1	116,964	Space Math Volume 1
2	60,574	Exploring Space Mathematics
3	44,124	Space Math Volume 5
4	30,258	Space Math Volume 2
5	28,335	Space Math Volume 3
6	23,420	Magnetic Math
7	18,753	Space Math Volume 4
8	17,970	Earth Math
9	17,728	Space Math Volume 6
10	14,756	Electromagnetic Math
11	12,289	Northern Lights
12	11,684	Algebra 2
13	10,223	Space Weather Math
14	9,520	The IMAGE Satellite – Introduction
15	9,323	Solar Math Volume 1
16	9,319	Black Hole Math
17	9,037	The IMAGE Satellite – Data
18	8,510	Solar Math Volume 2
19	7,478	Lunar Math
20	7,229	Tracking a Solar Storm

The rankings and 'shares' occupied by the problems and books indicate that visitors have a relatively narrow interest in math skills from unit conversion and scientific notation exercises, to problems involving distance, time, and speed. However, the most popular problems only account for 5% of all problems offered, so taken as a whole, visitor interests really do span the entire gamut of math abilities and topic areas from basic numeracy and fractions to advanced calculus. In terms of book-length products, it is clear that the annual collections of the previous years' individual problems are very popular. This reflects the fact that it is more convenient and efficient to download the past years' problems in one batch (about ten megabytes), rather than the 50–80 individual downloads. Moreover, the problem books all contain extensive explanatory text that provides additional science content information not found in the individual problems. It is of particular interest that, although the *Algebra 2* book was only available in 2011, its cumulative downloads and ranking already exceeds book products such as *Black Hole Math*, which have been available since 2008. Clearly the creation of this high-school-level math guide with over 200 space and astronomy application problems was a significant and ultimately wise decision, and one that will be replicated in an upcoming pre-algebra product for grades 6–8.

Assessments

Since 2008, the Technology For Learning Consortium has conducted teacher and student evaluations using online surveys and written/email surveys targeted toward workshop participants. Multiple methods of data collection were utilized to assist in the evaluation of the *SpaceMath@NASA* project during the period from 2008–2011. The *SpaceMath@NASA* website web statistics were analyzed for usage, teachers utilizing the math books were surveyed, students whose teachers were utilizing the *SpaceMath@NASA* problems in the classroom were surveyed, and conference attendees were surveyed and interviewed.

The majority of educators who download application problems from the Internet are getting the ones they utilize directly from the *SpaceMath@NASA* website. The website has also improved users' impression of NASA's support of STEM education in a significant way. Consistently, users praise *SpaceMath@NASA* as a valuable and accessible resource.

We learned that teachers plan to use *SpaceMath@NASA* problems to introduce a math concept or science concept, for group problem solving in class, and as curriculum extensions. They report that the problems they have looked at so far will improve the math skills featured in the problem, help them be more effective in teaching STEM topics, align well with what they teach, increase their students' interest in STEM, and encourage student exploration, discussion, and participation.

Students reported they would like to do more problems of this kind, that it was easier to do the math problems because of the science content, and that they felt inspired to learn more about the math. As one student shared, 'Definitely worth having and completing for a greater understanding of space science and mathematics'.

Most of the 300 educators who responded to the Annual Surveys since 2008 reported they are science teachers who use the problems to emphasize science content. About one-fifth of the educators described themselves as math teachers and an additional one-fifth of the respondents indicated that they were informal educators. More than half of the teachers reported using the problems a few times a month, and that their students are productively challenged by *SpaceMath@NASA* problems. In particular, the educators noted in the surveys that their students enjoy the application problems and the topics they presented. The students were also asking questions about the problems that demonstrated curiosity and interest. There was also considerable anecdotal evidence in the surveys that students improved their academic performance as a result of working these problems, and were actually looking forward to exploring new problems! Perhaps just as importantly for the longevity and impact of *SpaceMath@NASA* is that all of the surveyed teachers reported sharing the problems with at least one other educator during the year.

Conclusions

Amidst the hubris of modern education practice and theory, it is sometimes helpful to consider the simple principles that stimulate learning. As an astronomer who had matriculated through the conventional K-12 system in the United States, I recognized the sources of stimulation to my own learning process. Through *SpaceMath@NASA*, we have created a resource that brings students into contact with thrilling, and timely, instances of genuine scientific discovery. These moments, captured in NASA press releases, are then used to explore relevant mathematical topics. The use of Newspapers in Education has scored dramatic successes in other educational areas. We are hopeful that a similar approach using NASA press releases will ultimately foster the same kinds of successes in science and mathematics education.

Acknowledgments

This chapter was prepared through a NASA Science Mission Directorate ROSES/EPOESS Grant NNH10CC53C-EPO.

References

Abbott, W., & Nantz, K. (1994). History and economics: can students (and Professors) learn together? *College Teaching, 42*(1). Downloaded from www.jstor.org/pss/27558632 on September 25, 2011.

Ainley, M., & Ainley, J. (2011). A cultural perspective on the structure of student interest in science. *International Journal of Science Education, 33*(1), 51–71.

Nola Kortner, A. (1988). Using newspapers as effective teaching tools. Downloaded from www.eric.ed.gov/PDFS/ED300847.pdf on September 25, 2011.

Bandura, A. (1977). *Social Learning Theory*. New York: General learning press. Downloaded from http://psychology.about.com/od/developmentalpsychology/a/sociallearning.htm on September 25, 2011.

Bruder, P. (2006). Current Events: Newspapers in the classroom. *New Jersey Educational Review*, April 3. Downloaded from www.njea.org/teaching-and-learning/classroom-tools/toolbox/current-events-newspapers-in-the-classroom on September 25, 2011.

Cherry, K. (2011). An overview of Bandura's social learning theory. Downloaded from http://psychology.about.com/od/developmentalpsychology/a/sociallearning.htm on September 25, 2011.

DeRoche, E. F. (1991). Student interest in national news and its relation to school courses. Downloaded from http://nces.ed.gov/pubsearch/pubsinfo.asp?pubid=97970 on September 25, 2011.

Fensham, P. (2006). Student interest in science: The problem, possible solutions and constraints. Downloaded from www.acer.edu.au/documents/RC2006_Fensham.pdf on September 20, 2011.

Hazelwood, C. (2005). Science education in the era of no child left behind. Downloaded from www.ncrel.org/sdrs/areas/issues/content/cntareas/science/sc600.pdf on September 25, 2011.

Lord, M. (2006). No child left behind and science education. *Education Reform*. Education writers association. Downloaded from www.ewa.org/docs/scienceeducation2006.pdf on September 25, 2011.

National commission on mathematics and science teaching. (2001). Before it's too late. Downloaded from www2.ed.gov/inits/Math/glenn/toolate-execsum.html on September 25, 2011.

Newspapers in Education. (2011). Suggested math problems. Downloaded from www.press-enterprise.com/nie/math.html on September 25, 2011.

Odenwald, S. F. (2011). Space for math in the classroom: Searching for exoplanets. *Classroom Astronomer*, Fall. Downloaded from http://ClassroomAstronomer.ToTeachTheStars.Net on September 25, 2011.

Odenwald, S. F. (2012). NASA Press Releases: Exploring the math behind the news. *Astronomy Beat*, The astronomical society of the pacific (in press). Available at www.astrosociety.org/2012/.

Rhoades, L., & Rhoades, G. (1985). Using the daily newspaper to teach cognitive and affective skills. *Clearing House, 59*(4), 162–164.

Tugel, J. (2004). Time for science. *Alliance Access, 8*(2), 1–3.

Teaching with Contests. (2010). Downloaded from http://teachingwithcontests.com/?p=2166 on September 10, 2011.

Vockell, E. L., & Cusick, K. (1995). Teachers' attitudes toward using newspapers in the classroom. *Clearing House, 68*(6), 359–362. Downloaded from www.jstor.org/stable/30189108 on September 25, 2011.

Part VIII
Innovations in science education for developing innovators

25
Innovation in science, technology, engineering, and mathematics (STEM) disciplines
Implications for educational practices

David F. Feldon, Melissa D. Hurst, Christopher A. Rates, and Jennifer Elliott

UNIVERSITY OF VIRGINIA, USA AND UNIVERSITY OF CINCINNATI, USA

Summary: Innovation is fundamental to the execution of science and engineering. Scientists endeavor to generate new knowledge about the universe, and engineers apply that knowledge to generate effective solutions for new problems. Typically, the training of these individuals relies heavily on apprenticeship and trial-and-error experiences. However, more formalized and systematic approaches to science education at all levels of schooling can prepare students to enter their respective fields as innovators. This chapter reviews research from education, psychology, and the sociology of science that can inform educational practice. The manner in which students understand the nature and traditional boundaries of their disciplines, the strengths and limitations of their disciplinary tools, and the constraints imposed by specific problems can substantially influence the strategies they develop to solve novel problems.

Key words: Innovation, inquiry, knowledge transfer, problem-solving, science education, scientific creativity, STEM education.

Introduction

Innovation in science, technology, engineering, and mathematics (STEM) is considered a paramount driver of the modern economy (Carnevale, Smith, & Strohl, 2010). Consequently, in the United States and other countries, there exists an intense focus on the development of a highly skilled and creative scientific workforce to enhance national competitiveness in these areas (Augustine, 2005; Gereffi, Wadhwa, Rissing, & Ong, 2008). In the STEM context, innovation frequently manifests as one of several types:

1   A novel strategy is applied to successfully solve a previously intractable problem.
2   Knowledge—strategic, procedural, conceptual, or factual—from one field or discipline is

effectively applied to an existing, previously intractable problem in a different field or discipline.
3   A new conceptualization or framing of an existing, intractable problem provides better progress toward its resolution.
4   A new problem is identified whose resolution will enable significant theoretical advance.

Developing a workforce to perform such tasks requires a refocusing of educational practices to prepare students who not only understand scientific concepts, but also able to engage in STEM problem-solving. However, in the status quo, students receive relatively little instruction conducive to authentic scientific inquiry (Metz, 2008; Windschitl, 2004, 2009) or engineering practice (Brophy, Klein, Portsmore, & Rogers, 2008). Even at the doctoral level, many graduate students approach their dissertations feeling unprepared to conduct independent research (Golde & Dore, 2001; Lovitts, 2008).

This chapter examines the structure of the STEM disciplines in relation to research on creativity and problem-solving. Following these sections, it reviews the major modes of reasoning used by scientists and engineers as they innovate in response to novel problems. It concludes by reviewing some of the best validated educational practices to support students as they develop their disciplinary skills in a way that can facilitate their transfer and adaptation to new challenges.

Relevant distinctions between STEM disciplines

A discipline can be defined as a type of intellectual endeavor for which a group of scholars share a set of research questions, methods of inquiry, and an intellectual approach to solving problems (e.g., Chubin, 1983; Kuhn, 1962; Price, 1965). Disciplinary perspectives may differ regarding the use and value of logical, observational, or experimental evidence or the privileging of inductive (data-driven) or deductive (theory-driven) modes of reasoning (Bauer, 1992). Through these collective views, disciplines represent distinct "epistemic cultures" (Knorr-Cetina, 1997, p. 260) with their own normative practices, terminologies, conceptualizations, and essential skills.

Science uses empirical evidence and logical inference to identify generalizable principles that explain the mechanisms of natural phenomena. In this context, quality is judged by replicability (i.e., repeated demonstrations that a theory predicts empirical findings accurately) and parsimony (i.e., the ability of an empirically supported theory to explain the widest possible range of instances in the simplest way) (Blackburn, 1999).

Engineering also uses empirical evidence and logical inference, but to accomplish different goals: the development of targeted solutions for specific problems that are optimized in relation to a set of value-laden constraints (e.g., tradeoffs between reliability, cost, and speed) (Koen, 1985). Thus, generalized inferences about the operation of a system under all possible conditions are devalued in engineering contexts compared to highly specific inferences regarding the performance of a system under narrow—or even unique—conditions. For example, scientists might evaluate the reliability of a controllable event to determine the explanatory sufficiency of a certain theory, but engineers might evaluate the event's reliability to determine if the failure rate of the control mechanism was within acceptable parameters for application within a specific context. The outcome of the investigation would be a necessary but insufficient step toward accomplishing a scientific goal of developing or validating a theoretical account of causal mechanisms, but it could constitute an end point for an engineering goal of performance within specified constraints.

Scientific creativity

Creativity within a domain leads to productive and innovative solutions only when foundational knowledge is necessary but insufficient to yield progress (Csikszentmihalyi, 1996; Stokes, 2006). Scientists and engineers leverage their knowledge to identify the limits of existing knowledge, engage with problems at the boundaries of current knowledge, and critically evaluate the sufficiency of new solutions. A problem's constraints must preclude existing strategies from being effective for the context in which it is relevant to represent innovation. Thus, the most challenging problems in STEM fields are those that are fundamentally ill-structured with goals that are complex and indefinite (Metz, 2008). They "possess multiple solutions, solution paths, fewer parameters which are less manipulable, and contain uncertainty about which concepts, rules, and principles are necessary for the solution or how they are organized and which solution is best" (Jonassen, 1997, p. 65). However, not all important scientific achievements entail innovative problem-solving. In many cases, existing knowledge must only be applied to answer questions or meet needs that had not yet been raised.

For example, original developments of genome mapping techniques were highly innovative applications of science and engineering to explore the genetic composition of human beings in the 1970s (Maxam & Gilbert, 1977). The Human Genome Project began in 1990 and took 13 years to complete its mapping of ~23,000 human genes by systematically applying these methods (Human Genome Project Information, 2011). Early in the project, its means of isolating genes were not considered innovative, though it continued to generate new data vital to further research. Thus, the problem of how to map the human genome shifted from ill-structured at its outset (with unresolved issues about its methods and anticipated outcomes) to well-defined (with a fully predictable process yielding novel, but unsurprising results). The successful mapping of three billion chemical base pairs that compose human DNA was a major scientific accomplishment of great value for current scientific efforts, but its attainment was not the direct outcome of innovative practice within the project. Indeed, one of the originators of the project described it as "boring and repetitive" (DeLisi, 2008, p. 877).

Generating innovations

The process of generating creative solutions to problems is often described as a stochastic blind-variation-and-selective-retention process, in which various pieces of new and old information are randomly combined in working memory and conceptually tested against the needs of the problem being considered (Campbell, 1960; Simonton, 1999; Sweller, 2009; Sweller & Sweller, 2006). While this process is considered random, various factors can preemptively limit or expand the range of possible combinations. Variability in the ideas generated is related to an individual's knowledge base within the relevant domain (i.e., old information; Stokes, 2006). This knowledge base precludes the generation of new ideas that replicate known prior solutions or are not feasible based on knowledge of natural laws or task constraints. However, it also increases the number of information elements available to combine and the working memory capacity necessary to combine more of them in more complex ways (e.g., Sweller, 1988).

Simonton (2004) explores this core idea through his examination of four principal hypotheses about the genesis of scientific creativity: logic, genius, chance, and zeitgeist. By logic, he refers to the knowledge and skills that can be systematically acquired through training and study. This set of skills can be systematically applied to generate scientific progress. The genius hypothesis posits that certain individuals are possessed of greater innate ability, which allows them to solve problems through insight that elude most others. Chance holds that innovations occur at random. The zeitgeist perspective indicates that various cultural and societal pressures increase the likelihood of certain innovations occurring due to disseminated advances in knowledge and reasoning frameworks that become prevalent at a given point in history, as well as the universality of natural laws (Kantorovich, 1993). Using bibliographic data of scientific achievements, he argues that while each of these components can play a role in fostering innovation, chance is the driving factor that interacts with the others to best account for the patterns of creativity in the STEM fields.

For example, a strict logic hypothesis would predict that the best trained and most knowledgeable individuals would be the ones to come up with the most frequent innovations. However, no available indicators of those qualities predict differences in the number of publications reporting independent creative accomplishments, their impacts on others' work (i.e., citation rates), or the rate of innovative productivity. Similarly, although there are reasonable levels of variability in individuals' measured intelligence and other traits, they do not correlate significantly to measures of productive scholarly innovation (Ericsson & Lehmann, 1996). Certain individuals are consistently more productive and influential than others, but the impacts of their contributions are typically limited to a single field and do not deviate from established temporal patterns of innovation over the course of a career (Simonton, 1991). The zeitgeist hypothesis describes important contextual factors, but it does not preclude the possibility that the random combination of knowledge and observational elements within individuals could lead to multiple independent discoveries of the same solution. Indeed, as the levels of prior knowledge and specificity of task constraints increase, the chance of this occurring would increase as various scientists or engineers each investigated a diminishing combinatorial space (Simonton, 1987).

While the chance perspective fits with a number of observed patterns, it must be kept in mind that such patterns are only found among those who participate in scholarly publication. This group constitutes a very narrow subset of the general population due to its members' very high levels of training, so relevant achievements are not truly randomly distributed outside the group. However, the contributions to each of the other factors establishes a stronger foundation in which, as stated by Louis Pasteur (1854), "chance favors the prepared mind" with scholarly expertise, intellect, and socio-intellectual factors constituting that preparation.

Problem-solving as solution search in science and engineering

From a cognitive perspective, the search for answers to problems for which the correct solution or solution strategy is unknown is often referred to as the search of a problem space. The space is defined by the initial state and the goal state as the starting and ending points, and it is composed of all possible routes to move from the former to the latter (Newell & Simon, 1972). The number of possible routes is constrained by the knowledge of the individual attempting to solve the problem; strategies known to be ineffective or inapplicable within the constraints of the problem and/or disciplinary lens are disregarded preemptively.

In problems involving scientific reasoning, Klahr and Dunbar (1988) argue that two adjacent, mutually influential problem spaces must be searched—the *theoretical* space (what is the correct explanation or principle?) and the *methodological* space (what are the effective means of investigating its validity?).[1] To resolve the theoretical problem, scientists consider what is known about a given problem and what the most likely explanation or governing principle might be. Selection of this conjecture (hypothesis) narrows the possible range of investigative approaches (methodology), because it specifies what type of events must be observed to inform the veracity of the conjecture. Similarly, selection of an investigative approach inherently constrains the possible theoretical conclusions that can be drawn, because choices have been made about which aspects of the phenomenon will and will not be examined. In other words, what is known about a problem informs "where" solutions are sought, where one looks determines what one finds, and what is found in turn informs the conclusions drawn about both solutions and the constraints of the problem space. Thus, each navigation of one problem space results in incremental progress in determining an effective path through the other.

Given the vast number of possible steps within each problem space and the exponential increase in the size of the search space for each additional step, Klahr and Simon (2001) note that "much of the training in scientists is aimed at increasing the degree of well-definedness of problems in their domain" (p. 76). For example, the experimental design strategies used by expert scientists to answer posed research questions are optimized when they are working within their own area of specialty, because their specialized knowledge of prior research findings and common methods permit a more focused problem space search (Schraagen, 1993). The ability to narrow the search of problem spaces by leveraging the constraints of prior knowledge is essential for developing refined approaches to previously unsolved or novel problems as reflected by Sir Isaac Newton's oft-cited quote "If I have seen a little further, it is by standing on the shoulders of giants."

Engineers similarly navigate conjoined problem spaces when engaging in design thinking. However, instead of a theoretical problem space and a methodological problem space, they navigate a *framing* problem space and a *solution* problem space (Cross, 2004). In the framing space, the nature of the problem to be solved is determined in relation to the constraints of the situation presented, because a given problem might be conceptualized in multiple ways (e.g., to make two pieces fit, is the problem that one is too small or that the other is too large?). In the solution space, the potential mechanisms by which a problem as determined in the framing space might be solved are explored. Cross and Clayburn Cross (1998) observe that the most innovative design experts engage in the most thorough searches of the framing space and resist the tendency to terminate their search with previously established solutions. However, this extended framing search is facilitated by extensive prior knowledge of solution techniques, which constrains the concomitant search of the solution space to a manageable scope.

Common modes of reasoning in STEM disciplines

In both science and engineering, problem space search is facilitated by two prevalent reasoning strategies—logical deduction and analogical reasoning (Dunbar, 1997; Robinson, 1998). Logical deduction—also referred to as *strong methods* (Singley & Anderson, 1989) or *forward reasoning* (Chi, Feltovich, & Glaser, 1981)—leverages existing knowledge of relevant findings, principles, and theories to draw inferences about the solution to a problem or the necessary strategies to attain it. For example, graduate students in physics solve presented problems by rapidly identifying the relevant information available to answer a question and manipulating that information to yield the solution using known principles and strategies (Larkin, McDermott, Simon, & Simon, 1980a, 1980b). In contrast, undergraduates in introductory physics courses utilize inductive *weak method* heuristics that begin with determining what the appropriate type of solution would be and reason backwards to identify relevant given information and approaches that will generate the necessary outcome (Lovett & Anderson, 1996). Further, the development of expertise entails a progression from general heuristics to feedback-refined procedures that have integrated domain-specific knowledge (Anderson, 1987).

Innovation, however, entails the discovery of new strategies and solutions. When existing knowledge frameworks are not sufficient to solve a problem, scientists and engineers reason through analogy (Clement, 1988; Robinson, 1998). When the solution to a problem is not known, experts may consider previous experiences or parallel scenarios that relate directly to the issue at hand (e.g., material X is a member of a broader class of materials that generally possess certain properties, thus may be that it will behave similarly to another material from the same class). In Hmelo-Silver, Nagarajan, and Day (2002), participants with varying levels of expertise in clinical trial design used a simulation to demonstrate a hypothetical drug's suitability for medical use. Throughout multiple attempts, those participants with extensive prior experience (unlike novices) consistently used analogies to past experiences in their verbal protocols to reason abstractly about the process and the predicted outcomes.

More distant analogies may also be used where the underlying structure of one scenario can be linked to the structure of another from a different domain. For example, Rutherford conceptualized the structure of the atom as mirroring that of the solar system with electrons orbiting the nucleus like planets orbiting the sun (Dunbar, 1997). Although that particular model has not been borne out as a valid representation of atomic structure, the solar system analogy facilitated his extended reasoning about atomic interactions. Multiple studies of scientists thinking aloud while engaged in authentic scientific reasoning tasks find that the use of these more abstract analogies occur with significantly lower frequency than those that utilize other information from the same domain or discipline. At their most productive, these analogies are thoroughly tested against existing knowledge that has been empirically verified to determine their effectiveness for reformulating a problem (e.g., Clement, 1988; Dunbar, 2000; Trickett & Trafton, 2007).

Knowledge transfer for solving new problems

An important aspect of innovation in learning and educational practice is the concept of transfer. Transfer is defined as the application of acquired knowledge and skills to novel situations and problems (Barnett & Ceci, 2002; Perkins & Saloman, 1994). The extent of the departure from previously encountered situations and problems is generally characterized in terms of distance. New solutions or contexts with relatively small departures from an individual's previous experience employ *near transfer*, and those with substantial differences employ *far transfer* (Perkins & Saloman, 1994). For example, predicting the products of a chemical reaction which a chemist has not previously encountered would employ a great deal of previously acquired knowledge about other, similar reactions. Thus, making the prediction would involve only near transfer, as the problem is merely a new instantiation of a well-understood class of problems. In contrast, attempting to enhance the stability of a newly created compound whose properties are not yet fully understood would entail a greater transfer distance, because general principles could be applied from previously acquired knowledge, but specific strategies might not be applicable.

Although there are differing viewpoints regarding the criteria by which transfer is assessed, there are three general mechanisms through which transfer occurs: leveraging a thorough foundation of conceptual knowledge, analogical problem-solving, and the use of metacognitive strategies (Barnett & Ceci, 2002). Having extensive knowledge of a problem domain entails an understanding of the underlying (deep) structure of problems, which reflects the mechanisms and principles governing the interactions of problem elements. It also limits distraction by superficial features of problems that can hinder the recognition and application of successful solution strategies in new situations (Gick & Holyoak, 1980). Such maladaptive, or *negative transfer*, can occur when surface features are similar but deep structure differs. It can also occur when surface and deep structural features are similar enough to a previously solved problem that the learner deploys a suboptimal strategy rather than investing the effort to develop a newer, more efficient one (Halpern, Hansen, & Riefer, 1990).

When a thorough knowledge base is invoked, it permits the formation of appropriate and productive analogies between the novel problem and other situations from which viable solutions might be derived. However, the probability of identifying and successfully using an effective analogy is dependent on several features. As discussed above, an analogue must map effectively on to the deep structure of the new problem (Wharton et al., 1994). However, an effective mapping does not necessarily entail a perfect match. If the analogue is reasonably adaptable in its features, aspects can be modified to permit its application (Genter, Rattermann, & Forbus, 1993). In addition, sufficient working memory capacity, which is facilitated by extensive and well-organized knowledge (van Merriënboer, Kester, & Paas, 2006), must be available to permit the mapping and adaptation to occur (Keane, Ledgeway, & Duff, 1994).

Once an analogy is engaged, the efficacy of the strategies it yields must then be carefully monitored, assessed, and adapted if necessary to yield an ultimately effective solution (Brown, Kane, & Long, 1989; Dunbar & Fugelsang, 2005; Keane, 1996). The ability to exercise such metacognitive skills can be trained. When explicit training for strategies such as planning and monitoring incremental progress toward a goal, problem framing and assessing strategy effectiveness is provided, learners become more successful at solving novel problems (Halpern, 1998; Masui & De Corte, 1999; Sternberg, Ferrari, Clinkenbeard, & Grigorenko, 1996).

Instructional strategies to enhance capacity for innovation in STEM

Within the frameworks discussed above, there are a number of validated educational strategies that can be applied to foster the development of innovators in STEM disciplines. Collectively, they emphasize the need for students to engage in authentic problem-solving activities within science and engineering that jointly integrate relevant content and opportunities to apply it to conduct inquiry. Further, such experiences must occur frequently and engage with problems that vary in their focal area, emphasis, and constraints. They must also include explicit feedback on the effectiveness of the strategies employed by students and ongoing guidance to develop metacognitive skills.

Authentic problem-solving and inquiry

Science and engineering entail deeply entwined facets of both process and content knowledge (Klahr & Dunbar, 1988; Koen, 1985; Robinson, 1998). However, primary and secondary school science education has traditionally split apart content from process when teaching science (Metz, 2008; Windschitl, 2003). Often students memorize facts and are separately asked to do the process of science in a generic way or follow scripted procedural steps intended to yield a predetermined outcome. This is not authentic inquiry, and precludes the opportunity for scientific creativity, because it is "divorced from its disciplinary context" (Simonton, 2000, p. 155). Even at the undergraduate level, "cookbook labs" that require students to follow prescribed steps without engaging in scientific reasoning in order to replicate a known outcome are common (Buck, Bretz, & Towns, 2008). However, classroom-based inquiry and design that are situated within meaningful discussions of content can support the development both of knowledge of core science content and of the development of disciplinary reasoning (Minner, Levy, & Century, 2010; Kolodner, Gray, & Fasse, 2003; Raghavan, Sartoris, & Glaser, 1998). Without this interplay, students are left with "scientific knowledge without any hints about how it was generated or the nature of its tentative status" (Simonton, 2000, p. 141).

Further, there exists robust evidence that the direct teaching of strategies for inquiry is effective preparation for engaging in authentic disciplinary problem solving. For example, central features of much scientific inquiry include experimental design and the modeling of data for analysis. In the former, sources of variance are carefully controlled to permit valid inferences about the potential causes of phenomena. When direct instruction of strategies for the control of variables is presented to elementary school students along with probing questions about their reasoning, they are better able to design valid experiments than peers who only responded to the questions (Chen & Klahr, 1999). Further, out of four attempts, 77% of the students receiving explicit instruction on these skills were able to design at least three valid experiments, in contrast to only 23% of children who did not receive such explanations (Klahr & Nigam, 2004). In a follow-up task one week later, all of the participants who scored at least three out of four were assessed on their ability to critique presented experimental designs that were flawed in various ways. The two groups of successful students did not differ significantly in their ability to perform this transfer task, indicating that students who received direct instruction did not lose anything by not inferring effective strategies independently in place of receiving explicit procedural guidance.

Modeling skills are perhaps more challenging to acquire and necessitate sustained engagement with the practice of authentic inquiry in science and engineering (Lehrer & Schauble, 2006). In the process of posing questions for inquiry, students can be guided to think about the essential attributes or variables related to a problem and then how to construct meaningful measures and indicators to quantify them. Once data are collected, the conceptual foundation driving the measurement process can similarly support the structuring, representation, and analysis of the information to yield inferences which are both disciplinarily appropriate and meaningful to students as young as elementary school (Lehrer, Kim, & Schauble, 2007; Lehrer & Schauble, 2004).

Research on the development of adaptive expertise further suggests that instruction in strategic procedural knowledge is also effective for supporting transfer. More innovative practitioners acquire and apply both procedural and conceptual knowledge differently from those who are less adaptable (e.g., Besnard & Bastien-Toniazzo, 1999; Perkins & Grotzer, 1997). Frequently, the objection that is raised to teaching procedural knowledge for the purpose of preparing learners to perform adaptively is that it will lead to inappropriate rigidity that can hinder creativity (e.g., Ericsson, 1998, 2004). However studies of learning and transfer demonstrate that limited transfer of procedures to novel cues and circumstances does occur (for an extensive treatment of this issue, see Feldon [2007]).

Learning in this way not only promotes motivation and scientific creativity but even children as young as first grade are able to select and generate conclusive tests of simple hypotheses (Sodian, Zaitchik, & Carey, 1991). By providing active guidance to students while they engage in inquiry, teachers can constrain the scope of the problem spaces students need to navigate and allow them to learn to function competently within it (De Corte, 2003). Whereas open inquiry appears to allow students more control it disregards their limited working memory capacity, lack of schemas to place learned knowledge, and their inefficient heuristics (Kirschner, Sweller, & Clark, 2006). As students become more familiar with the tools and concepts of a discipline, the constraints on the problem space can be strategically changed to promote variability in both the types of problems solved and the range of solutions students have the opportunity to generate (Stokes, 2006). Rewarding novel solutions while maintaining rigor related to the application of disciplinary reasoning can both increase originality (Stokes, 2001) and facilitate learning (Stokes, Lai, Holtz, Rigsbee, & Cherrick, 2008).

Varied practice and contexts

Engaging in problem-solving with high levels of variability is necessary to evoke and foster innovative responses. However, variability also needs to be carefully balanced. Too much variability can be frustrating, because a student is unable to leverage knowledge gained from a previous experience to refine his or her solution search for a new problem. Likewise, too little variability can lead to boredom (Stokes, 1999, 2006).

One way of promoting growth in scientific creativity is by gradually increasing how ill-structured a problem is as a student gains scientific competence (Bell, Lederman, & Abd-El-Khalick, 2000; Bell, Smetana, & Binns, 2005). For example, an inquiry activity should contain a scientific question which is answered by analyzing results. Complexity in inquiry can be increased when the solution is unknown (Structured Inquiry), when the methods to find results and the solution are up to the student (Guided Inquiry), and finally when the question, method, and answer are unknown (Open Inquiry) (Bell et al., 2005). These constraints give students control and constrain their problem space as they learn.

Further, as students develop new procedures and strategies, it is important that they identify appropriate cues from problems' deep structures (e.g., Brown et al., 1989; Larkin et al., 1980b). Explicit instructional guidance that models the identification of the salient cues across a variety of appropriate problems will aid subsequent transfer attempts (Clark & Blake, 1997). Instruction should include structural aspects of problems with an emphasis on conceptual cues which students can use to guide retrieval in future situations (Halpern, 1998). The more learners practice in different context and use novelty in their practice exercises, the more successful they are likely to be in far transfer.

Specific feedback and reflection

Optimal feedback to support innovation entails guidance to learners that is both timely (Kluger & DeNisi, 1996) and elaborative in nature (Moreno, Reisslein, & Ozogul, 2009). For complex, challenging tasks that characterize the types of innovative STEM problem-solving discussed in this chapter, feedback provided immediately after performance is more beneficial than when it is delayed (Clariana, 1999; Mathan & Koedinger, 2002). However, the feedback offered consists of only simple information (e.g., a right/wrong evaluation or a presentation of a correct or effective response), it typically induces shallow learning, which emphasizes memorization and shallow assessment of strategies (Hattie & Timperley, 2007). In contrast, responses that require learners to engage thoughtfully and reflectively with their knowledge and reorganize it can be highly beneficial (Shute, 2008). Providing full explanations of the benefits and shortcomings to a problem's solution or solution strategy (i.e., why it was effective or ineffective) can enhance learners' abilities to successfully solve transfer tasks. For example, when users of a multimedia instructional program in botany received such information while solving the problems presented significantly outperformed their counterparts with less elaborate feedback on subsequent tasks asking them to design plants for specific climates or deduce the climate in which a plant was found given its features (Moreno, 2004).

An additional strategy that engages reflection and self-assessment to enhance transfer is self-explanation. The self-explanation effect occurs when learners verbalize their reasoning after completing a learning or problem-solving task (Chi, de Leeuw, Chiu, & Lavancher, 1994). Specifically, self-explanations that emphasize clarification, justification of decisions, and inferences are associated with more effective analogical problem-solving in subsequent problems (Neuman & Schwarz, 1998) and greater success in far transfer of problem-solving strategies in mathematics (Wong, Lawson, & Keeves, 2002). Self-explanation effects can also occur in teaching contexts, where advanced learners verbalize for less experienced students how to approach solving authentic problems in STEM disciplines (Feldon et al., 2011).

In conclusion

There exists an increasing demand for innovative problem-solvers in STEM disciplines. However, current educational practice is not optimally structured to train the future workforce to contribute to science and engineering creatively. As instructional practices more fully engage the differentiating qualities of these disciplines, provide opportunities for authentic problem-solving that integrate process skills and content knowledge, offer opportunities for creativity, and support the development of critical metacognitive skills, students will be better prepared to contribute to the development of new scientific knowledge and innovative technologies.

Note

1 Thagard (1998) advocated the existence of a third problem search space for selection of instruments used within experiments, given the key role that advances in instrumentation played in the understanding of bacteria's role in ulcer formation. However, it can be argued that instrument selection is encompassed in the attainment of sub-goals within the experimental design problem space.

References

Anderson, J. R. (1987). Skill acquisition: Compilation of weak-method problem situations. *Psychological Review, 94*(2), 192–210.

Augustine, N. R. (2005). *Rising Above the Gathering Storm: Energizing and Employing America for a Brighter Economic Future*. Washington, DC: National Academy Press.

Barnett, S. M., & Ceci, S. J. (2002). When and where do we apply what we learn? A taxonomy for far transfer. *Psychological Bulletin, 128*(4), 612–637.

Bauer, H. H. (1992). *Scientific Literacy and the Myth of the Scientific Method*. Chicago: University of Illinois Press.

Bell, R. L., Lederman, N. G., & Abd-El-Khalick, F. (2000). Developing and acting upon one's conception of the nature of science: A follow-up study. *Journal of Research in Science Teaching, 37*(6), 563–581.

Bell, R. L., Smetana, L., & Binns, I. (2005). Simplifying inquiry instruction: Assessing the inquiry level of classroom activities. *Science Teacher, 72*(7), 30–33.

Besnard, D., & Bastien-Toniazzo, M. (1999). Expert error in trouble-shooting: An exploratory study in electronics. *International Journal of Human-Computer Studies, 50*(5), 391–405.

Blackburn, S. (1999). *Think: A Compelling Introduction to Philosophy*. London: Oxford University Press.

Brophy, S., Klein, S., Portsmore, M., & Rogers, C. (2008). Advancing engineering education in P-12 classrooms. *Journal of Engineering Education, 97*(3), 369–387.

Brown, A. L., Kane, M. J., & Long, C. (1989). Analogical transfer in young children: Analogies as tools for communication and exposition. *Applied Cognitive Psychology, 3*(4), 275–293.

Buck, L. B., Bretz, S. L., & Towns, M. H. (2008). Characterizing the level of inquiry in the undergraduate laboratory. *Journal of College Science Teaching, 38*(1), 52–58.

Campbell, D. (1960). Blind variation and selective retention in creative thought as in other knowledge processes. *Psychological Review, 67*, 380–400.

Carnevale, A. P., Smith, N., & Strohl, J. (2010). *Help Wanted: Projections of Jobs and Education Requirements through 2018*. Washington, DC: Georgetown University Center on Education and the Workforce.

Chi, M. T., de Leeuw, N., Chiu, M., & Lavancher, C. (1994). Eliciting self-explanations improves understanding. *Cognitive Science, 18*, 439–477.

Chi, M. T., Feltovich, P. J., & Glaser, R. (1981). Categorization and representation of physics problems by experts and novices. *Cognitive Science, 5*, 121–152.

Chen, Z., & Klahr, D. (1999). All other things being equal: Acquisition and transfer of the control of variables strategy. *Child Development, 70*(5), 1098–1120.

Chubin, D. (1983). *Sociology of Sciences: An Annotated Bibliography on Invisible Colleges, 1972–1981*. New York: Garland.

Clariana, R. B. (1999). *Differential memory effects for immediate and delayed feedback: A delta rule explanation of feedback timing effects*. Paper presented at the annual meeting of the Association of Educational Communications and Technology Houston, TX: February, 1999.

Clark R. E., & Blake, S. (1997) Analyzing cognitive structures and processes to derive instructional methods for the transfer of problem solving expertise. In S. Dijkstra & N. M. Seel (Eds.), *Instructional Design Perspectives: Volume II, Solving Instructional Design Problems* (pp. 183–214). Oxford: Pergamon.

Clement, J. (1988). Observed methods for generating analogies in scientific problem solving. *Cognitive Science, 12*, 563–586.

Cross, N. (2004). Expertise in design: An overview. *Design Studies, 25*(5), 427–441.
Cross, N., & Clayburn Cross, A. (1998). Expertise in engineering design. *Research in Engineering Design, 10*(3), 141–149.
Csikszentmihalyi, M. (1996). *Creativity: Flow and the Psychology of Discovery and Invention.* New York: HarperCollins.
De Corte, E. (2003). Transfer as the productive use of acquired knowledge, skills, and motivations. *Current Directions in Psychological Science, 12*(4), 143–146.
DeLisi, C. (2008). Meetings that changed the world: Santa Fe 1986: Human genome baby steps. *Nature, 455*(7215), 876–877.
Dunbar, K. (1997). How scientists think: On-line creativity and conceptual change in science. In T. B. Ward, S. M. Smith, & J. Vaid (Eds.), *Conceptual Structures and Processes: Emergence, Discovery, and Change* (pp. 461–494). Washington, DC: American Psychological Association.
Dunbar, K. (2000). How scientists think in the real world: Implications for science education. *Journal of Applied Developmental Psychology, 21*(1), 49–58.
Dunbar, K., & Fugelsang, J. (2005). Scientific thinking and reasoning. In K. Holyoak & R. Morrison (Eds.), *Cambridge Handbook of Thinking and Reasoning* (pp. 705–725). New York: Cambridge University Press.
Ericsson, K. A. (1998). The scientific study of expert levels of performance: General implications for optimal learning and creativity. *High Ability Studies, 9*(1), 75–100.
Ericsson, K. A. (2004). Deliberate practice and the acquisition and maintenance of expert performance in medicine and related domains. *Academic Medicine, 79*(10 Suppl.), S70–S81.
Ericsson, K. A., & Lehmann, A. C. (1996). Expert and exceptional performance: Maximal adaptation to task constraints. *Annual Review of Psychology, 47,* 273–305.
Feldon, D. F. (2007). Implications of research on expertise for curriculum and pedagogy. *Educational Psychology Review, 19*(2), 91–110.
Feldon, D. F., Peugh, J., Timmerman, B. E., Maher, M. A., Hurst, M., Strickland, D., Gilmore, J. A., & Stiegelmeyer, C. (2011). Graduate students' teaching experiences improve their methodological research skills. *Science, 333*(6045), 1037–1039.
Genter, D., Rattermann, M. J., & Forbus, K. D. (1993). The roles of similarity in transfer. *Cognitive Psychology, 25*(4), 435–467.
Gereffi, G., Wadhwa, V., Rissing, B., & Ong, R. (2008). Getting the numbers right: International engineering education in the United States, China, and India. *Journal of Engineering Education, 97*(1), 13–25.
Gick, M., & Holyoak, K. (1980). Analogical problem solving. *Cognitive Psychology, 12,* 306–355.
Golde, C. M., & Dore, T. M. (2001). *At Cross Purposes: What the Experiences of Today's Doctoral Students Reveal about Doctoral Education.* Madison, WI: Wisconsin Center for Education Research.
Halpern, D. F. (1998). Teaching critical thinking for transfer across domains. *American Psychologist, 53*(4), 449–455.
Halpern, D. F., Hansen, C., & Riefer, D. (1990). Analogies as an aid to understanding and memory. *Journal of Educational Psychology, 82*(2), 298–305.
Hattie, J., & Timperley, H. (2007). The power of feedback. *Review of Educational Research, 77*(1), 81–112.
Hmelo-Silver, C. E., Nagarajan, A., and Day, R. S. (2002). "It's harder than we thought it would be": A comparative case study of expert–novice experimentation strategies. *Science Education, 86*(2), 219–243.
Human Genome Project Information. (2011). *About the Human Genome Project.* Accessed online at www.ornl.gov/sci/techresources/Human_Genome/project/about.shtml on January 31, 2012.
Jonassen, D. (1997). Instructional design models for well-structured and ill-structured problem-solving learning outcomes. *Educational Technology Research and Development, 45*(1), 65–94.
Kantorovich, A. (1993). *Scientific Discovery: Logic and Tinkering.* Albany, NY: State University of New York Press.
Keane, M. T. (1996). On adaptation in analogy: Tests of pragmatic-importance and adaptability in analogical problem solving. *Quarterly Journal of Experimental Psychology, 49A,* 1062–1085.
Keane, M. T., Ledgeway, K., & Duff, S. (1994). Constraints on analogical mapping: A comparison of three models. *Cognitive Science, 18*(3), 387–438.
Kirschner, P. A., Sweller, J., & Clark, R. E. (2006). Why minimal guidance during instruction does not work: An analysis of the failure of constructivist, discovery, problem-based, experiential, and inquiry-based teaching. *Educational Psychologist, 41*(2), 75–86.
Klahr, D., & Dunbar, K. (1988). Dual space search during scientific reasoning. *Cognitive Science, 12*(1), 1–55.
Klahr, D., & Nigam, M. (2004). The equivalence of learning paths in early science instruction: Effects of direct instruction and discovery learning. *Psychological Science, 15*(10), 661–667.

Klahr, D., & Simon, H. A. (2001). What have psychologists (and others) discovered about the process of scientific discovery? *Current Directions in Psychological Science, 10*(3), 75–79.

Kluger, A. N., & DeNisi, A. (1996). The effects of feedback interventions on performance: A historical review, a meta-analysis, and a preliminary feedback intervention theory. *Psychological Bulletin, 119*(2), 254–284.

Knorr-Cetina, K. (1997). What scientists do. In T. Ibáñez & L. Íñiguez (Eds.), *Critical Social Psychology* (pp. 260–272). London: Sage Publications.

Koen, B. V. (1985). *Definition of the Engineering Method*. Washington, DC: American Society for Engineering Education.

Kolodner, J. L., Gray, J. T., & Fasse, B. B. (2003). Promoting transfer through case-based reasoning: Rituals and practices in Learning by Design™ classrooms. *Cognitive Science Quarterly, 3*(2), 119–170.

Kuhn, T. S. (1962). *The Structure of Scientific Revolutions*. Chicago: University of Chicago Press.

Larkin, J., McDermott, J., Simon, D. P., & Simon, H. A. (1980a). Expert and novice performance in solving physics problems. *Science, 208*(4450), 1335–1342.

Larkin, J. H., McDermott, J., Simon, D. P., & Simon, H. A. (1980b). Models of competence in solving physics problems. *Cognitive Science, 4*(4), 317–345.

Lehrer, R., Kim, M., & Schauble, L. (2007). Supporting the development of conceptions of statistics by engaging students in measuring and modeling variability. *International Journal of Computers for Mathematical Learning, 12*(3), 195–216.

Lehrer, R., & Schauble, L. (2004). Modeling natural variation through distribution. *American Educational Research Journal, 41*(3), 635–679.

Lehrer, R., & Schauble, L. (2006). Cultivating model-based reasoning in science education. In K. Sawyer (Ed.), *Cambridge Handbook of the Learning Sciences* (pp. 371–388). Cambridge, MA: Cambridge University Press.

Lovett, M. C., & Anderson, J. R. (1996). History of success and current context in problem solving: Combined influences on operator selection. *Cognitive Psychology, 31*(2), 168–217.

Lovitts, B. E. (2008). The transition to independent research: Who makes it, who doesn't, and why. *Journal of Higher Education, 79*(3), 296–325.

Masui, C., & De Corte, E. (1999). Enhancing learning and problem solving skills: Orienting and self-judging, two powerful and trainable learning tools. *Learning and Instruction, 9*(6), 517–542.

Mathan, S. A., & Koedinger, K. R. (2002). An empirical assessment of comprehension fostering features in an intelligent tutoring system. In S. A. Cerri, G. Gouarderes & F. Paraguacu (Eds.), *Proceedings of the 6th International Conference in Intelligent Tutoring Systems* (pp. 330–343). New York: Springer-Verlag.

Maxam, A. M., & Gilbert, W. (1977). A new method for sequencing DNA. *Proceedings of the National Academy of Sciences of the United States of America, 74*, 560–564.

Metz, K. E. (2008). Narrowing the gulf between the practices of science and the elementary school science classroom. *Elementary School Journal, 109*(2), 138–161.

Minner, D. D., Levy, A. J., & Century, J. (2010). Inquiry-based science instruction. What is it and does it matter? Results from a research synthesis years 1984 to 2002. *Journal of Research in Science Teaching, 47*(4), 474–496.

Moreno, R. (2004). Decreasing cognitive load for novice students: Effects of explanatory versus corrective feedback in discovery-based multimedia. *Instructional Science, 32*(1/2), 99–113.

Moreno, R., Reisslein, M., & Ozogul, G. (2009). Optimizing worked-example instruction in electrical engineering: The role of fading and feedback during problem-solving practice. *Journal of Engineering Education, 98*, 83–92.

Neuman, Y., & Schwarz, B. (1998). Is self-explanation while solving problems helpful? The case of analogical problem-solving. *British Journal of Educational Psychology, 68*(1), 15–24.

Newell, A., & Simon, H. A. (1972). *Human Problem Solving*. Englewood Cliffs, NJ: Prentice-Hall.

Pasteur, L. (1854). Lecture, University of Lille. December 7, 1854.

Perkins, D. N., & Grotzner, T. A. (1997). Teaching intelligence. *American Psychologist, 52*(10), 1125–1133.

Perkins, D., & Saloman, G. (1994). Transfer of learning. In T. Husen & T. Postlethwaite (Eds.), *The International Encyclopedia of Education*, 2nd ed., vol. 11. Oxford: Elsevier Science Ltd.

Price, D. (1965). Networks of scientific papers. *Science, 149*(3683), 510–515.

Raghavan, K., Sartoris, M. L., & Glaser, R. (1998). Impact of the MARS curriculum: The mass unit. *Science Education, 83*(1), 53–91.

Robinson, J. A. (1998). Engineering thinking and rhetoric. *Journal of Engineering Education, 87*, 227–229.

Schraagen, J. (1993). How experts solve a novel problem in experimental design. *Cognitive Science, 17*(2), 285–305.

Shute, V. J. (2008). Focus on formative feedback. *Review of Educational Research, 78*(1), 153–189.

Simonton, D. K. (1987). Multiples, chance, genius, creativity, and zeitgeist. In D. N. Jackson & J. P. Rushton (Eds.), *Scientific Excellence: Origins and Assessment* (pp. 98–128). Beverly Hills, CA: Sage Publications.

Simonton, D. K. (1991). Career landmarks in science: Individual differences and interdisciplinary contrasts. *Developmental Psychology, 27*(1), 119–130.

Simonton, D. K. (1999). *Origins of Genius: Darwinian Perspectives on Creativity*. New York: Oxford University Press.

Simonton, D. K. (2000). Creativity: Cognitive, personal, developmental, and social aspects. *American Psychologist, 55*(1), 151.

Simonton, D. K. (2004). *Creativity in Science: Chance, Logic, Genius, and Zeitgeist*. New York: Cambridge University Press.

Singley, M. K., & Anderson, J. R. (1989). *Transfer of Cognitive Skill*. Cambridge, MA: Harvard University Press.

Sodian, B., Zaitchik, D., & Carey, S. (1991). Young children's differentiation of hypothetical beliefs from evidence. *Child Development, 62*(4), 753–766.

Sternberg, R. J., Ferrari, M., Clinkenbeard, P. M., & Grigorenko, E. (1996). Identification, instruction, and assessment of gifted children: A construction validation study of a triarchic model. *Gifted Child Quarterly, 40*(3), 129–137.

Stokes, P. D. (1999). Learned variability levels: Implications for creativity. *Creativity Research Journal, 12*(1), 37–45.

Stokes, P. D. (2001). Variability, constraints, and creativity: Shedding light on Claude Monet. *American Psychologist, 56*(4), 355.

Stokes, P. D. (2006). *Creativity from Constraints: The Psychology of Breakthrough*. New York: Springer Publishing Company.

Stokes, P. D., Lai, B., Holtz, D., Rigsbee, E., & Cherrick, D. (2008). Effects of practice on variability, effects of variability on transfer. *Journal of Experimental Psychology: Human Perception and Performance, 34*(3), 640–659.

Sweller, J. (1988). Cognitive load during problem solving: Effects on learning. *Cognitive Science, 12*(2), 257–285.

Sweller, J. (2009). Cognitive bases of human creativity. *Educational Psychology Review, 21*(1), 11–19.

Sweller, J., & Sweller, S. (2006). Natural information processing systems. *Evolutionary Psychology, 4*, 434–458.

Thagard, P. (1998). Ulcers and bacteria: I. Discovery and acceptance. *Studies in the History and Philosophy of Biology and Biomedical Sciences, 9*, 107–136.

Trickett, S. B., & Trafton, J. G. (2007). "What if…": the use of conceptual simulations in scientific reasoning. *Cognitive Science, 31*(5), 843–875.

van Merriënboer, J. J. G., Kester, L., & Paas, F. (2006). Teaching complex rather than simple tasks: Balancing intrinsic and germane load to enhance transfer of learning. *Applied Cognitive Psychology, 20*(3), 343–352.

Wharton, C. M., Holyoak, K. J., Downing, P. E., Lange, T. R., Wickens, T. D., & Melz, E. R. (1994). Below the surface: Analogical similarity and retrieval competition in reminding. *Cognitive Psychology, 26*(1), 64–101.

Windschitl, M. (2003). Inquiry projects in science teacher education: What can investigative experiences reveal about teacher thinking and eventual classroom practice? *Science Education, 87*(1), 112–143.

Windschitl, M. (2004). Caught in the cycle of reproducing folk theories of "Inquiry": How pre-service teachers continue the discourse and practices of an atheoretical scientific method. *Journal of Research in Science Teaching, 41*(5), 481–512.

Windschitl, M. (2009). *Cultivating 21st century skills in science learners: How systems of teacher preparation and professional development will have to evolve*. A paper presented to the National Academies of Science Workshop on 21st Century Skills. Washington, DC: February 5–6, 2009.

Wong, R. M. F., Lawson, M. J., & Keeves, J. (2002). The effects of self-explanation training on students' problem solving in high-school mathematics. *Learning and Instruction, 12*(2), 233–262.

26
The importance of informal learning in science for innovation education

Susan M. Stocklmayer and Bobby Cerini

AUSTRALIAN NATIONAL UNIVERSITY, AUSTRALIA

Summary: In this chapter, we contrast the roles of formal and informal learning in fostering the skills needed for innovative and creative thinking in science. Informal learning, whether it be at the hands of a mentor, in the environment of a science centre or a science club, or through a demanding hobby that requires an understanding of science, can be a powerful change agent in developing innovative skills. Formal education can, on the other hand, stifle such skills. Through specific case studies of innovators from the world of science, the authors show that there are common factors that may be identified and explored in order to bring more opportunities for innovative thinking into the formal science education sphere.

Key words: Informal learning, science, innovation, creativity, role models.

Introduction

Informal learning environments have long been recognised as fostering creativity and innovation in science, but it is only relatively recently that this learning has been investigated in a formal manner. In particular, the increase in the number of science centres from the1980s up to the present day (Durant, 2011) constitutes a tacit recognition of the importance of science play and "free choice learning" (Falk & Dierking, 2002).

The evidence from informal environments contrasts sharply, however, with research from the formal sector, where declining enrolments in senior science classes testify to deep-seated problems. Fensham (2008, p. 20) has commented: "It is urgent that educational policy makers address the lack of engagement that so many students experience in school science and technology education." These problems have their origins in a range of factors related to the way in which school science has traditionally been taught. Largely based on the textbook, with little room for exploration or for independent thought, school science has not reflected the way in which science is conducted in the "real world" outside school. Traditional teaching has been transmissive, with an emphasis on abstract, often difficult concepts that have little relevance for students. The lack of emphasis on creative, open-ended enquiry is a major problem. Students often assume science to be a discrete and well-established body of knowledge, echoing the comment said to have been made by Lord Kelvin in an address to the British Association in 1900: "There is nothing new to be discovered in physics now. All that remains is more and more precise measurement" (original source not known).

Wellington (1990, p. 248) summarised the characteristics of formal learning in science with adjectives such as: structured; compulsory; closed; teacher-led and teacher-centred; planned; and with few unintended outcomes. It is easy to see that students who prefer to be creative and innovative will be stifled by such a classroom.

Of course not all science classrooms are this constrained, but many are driven by the requirements of curricula grounded in the discoveries of nineteenth century science, which highlight the achievements and understandings of the Industrial Revolution. The accompanying textbooks largely ignore events of the last fifty years and are structured in a traditional, linear framework, which constitutes familiar and comfortable material for the teacher. It is rare that a science textbook conveys the challenges, frustrations and ultimate triumphs of scientific discovery. We acknowledge that these are generalisations, but the widespread nature of boring, rote-driven curricula has generally defied attempts at reform. "Inquiry-based teaching" has been a feature of recommended practice for almost a century and is now a specific requirement of many curricula around the world: nevertheless a cursory check of the Internet reveals many teacher education programmes that still define and explain what this actually means. From our own experience, introduction of student-centred inquiry-based learning is very difficult for those teachers experienced in the transmissive mode. Osborne (2006, p. 2) states that:

> Four decades after Schwab's (1962) argument that science should be taught as an *enquiry into enquiry*, and almost a century since John Dewey (in 1916) advocated that classroom learning be a student-centred process of enquiry, we still find ourselves struggling to achieve such practices in the science classroom.

Encouraging innovation is a widely recognised goal for education. The importance of early critical thinking, of a range of choices, of tinkering and play, are all widely acknowledged to be critical keys to producing the kind of innovators in science who will be important not only to the economy of their country, but to democracy. We have said (Rennie & Stocklmayer, 2003, p. 771) that the important goals for public awareness of science include a personal interest in contextually relevant science; a sense of "ownership" of science; understanding how to assess the impact of science on human lives; and a sense that individual knowledge and concerns are valued. All these are facilitated by an approach to education that celebrates the individual, encourages critical thinking and fosters innovative thought.

Informal learning in science

According to Crane *et al.* (1994, p. 3):

> Informal science learning refers to activities that occur outside the school setting, are not developed primarily for school use, are not developed to be part of an ongoing school curriculum, and are characterized as voluntary as opposed to mandatory participation as part of a credited school experience.

Contexts in which informal learning occurs might be structured, in the sense that a science centre or a science lecture might be organised to provide learning experiences in science. Rennie (2007) has classified structured learning environments into three groups: museums and similar institutions which have an educational purpose; community organisations, which are aimed at educating the public about matters usually relating to health and environment, and including after school programmes; and the media, which include print and electronic forms. In a significant review of informal environments, Bell et al. (2009) characterised museums, zoos and similar establishments as "designed settings". They differentiated learning in such settings from formal learning in several critical ways. First, learning tends to be fluid and sporadic. Second, the settings are experienced episodically. Third, participants may navigate freely. The personal, sociocultural and physical contexts, described by Falk and Dierking (2000) as essential components in understanding the museum experience, also affect learning outcomes in any designed setting and will determine choices to a marked degree.

On the other hand, the informal context may be much more casual, in the sense that learning through an unplanned experience or through a hobby (such as jewellery-making or keeping tropical fish) occurs when and if required. Research in this domain is scattered and limited: little is known, for example, about how people learn from television, or the media, or within the home. In respect of television, evaluations have been conducted of children's education programmes (see, for example, Fisch, 2004; Fisch et al., 1997) but these were controlled evaluations. Nevertheless they highlighted the importance of the social context in science learning. Information about adult learning is much more ephemeral. Bell et al. (2009, p. 252) reported:

> Much of this material is fugitive literature... for many of our queries respondents (both producers and researchers) were unsure as to whether their reports were public documents... almost all the reports we obtained were funded by the National Science Foundation.

Stocklmayer et al. (2010, p. 18) state:

> In general, evaluation of the effects of the media is difficult, but there are some reports that offer hints as to why the media experience is likely to facilitate learning.... These processes include the iterative nature of learning in informal contexts (Miller, 1998; Stocklmayer & Gilbert, 2002b), the importance of meaningful reminders (Stocklmayer & Gilbert, 2002a) and that learning takes time (Falk & Dierking, 2000).

People say that they are very interested in science: for example, in a survey of the Australian public (Lamberts et al., 2010) over 90 per cent of respondents were moderately or very interested in scientific and medical discoveries, health and environmental issues. Nevertheless, the nature of this interest and its effect on attitudes to science, technology and technical innovation are little understood. We shall return to this theme later in this chapter, as we recount the stories of innovative scientists who have reflected on their early experiences.

In summary, contexts for learning in the informal sector are:

> those out-of-school learning environments where (a) both attendance and involvement are voluntary or free-choice, rather than compulsory or coercive; (b) the curriculum, if any, and whether intended or not, has an underlying structure which is open, offers choices to learners and tends not to be transmissive; (c) the activities in which learners can be involved are non-evaluative and non-competitive, rather than assessed and graded; and (d) the social interaction is amongst groups likely to be heterogeneous with regard to age, rather than constrained between same-age peers and formalized with the teacher as the main adult.
>
> *(Stocklmayer et al., 2010, p. 10)*

Such contexts facilitate important attributes that enhance the learning experience. These include exploration, curiosity and surprise (see, for example, Sadler, 2006; Walker et al., in press). Bell et al.

Table 26.1 Factors affecting science learning for innovation education

Affective factors
 Providing for free choice
 Internally driven and challenging
 Encouraging wonder, delight and awe
 Entertaining, interesting, enjoyable

Factors related to learning science
 Holistic
 Useful, powerful and transferable knowledge

Factors related to learning about science
 Facilitating social and community interaction,
 Presenting science as messy, human and exploratory in nature, addressing real and current problems

Factors related to doing science
 Facilitating inquiry-based science
 Involving real projects with real outcomes

Source: adapted from Stocklmayer *et al.* (2010, p. 25).

(2009) also identified the qualities of excitement, delight and awe. McLean (1993, p. 93), commenting about interactive exhibits in the designed setting of a science centre, said that they provide opportunities to "gather evidence, select options, form conclusions, test skills, provide input and actually alter a situation based on that input". All of these attributes and opportunities will be familiar to anyone who is desirous of fostering innovation education, yet they are not readily found in formal settings. These factors were summarised by Stocklmayer *et al.* (2010) as falling into four categories. All are important to science learning, but those most likely to be relevant specifically to innovation education are shown in Table 26.1.

Many of these are clearly relevant to innovative approaches to science and need no explanation. To expand on a few, however: the holistic approach enables systems thinking, which is critical to effective innovation. Social and community interaction enables science to be seen in a social context, enhancing its relevance and enabling free thinking about technological development. The messy nature of science is critical to understanding that exploration may sometimes – but not always – lead to productive outcomes, but that all knowledge is important in formulating scientific theories which may, at a later date, be useful.

All of these factors have influenced innovative thinkers. In the next section, we will tell some of their stories and identify how and when these elements played a part.

Informal learning in the life of innovators

As previous chapters have noted, knowledge about the science and technology learning experiences of recognised innovators can help to illuminate possible future directions for innovation education. As Shavinina (Chapter 18) points out, creativity and high levels of innovation are not restricted to the "gifted and talented"; highly innovative adults may in fact have had quite unremarkable academic records as children, whilst precocious academic accomplishment by children is not a reliable predictor of their innovation as adults. This suggests that there are other factors at play and it is our argument that these are to be found outside the world of formal education, in the messy and holistic "real world" of experiential learning.

As part of our exploration into the informal learning experiences of highly innovative individuals, interviews were conducted over a two-year period with some of the world's most respected scientific, technological and communication innovators. They came from all walks of life, represented many different disciplines and, consistent with Shavinina's findings, they reported being of mixed ability in terms of their early education. Nevertheless, these are individuals who have been recognised as achieving something extraordinary. It is to their stories that we turn, in order to illustrate those other worlds of learning: the affective, holistic and "real", the open-ended, unsupervised and social. We present these stories here in abridged form, as brief interview extracts and further points for discussion.

Many of the innovative individuals who were interviewed reported social interactions that had a positive effect on their education, by influencing the values placed on learning and reinforcing and extending their existing knowledge. For example, geneticist and environmental campaigner David Suzuki describes how conversations with his father played a key role in consolidating his learning outside school:

> My father was my great mentor and hero. Every night after dinner he would say, "so what did you learn today at school?" I would have to try and remember the lessons and if I said something that he didn't understand, he would get me to explain it again.

This emphasis on the effective communication of science was clearly a powerful influence on Suzuki's later career as a public figure in environmental science. In keeping with findings about informal learning environments, his interaction with his father was cross-generational and inspirational. This is also consistent with other research showing that parents and other relatives have powerful and direct influences on an individual's career choices (Young, 1993). For innovators such as Emily Cummins, they also have a direct influence on the acquisition of practical skills:

> My grandad had a shed at the bottom of the garden and whenever we'd visit he'd invite us down to the shed where we could create and make things. From such a young age I was involved with the creative process and creative thinking. We'd have a go at coming up with ideas for trucks and what we could use for wheels, and as we got older we would learn to use tools and the different properties of things....We had amazing opportunities to be creative.... It was just this thinking out of the box, which was such a great skill to have as a kid.

Such anecodotes suggest an early investment in pathways to innovation. Researchers such as Grotevant (1987) have proposed that one's momentum along a particular career or life path is affected by the degree to which different aspects of one's life are intertwined with it. Certainly this seems to be the case with Cummins. A young inventor, her innovations include a sustainable refrigerator and water carrier for use in the developing world, both developed whilst still a student. Her approach to commercialisation is also innovative; she was named one of the Top Ten Outstanding Young People In The World by Junior Chamber International for her work distributing the technology directly into needy communities (JCI, 2010). In talks to high school students, Cummins attributes her successful career as an innovator to her efforts and experiences outside the classroom, rather than inside it:

> I wasn't the strongest at maths when I was at school, I wasn't very good at languages, I wasn't the best in the year by far; my invention was just an idea that I had and, having that passion, just wanting to make it work. I've researched specific areas to teach myself about engineering and design my products, and I've done it through my own research, not by studying engineering at school.

From these statements emerges a portrait of a young person whose desire to succeed is internally motivated and whose self-belief was encouraged by experiences outside school that were challenging, enjoyable and exploratory.

Cummins is far from alone when describing passion as a factor in pursuing particular tasks or subjects. Bill Fenical, a pioneer in the field of marine chemistry at the Scripps Institute in La Jolla, CA, reveals how his personal passions and training came together as a young man:

> When young children think about careers I don't know that they have adequate decision-making capacity, simply they're drawn to things. I grew up well away from the ocean, but I loved the water. My love of the water led me to be an amateur diver during high school. During college I realised that one of my loves was chemistry, but I was also biology major so I enjoyed biology and chemistry, and I never thought much about oceanography but continued to be very enthusiastic about the ocean. When I graduated with my degree in chemistry I took a job, somewhat naively, thinking that this was a perfect endpoint for a career in organic chemistry and I ended up in a position that I hated, it was terrible. I had little respect for the way in which things happened in these industries.... I asked myself some serious questions which were basically, what do I love to do, what can I do that is essentially my hobby and combine that with my training? So I realised that if I could do some type of organic chemistry in the ocean this would be for me.

Motivated to find a career that could combine his hobby and his training, Fenical went on to establish himself as a postdoctoral student looking at the chemistry of marine creatures. Within two years he began to find "interesting things" of the sort that would ultimately see him become a leader in the field of marine pharmachemistry.

Such examples suggest that both informal and formal learning experiences can work together to contribute to an individual's overall view of themselves and their competencies; while academic success offers its own rewards and can be highly motivating, personal success is at least equally important. If it were not, those self-proclaimed "average students" would not be leading innovators in so many fields.

One might surmise from the above examples that the presence of the right conversations – with others and with the self – is crucial to learning how things work and what things matter. Self-talk is known to play an important role in self-esteem and contributes to an individual's sense of identity in the world (Bandura, 1993; Seifert, 2004). Conversations with others about what is happening, and why, also form an important part of learning and are encouraged in informal environments, such as out-of-school activities and science centres and museums (Falk & Dierking, 2000). Within these environments, social and community interactions are also facilitated (Stocklmayer et al., 2010).

Such exploratory conversations are often restricted or absent in the classroom – especially conversations of the sort that allow personal, real-world experiences to be shared or that habituate the participants to valuing their own knowledge. In 2002, hundreds of teenagers across England participated in a nationwide review of the secondary school science curriculum. The review found that students experienced many frustrations with the curriculum, wanting it to be more relevant to everyday life, offer more opportunities for practical, hands-on learning in class and, significantly, provide opportunities for open-ended discussion and debate, rather than just transmission of facts and figures (Cerini et al., 2003).

The comments made by students paint a picture of a learning environment in which the curriculum is strongly geared towards passive learning and where individual interpretation and innovative thinking by students have little roles to play. Responsibility for making the connection between abstract concepts and real world practice therefore rests with the teacher, and even innovative students may struggle to engage with the content when the examples do not seem relevant to them. Tim Hunkin, an engineer, inventor and former host of the BBC's *The Secret Life of Machines*, makes this observation about the disconnection between his out-of-school experiences and his formal education, in which his own experience and curiosity seemed remote from what was being taught:

> At school, I had a sense they were telling me all this stupid stuff that I didn't need to know, while there were all these other interesting things out there in the world. I always had a feeling that there were these cool things out there, and it seemed natural for me to try and fill the gaps outside school. The careers advisor said that I wasn't bright enough to do science. But at the last minute I got good A levels, which they weren't expecting. It was partly due to one teacher, he was ex-army, a ballistics instructor, and while he taught us mechanics he peppered the lessons, which were basically mathematical, with these extraordinary stories of testing different munitions. And somehow he gave that sort of connection, between the sums and real life. He just had lots of stories to tell and he was not like somebody who had just learned it – rather, he knew it so well it was completely a part of him.

This recollection suggests a teacher who, in being able to locate the learning in a real-world context, rendered the knowledge useful, powerful and transferable.

Others have also reflected on the importance of ideas and how they are communicated. Sir Gustav Nossal, an immunologist at the University of Melbourne and former Director of the Walter and Eliza Hall Institute (a leading Australian medical research centre) speaks of his excitement at encountering stories from the real world of science:

> Often my eldest brother would come across from Adelaide to Sydney for the ANZAAS conference, not infrequently bringing one or two colleagues.... And then my Dad, who was great fun and loved an argument, would sit them down over dinner or after dinner and cross-examine each of them about what they were doing. One might be a physiologist, my brother was a biochemist, the third might be a bacteriologist, and that got them talking about their stuff. I was much younger, I was seven years younger, and that got me incredibly excited. Very, very excited. The other part that got me excited about science was… in first year medicine. It was a pretty harsh climate… and we weren't very well taught. The classes were too big, the number of professors were too small. In third year we started teaching each other. We were the sort of group of so-called clever kids, we found each other and we started reading our stuff up in the library and then giving each other mini-research seminars. Now that got me very excited. And that I think was the beginning of me wanting to do research myself.

These examples suggest that real-world stories and the discussion of them play an important role in helping learners to engage with abstract knowledge and how it relates to success as an innovator. It also suggests that would-be innovators learn to *imagine* themselves in the role, both before and whilst they are developing concrete skills in this area. For children, whose technical skills and abilities develop slowly over time and with repeated practice, the idea of being an innovator must frequently precede the reality of it.

Conversation and story-telling can thus be seen as playing important roles in capturing and inspiring the imagination, and supporting an individual's engagement with ideas of future self and identity. Contributing to this process are role models, who can both illustrate how future accomplishments may be achieved and also make ideas about the possible future self more tangible (Lockwood &

Kunda, 1997). Role models need not be immediately present – indeed, distant role models can appear through the media, film and in literature (Pace, 2008). An example comes from Nobel Laureate in Medicine, Barry Marshall, who recollects how ideas of individual success and practical innovation excited him as a child:

> When I was about 10 or 11 or 12, I started reading these biographical novels, they must have had a series of them at school and so I went right through them. Thomas Edison was pretty exciting because I could do things that he did, you know – I'd made a Morse code set, so that made an interesting story... and I read about the Mayo brothers, they were pretty exciting, they used to operate on their puppy and they grew up and started the Mayo Clinic.... And I read Brother Surgeons, which is a great story about these two young medical students who dug up bodies for the anatomists and were doing all this stuff in 18th century England, it was a pretty exciting story. I had those, and I really just wanted to be a GP.... I just thought, wow, I could cut and sew people up, do surgery, all that type of thing. And it was when I got into medicine that I realised there was a very interesting intellectual problem-solving side to it.

Another source of data comes from participants in the World Wide Day of Science, an initiative started by the University of New South Wales to record scientists' activities and aspirations. When asked to write about what got them interested in science, many report imagining themselves in the role of scientists from a young age, feeding into their sense of identity as a future innovator. One person, now a chemist, writes:

> Science has always been my hobby. For as long as I can remember I've asked "why?" and "how?". To help answer my own questions I used to develop theories and design a practical way to see if I was correct. When I was small these experiments were usually confined to learning about plants or food, although I once climbed onto the roof and photographed the movement of stars... I also enjoyed reading absolutely everything I could find including information on famous scientists and news of scientific discoveries reported in the newspapers. I felt proud, as though I was already part of this scientific community. It seemed that a scientific career would enable me to satisfy my own curiosity as well as perhaps make useful discoveries.
>
> *(UNSW, 2007)*

Excitement, enjoyment and wonder all feature strongly in these accounts. Such recollections also suggest that stories of achievement can motivate a desire for future success and bring focus to an individual's early attempts at innovation. As we have previously reported elsewhere, role models can play an important part in encouraging careers in science and technology (Priest, 2010). Indeed, a desire to inspire future generations is a key driver for innovators sharing their stories. As Cummins reports:

> I think it makes a difference to go and tell a story as someone who is not that much older, and specifically lay out exactly what I did, and tell them what I have learned from my whole experience. I think why it works so well is that young people relate to my story, it is quite straightforward, what's happening is quite clear cut and is quite achievable by other people. It is not so unrealistic: you know, I had an idea and I followed it through and I think that is a process young people relate to. I want people to believe they can do these things.

Yet as we have discussed, the curriculum has limited room for discussion or modelling of the process of innovation. Whilst some efforts have been made to mimic the process of knowledge production, for example through team-based enquiry and learning activities, within the traditional transmission model, conversations amongst learners tend to be limited (Chandler, 1994). But if the discussion of ideas cannot flourish in the classroom, then where does it have a place?

The answer currently lies in informal learning environments, which, in contrast to formal education, reward attempts at problem solving rather than displays of existing knowledge. Such environments encourage individuals to participate in the process of innovation by making observations and solving problems using an open-ended, trial-by-error approach, regardless of their level of educational attainment. This approach allows for multiple failures in the pursuit of an answer, and encourages original thinking in the pursuit of possible solutions. Science centres, with their interactive exhibits and open-ended learning experiences, are perhaps the best example of this. As scholars of informal learning point out (for example, Stocklmayer *et al.*, 2010) these are environments in which the affective and learning factors associated with science engagement are all present. Importantly, they allow for experiential learning – the making and doing of things, alone or in the company of others, which have messy, enjoyable and unpredictable outcomes.

Tim Hunkin describes the importance of such experiences to the innovation process:

> I think people forget that the sort of intuitive hands-on developing thing, which is how we got as far as we did in the mid twentieth century, is still very important for everything that's made or developed.... I've always liked making things. As a child I used to take things to bits and scavenge things from junkyards and make things out of it, like lots of kids do. There were places that I used to love to go, like the Science Museum, and there was also a comedian on TV called Michael Bentine who used to have crazy things that people had made on the show.

Hunkin's experience suggests that stories told on television can be highly memorable for the individual, when they reinforce existing interests and stimulate the imagination and the sense of awe. Other examples show that individuals may be motivated by experiences depicted in the media to seek out additional information and opportunities for learning. Chris Lintott, an astronomer at Oxford University and today also a presenter of the BBC's long-running *The Sky at Night* programme, recalls his experience watching the show as a child:

> I've always been an amateur astronomer as well as a professional one. I grew up in my local astronomical society, and as a member of the British Astronomical Association, contributing very bad sketches of the planet Jupiter and trying to take images. My school had a huge telescope that they built and they gave us the keys. And I came across Patrick (Moore). I was one of many kids who wrote to him. He came to my school when I was about 11 or something. And I wrote and I got a reply, and we kept up a vague correspondence. I once had a wonderful postcard back that said "Dear Chris. Yes. Patrick."

Lintott's childhood correspondence with Sir Patrick Moore is far from unusual. Countless others have been inspired by *The Sky at Night*, to the extent that they have actively corresponded with the presenter. Moore reports having received and responded to thousands of letters from young fans over the course of a continuous broadcasting career dating back to 1957. He also hosts visitors to his home observatory in southern England and, in Lintott's case, also facilitated opportunities to co-present *The Sky at Night*. For Lintott, interactions with Moore played a significant role in his future career, both in extending his interest and skills far beyond the possibilities of the classroom and ultimately in encouraging his future career path.

As we have already pointed out, the media can be an important source of education, encouraging not just passive learning but active knowledge seeking. In our study, others who, like Moore, regularly appear in the media have reported receiving unsolicited letters and emails from hundreds or even thousands of people seeking further information, assistance and advice. At least some of these correspondents later go on to emulate their heroes, achieving similar heights of ability and attainment. Other prominent science personalities reportedly inspired by Moore include the rock guitarist Brian May and the British Astronomer-Royal, Sir Martin Rees (Cerini, 2011). Similarly, Carl Sagan is cited as a huge source of inspiration for many. As astronomer Bryan Gaensler describes:

> Carl Sagan looms very big, not just because he was a really, really good scientist but also unbelievably eloquent. He had ways of communicating, he didn't just take the old hackneyed analogies that I and other people use over and over again, but he had these incredibly poetic ways of making you think about things in a way you hadn't thought about them before. And so I love the TV show that he produced in the early 80s, Cosmos. I love that a lot of people have gotten into astronomy because of that show. And I had the book, I used to read it over and over again, I've read all his books, just loved his turn of phrase and his insight, and the fact that he saw the spirituality of science as well.

Perhaps the strength of the media is its simple accessibility, which allows the user to move swiftly between stories, selecting the content that is most interesting to them. The user whose interest is piqued may concentrate deeply on the content, returning to it again by purchasing or recording private copies, or seeking out related stories that extend or develop the concepts. Thus it can be seen as a user-centric learning medium, shaped by public audiences through ratings and reviews but also used by them to advance their own knowledge in ways that suit themselves. This contrasts sharply with the realities of learning through the formal education process. As Cummins, who by secondary school age was already developing the skills needed to becoming an innovator, describes:

> When I got to secondary school our technology classes were absolutely boring... it put me off technology for a little while.... We all had to design exactly the same product, you know the same pencil holder... we could only choose the colour of plastic we could use... and that for me was absolutely boring. It was so straightforward to me and you weren't able to be creative.

Such comments are suggestive of a formal learning environment that is not geared towards capitalising on the creative skills of the learner and successfully transforming them into innovators. Hunkin describes the importance of asking open-ended questions and realising that there is an answer waiting to be discovered: "For whatever reason, I've sort of got a curiosity about the world around me and it's always very exciting when I realise 'oh! I don't know what that's made of' or 'I don't know how that works'."

Conclusions

Informal learning environments are to be found in many areas of our lives, from the structured environments offered within places such as science centres and out-of-school programmes to the unstructured environments of the home, television and bookshelf. As we have shown, learning by individuals about what innovation is and how it is practised occurs almost continuously throughout the course of one's life. It can arise in the company of others, through the telling of stories and the discussion of ideas, or in the solo hobbies and pastimes of the individual, through practical experimentation and experiences. In both cases, experiences of the world are mixed with and influenced by ideas about it, and vice versa. Critically, how individuals see themselves has an effect on what they do and where they are motivated to learn.

Formal learning environments may have limitations that prevent the individual from engaging with and practising innovation. As we have discussed, classrooms are constrained by a number of factors, including teaching traditions, assessment goals and the nature of the curriculum. Indeed one might argue that formal education is not designed to teach innovation but to prepare you for its possibility; perhaps, as Hunkin puts it, "the point of education is to get you to a state where you are curious enough to go on finding stuff out for yourself".

Informal learning appears to fill the gaps left by formal education – or indeed, the reverse may be true, for the majority of any learner's time is spent outside the classroom, in environments where learning is spontaneous, unstructured and personally directed. It is out in the "real world" that innovation appears to be modelled and practised to a greater or lesser extent. Largely engaged in the world outside school, future innovators find their interest fired by stories that stir their imagination, experiences that model the innovation process even when failure is involved and opportunities for further exploration, discussion, thought and learning.

What can formal education learn from this messy, uncontrolled world of experiential learning? The following suggestions emerge from the anecdotal evidence of our research subjects, some of whom have found voice here and all of whom have excelled in their respective fields.

First, school curricula inevitably have both strengths and weaknesses and, whilst curriculum change may be desirable, it is a slow and lengthy process that remains beyond the direct control of teachers. Nevertheless, the nature of the discussions that take place in classrooms can be remodelled, bringing them into line with previous calls to build enquiry-led curricula. To foster innovation, the connection between abstract concepts and practical outcomes must be strengthened by open-ended discussions that allow for the deeper exploration of ideas, and by the use of vivid real-life examples that are exciting to the listener.

The presence and awareness of inspirational figures and role models is a common theme for innovators. Importantly, these are encountered in a number of ways: not just through formal environments within the school or institution, but in the media, in literature, at home and within the wider community. Inspirational individuals have a role to play in bringing the stories and processes of innovation to life, in modelling innovative behaviour and in supporting and encouraging the aspirations of young people. Educators can capitalise on this, by sharing stories that reveal and celebrate the real world of innovation. The stories of innovators themselves should be more widely acknowledged and shared, and every opportunity taken to support the aspiration of students with examples of individuals who have embarked on ambitious journeys. These innovators need to be people with whom students can identify, true role models in every sense of the word.

Finally, as history shows, despite our most modern curricula and well-trained teachers, it remains difficult to integrate open-ended learning activities into the classroom. Yet inquiry-led, "messy" practical experiences that allow for individual exploration, decision-making and problem-solving are essential to the process of learning to innovate, and must be encouraged if we are to support innovation education. Central to this is a need to celebrate failure as a potentially positive experience, rather than depict it as a solely negative experience that will impinge future careers. Conversations *about* innovation, designed to open students' minds to an awareness of the value of their own knowledge and creative thinking, are critical to this process.

Certainly, within the competitive, assessment-driven environments of many schools, practical exploration that encourages repeated failure is not an attractive option. Yet the process of innovation remains dependent on valuing failure and encouraging persistence, ultimately rewarding the best solutions to emerge from multiple attempts at a problem. Somehow, we must learn to find new ways of celebrating this process, and embedding it into learning activities that deliver the unexpected.

As our examples have shown, it is through a combination of imagination, skills and persistence that innovators thrive outside the formal education system. Our challenge remains to connect the experiences offered in the world of informal learning with those offered within the classroom. One way to do this may be through closer integration between the two worlds, where innovation experiences encountered outside the classroom are anticipated and strengthened within it. Here there is a clear role for planned incursions and excursions to informal learning environments. Many institutions offer pre-visit planning resources and curriculum-linked learning materials for use in the classroom, and in-school visits from travelling programmes are also available. These clearly have value. It is important to recognise, however, that less structured experiences also have an important role to play, and consideration must be given to how film, literature, role models and practical learning opportunities can be accessed and integrated. A criticism sometimes levelled at informal environments is that people are "just having fun". As we have seen, the value of fun in fostering innovation should never be underestimated.

References

Bandura, A. (1993). Perceived self-efficacy in cognitive development and functioning, *Educational Psychologist, 28*, 117–148.

Bell, P., Lewenstein, B., Shouse, A. W., & Feder, M. A. (Eds.) (2009). *Learning Science in Informal Environments: People, Places, and Pursuits*. Washington, DC: National Academies Press.

Cerini, B. (2011). *Science Heroes in Popular Culture*. Paper presented at the 2011 Conference on Popular Culture in Australia and New Zealand, Auckland.

Cerini, B., Murray, I., & Reiss, M. (Eds.) (2003). *Student Review of the Science Curriculum: Major Findings*. London: National Endowment for Science, Technology and the Arts.

Chandler, D. (1994). *The Transmission Model of Communication*. Downloaded from www.aber.ac.uk/media/Documents/short/trans.html#I on November 1, 2011.

Crane, V., Nicholson, H., Chen, M., & Bitgood, S. (Eds.) (1994). *Informal Science Learning: What Research Says About Television, Science Museums, and Community-based Projects*. Dedham, MA: Research Communications Ltd.

Durant, G. (2011). *Building Capacity Across Africa: Next steps*. Paper presented at the sixth World Congress of Science Centres, Cape Town, September.

Falk, J. H., & Dierking, L. D. (2002). *Lessons Without Limit: How Free Choice Learning is Transforming Education*. Walnut Creek, CA: Altamira Press.

Falk, J. H., & Dierking, L. D. (2000). *Learning from Museums: Visitor Experiences and the Making of Meaning*. Walnut Creek, CA: Altamira Press.

Fensham, P. J. (2008). *Science Education Policy-making: Eleven Emerging Issues*. UNESCO. Downloaded from http://unesdoc.unesco.org/images/0015/001567/156700e.pdf on October 10, 2008.

Fisch, S. M. (2004). *Children's Learning from Educational Television*. Mahwah, NJ: Lawrence Erlbaum Associates.

Fisch, S. M., Yotive, W., Brown, S. K. M., Garner, M. S., & Chen, L. (1997). Science on Saturday morning: Children's perceptions of science in educational and non-educational cartoons. *Journal of Educational Media, 23*, 157–167.

Grotevant, H. D. (1987). Towards a process model of identity formation. *Journal of Adolescent Research, 2*, 203–222.

Junior Chamber International (2010). Press release: Emily Cummins of the United Kingdom selected as one of the 2010 JCI Ten Outstanding Young Persons of the World. Downloaded, from www.jci.cc/guests/w/press/pressreleases?newsid=16850 on October 30, 2011.

Lamberts, R., Grant, W. J., & Martin, A. (2010). *ANU Poll: Public Opinion About Science*. Canberra: Australian National University.

Lockwood, P., & Kunda, Z. (1997). Superstars and me: Predicting the impact of role models on the self. *Journal of Personality and Social Psychology, 73*, 91–103.

McLean, K. (1993). *Planning for People in Museum Exhibitions*. Washington, DC: Association of Science-Technology Centers.

Miller, J. D. (1998). The measurement of civic scientific literacy. *Public Understanding of Science, 7*, 203–223.

Osborne, J. (2006). *Towards a Science Education for All: The Role of Ideas, Evidence and Argument*. Paper presented at the 2006 ACER Research Conference: Boosting Science Learning – What Will It Take? Canberra, Australia.

Pace, S. (2008). The influence of models on the career aspirations of prospective tertiary students: A regional case study. *Studies in Learning Evaluation Innovation and Development, 5*, 10–18.

Priest, S. H. (Ed.) (2010). *Encyclopedia of Science and Technology Communication*, vols 1–2. Thousand Oaks, CA: Sage Publications Ltd.

Rennie, L. J. (2007). Learning science outside of school. In S. K. Abell & N. G. Lederman (Eds.), *Handbook of Research on Science Education* (pp. 125–167). Mahwah, NJ: Lawrence Erlbaum Associates.

Rennie, L. J., & Stocklmayer, S. M. (2003). The communication of science and technology: Past, present and future agendas. *International Journal of Science Education, 25*, 759–773.

Sadler, W. J. (2006). *Evaluating the Short and Long-term Impact of an Interactive Science Show*. Unpublished Master's thesis, Open University, Milton Keynes, UK.

Seifert, T. L. (2004). Understanding student motivation. *Educational Research, 46*, 137–149.

Stocklmayer, S. M., & Gilbert, J. K. (2002a). New experiences and old knowledge: Towards a model for the public awareness of science. *International Journal of Science Education, 24*, 835–858.

Stocklmayer, S., & Gilbert, J. K. (2002b). Informal chemical education. In J. K. Gilbert, O. De Jong, R. Justi, D. F. Treagust, & J. H. Van Driel (Eds.), *Chemical Education Towards Research-based Practice* (pp. 143–164). Dordrecht: Kluwer.

Stocklmayer, S. M., Rennie, L. J., & Gilbert, J. K. (2010). The roles of the formal and informal sectors in the provision of effective science education. *Studies in Science Education, 46*, 1–44.

UNSW (2007). Unpublished data from the World Wide Day In Science program. Downloaded from www.dayinscience.unsw.edu.au/index.html on September 15, 2009.

Walker, G. J., Stocklmayer, S. M., & Grant, W. J. (in press). Science theatre: Changing South African students' intended behaviour toward HIV AIDS. *International Journal of Science Education*.

Wellington, J. (1990). Formal and informal learning in science: The role of interactive science centres. *Physics Education, 25*, 247–252.

Young, R. A. (1993). *Parental Influence in the Career and Educational Development of Children and Adolescents: An Action Perspective*. Paper presented at the fifth Annual International Roundtable on Family, Community and School Partnerships, Atlanta, GA, April 12, 1993.

27
Designing an innovative approach to engage students in learning science
The evolving case of hybridized writing

Stephen M. Ritchie and Louisa Tomas
QUEENSLAND UNIVERSITY OF TECHNOLOGY, AUSTRALIA, AND JAMES COOK UNIVERSITY, AUSTRALIA

Summary: Innovations are usually attributed to ideas generated in the minds of individuals. As we reflect upon the evolving design of an online project to engage students in learning science through hybridized writing activities we propose a more distributed view of the process of innovative design. That is, our experience suggests ideas are generated in the activity of interacting with human and material resources that expand and constrain possibilities. This project is innovative in that it is a new educational response to the problem of disengagement of students in science, and has proven to be effective in changing classroom practice and improving students' scientific literacy. In this chapter, we identify the antecedents and trace the evolution of the project. This account illuminates the innovative design process, presents a summary of the evidence for the effectiveness of the project, and identifies future directions for further development and research.

Key words: Science learning, hybridized writing, case study, innovative approach

Curriculum innovation: an introduction

Curriculum initiatives in science education have been influenced heavily by psychological perspectives that focus on the individual learner as the unit of instruction and the unit of analysis in related research (see Bell, Cowie, & Jones, 2009; Fensham, 2009; Gunstone & Treagust, 2009). Accounting for the antecedents of research into alternative conceptions and conceptual change in Australia, for example, Gunstone and Treagust (2009) identified "in the head" perspectives that include information processing, "meaningful learning" and "cognitive structures" as the major theoretical influences (p. 187). Additional perspectives introduced over time have been more social, discursive, and dialectical (e.g., Bell et al., 2009), and these have reduced the hegemony of individualistic perspectives. In particular, many science educators "switched to the idea that knowing and knowedgeability are better thought of as cultural practices that are exhibited by practitioners belonging to various communities" (Roth & Lee, 2006, p. 27). The implication of these alternative perspectives for educators suggests that learning should no longer be viewed "as an asymptotic adaptation to the environment but a continuous transformation of individual and collective practices that themselves change the collectivity and its setting" (Roth, 1996a, p. 218).

In the relatively new field of innovation research in education, there is still a predominance of individualistic perspectives of innovation and the generation of ideas. For example, in promoting the possible benefits of high intellectual and creative educational multimedia technologies, Shavinina (2000; see also this volume) argued, "the human mind is a unique phenomenon in nature, and its development should be realized by individual means, which are appropriate for the increase of mental resources of a given person" (p. 193). This individualistic perspective assumes "that educational process is a psychological process" (p. 194) and "[a]ny learning, teaching, and training are based on the fundamental psychological mechanisms" (p. 194). Yet, Rogoff (1990) recognized that creative ideas are reformulations of existing ideas that occur "in the context of a community of thinkers (artists, inventors, scientists), where more than one person is working on the solution of a particular problem or within the particular genre of expression" (p. 198). Moreover, as Roth and Roychoudhury (1993) noted about the outcomes from ethnographic research of scientists' ideas and discoveries,

> there is ample evidence that scientific objects (knowledge) are co-constructed by groups of individuals who are unavoidably linked to their contexts. The source of an idea cannot be attributed to either individuals or environment, and has to be considered a social construct contingent on the local conditions.
>
> (p. 524)

Consistent with views of creative ideas and learning that decentre the individual as "a lone ranger in terms of innovating and pushing the boundaries of our field" (Tobin & Roth, 2002, p. 277), Tobin repositioned innovations as culturally embedded products that reflect social, cultural, and historical trajectories. For this reason, we wonder whether innovation should be studied in ways that are more cognizant of the generation of creative ideas within social contexts where the ownership of the ideas is not attributed exclusively to individuals, but is viewed as the product of interactions between individuals, and between an individual and collectives.

In this chapter, we continue to draw out the implications of developments in learning science in classrooms with the way educators should begin to think about innovation in education, particularly curriculum innovation. We do so by reflecting on the development and production of our innovative curricular and pedagogical moves to engage students in science activities. In particular, we focus on the case of engaging students in completing online *BioStories* that require them to transform scientific information into hybridized text that expresses technical content in everyday language within a given social context. Before we begin this reflective account of the design and evolution of the BioStories project, we exemplify research in science education that has highlighted the need to go beyond individualistic psychological perspectives of learning and the generation of new ideas.

Generating ideas with others

While numerous studies have employed non-individualistic sociocultural learning theories to understand the production of knowledge in science classrooms (e.g., Barab, Barnett, Yamagata-Lynch, Squire, & Keating, 2002; Roth, 1995; Varelas, Pappas, & Rife, 2005), arguably Wolff-Michael Roth has been the most prominent leader in this field. For this reason, we feature Roth's germinal contributions by summarizing his findings that support broadening perspectives to include those that decentre the individual in the process of generating ideas, learning, and innovation in curriculum design.

In a series of articles and book chapters, Roth (1995, 1996a, Roth, Tobin, & Ritchie, 2001) reported on the transformation of practices or learning that took place in a grade 4/5 class engaged in engineering activities that required students to design bridge structures using common household materials and tools. As students worked in pairs they appropriated structures seen in supplementary activities that included viewing films and visiting a museum. They also fixed materials (e.g., with a glue gun), strengthened (i.e., with triangular braces), and adorned (i.e., with a flag) structures in ways that at times were tried by other groups in the class. Interestingly, some students laid claim to specific ideas that then diffused elsewhere within the class without evidence of direct copying, whereas others deliberately copied ideas they had seen. Collectively, the class benefited from the dispersion of ideas and practices. Roth concluded that inventors could not be clearly identified when working in close proximity within a community characterized by high levels of interaction because focus on the same task/problem with access to similar resources can generate ideas almost simultaneously, as if there were *ideas in the air* (see Schoenfeld, 1989). Roth's investigation showed that a focus on the individual was only part of the evolution of the larger system or community, "and thus a partial view of learning" (1996a, p. 217).

Roth (1996b; Hwang, Roth, & Pozzer-Ardenghl, 2005; Roth, Hwang, Goulart, & Lee, 2005; Roth et al., 2001) continued the study of student learning in small groups with a grade 6/7 class that engaged in a unit on machines. An important theme that emerged from these studies was how knowledge was produced and reproduced in collaborative practice as small groups engaged in activities designed and taught by Roth. In these innovative design activities, collaborative practice was much more than the sum of individual actions. Just as individual subjectivities of the students were made available concretely to others through their actions, their interactions created additional possibilities for action (or room to manoeuvre) for the collective or collaborative group. For this reason, Hwang et al. (2005) argued that collaboration should be understood in terms of

> reading between the actions (in the same way we read between the lines). Such reading transcends reading between the lines in that actions are always concrete modes of acting human bodies extending beyond the words – actions speak louder than words.
>
> (p. 67)

We attempt to study the design and evolution of BioStories in terms of collaborative practice where we read between the actions, similar to the accounts of collaborative practice recorded in Ritchie (2007).

Background to the BioStories project

The online BioStories project required students to complete a sequence of story templates about biosecurity in such a way as to afford students opportunities to demonstrate their understanding of scientific/technical information, located in listed websites, within the provided storyline. More specifically, three writing tasks were developed, progressively fading out the scaffolding such that the final task was an open-written task. The first task involved completing a conversation between the two central characters (an expert with a novice) about an allocated biological incursion that has affected Australian ecosystems (e.g., fire ants); the second task required students to adopt the role of an expert by writing a story about the possible consequences of an incursion yet to affect Australia (e.g., varroa mite); and the final task was open ended because the students could choose to write their own story involving any of the ideas they had gleaned from their own or others' stories.

The technical information produced on government approved websites (e.g., www.dpi.qld.gov.au/4790_13505.htm) was directly linked to the resources page of the BioStories website. This ensured that accuracy and relevance of information could be checked prior to student use, and search time for relevant technical information by students was minimized. Students uploaded their stories to the BioStories webpage so that all participants could read and critique each other's stories, thus exposing them progressively to a wider range of information about different biological incursions and writing styles.

In writing BioStories students generate what is called hybridized text. Hybridized writing is a diversified approach to writing (Prain, 2006) that involves "crossing borders" between formal writing of technical information and everyday language conventions. For example, the following sample illustrates how technical information about a hawksbill turtle can be merged with the narrative storyline about solving a series of turtle killings on the beach in the language of the characters—in this case, a park ranger/father who is communicating with his children:

> Dad got there first. "It's a hawksbill. See its mouth. It's a bit like a beak. This is only the second hawksbill I've seen around here. No doubt about it, it's been stabbed to death. It's had time to dig the body pit and the egg chamber but not time to lay any eggs. Let's hope this wasn't its first batch."
> "What do you mean Dad?" asked Elisha.
>
> *(Year 4 Students, 2006, p. 15)*

The purpose of hybridized writing is principally to engage students in a learning activity with which students feel comfortable. Wellington and Osborne (2001, p. 76), for example, reasoned,

> if we wish to engage children with ideas in science, we should at least offer activities that initiate writing in science in a manner which is enjoyable. Using a familiar genre at least begins the process of helping children to express their thoughts in written language through being personally engaged.

Apart from making pragmatic sense that hybridized writing would engage students in science activities, there is growing empirical evidence to support this stance. Diversified writing tasks, including more imaginative writing, have been shown to assist students' learning processes, improve learning outcomes, have strong motivating effects and impact positively on students' attitudes and engagement (e.g., Hand & Prain, 1995; Prain & Hand, 1996, 1999). Specific to research related to BioStories, outcomes from a sequence of studies have shown that engaging students' imaginations in writing hybridized scientific narratives about biosecurity can enhance their scientific literacy while developing more positive attitudes toward science and science learning, and elicit positive emotional responses (Ritchie, Tomas, & Tones, 2011; Tomas, Ritchie, & Tones, 2011; Tomas & Ritchie, 2012). This research supports the inclusion of hybridized scientific genres in science curricula as a way of engaging students meaningfully in science learning (see also Ritchie & Tomas, 2012).

Reflective account[1] of the design and evolution of BioStories

The BioStories project arose from a fourth-grade classroom project conceptualized initially by Ritchie (i.e., the first listed author of this chapter), albeit with modifications from a previous project he had sponsored, and implemented by his colleague (Rigano) with a classroom teacher (Duane) in which an eco-mystery was co-authored by the class members and later published. Interestingly,

Tomas (i.e., the second listed author of this chapter) was employed casually towards the end of the project in the capacity as research assistant while she was a pre-service education student. The project was successful in that the children were engaged in the project throughout the year and they became fluent with their use of scientific concepts in everyday contexts (see Ritchie, Rigano, & Duane, 2008). Yet, one year was too long for most teachers to devote to such a demanding schedule with an overcrowded curriculum, and the cost of publishing books was prohibitive for schools. Additionally, the benefits of this project were constrained to the classroom participants. As Ritchie discussed these constraints with his then colleague, and to a lesser extent, Tomas, they wondered whether it was feasible to design a related project that addressed these constraints but yielded similar outcomes. Through these interactions, the first idea was to conceptualize a large-scale online writing competition configured in such a way that students or classes could upload stories on a given topic where the winner was determined by popular vote, in much the same way as popular television talent shows—thus reducing the need for a formal arrangement to appoint well-qualified judges. It might have been possible for Ritchie to initiate this idea "independently," but it emerged following interactions about the constraints experienced with the eco-mystery project and in cognizance of popular formats in the media. The issues of topic or format were yet to be determined, but they were hopeful that a national competition might attract funding from companies or philanthropists. After all, there had been national mathematics contests sponsored by a bank and other national contests that had received financial support. This was an important consideration because they wanted the project to be as much an outreach project as a research project—both requiring financial support for the necessary development and subsequent research.

In the first instance, information was sought from a known science communicator about possible further action. A subsequent meeting between the science communicator and Ritchie led to a list of additional contacts with other science communicators and government agencies that might be interested in supporting the project. Contact with the outreach coordinator from the Department of Primary Industries (DPI) proved the most fruitful meeting because the related topics of quarantine and biosecurity were identified as an area that had attracted education funding and advertising. Especially prominent was the successful *Quarantine Matters* television campaign that featured Steve Irwin—the legendary environmentalist better known as the "crocodile hunter." Through discussions with departmental officials, it seemed that biological incursions in periurban communities (i.e., semi-rural communities situated on the fringe of large municipalities) were particularly worrying. Interestingly, the science education literature had identified issues that frequently required moral and ethical decision making whereby science concepts could be learned within social contexts—known as scoioscientific issues (SSI) (see Bencze, this volume)—as valuable topics in the science curriculum to engage learners. The outcome from this meeting was that Ritchie would need to develop the concept further before any project could attract funding. He then began to write the first of the BioStories templates with the late Steve Irwin as the central character.

After drafting the first template called *Crikey*, Ritchie tried it out with his pre-service teacher education students. Actually, this draft was Part A of Crikey that focused on a conversation between Steve Irwin and a 13-year-old student about the damage done to natural and agricultural ecosystems by such incursions as fire ants, citrus canker, and tilapia. Around this time, and based on earlier success with the eco-mystery publication, Ritchie and Rigano were invited to provide a workshop for 250 middle-years students at the Brisbane Writers' Festival. With the assistance of Ritchie's volunteer pre-service science-teaching students, the workshop was conducted quite successfully. As student representatives read out their completed stories on stage, Ritchie noticed how attentive the audience was and how they might have benefitted from hearing about a range of incursions. This experience provided the impetus for Ritchie to draft Part B (i.e., incursions yet to arrive in

Australia) and Part C (i.e., open-ended story). These drafts were refined after further consultation and input from his colleague (i.e., Rigano).

The next challenge was to design a way to provide access to a wider audience online. The Institute for Sustainable Resources at Queensland University of Technology (QUT) provided seeding funds to develop a web page that would enable student access to relevant resources and become a repository for completed stories for all participants to view and comment. The latter feature of the online project enabled students to read stories written by children in other classes, thus dispersing ideas and writing practices more broadly (cf. Roth, 1996a). Ritchie employed a doctoral candidate to develop the web page. Conversations about the design brief and suggestions offered by the candidate led to a pilot version of the web page that was tried in several sixth-grade classes that had access to computers. The results from this trial were reported in Ritchie et al. (2011). Interest in a national writing competition faded in preference for designing innovative research as the results were analysed and the potential for making a significant contribution to the field of scientific literacy and socioscientific issues was realized.

After this trial, two important events occurred. First, Tomas became a doctoral candidate who was interested in extending the current trial for her project. Second, Ritchie won a grant from the auDA Foundation to support expansion of the project with grade 9 students. After revising the trial materials, we designed a small-scale quasi-experimental study to complement the trial study. The results from this study reinforced the earlier trial and both sets of results were published in the same article. This work was conducted concurrently with a larger-scale study sponsored by the auDA Foundation. The results from one school's implementation formed the basis of Tomas's doctoral thesis (see Tomas et al., 2011). Overall, there were several design changes contributed by Tomas and a new web-page designer (informed by their experiences of the grade 6 study) in ongoing discussions with Ritchie.

Upon completion of her doctoral thesis, Tomas reported a summary of her work in a professional magazine for teachers. This attracted interest from a school head of science who could see potential involving older students. With a small grant from her own university (i.e., Tomas had been appointed to James Cook University shortly before her thesis completion), we continued our collaboration. Up to this point, we had focused on conceptual and attitudinal outcomes of these writing projects. Due to Ritchie's engagement in a different project involving teachers' arousal of emotions during classroom interactions (e.g., Ritchie, Tobin, Hudson, Roth, & Mergard, 2011), students' emotions experienced in the moment became a new focus for our BioStories research. For the first time we (i.e., Tomas & Ritchie, 2012) used the Facial Action Coding System (Ekman & Friesen, 1978) to determine students' experienced emotions as they interacted with the BioStories webpage and completed the various writing tasks—a procedure that also has been explored more recently through computer facial recognition software (e.g., Ritchie et al., in press).

As attention focused on emotional arousal of students while they were engaged in the writing activities, Ritchie recognized potential for the design of a large-scale study that would explore the benefits of the emotional engagement of middle-years students in science activities that included but surpassed these writing activities. Two other school leaders also invited him to set up research programmes in their schools. These interactions and his collaborations with Tomas and other former doctoral students who were now colleagues at QUT, encouraged him to write a new grant that not only would lead to interesting outcomes, but also be a catalyst for the research careers for these early career researchers. With interest expressed by all parties, Ritchie immediately invited his long-term research collaborator from the USA (i.e., Tobin) to join the research team. Tomas recruited another school, and Ritchie wrote the grant in concert with the other Chief Investigators. This grant application was successful.

The plan was to replicate the BioStories activities on biosecurity in the first of three years of the new project as a form of professional development for the teachers that would provide leverage for the design of new activities in subsequent years. By the time the project team was ready to plan for the start of this grant, the new Australian Curriculum had been approved, but with a different content focus. The new constraints imposed externally on this occasion were that instead of an ecological unit that would suit a focus on biosecurity, new topics in the mandated eighth-grade curriculum included cells (with elaborations on tissues and organs) and energy. Possible solutions within the team were discussed. A particular focus was consideration of how the new topics could be converted into socioscientific issues that would structure classroom activities. The constraints afforded the team the opportunity for the generation of new ideas, and design innovative activities that extended the original plans before the start of the new grant. Ongoing interactions with researchers and school partners have transformed further the BioStories activities because both researchers and teachers have a vested interest in developing new activities that have potential for engaging students emotionally with substantive learning outcomes.

Even though the team is yet to evaluate the effectiveness of the new (writing) activities, the BioStories project has evolved over eight years and is likely to continue to be transformed as the co-researchers interact about curriculum design and research design/results issues. Over this period, Ritchie's participation was the one constant, but there were numerous other contributors whose input was essential for the particular direction the project took. These contributions were not limited to direct involvement by individuals, but also included external constraints by new curricular requirements and changing school structures and professional dynamics over time.

Implications for conceptualizing innovation

The preceding account was our attempt of mapping out the major developments and refinements of our BioStories project. We referred individually to researchers/developers to emphasize that these people had significant roles in the process, and our interactions with them contributed substantively to the development of the project. Even though we both feel it represents accurately the trajectory of the project, others could have different recollections of the design of the project from their different vantage points, possibly dependent on their commitment and enthusiasm for those stages of the project in which they participated. Nevertheless, during our interactions with these individuals and others, ideas were generated through and subsequent to the conversations. Ideas were not so much *in the air* as suggested by Schoenfeld (1989), but rather emergent from these interactions as they were related both simultaneously and subsequently to the experiences, the ideas of which we had become aware during other conversations, and constraints we experienced and identified along the way. Sometimes we, individually or collectively, would have *ah ha* moments of clarity from which ideas emerged for further consideration. These ideas would either be sustained (e.g., hybridized writing) or fade from prominence (e.g., national competition). Sustained ideas persisted throughout the project. Yet, as the project continues to evolve, these features have the potential to lose prominence, not because of any diminishing clarity, but rather because the possibilities of further refinements and/or research become exhausted, at least until interactions generate new possibilities.

From our experience, and from reading between our actions, individuals have been important to the project. They were important, not so much because they contributed unique or original ideas, but rather they created dynamic contexts through the sharing of resources; only the contributions of these individuals in our interactions could help produce the emergent ideas. The essence of this dialectical relationship between individual and collective within team interactions was captured nicely in Roth and Lee's (2006) conceptualization of action: "an individual concretely realizes an action (such as learning), the possibility of which exists at a collective (generalized) level. What the individual does defines and has repercussions for the collective such as affirming what are legitimate or illegitimate practices" (p. 28); and so it is with innovation in education, as illustrated through our account of the evolution of BioStories, specifically. The collective creates the possibilities for individuals to interact and generate ideas that are then realized through the actions of individuals. Tasks can be and were completed individually (i.e., division of labour), but in relation to the collective goals for the project. This *collaborative practice* involves,

> the actions of individual designers, who interact to bring separate individual ideas to bear on collective motives. Collaborative practice produces and reproduces not only the design artifact as a collective object but also the artifact designers themselves as a collective (designer) subject.
>
> *(Hwang et al., 2005, p. 54)*

Even though Hwang et al. were referring to student designers of engineering artefacts, we would argue, based on our experience, this practice was equally observable throughout the innovative curricular design of the BioStories project. Moreover, they claimed that intersubjectivity is achieved in collaborative practice through two dominant processes:

1. face-to-face discursive and bodily communication through which participants attend to and act on the same component of the object; and
2. division of labour within an activity where participants act on different parts of an object, and where they—even in the absence of talk—contribute to the production and reproduction of their shared actions (p. 55).

The individual-collective dialectic nature of the innovative actions in the latest developments of the project was particularly noticeable. The team was forced to redesign the activities because one of the teachers from the partner schools observed that biosecurity would no longer be a satisfactory topic in grade 8 with the Australian Curriculum. This new constraint set off a series of meetings between members of the research team to identify possible new topics to replace biosecurity that would align with the Australian Curriculum. Brainstorming was a process in which the team engaged initially. Some possibilities were considered further in terms of suitability as an SSI and writing task that would be engaging for young students. Two topics were identified as possibilities; organ transplants and retrieval of coal seam gas (i.e., fracking). Organ transplants would be a suitable focus for the cells theme while the fracking topic would be suitable for the energy theme. The recursive relationship between individual and collective was apparent throughout these actions.

While innovations are developed in response to particular needs or problems, constraints impact on the design process and innovative products. The constraint of time, and the need to make the project more accessible by more teachers, prompted Ritchie with others to design online writing activities for students that could be completed satisfactorily within a term rather than a year. The constraint of a new Australian Curriculum also created an opportunity to expand topics from biosecurity to organ and tissue transplantation and fracking. Constraints relating to computer access and availability for classes, quality of information or material resources, and different preferences from the partner schools, all will impact on how the project will continue to evolve. Rather than impediments to change, constraints can expand design opportunities for innovation (Roth et al., 2001).

Our reflective account appears to map out the project development chronologically. This artefact of our writing masks how several phases of the project occurred simultaneously, or at least much more disorderly. For example, the most fruitful meeting at the DPI was actually scheduled in between meetings with different science communicators. So rather than a clear resolution arising from the DPI meeting, tentative ideas from that meeting were "tested" at subsequent meetings until one of the messages from that meeting became the overall preferred course of action. Similarly, the various parts of Crikey were written sequentially, but each was written with a general image of how the three parts would relate to each other.

In conclusion, as the conceptualization of innovation is now considered, and perhaps how innovation in education should be studied, the literature on designing might provide a useful reference. The following points made in relation to designing engineering projects in schools (Roth et al., 2001), might apply equally to innovation in education. These points have been modified in terms of innovation in education, and were experienced at various stages through the BioStories curriculum innovation.

1 Innovation is not a cognitive activity that can be reduced to individual mental processes. Decisions taken by project developers cannot be understood apart from the tools, materials, artefacts, and constraints (p. 29).
2 Innovation does not occur in a vacuum; it occurs in a nexus of psychological, historical, sociological, and material conditions of the setting or context. Multiple individuals usually contribute to the development of the innovation at various stages throughout the process. The innovation is an assemblage arising from the conflux of several resources (pp. 29–30).
3 The emerging innovation provides a context for future innovative moves (p. 30).
4 Innovation is a non-linear and situated activity that attempts to bring order to messy situations (p. 135).

The study of innovation is likely to be as messy as other relational constructs. For example, the relational construct of collaboration has been studied by such varied means as personal-lived experience via self-reflective accounts and interview, and micro-analysis of video-recorded team meetings (see Ritchie, 2007). It is the latter method that offers the most exciting potential for gaining a deeper understanding of innovation, as it occurs in the moment. Innovative methods that are planned to apply in the ongoing study of engaging students in science (writing) activities, might also be useful in the study of innovation and collaborative practice (see Tobin & Ritchie, 2012). Important phenomena that could be studied during innovation could include those that were suggested by Ritchie for collaboration; namely, positive emotional energy, solidarity and trust, as well as emotional arousal and leadership.

Acknowledgments

We are grateful for the funding support from the auDA Foundation and the Australian Research Council (i.e. LP110200368) throughout the development and associated evolving program of research identified in this chapter.

Note

1 In this account "we" and "our" are reserved for Ritchie and Tomas. Individuals are identified by name when particular ideas and activities need to be distinguished between our activities or those of the project/research team.

References

Barab, S. A., Barnett, M., Yamagata-Lynch, L., Squire, K., & Keating, T. (2002). Using activity theory to understand the systemic tensions characterizing a technology-rich introductory astronomy course. *Mind, Culture, and Activity, 9*, 76–107.
Bell, B., Cowie, B., & Jones, A. (2009). Theorising learning in science. In S. M. Ritchie (Ed.), *The World of Science Education: Handbook of Research in Australasia* (pp. 85–105). Rotterdam, Netherlands: Sense Publishers.
Ekman, P., & Friesen, W. V. (1978). *The Facial Action Coding System*. Palo Alto, CA: Consulting Psychologists Press.
Fensham, P. J. (2009). The genesis of science education research in Australasia. In S. M. Ritchie (Ed.), *The World of Science Education: Handbook of Research in Australasia* (pp. 9–15). Rotterdam, Netherlands: Sense Publishers.
Gunstone, R. F., & Treagust, D. F. (2009). Conceptual change in science: Research at the forefront over the past three decades. In S. M. Ritchie (Ed.), *The World of Science Education: Handbook of Research in Australasia* (pp. 185–197). Rotterdam, Netherlands: Sense Publishers.
Hand, B., & Prain, V. (1995). *Teaching and Learning in Science: The Constructivist Classroom*. Sydney: Harcourt Brace.
Hwang, S. W., Roth, W.-M., & Pozzer-Ardenghi, L. (2005). Understanding collaborative practice. *Outlines. Critical Practice Studies, 1*, 50–79.
Prain, V. (2006). Learning from writing in secondary science: Some theoretical and practical implications. *International Journal of Science Education, 28*, 179–201.
Prain, V., & Hand, B. (1996). Writing to learn in the junior secondary science classroom: Issues arising from a case study. *International Journal of Science Education, 18*, 117–128.
Prain, V., & Hand, B. (1999). Student's perceptions of writing-to-learn in secondary school science. *Science Education, 83*, 151–162.
Ritchie, S. M. (Ed.) (2007). *Research Collaboration: Relationships and Praxis*. Rotterdam, Netherlands: Sense Publishers.
Ritchie, S. M., Rigano, D. L., & Duane, A. (2008). Writing an ecological mystery in class: Merging genres and learning science. *International Journal of Science Education, 30*, 143–166.
Ritchie, S. M., Tobin, K., Sandhu, M., Sandhu, S., Henderson, S., & Roth, W.-M. (in press). Emotional arousal of beginning physics teachers during extended experimental investigations. *Journal of Research in Science Teaching*.
Ritchie, S. M., Tobin, K., Hudson, P., Roth, W.-M., & Mergard, V. (2011). Reproducing successful rituals in bad times: Exploring emotional interactions of a new science teacher. *Science Education, 95*, 745–765.
Ritchie, S. M., & Tomas, L. (2012). Hybridized writing for scientific literacy: Pedagogy and evidence. In R. M. Gillies (Ed.), *Pedagogy: New Developments in the Learning Sciences* (pp. 213–226). Hauppauge: Nova Publishers.
Ritchie, S. M., Tomas, L., & Tones, M. (2011). Writing stories to enhance scientific literacy. *International Journal of Science Education, 33*, 685–707.
Rogoff, B. (1990). *Apprenticeship in Thinking: Cognitive Development in Social Context*. New York: Oxford University Press.
Roth, W.-M. (1995). Inventors, copycats, and everyone else: The emergence of shared resources and practices as defining aspects of classroom communities. *Science Education, 79*, 475–502.
Roth, W.-M. (1996a). Knowledge diffusion in a grade 4–5 classroom during a unit on civil engineering: An analysis of a classroom community in terms of its changing resources and practices. *Cognition and Instruction, 14*, 179–220.
Roth, W.-M. (1996b). Thinking with hands, eyes, and signs: Multimodal science talk in a grade 6/7 unit on simple machines. *Interactive Learning Environments, 4*, 170–187.
Roth, W.-M., Hwang, S., Goulart, M. I. M., & Lee, Y. J. (2005). *Participation, Learning, and Identity: Dialectical Perspectives*. Berlin: Lehmanns Media—Lob.de.
Roth, W.-M., & Lee, Y.-J. (2006). Contradictions in theorizing and implementing communities in education. *Educational Research Review, 1*, 27–40.
Roth, W.-M., & Roychoudhury, A. (1993). The concept map as a tool for the collaborative construction of knowledge: A microanalysis of high school physics students. *Journal of Research in Science Teaching, 30*, 503–534.
Roth, W.-M., Tobin, K., & Ritchie, S. (2001). *Re/constructing Elementary Science*. New York: Peter Lang.
Schoenfeld, A. H. (1989). Ideas in the air: Speculation on small group learning, environmental and cultural influences on cognition, and epistemology. *International Journal of Educational Research, 13*, 71–88.

Shavinina, L. V. (2000). High intellectual and creative educational multimedia technologies. *CyberPsychology and Behavior, 3*, 191–198.

Tobin, K., & Roth, W.-M. (2002). The contradictions in science education peer review and possibilities for change. *Research in Science Education, 32*, 269–280.

Tobin, K., & Ritchie, S. M. (2012). Multi-method, multi-theoretical, multi-level research in the learning sciences. *Asia-Pacific Education Researcher, 20*, 117–129.

Tomas, L., & Ritchie, S. M. (2012). Positive emotional responses to hybridised writing about a socioscientific issue. *Research in Science Education, 42*, 25–49.

Tomas, L., Ritchie, S. M., & Tones, M. (2011). Attitudinal impact of hybridized writing about a socioscientific issue. *Journal of Research in Science Teaching, 48*, 878–900.

Varelas, M., Pappas, C. C., & Rife, A. (2005). Dialogic inquiry in an urban second-grade classroom: How intertextuality shapes and is shaped by social interactions and scientific understandings. In R. Yerrick & W.-M. Roth (Eds.), *Establishing Scientific Classroom Discourse Communities: Multiple Voices of Teaching and Learning in Research* (pp. 139–168). Mahwah, NJ: Lawrence Erlbaum.

Wellington, J., & Osborne, J. (2001). *Language and Literacy in Science Education*. Buckingham, UK: Open University Press.

Year 4 Students (2006). *Ocean Action: An Adventure in Beachtown*. Townsville, Qld: James Cook University.

28
An integrated approach to the study of biology

Sergei Danilov and Olga Danilova

TARAS SHEVCHENKO NATIONAL UNIVERSITY OF KIEV AND KIEV NATIONAL UNIVERSITY OF ARCHITECTURE, UKRAINE

Avoid boring people.

(James Dewey Watson)

Summary: The use of computer based learning (CBL) systems in teaching biology provides a prime opportunity to implement educational technologies aimed at developing analytical and practical skills, promoting individualized orientation to education and a healthy way of life, providing equal access to knowledge and individualized knowledge testing and self-checking, and ecologizing education and creating foundations for nature-conservation activity. These technologies will contribute to the development of students' creative and innovation abilities.

Key words: Computer based learning (CBL) systems, biology, methodological apparatus, educational technologies.

Introduction

In a modern innovative society education should play a crucial role in the creation of a more humane world order. There is a manifest need to reorient teaching from the accumulation of a sum of knowledge toward shaping citizens capable of innovative work. The acquisition of certain competences – the desire to learn throughout one's life, an interest in scientific research and the practical use of biological knowledge – lay foundations both for personal social success and the stable development of society.

In its simplest interpretation, biology is the science of life and development of living things. Studying the subject "biology" in schools on a verbal level alone does not give a true picture of the objects and phenomena under study (Traytak, 2002). One way to solve this problem is to use new teaching technologies, and, in particular for biology, an integrated approach incorporating the use of information-communications technology.

To implement such an approach, a computer based learning system (CBL system) may be used. Such CBL systems are suitable for use with interactive whiteboards and for conducting lessons using inter-dialogic and active learning techniques. In an ideal situation an online system would be desirable whereby a team of researchers, university and pedagogical college students, and teachers would be able to analyze the students' work, answer their questions and explain new scientific advances and their practical application.

An innovative computer based learning system

The methodological apparatus of the system may contain the following elements:

- A course schedule based on the syllabus and possible methods of teaching it within the framework of the given topics.
- Texts on the topics (which could be in the form of games, using various techniques to stimulate interest in the material being studied), divided into two types: (1) core teaching materials in accordance with the syllabus, (2) extension materials, which analyze the way in which the material being studied links to the other natural sciences and the humanities; and practical applications of the scientific achievements of biology in agricultural production, medicine, psychology, ecology, the protection of the environment, biotechnology and in daily life.
- Illustrative material – diagrams, drawings, graphs, photographs, microscopic photographs, electron diffraction photographs, animation, video clips, interactive models of complex biological processes and the opportunity to design them.
- A set of readings, a reference book, a dictionary of terms with semantic explanations and etymology, a terminological index, and a set of procedures and virtual simulator for carrying out laboratory and practical work. The presence in the CBL system of classic readings and reference books gives the teacher and students, who are used to working with printed material, the opportunity to use their time effectively.
- A virtual trainer which includes several types of tests with different aims. (1) Knowledge testing tasks which stimulate interest and motivation to study the particular topic and set up the motivation for performing work and the logic for drawing conclusions from it. (2) Practice tests for students who are not sure of their knowledge. These students are invited to work in the practice area, where mistakes are shown during the performance of various test tasks, as is the time taken to complete the test. (3) Tests to monitor the students' progress. These should happen in every lesson and for each topic, and on completion of the course, at various levels of difficulty (to be selected by the student). The tests should take various forms: matching questions, completion questions, analysis of tables, diagrams, graphs and photographs, and solving practical problems.
- Tasks for independent work by the students, preparation for competitions and Olympiads, and discussions of current issues in biology. It is important to set tasks for independent work which is of a career-oriented nature, in order to familiarize students with the use of biological knowledge in various spheres of professional activity.
- The opportunity to connect to the Internet.
- Computer games using the study material.

We[1] have attempted to create CBL systems of this nature (Danilova, Danilov, & Shabanov, 2006; Makarchuk, 2006; Shabanov & Kozlenko, 2009). As yet, they do not fully meet our own requirements for a system of this type. Our experience of using these systems in the senior year of comprehensive secondary school showed their potential in implementing the following teaching technologies.

I. Procedural technology, supporting an active approach to learning. Lessons begin with an illustrated introduction comprising motivation to study and links to previous lessons or themes. The process of learning that follows uses the following procedure:

1 *Analysis* of modern scientific facts, introduced in the form of diagrams, tables, animation, graphs, video clips and experiments.
2 *Synthesis.* Forming inferences and conclusions to master the lesson content. In order to complete the synthesis, the students use test tasks which lead from one component of the lesson to another. The student uses the dictionary for this and may be given references to other lessons where there is information on or discussion of the problem being studied.

It may happen that the student is unable to manage the task and cannot draw the conclusions. In this case he or she can read the prepared instructional material in the set of readings included in the system. However, our aim is not to make the student simply learn from the instructional material, but to teach him or her to think, and therefore we encourage the student to work according to the procedures outlined above and gradually bring him or her to an understanding of the logic of learning and the skill of drawing conclusions. To interest the student who is not motivated to learn biology, we try to introduce into the lesson various "items of interest", game-playing interludes and even jokes. Emotional color means better memorizing of educational material. The following type of test is an example:

> Imagine that it is summer and you are walking barefoot on the grass. The grass seems cool because: A) it is cooled by the earth; B) your feet are poor conductors of heat; C) the blades of grass shade one another from the sun's rays; D) water is evaporated from the blades of grass.

The "Water and aqueous solutions" lesson demonstrates the procedure outlined. The first stage is analysis. The software allows the student the opportunity to work out his or her own daily water requirements; after this motivation he or she then needs to discover why water has such an important significance for sustaining life. The transition is made to studying the chemical and physical properties of water. The student analyzes animations and diagrams demonstrating the various states of water and its properties (viscosity, evaporation, capillary properties, surface tension and so on), carries out laboratory work testing the solubility in water of various substances found in the human body by comparison with their solubility in oil and alcohol. In order to highlight what is important about the analysis, test tasks are used, which lead from one component to the next. After analysis of all the material the student must carry out synthesis, showing what significance all the properties of water he or she has found have for sustaining life and the survival of the ecosystem. If some students need to have the work made easier, they may be set an interactive table where the properties of water are given in the right hand side, and the student selects on the left hand side the biological significance that each property has. This can be done several times, using a trial and error method, until the right answer is found. This allows for the best assimilation of the taught material.

   II. *The technology of developing practical skills to use acquired knowledge* in one's own activities. The practical value of the material being studied is demonstrated in every lesson, most often being taught in such a way that the student independently comes to conclusions about where, when and in what life situations this knowledge and these skills can be of use to him or her. The student gains the skill of working with biological objects by completing laboratory and practical work (Danilova, Danilov, & Goryanaya, 2007). We hold the opinion that they should be a logical constituent part of the study of certain topics, as material for analysis and synthesis. As well as a description of the process of laboratory work, students can see it carried out in the video clips. If the student is able to follow instructions independently, this may be omitted.

   During laboratory work, test tasks are also suggested for the students who do not manage to draw their own conclusions. These students can do the test tasks and independently evaluate their own performance. Other teachers prefer that students carry out such work in a separate laboratory practicum. There is a certain advantage in this approach, which is that the revision of learning material allows the student to gain better mastery of it. The end result of carrying out laboratory work should be conclusions, which can be drawn either through the test tasks or independently. This sort of approach sharpens the thinking process.

When studying historical materials, for example on the formulation of cell theory, it may be possible to invite some students to "walk the path" of the investigator – making preparations from oak bark and other biological objects using the methods of Hooke, Leeuwenhoek, Schleiden, Schwann and Purkyně, to attempt themselves to come to conclusions drawn by 19th century scientists. Excursions into history (which students don't like very much) will then become relevant to them, as a way of understanding the logic of scientific cognition. If it is not possible to make the preparations, pre-prepared microslides can be used to the same end. When studying the structure of the DNA molecule, students could manipulate models of nucleotides to build the nucleic acid molecules on the basic principle of complementarity. The result is that the students replicate, so to speak, the scientists' logic that unlocked the structure of DNA (Watson and Crick's model). In addition, they understand the principle of complementarity and the interaction of various biomolecules and viruses with the host cell and other processes.

The CBL system gives the student the opportunity independently to learn how to make a number of microslides, set up experiments and construct models. For example, after determining the chemical composition of membranes, the system gives the student the chance to construct a model of membrane structure from lipids, proteins and carbohydrates. The polytechnic orientation of its content provides familiarity with the use of fertilizers and the conservation of soil fertility, methods of artificially obtaining essential human biological molecules, biotechnology methods, methods of diagnosing illness in plants and animals, indication methods in ecology, geology and more. For example, under the rubric "That's interesting" a task may be suggested that can be carried out at school and in some cases at home. On a glass slide place a drop of milk and examine it under a microscope. Observe that the milk is a solution of fat and water, or an emulsion. The manufacture of food products often involves the use of emulsifiers, which stabilize them in the state of an emulsion (ice cream, mayonnaise and so on). Try to stabilize an emulsion with a soda solution. Put 0.5 ml of vegetable oil and 2–3 ml of water into a test tube and mix well. You will get an opaque emulsion. Let the test tube stand for five minutes. The emulsion will separate into oil at the top and water at the bottom. In another test tube instead of water add a 1% soda solution (Na_2CO_3) to the oil. Let the test tube stand for five minutes. Has the emulsion become more stable? Here is another simple task. By using an iodine reaction it is possible to detect the starch present in any given food product; for example in salami, honey and so on.

III. The technology of promoting a healthy way of life. We moved this point into its own separate technology as a result of our work experience. There is so much said on this topic nowadays that the majority of students simply don't take in valuable information, so our task is to teach this issue in such a way so as not to impose a point of view or moralize. The student should draw his or her own conclusions. This idea should run through all topics without any special emphasis. For example, when discussing the level of organization of living things, we introduce microscopic photography of liver cells from a healthy individual and from an individual who is suffering from alcoholism. During discussion of the effect of hormones on cells, attention can be paid to how nicotine acts on proteins that transmit information to the cell. During study of the topic "Viruses", useful advice could be given about when and from where it is necessary to seek help in the event of unsafe behavior and risk of infection with HIV/AIDS or other pathogens (Strashko, Zhivotovskaya, & Grechishkina, 2006). An example, which vividly demonstrates methods of studying cells in medicine, is the comparison of nervous system cells from a rabid dog and a healthy one. At the same time as studying this, emphasis can be given to preventing the disease in humans, what to do in case of bites and the laws concerning keeping domestic animals.

IV. The technology of an individualized orientation to learning. The CBL system gives each student the opportunity to choose his or her own mode and trajectory of learning, but the teacher should tailor for particular psychological features of children of different ages. The extension material can be used with children who display a high level of creativity and are interested in studying biology. The reference book, readings and other materials included in the system, as well as work on the Internet, provide the opportunity to write short reports, reviews and so on. The CBL system teaches students how to work with information, structure it and incorporate it into their knowledge system. The tasks can be done as individual or group work. Children can be incentivized by forming groups according to interest. For example, one group could be made up of students who play sports. Another could be for those interested in cosmetics, and so on. There will undoubtedly be children in the class who follow mass media or Internet news items. The discussion of news involving biology always provokes lively interest and provides an opportunity to revise what has previously been learnt to interpret the facts, or to become a preamble to an upcoming topic. This kind of work imparts a vivid emotional tint to learning. For example, it has been reported that water had been discovered on the moon, in the form of ice. Students could then discuss how to interpret this fact, all the more if they have already studied the properties of water and its significance for life at an earlier stage. Students who don't remember the taught material very well can return here to the subject and take part in the discussion. For students who are interested in the natural sciences links may be explored in the second part of the lesson or extension (to use an example from the same theme, "Water and aqueous solutions") examining issues such as: water on our planet and other cosmic objects; water supply on earth and associated ecological problems; characteristics of hydrogen bonding; anomalous properties of water in comparison to other liquids; osmosis; water dissociation; pH; determining pH; buffering capacity of aqueous solutions; characteristics of popular drinks and a discussion of their safety. Such integration allows use of the knowledge gained in any sphere of activity in future.

Using the CBL system in the educational process cultivates in the student the development of observation skills (comparing karyograms of males and females; healthy individuals and individuals with hereditary diseases), the ability to carry out delicate and meticulous work, and the development of a scientific worldview. It assists in self-appraisal through self-observation of one's own physical and emotional state. While studying muscular tissue, observations can be carried out through somatoscopy – visual examination – for example, to detect the level of musculature development by external contours; the ability to carry out classificatory work, for example, in the latter case to establish reasons for the development of weak or good muscle. Developing the ability to identify goals means learning how to study and practice and developing one's own capacity to work. The use of a large quantity of illustrated material, video clips and reproductions of pictures plays a role in cultivating aesthetic sense.

V. The technology of providing equal access to knowledge. There are students who do not take in lecture material and text from textbooks or cannot independently draw conclusions. Other issues also affect students' access to knowledge. The availability of various different forms of study material gives everyone the opportunity to learn; the teacher may help to select the form of work that develops the abilities of each student.

VI. The technology of individualized knowledge-testing and self-checking. In evaluating students' educational attainments it is important to emphasize not the memorization of ready-made facts but how knowledge is gained and interpreted in various situations and whether students are able to activate their knowledge: using it in real life, acting appropriately and taking independent decisions. Another component of evaluating of a student's educational attainment is the ability to work with tests (Danilova & Danilov, 2002). The variety and quantity of tests give the teacher the opportunity to select tests taking into account students' individual needs. Our main task is to improve the students' level of self-evaluation, not to kill their creativity, but to develop their ability to solve any problem. In our system, there are sets of answers for each lesson, as well as for overall testing.

The importance of correct self-evaluation plays a decisive role in personality development. Both parents and teachers in search of giftedness in children assign an essential role to evaluating the student's attainments. Clearly, the level of development of an individual's talents does not guarantee success in any given area of human endeavor, but only indicates the possibility of attaining that success. To do well in any field, in addition to having a set of talents, the individual must also possess a certain amount of knowledge, skill and ability, as well as an appropriate level of self-appraisal. Here we might introduce an example.

Having assisted with the running of Olympiads of various levels, including International Biology Olympiads (IBO), we came across a case where a well-educated student with a good memory failed the practical round of an international competition, and fell into a state verging on prostration (a blank look, muscle spasms, lack of appetite and so on). After talking to him, it became clear that the student had been undertaking training for gifted children and had formed an attitude of unconditional victory, strongly encouraged by his teachers and parents. Speaking to him about the fact that the unexpected sometimes happens, he began to seek an answer to the question of why things had happened that way. We were very surprised when the first answer he sought was checking for possible inaccuracies in the translations of the tests and possible hints in those translations. Most surprisingly, when he was convinced that there had been no violation in the conduct of the work, he moved on to an analysis of the situation. The student then realized that his high achievements in accumulating knowledge could not be implemented if there was no skill of analysis and ability to assemble and interpret scientific facts or sufficient procedural training. The psychological shock passed, the student received one of the lesser prizes, lowered his level of self-appraisal and realized how to work, went on to university and success in later life. The CBL system gives the opportunity to prepare for biology Olympiads at various levels (Danilova, Zadorozhniy, Shabanov, & Danilov, 2006; Vashchenko, Danilova, Makarchuk, & Motuzniy, 2002).

VII. The technology of ecologizing education and creating foundations for nature-conservation activity. In studying each theme in biology there is the opportunity to direct students' attention to the influence of various external environmental factors on the processes of metabolism, different cell structures and the life processes in general. In our experience, it is best not to make these digressions explicit every time, so as not to overload the students with information, but to teach them to analyze the problem and draw their own conclusions independently. For example, in the "Water and aqueous solutions" topic we suggest the following test task. Indicate the optimal structure of soil for cultivating plants in agricultural production: (A) large clumps; (B) dust-like; (C) fine-grained. In the process of discussing the answer, attention is directed to the preservation of soil structure with respect to various methods of cultivation. On the basis of their knowledge of biology, chemistry and physics the students may be encouraged to identify the environmental conditions of their own homes, school, classroom, sports complex, district and region and to establish relations of cause and effect on their own activities and health.

Our experience of compiling the CBL system and its application in schools and a number of university courses has shown the potential for using the most diverse pedagogical methods and different didactic techniques, and for elevating the cognitive level of the educational material. This is facilitated by comparing various objectives and the presentation of a large number of tasks and exercises as a means of stimulating students' intellectual activity. The process of compiling these CBL systems is fairly labor intensive and expensive. Our experience has not been entirely successful; we have not finished the advertised games using educational material or online links. Only the study center has been completed, including:

- a classroom in which there are 30 lessons;
- a laboratory containing video clips showing how to carry out laboratory work;
- a practice area with a large selection of test tasks, knowledge tests, suggested answers and statistics;
- a library containing the set of readings, fully illustrated texts which do not duplicate the classroom materials, a reference book with material complementing the readings and a selection of materials required to complete practical work, a dictionary of biological terms with semantic explanations and etymology, an index of terms, a portrait gallery, and bibliography.

In addition, a number of print manuals have been published which complement the electronic version.

Conclusion

We were not entirely successful in creating the full textbook kit and providing links to the Internet, although many people were interested in this work. Teachers have rated the CBL system highly; it makes their task easier, saves time and helps to diversify teaching methods. There was no state funding for the project and there was insufficient finance from sponsorship for its continuation. However, if we were to calculate expenditure on printing and reprinting textbooks, reference books, books of biology readings, collections of tests and tasks, methodological guides for teachers, visual aids and other material supporting the educational process, it would be clear that those are considerably more expensive than the creation of a corresponding electronic system. Using such materials is also less rational because the opportunities for teaching and appraisal are narrower and a relatively large expenditure of time is required for the development of competences necessary to the modern individual. The CBL system provides the opportunity to master a large volume of material in a relatively short amount of time without being overloaded, since it is possible to choose the mode of learning suitable to the specific student, to introduce diversity, diversions, switch elements around and so on. Our experience shows that only with the help of a computer based learning system can the full kit of teaching material be created in order to support the application of various teaching technologies. The integration of computer technology with a pedagogical system for organizing educational activity allows for a substantial increase in educational opportunities for students, puts choice into practice and enables individual trajectories in an open educational space.

Acknowledgments

We are very grateful to Mary Bailes for her excellent translation of our chapter from Russian into English.

Note

1 The work was carried out in collaboration with D. Shabanov, N. Makarchuk, V. Kramarenko and many others.

References

Danilova, O. V., & Danilov, S. A. (2002). *A Collection of State Examination Tasks in Biology: 11th Grade*. Kiev Tsentr uchebno-metodicheskoy literature. (In Ukrainian.)

Danilova, O. V., Danilov, S. A., & Goryanaya, L. G. (2007). *Laboratory and Practical Work. Biology. 10th Grade: Workbook*. Kiev: Osnova. (In Russian and Ukrainian.)

Danilova, O. V., Danilov, S. A., & Shabanov, D. A. (2006). *Computer Based Learning System. Educational Aid: "Biology, Molecules, Cells, Organisms"*. Kiev: Xedex ITBS. (In Russian and Ukranian.)

Danilova, O. V., Zadorozhniy, K. M., Shabanov, D. A., & Danilov S. A. (2006). *Olympiads in Biology for Schoolchildren*. Khar'kov: Osnova Publishing Group (In Ukrainian.)

Makarchuk, N. E. (Ed.) (2006). *Computer Based Learning System. General Physiology. Stannius Ligatures*. Kiev: Xedex ITBS. (In Russian and English.)

Shabanov, D. A., & Kozlenko, A. S. (2009). *Ecology: Construction of the Biosphere*. (Grades 10 and 11). An innovative teaching methodology complex. An integrated collection of digital educational resources. Moscow: Rossiya. (In Russian.)

Strashko, S. V., Zhivotovskaya, L. A., & Grechishkina, O. D. (2006). *Social-Educational Training on Forming Incentives for a Healthy Way of Life and HIV AIDS Prevention*. Kiev: Obrazovanie Ukrainy. (In Ukrainian.)

Traytak, D. I. (2002). *Methodological Problems in Teaching Biology: Works by Members of the International Academy of Educational Sciences*. Moscow: Mnemozina. (In Russian.)

Vashchenko, L. S., Danilova, O. V., Makarchuk, M., & Motuzniy, V. O. (2002). *Biology Olympiads for Schoolchildren: An Educational Methodological Handbook*. Kiev: Geneza. (In Ukrainian.)

29
Socioscientific innovation for the common good

John Lawrence Bencze
UNIVERSITY OF TORONTO, CANADA

Summary: While relishing traditions and group belonging, humans also strive to innovate—to actualize desired changes. In many contexts throughout the world, however, innovation has been increasingly limited to a few powerful people and groups. Much of the world seems highly influenced by neoliberal capitalism, in which financiers and corporations are the main controllers of innovation—using it to encourage consumption of for-profit products and services that often prioritize image over quality. This zeitgeist of hyper-consumerism appears to be leading to many problems for the wellbeing of individuals, societies and environments. Given the important roles of fields science and engineering in capitalist activities, school science can make significant contributions to generating societies that may effectively address such problems. As argued here, students can gain significant expertise and confidence for carrying out research-informed actions to address so-called 'socio-scientific issues.' In doing so, more citizens could be innovating for the common good.

Key words: Neoliberalism, consumerism, semiotics, science and technology, science education, inquiry, socioscientific issues, innovation, social activism.

Introduction

In most societies, there are balances between maintenance of traditions and achievement of particular ends that individuals or groups agree are desirable. People want to maintain certain cultural practices, such as celebration of Christmas, Hanukkah, etc., while other societal members may want to revise them. Curiously, we want to belong, but we also want to establish unique identities within our groups. In the last thirty or so years, however, there appears to have been a shift in many societies toward an emphasis on continual renewal of personal identities. In that time, societies around the globe have, to a great extent, been increasingly influenced by an intense renewal of economic liberalism; that is, by *neoliberal* capitalism. Under this ethic, in direct contrast to traditional economic liberalism, strategic intervention by governing bodies is encouraged in order to promote maximum profit generation by owners of capital. A key strategic element of this agenda appears to be an emphasis on repeating cycles of consumption of continually revised for-profit products and services that, to a great extent, are attractive to consumers on the basis of new identities they may bestow upon them. To encourage such repeating consumption/re-identification, neoliberal capitalists—often with assistance from creative professionals—need to be continually *innovative*; that is, to bring about substantial changes in products and services and their inherent identity possibilities. Among the creative professions assisting with this process are fields of science and technology—and, given its role in generation of such professionals, science education. Accordingly, this chapter provides analyses of neoliberal-influenced science and engineering and, in light of resulting issues, makes suggestions for science education reform in ways that serve interests of all members of societies and their living and non-living environments.

Innovation in neoliberal times

The neoliberal capitalist zeitgeist

From the general perspective of social epistemology (Fuller, 2002) and in light of actor network theory (Latour, 2005), more particularly, each of us influences and is influenced by a multitude of 'actants' (animate and inanimate agents) in a complex web of interactions. For example, each of us owes, to a great extent, a debt of gratitude to those who have come before us and conditions—such as parts of our social and physical infrastructure—created by them. Alperovitz and Daly (2008) suggest, for instance, that someone like Bill Gates (former CEO of Microsoft Corp. and one of the richest people on Earth) would not have been able to achieve his great wealth without previous development, for example, of fundamental physics knowledge, electrical systems, computer hardware and software engineering, legal systems such as copyright laws, and surface and air transportation networks. With his same level of intelligence, motivation, etc., in an earlier era, he may not have become as rich as has, actually, occurred. Among the actants that may influence people, many suggest that a collectivity of them associated with a general tendency toward *economization* is most powerful in many societal contexts (e.g., Bakan, 2004; Barber, 2007; Dobbin, 1998; Gabbard, 2000; Harvey, 2010; McMurtry, 1999; Reich, 2007; Wood, 2005). Gabbard (2000) argues, for instance, that *economization* 'subordinates all [globally] ... forms of social interaction to economic logic and transforms nonmaterial needs ... into commodities' (p. xvii). In other words, it seems that many of our thoughts and actions are guided by a general affinity for capitalist economic exchanges.

The veritable economized zeitgeist that appears to characterize many societies has, apparently, undergone dramatic changes in the last few decades. Generally, it appears to have *intensified*, becoming more globalized and strategic. After World War II, many of the countries involved instituted large-scale spending programs to increase the living conditions of their populations. During this time of recovery, governments spent considerable fractions of their wealth on such social benefits as health care, education, housing and general infrastructure (e.g., roads); and, they enacted laws to protect workers. To accomplish such developments, increases in general taxation were implemented. This, however, apparently led to reductions in income of the richest members of societies who, consequently, placed pressure on governments to re-establish traditional economic *liberalism*; that is, policies enabling capitalists to pursue, with little interference from government, individual economic self-interests. According to McQuaig and Brooks (2010), such policies contributed to a return to the pre-war situation in which the richest 1% of the population controlled nearly one-quarter (instead of only about 10% in the post-war recovery period) of the society's wealth. To accomplish such dramatic shifts in wealth distribution, it is apparent that a new kind of economic liberalism, called *neo*liberalism, was promoted. Under this ethic, strategic intervention by governments and newly formed supranational[1] organizations like the World Bank and World Trade Organization is promoted—mainly in ways that enable capitalists to maximize profit (Bakan, 2004; McMurtry, 1999; Wood, 2005). Neoliberal policies include: privatization of public services, such as energy and transportation systems; reductions in social spending, including for health care and education; easing of regulations governing international trade of goods and services and workers; easing of laws governing labour and environmental protections; and, income tax reductions, particularly for rich individuals and corporations (Harvey, 2010; McMurtry, 1999).

Although overt force has been used, as in Pinochet's Chile, to enable neoliberal policies to be accepted by populations, it is common for capitalists to use a more subtle/subliminal technique, such as neoliberal *governmentality* (Foucault, 1991). Through innovative control of technology-enhanced media organizations (e.g., television, films, internet-based gaming), capitalists have been able to convince people to enact many neoliberal principles such as: *excellence, efficiency, standardization, competition, privatization, individual responsibility* and *commodification*—while believing that they are self-governing (Larner, 2000). Apparently enhancing this effect is citizens' use of hand-held video recording devices. Like the classic panopticon prison described by Foucault (1991), in which inmates self-regulate their behaviour because of the often misled assumption that they are being perpetually watched, citizens increasingly seem to be engaged in self-surveillance—emulating official watchfulness they more frequently experience through, for example, the presence of video surveillance equipment in public and private spaces (Dennis, 2008). Such seems to be the power of neoliberal capitalism.

Neoliberalism-influenced science and technology

Fundamental to the functioning of neoliberal capitalist systems are fields of science and technology[2]—which supply, for instance, computer systems, forms of transportation, manufacturing machinery, for business operations; and which have key roles in generation of many or most of the saleable products and services. Given their importance to neoliberal capitalists, it seems logical to assume that fields of science and technology may be aligned with neoliberal perspectives and preferred practices. There is, indeed, considerable evidence along these lines (Angell, 2004; Krimsky, 2003; Ziman, 2000). Of particular importance in relationships between neoliberal capitalists and fields of science and technology is frequent *innovation*; that is, creation, marketing and distribution of new for-profit products and services.

Many for-profit innovations involving fields of science and technology may be considered positive, often by proponents of business–science partnerships—who tend to claim that 'scientists who can turn ideas into profits are the ones who are contributing to a better world' (Krimsky, 2003, p. 2). There are, indeed, many clearly positive advancements when companies associate themselves with scientists and/or engineers—such as those in the medical technologies field, one example being hip replacement surgery (Weinstein, 2007). Such associations are, apparently, often necessary because of the costs of technical equipment for conduct of science and technology—including, for example, for measurement equipment (e.g., a synchrotron) and computer software (e.g., for complex statistical analyses); and, consequently, can often benefit from private sector financing (Dzisah, 2007; Ziman, 2000).

Although there are, undoubtedly, many benefits of capitalist relations with scientists and engineers, there are thought to be numerous negative side-effects of such partnerships and influences. In broad terms, there are concerns that the focus of research and development may be shifted toward benefits of the relatively few wealthy individuals and companies that, as a 'super-entity,' dominate societies (Vitali, Glattfelder, & Battison, 2011). A major repercussion of this is that university-based scientists, who often may want to focus on research that is intellectually interesting to them and not, necessarily, focused on particular marketable products and services, apparently have—to a significant extent—been transformed into 'incubators for generating wealth and intellectual property while significantly compromising their virtue and public interest roles' (Krimsky, 2003, p. 24). A university-based scientist lamented, for instance, that private and public funding agencies increasingly seek to reshape the purpose of research to define curiosity-based scholarship right out of the equation. If the results of our labours are not judged market-worthy, they are deemed to be of minor importance.

(Axelrod, 2000, p. 201)

A good example of this is that pharmaceutical companies often appear to avoid innovation for diseases and conditions that affect few people (a small market) and in places in the world ('markets') where markets may be large but few people would be able to pay for medications (Angell, 2004).

In terms of marketable products and services, there has, apparently, been a dramatic shift in capitalists' focus with the advent of neoliberalism. To a great extent, there is less focus on innovation in *physical* products and services—such as for housing, transportation, nutrition and health services—for people who *need* them and more emphasis on innovation leading to repeating cycles of consumer *desires* in people who have few needs (Barber, 2007; McMurtry, 1999; Usher, 2010). This re-invention in capitalist focus is linked to what have been called *knowledge* economies/societies, in which emphasis is on innovation of marketable ideas, concepts, perspectives, etc. (Gabbard, 2000). This can be understood in terms of the general framework for relationships between fields of science and technology/engineering given in Figure 29.1.[3] With regards to it, although the degree to which fields of science and technology interact with each other varies (Gardner, 1999), they often are linked in the sense of the schema in Figure 29.1. Although 'science' is traditionally considered to focus on 'World → Sign' translations (e.g., developing a chemical equation to describe a particular chemical reaction) and 'technology' is frequently associated with 'Sign → World' translations (e.g., using a chemical equation to guide a chemical industry process), both fields are thought to engage in reciprocal World ↔ Sign translations (Roth, 2001). In that sense, the distinctness of 'science' and 'technology' is called into question; and, consequently, it may be better to think of them as one merged entity, perhaps called 'technoscience' (Sismondo, 2008). This merged field is considered *material-semiotic*; meaning that it has both physical and symbolic components. Neoliberal capitalism has, apparently, made significant use—subconsciously or otherwise—of innovations in material-semiotic entities that fields of technoscience help generate. With its emphasis on consumerism, however, the focus tends to be on the semiotic (Sign) aspect of the relationship.

On the one hand, it is, arguably, unavoidable for Signs to somewhat misrepresent (reify) the World. Translating from the World to the Sign may involve numerous translations from one ontological entity to another, such as from a picture of a plant, to a drawing of it, to a map of a garden, etc. With each translation, there may be *ontological gaps* (refer to Figure 29.1); that is, inaccuracies

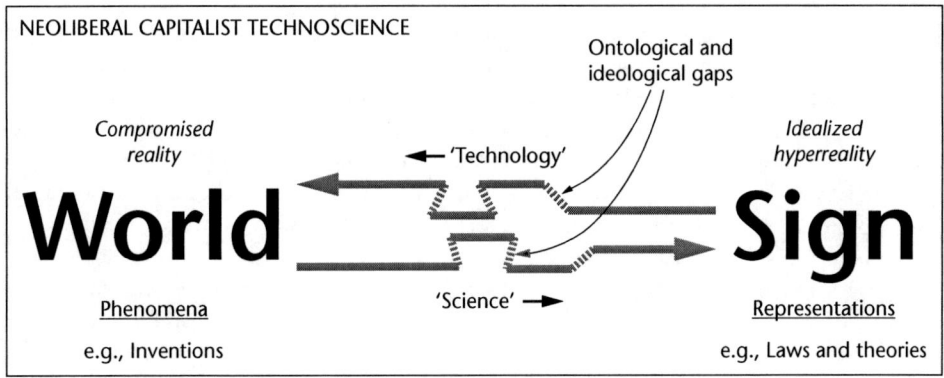

Figure 29.1 Neoliberalism-influenced technoscience.

in representation of details of each previous stage. The more translations, the greater the possibility for misrepresentation (Pozzer & Roth, 2003). There are suggestions, however, that companies are not satisfied with these apparently natural inefficiencies in representation. According to several authors (e.g., Barber, 2007; Norris, 2011; Usher, 2010) who have drawn on Baudrillard's (1998) analysis of consumer societies, companies have, in coordination with their intense emphasis on consumer advertising in various media environments (e.g., television, movies, internet sites), supported innovations in brand identities that are associated with for-profit products and services. When we purchase clothing for instance, we are acquiring not just physical objects but also semiotic messages associated with logos/brands that communicate to others something about us (providing us with, at least a contribution to, an *identity*), such as that we are 'rich,' 'relaxed,' 'stylish,' etc. (Klein, 2000). These identities may not, however, be generated from the consumer products and services as much as *applied* to them. They need not actually be derived from the product/service as long as people associate them with those commodities. With reference to Figure 29.1, some of the gaps in the translations may be *ideological*; that is, created for particular purposes. In creating such links to commodities, Baudrillard (1998) suggested that capitalists, with assistance from marketers, create a *hyperreal* state in people's minds; that is, they are unable to distinguish the created abstractions (Signs) from the real commodities (World) they are purported to represent. This can, and may very well be, misleading. Nevertheless, it can be very useful to capitalists. Latour (1987) suggested that the more abstracted from reality is a (purported) 'representation' the more amenable it is to manipulation by its creators. For capitalists, this aligns with an essential aspect of neoliberalism in knowledge economies; that is, with the desire to encourage people to *repeatedly* purchase purportedly 'new and improved' products and services. This can be very effective, according to Baudrillard (1998), in part because people have an innate need to reproduce and, perhaps, to delay death.

Although consumerism keeps many of us working in various ways, for example, there is much concern about possible adverse effects it is having on individuals, societies and environments. Given its use of hyperreal abstractions, people are critical of capitalists' control over consumers' identities, thoughts and actions, which seem subject to manipulation on a regular basis through the material-semiotic entities they regularly purchase. Among the many issues in this regard, several scholars have called into question capitalists' innovations and marketing to children (e.g., Bakan, 2011; Barber, 2007; Norris, 2011; Usher, 2010). In *Consumed*, for example, Barber (2007) suggests that a veritably 'religious' attachment to consumption of goods and services is enhanced by *infantilization*; that is, a double-barreled process to regress adults to and inhibit children from developing past child-like personal possessiveness. He claims that infantilization features at least three major techniques, including: *easiness* (e.g., watching instead of doing); *simplicity* (e.g., viewing images vs. reading); and, *speediness* (e.g., rapidly changing focus, instead of extended involvement). Moreover, he suggests that consumer encouragements/enticements are: *ubiquitous* (everywhere); *omnipresent* (always there); *addictive* (creates reinforcements); *self-replicating* (spreads, 'virally'); and, *omnilegitimate* (self-promotional). About similar issues, Bakan (2011) writes that societies seem to be in a *crisis*, in which there is 'erosion and sometimes outright destruction of our capacity to protect children from economic activities that might cause them harm ... [which he says] is arguably the most chilling effect of the turn to neoliberalism' (p. 10). Among the many specific concerns he raises in his book, *Childhood Under Siege*, is that it 'may seem diabolical—manipulating kids' emotions so as to "addict" and therefore monetize them—but that is the accepted strategy ... throughout the gaming and virtual world industries' (p. 22).

A principle underlying much neoliberal-informed innovation is that corporations, which embody many of the ideological characteristics of neoliberalism (Bakan, 2004; Dobbin, 1998; McMurtry, 1999), are legally sanctioned to maximize profits by minimizing their costs, such as by minimizing labour and materials expenditures. Associated with this right is companies' legal ability to *externalize* their costs; that is, to arrange for others to pay some costs of companies' operations. When a worker's wages are lowered, with similar or even greater labour responsibilities, s/he is, in effect, subsidizing the company's operations. This right has, however, also led capitalists to compromise the quality of their products and services which, in turn, seems to have contributed to many social and environmental ills. In *The Story of Stuff*, for example, Leonard (2010) documents how manufacturers of many materials-based products incorporate many toxic chemicals into their products, but without testing their health and environmental side-effects, and then, through *planned* (engineered self-destruction) and *perceived* (through advertising innovations) obsolescence, convince consumers to regularly dispose of their recently purchased products in favour of the 'new and improved' ones on offer. Disposal of many toxin-laden products in environments is contributing to numerous human health problems, with rates of cancer being a major concern. Other authors cite similar health and environment concerns associated with companies' tendencies to promote consumption of innovatively compromised products, including industries for production and marketing of cigarettes (Barnes, Hammond, & Glantz, 2006), pesticides (Hileman, 1998), genetically modified foods, etc. (Kleinman, 2003; Krimsky, 2003) and pharmaceuticals (Angell, 2004). Very often, companies responsible for these personal, social and/or environmental problems do not bear significant proportions of the costs associated with them. Such costs have been externalized out of companies.

Ultimately, questions of ethics pervade innovations of for-profit material-semiotic entities. In light of the above information and arguments, it appears that neoliberalism-influenced innovation in knowledge economies promotes repeated consumption and disposal cycles of goods and services that are, to a great extent, compromising the wellbeing of individuals, societies and environments. With reference to the schema in Figure 29.1, for example, it seems that there is an emphasis on selling idealized abstractions that, in effect, *occlude* compromised products and services. This seems highly deceptive. Add to this, however, that companies routinely externalize many of their costs and such innovation seems quite problematic. Bakan (2004), for example, suggests that 'the corporation [often actualizing neoliberal ideals] … is an externalizing machine, in the same way that a shark is a killing machine' and that this makes it 'potentially very, very damaging to society' (p. 20). In that light, he (Bakan, 2004) suggests that corporations—which are legally considered to be individuals—share many characteristics of psychopaths, who are extremely self-interested and lack empathy for others. In support of this claim, Babiak and Hare (2006) found that, largely because of the competitive nature of corporate life, many senior executives exhibit such self-obsessed behaviours. This is a worrisome aspect of the turn toward neoliberalism and associated innovations, including those related to marketing of problematic products and services masked by overly positive semiotic messages that appear to pre-occupy many citizens' thoughts and actions.

Toward socioscientific innovation for the common good

Preamble

Given its role in educating students for careers in fields of science and technology and for general citizenship, it seems reasonable to assume that school science may play significant roles in preparing students for life in a world apparently dominated by neoliberal capitalism. There is, indeed, much support for this claim. Carter (2005, 2008), for instance, has contributed greatly to this discussion in her treatment of influences of globalization on fields of science and science education. My work parallels these discussions, offering specific suggestions about ways in which science education is affected by and contributes to neoliberal capitalism (Bencze, 2008, 2010). Although space limitations preclude a detailed treatment here of such effects, it is apparent that neoliberal-influenced science education serves to generate citizens amenable to the different aspects of the technoscience model in

J.L. Bencze

Figure 29.1. Briefly, it seems that it generates: a relatively small cohort of *knowledge producers*, including potential technoscientists (and other professionals) who can conduct innovative work to generate for-profit material-semiotic entities on behalf of financiers and corporations; and, much larger groups of *knowledge consumers*, who are prepared to serve capitalists as compliant workers and enthusiastic and unquestioning purchasers of for-profit material-semiotic entities. Such a role for science education seems undemocratic, more aligned to serving the world's economic elite than the students to be educated. Moreover, as a veritable contributor to consumerism, science education may be considered complicit in many of the personal, social and environmental problems associated with fields of science and technology that humanity is currently facing. Consequently, it seems essential that science education undergo radical transformations in order for it to more effectively serve students in its care societies, more generally, and environments.

Since 2006, I have led an educational research and development project based on the curricular and pedagogical framework provided in Figure 29.2. Although it has been used mainly for studies relating to school science education (especially for grades 7–12), it can be applied to many other educational contexts. We have, for instance, had some successes in using it in the context of university-based science teacher education (Bencze & Sperling, 2012). Educators in various other contexts may, therefore, find the illustrated summary below useful in their work.

Proactively seeking socioscientific innovation for the common good

Although school science education appears to be a major actant in support of the neoliberal agenda, there also are movements to help reform it in ways conducive to personal, social and environmental

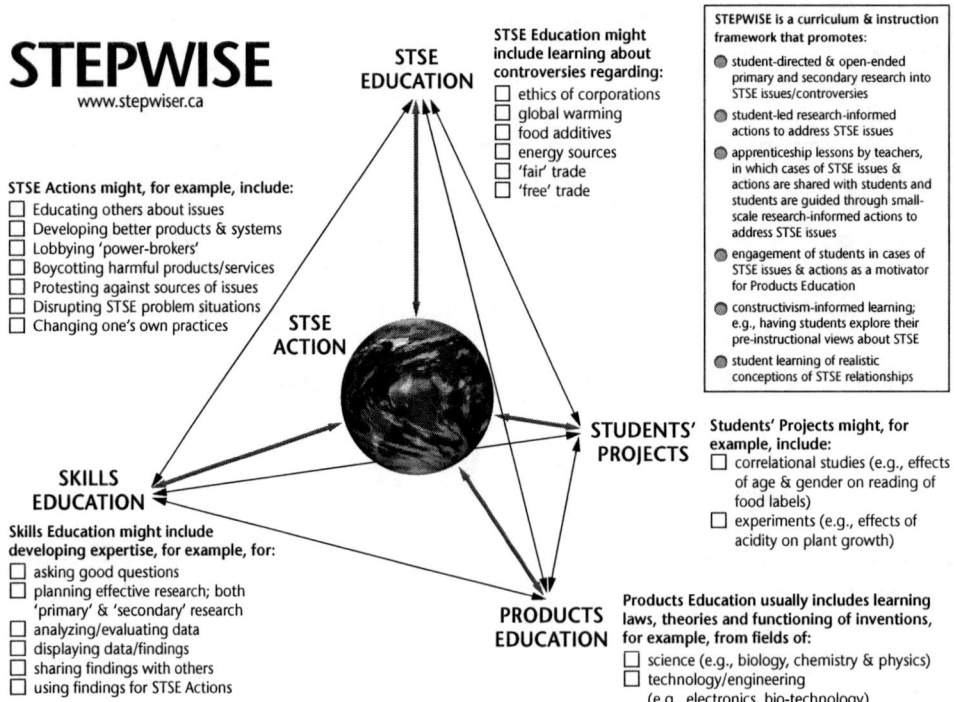

Figure 29.2 A framework for activist science and technology education.

wellbeing. For at least the last four decades, for instance, there has been significant scholarly work, primarily in the UK and Canada, promoting education in relationships among fields of science and technology and societies and environments (STSE)—including in terms of potential problematic relationships of those types (Pedretti & Nazir, 2011). A nearly parallel body of work, primarily emanating from within the USA and now extending more globally, has promoted education relating to *socioscientific issues* (SSIs) (Sadler, 2011). This work has made significant contributions to efforts to enlighten students and encourage them to establish their personal views about such controversial issues as those pertaining to cell phone uses (e.g., Pouliot, 2008), climate change (Simonneaux & Simonneaux, 2009) and genetics issues (Dawson & Venville, 2010). Much less emphasis has been placed, however, on encouraging and enabling students to take actions to address such issues (Hodson, 2003). Nevertheless, many scholars are urging such a focus—for at least the reasons that many of the issues pose significant potential threats to the wellbeing of individuals, societies and environments, and because of the apparent deep hold that neoliberal capitalists appear to have over numerous actors associated with such issues (Hodson, 2003, 2011; Santos, 2009).

To promote development of societies amenable to activism for addressing social and environmental issues associated with science and technology, the framework in Figure 29.2 was conceived. It is based on a number of well-supported educational and philosophical principles chosen to address many of the concerns identified in the first half of this chapter. Prime among these is *communitarianism*; the view that, because all actants are connected in actor networks, our positive actions for others may also benefit us (Fuller, 2002). For this reason, along with the severity of SSIs, 'STSE Actions' are placed in the center of the tetrahedral model in Figure 29.2. In theory, students may draw on their pre-instructional cultural and social (and other) capital (Bourdieu, 1986) to develop and enact plans of action. This may be particularly the case for advantaged students, whose home lives may provide them with considerable cultural and social capital (Hodson, 2011), but others may require intervention—such as instruction regarding the elements around the periphery of the tetrahedron in Figure 29.2. Having said that, it may be that all students require some intervention/instruction—in the sense that science education has, to a great extent, been limited to instruction in 'products,' such as laws and theories, of fields of professional science and technology. About this, Bell (2006), for example, writes: 'In the typical classroom, instruction has focused almost exclusively on the well-established products of science and cookbook approaches to laboratory exercises, using authoritarian teaching modes' (p. 430). Consequently, fields of professional science often are portrayed as highly logical and efficient in achieving truths. In a study of textbooks' historical accounts, for example, Allchin (2003) concluded that the following idealized images about scientists and their work were promoted: *Monumentality*: scientists appear as 'larger-than-life' heroic figures, often working alone and, moreover, their work is seen as very difficult, but very important; *Idealization*: false starts, complexities and biases are absent; *Affective drama*: the excitement and emotional elation of 'discovery' are exaggerated; *Explanatory and justificatory narrative*: conclusions of [individual investigations] are seen as final/unchanging and correct (pp. 341–347). In other words, school science may create a hyper-real state in students' minds—with them unable to distinguish between such idealized images and professional science. For a democratic education, students need to be exposed to more realistic conceptions about science and scientists. Indeed, given the hold that neoliberal capitalism appears to have over school science, this may be require 'revolutionary' *conscientization*; that is, exposure to power relations and issues with regards to fields of science (McLaren, 2000). In light of the arguments above, for example, they may need—as part of their 'STSE Education' in Figure 29.2—exposure to ways in which neoliberal capitalists influence fields of technoscience (Carter, 2005, 2008; Bencze, 2008, 2010). Included in this education could be efforts to enlighten students about neoliberal capitalists' use of innovations in creating idealized Signs that may occlude a compromised World (refer to Figure 29.1 and corresponding discussions, above).

As discussed above, although crucial for democratic awareness, enlightening students about problematic power relations regarding fields of technoscience is, likely, insufficient. They need to gain expertise and confidence for developing and implementing innovative plans of action to address socioscientific issues (Hodson, 2003, 2011; Santos, 2009). However, as Hodson (2003) advised: "It is almost always much *easier* to *proclaim* that one cares about an issue than to *do* something about it!" (p. 657; emphases added). Students may need to be personally motivated about an issue to act on it. Accordingly, the model in Figure 29.2 encourages students to be engaged in 'Students' Projects'; that is, self-directed and open-ended (no pre-determined conclusions) science inquiry projects. In doing so, they would, in effect, be engaged in self-determined reciprocal World↔Sign relationships (Figure 29.1), a situation that Wenger's (1998) work on communities-of-practice suggests may help students to develop deep and committed learning. Evidence suggests that such student-led research may, indeed, motivate students to take actions on SSIs important to them (Bencze, Sperling, & Carter, 2012). Having concluded, for instance, that there appeared to be a correlation between the number of fast food meals consumed per week and teenagers' average heart rates, a group of students then used data from that study to develop an innovative pamphlet that asked readers to problematize eating at fast food restaurants (Bencze et al., 2012).

In working with students, it is apparent that many of them often are reluctant to self-direct research-informed activism without first having been engaged in an *apprenticeship*—in which, as recommended by Hennessy (1993), teachers provide students with models of research-informed activism and guide them through small-scale research-informed activism projects (Bencze et al., 2012; Hodson, 2011). For example, teachers might provide students with a case documentary illustrating an activist video posted to YouTube™ that features research indicating that companies may misleadingly label food products as 'trans-fat[4] free,' while still using hydrogenated vegetable oils in the product. In coordination with this case method, students might then be guided through a study possibly correlating gender and age ranges with awareness of trans-fats labeling and then use their findings in innovative production of a PowerPoint™ presentation to peers in their school. Similarly, they could learn about how clothing advertisements promote certain styles and images of prestige (Signs) that may mask that the cotton from which it is made may be heavily laden with pesticide residues—as well as being produced by poor people under degraded working conditions (Vasil, 2007). Students could then be guided in secondary research about the cotton industry and a study of student peers' clothing choices and reasons—findings of which they could use to launch an innovative campaign to challenge other students to consider their clothing/fashion choices. Such apprenticeship activities combine 'Skills Education' and 'STSE Education' with 'STSE Actions' from the framework in Figure 29.2. In other words, students may develop innovative abilities—such as for experiment and correlational design and activism product development—in critical STSE contexts. Of particular focus, given the discussion above, is that students be enlightened about the role of technosciences in innovations leading to idealized hyperreal abstractions occluding compromised for-profit products and services. Using such critical STSE cases for skill development—for innovative investigations and actions—may, eventually, lead them to conduct student-led research-informed activism projects dealing with such critical issues. In that vein, it is, indeed, critical that students do, ultimately, conduct student-led and open-ended science investigations into SSIs and later use their results for developing and implementing innovative plans of action to address such issues. It is only when students gain full control over decisions on such investigations and actions that they develop deep attachments to them (Wenger, 1998) and, moreover, to authentic innovative research and actions—in that they are linked to particular instances of their praxis (McLaren, 2000), something others cannot plan. Such innovation may help contribute to the common good.

Summary and conclusions

Innovation—to create and implement phenomena that had not existed—can be a wonderful thing. It can bring recognition to innovators and happiness to others. In many places in the world, however, human ingenuity appears to often been usurped by private enterprise—which, under an ethic of intensified economic liberalism, has been given 'free rein' to orient societies toward enthusiastic and unquestioning cycles of consumption of commodities. Excessive consumerism appears to be associated with significant challenges to the wellbeing of many individuals, societies and environments.

A key actant in promotion of consumerism appears to be fields of technoscience that, for example, help innovate idealized representations of commodities that seem to create consumer desire while simultaneously masking potential shortcomings of those for-profit goods and services. Many fields of technoscience appear to be overly influenced, if not controlled, by private enterprise. Convincing capitalists to at least moderate this orientation toward consumerism may not be easy. However, given its role in identifying and educating potential technoscientists and, also, in conditioning large fractions of societies into habits of consumption, science education may be an essential agent addressing problems associated with consumerism. In this chapter, a framework for science (and technology) education has been presented that may help address such concerns, particularly in ways in which it encourages and enables students/citizens to *self-direct* innovative inquiry-informed activist projects to address personal, social and environmental issues associated with fields of technoscience. Ceding more control of innovation, particularly with a positive ethical bent, may help to reverse consumerist tendencies and contribute to a more sustainable world.

Notes

1 Supranational organizations, which arose after World War II, are independent from nation states/governments and are, generally, mainly beholding to powerful capitalists (including corporations).
2 'Technology' is a broad term, encompassing more technical activities such as those carried-out by technicians; but, it also encompasses more theoretical and creative activities of engineers.
3 I suggest that, as with any representation, aspects of the phenomena (here, fields of science and technology) to be represented may be omitted or misconstrued. This is, perhaps, ironic, given that the central argument here is that commercialized fields of technoscience are currently generating profit largely on the basis of constructed images associated with their products and services. Nevertheless, the representation in Figure 29.1 may be helpful as a focus for discussion and debate.
4 The term 'trans-fat' is the short form for trans fatty acid, which is a fatty substance in solid form made by adding hydrogen to vegetable oils (which are normally liquid at room temperature). Trans-fats have been widely linked to cardio-vascular illnesses, like heart disease. Hydrogenated vegetable oils are trans-fats.

References

Allchin, D. (2003). Scientific myth-conceptions. *Science Education, 87*(3), 329–351.
Alperovitz, G., & Daly, L. (2008). *Unjust Deserts: How the Rich are Taking Our Common Inheritance*. New York: New Press.
Angell, M. (2004). *The Truth About the Drug Companies: How They Deceive Us and What to Do About It*. New York: Random House.
Axelrod, P. (2000). What is to be done? Envisioning the university's future. In J. L. Turk (Ed.), *The Corporate Campus: Commercialization and the Dangers to Canada's Colleges and Universities* (pp. 201–208). Toronto: James Lorimer.
Babiak, P., & Hare, R. D. (2006). *Snakes in Suits: Why Psychopaths Go to Work*. New York: HarperCollins.
Bakan, J. (2004). *The Corporation: The Pathological Pursuit of Profit and Power*. Toronto: Viking.
Bakan, J. (2011). *Childhood Under Siege: How Big Business Targets Children*. Toronto: Allen Lane.
Barber, B. R. (2007). *Consumed: How Markets Corrupt Children, Infantilize Adults, and Swallow Citizens Whole*. New York: Norton.

Barnes, R. L., Hammond, S. K., & Glantz, S. A. (2006). The tobacco industry's role in the 16 Cities Study of Secondhand Tobacco Smoke: Do the data support the stated conclusions? *Environmental Health Perspectives, 114*(12), 1890–1897.

Baudrillard, J. (1998). *The Consumer Society*. London: Sage.

Bell, R. L. (2006). Perusing Pandora's Box: Exploring the what, when, and how of nature of science instruction. In L. B. Flick & N. G. Lederman (Eds.), *Scientific Inquiry and Nature of Science: Implications for Teaching, Learning, and Teacher Education* (pp. 427–446). Dordrecht: Springer.

Bencze, J. L. (2008). Private profit, science and science education: Critical problems and possibilities for action. *Canadian Journal of Science, Mathematics and Technology Education, 8*(4), 297–312.

Bencze, J. L. (2010). Exposing and deposing hyper-economized school science. *Cultural Studies of Science Education, 5*(2), 293–303.

Bencze, J. L., & Sperling, E. R. (2012). Student-teachers as advocates for student-led research-informed socioscientific activism. *Canadian Journal of Science, Mathematics and Technology Education, 12*(1), 62–85.

Bencze, L., Sperling, E., & Carter, L. (2012). Students' research-informed socioscientific activism: Re/visions for a sustainable future. *Research in Science Education, 42*(1), 129–148.

Bourdieu, P. (1986). The forms of capital. In J. G. Richardson (Ed.), *The Handbook of Theory: Research for the Sociology of Education* (pp. 241–258). New York: Greenwood Press.

Carter, L. (2005). Globalisation and science education: Rethinking science education reforms. *Journal of Research in Science Teaching, 42*(5), 561–580.

Carter, L. (2008). Globalisation and science education: The implications for science in the new economy. *Journal of Research in Science Teaching, 45*(5), 617–633.

Dawson, V. M., & Venville, G. (2010). Teaching strategies for developing students' argumentation skills about socioscientific issues in high school genetics. *Research in Science Education, 40*(2), 133–148.

Dennis, K. (2008). Keeping a close watch: The rise of self-surveillance and the threat of digital exposure. *Sociological Review, 56*(3), 347–357.

Dobbin, M. (1998). *The Myth of the Good Corporate Citizen: Democracy Under the Rule of Big Business*. Toronto: Stoddart.

Dzisah, J. (2007). Institutional transformations in the regime of knowledge production: The university as a catalyst for the science-based knowledge economy. *Asian Journal of Social Science, 35*(1), 126–140.

Foucault, M. (1977). *Discipline and Punish: The Birth of the Prison*. New York: Pantheon Books.

Foucault, M. (1991). Governmentality. In G. Burchell, C. Gordon & P. Miller (Eds.), *The Foucault Effect: Studies in Governmentality* (pp. 87–104). Hemel Hempstead, UK: Harvester Wheatsheaf.

Fuller, S. (2002). *Social Epistemology*, 2nd ed. Bloomington, IN: Indiana University Press.

Gabbard, D. A. (Ed.) (2000). *Knowledge and Power in the Global Economy: Politics and the Rhetoric of School Reform*. Mahwah, NJ: Lawrence-Erlbaum.

Gardner, P. L. (1999). The representation of science–technology relationships in Canadian physics textbooks. *International Journal of Science Education, 21*(3), 329–347.

Harvey, D. (2010). *The Enigma of Capital, and the Crises of Capitalism*. London: Oxford University Press.

Hennessy, S. (1993). Situated cognition and cognitive apprenticeship: Implications for classroom learning. *Studies in Science Education, 22*(1), 1–41.

Hileman, B. (1998). Industry's privacy rights: Is science shortchanged? *Chemical and Engineering News, 76*(17), 36.

Hodson, D. (2003). Time for action: Science education for an alternative future. *International Journal of Science Education, 25*(6), 645–670.

Hodson, D. (2011). *Looking to the Future: Building a Curriculum for Social Activism*. Rotterdam: Sense.

Klein, N. (2000). *No Logo: Taking Aim at the Brand Bullies*. Toronto: Vintage.

Kleinman, D. L. (2003). *Impure Cultures: University Biology and the World of Commerce*. Madison, WI: University of Wisconsin Press.

Krimsky, S. (2003). *Science in the Private Interest: Has the Lure of Profits Corrupted Biomedical Research?* Lanham, MD: Rowman & Littlefield.

Larner, W. (2000). Neo-liberalism: Policy, ideology, governmentality. *Studies in Political Economy, 63*, 5–26.

Latour, B. (1987). *Science in Action: How to Follow Scientists and Engineers Through Society*. Milton Keynes, UK: Open University Press.

Latour, B. (2005). *Reassembling the Social: An Introduction to Actor-Network-Theory*. Oxford, UK: Oxford University Press.

Leonard, A. (2010). *The Story of Stuff: How Our Obsession with Stuff is Trashing the Planet, our Communities, and our Health—and A Vision for Change*. New York: Free Press.

McLaren, P. (2000). *Che Guevara, Paulo Freire, and the Pedagogy of the Revolution*. Lanham, MD: Rowman & Littlefield.
McMurtry, J. (1999). *The Cancer Stage of Capitalism*. London: Pluto.
McQuaig, L., & Brooks, N. (2010). *The Trouble with Billionaires*. Toronto: Viking.
Norris, T. (2011). *Consuming Schools: Commercialism and the End of Politics*. Toronto: University of Toronto Press.
Pedretti, E., & Nazir, J. (2011). Currents in STSE education: Mapping a complex field, 40 years on. *Science Education, 95*(4), 601–626.
Pouliot, C. (2008). Students' inventory of social actors concerned by the controversy surrounding cellular telephones: A case study. *Science Education, 92*(3), 543–559.
Pozzer, L. L., & Roth, W.-M. (2003). Prevalence, function, and structure of photographs in high school biology textbooks. *Journal of Research in Science Teaching, 40*(10), 1089–1114.
Reich, R. B. (2007). *Supercapitalism: The Transformation of Business, Democracy, and Everyday Life*. New York: Knopf.
Roth, W.-M. (2001). Learning science through technological design. *Journal of Research in Science Teaching, 38*(7), 768–790.
Sadler, T. (Ed.) (2011). *Socio-Scientific Issues in the Classroom: Teaching, Learning and Trends*. Dordrecht: Springer.
Santos, W. L. P. dos (2009). Scientific literacy: A Freirean perspective as a radical view of humanistic science education. *Science Education, 93*(2), 361–382.
Simonneaux, L., & Simonneaux, J. (2009). Students' socio-scientific reasoning on controversies from the viewpoint of education for sustainable development. *Cultural Studies of Science Education, 4*(3), 657–687.
Sismondo, S. (2008). Science and technology studies and an engaged program. In E. J. Hackett, O. Amsterdamska, M. Lynch & J. Wajcman (Eds.), *The Handbook of Science and Technology Studies*, 3rd ed. (pp. 13–31). Cambridge, MA: MIT Press.
Usher, R. (2010). Consuming learning. In J. A. Sandlin & P. McLaren (Eds.), *Critical Pedagogies of Consumption: Living and Learning in the Shadow of the "Shopocalypse"* (pp. 36–46). New York: Routledge.
Vasil, A. (2007). *Ecoholic: Your Guide to the Most Environmentally Friendly Information, Products and Services in Canada*. Toronto: Vintage.
Vitali, S., Glattfelder, J. B., & Battison, S. (2011). The network of global corporate control. *PLoS ONE 6*(10); available at: www.plosone.org/article/info%3Adoi%2F10.1371%2Fjournal.pone.0025995.
Weinstein, J. N. (2007). Threats to scientific advancement in clinical practice. *SPINE, 32*(11S), S58–S62.
Wenger, E. (1998). *Communities of Practice: Learning, Meaning, and Identity*. New York: Cambridge University Press.
Wood, E. M. (2005). *Empire of Capital*, 2nd ed. London: Verso.
Ziman, J. (2000). *Real Science: What It Is, and What It Means*. Cambridge, UK: Cambridge University Press.

Part IX
How does technology education contribute to innovation education?

30

The role and place of science and technology education in developing innovation education

Alister Jones and Cathy Buntting

UNIVERSITY OF WAIKATO, NEW ZEALAND

Summary: An innovative future-focused culture is needed to address global, regional and local needs and opportunities for sustainable social, environment and economic well-being. Science and technology are likely to underpin many of the responses to these challenges. However it is not clear how traditional forms of science and technology education will produce future citizens who are capable of contributing meaningfully within this space. Global moves to expand the notion of science and technology education for citizenship as well as pre-professional training offer scope to address this challenge. However, despite the rhetoric, little is understood how expanding science and technology education in this way might contribute to the development of a future-focused innovative culture. This chapter explores the nature of science and nature of technology and potential links to innovation education as well as ways in which innovation education might be fostered through science education and perhaps more particularly technology education.

Key words: Science education, technology education, innovation education, futures thinking, assessment, communities of practice, formal education

Introduction

There is much rhetoric internationally of science and technology education being an essential part of the development of a globally networked knowledge economy. Part of the response to this has been an increased emphasis internationally on the development of a scientifically and technologically literate citizenship and workforce. Despite the rhetoric, however, little is understood about the different ways in which science and technology education might specifically contribute to the development of a future-focused innovative culture. We know, for example, that science and technology are likely to underpin many of the world's responses to global – as well as niche – issues and opportunities, but it is still far from clear how the innovative potential of individuals and groups can be fostered within the boundaries of formal education.

This chapter will explore the role and place of science and technology education as a potential space in which to support innovative culture and thinking in schools. First, we will review the connections between science and technology education and innovation education, arguing that technology education seems more likely to offer more robust opportunities for innovation education than science. Second, we will consider some of the factors influencing the expression of innovation education in technology, and emergent approaches to foster innovation education within technology and science. These include developing strong connections between the work of science, technology and innovation with science and technology classrooms, and the place of futures thinking in both technology and science education. Third, we will explore the need for authentic assessment if innovation education is to be valued and become embedded in teaching and learning programs. Our intention throughout the chapter is to provide a broad scope of the possibilities for supporting innovation within science and technology education, with a particular focus on the latter.

The nature of science and the nature of technology: links to innovation education

Curriculum reforms since the 1990s have expanded science and technology education internationally to include notions of the nature of science and the nature of technology. This has been driven largely by recognition that students live in an increasingly scientific and technological world and that they should be educated to participate knowledgeably and responsibly within this environment. Including human aspects of science in science education is also seen as a way of re-engaging a student audience increasingly disinterested in courses focused solely on conceptual development. In technology, the move has extended technical- and skills-based programmes to include technological knowledge, or 'knowing that', as well as technological practice ('knowing how'). The broadening of curricula in science and technology has also opened up opportunities for innovation education to be considered and explored in a school setting although these spaces are not yet clearly articulated.

Science and technology currently tend to sit as two discrete learning areas within school curricula around the world, with different goals and objectives. This is premised on understandings of the differences between the nature of science and the nature of technology as distinct disciplines. Science, on the one hand, is primarily focused on providing coherent frameworks to understand the physical world. Scientists generate and test ideas by collecting and interpreting evidence and presenting them for peer review. Technology, on the other hand, is an intervening in the world to develop products and systems that address human-identified needs and opportunities. This may draw strongly on scientific knowledge, such as in biotechnology; or be more mathematically and logic-driven, such as in information and communication technologies (ICTs). The different purposes of science and technology mean, however, that technological knowledge is often situated and contextualized, whereas scientific knowledge is more generalizable (Jones, 2012).

Science education through its mainstream curriculum programmes has historically focused primarily on conceptual development and not linked strongly with notions of innovation education. However, recent curriculum reforms place an increasing emphasis on linking science with applied science and technological outcomes in the marketplace. For example, technological contexts are increasingly used in science classrooms to demonstrate how science can be used to solve human problems (Jones, 2012). Introducing technological applications in science classrooms can also provide opportunities for enhancing student autonomy, logical thinking and creativity (Norton, McRobbie, & Ginns, 2007). This approach, of using technological problem solving in science, can therefore enhance students' problem solving and strategic thinking skills, but it does not necessarily deepen their understanding of technology and/or innovation.

An alternative strategy for building innovation into science education is to create opportunities for genuine curriculum integration where space is created to include other knowledge bases and practices. However, Rennie, Venville, and Wallace (2011) point out the challenges associated with assessing such initiatives and the need 'to analyse the multiplicities of learning that lie beyond the learning of discipline-based content and processes' (p. 158). To illustrate this, they use the example of a Grade 9 class whose science, mathematics and technology teachers worked together over a term to teach concepts integral to the design, building and testing of a solar-powered boat, the series of lessons culminating in the testing of boats constructed by the 13- and 14-year old students. Significantly, Rennie et al. demonstrate how measuring the impact of the teaching programme on student learning relies heavily on what is valued – identified as the theoretical lens that is used to measure the learning. Thus, from an integrated perspective, learning was considered to have been enhanced when compared with what students might have learnt in the individual subjects. From a disciplinary (science) perspective, however, learning was seen to be limited because, for example, students' understanding of 'electrical current' was scientifically incorrect. In order for more non-traditional learning outcomes, such as innovation, to therefore be valued in integrated teaching approaches, cognizance needs to be taken of the ways in which science is often still accorded higher status than technology. The ways in which teachers' and students' understandings of the relationships between science and technology influence learning outcomes also need to be considered (Jones, 2012).

In contrast to many traditional school science programmes – and even integrated initiatives – science competitions and award schemes offer students opportunities to pursue projects of individual interest and relevance. CREST (Creativity Education in Science and Technology), as one example, is an international awards scheme designed to encourage students to be innovative, creative and to problem solve in science, technology and environmental studies. In New Zealand, CREST projects can also help provide evidence for high-stakes science and technology assessments at the senior secondary level (www.royalsociety.org.nz/programmes/awards/crest/). The work generally represents a long-term investigation, often of at least a year, and is innovative and at an advanced level with an emphasis on the processes undertaken rather than the final outcome. For example, Seok Jun Bing and Alec Wang, two students in their final year of schooling were awarded a Gold CREST in 2011 for their project 'Carbon dioxide sequestration from internal combustion engines'. Using knowledge in chemistry, engineering and fabrication they researched, designed and developed a concept model of a device capable of reducing the CO_2 concentration in car exhaust fumes: a sodium hydroxide filter that sequesters CO_2 in the form of sodium carbonate.

CREST and similar schemes internationally represent an alternative approach to the science education that the majority of school students experience. Technology education, on other hand, generally includes even more creative components. In comparison with most contemporary forms of classroom science, technology education is therefore likely to be more directly aligned with possibilities for innovation education in schools. This has been enhanced by the move in many countries to broaden the traditional areas of craft and skill development to include aspects of design and notions of technological literacy for all. Key in this is a consideration of the philosophy of technology, which has developed into four main areas of interest: technology as artefacts, as knowledge, as activities and as an aspect of humanity (Mitcham, 1994). Each is relevant to technology education in that it explores a different aspect of what technology is. Thus, the dual nature of artefacts – as having both physical and functional properties – is given value, as is technology as an area of knowledge distinct from other knowledges (including science); the role of design in technological activity; and how technology is shaped by, and also shapes, humans and human culture and society (Jones, Buntting, & de Vries, 2011). In other words, technology is not only about artefacts but how and why those artefacts are developed and the impact they might have on people and our world (Jones, Cowie, & Moreland, 2010). These aspects all seem to also be important in considering how innovation and innovation education might be fostered.

Technology education, we suggest, therefore offers a robust site for innovation education in cases where the curriculum reflects the breadth of technological practice. This is significant given current international recognition of the need for innovation and a knowledge economy. However, school reform has been slower in reflecting this need and there is often a disconnect between the public rhetoric and education policy in that governments want a knowledge economy but do not go far enough in curriculum and school changes, including changes in teacher education and teacher professional learning.

Fostering innovation education in technology education

For technology education to offer students diverse opportunities to develop as innovators, a number of aspects need to be addressed. These include the precarious positioning of technology within many school curricula, the ongoing historical influences of technology as a skills-based handicraft subject and the difficulties in creating authentic learning experiences and linking with appropriate communities of practice.

Technology as a curriculum area

Although the 1990s saw an international move to develop technology curricula, its position is still fragile in many school jurisdictions, with national and school policy frameworks struggling to balance literacy and numeracy initiatives alongside other competing subject areas. Technology education is therefore at a critical juncture in its development as a field of teaching and learning in the compulsory school sector (see Jones & de Vries, 2009). In addition, the broad definitions of technology education as problem solving, design and creativity used in the past sometimes belie the existence of a more extensive knowledge base. Thus, the translation of the historical and philosophical aspects of technology for technology education continues to be crucial in the ongoing development of the field. A robust understanding of technology is also necessary if technology is to retain its own identity rather than being perceived as simply a subset of other more-established disciplines such as science. This is perhaps most pertinent to more advanced technologies, such as biotechnology and nanotechnology, where the scientific and technological aspects are tightly interwoven (Buntting & Jones, 2009).

Maintaining technology as a distinct domain rather than subsuming it within science helps to retain the distinctions between the nature of science and the nature of technology. The resurgence of STEM (science, technology, engineering and mathematics) in countries such as the USA and UK poses a threat to this unless the unique aspects of technology can somehow be embedded within any changes. Similarly, with STS (science-technology-society) approaches to science education (e.g., Solomon & Aikenhead, 1994; Yager, 1996) it is critical that the nature of technology be understood by teachers and taught alongside the nature of science if students' understanding of technology – and innovation – is to be enhanced. As Gardner (1995) notes, if technology is perceived as simply applied science, then economic, social, personal and environmental needs and constraints are often ignored. This will therefore limit students' learning of technological (and innovation) concepts (Jones, 1997).

In addition to pressures on the place of technology education within school curricula, the potential it offers for fostering innovation education can be limited by teacher and student expectations, teacher knowledge and a narrowing of the intended curriculum to fit with past practice. For example, the concepts that teachers and students hold of technology and what it means to learn in technology impact on their classroom experiences. To illustrate this, Jones (1994) relates an incident where a group of students made a waterproof container more 'technological' by adding flashing lights, their narrow concept of technology (as something 'hi-tech') constraining their practice. Similarly, when teacher understanding of technological capability is limited there is a tendency for the teacher to focus on the actual production of a product rather than the thinking skills, creativity, processes, issues and key learning involved (Jones & Compton, 1998).

Teacher subject subculture is a powerful determinant of teacher views and ways of acting (Goodson, 1985). These subcultures often represent reasonably consistent views about the role of the teacher, the nature of their subject and the way it should be taught, and expectations of the students' learning. This was perhaps particularly pertinent when technology was first introduced as a learning area in school curricula, where teachers were likely to have backgrounds in subjects like science, or home economics (Jones & Carr, 1992; Paechter, 1995). Nearly 20 years on, many of the vestiges of these historical subcultures continue to hold sway. For example, a teacher with a long career in teaching technical skills in a wood or metal workshop who continues to value the development of particular technical skills over other technological or innovative aspects is likely to be influenced by this when choosing teaching, learning and assessment tasks.

Since existing concepts affect future predictions, explanations and perceptions, students' and teachers' concepts of technology, including concepts of technological knowledge and processes, will also impact on student learning about technology and innovation (Jones, 2002). To address this, many countries support a range of strategies to broaden the experiences of both teachers and students. One important approach is to provide meaningful teacher and student exemplars. In New Zealand, this purpose is served by Techlink (www.techlink.org.nz), a website dedicated to showcasing examples of contemporary teaching and learning in technology education. For example, the section 'Student showcase' celebrates some of the innovative work being carried out by students. This includes a description of 16-year-old Kyle van de Pas' invention of 'Didy-go' for a Year 12 course in materials technology. Kyle's invention is a portable plastic bag with a spray nozzle attached (a 'spraybag') that can be used by mountain bikers and trampers to clean their equipment and kill didymo, a microscopic pest that has taken hold in New Zealand's alpine aquatic environments. This project therefore demonstrates both to teachers and to students the type of innovative undertaking that is possible. The narrative describing Kyle's liaison with industry mentors also highlights the value of being able to access relevant expertise.

Linking classrooms with the community of practice

Establishing links between schools and technologists is therefore another approach to broadening the knowledge base of technology teachers and their students. However, establishing and maintaining school–industry links can be challenging. For example, it takes a significant amount of time for teachers to locate relevant expertise, make contact, establish a meaningful relationship and work together to enhance students' learning. Moreland, France, Cowie and Milne (2004), for instance, present a case study where a secondary science and secondary technology teacher made individual approaches to contacts within the forensics community and were then 'referred along the chain to somebody who finally volunteered to help' (p. 185). The outcome of one of these approaches was that a newly graduated forensic scientist visited the school to talk to the science and technology classes. Not surprisingly, this was considered by students and teachers to be one of the high points of the classroom programme. However, both teachers 'were acutely aware of the imposition their demands had on these people' (p. 185). This telling example suggests that while contact with the community of practice significantly enhanced the learning opportunities for both the teachers and their students, it was not an easy process and required commitment and persistence on the part of the teacher. There was also a degree of hesitation about asking for expert involvement (Buntting & Jones, 2012). Similar barriers to establishing such relationships were reported by Slatter and France (2011), even though accessing the community of practice is considered key to enacting the New Zealand technology curriculum (Ministry of Education, 2007).

One approach in New Zealand to ameliorate at least some of the challenges teachers and students experience in identifying and contacting people who may (or may not) have the appropriate expertise has been the establishment of Futureintech (www.futureintech.org.nz), a government-funded initiative of the Institute of Professional Engineers New Zealand. Central to this work is the employment of eight regional facilitators charged with linking schools with local industries, including the recruitment of industry-based 'ambassadors' who are trained by the facilitators to work in schools, supporting innovation projects and providing a real-world perspective. An alternative approach is to use an online portal to connect the two communities, as has been done with the New Zealand Biotechnology Learning Hub.

The New Zealand Biotechnology Learning Hub (www.biotechlearn.org.nz) is an online portal developed to make the work of New Zealand biotechnologists more accessible and relevant to New Zealand school students (Buntting & Jones, 2012). It was initiated by the Ministry of Research, Science and Technology (now the Ministry of Business, Innovation and Employment) in response to the call from the biotechnology community for opportunities to interact with the education sector in a sustainable way. Many of these industries – which include large government-funded organizations as well as private companies, some with fewer than ten staff – valued opportunities to engage with students and teachers for such reasons as offering insights into contemporary practice and providing up-to-date knowledge; enhancing understanding about the nature of science, technology and innovation; and illustrating the multiple career opportunities and pathways that are possible. However, resourcing constraints including time were reported by industry representatives as impacting significantly on the successful establishment and maintenance of meaningful school–industry relationships.

The Biotechnology Learning Hub, by using a free-to-access Internet-based portal, seeks to make the biotechnology community more visible and accessible to a school audience. Content is presented largely as 'focus stories', in-depth case studies of modern biotechnological practice and innovation. These have the dual purpose of reflecting the situated nature of biotechnological innovation, and providing a hook for the development of engaging classroom programmes. Information is provided in multimodal form, including text, video clips, animations and interactive activities, and can be used as teacher background and/or for student investigations. Examples of learning activities are also provided to demonstrate how teachers can transform the biotechnology into relevant classroom experiences.

A key strength of the project is therefore the development of content specifically for educational purposes. Real-life contexts are selected to reflect New Zealand innovation, as well as what is likely to be of interest and relevance to students. Within this, the human aspects of science, technology and innovation are emphasized, including the people and stories of science, technology and innovation, as well as the ways in which science, technology and innovation relate to everyday life. Also related to innovation is the Hub's commitment to represent not only the research and development (or design) phase, but the routes whereby ideas are generated and come to being realized in the marketplace. For example, a teacher of a Year 8 (13 year olds) technology class effectively used the Biotechnology Learning Hub during a lesson designed to stimulate students' thinking about how ideas for new or improved products develop. One example that she accessed from the Hub was the story of Potatopak, a New Zealand company that makes biodegradable products. In short video clips that the teacher showed to the class the founder of the company, Richard Williams, explains where some of the ideas originated as well as the importance of market research and prototype development. The teacher used these to stimulate discussion about how the innovation of other products has been inspired by societal and environmental needs.

Another focus story on the Biotechnology Learning Hub, 'Wool innovations' (www.biotechlearn. org.nz/focus_stories/wool_innovations), was developed in conjunction with a large, government-funded research company and portrays the development of innovative fabrics such as Natural Easy Care

Wool and stab- and flame-resistant fabrics. The story outlines the impact of lighter-weight, synthetic fibres on the wool industry and the need to develop new wool fabrics to meet changing consumer needs, as well as the advantages of wool over synthetics in terms of growing consumer demand for renewable resources harvested via sustainable practices. The significance of new developments, for example in spinning technology, in fostering innovation is highlighted, as is creating and using new market opportunities. In the sample unit plan, 'New opportunities for protective wear', teachers are provided with ideas for a classroom programme in which students consider the performance properties of the new stab- and flame-resistant fabric, identify and question a potential market, and design an appropriate garment that they then present to the stakeholder group. The suggested focus of student learning in this example is on understanding factors that influence the development of new technologies, and thinking critically about societal influences on marketing new products.

These examples demonstrate the focus of the Biotechnology Learning Hub's content on innovation, from research and development through to marketing, presented in ways that are intended to give it coherence within an educational setting. They also demonstrate how the portal provides the school community with virtual access to the biotechnological community in ways that have hitherto been difficult to establish and maintain. The enthusiasm with which the biotechnology community has embraced the endeavour is a key to its initial success and continued development. Recognition of the value of this model in making contemporary innovation available to school audiences has also led to the development of a sister site, the Science Learning Hub (www.sciencelearn.org.nz), which showcases the place of science research and innovation in contemporary society.

Both Hubs also promote a futures focus as a significant driver for much scientific and technological innovation, and incorporating such thinking into science and technology classrooms seems another important strategy to foster students' innovative and evaluative potential.

The case for futures thinking

Futures thinking involves a structured exploration into how society and its physical and cultural environment could be shaped in the future, usually through developing possible, probable and preferable scenarios (Jones et al., 2012). It is related to innovation, which develops out of a vision of what alternatives might be possible, and draws on the following principles: the future world will likely differ in many respects from the present world; the future is not fixed, but consists of a variety of alternatives; people are responsible for choosing between alternatives; and small changes can become major changes over time (Cornish, 1977). In relation to innovation education, futures thinking has the potential to empower individuals and communities to envisage, value and work towards alternative futures. An overarching methodology of futures studies is the development of scenario models: 'pictures of future worlds that describe a *possibility space* – a set of plausible futures that span a range of conceivable outcomes' (Eames, Berkhout, Hertin, & Hawkins, 2000, p. 4, emphasis in orginal). They can be either exploratory – where the thinking moves from the present towards futures that could conceivably evolve from the present – or strategic, moving from an envisaged desirable future back to the present (Coates, 1996).

A conceptual framework developed by Jones et al. (2012) to scaffold students in developing and evaluating alternative scenarios is based on five key components:

- understanding the current situation,
- analysing relevant trends,
- identifying drivers,
- exploring possible and probable futures and
- selecting preferable futures.

Each component is explored at a personal, local, national and global level in order to encourage students to think beyond how the issue affects them personally, emphasizing both the critical role of the social context in futures thinking and the existence of multiple perspectives. The aim of this framework is to provide teachers and students with an analytical tool that they can use to develop and evaluate future scenarios. In doing so, space is provided not only for selecting preferred scenarios, but also for identifying ways in which these preferred futures might be pursued by individuals and communities. Opportunities for innovation – indeed the *need* for innovation – in this are likely to become apparent.

For example, the teacher of a Year 10 (14 year old) class used the futures thinking framework to design a six-lesson sequence that culminated in student groups promoting the development of a future food they had designed (Jones et al., 2012). Students' presentations needed to address the need or opportunity driving the development of the future food, the scientific techniques that would be required for its development and the potential risks and benefits associated with its development. Following the group presentations, the teacher facilitated a whole-class discussion about factors that would shape the development of future foods (e.g., new technologies, public support for new technologies and needs, such as feeding a growing population). This helped the students to link their presentations with the overall aim of developing futures thinking skills by highlighting the central role of trends and drivers.

The future foods proposed by the students suggested that they were able to identify a need (nutritional, environmental) and propose a solution. The diversity in both the student-designed foods and their evaluation of the specialized foods in an earlier activity powerfully demonstrated the influence of values in decision making. This suggests opportunities for the teacher to carefully guide students to identify and respectfully evaluate alternative perspectives. Although student presentations demonstrated limited exploration of wider environmental issues, such as environmental sustainability of food production and transport processes, opportunities for innovative solutions could have been created in this space. This case study demonstrates the rich potential offered by incorporating futures studies into a science or technology programme for enhancing students' critical thinking and opening up opportunities to explore areas of need and opportunity – leading, ultimately, to the proposal and evaluation of innovative products and solutions.

Authentic assessment

If technology and/or science education is to truly offer space for innovation education and the fostering of students' innovative potential, then this needs to be appropriately reflected in assessment. As Fensham (2006) points out, the nature of what is assessed in formal education determines what teachers and students recognize as knowledge of worth. Reflecting the aims of innovation education and whether or not these have been achieved using traditional assessment strategies in subject areas like science – particularly pen and paper tests – is obviously problematic. Technology education, however, with its historical underpinnings in craft- and technical-type subjects, has already developed alternative assessment strategies such as the assessment of artefacts and/or skills.

Grounding discussion of assessment within a socio-cultural view of innovation education brings to the fore the complexity and impact of interactions between people, ideas, tools and settings over time (Wertsch, 1998) and has implications for allocating importance to both the situated environment and the artefacts and tools in that environment (Cowie & Moreland, 2009). Seen this way, assessment in innovation education is a situated social and cultural activity that cannot be separated completely from the classroom or from ongoing student and teacher interaction. This recognizes the need for the teacher to be knowledgeable of and involved in the assessment process. However, as Jones et al. (2010) point out, teachers cannot design and evaluate valid assessment tasks or interact formatively unless they have a clear sense of the ideas of the subject. Teachers therefore need to have a comprehensive understanding of the nature of the discipline, its organizing concepts, mediational tools, cultural values, and symbolic and language systems. This raises several challenges for innovation education. First, teachers need to have this knowledge. Second, they need to be trusted to make valid assessment judgments. While this is usually the case at the elementary and junior high school level in the high-stakes environment of the higher end of schooling in many countries inter-teacher reliability is often more contentious.

A key aspect of effective assessment in innovation education, as has already been recognized for technology education more generally (Jones et al., 2010), is that it needs to focus on the multifaceted and multimodal nature of innovation process. As such, effective assessment should accommodate multiple modes such as drawing and modelling, not just writing or talk, to communicate and develop ideas, in order to reflect authentic innovation practice. The long-term nature of many technology and innovation education tasks also poses particular issues for assessment. One way to help with building connections, continuity and coherency is to think about these aspects when planning (Cowie, Moreland, Otrel-Cass, & Jones, 2008). Portfolios offer a powerful example of a means to provide connected evidence of learning for formative and summative assessment purposes, and ICTs are increasingly being used for developing e-portfolios to document learning and achievement in a multimodal way by including written work, photographs, video, audio and other digital media. These allow both teachers and students to document the process and products of learning as well as teacher feedback and student responses to this. This type of e-portfolio can be used for national assessment and moderation processes, as well as providing teachers with an opportunity to track students' conceptual development over time (Kimbell, 2009). In addition, digital evidence is easy to share and easy to search. Teachers and schools are able to ask probing questions about the nature of their students' learning and to consider the implications for their teaching and learning programmes, providing professional opportunities for teachers to expand their teaching repertoire and enhance students' development.

The road ahead: concluding comments

The expansion of science and technology school curricula to include broader aspects of the nature of each discipline has created space for innovation education to become better embedded in general education. Technology education, in particular, offers powerful opportunities to foster students' innovative potential. However, our adherence to the past limits the possibilities of the future in relation to the learning of today for tomorrow. In particular, teacher conceptualization of the nature of innovation and how this relates to technology in both the intended and experienced curriculum is critical.

Teachers' understandings of science, technology and innovation are complex and influenced by a range of factors, including teachers' background experiences, the subject subculture, how willing teachers are to critically evaluate and even change their own conceptual understandings, and the level of support given to teachers during any change process. This highlights the need for more research into teacher knowledges, including teacher pedagogical content knowledge in and of innovation education and how this can be enhanced. The role and place of school–industry connections in supporting students' innovation projects in ways that are meaningful and sustainable is another key avenue for further research. Even though findings may be localized to the setting in which the research is carried out, key messages can be re-contextualized and trialled in different locations. Futures thinking as an emergent area of formal education also seems very relevant to innovation education and is worthy of further investigation. As demonstrated in this chapter, early evidence suggests that a futures thinking framework has the potential to support the development of the teaching and learning of innovation.

A consideration of the ICT-rich environment in which schooling now can take place is another critical dimension impacting on teaching and learning, including how students are introduced to innovation as an area of study. The advent of sophisticated, affordable ICTs has had a profound effect on society and within this context, knowing and learning in science and technology are as much to do with access, participation and knowing how as with the acquisition of skills and knowing that. While an increasing number of studies indicate a positive role for ICTs in enhancing student engagement and motivation, greater research and analysis is needed to identify how specific ICTs can be used to foster student learning in science, technology and innovation.

While this chapter has highlighted the potential for technology education, and science to a lesser degree, to create space for innovation education to be explicitly incorporated in students' general education experiences, the opportunities are unlikely to be realized unless this is strategically pursued at the national and school level. This seems critical since innovation education has the potential not only to engage students in science and technology as school subjects, it also is an area of study in its own right that is important not only for individual development but also community well-being.

References

Buntting, C., & Jones, A. (2009). Unpacking the interface between science, technology and the environment: Biotechnology as an example. In A. Jones & M. de Vries (Eds.), *Handbook of Research and Development in Technology Education* (pp. 275–285). Rotterdam, Netherlands/Boston, MA/Taipei: Sense.

Buntting, C., & Jones, A. (2012). Expanding the capacity for connection: The New Zealand Biotechnology Learning Hub. In B. France & V. Compton (Eds.), *Bringing Communities Together: Connecting Learners with Scientists or Technologists* (pp. 101–112). Rotterdam, Netherlands: Sense.

Coates, J. (1996). An overview of futures methods. In R. Slaughter (Ed.), *The Knowledge Base of Futures Studies, Volume 2: Organisations, Practices, Products* (pp. 57–75). Hawthorn, Victoria: DDM Media Group.

Cornish, E. (1977). *The Study of the Future: An Introduction to the Art and Science of Understanding and Shaping Tomorrow's World*. Bethesda, MD: World Futures Society.

Cowie, B., & Moreland, J. (2009). Methodological considerations in studying classroom interactions in technology education. In A. Jones & M. de Vries (Eds.), *International Handbook of Research and Development in Technology Education* (pp. 625–635). Rotterdam, Netherlands: Sense.

Cowie, B., Moreland, J., Otrel-Cass, K., & Jones, A. (2008). Making connections in the teaching and learning of science and technology. *Set: Research Information for Teachers, 3*, 42–45.

Eames, M., Berkhout, F., Hertin, J., & Hawkins, R. (2000). *E-topia? Contextual Scenarios for Digital Futures. Final Report*. Brighton, UK: Science and Technology Policy Research, University of Sussex. Retrieved November 1, 2011, from www.sussex.ac.uk/Units/spru/publications/reports/etopia/etopia.pdf.

Fensham, P. J. (2006). Foreword. In B. H. W. Yung (Ed.), *Assessment Reform in Science: Fairness and Fear* (p. i). Dordrecht, Netherlands: Springer.

Gardner, P. (1995). The relationship between technology and science: Some historical and philosophical reflections. Part 2. *International Journal of Technology and Design Education, 5*(1), 1–33.

Goodson, I. F. (1985). Subjects for study. In I. F. Goodson (Ed.), *Social Histories of the Secondary Curriculum* (pp. 9–18). Lewes, UK: Falmer.

Jones, A. (1994). Technological problem solving in two science classrooms. *Research in Science Education, 24*(1), 182–190.

Jones, A. (1997). Recent research in student learning of technological concepts and processes. *International Journal of Technology and Design Education, 7*(1–2), 83–96.

Jones, A. (2002). Research in learning technological concepts and processes. In G. Owen-Jackson (Ed.), *Teaching Design and Technology in Secondary Schools* (pp. 79–90). London: RoutledgeFalmer.

Jones, A. (2012). Technology in science education: Context, contestation, and connection. In B. J. Fraser, K. Tobin & C. McRobbie (Eds.), *Second International Handbook of Science Education* (pp. 811–821). New York: Springer.

Jones, A., & Carr, M. (1992). Teachers' perceptions of technology education: Implications for curriculum innovation. *Research in Science Education, 22*(1), 230–239.

Jones, A., & Compton, V. (1998). Towards a model for teacher development in technology education: From research to practice. *International Journal of Technology and Design Education, 8*(1), 51–65.

Jones, A., & de Vries, M. (Eds.) (2009). *International Handbook of Research and Development in Technology Education*. Rotterdam, Netherlands: Sense.

Jones, A., Buntting, C., & de Vries, M. (2011). The developing field of technology education: A review to look forward. *International Journal of Technology and Design Education,* June, doi: 10.1007/s10798-011-9174-4.

Jones, A., Buntting, C., Hipkins, R., McKim, A., Conner, L., & Saunders, K. (2011). Developing students' futures thinking in science education. *Research in Science Education, 42*(4), 687–708.

Jones, A., Cowie, B., & Moreland, J. (2010). Assessment in schools: Technology education and ICT. In P. Peterson, E. Baker & B. McGaw (Eds.), *International Encyclopedia of Education, Vol. 3* (pp. 311–315). Amsterdam, Netherlands: Elsevier.

Kimbell, R. (2009). Performance portfolios ... problems, potential and policy. In A. Jones & M. de Vries (Eds.), *International Handbook of Research and Development in Technology Education* (pp. 509–522). Rotterdam, Netherlands: Sense.

Ministry of Education. (2007). *The New Zealand Curriculum*. Wellington, New Zealand: Learning Media.

Mitcham, C. (1994). *Thinking Through Technology. The Path Between Engineering and Philosophy*. Chicago: Chicago University.

Moreland, J., France, B., Cowie, B., & Milne, L. (2004). Case studies of biotechnology in the classroom. In A. Jones (Ed.), *Biotechnology in the New Zealand Curriculum: Final Research Report to the Ministry of Research, Science and Technology* (pp. 94–202). Hamilton, New Zealand: Centre for Science and Technology Education Research, University of Waikato.

Norton, S. J., McRobbie, C. J., & Ginns, I. S. (2007). Problem solving in a middle school robotics design classroom. *Research in Science Education, 37*(3), 261–277.

Paechter, C. (1995). Subcultural retreat: Negotiating the design and technology curriculum. *British Educational Research Journal, 21*(1), 75–87.

Rennie, L. J., Venville, G., & Wallace, J. (2011). Learning science in an integrated classroom: Finding balance through theoretical triangulation. *Journal of Curriculum Studies, 43*(2), 139–162.

Slatter, W., & France, B. (2011). The teacher-community of practice-student interaction in the New Zealand technology classroom. *International Journal of Technology and Design Education, 21*(2), 149–160.

Solomon, J., & Aikenhead, G. (Eds.) (1994). *STS Education: International Perspectives on Reform*. New York: Teachers College.

Wertsch, J. (1998). *Mind as Action*. New York: Oxford University.

Yager, R. E. (Ed.) (1996). *Science/Technology/Society as Reform in Science Education*. Albany, NY: State University of New York.

31

Nurturing innovation through online learning

Patricia Wallace

JOHNS HOPKINS UNIVERSITY, USA

Summary: With rapidly growing participation in online learning, educators are developing strategies that expand upon early attempts to replicate the traditional classroom environment, or digitize print-based texts, quizzes, and other resources. As the technologies advance, opportunities arise to tap the distinctive capabilities of these environments for nurturing critical thinking, creativity, and innovation. While some characteristics of the online world might appear to hinder these efforts, others may actually help create an environment that promotes attitudes and behaviors thought to be related to innovation, such as risk taking and willingness to challenge or defy norms. This chapter examines the evolution of online learning environments, their psychological, social, and cognitive characteristics, and how these environments may contribute to innovation education.

Key words: Innovation, online learning, distance education, innovation education, online education, distance learning, online courses, gifted education.

Introduction

The potential of online environments for education has received considerable attention from researchers, educators, and policy makers, particularly as the programs continue to grow and expand. Much of the attention focuses on whether online learning can achieve learning outcomes comparable to traditional classroom-based instruction. While these environments are quite varied, many of them share characteristics that may play an important, even if inadvertent, role in nurturing attitudes and behaviors thought to be important for innovation. This chapter explores the evolution of these environments, their psychological, social, and cognitive characteristics, and their potential for innovation education.

Terms like online learning, distance education, virtual learning communities, e-learning, tele-learning, distributed learning, virtual education, and distance learning have all been used to describe a very wide variety of technology-based environments in which students learn (Garrison & Shale, 1987; Simonson, Smaldino, Albright, & Zvacek, 2012; Wallace, 2004). The enormous variety in itself has caused confusion in the field, and has led to a situation in which people hold very different perceptions about what is actually involved. Some envision a remote student taking an impersonal, self-paced course with no interaction with other students or a teacher. For others, the terms conjure up the image of a synchronous web meeting, in which all the students gather each week to hear a traditional lecture, delivered via a web cam. The labels themselves are often used interchangeably and do not distinguish among the many different types of environments. Furthermore, programs identified by one of the terms might actually be quite different from one another, drawing on various technologies and pedagogical approaches, and producing different outcomes. To reduce the confusion, it is helpful to simply describe several examples of these environments, all of which share a key ingredient: the teacher and students are geographically separated.

Generations of online learning environments

Although the term "distance education" now implies a strong technology base, the first generation of educational programming in which there was geographical distance between teacher and student began in the 19th century, as correspondence study. One very early version started with an advertisement in a Swedish newspaper in 1833, offering people the opportunity to study composition by postal mail (Holmberg, 1995). Correspondence programs sprang up in Germany, Britain, Japan, the United States, and other countries as a way to reach students who were unable to join classrooms. They continued to expand throughout the early 20th century, adding audio recordings, lab kits, radio transmissions, and other tools to enrich the learning experience.

The explosion of information and communications technologies that began in the 1960s transformed the landscape for distance education models, bringing in waves of ground-breaking, disruptive innovations that underpinned the growing diversity in these environments. For example, instructional television began to grow with the advent of satellite technology, and governments funded many programs to bring high quality educational programming to remote areas. Interactive video networks were introduced in the 1970s to connect classrooms in different campuses or schools. A model used in many school systems involved the installation of video cameras and monitors in a classroom in several schools, which are then connected via a private network. The teacher, who is physically present in one of the classrooms, can see the students in the remote classrooms on monitors, and the geographically remote students can see the teacher on their monitors. School districts deployed this model to expand options for students, particularly for courses that were rarely offered because of the scarcity of qualified teachers (Wilson, Litle, Coleman, & Gallagher, 1997). Interactive videoconferencing in settings like these is an attempt to extend the traditional classroom experience using technology, typically emphasizing lectures and synchronous interaction between students and teachers.

By the 1990s, the Internet became the catalyst that generated remarkable innovations in information and communications technology, and a plethora of new possibilities for online learning environments. The exponential growth in the availability of online resources transformed the remote student's access to materials and communication tools. As online resources multiplied, students gained access to online library databases with a wealth of primary sources, statistical data, e-journals, online newspapers, e-books, art collections, museum tours, interviews, and video lectures. Instead of relying on a textbook and course notes, the remote student could draw upon a broad array of high quality resources. Indeed, a growing challenge for online learning environments is the sheer volume of material available, and the need for teachers to help students develop information literacy skills to judge quality and accuracy. Interactive resources also expanded, as educators created computer-graded quizzes, games, simulations, and polls.

A major feature of these online learning environments was the addition of viable, alternative communications technologies. Email became widespread in the late 1970s, and was followed quickly by technologies such as asynchronous discussion forums, interactive whiteboards, shared team workspaces, live text-based chats, instant messaging, two-way interactive video, group videoconferencing, virtual classrooms, and immersive virtual worlds.

The next generation of online learning environments will further enrich their capabilities. One important trend is mobility, which is especially reliant on the growing 4G cellular networks and wifi access points (Wallace, 2011). Users with specially equipped smartphones, tablets, and laptop computers can switch between the two technologies, depending on where they are and what services are available. These devices add the vast potential of "m-learning" to online learning environments, with features such as cameras, camcorders, and location awareness using global positioning satellites (GPS), in addition to their smaller size and lower cost.

A second trend involves a broader integration of Web 2.0 features into the online learning environments, including social networking tools, blogs, microblogging, wikis, and others. These features add interactive, student-centered capabilities that also support more collaboration and engagement. Social networks and online blogging platforms, in particular, can offer opportunities for building communities of learners that extend well beyond a single, structured online course (Abedin, 2011; Birch, Blackburn, Brody, & Wallace, 2011; Toikkanen & Lipponen, 2011).

A third significant trend is immersive education, which involves virtual worlds, virtual reality, commercial game and simulation technologies, and other kinds of rich digital media that support a sense of physically "being there." This trend is partly gaining momentum because of technological advances in the online multiplayer game industry, which attracts millions of players who interact with one another using avatars, inside vivid, 3D game worlds. The trend could transform online learning environments so that they support rich, avatar-based social interactions. Immersive education can also take advantage of virtual reality, a term used to describe what people experience when the sensory input is simulated rather than real. With the help of head-mounted gear, special gloves equipped with sensors, and other equipment, the student can be immersed in any environment, from the inside of a living cell to a landscape on Pluto. Strategies for adapting these emerging technologies to education are evolving quickly (de Freitas, Rebolledo-Mendez, Liarokapis, Magoulas, & Poulovassilis, 2010).

A fourth trend involves the growing expertise about how to innovatively exploit online environments to create novel and unique pedagogical approaches and learning resources. Initially, most educators who ventured into this space made few pedagogical changes, either in teaching styles or in the kinds of resources offered to students. In a sense, they were using technology, such as interactive video, to support a teaching tradition that dates back to Europe's first university in 11th century Bologna. Some adaptations offered small departures from this tradition, such as the digitized lecture notes and handouts, computer-graded quizzes, videos of the teacher's classroom lectures, and Powerpoint slides. Publishers converted their textbooks to "e-books," with some additional features, such as full text searching, hyperlinks, or online note taking.

However, online environments offer far more potential to create high intellectual and creative educational multimedia technologies (HICEMT) (Shavinina & Ponomarev, 2003). Computer-based simulations, for example, can support a wide variety of active, experiential learning scenarios, with some that are not feasible or even possible in a physical setting. In role-play simulations, a student might take over a new business and make strategic decisions that lead to its success or failure. Flight simulators offer novice pilots a safe place to practice necessary skills. Immersive environments can be created to simulate hazardous environments, so students can learn through simulated experience.

Characteristics of online learning environments

While there is considerable variety in online learning environments, many of them share certain underlying cognitive, social, and psychological characteristics that distinguish them from face-to-face settings. One key example is that the primary communication medium in a face-to-face class is speaking, while for online environments, much of the communication, though not all, involves keyboard, mouse, and touchscreen. Online learners use email, blogs, text messages, instant messages, asynchronous discussion forums, social network wall postings, interactive whiteboard, and text-based chat rooms to communicate with their instructors and classmates, both asynchronously and synchronously. This computer-mediated communication (CMC) results in reduced media richness with fewer communication channels. In particular, the nuances of nonverbal communication are often absent from the online learning environment. Even when the class uses videoconferencing, the small images have limited resolution and thus diminish the psychological and social impact of physical presence.

Physical distance is another factor that distinguishes online environments from a classroom. Students may not only be geographically separated from their instructors and classmates, but from all other people as they focus on the computer screen in front of them. The fact that distance is not relevant when creating and joining classes also means that online learning can expose students to greater diversity within each class, with students coming from quite varied backgrounds in terms of national origin, culture, ethnicity, socio-economic status, and age. In some settings, online students have opportunities to join classes with diversity that would not be possible in a face-to-face classroom. For example, Saudi Arabian girls and boys freely enroll in Advanced Placement courses offered through the CTY*Online* program at the Johns Hopkins University Center for Talented Youth. The online classes include girls and boys from around the world, but face-to-face classes in Saudi Arabia are segregated by gender.

A related phenomenon that enhances exposure to diversity occurs when students take more than one class. Even in large brick and mortar schools, students will meet many of the same classmates in their courses as they progress. Online, however, students have more opportunities to encounter new classmates.

The composition and size of the audience can be more difficult to judge online, and easy to over- or underestimate. For example, an instructor might send a student an email, adding a blind copy to a teaching assistant. Or a student might choose to forward a private message from the instructor to the whole class, or post a link to it on a Twitter account with hundreds of followers. Unlike an unrecorded spoken conversation in which the parties present can see exactly who is listening, electronic communications are more difficult to control in terms of the audience. Indeed, people often click the "reply all" button inadvertently, and send messages meant for one party to entire groups.

While many online learning environments are private "islands" based on learning management systems such as Moodle or Blackboard, educators are increasingly using more publicly available tools, such as Google docs, YouTube, Facebook, Second Life, wikis, or open class websites. In those cases, communications and resources that are intended for class viewing may be open to the public, creating some uncertainty regarding the privacy of in-class discussions, and who exactly is viewing students' contributions. These features add to the difficulty of clarifying who the audience is.

The online learning environment offers very easy access to additional resources and communication tools, as well as a sense of control over their use. A student viewing a video lecture on mitosis, for instance, can pause to look up an unfamiliar word, rewind it to repeat sections, or use a search engine to obtain resources that clarify the concept. In a face-to-face classroom, a student who does not understand a segment of the lecture might ask the teacher to slow down, but such public requests in front of classmates carry a psychological cost.

The ease of access has a downside as well. Having such vast resources within easy reach can tempt students to do more unrelated multitasking – posting Facebook wall messages, playing games, chatting with friends, or surfing the web. It also facilitates plagiarism, with simple cut and paste operations. Indeed, students judge ethical violations such as plagiarism as more acceptable when the Internet and computers are involved, compared to situations in which the violation does not involve any technology (Molnar, Kletke, & Chongwatpol, 2008).

Another potentially important feature of online environments may be a heightened sense of anonymity. Anonymity is a relative term, and complete anonymity is becoming increasingly

433

difficult to achieve in any online environment. Nevertheless, people tend to feel less personally identifiable online. In open and public areas on the Internet, such as the comments sections of news sites or discussion forums for sports, individuals typically create accounts tied to valid email addresses, but use pseudonyms to comment. In online learning environments, non-anonymous postings are more common, but instructors can include activities that involve anonymous contributions.

Finally, online learning environments offer students opportunities to manage the impressions they make on others in ways that differ from a face-to-face classroom. They can carefully select an attractive photo to upload, or choose a fanciful avatar. They can edit their words before posting in a discussion, and use a spell checker and a thesaurus to improve the quality of their contributions to the class discussions. In a virtual world, they may have the freedom to craft their own avatars to represent themselves. Even if the class uses webcams, the student has more control over the impression the image makes compared to the situation in a live classroom.

How do these characteristics influence the online learning experience, and how can they be tapped to support innovation education? The next section examines these questions, pointing out specific features that may promote behaviors and conditions that are hypothesized to be important for innovation (Denning, 2012; Shavinina, 2003; Sternberg, 2005).

Online learning environments and innovation education

The characteristics of online learning environments described above, along with the increasingly sophisticated technological advances, are already being tapped to promote positive learning outcomes among students (Means, Toyama, Murphy, Bakia, & Jones, 2009). But they may also be leveraged to promote attitudes and experiences that are thought to nurture innovation in young people. Examples include risk taking, a willingness to defy group norms, self-esteem and self-confidence, broad exposure to diverse people and ideas, and active learning.

Risk taking

Consider, for example, the greater physical distance between teachers and learners common to online learning environments. This distance, combined with the fact that each student is physically located in his or her own home or other familiar environment, can promote a sense of safety that may support more risk-taking behavior. Risk taking is frequently mentioned as an important characteristic of innovators, particularly calculated risk taking that has some chance to pay off (Sternberg, 2005). Online, students are less exposed to the nonverbal nuances that classmates and teachers might convey when the student makes a mistake that would cause embarrassment.

The online learning environment appears to foster greater risk taking in people who are especially shy or quiet in face-to-face classes. In a study of students engaged in online learning, several offered comments about this phenomenon (Sullivan, 2001, p. 811):

> "Yes, I feel more at ease and can take time in forming my answers online. I feel put on the spot in the classroom."

> "It's easier to participate because you're not worried what people will think about you because they don't know who you are."

> "I feel that I was more inclined to participate and express myself in the online format, as opposed to the classroom situation, where I feel self-conscious about raising my hand."

Online discussion groups typically show more equal participation by group members compared to face-to-face groups (Sproull & Kiesler, 1991; Wallace, 1999; Walther, 1996). The latter are often dominated by individuals with higher status or social power, a factor that can stifle contributions from less powerful members. Thus the environment may also be supporting innovation by encouraging the fluent generation of ideas from group members who might not have spoken up in a face-to-face setting. Indeed, the development of group decision support software (GDSS) for business was partly driven by this status equalization effect (DeSanctis et al., 2008). Initially, this software was used for computer-supported face-to-face groups, in which each member is equipped with a computer but all members are in the same room. A facilitator structures the discussion, and participants type their comments, which appear on a large screen for all to see without attribution. In addition to more rational decision making, an important goal of such relatively anonymous environments is to level corporate hierarchies and promote the generation of novel ideas that might remain unspoken otherwise.

Willingness to defy norms and nonconformity

A willingness to defy group norms is another attribute often associated with innovators (Sternberg, 2005). Norms unfold differently in online environments compared to face-to-face settings. In a brick and mortar classroom, students have acquired a deep understanding of the norms simply by having been in school for years. Even young students have absorbed norms such as raising their hands to ask a question, asking permission to leave the room, or not cutting in line. A new student would observe classmates carefully, seeking to fit in as quickly as possible by detecting whether the norms the student understands from a previous school are the same in the new one. Nonverbal cues are especially revealing and effective. A smile, a nod of the head, a stifled laugh, or an arched brow can all provide meaningful cues that both reveal and enforce group norms.

Online, however, norms are a work in progress and more difficult to ascertain. They may not even exist, leaving students opportunities to make their own choices, or develop very local norms for a particular class. A student entering his first online course, for example, would have little knowledge about norms pertaining to how students should introduce themselves. The student could choose a formal style, or an informal one, typing in all lower case and adding emoticons. He might upload a video, or write a lengthy and personal biography of his family history. The remaining students may follow that lead and assume the first student's post reveals a norm, or they might choose a new style.

Early studies that compared online discussions to comparable face-to-face discussions also show differences in group dynamics (Hiltz & Turoff, 1978; Walther, 1996). For example, an analysis of the utterances of students interacting face-to-face showed that the students expressed more agreement with one another, often with a simple "uh-huh" to show understanding and build consensus. The groups conducting the same discussion online made such utterances rarely, and instead made more remarks that expressed disagreement. Norms that apply to group dynamics and consensus building are weaker online, making it easier for students to defy or ignore them.

To a large extent, the willingness to defy norms is related to nonconformity, and online environments provide more leeway for nonconformists compared to face-to-face settings. A classic series of studies on conformity conducted by Solomon Asch demonstrated how conformity unfolds in face-to-face group settings, even when group pressure is nonexistent (Asch, 1955). Asch invited subjects into a room in which four confederates were already sitting, and asked each of them to make perceptual judgments about the comparative length of lines on large cards presented one at a time. The confederates were instructed to give the same wrong answer on certain turns, so that the subject, who went last, heard one person after another announce what was clearly an incorrect judgment about which line was longest. More than one-third of the subjects went along with incorrect judgments, even though there was no particular consequence or pressure to conform.

A similar study was conducted online, with the subject sitting in front of a computer that was purportedly networked to the computers of the other four "subjects" in the five-person group. For each trial, a set of four lines appeared on screen, and then the responses of the confederates were displayed, one at a time, showing the subject each person's judgment about which line was longest. In this setting, conformity dropped considerably. While Asch found that only 25% of the subjects in the face-to-face setting refused to conform on any trial, over two-thirds of the subjects in the online replication made no incorrect judgments (Smilowitz, Compton, & Flint, 1988).

Networking and exposure to diverse people and ideas

Wide exposure to different ideas and people is thought to be an important underpinning for innovation (Dyer, Gregersen, & Christensen, 2009). Indeed, the explosion of creativity that occurred in Florence during the Renaissance has been described as the "Medici effect," because the catalyst was arguably the Medici family's drive to bring together people from many different backgrounds – from those seeking to resurrect Greco-Roman learning to sculptors, poets, scientists, and architects (Johansson, 2004). Studies of Silicon Valley highlight the importance of its cultural mix and global connections in encouraging innovation in high tech industries, and also how the social networks that arise continue to support innovation for the migrant entrepreneurs who return to their home countries to set up businesses there (Saxenian, 2006).

Peter Denning points out that "sensing" is the first step in innovation, in which innovators must locate and articulate a new possibility, often found by observing disharmonies or incongruous events (Denning, 2012). Because an online learning environment can accommodate people from all over the world, it offers the potential to provide this diversity and exposure to new ideas. The ease of access to rich Internet resources adds further to the environment's potential to expose individuals to novelty. These features, combined with the enhanced risk taking discussed earlier, can provide a place in which disharmonies emerge more easily, and associations among them freely generated. Innovation, however, is not just about generating ideas, and Denning identifies many other steps to a successful adoption of an innovation. These include making a compelling case for the innovation, gaining a commitment from potential adopters to consider it, managing the adoption process, and others. Most people do not have all these skills, but the online environment and its support for social networks may help people who have some of them identify complementary players who can fulfill the other roles.

Self-esteem, self-confidence, and courage

Courage is cited as another critical component of innovation, and this attribute may spring from enhanced self-esteem and self-confidence (Shavinina, 2009; Sternberg, 2005). Some studies find that use of computers and the Internet itself enhances measures of self-esteem among young people, though the relationships are complex (Jackson, von Eye, Fitzgerald, Zhao, & Witt, 2010; Johnson, 2011). For example, the context in which the technologies are used is an important mediating variable. Game playing tends to be negatively related to measures of self-esteem, as does excessive Internet use.

An important aspect of online learning environments is the student's ability to manage his or her own persona and the impression it makes in novel and creative ways, a feature that may also contribute to self-confidence. Students can craft a persona that highlights their most positive attributes, and they can interact with others, unhindered by stereotypes that bias impressions in person. One student reported, "Everyone gets a chance to talk in a distance learning course ... It really helps build your self-esteem and confidence, in not only your writing ability, but also your deeper understanding of the material" (Sullivan, 2001, p. 811).

Immersive environments offer even more potential for students to craft personas that boost self-esteem and self-confidence. Self-perception theory, first proposed in the 1970s, suggests that people's attitudes and beliefs about themselves arise partly from their observations of themselves, as though they are observing themselves from outside (Bem, 1972). In face-to-face settings, this means that the person's self-concept will be influenced by how they think others judge their appearance or behavior. For example, if people wearing black clothing are generally judged to be more aggressive, then if the person wears black, that person will begin to hold the belief that he is more aggressive. Studies in which subjects don various costumes or uniforms that carry certain meanings have supported the hypothesis that self-perception is malleable and plays a role in attitudes.

Thus, when a student chooses an avatar for an online class, the details matter. In one study, for instance, the subjects were given avatars that were pretested and judged as either unattractive or attractive. Then, each one interacted in an immersive environment with a confederate of the researchers, but the confederate only saw the same average-looking avatar for all subjects, not the one that the subject actually thought was in place. The subjects who thought they had an attractive avatar displayed more extroversion and disclosed more personal information to the confederate (Yee & Bailenson, 2007). In a follow-up study, subjects were given either short or tall avatars, and then they negotiated with the confederate to split up a pool of money. Subjects who thought their avatar was considerably taller than the confederate behaved more confidently, by rejecting unfair offers or demanding other concessions in the negotiation. Interestingly, the heightened self-confidence extended beyond the virtual negotiation, into a face-to-face negotiation. Whether the avatar one chooses will positively influence courage is unknown, but it does appear to affect self-confidence (Yee, Bailenson, & Ducheneaut, 2009).

The individualized attention that online teachers can provide also has positive effects on self-confidence. Because the teacher typically spends much less time lecturing, and also because the course management system maintains detailed records of student activity and progress, teachers can offer very precise feedback combined with thoughtful encouragement and constructive criticism. A parent (personal communication, September 5, 2011) whose 6th grade daughter took a CTY-Online writing course wrote, "[The instructor] always had kind yet constructive comments for our daughter. Our daughter started the class doubting her writing ability, and over a 3 month period, really started to enjoy the process and gain confidence."

Active learning and higher order thinking skills

Nurturing innovation requires pedagogical approaches that promote higher order thinking skills, and that draw on evidence-based best practices that are known to foster such skills, such as the use of active learning strategies (Handelsman et al., 2004). For example, online discussion forums offer students opportunities to collaboratively explore issues in depth, without the limitation of a 50-minute class period, or the competition among students for speaking time. Although computer-supported collaborative work (CSCW) might be hindered by the lack of physical presence, the environment offers other advantages that may support the acquisition of higher order thinking skills, particularly when active learning strategies such as problem-based learning, inquiry-based learning, experiential learning, or case studies are applied (Murphrey, 2010). In one study, for instance, 138 students ranging in age from 12 to 17 enrolled in an online environmental science course that stressed a case-based approach, offered by the University of Virginia. Analysis of the discussion showed that out of 3,700 posts, 1,450 (39%) were coded as examples of higher level thinking, because they combined concepts, made connections across the cases, drew on knowledge from previous courses in subjects such as physics or history, or pushed their classmates to analyze the cases more deeply (Missett, Reed, Scot, Callahan, & Slade, 2010).

Conclusions

The rapid growth of online learning is a significant trend that may offer favorable opportunities to promote innovation education. A key reason for this growth is the opportunity to expand access to courses that might not be available locally. High schools in rural areas, for example, are able to tap into online learning to offer Advanced Placement courses that would otherwise be unavailable because there is no teacher qualified to teach them or because the number of students interested is too small to support a class (Picciano & Seaman, 2009). Middle school students can take courses in subjects rarely offered at this level, such as computer science.

Another reason for the growth in online learning is that these courses can provide individualized learning to meet the needs of certain populations of students whose needs are not well served in traditional classrooms. Examples include gifted and talented students, students with disabilities who are either unable to attend class or who are not benefitting from regular instruction, students who travel a great deal, or students in juvenile detention. An educator from one detention center commented, "We are a secure care facility with accredited high school programs. Our students love the online and blended courses. They experience success here and develop a sense of hope of attending a postsecondary school upon release" (Picciano & Seaman, 2009, p. 13).

Especially promising for innovation education is the success of online learning programs for high ability students (Miller, Adams, & Olszewski-Kubilius, 2007; Olszewski-Kubilius & Lee, 2004; Wallace, 2005, 2009). Indeed, a case has been made that out-of-school, supplementary programs tailored to gifted students, including advanced online courses, are a key factor in the remarkable success the United States enjoys in terms of Nobel Prizes in science and economics, Fields Medals for achievement in mathematics, and other top honors (Schrag, 2010).

In the CTY*Online* program at Johns Hopkins University Center for Talented Youth (www.cty.jhu.edu/ctyonline), academically advanced PreK-12 students can enroll in online courses well above their grade level in math, science, computer science, foreign languages, language arts, and other subjects. Students eager for advanced challenges can accelerate and enrich their school curriculum, pursue advanced coursework during the summer or holiday breaks, or arrange with their schools for credit and/or placement.

One example of a CTY*Online* course that especially stresses innovation education is "Inventions in Engineering," for students in grades 3 to 5. Students learn about the scientific process and the fundamentals of engineering, and also complete independent invention projects using materials such as paper clips, foil, grocery bags, string, and broken appliances. They upload photos and videos of their creations, and, in the online forums, they discuss how other students use the same materials to achieve different results. During the course, students also maintain "idea logs," a diary that helps them practice documenting innovative ideas that occur to them.

Another CTY*Online* course called "Scratch Programming" engages elementary and middle school students in computer programming using a Scratch, a graphical programming language developed at MIT. While learning programming fundamentals, students begin to develop ideas for their individual projects which often involve creating innovative games of their own design that they can share with family and friends.

For schools, the use of online programs that target gifted students offers several benefits. The programs eliminate the need to hire specialized teachers for a handful of students, reduce the pressure to skip a grade, and leave the classroom teacher with more time to spend with struggling students. At the same time, the programs stress highly individualized coursework that provides appropriately advanced levels of challenge to this student population. One parent (personal communication, September 10, 2011) of a gifted student enrolled in the CTY *Online* program pointed out,

> After the [final virtual class] session, he said to me "that was hard Mom". He welcomed the challenge, and his comment has resonated with me since. He has been in the public school system for 8 years now, and I have never heard those words.

Some educators perceive online learning as inferior to traditional classroom-based education, but acceptable when other options are unavailable. However, the research shows that in general, online learning is at least as effective as face-to-face classes in terms of learning outcomes (Means et al., 2009). The studies do report mixed results in many cases, especially due to the enormous diversity and rapid evolution of online learning environments described in this chapter. Evidence for their effectiveness should mount as researchers continue to identify best practices, providing guidance to educators about incorporating clear learning goals, active learning, authentic problems, multimodal resources, student-centered activities, collaboration, and highly individualized feedback (Sitzmann, Kraiger, Stewart, & Wisher, 2006; Thomson, 2010).

Future research should certainly focus on best practices as they pertain to learning outcomes. But it should also build knowledge about how the characteristics of online learning environments can promote attitudes and behaviors that support innovation education. This chapter identified several aspects that make them promising platforms for fostering innovation and nurturing tomorrow's innovators.

References

Abedin, B. (2011). Web 2.0 and online learning and teaching: A preliminary benchmarking study. *Asian Social Science, 7*(11), 5–12.

Asch, S. (1955). Opinions and social pressure. *Scientific American* (November), 31–35.

Bem, D. (1972). Self perception theory. In L. Berkowitz (Ed.), *Advances in Experimental Social Psychology*, vol. 6 (pp. 2–57). New York: Academic Press.

Birch, K., Blackburn, C., Brody, L., & Wallace, P. (2011). An online community for students who love STEM. *Science, 334*(6055), 467–468.

de Freitas, S., Rebolledo-Mendez, G., Liarokapis, F., Magoulas, G., & Poulovassilis, A. (2010). Learning as immersive experiences: Using the four-dimensional framework for designing and evaluating immersive learning experiences in a virtual world. *British Journal of Educational Technology, 41*(1), 69–85.

Denning, P. J. (2012). Innovating the future: From ideas to adoption. *Futurist, 46*(1), 40–45.

DeSanctis, G., Poole, M., Zigurs, I., DeSharnais, G., D'Onofrio, M., Gallupe, et al. (2008). The Minnesota GDSS research project: Group support systems, group processes, and outcomes. *Journal of the Association for Information Systems, 9*(10), Special Issue, 551–608. Retrieved from ABI/INFORM Global database.

Dyer, J. H., Gregersen, H. B., & Christensen, C. M. (2009). The innovator's DNA. *Harvard Business Review, 87*(12), 60–67.

Garrison, D. R., & Shale, D. (1987). Mapping the boundaries of distance education: Problems in defining the field. *American Journal of Distance Education, 1*(1), 4–13.

Handelsman, J., Ebert-May, D., Beichner, R., Bruns, P., Chang, A., DeHaan, R., et al. (2004). Scientific teaching. *Science, 304*(5670), 521–522.

Hiltz, S. R., & Turoff, M. (1978). *The Network Nation: Human Communication Via Computer.* Cambridge, MA: MIT Press.

Holmberg, B. (1995). The evolution of the character and practice of distance education. *Open Learning* (June), 47–53.

Jackson, L. A., von Eye, A., Fitzgerald, H. E., Zhao, Y., & Witt, E. A. (2010). Self-concept, self-esteem, gender, race and information technology use. *Computers in Human Behavior, 26*(3), 323–328.

Johansson, F. (2004). *Medici Effect: Breakthrough Insights at the Intersection of Ideas, Concepts, and Cultures.* Cambridge, MA: Harvard Business School Press Books.

Johnson, G. M. (2011). Self-esteem and use of the Internet among young school-age children. *International Journal of Psychological Studies, 3*(2), 48–53.

Means, B., Toyama, Y., Murphy, R., Bakia, M., & Jones, K. (2009). *Evaluation of Evidence-based Practices in Online Learning: A Meta-analysis and Review of Online Learning Studies.* Washington, DC: U.S. Department of Education.

Miller, K. A., Adams, C. M., & Olszewski-Kubilius, P. (2007). Distance learning and gifted students. In J. L. VanTassel-Baska (Ed.), *Serving Gifted Learners Beyond the Traditional Classroom* (pp. 169–188). Waco, TX: Prufrock Press.

Missett, T. C., Reed, C. B., Scot, T. P., Callahan, C. M., & Slade, M. (2010). Describing learning in an advanced online case-based course in environmental science. *Journal of Advanced Academics, 22*(1), 10–50.

Molnar, K., Kletke, M., & Chongwatpol, J. (2008). Ethics vs. IT ethics: Do undergraduate students perceive a difference? *Journal of Business Ethics, 83*(4), 657–671.

Murphrey, T. P. (2010). A case study of eLearning: Using technology to create and facilitate experiential learning. *Quarterly Review of Distance Education, 11*(4), 211–221.

Olszewski-Kubilius, P., & Lee, S. Y. (2004). Gifted adolescents' talent development through distance learning. *Journal for the Education of the Gifted, 28*(1), 7–35.

Picciano, A. G., & Seaman, J. (2009). *K-12 Online Learning: A 2008 Follow-Up of the Survey of U.S. School District Administrators.* Needham, MA: The Sloan Consortium.

Saxenian, A. (2006). *The New Argonauts: Regional Advantage in a Global Economy.* Cambridge, MA: Harvard University Press.

Schrag, F. K. (2010). Nurturing talent: How the U.S. succeeds. *Education Week, 29*(25), 22–23.

Shavinina, L. V. (2003). On the nature of individual innovation. In L. V. Shavinina (Ed.), *The International Handbook on Innovation* (pp. 31–43). Oxford, UK: Elsevier.

Shavinina, L. V. (2009). Innovation education for the gifted: A new direction for gifted education. In L. V. Shavinina (Ed.), *International Handbook of Gifted Education* (pp. 1257–1268). New York: Springer.

Shavinina, L. V., & Ponomarev, E. A. (2003). Developing innovative ideas through high intellectual and creative educational multimedia technologies. In L. V. Shavinina (Ed.), *Handbook of Innovation* (pp. 401–420). Oxford, UK: Elsevier.

Simonson, M., Smaldino, S., Albright, M., & Zvacek, S. (2012). *Teaching and Learning at a Distance: Foundations of Distance Education*, 5th ed. Boston, MA: Pearson Education, Inc.

Sitzmann, T., Kraiger, K., Stewart, D., & Wisher, R. (2006). The comparative effectiveness of web-based and classroom instruction: A meta-analysis. *Personnel Psychology, 59*(3), 623–664.

Smilowitz, M., Compton, D. D., & Flint, L. (1988). The effects of computer-mediated communication on an individual's judgment: A study based on the methods of Asch's social influence experiment. *Computers in Human Behavior, 4*(4), 311–321.

Sproull, L., & Kiesler, S. (1991). *Connections: New Ways of Working in the Networked Organization.* Cambridge, MA: MIT Press.

Sternberg, R. J. (2005). Innovation in the new millenium. *Innovation, 5*(3), 30–31.

Sullivan, P. (2001). Gender differences and the online classroom: Male and female college students evaluate their experiences. *Community College Journal of Research and Practice, 25*(10), 805–818.

Thomson, D. L. (2010). Beyond the classroom walls: Teachers' and students' perspectives on how online learning can meet the needs of gifted students. *Journal of Advanced Academics, 21*(4), 662–712.

Toikkanen, T., & Lipponen, L. (2011). The applicability of social network analysis to the study of networked learning. *Interactive Learning Environments, 19*(4), 365–379.

Wallace, P. (1999). *The Psychology of the Internet.* Cambridge, UK: Cambridge University Press.

Wallace, P. (2004). *The Internet in the Workplace: How New Technologies Transform Work.* Cambridge, UK: Cambridge University Press.

Wallace, P. (2005). Distance education for gifted students: Leveraging technology to expand academic programs. *High Ability Studies, 16*(1), 77–86.

Wallace, P. (2009). Distance learning for gifted students: Outcomes for elementary, middle, and high school aged students. *Journal for the Education of the Gifted, 32*(3), 295–320.

Wallace, P. (2011). M-learning: Promises, perils, and challenges for K-12 education. *New Horizons for Learning, 9*(1). Retrieved from http://education.jhu.edu/newhorizons/Journals/Winter2011/Wallace.

Walther, J. B. (1996). Computer-mediated communication: Impersonal, interpersonal, and hyperpersonal interaction. *Communication Research, 23*(1), 3–43.

Wilson, V., Litle, J., Coleman, M. R., & Gallagher, J. (1997). Distance learning: One school's experience on the information highway. *Journal of Secondary Gifted Education, 9*(2), 89–100.

Yee, N., & Bailenson, J. (2007). The Proteus Effect: The effect of transformed self-representation on behavior. *Human Communication Research, 33*(3), 271–290.

Yee, N., Bailenson, J. N., & Ducheneaut, N. (2009). The Proteus Effect: Implications of transformed digital self-representation on online and offline behavior. *Communication Research, 36*(2), 285–312.

32

E-learning as educational innovation in universities

Two case studies

Lorraine Carter and Vince Salyers

NIPISSING UNIVERSITY, CANADA, AND MOUNT ROYAL UNIVERSITY, CANADA

Summary: This chapter is a response to the trend among Canadian universities to embrace innovative and technologically based approaches to education. Using a case study approach, we explore innovation in universities from two perspectives: that of a university community as a whole engaged in embracing innovative and flexible ways of teaching and learning and that of teachers and students engaged in Internet-based learning in a professional program of study. A review of relevant terminology and literature offers important context. Issues related to change management and university culture are explored as well as strategies that may assist stakeholders as they adopt new and innovative approaches. Recommendations to assist learning leaders in universities as they plan for and implement innovative teaching and learning strategies are also provided.

Key words: E-learning, innovation, ICARE, blended learning, face-to-face learning.

Introduction

> One doesn't discover new lands without losing sight of the shore for a very long time.
> *(Andre Gide, French critic, essayist, and novelist (1869–1951))*

In Canada, like elsewhere in the world, university administrators and educators are challenged to enable a shift from more traditional teaching and learning methods to more innovative approaches. They are also required to meet the specialized needs of faculty and students as these methods are established in practice. In part, these trends are the result of the need to cultivate a competitive university marketplace grounded in flexible, accessible user-centric learning experiences (Carter et al., 2012; Kearns, Shoaf, & Summey, 2004; Reeves & Reeves, 2008; Ryan, Carlton, & Ali, 2004; Salyers, 2005; Thiele, 2003; Weber & Lennon, 2007). As an example, Canadian nursing schools are actively engaged in responding to the need for nurses around the world through compressed programs that use online teaching and learning methods. Additionally, most students at Canadian universities have "grown up digital" (Tapscott, 2008a) and, therefore, see technology-supported learning to be a natural component of life in the 21st century. They expect their universities to embrace innovation in the same way they have in their daily lives.

In this chapter, educational innovation in universities is considered through a case study approach that challenges us to reflect on educational innovation from two perspectives: the perspective of an entire university community engaged in embracing innovative and flexible ways of teaching and learning and that of teachers and students engaged in Internet-based learning in a professional program of study. In the first case, the development of the Centre for Flexible Teaching and Learning at Nipissing University in North Bay, Ontario, Canada during its first year of development is examined. Emphasis is placed on issues of change management and leadership. The use of a teaching and learning framework called Introduction, Connect, Apply, Reflect, and Extend (ICARE) in various nursing courses and programs will also be discussed. In this case study, educational innovation is viewed at the front line. Lessons derived from it might easily be extended beyond nursing classrooms.

In addition to describing these two cases, we will share observations that may assist other learning leaders immersed in the politic and practicalities of educational innovation at their universities. The key takeaways pertain to leaders who will own the change experience, faculty buy-in as achieved through a cross-section of strategies over time, and a commitment to change based on evidence.

Educational innovation in context

While not intended to be an exhaustive discussion of the various forms for innovation within educational contexts, we believe that a brief consideration of asynchronous, synchronous, blended, and online educational innovations is warranted. Not to provide this would muddy all further discussion.

Asynchronous learning environments. Over the past 20 years, the utilization of asynchronous and synchronous teaching and learning strategies has emerged. Prior to the development of learning management systems such as WebCT, Blackboard, and Desire2Learn in the 1990s, a number of independent companies developed websites that hosted course materials that students and faculty could access at their leisure, any time, any place. These early asynchronous environments functioned as repositories of information associated with particular courses and frequently included tools such as email and discussion boards (Hrastinski, 2008).

Synchronous learning environments. With the advent of educational technologies such as videoconferencing, chat, podcasts, wikis, blogs, and, more recently, social networking spaces, synchronous and "real time" teaching and learning experiences have provided for greater flexibility and more variety in teaching and learning environments (Hrastinski, 2008).

Blended classroom educational approaches. Blended educational approaches, also known as web-enhanced approaches, have been around for as long as learning management systems have existed. Blended approaches combine the best of face-to-face, asynchronous, and asynchronous learning environments (Bonk & Graham, 2006; Ellis, Goodyear, Prosser, & O'Hara, 2006; So & Bonk, 2010).

Fully online learning environments. Often regarded to be distance learning, the fully online learning environment removes the need for face-to-face and blended learning strategies from the teaching and learning experience. In other words, course work may be completed at a geographic distance from the institution and/or brick and mortar classroom. In other words, there is no face-to-face engagement of faculty and students. Frequently cited challenges of online learning environments include social isolation, poor retention rates, and an increased need for student motivation and self-directed learning (Brown, 1996; Frankola, 2001; Golladay, Prybutok, & Huff, 2000; Laine, 2003; Ryan, 2001; Serwatka, 2003; Smart & Cappel, 2006).

Special considerations. In addition to the teaching and learning modalities just described, there are other a number of items to consider in any discussion of educational innovation in higher education. For instance, there can be confusion about the pedagogies used to scaffold effective online learning environments; there are also differences between disciplines and professions. For example, the critical thinking activities often embedded in nursing education environments may not be same the as those used in the arts and humanities. It is further important to develop e-learning environments in which the contrasting learning preferences of students are considered. Although this is not an easy task, it may be mitigated through activities that involve multiple learning styles and promote general student engagement. The increasing use of web analytics will make contributions to our understanding of this need and possible ways of responding to it.

Finally, in e-learning settings, educators need to evaluate their underlying assumptions about the technological readiness and savvy of their students. Research has demonstrated that students who self-identify as possessing adequate skill with technology tend to perform better in and be more satisfied with blended and online learning environments than those with lesser skill with technology (Gefen, Karahanna, & Straub, 2003; Henry & Stone, 1994; Martins & Kellermanns, 2004; Salyers, 2005; Stoel & Lee, 2003; Wober & Gretzel, 2000).

Setting the stage for two case studies

As preface, having a fundamental understanding of our perspectives on educational innovation in the context of university education is vital. In addition to presenting some thoughts on the concept of innovation in education, we will provide insight into the terms e-learning and flexible learning.

Innovation. The words innovation and novelty derive from the same Latin root, both suggesting something new and, ideally, improved. An early definition of the verb to innovate states that "the desire to innovate moveth all troublesome men"; stated in more modern terms, the act of innovation can stir strong emotions among stakeholders (Ellis, 2005, p. 13).

In education, innovation is highly regarded for at least two reasons. First, throughout history, education has been expected to be on the leading edge. Second is that, as societal needs, demands, and expectations change, so too must education.

E-learning. Reflecting on educational innovation in North America over the last 20 years, a number of practices come to mind: online education, blended education, technology-supported education, and e-learning. While the first three of these terms can be considered in their own right, each is a subset of item four, namely, e-learning.

In the two cases we will discuss shortly, e-learning is understood in the following way: e-learning is an integration of pedagogy, content, and e-technologies within teaching and learning processes. As such, e-learning can include

1. face-to-face (f2f) classrooms in which information technologies (e.g., learning management systems, video-conferencing and web-conferencing, mobile devices) are used;
2. blended and web-enhanced learning environments;
3. fully online learning environments.

E-learning is also understood as an experience that can occur synchronously or asynchronously, or as a blend of the two.

Flexible learning. While the term e-learning is relatively new, the term flexible learning is even more recent. In its essence, flexible learning is about facilitating accessible, appropriate, and effective learning for all students, irrespective of students' geographic locations in relation to the university.

Flexible learning opportunities are designed with awareness of students' busy lives and their need to balance study, work, and family responsibilities. It is also about alternate educational models, partnerships, and evidence-informed teaching and learning practices (www.csu.edu.au/division/landt/index.htm; www.nipissingu.ca/cftl).

While e-learning is typically a major component of flexible learning particularly when programs serve professional and working learners, flexible learning is more encompassing than e-learning and may or may not include extensive use of educational technology.

Flexible learning in action. As an example of flexible learning, consider the following scenario. A professor may decide that the best way to inspire students in a course on medieval literature is to use well prepared lectures and an Internet-based library assignment. Imagine that this teacher is an authority in the field, a good explainer, an interesting lecturer, and an enthusiastic dramatist when reading passages out loud. The instructor revisits ideas that may not be clear to students and tries different strategies to facilitate student learning.

While there is minimal technology here except for the students' library work, in all likelihood, the students will come to understand and enjoy Chaucer's *Canterbury Tales* a great deal. Use of technology then is neither indicative of flexible learning nor of good practice. In fact, flexible learning is something much more sophisticated; it is something grounded in evidence-informed decisions.

A further example of flexible learning is the delivery of a nursing program in a compressed format where the emphasis is experiential learning rather than transmissive techniques. In this kind of program, students might study using a learning management system, e-portfolios, and iPads which they take into the clinical setting. These experiences could, in turn, be complemented by intensive classes that use narrative inquiry as the learning model.

Although this situation uses technology in the pursuit of learning, technology should be used only when there have been pedagogical discussions about its appropriateness in specific learning contexts. Moreover, in a program such as this, faculty and program planners need to be flexible to make changes if one or more strategy is not working as anticipated.

In both instances, the learning experience is based on informed choices about the subject matter, the students' learning needs and preferences, relevant demographic information, the instructor's expertise, and a commitment to student engagement. At the same time, in today's wired world, the Internet is often regarded to be the nexus of flexible teaching and learning. The reasons for this are several: the Internet can and does support synchronous and asynchronous learning in a wide cross-section of disciplines with greater and greater sophistication. Indeed, this should come as no surprise given how the Internet is critical to how we bank, plan vacations, listen to music, talk to friends and children around the world. In many instances, these experiences are transformative.

Case study 1: Flexible learning at Nipissing University

About Nipissing University

Economies of scale and geographic limitations are no strangers to small to mid-sized communities from which university-age students often migrate to major population centers with large universities. In Ontario, Canada, these larger university communities could include Toronto, the capital city of the province of Ontario, and Ottawa, the capital city of Canada. Located in North Bay, Ontario, Nipissing University knows these circumstances too well.

The roots of Nipissing University reach back more than 100 years when the North Bay Normal School was established as an institution dedicated to the preparation of elementary and secondary school teachers. Much later, in 1967, Nipissing University College was formed as an affiliate of Laurentian University in Sudbury, Ontario. Sudbury is a city located approximately 80 miles west of North Bay.

Later, in 1992, Nipissing University received its charter as an independent university. Today, Nipissing continues to be a primarily undergraduate university with a reputation for excellence in teacher education, the arts, science, and professional programs including business and nursing. Students experience a high quality academic environment that is student focused and based on interactive teaching practices, proven approaches to learning, and a growing research culture (www.nipissingu.ca). In addition to its undergraduate offerings, Nipissing now offers a number of Master's degrees and a new PhD in Educational Sustainability.

While commitment to high levels of faculty–student interaction in a small collegial setting is Nipissing's strong suit, Nipissing has recently felt the aforementioned pull of bigger schools in southern locations as well as the effect of a decreased market for Ontario teachers. Because of this, Nipissing is embracing an approach to university education that builds on its 100-year history in education but in new and innovative ways: it is championing flexible learning as a form of educational innovation. Flexible learning has been identified as a key goal in Nipissing's strategic plan for 2010–2015. While flexible learning is something bigger than e-learning, e-learning and technology-supported learning are major components of Nipissing's brand of flexible learning.

Rationale for flexible learning at Nipissing University

The rationale behind Nipissing's decision to champion flexible learning includes three components:

1. existing expertise in education as well as early integration of educational technology into teaching and learning in undergraduate programs;
2. changing requirements of university students for time- and place-independent courses and programs;
3. the blurring of lines between and among face-to-face learning situations, online and distance education, and blended learning situations.

Each of the above are explored in the following paragraphs. Relevant examples are provided.

Educational expertise and early integration. Nipissing University holds a strong and positive reputation among its students. As support of this idea, during the preparation of this chapter, Canada's national newspaper called the *Globe and Mail* released the Canadian University Report. The data presented in the report comes from a 100-question survey filled out by 33,000 undergraduate students across Canada.

In the report, Nipissing was first or tied for first among all participating small universities (enrollment between 4,000 and 10,000 students) in seven categories. In categories focused on teaching and learning, Nipissing did particularly well with the following scores: Quality of teaching and learning (A–), Instructors' teaching style (A–), Student–faculty interaction (A), and Class size (A+). Additionally, as indicated earlier, Nipissing's history in education is long standing—over 100 years.

Despite its small size, Nipissing is technically sophisticated. Examples of its technical panache include early adoption of e-learning by faculty and administrators in the Faculty of Applied and Professional Programs, growing registration numbers in online courses, wireless connectivity on its three campuses, use of Gmail as the university email system, and leading edge use of mobile technologies in classrooms. Throughout Ontario, Nipissing is a leader in the deployment, use of, and research of Blackberries, iPads, and other mobile devices in teaching and learning contexts. One further example of Nipissing's leadership in educational technology is the iTeach Laptop Learning Program which was designed to provide pre-service teacher candidates the skills and knowledge they need to choose and implement appropriate software and hardware for effective teaching, learning, and student success in K-12 classrooms. As part of the program, faculty members are trained in the application of educational technology while teacher candidates learn how educational technologies can enhance the learning experience of students in K-12 classrooms and how to assess, use, and integrate high-quality software resources and integrate hardware such as SMART Boards, Promethean Activboards, digital documents, and video cameras into the classroom experience.

Responding to the student market. There is no disputing the evidence that today's university students—the so-called millennials—have grown up with technology as part of their everyday lives. The millennials (also called the digital generation, net generation, digital natives) are people born from 1982 through to the 2000s.

Research abounds with contrary opinions about the effect of technology on learning and student engagement. Some experts argue that the millennials' exposure to technology has affected how they learn and even how their brains work (Tapscott, 2008a, 2008b). Others suggest that there is no credible evidence that today's generation of university students learn differently from any other generation (Bullen, 2011).

Regardless of the conflicting positions about how millennials learn, there is one indisputable reality—that they use technology more than any generation before them and that they display strong interest in integrating technology with their learning. One example of this is the portable electronic devices and wireless technologies which have evolved significantly over the past decade and created drastic changes in our lives which are characterized by travel and, therefore, a need for mobility (Beckmann, 2010; El-Hussein & Cronje, 2010). In total, today's mobile market includes nearly four billion subscribers, a number which continues to escalate (Johnson, Levine, & Smith, 2010). Billions of new innovative mobile phones are produced every year, with the fastest growing sector of the market belonging to Smartphones; that is, devices which incorporate the benefits of a computer within a small handheld unit and which can be taken virtually anywhere (Johnson et al.).

While pedagogical change tends to occur slowly, not to do so at this time represents risk (Margaryan, Littlejohn, & Vojt, 2011; Parson, Ready, Wood, & Senior, 2009). The most significant risk is that of disengaged learners, especially in those settings where there is a disconnect between student lifestyle in the digital age and methods for learning (Ontario Public School Boards' Association, 2010). There is substantive and negative possibility of unmotivated learners. While the incorporation of online materials within a course may lead to decreased lecture attendance and, hence, insecurity among professors about their roles and ultimately their jobs, students in studies by Beckmann (2010), Margaryan et al. (2011), and Parson et al. (2009) continue to express that the professor will always be irreplaceable and integral to their learning experiences.

In addition to the fact that today's university students have grown up with technology, there are two other reasons for Nipissing to endorse flexible and e-learning. The first is that the majority of university students need to work while they attend university; this is the reality of younger, straight from high school university students and adult learners who return to university to upgrade or to earn a professional credential. Younger learners may hold more than one part-time job in order to cover the rising cost of university education in Canada. At the time of writing, at Nipissing University, the first year tuition cost was $5,148.00. Providing educational opportunities that offer some flexibility around place and time then is a good business decision if universities are serious about retaining their students.

Adult learners require flexibility and access to professional learning that often looks different from face-to-face classes. Nursing education in Ontario is a prime example. E-learning for the purpose of nursing education has existed in Ontario for last the ten years as Ontario anticipated and then responded to the January 1, 2005 entry to practice requirement of baccalaureate education in nursing (Carter, 2008; Carter & Rukholm, 2008; Dickieson, Carter, & Walsh, 2008; Carter et al., 2006; Carter, 2003; Carter & Rukholm, 2002).

Finally, there is no question that Canadian universities are becoming increasingly involved in international education in Asia, Africa, and elsewhere. For Nipissing University not to be engaged in alternate forms of education would be short sighted. Flexible and accessible educational opportunities will most certainly make a difference to the urgent call for higher education around the world.

Blurred lines between and among. With the advent of Internet-supported education, the lines between and among face-to-face learning situations, online and distance education, and blended learning situations have become increasingly blurred. Initially, online and other technology-supported forms of teaching and learning were associated with distance education. For some who might call themselves educational purists, early online education held the status of poor cousin to face-to-face learning. This sense of second class citizenship was bred from within the practice field through a kind of identity crisis. For example, in 1990, Shale remarked, "Distance education is beset with a remarkable paradox—it has asserted its existence but it cannot define itself" (p. 333).

If distance educators were unable to define distance education in 1990, achieving a distinctive identity today is even more daunting. Today the notion of distance education has become increasingly complex due to the evolution of highly sophisticated educational technologies, changing understandings of time and space, and increased globalization.

Some might argue that today the angst is being felt not so much by distance and e-learning practitioners but by those who continue to hold to more traditional ways of thinking about teaching and learning. Because of the many and rapid advancements in educational technology, there is tremendous pressure facing institutions to become more flexible in relation to time, place, pace, and learner entry into programs. Bates (2008) points out that, today, distance is more psychological than anything.

A final comment on Nipissing University's decision to champion flexible learning is the simple reality that, with the requisite planning and supports, it works. Effective teachers know that learning is best enabled when there is variety and change in the learning strategies. Regarding e-learning as a form of flexible learning, it has existed in Canada for approximately a decade, and considerable evidence has been gathered about its efficacy in facilitating excellence in education. The message is that there is no statistically significant difference between learning achieved online and learning achieved in a face-to-face classroom.

The development of the Centre for Flexible Teaching and Learning: year one

While still in its early days, the Centre for Flexible Teaching and Learning has grown and developed substantively. In less than 12 months, it has evolved from a loosely affiliated staff relocated from various sectors of the university to a dynamic team that delivers a robust program of faculty, staff, and student development; supports a fellowship program that recognizes faculty who have made important contributions to the Scholarship of Teaching and Learning (SoTL) at Nipissing; has consolidated policies and processes around online course development and delivery; provides support to all faculty, staff, and students engaged in the use of educational technologies; and works collaboratively with faculty and others on wide ranging innovative educational projects. While the deliverables and accomplishments of the Centre for Flexible Teaching and Learning in under a year are impressive, the road to innovation has been circuitous and potholed. Educational innovation is not something real and tangible without significant naysaying and, at times, obstructionism. Leadership is an absolute requirement.

The connection between innovation and leadership in institutions including universities cannot be underestimated. In the context of the Centre for Flexible Teaching and Learning, such leadership has been necessary on at least three levels: that of the senior persons in the Centre (Director and Manager), the Deans, and the senior executive team of the university. The reason for this layered approach is that, when innovation occurs within an institution, the process requires the expertise and passion of those who understand the bigger picture and who will persevere when blockages present. Without an experienced Director with strong academic and administrative credentials, there is little hope that a project of this scope will grow and mature. Leadership among the community of the Deans is required in order to build positive relationships with faculty and promote understanding of the need for and benefits of flexible learning. The Deans also provide leadership relative to course and program development. Executive leadership is required for a number of reasons: to ensure that the human and technical resources for success are in place; to provide support to those actualizing the innovation; and to share successes and accomplishments within their larger networks.

Despite Nipissing's readiness for educational innovation on several fronts, on others, there has been some trepidation. Three principal fears have emerged: a generalized fear about something new; an almost visceral fear that e-learning will replace more traditional forms of teaching and learning; and reticence of those with a background in education to believe that others can assist and otherwise support their development as educators. Because Nipissing University is largely defined by its teacher education identity, the latter issue is uniquely complex. Interestingly, the faculty's reticence to embrace and use the supports of the Centre conflicts with repeated evidence that university professors declare interest in teaching supports (Carter & Brockerhoff-Macdonald, 2011).

What's next?

Thus far, we have provided context about the concept of educational innovation in universities and presented the case of Nipissing University as a study of educational innovation at a macro level. What follows is an exploration of educational innovation in a very specific context—nursing education. The need for delivering nursing education in innovative ways including e-learning and an up-close look at a pedagogical framework used to help nursing teachers and learners make sense of their experiences, despite the involved media, will also be explored. In the last section of the chapter, we will offer some observations about educational innovation in universities in general and present a number of recommendations for those considering the development and implementation of innovative approaches to teaching and learning in higher education.

Case study 2: a pedagogical approach (ICARE)

The need for innovative approaches to teaching and learning in nursing education

Statistics show that student numbers in post-secondary/tertiary programs are growing with students distributed across multiple campuses and in geographically dispersed locations across the globe (Altbach, Reisberg, & Rumbley, 2009; AUCC, 2011; Kim & Bonk, 2006; ICDE, 2009). Design of multiple course, subject, and degree offerings via flexible and distance learning modes has led to rapid growth in the use of e-learning technologies.

As technology and e-learning strategies continue to evolve, schools of nursing are using e-learning technologies to deliver courses and even entire curricula through a combination of face-to-face, web-enhanced, and fully online strategies. Challenges associated with course delivery include geographic and technological barriers; insufficient instructional design; inconsistent, inadequate, and/or unreliable support infrastructure; and varying degrees of faculty and student experience with learning management systems (Salyers, Carter, Barrett, & Williams, 2010). While the need for innovative delivery of nursing courses and programs is clear, it is also evident that nurse-learners and their instructors require, in addition to technical support, pedagogical support to navigate differing learning media and settings. The ICARE model discussed in the following paragraphs is presented as a system of organizing the teaching and learning experience and is particularly valuable when e-learning is involved.

What is the ICARE model?

The ICARE model is a pedagogical framework developed by staff and faculty at San Diego State University in 1997 to structure and organize course modules, modules being natural sub-sections of courses (Salyers, 2005; Salyers et al., 2010). The five steps of ICARE (Introduction, Connect, Apply, Reflect, and Extend) are repeated in each module of a course and can be used in face-to-face, blended, and fully online learning environments. More information regarding ICARE is presented by Hoffman and Ritchie (1998, 2005).

In the "Introduction" section of any ICARE module, context is provided. For example, learning objectives and reading assignments are presented. The "Connect" (or Content) section might provide lecture material and information to be discussed in other ICARE sections. In the "Apply" section, students might be required to write a short paper or complete a self-assessment in the form of a quiz which requires synthesis and application of ideas presented in the module. In the "Reflect" section, students might be asked to reflect on newly developed skills and knowledge (e.g., lessons learned). The "Extend" section might be structured around evidence-based articles associated with concepts presented in the module and "real world" applications (Salyers, 2005; Salyers et al., 2010).

Where has ICARE been used?

The ICARE framework has been implemented in two schools of nursing—one American and one Canadian. In both universities, the majority of programs offered at the undergraduate and graduate levels provided face-to-face, web-enhanced, and fully online learning experiences. Both institutions made a conscious decision to utilize the ICARE framework across the curricula of their various nursing programs.

The University of San Diego (United States)

The University of San Diego (California) is a private Roman Catholic institution founded in 1949. It has a total undergraduate enrollment of 5,388, its setting is urban, and the campus size is 180 acres. At the Hahn School of Nursing and Health Science, pre-licensure, undergraduate, and graduate nursing programs that include national certifications are offered. Courses for all of these programs use face-to-face, web-enhanced (blended), and fully online formats.

The University of Northern British Columbia (Canada)

The University of Northern British Columbia has four campuses in British Columbia. The main campus is located approximately ten hours away from Vancouver by car. Three regional campuses are located throughout British Columbia in rural and remote areas of the province. The university has a student population of nearly 4,200.

At the undergraduate level, the School of Nursing offers a Bachelor's degree in nursing (BScN) in partnership with two regional colleges, a post-diploma BScN, and a rural nursing certificate. At the graduate level, two options, the Master of Science in Nursing (Family Nurse Practitioner) and the Master of Science in Nursing (thesis stream), are offered. Total enrollment across all programs and campuses is approximately 650 students. Courses are delivered in face-to-face, web-enhanced (blended), and fully online formats at all campuses.

Contextual similarities between the two schools of nursing

Based on student surveys and anecdotal feedback provided by faculty at both institutions, a number of challenges about how courses are offered were identified. The first challenge related to faculty experience and expertise with e-learning formats. Faculty were inconsistent in their delivery of courses (e.g., one faculty might deliver his or her course using a face-to-face format while another might use a web-enhanced or fully online format). Some faculty were avid users of Blackboard, Moodle, or Desire2Learn and thus provided students with a variety of online experiences including discussion board activities, online quizzes and examinations, links to online resources, and so forth. Other faculty used the learning management system strictly to host course syllabi.

A second challenge was the variation among students' abilities to navigate their courses and to experience meaningful learning. Students frequently cited difficulties in finding course materials and general navigational issues for courses hosted online. In these instances, the students experiences were grounded in technology rather than learning.

Inadequate design of courses was cited as a third problem. Both the anecdotal and research evidence suggest that online courses require careful design in order to support both students and teachers. While design is related to issues of navigation as noted above, its purpose is much broader than technical decisions. It is particularly important in relation to the teaching and learning principles or pedagogy that supports the course.

While many technological, geographic, and other variables affected student and faculty satisfaction with their teaching and learning experiences, the three challenges previously discussed were identified as highest priority for improvement or change. In order to address some of these issues, the ICARE pedagogical framework was piloted to determine its effectiveness in mitigating some of the challenges in both institutions.

What does the research on ICARE tell us?

In their research on the ICARE framework, Salyers et al. (2010) found evidence to support use of the ICARE framework in structuring quality-based and satisfying courses from student and faculty perspectives. The study also included instructional design recommendations for the implementation of the framework in all programs.

Previous research based on the ICARE model found no differences in technical ability, learning styles, learning outcomes, and course satisfaction for graduate nursing students enrolled in face-to-face and web-enhanced sections of a course that used the ICARE framework (Salyers, 2005). Overall, students in web-enhanced sections were more satisfied with the course and reported advantages such as greater flexibility in scheduling, less travel, and greater independence and self-pacing in relation to content than their counterparts in face-to-face classes (Dimitrova, Mimirinis, & Murphy, 2004; Salyers, 2005; Salyers et al., 2010).

Reflections and recommendations

Barriers to building capacity in e-learning and flexible learning in general have been identified by many researchers: as the literature indicates, these barriers include inadequacy of infrastructure, inaccessibility of equipment, telecommunication costs, a lack of staff with appropriate skills, need for faculty training and support, administrative supports, and technological supports for students (Aczel, Peake, & Hardy, 2008). While each of these can be addressed in the planning and execution of a specific educational innovation, the cases presented in this chapter suggest that there are at least three other variables that require considerable attention if an innovation is to be successful. These variables are leaders who will own the change experience, faculty buy-in, and a commitment to change based on evidence.

University culture is fraught with an undeniable politic and, therefore, the path of least resistance for those with new ideas can be the status quo. In short, it is tough to deal with the naysaying and obstructionism and to move forward with actualization. This is particularly true when the intended innovation challenges university professors in their professional domain: teaching and learning. Thus, when an institution or group commits to the implementation of an educational innovation, the leaders must be prepared for considerable resistance and for owning the process in all its manifestations. While the change management and organizational behavior literature recommends, as possible, the hiring of a change agent to assist in the innovation process, fiscal realities seldom permit this at universities. Instead, it is incumbent upon the leaders of the innovation to support each other and to be highly skilled in articulating the benefits and opportunities of the proposed change. The Centre for Flexible Teaching and Learning will only reach its potential through sustained tenacity on the part of its three-part leadership team. Notably, implementation of the ICARE model occurred at the involved universities due to leadership's commitment to better experiences for students and instructors.

Educational innovation at universities will be successful only if there is faculty buy-in. This is not to suggest that full faculty buy-in will ever exist before an innovation can be deployed. While such support is desired, a more realistic scenario is one in which buy-in develops incrementally. Not only will buy-in be incremental, how it looks will vary from professor to professor. In the case of the Centre for Flexible Teaching and Learning, buy-in strategies include practical and highly relevant mentoring sessions on how to use educational technologies, discussions about using new models of teaching in different disciplines, participation in research and SoTL activity where the focus is educational innovation, and so forth. Moreover, since the wins may occur on a one by one basis, the leadership team must be patient with the process. Champions from within the faculty community will, of course, be valuable in the buy-in process. Buy-in is equally important in more specific settings such as the use of ICARE since resistance to use ICARE would have led to more of the same: dissatisfaction with the teaching and learning experience.

Last is that evidence is the innovator's ally. In the case of ICARE, its implementation was grounded in evidence of existing problems requiring a solution. In the Centre for Flexible Teaching and Learning, evidence needs to be used at all points: to document the need for innovative approaches to teaching and learning by universities; to point out the benefits and opportunities of innovative models and approaches; and to discover appropriate supports for professors in their commitment to educational excellence.

Conclusion

In conclusion, the evidence is clear: universities are being challenged to embrace new and innovative ways of thinking about teaching and learning or, very simply, they will be left behind. Universities that think and act innovatively are more likely to have engaged students including students who enter university directly from high school as well as working adults who return to university later for personal and professional reasons than their counterparts. Universities that see innovation as part of the present and the future will have a competitive advantage.

While the call to innovation is real, equally real are the obstacles to successful implementation. Although some of these obstacles are due to simple resistance to change in all its iterations, a good number of them stem from university culture and thinking grounded in old ways and fears. As stated earlier, some faculty fear that e-learning will replace more traditional forms of teaching and learning and possibly even their jobs. Rather than participate, they dig in and hold on to past practices, stating that they are the stewards of their classrooms and therefore know what is best. Contrarily, they may be ill equipped from a skill point of view and refuse to commit the time required to construct experiences that facilitate meaningful learning in our increasingly technologically defined world.

Based on this kind of pushback, the only appropriate response is that of courageous and tenacious leadership. Such leadership needs to occur throughout the university and must be grounded in evidence—evidence of why such change is required, evidence of the benefits for students and teachers. In the case of ICARE, nursing leaders saw evidence that this pedagogical framework facilitates learner and professorial competence and confidence. Thus, ICARE was implemented across all nursing programs.

To go forward is never easy. However, when educational leaders, practitioners, and innovators agree on specific goals and pool their energies, the hurdles become less daunting and the goals more readily achievable. In the tumult of 21st century education, the stakes are high with innovation being central to our successes and challenges. Educators of all stripes are, therefore, encouraged to put timidity behind them and to find colleagues who are equally committed to innovative ways of facilitating educational access and excellence.

References

Aczel, J. C., Peake, S. R., & Hardy, P. (2008). Designing capacity-building in e-learning expertise: Challenges and strategies. *Computers and Education, 50*(2), 499–510.

Altbach, P., Reisberg, L., & Rumbley, L. (2009). Trends in global higher education: Tracking an academic revolution. A Report prepared for the UNESCO 2009 World Conference on Higher Education. Retrieved November 8, 2012 from www.uis.unesco.org/Library/Documents/trends-global-higher-education-2009-world-conference-en.pdf.

AUCC (2011). *Trends in Higher Education. Volume 1: Enrolment*. Retrieved November 8, 2012 from www.aucc.ca/wp-content/uploads/2011/05/trends-2011-vol1-enrolment-e.pdf.

Bates, T. (2008). What is distance education? Retrieved November, 8 2012 from http://chronicle.com/article/America-FallingLongtime-D/48683/.

Beckmann, E. A. (2010). Learners on the move: Mobile modalities in development studies. *Distance Education, 31*(2), 159–173.

Bonk, C., & Graham, C. (2006). *The Handbook of Blended Learning: Global Perspectives, Local Design*. San Francisco: Pfeiffer Publishing.

Brown, K. M. (1996). The role of internal and external factors in the discontinuation of off-campus students. *Distance Education, 17*(1), 44–71.

Bullen, M. (2011). Separating fact from fiction in the digital generation discourse. Online presentation to IT Silgo, Ireland, January 19, 2011.

Bullen, M., Morgan, T., & Qayyum, A. (2011). Digital learners in higher education: Generation is not the issue. *Canadian Journal of Learning and Technology, 37*(1).

Carter, L. (2003). Distance wise: Meeting the learning needs of post-RN nurses in Ontario. *Canadian Nurse, 99*(10), 24–27.

Carter, L. (2008). Critical thinking dispositions in online nursing education. *Journal of Distance Education/Revue de l'Éducation à Distance, 22*(3), 89–144.

Carter, L., & Brockerhoff-Macdonald, B. (2011). The continuing education of faculty as teachers at a mid-sized Ontario university. *Canadian Journal for the Scholarship of Teaching and Learning, 2*(1). Retrieved November 15, 2011 from: http://ir.lib.uwo.ca/cjsotl_rcacea/vol. 2/iss1/4.

Carter, L., & Rukholm, E. (2002). Online scholarly discourse: Lessons learned for continuing and nurse educators. *Canadian Journal of University Continuing Education, 28*(2), 33–50.

Carter, L., & Rukholm, E. (2008). A study of critical thinking, teacher–student interaction, and discipline-specific writing in an online educational setting for registered nurses. *Journal of Continuing Education in Nursing, 39*(3), 133–138.

Carter, L., Rukholm, E., Mossey, S., Viverais-Dresler, G., Bakker, D., & Sheehan, C. (2006). Critical thinking in online nursing education setting: Raising the bar. *Canadian Journal of University Continuing Education, 32*(1), 23–46.

Carter, L., Salyers, V., Page, A., Williams, L., Hofsink, C., & Albl, L. (2012). Highly relevant mentoring (HRM) as a faculty development model for web-based instruction. *Canadian Journal of Learning and Technology, 38*(1).

Dickieson, P., Carter, L., & Walsh, M. (2008). Integrative thinking and learning in undergraduate nursing education: Three strategies. *International Journal of Nursing Education Scholarship, 5*(1), Article 39.

Dimitrova, M., Mimirinis, M., & Murphy, A. (2004). Evaluating the flexibility of a pedagogical framework for e-learning. Retrieved November 20, 2011 from: www.computer.org/portal/web/csdl/doi/10.1109/ICALT.2004.1357422.

El-Hussein, M. O., & Cronje, J. C. (2010). Defining mobile learning in the higher education landscape. *Journal of Educational Technology and Society, 13*(3), 12–21. Retrieved November 10, 2012 from www.ifets.info/journals/13_3/3.pdf.

Ellis, A. (2005). *Research on Educational Innovations*, 4th ed. Larchmont, NY: Eye On Education.

Ellis, R. A., Goodyear, P., Prosser, M., & O'Hara, A. (2006). How and what university students learn through online and face-to-face discussion: Conceptions, intentions, and approaches. *Journal of Computer Assisted Learning, 22*(4), 244–256.

Frankola, K. (2001). Why online learners drop out. *Workforce, 10*, 52–60.

Gefen, D., Karahanna, E., & Straub, D. W. (2003). Inexperience and experience with online stores: The importance of TAM and trust. *IEEE Transactions on Engineering Management, 50*(3), 307–321.

Golladay, R., Prybutok, V., & Huff, R. (2000). Critical success factors for the online learner. *Journal of Computer Information Systems, 40*(4), 69–71.

Henry, J. W., & Stone, R. W. (1994). A structural equation model of end-user satisfaction with a computer-based medical information system. *Information Resources Management Journal, 7*(3), 21–34.

Hoffman, B., & Ritchie, D. C. (1998). *Teaching and Learning Online: Tools, Templates, and Training*. Washington, DC: Society for Information Technology and Teacher Education.

Hoffman, B., & Ritchie, D. C. (2005). Teaching and learning online: Tools, templates, and training. In J. Willis, D. Willis, & J. Price (Eds.), *Technology and Teacher Education Annual, 1998* [CD ROM]. Charlottesville, VA: AACE.

Hrastinski, S. (2008). Asynchronous and synchronous e-learning: A study of asynchronous and synchronous e-learning methods discovered that each supports different purposes. Retrieved November 20, 2011 from: www.educause.edu/EDUCAUSE+Quarterly/EDUCAUSEQuarterlyMagazineVolum/AsynchronousandSynchronousELea/163445.

ICDE (2009). Global trends in higher education, adult and distance learning. Retrieved November 8, 2012 from www.icde.org/filestore/Resources/Reports/FINALICDEENVIRNOMENTALSCAN05.02.pdf.

Johnson, L., Levine, A., Smith, R., & Stone, S. (2010). *The 2010 Horizon Report*. Austin, TX: New Media Consortium.

Kearns, L., Shoaf, J., & Summey, M. (2004). Performance and satisfaction of second-degree BSN students in web-based and traditional course delivery environments. *Journal of Nursing Education, 43*(6), 280–284.

Kim, K., & Bonk, C. (2006). The future of online teaching and learning in higher education: The survey says … *EDUCAUSE Quarterly, 29*(4). Retrieved November 20, 2011 from: www.educause.edu/EDUCAUSE+Quarterly/EDUCAUSEQuarterlyMagazineVolum/TheFutureofOnlineTeachingandLe/157426.

Laine, L. (2003). Is e-learning effective for IT training? *T +D, 57*(6), 55–60.

Margaryan, A., Littlejohn, A., & Vojt, G. (2011). Are digital natives a myth or reality? University students' use of digital technologies. *Computers and Education, 56*(2), 429–440.

Martins, L. L., & Kellermanns, F. W. (2004). A model of business school students' acceptance of a web-based course management system. *Academy of Management Learning and Education, 3*(1), 7–39.

Ontario Public School Boards' Association (2010). What if? Technology in the 21st century classroom: An OPSBA discussion paper. Retrieved on November 8, 2012 from www.opsba.org/files/WhatIf.pdf.

Parson, V., Reddy, P., Wood, J., & Senior, C. (2009). Educating an iPod generation: Undergraduate attitudes, experiences and understanding of vodcast and podcast use. *Learning, Media and Technology, 34*(3), 215–228.

Reeves, P., & Reeves, T. (2008). Design considerations for online learning in health and social work education. *Learning in Health and Social Care, 7*(1), 46–58.

Ryan, M., Carlton, K., & Ali, N. (2004). Reflections on the role of faculty in distance learning and changing pedagogies. *Nursing Education Perspectives, 25*(2), 73–80.

Ryan, S. (2001). Is online learning right for you? *American Agent and Broker, 73*(6), 54–58.

Salyers, V. (2005). Web-enhanced and face-to-face classroom instructional methods: Effects on course outcomes and student satisfaction. *International Journal of Nursing Education Scholarship, 2*(1), Article 29, 1–13.

Salyers, V., Carter, L., Barrett, P., & Williams, L. (2010). Evaluating student and faculty satisfaction with a pedagogical framework. *Journal of Distance Education/Revue de l'Éducation à Distance, 24*(3).

Serwatka, J. (2003). Assessment in on-line CIS courses. *Journal of Computer Information Systems, 43*(3), 16–20.

Shale, D. (1990). Towards a reconceptualization of distance education. In M. G. Moore (Ed.), *Contemporary Issues in American Distance Education* (pp. 333–343). Oxford: Pergamon.

Smart, K., & Cappel, J. (2006). Students' perceptions of online learning: A comparative study. *Journal of Information Technology Education, 5*, 201–219.

So, H., & Bonk, C. (2010). Examining the roles of blended learning approaches in computer-supported collaborative learning (CSCL) environments: A delphi study. *Educational Technology and Society, 13*(3), 189–200.

Stoel, L., & Lee, K. H. (2003). Modeling the effect of experience on student acceptance of Web-based courseware. *Internet Research, 13*(5), 364–374.

Tapscott, D. (2008a). *Grown Up Digital: How the Net Generation is Changing Your World*. Columbus, OH: McGraw-Hill.

Tapscott, D. (2008b). How digital technology has changed the brain. *Business Week*. Retrieved August 2, 2011 from: www.businessweek.com/technology/content/nov2008/tc2008117_034517.htm.

Thiele, J. (2003). Learning patterns of online students. *Journal of Nursing Education, 42*(8), 364–366.

Weber, J., & Lennon, R. (2007). Multi-course comparison of traditional versus web-based course delivery systems. *Journal of Educators Online, 4*(2), 1–19.

Wober, K., & Gretzel, U. (2000). Tourism managers' adoption of marketing decision support systems. *Journal of Travel Research, 39*(2), 172–182.

33

Developing an understanding of the pedagogy of using a Virtual Reality Learning Environment (VRLE) to support innovation education

Gisli Thorsteinsson

UNIVERSITY OF ICELAND, ICELAND

Summary: Innovation Education (IE) is a new subject area in Icelandic schools. The aim is to train students to identify needs and problems in their environment and to find solutions: a process of ideation. This activity has been classroom based but now a specific Virtual Reality Learning Environment technology (VRLE) has been created to support ideation. This technology supports online communications between students and teacher and enables them to develop drawings and descriptions of solutions. The VRLE is Internet connected and the students work online with their ideas in real time. As this learning environment is new, it was important to evaluate and explore its use and value in supporting ideation in the context of IE. The author has run a series of case studies to identify the pedagogical issues of using the new VRLE to support ideation within IE.

Key words: Innovation Education, ideation, case study lessons, Virtual Reality Learning Environment, pedagogy, ideation process.

Introduction

Innovation Education (IE) in Iceland aims to train students to identify needs and problems in their environment and to find solutions: this is referred to as the process of ideation. This chapter explores the contexts of teaching and learning, incorporating the VRLE with IE to support the students' work. There is a focus on blended learning, as the VRLE is used in conjunction with conventional classroom-based activity.

The work employed the grounded theory (Glaser & Strauss, 1967) perspective, in order to observe the complex social/educational activity relating to this real-life learning context. It was intended to build understanding (grounded theory), rather than an attempt to establish cause and effect. The author intended to observe, describe and interpret settings as sources of data and the main aim was to 'gain a greater understanding of the use of the VRLE, in supporting students' work in conventional Innovation Education classes within Icelandic schools'.

The overall research question was: 'How does the use of the VRLE affect teacher's pedagogy and the students' work, in conventional Innovation Education in Iceland?'

Research tools and multiple data sources were selected, in order to gather triangulated data. Specifically, a series of case studies were employed in an Icelandic elementary school, featuring groups of volunteers from the seventh class (age 12).

The main findings highlighted the significant categories and issues relating to the impact of the VRLE in the contexts of teaching and learning, which were:

1. the teacher and his/her approach to his/her work;
2. use of homework;
3. use of the VRLE;
4. Innovation Education and idea generation;
5. drawing;
6. values.

The findings indicated that a teacher's approach to his or her work is significant; s/he must be able to alternate between various roles during lessons. Such roles include computer administrator, instructor and a facilitator, in encouraging students to become self-sufficient and autonomous. The students' homework enabled them to generate the content of the course and make meaning of their work, while the VRLE facilitated students' collaboration and cooperation during ideation work. Training students in the use of the CAD programme, drawing and the VRLE appeared to be important in enabling their ideation work. However, the students also learned through their own practice.

The chapter introduces the Icelandic IE and describes the background of the VRLE. The overall aims, objectives and research questions are stated. Subsequently idea generation is defined and a specific pedagogical model for idea generation demonstrated. The research methods are explained and findings reported. Subsequently these findings are discussed and conclusion drawn.

Using a Virtual Reality Learning Environment for innovation education

Innovation Education originated in Iceland in 1991 (see Chapters 2 and 19, this volume). It was developed within Design and Craft lessons and was closely linked to the principles of Nordic Sloyd (Olafsson & Thorsteinsson, 2009), in that it also aimed to educate children holistically, via a carefully structured system. In the case of Sloyd, such a carefully structured system was handicraft and, with regards to IE, the system refers to ideation skills within the context of innovation. IE focused on the conceptual work of students, searching for needs and problems in their own environments, generating appropriate solutions or applying and developing known solutions (Thorsteinsson & Denton, 2003; Gunnarsdottir, 2001a). While IE had its roots in Design and Craft, it was aimed at general education and, in 1999, IE was developed into a new subject within the Icelandic National Curriculum 1999 (Ministry of Education, 1999). In 2006, it became a cross-curricular element of the National Curriculum (2006).

In 1996, the Iceland University of Education coordinated the three-year EU funded *Practical use of Information Technology (IT)* and *Open and Distance Learning (ODL) in Innovation Education* (InnoEd, 2011). This project combined computer-based technologies and ODL with original IE concepts, in order to develop new ways of supporting student ideation work in IE classes. A major output of the *InnoEd* project was the development of a specific 'Virtual Reality Learning Environment' (VRLE), in which students could communicate, interact, and develop and host their IE work.

As the relationship between the VRLE and the IE process was new and potentially complex it became important to research student work and teaching approaches (pedagogy) relating to this context and existing learning theories. The curriculum development work moved into the research phase, using research strategies and methodologies appropriate for exploring and describing the issues relating to the new learning and teaching context. This built on the work of Gunnarsdottir (this volume), who outlined the pedagogy for early IE work, prior to the introduction of the VRLE (Gunnarsdottir, 2001a; Thorsteinsson, 1998).

The pedagogy of Innovation Education

Ideation is at the core of the IE pedagogical framework and the emphasis is to develop students' ideation skills through the innovation process (Gunnarsdottir, 2001a). IE is based on conceptual work that involves searching for needs and problems in students' environments and finding appropriate solutions, or applying and developing known solutions (Thorsteinsson & Denton, 2003). The aim is to develop students' innovativeness and make them better equipped to deal with their world and take an active part in society.

Innovation relates to the usefulness of ideas and/or how they can be implemented as solutions to the many problems encountered in daily life. In Innovation Education, students use appropriate knowledge and information from different sources to find solutions to the problems or opportunities identified: this mirrors Vygotsky's (1978) zone of proximal development. Students work with their own concepts, but learn to work through the innovation process (see Figure 33.1) to bring their ideas into being and gain what is now known as *Creative Relevant Skills* (see Chapter 2 and Gunnarsdottir, 2001b).

The Thorsteinsson initial pedagogical model for IE

Gunnarsdottir (this volume) gained an understanding of how students learnt in IE classes prior to the introduction of the VRLE. She looked at how students learned through their social activities

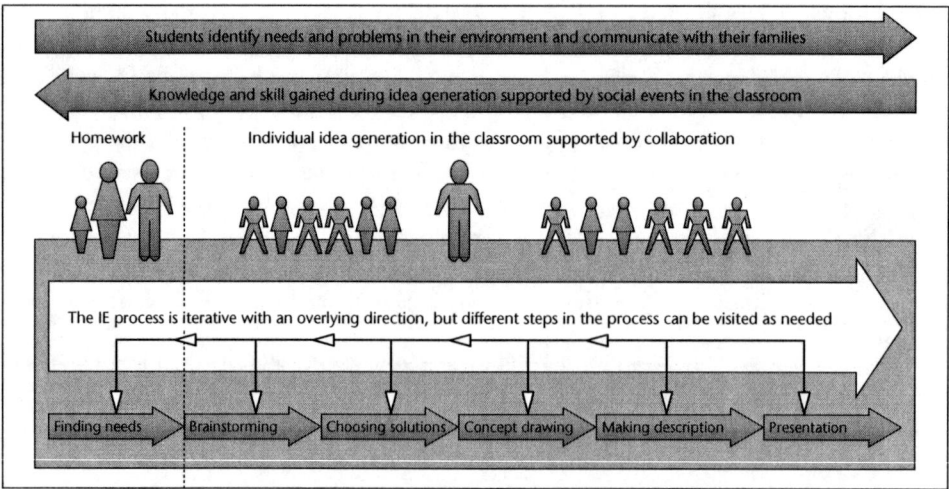

Figure 33.1 The basic pedagogical model of the IE innovation process. The model illustrates innovation as a 'process', with appropriate feedback loops and options.

during ideation in IE and put forward a pedagogical model of teaching and learning in *Innovation Education* (Gunnarsdottir, 2001a). Based on Gunnarsdottir's work (this volume) and Thorsteinsson's description of the innovation process in IE, Thorsteinsson put forward an initial model (Thorsteinsson & Denton, 2003; as well as Figure 33.1).

The model illustrates the way students work through the innovation process in Innovation Education classes and is based on a series of steps, iterations and relationships, with the overlying direction leading from 'finding needs' to 'presentation of solutions'. Students employ ideational skills at all stages and learn through the innovation process within the overall IE pedagogical framework (Ministry of Education, 1999, 2006) which is managed by the teacher. The process is as follows:

1. finding needs;
2. brainstorming;
3. creating and choosing initial solutions;
4. concept drawing or modelling, in order to develop the technical solution;
5. creating a description of the solution, in addition to the drawing;
6. presentation.

Innovation and ideation

According to the *Oxford English Dictionaries Online* (2011), innovation is the action of innovating; the introduction of novelties and the alteration of what is established by the introduction of new elements or forms. Innovation is a form of problem solving that begins with the feeling that change is needed and ends with the successful implementation of an idea (Smith, 2003). Innovation includes the generation of ideas, alternatives and possibilities (Smith, 1998).

Rogers (2003) states that:

> Innovation is an idea, practice or object that is perceived as new by an individual or other unit of adoption. It matters little whether the idea is objectively new, as measured by the lapse of time since its first use or discovery. The perceived newness of the idea for the individual determines his or her reaction to it: if the idea seems new to the individual, it is an innovation.

In IE the novelty in a student's work has an individual meaning that is concerned with the individual's ability to deal with their world by calling upon their creative talents on a daily basis (Thorsteinsson & Denton, 2003). Thus, innovation is different from creativity, although both share elements of meaning. Innovation is the application of new ideas (Rogers, 2003), whereas creativity, in contrast, is the generation and articulation of new ideas.

Ideation and ideation skills are important parts of the IE pedagogy, as they enable students to go through the innovation process. *Ideation* is a concept derived from Guilford (1950) (Thompson, 2008) and is used to describe the pattern of interactions that arise when a person works on and produces an idea. As the *Oxford English Dictionaries Online* (2011) states, ideation is the formation of ideas or mental images of things not present to the senses. According to the *Webster's Dictionary* (2005), ideation is described as 'the faculty or capacity of the mind for forming ideas; the exercise of this capacity; the act of the mind by which objects of sense are apprehended and retained as objects of thought' (p. 725). Ideation is important during several phases of innovative problem solving, including the generation of ideas, in terms of solving problems, and the development of solutions to such problems (Clapham, 2003; Doolittle, 1995).

Ideation and its role in building innovativeness through general education

The main emphasis of the pedagogy of IE is to make students better equipped to deal with their world and take an active part in society through innovation (Gunnarsdottir, 2001a; Thorsteinsson & Denton, 2003). The ideational skills developed during IE aim to encourage this aspect of students' development and thus strengthen the ability of future societies, in terms of innovation and development (Ministry of Education, 1999).

Ethical maturation is an important element of IE, in the context of developing ideation skills. Ethics is the ethical judgement of an individual. It supports an individual's responsibility to take part in and help shape society (Thorsteinsson, 1996). Ethics develop through a student's innovation work as they are working with real world problems. Students augment their ethical maturity and ability to utilize their innovativeness. This enables them to move in a positive direction, believe in their future and feel themselves to be an integral and independent person.

In IE, students are introduced to a process of innovation that focuses on the 'front-end' of the design process; i.e. problem and need identification, initial concept generation, the development of basic solutions using simple models and descriptions with images or multimedia content (Thorsteinsson & Denton, 2003) (ideation skills are central to the formation of ideas in this process). The Icelandic National Curriculum takes the position that everyone can be innovative and that it is possible to introduce classroom activities that develop ideation. Innovation Education is integrated into regular ordinary schoolwork and taught by non-specialist teachers, who aim to:

1. stimulate and develop innovativeness in students and teach them certain approaches and processes, from concept through to realisation;
2. teach individuals to be innovative in daily life, so that they become better equipped to adapt their environment;
3. encourage and develop students' initiative and strengthen their self-image;
4. make students aware of the ethical values of 'objects', while teaching ways in which to improve their environment (Thorsteinsson, 1998, p. 143).

Using a VRLE to support innovation education

The original idea behind the VRLE was to find a new way of supporting students' ideation work, using information and computer technology (Thorsteinsson, Denton, Page, & Yokoyama, 2005). The specific VRLE was designed to enhance ideation via collaborative learning support and thus offers individual and social educational opportunities. The main output of the project was an online VRLE, linked to a database: this VRLE was developed as a combination of the managed learning environment (MLE) and the virtual reality environment (VRE). The MLE provided the framework for teachers to manage student learning, while the VRE provided a simple virtual environment that enabled students to meet and communicate through a number of means, such as voice, text, drawings, photographs and presentations. The database enabled these ideas to be shared and recorded and these, as a whole, represented the VRLE.

The VRLE is potentially a tool for experiential learning, as it provides various dynamic and rapid ways to see, experience and generate ideas and information. The VRLE can be used as a tool for problem solving and communicating ideas and includes the possibility of promoting a high degree of interactivity and immersion (Ogle, 2002; Bricken, 1991; Johnson, Moher, Choll, Lin, & Kim, 2002; Jonassen, 1999; McLellan, 1996; Winn, Windschiti, & Thomson-Bulldis, 1999; Osberg, 1993). The VRLE is interactive in two ways: first, a user interacts with data in the database within the VRLE and also beyond; for example, via the World Wide Web (www). Second, it

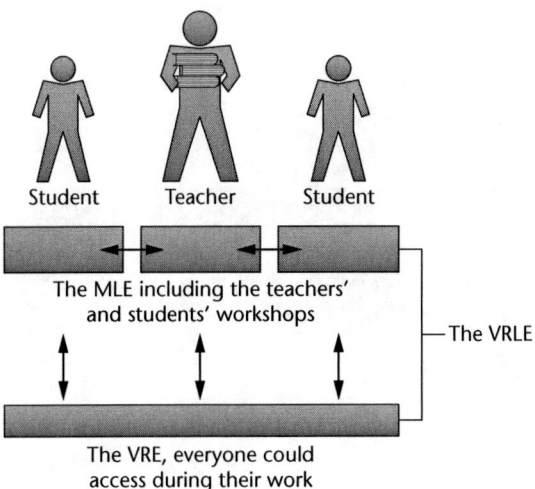

Figure 33.2 The VRLE is a combination of a managed learning environment and a virtual reality environment.

allows the interaction of a number of students and staff within the VRLE, using a range of modes including speech, drawing and writing. Students could be from the same class or in other schools or countries, accessing the VRLE via the WWW.

Using the VRLE within the classroom context offers multi-modal communication through the Internet to the world outside and this would be expected to influence students' learning experiences (see Figure 33.2).

Extraneous interruptions while using the VRLE are avoided by the use of a user name and password (it is important to prevent anyone interfering with the scaffolding of the lesson for malicious reasons). As before, the student brings needs and problems identified at home into the school and works there, supported by the VRLE.

The main reasons for students using the VRLE in IE classes were:

- to offer another mode of working together, in terms of ideas, sharing problems, solving such problems and developing solutions;

Figure 33.3 The teacher and students using the MLE in the conventional classroom and the VRE in the virtual classroom.

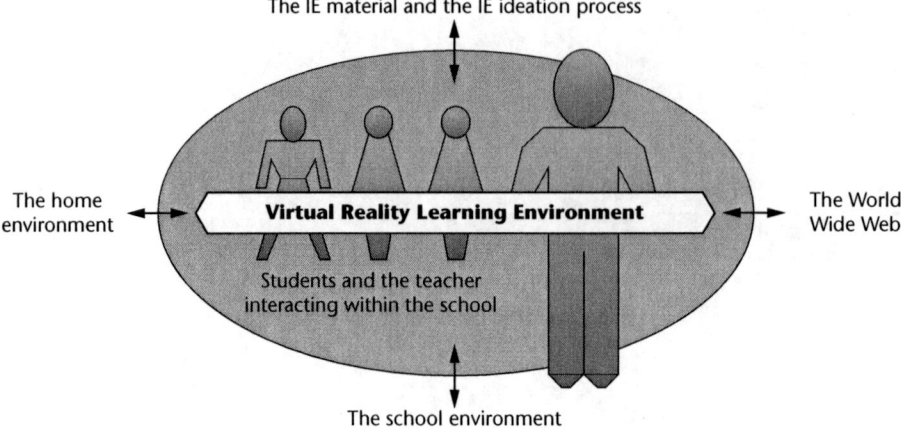

Figure 33.4 The VRLE offers different dimensions of communication.

- to enable students to meet each other and their teacher online;
- to facilitate easy communication inside virtual 3D spaces, where students and teachers could meet in real time, share information and work together with ideas;
- to provide the opportunity to develop certain skills within the innovation process (i.e. brainstorming, drawing and discussion);
- to allow students the opportunity to give online presentations;
- to set up virtual exhibitions;
- to enable virtual meetings between participants.

The focus and aims of the research project

As the use of the VRLE for IE was new and the learning and teaching context complex and dynamic, the focus became the exploration of the use of the VRLE to support student ideation work and the pedagogy developed within the context of IE in Iceland. The intention was to identify the issues involved, to use literature and fieldwork to understand how these issues were related and, eventually, to be able to prepare a map of directions for further research.

While the VRLE has the potential to enable open and distance learning in IE work, in terms of cooperation between students and teachers across continents, it was decided that this would be too large a dimension for this research. Thus, the focus is on the use of the VRLE within the conventional classroom context, as a logical precursor to future work and looking at ODL work.

The project's main aim was to: *gain a greater understanding of the use of the VRLE in supporting students' work in conventional Innovation Education classes within Icelandic schools.*

The main objectives were:

a   *to develop an understanding of the pedagogy needed to use the VRLE in this context;*
b   *to upgrade the Icelandic Innovation Education pedagogical model, in terms of supporting the VRLE.*

The following overall research question was thus formed to guide the research:

> *How does the use of the VRLE affect teacher's pedagogy and student work in conventional Innovation Education in Iceland?*

Research methodology development

The author ran a series of case study lessons during a two year period. A research plan was established, on the aim and research question. Students from class 7 in an Icelandic elementary school volunteered for the research. The teacher set up email accounts and registered them to the VRLE; he also took digital photographs, in order to enable the students to personalise their VRLE workshops.

As the research took place in a complex social/educational context, grounded theory (Glaser & Strauss, 1967) principles were used as a way of observing, describing and interpreting settings as sources of data (grounded theory is a principle based on the systematic building of theory, using qualitative or/and quantitative data). The key points in the data are marked with a series of codes, which are then grouped into emerging conceptual categories. These categories are related to each other as a theoretical explanation of the action(s) that continually resolve the main concerns of the participants within a substantive area (Denzin & Lincoln, 1994).

Grounded theory focuses on obtaining an abstract analytical schema of a phenomenon that relates to a particular situation (Creswell, 1998). However, Strauss and Corbin (1998) explicitly pointed out that the value of grounded theory lies in its ability not only to generate the theory, but also to ground that theory in data. This inductive method is particularly helpful in identifying patterns of behaviour or thought in a particular group of people, as in this study.

In the case study series, different types of qualitative data were collected in the form of interviews with the participating teacher and students, classroom observations, video recordings of students' activity when using the VRLE, screen video recordings, student work samples and the teacher's and researcher's logbooks. These multiple perspectives helped the researcher to 'validate and crosscheck findings' (Patton, 1990, p. 244) and offered a good degree of triangulation (Denzin, 1984; Cohen, Manion, & Morrison, 2005).

General findings

Throughout the research the VRLE worked well in general; it was stable and easy to register the students. However, dealing with the VRLE technology might have been more difficult for a teacher without strong information technology skills. Due to good computer literacy, students learned to use the VRLE through direct experience. Using the VRLE network inside the classroom made it possible for them to learn from one another both face-to-face and online. They also had some instruction by the teacher. They quickly became self-reliant but the teacher considered they needed more concrete learning material and a traditional instructional phase.

The teacher's role was to help students to understand IE and the innovation process. Training them via the VRLE was beneficial for their idea generation. Students normally quickly understood the innovation process and were able to identify needs and problems in their own environment. Identifying problems and need at home played a significant role in the first stages of the innovation process that took place at home. This was intended to trigger idea generation in lessons, helping students to generate the content of the course, make them self-directed and give their work a personal meaning.

Students usually defined their findings spontaneously and tended to record solutions in the notebook instead of needs and problems. However, the teacher was able to help them to define needs rather than solutions by discussions while they worked inside the VRLE without imposing his own value judgements.

The VRLE directed students' idea generation as it was structured upon the idea generation process. The VRLE facility for sharing needs, solutions and to brainstorm during work was identified as beneficial. Students frequently shared needs and problems with each other, both face-to-face and online. There was a balance between needs identified at home and at school. However, most ideas were generated when students were working collaboratively inside the VRLE.

Discussion and conclusions

The VRLE was structured upon the IE innovation process and included a facility to share needs and solutions, with the opportunity to brainstorm (Thorsteinsson & Denton, 2003). It can thus be seen as an interactive, collaborative, learning tool that supports idea generation. Students often shared needs and solutions inside the VRLE; they usually came up with many ideas when working inside the MLE, but, subsequently, typically worked cooperatively inside the VRE on one idea chosen from the ideas presented by the group (Thorsteinsson, 2009).

Students in the case studies were generally self-reliant and often worked individually inside the MLE part of the VRLE; however they also worked collaboratively inside the VRE and this collaboration supported individually based idea generation, as it enabled students to help each other (Dennis & Valacich, 1993). Students were also able to access separate virtual whiteboards, but communicated their ideas inside the VRE and the classroom at the same time. However, students were still less productive and fewer ideas were generated, as this was time consuming (see accordingly Taylor, Berry, & Block, 1958; Paulus, Larey, & Ortega, 1995).

Being able to play inside the VRE, when working in the MLE, was a form of informal 'edutainment', which supported collaboration and the generation of skills (Rieber, 2001; O'Quin & Derks, 1999). A light-hearted spirit in lessons appeared to positively influence idea generation, supporting the position of O'Quin and Derks (1999). However, further research would be needed, with regards to differentiating between the effects of collaboration and the 'light hearted spirit' of idea generation.

The VRLE guided the students' work, gave structure and reflected the role of the computer as a tutor, tutee and tool (as in Blom & Monk, 2003; Taylor, 1980) and enabled both CSCL and CSCW (Thorsteinsson & Denton, 2008; Thorsteinsson, Page, & Niculescu, 2010b). The VRLE worked as a tool students used to enable their work. It included help pages and was structured on the innovation process. This structure and help pages guided and directed students during their work and was therefore a form of tutee.

During the research, students had no major problems in using the VRLE and quickly became self-reliant (Thorsteinsson & Denton, 2006). Their confidence and IT ability enabled them to start using the VRLE easily. However, the case studies showed that additional training was needed for the hardware and the VRLE. The teacher also considered students needed training in using the VRLE for cooperative idea generation (Thorsteinsson, Page, & Niculescu, 2010a). This involved use of avatars and CAD. These are all areas which need subsequent specific exploration.

Social presence was an important aspect of using the VRLE and enabled a community of learners to grow as Hamburg, Lindecke and Thij (2003), Thorsteinsson and Page (2007) and Hauber, Regenbrecht, Hills, Cockburn and Billinghurst (2005) had all indicated. Playing informally in the VRLE was shown to promote the students' skills and confidence in using the VRLE, and familiarity with each other (Prensky, 2005; Hussain et al., 2003). The case studies indicated that being physically together and being able to speak to the teacher both inside the classroom and over the Internet at the same time appeared to assist students' learning, probably via having multiple modes of communication (Loiselle, St-Louis, & Dupuy-Walker, 1998; Schrum & Berenfeld, 1997; Thurlow, Lengel, & Tomic, 2004; Romiszowski & Mason, 1996). The capability of students personalising the interface of their virtual workshops appeared to be important in relation to increasing their perception of relevance and ownership of the VRLE, echoing Oulasvirta and Blom (2008) and Blom and Monk (2003).

The research identified the importance of the IE teacher, in both roles of traditional instructor and facilitator and adapting to the role of facilitator rather than instructor echoes the social constructivist approach (Bauersfeld, 1995; Gunnarsdottir, 2001a; Jónsdóttir, 2005). However, the teacher appeared to feel uncomfortable during the case studies, especially when operating primarily as a facilitator, and he asserted that he wanted to incorporate more formal taught sections within the course, in terms of the use of the VRLE and the use of the input devices for drawing. He believed that training students, providing them with knowledge (i.e. instructor role) and personal experience (facilitator role) were beneficial in idea generation and he also considered that students should learn to draw, in order to speed up their work. The structure of the VRLE supported the teacher in his role as facilitator, as indicated by Bonk and Cunningham (1998) and Heinze (2008). Furthermore, it appeared to enable individuality, collaboration and cooperation and supported students' autonomy.

It was the teacher's role to help students to understand IE and the innovation process (Gunnarsdottir, 2001b) both with and without the VRLE (Thorsteinsson & Denton, 2008). They quickly became familiar with the innovation process in so far as bringing basic ideas to school to act as start points for effective collaborative idea development. However, it was evident that students in the case studies did not understand the fine differences between problems, opportunities, needs and initial ideas.

Training students in idea generation via the VRLE and in the classroom appeared to be encouraging self-reliance and independence and appeared to be beneficial for idea generation. It furthermore gave the teacher a little more freedom to stand back and observe the group carefully. This supported him in adopting the role of a facilitator to a greater extent (Thorsteinsson & Denton, 2008).

Students in the case studies were generally self-reliant and worked most often individually inside the MLE part of the VRLE, but also collaboratively inside the VRE at the same time. This collaboration was supportive for individually based idea generation (see also Dennis & Valacich, 1993). However, students were still less productive and fewer ideas were generated as it was time consuming (as with Taylor et al., 1958 and Paulus et al., 1995).

The approach reflected blended learning, where the teacher's role was based on multiple teaching methods that supported idea generation inside the VRLE (Page, Thorsteinsson, & Niculescu, 2008; Thorsteinsson & Denton, 2008; Worthington, 2008). Students were thus able to use various learning resources during idea generation, such as their inventor's notebook, the Internet, tutorials inside the VRLE, written material from the teacher and information gained during collaboration, both face-to-face and online.

Courses were built on combinations of conventional IE learning with the VRLE, using different didactic methods and delivery formats (Whitelock & Jelfs, 2003; Kerres & De Witt, 2003) and the teacher's role was to enable idea generation inside the VRLE through appropriate teaching methods. Lessons were thus based on the following basic structure:

a   introduction;
b   basic training;
c   students reporting needs and problems;
d   brainstorming sessions;
e   students developing solutions inside the VRLE, both as individuals and in groups;
f   summarising lessons and looking forward to/preparing for the next lesson in the series.

Due to the ability of the VRLE to enhance individual work, collaboration and cooperation, the teacher was able to employ different teaching methods (Thorsteinsson & Denton, 2008) and the

MLE was used to encourage individual idea generation, supported by the students' collaboration inside the VRE. However, the VRE was used for cooperative work and to support collaboration during individual idea generation inside the MLE.

The VRLE provided the teacher and students with an electronic form of individual and group learning support (Loiselle et al., 1998; Schrum & Berenfeld, 1997) and, in addition to enabling face-to-face collaboration, the VRLE supported communication via computer-supported media, which further enabled students to interact with each other (Thurlow et al., 2004; Wolz et al., 1997). The VRLE also facilitated the students' access to information within the conventional classroom and enabled multi-modal communication (Gilbert & Dabbagh, 2005; Gabriel, 2004).

Collaboration played an important role, both at home, in the classroom and inside the VRLE to facilitate idea generation, supporting the position of Hamburg et al. (2003). Collaboration between students is an important factor, with regards to the facilitation of their learning (Heinze & Procter, 2005), and the value of the VRLE, in enabling and supporting collaboration, mirrors Vygotsky's concept of the Zone of Proximal Development (1978). A logical follow up of this research would be to establish collaborative IE work via distance learning; for example, children could collaborate with a child in a different location, via the VRLE. This would enable further understanding of the role of multimodal communication in this context, as the ability to switch instantly from communication via the VRE to face-to-face communication would no longer be available.

Upgrading the IE pedagogical model

The original pedagogical model of the IE innovation process (see Figure 33.1) was developed prior to the introduction of the VRLE and has been useful in enabling the discussion of pedagogical characteristics of IE. However, to enrich the understanding of the emerging pedagogy, in terms of the VRLE, the model in Figure 33.5 was designed: this demonstrates the IE pedagogy as it appeared during the research. Figure 33.5 shows how students learn through idea generation and how learners' interactions between home, the classroom and the VRLE are fundamental to this process.

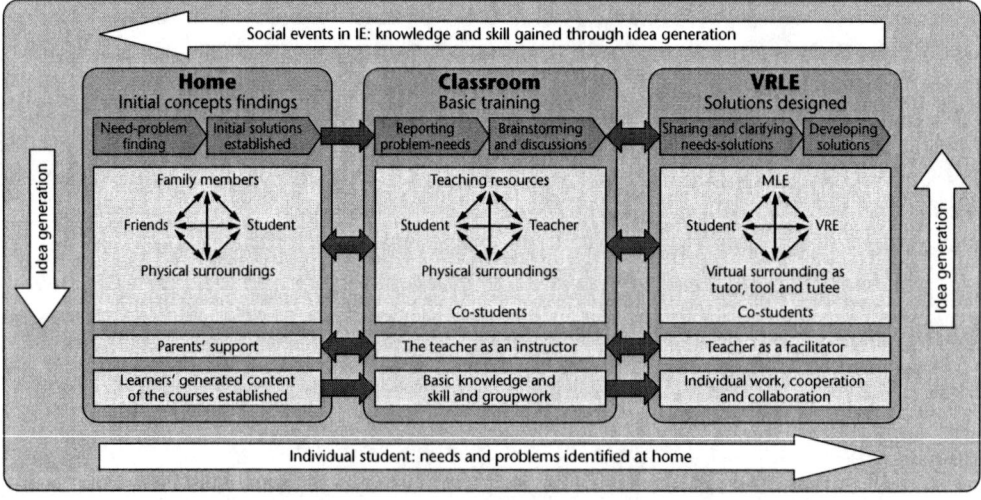

Figure 33.5 The pedagogical model for IE incorporating the VRLE.

Ideation skills are still employed at all stages and innovation relates to the usefulness of ideas and/or how they can be implemented as solutions to many problems encountered in everyday life. Students learn through the cycles of the innovation process, supported by the collaboration amongst individuals, as a group, and by the teacher. Both individual and social events are important in the process of idea generation and the teacher plays a fundamental role in both training and the facilitation of learning.

Chapter summary

Innovation Education (IE) in Iceland aims to train students to identify needs and problems in their environment and to find solutions: a process of ideation. This chapter presents work which explored this context for teaching and learning using a VRLE with IE to support development of students' ideation skills. It focuses on blended learning in that the VRLE is used in parallel with conventional classroom based activity.

The work took a Grounded Theory (Glaser & Strauss, 1967) perspective as a way of observing the complex social/educational activity relating to this live learning context. It was intended to build understanding (grounded theory) rather than attempt to establish cause and effect. The author intended to observe, describe and interpret settings as sources of data. The main aim was to: *gain a greater understanding of the use of the VRLE in supporting students' work in conventional Innovation Education classes within Icelandic schools.*

The overall research question was: *How does the use of the VRLE affect teacher's pedagogy and student work in conventional Innovation Education in Iceland?*

Research tools and multiple data sources were selected to gather triangulated data. Specifically a series of case studies in an Icelandic elementary school with groups of volunteers from the seventh class (age 12) were used.

Main findings demonstrate the significant categories and issues relating to the impact of the VRLE on the context of teaching and learning:

1   the teacher and his/her approach to his work;
2   use of homework;
3   use of the VRLE;
4   Innovation Education and idea generation;
5   drawing;
6   values.

The findings show the teacher's approach to his/her work as significant. The students' use of homework is important as it enables students to generate the content of the course and makes a meaning of their work. The VRLE enables students' collaboration and cooperation during their ideation work. The teacher has to change between different role during the lessons, including being an instructor and a facilitator, in order to support student self-sufficiency and autonomy. Training in using CAD, drawing and the VRLE appeared to be important. However, the students also learned through their own experience.

The works' main contribution to knowledge is understanding the context of teaching and learning using the VRLE for developing students' ideation skills in IE. It furthermore, contributes to a practical use of IE in education. The work is an exemplar of a qualitative approach based on case study methodology and grounded theory in an Innovation Educational context.

To enable further progress and development in Innovation Education we have to be aware of technological developments in both hard and software which can be employed in the pedagogical context.

The pedagogical understanding of using the VRLE for Ideation has to be developed further. The educational efficacy of using the VRLE in schools requires the development of meaningful forms of such learning support. The basis of the technology is already part of the daily lives of young people, but to date less advanced in general education.

References

Bauersfeld, H. (1995). The structuring of the structures: Development and function of mathematizing as a social practice. In L. P. Steffe & J. Gale (Eds.), *Constructivism in Education*. Hillsdale, NJ: Lawrence Erlbaum Associates Publishers.

Blom, J. O., & Monk, A. F. (2003). A theory of personalisation of appearance: Why users personalise their PCs and mobile phones. *Human–Computer Interaction, 18*(3), 193–228.

Bonk, C. J., & Cunningham, D. J. (1998). Searching for learner-centred, constructivist, and socio-cultural components of collaborative educational learning tools. In C. J. Bonk & K. S. King (Eds.), *Electronic Collaborators: Learner-Centred Technologies for Literacy, Apprenticeship, and Discourse* (pp. 25–50). Mahwah, NJ: Erlbaum.

Bricken, M. (1991). Virtual reality learning environments: Potentials and challenges. *Computer Graphics, 25*(3), 178–184.

Clapham, M. M. (2003). The development of innovative ideas through creativity training. In L. V. Shavinina (Ed.), *The International Handbook on Innovation* (pp. 366–376). Oxford: Elsevier Science.

Cohen, L., Manion, L., & Morrison, K. (2005). *Research Methods in Education*, 5th ed. London: Taylor & Francis e-Library.

Creswell, J. W. (1998). *Qualitative Inquiry and Research Design: Choosing Among Five Traditions*. Thousand Oaks, CA: Sage.

Dennis, A. R., & Valachich, J. S. (1993). Computer brainstorming: More heads are better than one. *Journal of Applied Psychology, 78*(1), 531–537.

Denzin, N. K. (1984). *The Research Act*. Englewood Cliffs, NJ: Prentice Hall.

Denzin, N. K., & Lincoln, Y. S. (Eds.) (1994). *Handbook of Qualitative Research*. Thousand Oaks, CA: Sage Publications, Inc.

Doolittle, P. E. (1995). Understanding cooperative learning through Vygotsky's zone of proximal development. Downloaded from www.ebscohostProquest. innopac.wits.ac.za (August 1, 2009).

Gabriel, M. A. (2004). Learning together: Exploring group interactions online. *Journal of Distance Education, 19*(1), 54–72.

Gilbert, P. K., & Dabbagh, N. (2005). How to structure online discussions for meaningful discourse: A case study. *British Journal of Educational Technology, 36*(1), 5–18.

Glaser, B. G., & Strauss, A. L. (1967). *The Discovery of Grounded Theory: Strategies for Qualitative Research*. New York: Aldine Publishing Company.

Guilford, J. P. (1950). Creativity. *American Psychologist, 5*(9), 444–454.

Gunnarsdottir, R. (2001a). Innovation education: Defining the phenomenon. Unpublished doctoral thesis. Leeds: University of Leeds.

Gunnarsdottir, R. (2001b). Research in innovation education: Socio-cultural methods for research and analysis for defining educational phenomenon. Visions on Sloyd and Sloyd Education. *Techne Series: Research in Sloyd Education and Crafts Science B, 10*(1), 65–104. Finland Vasa: Åbo Akademi University.

Hamburg, I., Lindecke, C., & Thij, H. T. (2003). Social aspects of e-learning and blending learning methods. *In Proceedings of the 4th European conference E-comm-line*. Bucharest.

Hauber, J., Regenbrecht, H., Hills, A., Cockburn, A., & Billinghurst, M. (2005) *Social Presence in Two- and Three-Dimensional Videoconferencing*. London: Proceedings Presence Workshop.

Heinze, A. (2008). Blended learning: An interpretive action research study. PhD thesis. Salford: University of Salford.

Heinze, A., & Procter, C. (2005). *Communication: A Challenge and an Enabler for Facilitating Blended Learning Community. Socio-Cultural Theory in Educational Research and Practice*. Manchester: University of Manchester.

Hussain, H., Embi, Z. C., & Hashim, S. (2003). A conceptualized framework for edutainment. *Informing Science*: InSite – Where Parallels Intersect, pp. 1077–1083.

InnoEd (2011). The InnoEd website. Downloaded from www.innoed.is/frame.asp?page=main%2Easp&PageID=4 (April 5, 2006).

Johnson, A., Moher, T., Choo, Y., Lin, Y. J., & Kim, J. (2002). Augmenting elementary school education with VR. *IEEE Computer Graphics and Applications,* March/April, 6–9.

Jonassen, D. H. (1999). Constructing learning environments on the web: Engaging students in meaningful learning. *EdTech 99: Educational Technology Conference and Exhibition 1999: Thinking Schools, Learning Nation*.

Jónsdóttir, S. R. (2005). Ny namsgrein verður til: Nyskopunarmennt í grunnskola. Unpublished MA. Reykjavik: University of Iceland.

Kerres, M., & de Witt, C. (2003). A didactical framework for the design of blended learning arrangements. *Journal of Educational Media, 28*(2–3), 101–113.

Loiselle, J., St-Louis, M., & Dupuy-Walker, L. (1998). Giving professional help to preservice teachers through computer-mediated communication. Paper presented at the Annual Meeting of the Association of Teacher Educators, Dallas, TX. (ERIC Document Reproduction No. ED 418 087.)

McLellan, H. (1996). Virtual reality. In D. Jonassen (Ed.), *Handbook of Research for Educational Communications and Technology* (pp. 457–487). Boston, MA: Kluwer-Nijhoff Publishing.

Ministry of Education. (1999). *The Icelandic National Curriculum 1999*. Reykjavik: Icelandic Ministry of Education.

Ministry of Education. (2006). *The Icelandic National Curriculum 2006*. Reykjavik: Icelandic Ministry of Education.

O'Quin, K., & Derks, P. (1999). Humor. In M. A. Runco & S. R. Pritzker (Eds.), *Encyclopedia of Creativity* (pp. 845–853). San Diego, CA: Elsevier Academic Press.

Ogle, T. (2002). The effects of virtual environments on recall in participants of differing levels of field dependence. PhD Dissertation. Virginia Polytechnic and State University, Blacksburg, VA.

Olafsson, B., & Thorsteinsson, G. (2009). Design and craft education in Iceland, pedagogical background and development: A literature review. *Design and Technology Education: An International Journal, 14*(2), 10–24.

Osberg, K. M. (1993). Virtual reality and education: A look at both sides of the sword. Downloaded from www.hitl.washington.edu/publications/r-93-7/ (August 12, 2009).

Oulasvirta, A., & Blom, J. (2008). Motivations in personalisation behaviour. *Interacting with Computers, 20*(1), 1–16.

Oxford English Dictionary Online. (2011). Downloaded from www.oed.com/ (April 5, 2011).

Page, T., Thorsteinsson, G., & Niculescu, A. (2008). A blended learning approach to enhancing innovation. *Studies in Informatics and Control, 17*(3), 297–311.

Patton, M. Q. (1990). *Qualitative Evaluation and Research Methods*, 2nd ed. Newbury Park, CA: Sage.

Paulus, P. B., Larey, T. S., & Ortega, A. H. (1995). Performance and perceptions of brainstormers in an organizational setting. *Basic and Applied Social Psychology, 17*(1–2), 249–265.

Prensky, M. (2005). Engage me or enrage me: What today's learners demand. *Educause Review, 40*(5), 60–65.

Rieber, L. P. (2001). Designing learning environments that excite serious play. Paper presented at the annual meeting of the Australasian Society for Computers in Learning in Tertiary Education. Australian Society for Computers in Learning in Tertiary Education, Melbourne, Australia.

Rogers, E. (2003). *The Diffusion of Innovations*, 3rd ed. New York: The Free Press.

Romiszowski, A. J., & Mason, R. (1996). Computer-mediated communication. In D. H. Jonassen (Ed.), *Handbook of Research for Educational Communications and Technology* (pp. 438–456). New York: Simon & Schuster Macmillan.

Schrum, L., & Berenfeld, B. (1997). *Teaching and Learning in the Information Age: A Guide to Educational Telecommunications*. Needham Heights: Allyn & Bacon.

Smith, G. F. (1998). Idea-generation techniques: A formulary of active ingredients. *Journal of Creative Behaviour, 32*(2), 107–133.

Smith, G. F. (2003). Towards a logic of innovation. In L. V. Shavinina (Ed.), *The International Handbook on Innovation* (pp. 347–365). Oxford: Elsevier Science.

Strauss, A., & Corbin, J. (1998). *Basics of Qualitative Research: Techniques and Procedures for Developing Grounded Theory*. Thousand Oaks, CA: Sage Publications.

Taylor, D. W., Berry, P. C., & Block, C. H. (1958). Does group participation when using brainstorming facilitate or inhibit creative thinking? *Administrative Science Quarterly, 3*(1), 23–47.

Taylor, R. P. (1980). Introduction. In R. P. Taylor (Ed.), *The Computer in the School: Tutor, Tool, Tutee* (pp. 1–10). New York: Teachers College Press.

Thompson, L. (2008). *Making the Team: A Guide for Managers*, 3rd ed. Upper Saddle River, NJ: Prentice Hall.

Thorsteinsson, G. (1996). Nyskapelse i islandske grunnskoler. *Håndarbejde i skolen, 1*, 22–23.

Thorsteinsson, G. (1998). Innovation in the elementary school. *Uppeldi, 6*(1), 14–148.

Thorsteinsson, G. (2009). Co-operative learning in a virtual reality environment (VRE) through idea generation: A pilot study from Iceland. In E. W. L. Norman & D. Spendlove (Eds.), *The Design and Technology*

Association Education and International Conference Book, 2009, The D&T – A Platform for Success (p. 115). Wellesbourne: Design and Technology Association.

Thorsteinsson, G., & Denton, H. (2003). The development of innovation education in Iceland: A pathway to modern pedagogy and potential value in the UK. *Journal of Design and Technology Education, 8*(3), 172–179.

Thorsteinsson, G., & Denton, H. G. (2006). Ideation in a virtual learning environment: A pilot project from iceland in innovation education. In E. W. L. Norman, D. Spendlove & G. Owen-Jackson (Eds.), *The Design and Technology Association International Research Conference Book 2006* (pp. 155–164). Telford, Wellesbourne, UK: The Design and Technology Association.

Thorsteinsson, G., & Denton, H. G. (2008). Developing an understanding of the pedagogy of using a Virtual Reality Learning Environment (VRLE) to support Innovation Education (IE) in Iceland: A literature survey. *Design and Technology Education: An International Journal, 13*(2), 15–26.

Thorsteinsson, G., Denton, H. G., Page, T., & Yokoyama, E. (2005). Innovation education within the technology curriculum in Iceland. *Bulletin of Institute of Vocational and Technical Education, 5*(1), 1–9. Graduate School of Education and Human Development. Japan: Nagoya University.

Thorsteinsson, G., & Page, T. (2007). Pedagogic development of computer applications and learning tools in design and technology education. *Educatia 21*(4), 97–116.

Thorsteinsson, G., Page, T., & Niculescu, A. (2010a). Adoption of ICT in supporting ideation skills in conventional classroom settings. *Informatics and Control, 19*(3), 309–318.

Thorsteinsson, G., Page, T., & Niculescu, A. (2010b). Using virtual reality for developing design communication. *Journal of Studies in Information and Control, 19*(1), 93–106.

Thurlow, C., Lengel, L., & Tomic, A. (2004). *Computer Mediated Communication: Social Interaction and the Internet.* London: Sage.

Vygotsky, L. (1978). *Thought and Language.* Cambridge, MA: MIT Press.

Webster's Revised Unabridged Dictionary. (2005). Downloaded from http://machaut.uchicago.edu/?resource=Webster (April 5, 2009).

Whitelock, D., & Jelfs, A. (2003). Editorial: Journal of Educational Media special issue on blended learning. *Journal of Educational Media, 28*(2–3), 99–100.

Winn, W., Windschiti, M., & Thomson-Bulldis, A. (1999). Learning science in virtual environments: A theoretical framework and research agenda. Paper presented at the Annual Meeting of the American Educational Research Association, Montreal, Canada. Downloaded from http://faculty.washington.edu/billwin/aera99.htm (April 5, 2009).

Woltz, U., Walker, H., Palme, J., Anderson, P., Chen, Z., Dunne, J., Karlsson, G., Laribi, A., Mannikko, S., & Spielvogel, R. (1997). Computer-mediated communication in collaborative educational settings (report of the ITiCSE '97 working group on CMC in collaborative educational settings). *The supplemental proceedings of the conference on Integrating technology into computer science education: working group reports and supplemental proceedings* (pp. 51–69). New York: ACM.

Worthington, T. (2008). Blended learning: Using a learning management system live in the classroom. *Australian National University.* Downloaded from www.tomw.net.au/blog/labels/ANU.html (May 13, 2009).

Part X
Innovation management, entrepreneurship, and innovation education

34
Creating an innovative and entrepreneurial collegiate academic program

Lynn A. Fish and Ji-Hee Kim

CANISIUS COLLEGE, USA

Summary: Starting and growing an entrepreneurship program is an arduous and rewarding task that requires process development, communication skills, real-world experiences and business planning; empowering students, faculty and administrators; networking and partnerships; leveraging current institutional programs; growth to other academic, local and international arenas; and financial and top management support. The successful program which emphasizes creativity, innovation, interpersonal skills and entrepreneurial leadership, aligns learning goals with curricular activities, and embedded experiences within – and outside – the program. The integrated curriculum provides students with a broad and necessary business background while developing entrepreneurial skills and insights. Students discover their innate entrepreneurial potential; develop processes, tools and perspectives to capitalize on that potential; learn to identify and evaluate business opportunities, acquire capital and other resources, and to start, develop, grow, operate and harvest a business. The thriving program continues to grow to other college academic areas, locally and internationally.

Key words: Entrepreneurship collegiate program, innovation, creativity, experiential entrepreneurship, interdisciplinary, entrepreneurial leadership.

Introduction

The Canisius College entrepreneurship major is a multi-faceted educational experience with a focus on individual innovation, creativity and entrepreneurial development in and out of the classroom at the collegiate level. The underlying concept students develop through the program is "innovation and creativity is key to success in any business." Critical aspects to creativity and innovation development include: developing the process, techniques and tools; motivation; developing student presentation skills and real-world, practical experiences; assisting in business plan development; and empowerment through individual experiences. These aspects are supplemented by outside support from business leaders, entrepreneurs and grants with a key factor being leveraging organizational campus communities to accelerate growth. Within a relatively short time frame, the successful entrepreneurship program grew within the College, regionally and internationally.

Literature review

The current entrepreneurship program is the brainchild of Director, Dr. Ji-Hee Kim. Her process to develop the program came from benchmarking other entrepreneurship programs, attending various conferences and faculty workshops. Various papers, conferences and workshops assisted in the curriculum development (specifically, Morris, Kuratoko, Seeling, Sause, & White, 2004; Academy for Entrepreneurial Leadership, 2007; Besant & Tidd, 2007; Brockhaus, Hills, Klandt, & Welsch, 2001). The resources feature an entrepreneurship-centered model, discuss resource issues, funding sources, challenges and obstacles, benchmark programs, aspiration reference points and goals, strategies for sustainability and infrastructure, mission and measurable goals, faculty involvement and curriculum responsibilities, continuous innovation and top quality collaboration for a successful entrepreneurship program.

Entrepreneurship major at Richard J. Wehle School of Business, Canisius College

At some point in time, every business that exists was someone's innovative and creative brainchild. Teaching students to develop their own creativity and innovation can foster this individual growth. In fact, one of the greatest opportunities for personal wealth and fulfillment comes from starting a business, growing a fledgling business or participating in entrepreneurial ventures within current corporations (also known as intrapreneurship). Small businesses remain a critical component of today's economy and critical to developing new technology, products and services.

History of program

The entrepreneurship major became a recognized, differentiable major at several colleges in the United States during the later part of the 20th century. Entrepreneurship involves the skills necessary to identify and evaluate opportunities, acquire capital and resources, and to start, develop, operate and, for some, harvest a business. In the 1990s, under his new venture and family business programs, Dr. Alan Weinstein recognized the potential for an entrepreneurship major at Canisius College. Along with several of his colleagues, including Dr. Steven Molloy, Dr. Guy Gessner and Dr. Richard Wall, Dr. Weinstein developed an entrepreneurship major at Canisius College. The program attracted a few students each year during its onset.

In October 2005, the Richard J. Wehle School of Business at Canisius College under its strategic plan positioned the revitalization of its entrepreneurship program as a high priority. Simultaneously, by 2004, the Coleman Foundation, as part of its 1998 EAEG program, offered a grant to develop an undergraduate internship program that placed students in community economic revitalization roles. As part of these revitalization efforts, in the fall of 2006, Canisius hired Dr. Ji-Hee Kim as the lead entrepreneurship faculty member and Director of the Entrepreneurship major. Upon arriving at Canisius, Dr. Kim applied for and received the Coleman Grant to support the "Support Tomorrow's Entrepreneurs: Entrepreneurship Education at Canisius College – Dreams>Passion>Vision>Opportunity>Success Program" (A Fast Start at Canisius College, 2008), and the Coleman Foundation Faculty Entrepreneurship Fellows Program Grants (2010–2011, 2011–2012) to support faculty members from disciplines outside entrepreneurship and the business school (Kim Awarded Coleman Faculty Entrepreneurship Grant, 2010; Kim Uses Grant to Expand Entrepreneurial Education, 2011). (We discuss these grants below.) Under Dr. Kim's tutelage, the entrepreneurship major expanded and grew.

We continue by outlining the learning goals for the entrepreneurship major at Canisius College, the curriculum along with a discussion of several key learning activities embedded within the program, and critical outside activities that foster entrepreneurship, innovation and creativity, before moving on to discuss the critical role the outside financial support played in program development and the recent extension to other College academic areas.

Learning goals

The Canisius College program is specifically designed to assist students in creating their own businesses. The entrepreneurship major specifically prepares students to:

- acquire an existing business or franchise, or start up a new venture;
- manage an existing family business for growth;
- engage in intrapreneurship (that is, develop new products or programs or evaluate and pursue potential merger or acquisition candidates within a mature corporation).

As outlined below, the Canisius College entrepreneurship major offers an integrated curriculum, drawing upon management, marketing and finance to provide students with the broad and necessary business background while developing entrepreneurial skills and insights. The entrepreneurship major is built upon a solid Association to Advance Collegiate Schools of Business (AACSB) International business and liberal arts core but tailored to entrepreneurship needs.

Specifically taught and tested entrepreneurship major learning goals are as shown in Table 34.1.

Table 34.1 Entrepreneurship major learning goals and objectives

Goals	Objectives
Entrepreneurship majors will apply a working knowledge of the principles of entrepreneurship to analysis and problem solving	Identify and apply the elements of entrepreneurship and to entrepreneurial processes Recognize the importance of entrepreneurship and identify the profile of entrepreneurs and their role in economic growth Use the entrepreneurial mind-set and behave responsibly and ethically in their roles as entrepreneurs
Entrepreneurship majors will be able to create and start new ventures	Creatively analyze the business environment, opportunity recognition and the business idea generation process Know how to acquire necessary resources and organizational matters of new venture creation process Write a business plan that creates and starts a new venture
Entrepreneurship majors will manage, grow and harvest a new venture	Apply a strategy for growth and manage the implications of growth Use capital budgeting that includes cost of capital, leverage and dividend policy in a financial management context, accessing resources for growth from external sources Appropriately end a venture
Entrepreneurship majors will understand entrepreneurship in the global marketplace	Know the relationship between domestic entrepreneurship and international entrepreneurship, based on economic, political, legal and cultural systems Know how to adapt domestic entrepreneurship to the global market

Entrepreneurship curriculum

Through the curriculum, enhanced through external activities (as we discuss below), students discover their innate entrepreneurial potential and develop processes, tools and perspectives to capitalize on that potential. The curriculum emphasizes creativity, innovation, interpersonal skills and entrepreneurial leadership through each course and embedded exercises along with opportunities for individual growth outside the classroom. Students learn to identify and evaluate business opportunities, acquire capital and other resources, and to start, develop, grow, operate and harvest a business.

As outlined in Table 34.2, the entrepreneurship major consists of 21 (or 22) courses (depending upon the student's mathematics ability). Each student is required to take seven courses (three credit hours each) specific to the entrepreneurship curriculum.

ENT101: Experiential Entrepreneurship

In the freshman year, through the foundation course ENT101 Experiential Entrepreneurship: Creativity, Innovation, Opportunity and Idea Generation, students are broadly introduced to entrepreneurship and begin to develop their individual creativity and innovation, to recognize business opportunities and generate entrepreneurial ideas. In the teaching brief "From Nothing to Something: An Entrepreneurial Exercise," we detail the pedagogy and activities embedded in ENT101

Table 34.2 Entrepreneurship major course requirements

Common body of business knowledge	Entrepreneurship major curriculum
MAT 105 Finite Mathematics *and* MAT106 Calculus for the Non-Sciences – *or* – MAT111 Calculus I – *or* – MAT115 Calculus for Business I	ENT101 Experiential Entrepreneurship: Creativity, Innovation, Opportunity and Idea Generation
ACC201 Financial Accounting	ENT201 Introduction to Entrepreneurship
ACC202 Managerial Accounting	ENT401 Small Business Management and Entrepreneurship
ECO101 Principles of Macroeconomics	ENT402 New Venture Creation
ECO102 Principles of Microeconomics	ENT498 Practicum in Entrepreneurship(0 credit; required)
ECO255 Business Statistics I	*Entrepreneurship electives: 3 courses*
ECO256 Business Statistics II	ENT310 International Entrepreneurship Practicum
FIN201 Introduction to Finance	ENT311 Entrepreneurship and Family Business
ISB101 Management Technology	ENT312 International Entrepreneurship
MGT325 Operations Analysis for Business	ENT314 Social Entrepreneurship and Not-for-Profit Sector Enterprise
	ENT316 Entrepreneurial Finance
MGT370 Managerial Environment	ENT411 Entrepreneurial Leadership
MKT201 Principles of Marketing (ENT section)	ENT412 Real Estate and Entrepreneurship
	ENT414 Franchising and Entrepreneurship
	ENT 496 Internship in Entrepreneurship
	ENT 497 Entrepreneurial Summer Internship in Korea
	ABR 496 Contemporary Business and Culture in Korea

to teach students a creative innovative *process* from recognition of a unique business opportunity through actual product development (Kim & Fish, 2010). In the teaching brief, we outline the semester-long experience whereby students develop from "nothing" – but junk and their business knowledge – and deliver "something" to the business world. Students learn a process to develop their individual creativity and innovation – and not a *cookbook* solution. Additional course aspects encourage creativity and innovation including the "Bug Reports and Two Cools" exercises (Kim & Fish, 2009), the "expanding on lessons from Hollywood: Using Hollywood Video-clips" and the "business idea pitch and empire creativity competition" exercises. In the "Bug Reports and Two Cools" exercises students learn to generate and improve upon their ideas from "best" and "worst" perspectives, and, through the exercise, a process of idea generation and development (Kim & Fish, 2009). In the "expanding on lessons from Hollywood" exercise, students utilize the original Hollywood experience (Schindehutte, 2006) but with a variation. Specifically, students seek their *own* visual materials, which further motivates them. Students utilize these visual aids to develop their technological and advertising skills. Through the "business idea pitch and empire creativity competition," students learn the important skills to present themselves and their creative idea in two minutes (i.e., the time it takes to ride up an elevator). Critical presentation skills – particularly important to grabbing investors' attention – are enhanced through the class and campus-wide competition. The course concludes with the "creative individual final project: entrepreneurial personal mission statement and reflection" whereby each student reflects and presents a visual description – not a written one – on his or her semester learning, personal mission and value statement. Students use their creativity to demonstrate what they learned over the entire semester, with Powerpoint presentations and written reports as unacceptable. Thus, through these activities – and many more (Kim & Fish, 2010) – students learn a process to generate, grow, and present creative and innovative ideas.

ENT201: Introduction to Entrepreneurship

The sophomore-level entrepreneurship course examines the nature and role of entrepreneurship in society. Students investigate the entrepreneurial process in a variety of context, explore issues surrounding new venture creation, business economics, resource and marketing requirements, deal structures and technology issues, the entrepreneurial perspective, entrepreneurial mindset and lifecycle. The process is taught with the lifecycle in mind – from idea to opportunity, to the business plan, to funding the venture, from the funding to launching, growing and ending the new venture. Continuing their "hands on" education, students engage in real entrepreneurial projects. In the "Class Venture: Group Business Plan and Funding the Venture" exercises, students learn important knowledge to write a profitable business plan and capital acquisition. After brainstorming a group business idea, student teams complete their own group business plan. Students learn a process to find potential investors as they present their business plan to bankers and investors for feedback.

ENT401/ENT402: Small Business Management and Entrepreneurship/New Venture Creation

Through a senior year-long experience (ENT401 and ENT402), students work with entrepreneurs and develop business plans for their future company. In the fall semester, students develop their problem-solving skills through focusing on operating small enterprises' strategy, new venture opportunities, creative marketing, innovative management, entrepreneurial financial management and legal requirements. In ENT401, students – under faculty guidance – work with local, small businesses and their owner to complete a comprehensive "Small Business Analysis and Consulting" project. Students gain valuable experience and insight into both success and failure through working with entrepreneurs in a small business enterprise. Student teams visit the business and analyze key functional areas. Through their analysis, teams offer recommendations for improvement.

ENT402 is designed to help students learn how to write a feasible business plan for a new venture. The plan outlines the blueprint that guides an entrepreneur through a complex set of variables and communicates to others the various plans. The capstone course, designed in collaboration with 14 local entrepreneurs under "Entrepreneurs on Campus," provides an opportunity for students to analyze real-world cases for startup and business growth. (The off-campus participants started, acquired or took over an entrepreneurial company.) Students learn from this mentorship experience, which adds to their own business plans. Student business plans are entered into the Canisius College Business Plan Competition, which includes a financial reward for the top three plans.

ENT 498: Practicum in Entrepreneurship

In addition to the required courses and three additional electives of the students' choice (Table 34.1), students must actively participate in the Canisius Entrepreneurs Organization and Canisius College Women's Business Center.

Additional comments

Entrepreneurship students take specific entrepreneurship sections that focus on entrepreneurship issues for MKT201 Introduction to Marketing, and FIN201 Introduction to Finance. Entrepreneurship majors are encouraged to seek a dual major, which can be done with management, marketing, science and technology fields.

Extending entrepreneurial learning outside the classroom

Outside activities enforce classroom activities and are critical to individual growth. We turn now to discuss the various outside activities in the program and their impact on student learning.

Annual empire creativity competition

Every fall since 2007 – and as a follow-up to the ENT101 business pitch development – as individuals or teams, students pitch their business ideas to a panel of judges with a chance to win cash prizes. The pitch ideas can be at any stage in development, and the competition is open to all Canisius College students regardless of major. A monetary prize, from the Coleman Foundation grant (2007), Canisius Undergraduate Students Association and Entrepreneurship Major (since 2008), is awarded to the winner. In its first year alone, 48 students participated. Participation continues to increase each year. The winner goes on to compete against 60 campus champions from various universities across the nation at the annual Collegiate Entrepreneurs' Organization conference's elevator pitch competition.

Can Do Society

In 2008, Dr. Kim established the Can Do Society to connect local professional and community leaders with students. An organization of Canisius College students, alumni and mentors, the Can

Do Society is committed to excellence, professionalism, personal growth along with positively impacting the community. Leaders serve as mentors to the students, assisting in developing an entrepreneurial attitude and skills. The Can Do Society organizes an annual Can Do Seminar focused on individual development with such titles as "Are you a 'Can Do' leader? We count on you (2010)" presented with Seth Godin, and "Who Am I and Where Am I going? (2011)" presented with Jeff Hoffman (Founder of priceline.com). These presentations serve as an additional support to encourage entrepreneurship in all Canisius students.

Canisius Entrepreneurs' Organization (CEO)

In 2007, Dr. Kim initiated the Collegiate Entrepreneurs' Organization (CEO) as an on-campus student group. Student members develop entrepreneurial leadership and professional communication skills through teamwork as well as entrepreneurial knowledge by planning and implementing educational outreach, and experiential entrepreneurship projects. CEO informs, supports and inspires students to be entrepreneurial and seek opportunities through creativity, innovation and enterprise creation. Guided by faculty advisors, CEO teams are supported by businesses and not-for-profit community organizations. The Canisius CEO group annually travels to the Collegiate Entrepreneurs' Organization conference. As a demonstration of its success toward its mission, the Canisius Chapter won awards in every year it has competed: 2007 Top 10 Business Pitch Category, 2008 Best Chapter Award in the E-Diffusion category which promotes entrepreneurship to non-business students (CEO Chapter Wins Awards at National Competition, 2008), 2009 Best Chapter Award in Teaching Entrepreneurship (CEO Chapter Wins Top Honors at Competition, 2009), 2010 second place in the Best Marketing Plan Category (CEO Recognized at National Competition, 2010) and, recently, 2011 Best Chapter Award in Teaching Entrepreneurship (CEO Wins Two Awards at National Competition, 2011). Since 2007, Dr. Kim won the CEO Best Chapter Advisor twice (2008 and 2011). Student projects include marketing, youth, social or international entrepreneurship, new venture creation, fundraising, in-person networking and entrepreneurial leadership, and starting a chapter business.

Career opportunities

Paramount to the business writing is the business plan. Students are encouraged to write business plans for their actual business, and to start and run them concurrent to the degree. To date, several companies started through this program – a car detailing firm, a landscaping service, a promotional agency and a retail store. Most – if not all – entrepreneurship majors start and run their own company upon program completion. As we discussed above, the creative processes and tools to accomplish this are taught through the curriculum.

Entrepreneurs on campus

In 2009, the curriculum, with support from Professor Emeritus Dr. Alan Weinstein, expanded to include Entrepreneurs on Campus. The group's mission is to guide and support the Canisius' entrepreneurship program to excel and become one of the top programs in the country.

The group's vision is to create a strong entrepreneurial presence at Canisius College, to differentiate Canisius College as a leader in entrepreneurship education and to create the next generation of Western New York (WNY) entrepreneurs. To accomplish these objectives, they support students' initiatives, provide advice and financial assistance, and interact directly with students.

Internships

A Canisius College business education emphasizes learning through experience. During the senior level, year-long program, students work hand-in-hand with a Women's Business Center (WBC) member. The WBC member is typically the CEO and innovator for the program. It's a win–win situation as many of the WBC do not have business backgrounds and benefit from the student's learning, while the student learns the real-world issues associated with business start-ups and importance of a valuable business network. Thus, while the emphasis is on the experience, students learn how to overcome the issues with making their creative and innovative ideas a reality.

MyLinkFace

MyLinkFace, the brainchild of Canisius entrepreneurship students, provides a virtual classroom as an outreach program. The student-initiated, student-run, international, non-profit, social venture takes distance learning to an extreme! Established in 2008, the classroom is the outcome of Dr. Kim's social entrepreneurship course, which challenged students to identify and bring a positive solution to a social issue. Through MyLinkFace, entrepreneurship majors manage the venture, and education and language students teach Korean students English. Through Skype and a dynamic, interactive online multimedia environment, Canisius College students and certified native English speaking teachers develop conversation skills, target and support foreign students' English learning needs. Currently, MyLinkFace delivers customized, one-on-one conversational English lessons to an average of 80 international clients each semester. Language and educational students gain direct experience, while entrepreneurship students learn everything about running a new venture. Current MyLinkFace goals include obtaining additional financial support, sponsorships and growing the program.

NexBizSolutions

Local small business owners participate in the bi-weekly CEO Market Place that promotes small businesses products and services to the Canisius College market. The CEO Market Place is instrumental in advertising, targeting and retailing small business products and services to the College community.

Women's Business Center

Entrepreneurship students are paired with Women's Business Center (WBC) members in the ENT401 Small Business Management and Entrepreneurship course (as already discussed). This experience is valuable to both the student and the mentor. Students experience real-world issues in making a product or service a reality, develop their own business network and gain practical experience. WBC members gain student business knowledge. It's a win–win relationship for both!

Extending entrepreneurship and innovation to other academic areas

The fast growth of the entrepreneurship major since 2006 could not be accomplished without the infusion of outside funding. Dr. Kim successfully applied for and received two grants from the Coleman Foundation. We continue by reviewing the outcomes from the first grant and the current status of the second grant, which focuses on extending the entrepreneurship program outside the Richard J. Wehle School of Business.

Support tomorrow's entrepreneurs: entrepreneurship education at Canisius College

As previously mentioned, in 2008, Dr. Kim received one of 20 nationally awarded grants awarded by the Coleman Foundation as part of its 1998 EAEG program to develop internship sponsored community economic revitalization (A Fast Start at Canisius College, 2008). Through the $73,000 grant, Dr. Kim developed courses that reinvigorated the entrepreneurship major, created CEO and promoted faculty research in entrepreneurship and community development, specifically the Empire Creativity Competition. (Future plans include creating an interdisciplinary entrepreneurship minor.) Specific courses developed included ENT101 Experiential Entrepreneurship: Creativity, Innovation, Opportunity and Idea Generation, MKT201 Introduction to Marketing – entrepreneurship section, ENT312 International Entrepreneurship and ENT411 Entrepreneurial Leadership. Canisius boosts small class-sizes averaging 17; however, in its first year, 105 students participated in these four classes. Since 2006, enrollment in entrepreneurship courses increased steadily from 54 to 239 today, as shown in Table 34.3. The CEO club became one of the fastest growing clubs on US campuses with over 60 members in its first year, of which 16 attended the 2007 CEO National Conference.

A critical component to the accelerate growth was Dr. Kim's strategy to leverage existing campus organizations. Specifically, Dr. Kim engaged four organizations: the Center for Teaching Excellence, the Office of Campus Programming and Leadership Development, the Career Center, and the Office of Undergraduate Admissions. She worked closely with each organization to nurture and empower them as champions and partners for entrepreneurial program growth on campus, which resulted in faster growth than anticipated. Specifically:

- The *Center for Teaching Excellence* (CTE), a resource for faculty development, assisted in identifying and recruiting faculty to support entrepreneurship activities.
- The *Office of Campus Programming & Leadership Development* (CPLD) promoted Dr. Kim's undergraduate extra-curricular activities, and co-sponsored the creativity competition. CPLD's function on campus is to encourage and promote student involvement and enhances learning through co-curricular programming and leadership development opportunities. Since its inception, CPLD assists CEO with annual funding through the Undergraduate Student Association (USA).
- The *Career Center* supported the efforts on various levels. First, the Career Center encouraged students to pursue entrepreneurial career opportunities. Second, the Center co-sponsored career workshops with entrepreneurs, alumni networking events and an alumni entrepreneur speaker series with the entrepreneurship major program and CEO club.
- The *Office of Undergraduate Admissions* continues to promote and recruit local high school students for Dr. Kim's Youth Entrepreneurship initiatives, as discussed below.

Table 34.3 Student enrollment and attendance at CEO National Conference since 2006

Year	Entrepreneurship enrollment	Student attendance at CEO National Conference
2006	54	–
2007	161	16
2008	119	20
2009	171	25
2010	144	12
2011	239	16

Another key lesson from the grant involved the difficulties and constant work to get senior administrative support for growing an entrepreneurship program. Dr. Kim worked countless hours with administration to gain their support – even with a fast-growing program. Key to her success in this area was her ability to attract alumni donors and a subsequent $1 million gift from a local philanthropist. Dr. Kim attracted donors through the "Entrepreneurs on Campus" initiatives which brought supportive entrepreneurs into the classroom (as previously discussed) and demonstrated to administrators that donors were willing to support the program. At its inception in 2009, ten entrepreneurs participated in the program. Currently, there are 14 official members in Entrepreneurs on Campus, with more than 30 supporting classroom activities. The $1 million gift (Million Dollar Gift from Peter & Elizabeth Tower, 2007), established a professorship to foster entrepreneurship, innovation and economic development in WNY and provided educational programming at the graduate and undergraduate levels to encourage and prepare students to accomplish these goals. Dr. Patricia Hutton, the first recipient of the Peter Tower Professorship, used the funds to support the Canisius College chapter of Students in Free Enterprise (SIFE) in over a dozen local projects in the region, fund a global outreach project whereby SIFE students promote education and artwork of school children in Mexico, host an annual market for women entrepreneurs to sell their products on campus, provide for a WNY Youth Business Leadership Conference through collaboration with CEO, and start a student-run business, QuadGear™, to provide hands-on entrepreneurial education. Additionally, projects under her direction include teaching financial literacy in the Buffalo public schools and promoting ethical practices among school-age children. Also, the WNY Prosperity Scholarship, established through the Prentice Family Foundation in 2010 and continued through 2012, provides student scholarship assistance and credit-bearing internships. In return, students commit to work in WNY for at least two years after attaining their degrees. The Prosperity Scholarship funds 12-week, 360-hour summer internship experiences for each of its ten recipients.

Coleman Faculty Entrepreneurship Fellows Program

In 2010, Dr. Kim received a second Coleman Foundation grant, a $15,000 Faculty Entrepreneurship Fellows Program Grant, to expand entrepreneurial educational opportunities for non-business faculty and students in the 2010–2011 academic year (Kim Awarded Coleman Faculty Entrepreneurship Grant, 2010). Specific areas addressed by the second grant include developing entrepreneurial skills in the MexiCanisius program under the direction of Dr. Julia Wescott, in adolescent education under Dr. Betsy DelleBovi (which supports the MyLinkFace describe above) and in the digital media art program under Mr. Jamie O'Neil.

Recently, Dr. Kim received a third Coleman grant for the 2011–2012 academic year to expand entrepreneurial education opportunities for non-business faculty and students (Kim Uses Grant to Expand Entrepreneurial Education, 2011). The $18,000, one-year grant to appoint three Canisius College faculty members to serve as Coleman Faculty Fellows, in addition to the original three. The Coleman Faculty Fellows' charge is to expand entrepreneurial education opportunities for non-business faculty and students, inspiring students to gain self-employment skills and experiences. Specific areas assisted through the third grant include developing entrepreneurship skills in the modern language-German program under Dr. Peter Böhm, in the field of health and exercise under Dr. Dennis Koch and in digital media arts and computational science under Dr. William Sack.

Additional outreach activities

Additional outreach activities focus on growing the entrepreneurship program in several directions: toward the high school level and globally. Over the 2009 and 2010 summers, with Dr. Coral Snodgrass, an international entrepreneurship, two-day session was held at Canisius College to promote international and entrepreneurial learning. Entitled "International Entrepreneurship Boot Camp," 30 local, high school students attended the international entrepreneurship workshop. The students met local entrepreneurs who run international businesses and leaders from community organizations, such as the World Trade Center and International Institute of Buffalo. Student teams performed market research and strategy development on global products. Recently, 118 teens from 15 countries attended the College over the summer for the Students for the Advancement of Global Entrepereneurship (SAGE) World Cup (Global Student Entrepreneurship Tournament at Canisius, 2011). Teens compete upon creativity and impact of their socially responsible businesses.

Developing creativity and innovation at the collegiate level: a summary

In summary, the Canisius College entrepreneurship major is a vibrant, growing program that is developing creativity, innovation, intrapersonal skills and entrepreneurial leadership in our students as well as faculty, administrators and others. What was once a small, underdeveloped program is now a model for learning at the Collegiate level. Specifically, the program grew from nine graduating students in the mid-1990s, to 17 in 2005 and 49 in 2010 (over 288% increase over 2005). As shown in Table 34.4 (2008–2009 results), students are achieving at least a 65% on all learning objectives, with mastery (85%) on most. The program was weakest in the capital budgeting and legal system issues in the global marketplace. As a result, two specific actions were taken at that time: future work to improve capital budgeting knowledge for entrepreneurship majors (specifically designated FIN201), a new course ENT 316 (Entrepreneurial Finance) is being developed for spring 2012 and a modification to the curriculum regarding global awareness in legal systems introduced in the international entrepreneurship course. Individual testimonials on learning include such comments as:

> Canisius entrepreneurship has offered me an abundance of new knowledge which I am sure I will find extremely useful in my life after college to be a successful entrepreneur. Our classes covered all of the topics which someone would need to be a successful entrepreneur in the busy and fast paced business world of today. Some of most important lessons which I learn did not come from the textbook but rather from our creative and innovative projects and hands on experiences.
>
> *(Student, Class of 2011)*

Hence, the entrepreneurship program is reaching its goals to foster creativity and innovation within the business school. Additionally, through the Coleman Fellows Foundation, 121 students are currently receiving entrepreneurial instruction within their non-business areas. Through the Can Do Society, CEO organization, MyLinkFace and SIFE, various people throughout the world are reaping the benefits of the entrepreneurship major. Additionally, countless new ventures are started each year – and continue to function – due to the entrepreneurship major.

What are the critical components to developing student creativity and innovation at the Collegiate level? Given this program, we feel the key components include:

Table 34.4 Assurance of learning results 2008–2009

Goal	Objectives	% attaining goal
Entrepreneurship majors will apply a working knowledge of the principles of entrepreneurship to analysis and problem solving	Identify and apply the elements of entrepreneurship and to entrepreneurial processes	100
	Recognize the importance of entrepreneurship and identify the profile of entrepreneurs and their role in economic growth	100
	Use the entrepreneurial mind-set and behave responsibly and ethically in their roles as entrepreneurs	100
Entrepreneurship majors will be able to create and start new ventures	Creatively analyze the business environment, opportunity recognition and the business idea generation process	83.3
	Know how to acquire necessary resources and organizational matters of new venture creation process	83.3
	Write a business plan that creates and starts a new venture	100
Entrepreneurship majors will manage, grow and harvest a new venture	Apply a strategy for growth and manage the implications of growth	83.3
	Use capital budgeting that includes cost of capital, leverage and dividend policy in a financial management context, accessing resources for growth from external sources	66.7
	Appropriately end a venture	83.3
Entrepreneurship majors will understand entrepreneurship in the global marketplace	Know the relationship between domestic entrepreneurship and international entrepreneurship, based on economic, political, legal and cultural systems	83.3
	Know how to adapt domestic entrepreneurship to the global market	83.3

- *Process development.* Students develop processes – not cookbook, theory driven templates – to develop new business ventures, products and services. Our experiences with the "From Nothing to Something: An Experiential Exercise" (Kim & Fish, 2010), "'Bug Reports' and 'Too Cools': Experiential Entrepreneurship Exercises to Develop Students' Creative, Innovative, and Technological Abilities" (Kim & Fish, 2009) directly address these processes that are critical to developing these skills.
- *Communication skills.* Presentation and writing skills are paramount to planning and obtaining necessary financial, marketing and operational resources. The major addresses presentation skills throughout each of the required courses and culminates in the writing of the business plan in the year-long senior experience.
- *Real-world experiences.* Lifelong student learning versus rout memorization occurs in the program through interaction with entrepreneurs through consulting, attending presentations, internships and, for most, through starting their own company.
- *Business planning.* Successful business start-ups require thoughtful, deep thinking regarding the management of the new venture, product or service. The program culminates in students taking a creative thought and transposing it into a well-thought out written description of how to bring that idea to the marketplace – and continue it.
- *Empowerment.* Students, faculty and administrators are empowered to think of ideas, motivated to bring them to the world around them. For example, the Hollywood exercise with Canisius College's variation provides motivation to the students to use technology to develop their ideas. The Coleman Fellowships provide six instructors' motivation to bring entrepreneurial ideas to other non-traditional business areas. In collaborating with additional administrative functions, other collegiate functions note the positive benefits to the entrepreneurship program and assist in fostering an entrepreneurial knowledge and growth.
- *Grow.* Like the skills that it teaches, the entrepreneurship program uses a process to "think outside the box" and address areas that may never have considered entrepreneurial skills. The Canisius College entrepreneurship program extends outside the Richard J. Wehle School of Business to other academic programs (through such areas as the Coleman Fellows Program), to other WNY organizations (through the Peter Tower Professorship) and to other international areas (such as MyLinkFace and MexiCanisius programs).
- *Leverage existing institutions.* As learned in the first Coleman grant, it is important to interact with other academic functions on the campus for support and growth. Other areas provided financial support, promotion to local high schools and other academic areas, teaching support, and entrepreneurship contacts in the community and alumni.
- *Top management support.* Like all new programs, it is imperative to gain top administrator support. This is not an easy task and requires countless hours demonstrating the increase in enrollment, positive societal impact of activities and increase in the bottom-line. Our attainment of this goal can be summarized by a recent statement made by Canisius College President John J. Hurley,

New ventures and small businesses are a rapidly growing component of the American economy. At Canisius, we foster the study of entrepreneurship not only through our academic major in the field but also through internships and scholarship programs; popular student-run business ventures such as MyLinkFace and QuadGear; as well as the student chapters of Students in Free Enterprise (SIFE), the Canisius Entrepreneurs' Organization (CEO) and the Can Do Society.
(Global Student Entrepreneurship Tournament at Canisius, 2011).

Creativity and innovation skill development at the collegiate level requires more than just learning the theories. It involves active involvement from the campus to develop a rich experiential learning environment focused on service to others on campus, in the surrounding community and around the world.

References

A fast start at Canisius College. (2008). The Coleman Foundation blog, June 9. Downloaded from http://colemanfoundation.typepad.com/cfi_blog/2008/06/canisius-college-starts-fast.html on November 13, 2011.

Academy for entrepreneurial leadership, university of Illinois, Urbana-champaign, 2007 USABE annual conference (A joint SBI (Small Business Institute) and USASBE (United States association for small business and entrepreneurship)), Orlando, Florida, January 10–14, 2007.

Bessant, J., & Tidd, J. (2007). *Innovation and Entrepreneurship*. Chichester, UK: John Wiley & Sons Ltd.

Brockhaus, R. H., Hills, G. E., Klandt, H., & Welsch, H. P. (2001). *Entrepreneurship Education: A Global View*, Burlington, VT: Ashgate Publishing Limited.

CEO chapter wins awards at national competition. (2008). Date released: November 21, 2008. Downloaded from www.canisius.edu/newsevents/display_story.asp?iNewsID=6141 on November 13, 2011.

CEO chapter wins top honors at competition. (2009). Date released: October 29, 2009. Downloaded from www.canisius.edu/newsevents/display_story.asp?iNewsID=6331 on November 14, 2011.

CEO recognized at national competition. (2010). Date released: November 15, 2010. Downloaded from www.canisius.edu/newsevents/display_story.asp?iNewsID=6509 on November 14, 2011.

CEO wins two awards at national competition. (2011). Date released: November 2, 2011. Downloaded from www.canisius.edu/newsevents/display_story.asp?iNewsID=6712 on November 14, 2011.

First tower professorship awarded to Hutton. (Fall 2007). *Canisius College Magazine*, 11. Downloaded from www.canisius.edu/alumni/magazine/fall07/faculty_notes.pdf on November 14, 2011.

Global student entrepreneurship tournament at Canisius. (2011). Date released: July 20, 2011. Downloaded from www.canisius.edu/newsevents/display_story.asp?iNewsID=6647 on November. 14, 2011.

Kim awarded Coleman faculty entrepreneurship grant. (2010). Date released: August 30, 2010. Downloaded from www.canisius.edu/newsevents/display_story.asp?iNewsID=6462&strBack=default.asp on November 14, 2011.

Kim uses grant to expand entrepreneurial education. (2011). Date released: September 21, 2011. Downloaded from www.canisius.edu/newsevents/display_story.asp?iNewsID=6676 on November 14, 2011.

Kim, J., & Fish, L. A. (2010). From nothing to something: An experiential entrepreneurship exercise. *Decision Sciences Journal of Innovative Education*, 8(1), 241–255.

Kim, J., & Fish, L. A. (2009). "Bug reports" and "too cools": Experiential entrepreneurship exercises to develop students' creative, innovative, and technological abilities. *Business Education Innovation Journal*, 1(2), 13–21.

Million dollar gift from Peter, Elizabeth Tower. (2007). Date released: May 14, 2007. Downloaded from www.canisius.edu/newsevents/display_story.asp?iNewsID=4882&strBack=default.asp on November 14, 2011.

Morris, M., Kuratko, D., Seeling, T., Sause, H., & White, R. (2004). Models of entrepreneurship centers: Emerging issues and approaches. 2004 National Consortium of Entrepreneurship Centers Conference, Syracuse, NY.

Schindehutte, M. (2006). Lessons from Hollywood workshop, 2006 United States Association for Small Business and Entrepreneurship (USASBE) Annual Conference, Tucson, Arizona, January 12–15, 2006.

35

Educating the innovation managers of the Web 2.0 age

A problem-based learning approach to user innovation training programs

Peter Keinz and Reinhard Prügl

VIENNA UNIVERSITY OF ECONOMICS AND BUSINESS, AUSTRIA, AND
ZEPPELIN UNIVERSITY FRIEDRICHSHAFEN, GERMANY

Summary: In this chapter we report and reflect on the structure, outcomes, and success factors of two problem-based learning course formats in the field of user innovation education. These courses have been designed to enable prospective innovation managers to apply two different user innovation methods (lead user method and technological competence leveraging) in a real-life setting and to think and learn about the consequences of a new paradigm where instead of acting as an innovator itself, the company becomes a "facilitator" being responsible to trigger, support, and manage innovative activities of individuals from outside the company.

Key words: User innovation, lead user, technological competence leveraging, problem-based learning, innovation education.

Introduction: a new paradigm in innovation management

Latest advancements in information and communication technologies – especially the development of the Web 2.0 – have substantially changed the way we interact with each other. Today, we not only communicate with members of our personal, real-life networks consisting of relatives, friends, and colleagues: in addition, most of us also somehow interact with a large number of unknown peers online. The major motive of connecting with the anonymous crowd of Internet users is to get information that cannot be accessed within our personal networks. Interaction on the Internet is often centered on an object or topic of joint interest, e.g., hobbies or products. For example, people planning to purchase a certain product that they do not know well may reduce uncertainty by surfing through the Internet and searching for reviews and opinions on the specific object of desire. In many cases, there are so-called user communities which are informally organized networks of people with a shared interest in a specific product. Members of such user communities usually freely reveal their experiences and opinions about the product; they discuss its major flaws as well as ideas of how to fix them. Some users even develop solutions to resolve the most important issues with a product and ask their user community peers for feedback and support in their innovative activities.

Obviously, this kind of interaction between users is of potentially high value for the focal producer firm. Observing discussions and collaborative innovation activities among users might sharpen the producer's understanding of user needs, and may yield in ideas for new product development. Actually, researchers and practitioners alike are about to realize the value of company-external individuals and organizations (especially users) as sources of innovation. For example, von Hippel (2005) calls for "democratization" of corporate innovation processes. He argues that users should be integrated systematically into corporate innovation processes in order to improve their effectiveness. It is argued that users are usually better capable of coming-up with commercially promising product ideas and concepts than the members of the internal Research and Development (later referred to as R&D) department because of their use experience and needs-related information. By definition, users know the product from a use perspective and therefore possess specific know-how about its main features and functions, the benefits it delivers, as well as its major flaws in the actual use situation. They have a very clear picture of how the product should look like in order to perfectly satisfy their needs. Naturally, manufacturers are lacking this kind of use specific information which reduces their ability to improve their products with regard to the users' needs. Thus, they have to conduct market research in order to gain insights into what their users want. However, market research is a costly and time-consuming task that does not always lead to success: needs-related information often is what literature refers to as "sticky information" – it cannot be expressed and transferred to the producer firm easily. Flop rates of new products of up to 90% in some industries reflect the difficulties in collecting valid information on user needs by traditional market research. As a consequence, more and more companies are trying to complement their traditional, market research-based new product development and innovation routines by open and user innovation approaches. For example, industry leaders like IBM, Procter & Gamble, and 3M have opened-up their innovation processes and launched systematic activities to tackle the creative potential of external individuals and organizations. The underlying idea of all of these approaches is to integrate external users and/or their ideas and solutions into corporate innovation activities, especially new product development projects. These user innovation approaches prove to be highly successful: innovative ideas and concepts by users tend to have a higher degree of newness and originality and are more promising with regard to potential sales revenues compared to internally developed innovations (Lilien, Morrison, Searls, Sonnack, & von Hippel, 2002).

User innovation approaches deliver a relatively high value proposition. However, they also pose quite some challenges on a company. Integrating users into corporate innovation processes goes along with a shift from the manufacturer-active to the customer-active innovation paradigm (von Hippel, 2005). As soon as a company decides to outsource at least some of its innovation activities instead of conducting market research and developing corresponding products completely on its own, the whole game play changes: users are promoted from passive consumers to active innovators; in turn, the role of the focal producer firm also changes. Instead of acting as an innovator itself, the company becomes a "facilitator" being responsible to trigger, support, and manage innovative activities of external users. In the customer-active paradigm, the manufacturer is concerned with providing users and other external stakeholders with the know-how, skills, and means (like toolkits for user innovation and design) necessary to perform innovation-related tasks. Associated with this change of the functional role in innovation processes as well as the new tasks and responsibilities, the following questions arise: Which methods and instruments to use in order to implement users into standard corporate innovation processes? How to find users capable of participating in corporate innovation processes? How to motivate external stakeholders to put energy and effort into corporate innovation tasks? How to align innovative activities of external stakeholders with the corporate strategy? What about intellectual property (IP) rights of user-generated ideas and solutions? How to overcome company-internal barriers (e.g., employees' fear of loss of control, or the not-invented-here syndrome) against opening up the organization's core business processes? Answering these and similar questions is vital to the success of user innovation approaches. Thus, innovation managers employing user innovation approaches will have to address these topics and help transforming their organizations into "facilitators" of user innovation.

In the following sections, we will report and reflect on the structure, outcomes, and success factors of two problem-based learning course formats in the field of user innovation education that have been developed at the Institute for Entrepreneurship and Innovation of the WU – Vienna University of Economics and Business. These courses have been designed to enable prospective innovation managers to apply two different user innovation methods in a real-life setting and to think about the questions raised above.

Common user innovation approaches and how to teach them

Systematically implementing external users into a company's core business processes may take different forms: depending on a company's goals and strategy, it can choose among various user innovation approaches that differ from each other with respect to the number of external individuals participating into the corporate innovation processes as well as the sustainability of the activity (Keinz, Lettl, & Hienerth, 2012). Out of the vast arsenal of different user innovation approaches (as there are the lead user method, the user community approach, toolkits for user innovation and design, etc.), there are two user innovation methods that are especially suitable for teaching them within a class at university: the lead user method (von Hippel, 2005) and the user community-based technological competence leveraging method (Keinz & Prügl, 2010). Both approaches can be designed as projects with a duration of four to six months which makes them applicable at universities with their rather inflexible semester rhythm. Furthermore and even more important, both methods can easily be taught applying a problem-based learning approach (in the following, we refer to the problem-based learning approach as PBL approach). The PBL pedagogy usually takes the form of student-centered learning, this is, students learn in small teams with the teacher acting as a facilitator only. The students obtain new knowledge through means of self-directed learning which is triggered by a real-life problem that stimulates the development and use of problem solving skills, resulting in a learning experience (Barrows, 1996). PBL has proved to be pedagogically highly effective in teaching content but also in conveying softer skills such as epistemic practices, self-directed learning, and collaboration to the learners (Hmelo-Silver, Duncan, & Chinn, 2007). Especially these soft skills are a very important take-away for future innovation managers that need to be capable of creatively developing problem solving strategies and interacting with different kinds of company-external stakeholders.

"Sources of Innovation" – a problem-based learning course format based on the lead user method

The idea of the course format "Sources of Innovation" is to enable business students to apply the lead user method (1) in a problem-based learning environment, and (2) in cooperation with a real-life project partner. The objectives of this course format are to equip the students with the methodological know-how and skills to conduct lead user projects on their own later on in their jobs as innovation managers. At the Institute for Entrepreneurship and Innovation of the WU – Vienna University of Economics and Business, there have been conducted approximately 20 lead user projects during the past ten years. We will report on the experiences gained throughout these projects in the following paragraphs.

What is the lead user method?

The lead user method was developed by Eric von Hippel at the Massachusetts Institute of Technology. It is a four-step process template helping organizations to systematically identify, and in a subsequent step, involve so-called lead users into corporate innovation processes for the purpose of generating radical innovations. Lead users are highly progressive users that have been found to be valuable contributors to innovation processes because of two very important characteristics. First, lead users are ahead of an important market trend (referred to as "trend leadership"). They usually perceive certain needs prior to the mass market. As existing products do not fit with their requirements and needs, lead users are highly motivated to invent and develop customized solutions themselves to satisfy their unmet needs (literature refers to this aspect as "high expected benefit from an innovation"). In the course of their innovative activities, lead users draw on their specific personal and professional know-how and skills that usually differ from the know-how and skills disposable in corporate R&D departments. Thus, lead users often come up with solutions that are based on completely different technologies or solution principles than in-house generated solutions. Despite their high degree of novelty and radicalness, lead user generated solutions have proven to be highly promising from a commercial point of view. Lead user concepts potentially become attractive to large market segments later on, as lead users literally anticipate future needs of the mass market.

The lead user method aims at identifying lead users and integrating them into corporate R&D projects. Because of the high value proposition of lead user generated concepts and solutions, an increasing number of companies employs the lead user method as described by von Hippel (1986). The typical lead user project consists of four interrelated steps. In a first step, the company has to define a search field; search fields might be technological problems that could not be solved within the company over a longer period of time as well as a call for an innovative solution for a specific problem the customers face. In a next step, the most important trends (yet unsatisfied user needs) within the specific search field have to be identified; the subsequent third step is dedicated to finding individuals possessing a trend leadership position for the most important trends. Ultimately, the lead users identified are invited to participate in a workshop together with a cross-functional project team consisting of employees of the focal company. The goal of these workshops is to generate a relatively small number of concrete and radical new product concepts (Herstatt & von Hippel, 1992; Lüthje & Herstatt, 2004; von Hippel, 1986).

What does the course format look like and what are the biggest challenges in running it?

The course format "Sources of Innovation" is designed as a project with a time scope of four months. Due to its nature as a PBL course, the number of students that can be enrolled to the class is limited with 15 participants. Of course, the teacher could run several lead user projects in parallel in order to increase the number of participants. However, as these types of project are quite demanding not only for the students but for the teacher as well, we would recommend not to conduct more than two lead user projects at once, resulting in a maximal number of participants of 30 students. Basically, the course's structure follows the logic of the lead user method as described above. However, due to the fact that it is a course, it consists of five instead of only four phases (the four phases of the lead user method as well as a project closing and documentation phase). Each of these phases is associated with specific tasks and challenges that are summarized in Figure 35.1.

In the first step, the concrete search field has to be specified. This is a very crucial task as the search field definition heavily influences the work on the subsequent stages as well as the success of

Educating the innovation managers

	Step I: Definition of the search field	Step II: Identification of important trends	Step III: Identification of lead-users	Step IV: Accomplishment of lead-user workshop	Step V: Documentation of the project
Tasks for students	• Understanding the project partner's problem • Challenging the project partner's search field definition	• Gathering of market data by drawing on primary and secondary data • Evaluation of market trends	• Development of a search strategy aiming at the identification of lead-users • Development of a scheme for evaluating lead-userness	• Planning the workshop • Composition of teams • Moderation and documentation of break-out session	• Summary of project and outcomes
Tasks for teachers	• Definition of search field • Composition of the partner organization's project team • Theoretical introduction to the lead user method	• Provision of feedback on the students' market research strategy • Provision of feedback on interview guideline	• Inputs on social search techniques (pyramiding and broadcast search) • Feedback on search processes and lead-users identified	• Provision of trainings on creativity techniques • Trainings fostering presentation skills • Moderation and documentation of plenary sessions	• Provision of templates of a written report and a final project presentation
Biggest challenge(s)	• Definition of an adequate search field	• Identification of most important market trends • Overcoming persistence to conducting interviews with users	• Evaluation of the potential participants' lead-userness	• Unforeseeable events in the course of the lead-user workshop (like discussion of IP rights issues, a lack of creativity among participants)	
Success factors	• Use of a template supporting the definition of the search field • Kick-off on method (with mini cases)	• Special training on how to conduct interviews	• Special training on evaluation of trend leadership and benefit from an innovation of potential workshop participants	• Detailed briefing of participants prior to the lead-user workshop	

Figure 35.1 Tasks and challenges of the lead user method.

the whole project. If the search field is defined to narrow (i.e., the project partner defines a very specific technological problem and also specifies how a solution should look in terms of the technology applied to solve the problem), there won't be enough space to come-up with really novel solutions with might lead to dissatisfaction of the students as well as the lead users involved. On the other hand, if a search field is too open, it will be difficult for the students to identify the most important market trends and corresponding lead users; in the end, the solutions generated might be too much out of the box and not really feasible from a technical and/or corporate strategy perspective, preventing the project partner from taking up the ideas and concepts generated. Defining an adequate search field is a complex task and probably the most challenging activity within this step. The following questions should be clarified prior to the project: Is there a real problem to be solved? What is the exact problem? Is it of relevance? Are there any types of solutions that should be excluded from the search? What is the time scope of the innovation, in other words, are concepts preferred that can be transformed in a marketable product almost immediately? How big is the budget for developing a first concept? Defining the search field calls for a high level of experience, thus this task should be fulfilled by the teacher prior to the start of the course. As said earlier, a well-defined search field is a prerequisite for a manageable and potentially successful lead user project. Another important prerequisite is a highly committed, cross-functional project team of representatives of the partner organization. The team members need to be interested in the project, they need to be willing to collaborate and to meet with students on a regularly basis in addition to their day-to-day job activities, and they need to be open to inputs from external users without a real "expert" status. Another task extremely important in this first phase of the project is to equip the students with the knowledge and know-how necessary to conduct the later stages of the lead user project. The students need to get an idea of the lead user method's logic, how it works, and which activities have to be conducted at what point of time. A good way of teaching them the lead user method is to conduct a one-day kick-off meeting at the beginning of the term. The kick-off meeting covers an overview about the method (given by the lecturer), a presentation of the search field (given by the project partner), as well as some mini cases and group assignments. Those assignments should address topics, problems, and questions that might arise during the lead user project in the later stages (e.g., the design of a social search strategy to identify potential lead users). Past lead user projects provide a rich source of such mini cases; furthermore, examples of successful lead user projects increase the students' motivation as well as their trust into the power of the method, which are both important success factors during the project. The tasks to be carried out by the students in this first project phase comprise the development of an understanding of the project partner's problem that is to be solved by an innovative concept. Furthermore, the students should critically challenge the search field presentation provided by the project partner with regard to its appropriateness for the lead user method.

The second step of the lead user project is dedicated to identifying the most important trends, that is unsatisfied user needs, within the search fields. In this phase, the students have to conduct a detailed market research using primary and secondary data in order to find out about unmet needs within the search field. They have to develop an action plan for gathering and evaluating relevant data. In this phase, the role of the teacher is the one of a coach challenging the student team's search strategy and triggering the discussions about adequate criteria for evaluating the trends identified. Usually, this phase takes approximately three to four weeks. At the end of this phase, the student team presents its findings and insights regarding the most important trends to the project partner who then decides which trends to follow in the subsequent steps. The biggest challenge in this phase certainly is to evaluate the trends. Students should try to identify the most important market trends (unmet needs) in terms of their relevance to potential customers as well as the prospective market size for a corresponding solution. In order to get the data needed, students will have to conduct a large number of interviews with users and customers. Many undergraduate students (at least in Europe) do not feel too comfortable with contacting and interviewing people unknown to them and asking them for information and thus try to get the information via secondary data. However, as this is not sufficient in many cases, the teacher has to support the students in this phase by providing them with special interview training. Based on the insights gained in these training sessions, the students are asked to develop a guideline for the interviews that is fed back by the teacher.

Step three is concerned with identifying lead users that are (1) ahead of the chosen market trends and (2) willing to participate in a workshop hosted jointly by the project partner and the university. Again, students will have to conduct lots of interviews and other search processes in order to identify real lead users. The teacher might support this phase by referring the students to literature addressing social search techniques like pyramiding (e.g., von Hippel, Franke, & Prügl, 2009) and/or broadcasting (e.g., Jeppesen & Lakhani, 2010). One of the biggest challenges in this phase is to evaluate the lead userness (consisting of trend leadership and an expected benefit from an innovation) of the interviewees. On the search for individuals leading a general market trend, students often find individuals that are "experts" rather than "lead users." The main difference between these two groups is that lead users do have use experience (thus, they know the current solution with all its advantages and bugs) and may benefit themselves from an innovative solution within the search field. In contrast, experts do not necessarily have use experience and are unlikely to benefit from a new solution as they do not have any unsatisfied needs with regard to existing solutions; they are ahead of a trend because of professional reasons and might be used to selling their specific expertise (like consultants and/or academics). Experience in more than 20 lead user workshops has shown that inviting experts to these events is counterproductive as they often negatively affect the climate at the lead user workshop. Very often, experts try to reduce the other participants' willingness to freely reveal needs-based information and exchange ideas, only in order to increase their own value as a potential partner to the project partner after the end of the lead user project. In order to sort out the lead users from the group of average users and experts, the students have to develop a valid and reliable instrument for measuring the interviewees' lead userness. Again, in this phase the role of the teacher is restricted to the one of a sparring partner providing feedback on the search processes, the instruments to choose appropriate participants for the lead user workshop as well as on the potential lead users themselves.

The fourth phase of the project is concerned with organizing and conducting the lead user workshop. A lead user workshop usually runs for 2.5 days and is organized as a sequence of break-out and plenary sessions. After an introduction and the presentation of the search field, the previously assigned composed teams (consisting of three to four lead users as well as one representative of the project partner company each) work on their concepts in parallel break-out sessions. In between these break-out sessions, there are plenary sessions in which all teams present their preliminary ideas and concepts and collect feedback by the other teams to further improve their developments. The students are fully responsible for managing the whole workshop, they have to moderate and document the break-out sessions, they have to arrange a social program, and they are in charge of all the *ex-ante* workshop preparations (finding an adequate facility, composing the teams, inviting the lead users, etc.). This phase definitely is the highlight of each lead user project. Students perceive it as highly rewarding to participate as facilitators in the generation of highly radical innovations that usually display high values to the project partner organization. In this phase, the teacher has to secure the preparedness of the students to conduct all activities necessary. For example, the students will need to get special training in presenting, in guiding and moderating brain storming sessions and other creativity techniques, in composing teams with regard to different factors as personal characteristics of workshop participants, as well as in business etiquette. Challenges in the workshop phase are to deal with upcoming questions of IP rights issues (which should be addressed by the teacher in accordance with the project partner) and to react to other unforeseeable developments (like a lack of creativity of workshop participants or company members influencing the direction of the developments and overruling their team members because of strategic considerations). These potential problems can be reduced by thoroughly briefing all workshop participants, the lead users as well as the project partner representatives.

Phase five is the project closing and documentation phase. Here, the students have to summarize the project and the outcomes achieved. They have to compile a comprehensive written report and to present the concepts generated throughout the workshop to the project partner organization's top management. The teacher might support this phase by providing the students with best-practice examples and templates.

"New Business Development" – a problem-based learning course format based on the user community-based technological competence leveraging method

The course format "New Business Development" was developed at the Institute for Entrepreneurship and Innovation of the WU – Vienna University of Economics and Business in response to an increased interest in technology-push innovation methods. Maybe induced by the upcoming, worldwide economic crisis in 2009, more and more companies cut their internal R&D budgets and decided to pursue technology-push instead of market-pull innovation strategies. The trend was and in many industries still is to leverage already existing technological competences by systematically transferring them to and exploiting them in completely new industries. The main objectives of such innovation strategies are to increase the return on investment in past R&D investments, to reduce the strategic dependence on a certain market by entering new industries, and/or to gain economies of scale increasing the productivity of manufacturing facilities.

Irrespective of the motive for pursuing a technology-push innovation strategy, the tasks involved (i.e., the identification of new applications in new industries, the evaluation of these new market opportunities, the design of an adequate strategy for entering these new and unknown markets) often are very demanding for technology-driven organizations because of various reasons, like for example the so-called local search bias (Rosenkopf & Nerkar, 2001). The course format "New Business Development" was invented in order to provide business students with the opportunity to apply the "user community-based technological competence leveraging method" (also referred to as ISAA method)[1] to a real-life project. The course format offers a problem-based learning environment to the students and also involves technology-driven organizations looking for additional application fields for a technology or a product as project partners.

What is the user community-based technological competence leveraging method?

User community-based technological competence leveraging denotes a methodological approach to systematically searching for and evaluating alternative fields of application for existing technologies. This method helps to overcome local search biases by integrating user communities without having to reveal the full functionality of the underlying technological solution. Thus, it is a method helping to conduct technology-push innovation strategies: starting point is a solution (a specific technology or product) that is looking for a (new) problem (new applications) to be solved. In this sense, the user community-based technological competence leveraging method is quite different from the lead user method in which new solutions for existing problems are searched for. Another difference between these two methods is the type of users involved. In contrast to the lead user method, technological competence leveraging (also referred to as TCL in the following) does not call for the integration of lead users into the innovation process. It is rather searched for average users that realize the technology as a solution to one of their specific problems, thereby pointing the inventor/technology owner to potential fields of application.

User community-based technological competence leveraging consists of four steps. In a first step, the technology needs to be analyzed with respect to the benefits it delivers. Inventors often describe their developments along their main technological features. However, in order to find alternative applications a technology needs to be described along the benefits it delivers to its users or its problem solving competence. To identify the benefits of a technology from a user perspective, current users have to be involved. The benefits revealed within interviews with current users prove helpful in the second phase, which is the systematic search for people that face problems similar to those of the technology's current users and that might also benefit from the focal technological solution. Step three is concerned with the evaluation of the different fields of application identified, and in step four actionable commercialization strategies for the most promising fields of application are crafted (Keinz & Prügl, 2010).

What does the course format look like and what are the biggest challenges in running it?

The course format "New Business Development" is designed similarly to the course "Sources of Innovation." The time scope again is four months, which is sufficient to conduct the five phases of the project of a typical TCL project (the four phases of the method as well as a project closing and documentation phase, see Figure 35.2). In practice, a group size of up to 15 students per project has proven to be manageable, for the teacher as well as for the students. In order to increase the number of participants that can be enrolled to the course, the teacher could run several TCL projects in parallel. However, as these projects tend to be very demanding in terms of coaching activities, we would suggest conducting a maximum of two projects at once.

Educating the innovation managers

	Step I: Identify the technology's use benefits	Step II: Search for alternative application areas	Step III: Analyze potential	Step IV: Assemble an actionable commercialization strategy	Step V: Documentation of the project
Tasks for students	• Understanding the project partner's technology • Challenging the project partner's technology	• Gathering of application ideas by drawing on primary and secondary data • Development of a search strategy	• Development of a scheme for evaluating identified application areas • Gathering additional data for evaluation	• Assemble commercialization strategy based on data from primary and secondary sources in every chosen field of application	• Summary of project and outcomes
Tasks for teachers	• Definition of search field • Composition of the partner organization's project team • Theoretical introduction to the method	• Provision of feedback on the students' search strategy and interview guideline • Inputs on social search techniques	• Feedback on evaluation processes • Preparation session for project partner meeting	• Provision of trainings on strategic options for technology commercialization • Supporting preparation of presentation of results	• Provision of templates of a written report and a final project presentation
Biggest challenge(s)	• Definition of an adequate search scope and search specification	• Identification of different applications • Overcoming persistence to conducting interviews with users	• Evaluation of the potential application areas	• Soundness of strategic reasoning	
Success factors	• Identification of unique benefit dimensions • Kick-off on method (with mini cases)	• Special training on how to conduct interviews	• Special training on evaluation of application areas and preparation of evaluation workshop with project partner	• Detailed briefing of participants regarding strategic options • Intense coaching	

Figure 35.2 Tasks and challenges of the user community-based technological competence leveraging method.

The first step involves an in-depth exploration of the technology. The goal is to fully understand it from a benefits perspective, that is, to identify its unique benefit dimensions. One starting point here is to analyze the situation in which the technology is used in the originally intended application (where possible). Employing this approach does not necessarily require the search team to know all the functional details of how the underlying technology works. What really matters is to get an idea of how the technology creates value for its potential users. In this step, the following questions were identified as central, and discussion within the search team revolved around these issues: What benefit can a (potential) user draw from the technology? Which problems does the technology solve? Which needs does the technology potentially fulfill? Having understood how current users benefit from the technology, the search team then translates those insights into an abstract problem-based search definition (i.e., an abstract schema). If this schema is defined too narrow, there won't be enough space to come-up with really novel application areas which might lead to dissatisfaction of the students as well as the project partner involved. On the other hand, if a search specification is too abstract, in the end, the solutions generated might be too much out of the box and not really feasible from a technical and/or corporate strategy perspective, preventing the project partner from seriously considering the application areas discovered. Another important prerequisite is a highly committed, cross-functional project team of representatives of the partner organization. The team members need to be interested in the project, they need to be willing to collaborate and to meet with students on a regularly basis in addition to their day-to-day job activities, and they need to be open to inputs from external users. Another task extremely important in this first phase of the project is to equip the students with the knowledge and know-how necessary to conduct the later stages of the project. The students need to get an idea of the method's logic, how it works, and which activities have to be conducted at what point of time. A good way of teaching them the user community-based technological competence leveraging method is to conduct a one-day kick-off meeting at the beginning of the term. The kick-off meeting covers an overview about the method (given by the lecturer), a presentation of the technological solution (given by the project partner), and some mini cases and group assignments. Those assignments should address topics, problems, and questions that might arise during the project in the later stages (e.g., the design of a social search strategy to identify potential application areas). Past projects provide a rich source of such mini cases; furthermore, examples of successful technological competence leveraging projects increase the students' motivation as well as their trust into the power of the method, which are both important success factors during the project. The tasks to be carried out by the students in this first project phase comprise the development of an understanding of the project partner's technological solution and its application boundaries. Furthermore, the students should critically challenge the presentation provided by the project partner with regard to its appropriateness for technological competence leveraging.

The second step of the technological competence leveraging project is dedicated to identifying the alternative areas of application. In this phase of the user community-based approach to technological competence leveraging, the main objective is to identify as many concrete applications for the existing technology as possible. However, instead of looking directly for other applications (based on a technological view), in our case this is done by systematically searching for people with high levels of knowledge or experience regarding problems linked to the benefits defined in the first step. In order to identify a wide range of different possible applications, it seems necessary to introduce the technology's use benefits (the problem-based search specification), formulated as an abstract schema, to a large number of people from as many different fields and professions as possible. One imperative is to ensure that the problem-based description of the technology is presented to the interviewees in an easy-to-understand format, accompanied by figures or videos showing the technology in use when solving a specific problem (if possible). The interviewees are then asked (1) whether they are familiar with situations in their professional lives in which they regularly face those problems or where those problems play a central role, (2) whether those situations really matter to them (level of eagerness for a solution), (3) whether they think a solution to this problem would also be beneficial to others facing the same problem, and (4) whether they know other persons who faced a similar or somehow related problem and could therefore benefit from a solution. As this is one of the most critical activities within the proposed framework, a great deal of cognitive effort has to go into the preparation of and the actual search for adequate interviewees. In this phase, the role of the teacher is the one of a coach challenging the student teams search strategy and triggering the discussions about adequate criteria for evaluating the application areas identified. In order to get the data needed, students will have to conduct a large number of interviews with users and customers. The teacher might support this phase by referring the students to literature addressing social search techniques like pyramiding (e.g., von Hippel et al., 2009) and/or broadcasting (e.g., Jeppesen & Lakhani, 2010). Many undergraduate students (at least in Europe) do not feel too comfortable with contacting and interviewing people unknown to them and asking them for information and thus try to get the information via secondary data. However, as this is not sufficient in many cases, the teacher has to support the students in this phase by providing them with special interview training. Based on the insights gained in these training sessions, the students are asked to develop a guideline for the interviews that is feed backed by the teacher.

Step three is concerned with analysis of the potential application areas. Once the potential fields of application have been identified, the next step is to evaluate their commercial attractiveness. Due to limited resources, it is necessary to identify the most promising applications in order to decide which markets to enter first. Our experience revealed a two-stage process in the analysis of different fields of application as the most effective modus operandi. The first stage of this analysis is intended to support the searching entity (e.g., a start-up team) in quickly rejecting those applications that do not satisfy the basic criteria for successful commercialization of the technology. In this stage, all potential fields of application are rated with respect to their fit with the company's strategy as well as the number of use benefits relevant in the particular application and their relative importance. Generally speaking, the more benefits are highly relevant within a certain field of application and the better that field corresponds to the strategic alignment of the company, the higher the potential success of the ensuing commercialization effort will be. At the end of this phase, the student team presents its findings and insights regarding the most promising application areas to the project partner which then decides which ones to follow in the subsequent steps. The biggest challenge in this phase certainly is to find the most promising application areas. Students should try to identify the most important application areas (unmet needs) in terms of their relevance to potential customers as well as the strategic fit of a corresponding market opportunity. Again, in this phase the role of the teacher is restricted to the one of a sparring partner providing feedback on the evaluation processes.

The fourth and fifth phase of the project are about drafting an actionable commercialization strategy and project closing and documentation. Although it should be clear from the previous steps which new markets are to be tackled first, the important question of *how* to commercialize the technology is still left unanswered. Thus, in the fourth step the team dedicates resources to developing what might be called "commercialization proposals." For each field of application, the team comes up with a brief report describing (1) the concrete problem which can be solved by the given technology in the specific field of application, (2) the market potential and future growth rates, (3) substitutes and products from competitors (where applicable), (4) potential user companies or distributors, as well as (5) strategic options regarding the commercialization mode (ranging from creating a new company to commercializing the technology to licensing the technology to established players or start-ups). Finally, phase five is the project closing and documentation phase. Here, the students have to summarize the project and the outcomes achieved. They have to compile a comprehensive written report and to present the final results to the project partner organization's top management. The teacher might support this phase by providing the students with best-practice examples and templates.

Success factors in teaching user innovation approaches in a problem-based learning environment: concluding remarks

In the sections above we reported and reflected on the structure, outcomes, and success factors of two problem-based learning course formats in the field of user innovation education. These courses have been designed to enable prospective innovation managers to apply two different user innovation methods in a real-life setting and to think about the questions outlined above. Taken together, there can be seen the following success factors that enhance learning opportunities in these approaches: (1) real-life problems, (2) real-life project partners, (3) diverse and mixed project teams consisting of students, project partners and a supervisor, (4) support and input "as the problems arise" throughout the entire course (provided by supervisors, coaches, and project partners), and (5) kick-off meetings in order to "lay the ground" for these learning endeavors increasing the knowledge as well as the motivation of the participating students.

Note

1 In the practical realm, the resulting user-driven search approach to technological competence leveraging has become known as the "ISAA method." ISAA can be interpreted as an acronym for Iterative Search for Alternative Applications and also reflects the four steps involved (Identify, Search, Analyze, Assemble).

References

Barrows, H. S. (1996). Problem-based learning in medicine and beyond: A brief overview. *New Directions for Teaching and Learning,* (68), 3–12.

Herstatt, C., & von Hippel, E. (1992). From experience: Developing new product concepts via the lead user method: A case study in a "low-tech" field. *Journal of Product Innovation Management, 9*(3), 213–221.

Hmelo-Silver, C. E., Duncan, R. G., & Chinn, C. A. (2007). Scaffolding and achievement in problem-based and inquiry learning: A response to Kirschner, Sweller, and Clark (2006). *Educational Psychologist, 42*(2), 99–107.

Jeppesen, L. B., & Lakhani, K. R. (2010). Marginality and problem-solving effectiveness in broadcast search. *Organization Science, 21*(5), 1016–1033.

Keinz, P., Lettl, C., & Hienerth, C. (2012). *Designing the Organization for User-driven Innovation.* Working Paper. WU Vienna University of Economics and Business, Vienna.

Keinz, P., & Prügl, R. (2010). A user community-based approach to leveraging technological competences: An exploratory case study of a technology start-up from MIT. *Creativity and Innovation Management, 19*(3), 269–289.

Lilien, G. L., Morrison, P. D., Searls, K., Sonnack, M., & von Hippel, E. (2002). Performance assessment of the lead user idea-generation process for new product development. *Management Science, 48*(8), 1042–1059.

Lüthje, C., & Herstatt, C. (2004). The lead user method: an outline of empirical findings and issues for future research. *R&D Management, 34*(5), 553–568.

Rosenkopf, L., & Nerkar, A. (2001). Beyond local search: Boundary-spanning, exploration, and impact in the optical disk industry. *Strategic Management Journal, 22*(4), 287–306.

von Hippel, E. (1986). Lead users: A source of novel product concepts. *Management Science, 32*(7), 791–805.

von Hippel, E. (2005). *Democratizing Innovation.* Cambridge, MA: MIT Press.

von Hippel, E., Franke, N., & Prügl, R. (2009). Pyramiding: Efficient search for rare subjects. *Research Policy, 38*(9), 1397–1406.

36

What can innovation education learn from innovators with longstanding records of breakthrough innovations?

Larisa V. Shavinina

UNIVERSITÉ DU QUÉBEC EN OUTAOUAIS, CANADA

Summary: This chapter presents preliminary findings from the study of the phenomenon of individual innovation in the case of outstanding innovators with longstanding records of breakthrough innovations. In sharp contrast to the conventional wisdom of innovation science emphasizing that (1) innovation is a team sport, and (2) people are good either in generating ideas (i.e., creativity) or in their implementing into practice (i.e., innovation), just to mention a few dogmas, there is a rare group of individual innovators. They possess a unique ability to both generate great ideas and to implement them into practice in the form of new products, services, and processes by putting into place all the necessary organizational, human, and "environmental" structures. This is the phenomenon of individual innovation. The findings will be presented from the point of view of innovation education: what can be learned from famous innovators with longstanding records of breakthrough innovations.

Key words: Innovation, innovators, breakthrough innovations, the phenomenon of individual innovation, innovation education.

Introduction

> For he who innovates ... it ought to be remembered that there is nothing more difficult to take in hand, more perilous to conduct, or more uncertain in its success, than to take the lead in the introduction of a new order of things.
>
> *(Niccolò Machiavelli)*

Today innovation is the cornerstone of economic prosperity, scientific discovery, technological invention, and cultural vibrancy. As the Premier Minister of Québec Jean Charest emphasized launching Québec's innovation strategy, "Innovation has been critical to the economic development of modern societies. ... Our prosperity in the future depends on it" (Charest, 2006). As governments launch innovation strategies and design innovation policies in an attempt to make their economies the innovation-based economies and to transform their societies into innovative societies, little is known what should be done exactly and how. This is why those strategies and policies are rarely successful. Financial investments in research and development (R&D) of new technologies are essential, but this is not the whole story. Despite the ever-increasing importance of innovation in society, one should acknowledge that innovation does not happen very often. During many years of my work on the bestselling *International Handbook on Innovation* (the first book of this type that is considered the beginning of innovation science; Shavinina, 2003) I found that something important is missing: individual innovation and how to develop it in everyone in society. It has never been studied. Societies can progress today only by innovations and every effort should be made to develop potential innovators.

In direct contrast to the traditional practice of innovation science emphasizing that (a) innovation is a team sport, and (b) people are good either in generating ideas (i.e., creativity) or in implementing them into practice (i.e., innovation), just to note some dogmas, I found that there is a rare group of individual innovators. They possess an excellent ability to both generate great ideas and to implement them into practice in the form of new products, services, and processes by putting into place all the necessary organizational (e.g., creating a research laboratory as Thomas Edison did, or co-founding a company as Steven Jobs did, or setting up many new companies as Richard Branson did), human (i.e., hiring the best talent), and "environmental" (e.g., changing the dominant management culture as Akio Morita did when he almost rejected the traditional Japanese way of doing business at *Sony*) structures. This is what the phenomenon of individual innovation is all about.

The overall objective of this chapter is to present the preliminary findings from the study of this phenomenon in the case of prominent innovators with longstanding records of breakthrough innovations such as Richard Branson, Jeff Bezos, Warren Buffet, Michael Dell, Bill Gates, Stelios Haji-Ioannou, Steven Jobs, Herbert Kelleher, Mike Lazaridis, Akio Morita, Robert Noyce, Anita Roddick, Sam Walton, and others. Specifically, the findings will be presented from the point of view of innovation education. That is, what today's children can learn from outstanding innovators who constantly make innovation happen. (An in-depth developmental version of the phenomenon of the individual innovation appears in Chapter 16 of this volume, *The trajectory of early development of prominent innovators: entrepreneurial giftedness in childhood*.)

From the methodological point of view, this is the idiographic exploratory investigation: qualitative analysis of case studies (i.e., micro-unit analysis of each case of individual innovation). The case study method fits well to research on people distinguished by their rarity that is the case of great innovators. This method allows analyzing their specific talents (e.g., wisdom or intuition) in detail. As each of the above-listed prominent innovators represents a significant sample, the given research uses a specific sampling procedure: *significant samples* (Simonton, 1999). That is, rather than sample randomly from the entire population of great innovators, the presented study concentrates on the best cases of outstanding innovation achievement relying on the available (auto) biographical sources, interviews, and corporate information. It is practically impossible to present findings with respect to all studied innovators in one chapter. Therefore, when each component of the phenomenon of individual innovation is considered below, examples are drawn from only some of the above-mentioned innovators.

Preliminary research demonstrates that individual innovators are characterized by a rare combination of highly developed creative abilities, applied wisdom, practical intuition, managerial talent, entrepreneurial giftedness, excellence, unique objective vision, persistence, and courage, which function jointly and compensate for one another. While persistence and courage of innovators are understandable, nothing is known about the unique combination of the other elements. No one examined wisdom, intuition, managerial talent, excellence, unique objective vision, and compensatory mechanisms of innovators. The goal of this research project supported by a special grant from the *Fonds québécois de la recherche sur la société et la culture* (FQRSC) is to understand their essence and joint functioning in outstanding individual innovation. This is a new, groundbreaking research that was undertaken for the first time with a great potential to advance knowledge in innovation science, business, management, education, and economy, as well as managerial practice in virtually all industries.

Initial exploration shows that innovators during their childhood encompassed a wide range of abilities, including the gifted (e.g., Jeff Bezos, Warren Buffett, Bill Gates, and Sam Walton, just to mention a few), underachievers (e.g., Richard Branson), and children without any special talents. Their divergent trajectories of innovation talent development ultimately led to the same result: amazing innovations, which testify to the outstanding minds of those who made them. Eventually, all the trajectories led to the same point: zenith in business. This study aims to understand how and why this happened, and what lessons can be derived for the education of today's youth–tomorrow innovators. The discovery of the principles involved in the talent development of outstanding innovators will allow educators to accordingly improve, develop, modify, and transcend areas in the current curriculum in an attempt to cultivate great innovation talent in future generations (Shavinina, 2012a).

The phenomenon of individual innovation: when unique objective vision, creativity, wisdom, intuition, managerial talent, entrepreneurial giftedness, excellence, and courage come together, which function jointly and compensate for one another

Unique vision and the objectivization of cognition

The basis of the phenomenon of individual innovation is innovators' unique cognitive experience that expresses itself in their specific type of representations of reality or in their *unique vision*. It means that innovators see, understand, and interpret everything that is going on around them differently from other people (Shavinina, 2003, 2007). Innovators' unique vision is based on their unusual sensitivity (Shavinina, 2009). This is why they are distinguished by a unique creativity, powerful intuition, great wisdom, entrepreneurial giftedness, managerial talent, excellence, and courage. This explains why numerous leadership researchers and innovation leaders have pointed out that "vision" is a central facet of leadership. Cases of the unique vision of great innovators and how many important business decisions were made based on their unrepeatable points of view are impressive. For example, *Xerox*'s brilliant scientists and engineers at its Palo Alto research center (PARC) invented the very first computer and a lot of related things. However, neither they nor Xerox top management knew what to do with those inventions. It was Steve Jobs, who during his visit to PARC, *saw* a prototype of the Macintosh and immediately realized the future of computing. As he recalled in 1996,

> When I went to Xerox PARC in 1979, I *saw* a very rudimentary graphical user interface. It was not complete. It was not quite right. But *within 10 minutes, it was obvious* that every computer in the world would work this way someday.
>
> *(quoted in Bennis & Biederman, 1997, pp. 79–80; italics added)*

The unique vision of innovators displays itself in every facet of their professional activity. For example, Richard Branson's unique vision was evident since his first business venture, the *Student* magazine, and was reflected in the *Virgin* brand, especially in his new model of business expansion via branded venture capital. The origins of this model—admired by the management scholars and practitioners alike and considered as the future of business—go back to the *Student*.

> I began to think of ways to develop the magazine, and the Student name, in other directions: a Student conference, a Student travel company, a Student accommodation agency. I didn't just see Student as an end in itself, a noun. I saw it as the beginning of a whole range of services.... In this way I was a little removed from the rest of my friends, who concentrated exclusively on the magazine.
>
> *(Branson, 2002, p. 64)*

It is safe to say today that this remoteness or the availability of vision was something that distinguished Richard out of crowd and eventually brought *Virgin* to a success.

One aspect of the unique "vision" of great innovators is related to the *objectivization of cognition*. It means that innovators see, understand, and interpret everything in a very *objective* way. To re-phrase Marina Kholodnaya, I would say that the value of innovators in society should be seen not only in that they make breakthrough innovations possible, but mainly in "their *ability to see everything in an objective manner:* they see the world as it was, as it is, and as it will be in its objective reality" (Kholodnaya, 1990, p. 128; italics added). This is why many innovators—Jeff Bezos, Richard Branson, Michael Dell, Steven Jobs, Mike Lazaridis, Akio Morita, just to mention a few—were and are exceptional in "sensing" or seeing new market opportunities, which turn out to be highly profitable for their companies. For example, Richard Branson upgraded his fleet at *Virgin Atlantic* during an economic down-turn at rock-bottom prices, because he *saw* this as the best time for upgrading when other airlines were not buying new aircrafts. He is convinced that managers always have to have a long-term vision. It does not matter what the current state of their business is, they have to see ahead. "However tight things are, you still need to have the big picture at the forefront of your mind" (Branson, 2002, p. 492).

The ability of famous innovators to see any aspect of the surrounding reality from an objective point of view is very important in business. Innovators are able to objectively see either hidden consumers' needs, or potential developments in these needs, or changes in technology. The example of Akio Morita—when he *saw* a great market opportunity for Walkman when nobody else could see and marketing research demonstrated just the opposite—is quite appropriate here.

Tomorrow's innovators can learn the importance of a unique objective vision from such examples, which should be a key part of innovation education.

Creativity

Creativity is the first step in innovation process, because every innovation originates from new ideas. There is a consensus today that innovation refers to the implementation of ideas into practice in the form of new products, processes, or services, whereas creativity means the generation of novel, original, and appropriate ideas (Shavinina, 2003). Innovation is essentially about the *implementation* of ideas. Creators thus generate ideas, while innovators implement them into practice. There is, however, a unique group of people who are both able to generate ideas and to implement them into practice (Shavinina, 2007). They are individual innovators.

The preliminary findings of this study demonstrate that prominent innovators with longstanding records of breakthrough innovations are always full of new ideas and open to creative ideas of other people. They have developed unique approaches to creativity. Conventional creativity training does not contain anything similar. This is because innovators were in such business situations where traditional methods of creativity did not work. They went ahead, invented their own approaches, and came up with great ideas, the implementation of which led to multimillion dollar profits for their companies. Examples are abound. For instance, Richard Branson's philosophy of business is based on creativity: "I have never gone into any business purely to make money. ... A business has to ... exercise your creative instincts" (Branson, 2002, p. 58).

According to Akio Morita,

> it is very important to know how to unleash people's inborn creativity. My concept is that anybody has creative ability, but very few people know how to use it. My solution to the problem of unleashing creativity is always to set up a target.
>
> (Morita, 1987, p. 164)

> When an engineer or scientist is given a clear target, he will struggle to reach it. But without having a target—if your company or organization just gives him a lot of money and says "Invent something"—you cannot expect success.
>
> (p. 166)

"In industry, we must have the theoretical background, and we must have the pure research that precedes development of new things, but I have learned that only if we have a clear goal can we concentrate our efforts" (p. 166).

It should be noted, however, that when Morita and Ibuka founded *Sony* and even before they took stock of their abilities, the number of people they had, and what their talents and expertise were, Ibuka said, "Let's make a tape recorder." So, they set up a target even before they knew what the tape was made of or how it was coated—or even what it looked like. They decided to make innovative products; this was their global target (Shavinina, 2012b). Sony's typical reaction when they learned of some new development or came across a phenomenon was invariably "How can we use this? What can we make with it? How can it be used to produce a useful product?" For example, when video recording was being used in America by the major broadcasting stations, Morita and Ibuka thought people should have the same capabilities in their homes. The big commercial grade TV video tape machines were not appropriate for home use because they were cumbersome and very expensive. Therefore, *Sony* set up a special target to bring this machine into the home. As *Sony* engineers devised each new model, it seemed more and more incredible to them that they could make it so small.

> Yet it was not small enough for Ibuka. But nobody knew exactly where we were headed until he tossed that paperback book onto the conference table and said that was the target, a videocassette the size of the book that could hold at least one hour of color program. That focused all the development. It wasn't just a matter of making a small cassette—a whole new concept of recording and reading the tape had to be devised.
>
> (Morita, 1987, p. 167)

From Akio Morita's point of view, it is possible to have a good idea, a fine invention, but still miss the boat. He gave a good example of how *Texas Instruments* pioneered a small radio, but did not see its future and, therefore, gave up.

> So product planning, which means deciding how to use technology in a given product, demands creativity. And once you have a good product it is important to use creativity in marketing it. Only with these three kinds of creativity—technology, product planning, and marketing—can the public receive the benefit of a new technology. And without an organization that can work together, sometimes over a very long period, it is difficult to see new projects to fruition.
>
> (Morita, 1987, pp. 169–170)

Renowned innovators are very good in creating a fun working environment, where creative ideas flourish. "Fun is at the core of the way I like to do business and it has been the key to everything I've done from the outset. More than any other element, fun is the secret of *Virgin's* success" (Branson, 2002, p. 489). Very often, Richard Branson has organized parties for his staff to have fun and, moreover, he entertained his employees. "I have always enjoyed parties, and I love throwing the *Virgin* staff together. It's an important part of life at *Virgin*" (Branson, 2002, p. 120).

Herbert Kelleher, the founder and former Chairman of *Southwest Airlines*, who has also organized fun parties for his staff and entertained them, has an interesting approach to creativity.

> The rule at *Southwest* is, if somebody has an idea, you read it quickly and you respond instantaneously. You may say no, but you give a lot of reasons why you're saying no, or you may say we're going to experiment with it in the field, see if it works.
>
> *(quoted in Krames, 2003, p. 173)*

Approaches to and methods of creativity of prominent innovators are of great benefit to every teacher wishing to develop creative abilities of his or her students. These findings were transformed into a set of practical techniques for a special workshop on creativity entitled, *From New Ideas to Success and Prosperity: Managing for Creativity* (www.innocrex.com). This is a real breakthrough in creativity training: learning from innovators with longstanding records of breakthrough innovations. Such learning will be a great asset for today's children–tomorrow's innovators and should, therefore, be an important element of innovation education.

Wisdom

Wisdom involves an individual's ability to balance the interests of all parties involved in a way that serves everyone's needs and well-being over the long and short terms. Wisdom has always been important to everyone because it is related to prosperity (e.g., Herbert Kelleher's wisdom and success of *Southwest Airlines*) or disaster of organizations (e.g., *Enron* executives). It is linked with success in any area of human endeavor whether in public service, science, and/or business. Wisdom becomes even more important today because of ever increasing complexity of tasks (e.g., environmental problems, security related issues, economic recession, just to mention a few) and, consequently, decisions. Information explosion and new and smart technologies also require wisdom. Conflicting challenges of multiple "dualisms"—(a) working productively in the present while innovating fruitfully for the future, (b) being successful but not becoming a victim of that success, and (c) acting under conditions of uncertainty while having a clear vision of the future—require wisdom as well.

When it comes to wisdom, the findings of the given study demonstrate that this is the most critical facet of individual innovation. Creative abilities, practical intuition, entrepreneurial giftedness, managerial talent, excellence, and courage—based on a unique objective vision—are important, as well as highly developed compensatory mechanisms. But ultimately it is innovators' wisdom that makes the difference between great, continued success and death in business. The recent economic recession clearly demonstrated this. It means that those companies were doing well, which had either wise CEOs and Presidents (e.g., *Virgin* and *EasyGroup*) or highly innovative products (e.g., Blackberry, iPod, and iPhone, just to mention a few) or both. As Richard Branson concluded, "Fortunes are made out of recessions."

Through an economic downturn wisdom means that great innovators anticipate economic troubles and do whatever possible to protect their companies. This is a unique management style. As Herbert Kelleher puts it, "In good times, manage as though bad times are just around the corner because they are sure to come." Wise managers are prepared for a recession (e.g., long-term investment strategy of Warren Buffett) and prepare their staff for it (e.g., increasing training budget at

Southwest Airlines). Other examples of wisdom-related performance of distinguished innovators are as follows:

- Innovators with longstanding records of breakthrough innovations have cash: the "save for a rainy day" policy (e.g., *Southwest Airlines, easyGroup, Virgin*). "The winners will have cash and the losers will not," pointed out Stelios Haji-Ioannou, the founder of *easyGroup* during the last economic recession.
- They will not only survive during recessions, they take advantage of opportunities and continue to grow and expand. For instance, Richard Branson discussed the opening of a new airline in Brazil or Russia in 2009 and predicted that "there are a lot of Richard Bransons that will come out of the next 3 or 4 years."
- They do long-term planning and have a long-term vision (e.g., Richard Branson, Japanese companies).

Prominent innovators are wise in every facet of their business life. For example, Herb Kelleher demonstrated this in its approach to managing people:

> Communication from the heart is more important than communication from the head, and informal communication is just as important as formal communication (e.g., It's a pleasure to see you. I was delighted to hear about your new baby. Did you find that hat in a trash can?). Moreover, communicating goals, ideas, emotions, inspiration, and love is just as important as communicating facts and figures.
>
> *(quoted in Krames, 2003, p. 191)*

The wisdom of outstanding innovators is based on their high standards of ethical behavior in professional life, as well as on the ability to take into account interests of all team members and act in the interests of everyone. High ethical standards lead to moral responsibility that is a true sign of wisdom. There are many examples of such wisdom. In the case of Branson the most sound of them is when the Customs and Excise officials arrested the twenty-year-old Richard because of the illegal profit, which *Virgin* had made from avoiding the purchase tax, and he spent one night in prison.

> I had never really focused on what a good name truly meant before, but that night in prison made me understand… I vowed to myself that I would never again do anything that would cause me to be imprisoned, or indeed do any kind of business deal by which I would ever have cause to be embarrassed. In the many different business worlds I have inhabited since that night in prison, there have been times when I could have succumbed to some form of bribe, or could have had my way by offering one. But ever since that night in Dover prison I have never been tempted to break my vow.
>
> *(Branson, 2002, pp. 100–101)*

It is interesting to note that high standards of ethical behavior—and wisdom as a result—originate from childhood. Thus, it was parents who set high standards of ethical behavior for Richard Branson.

> My parents had always drummed into me that all you have in life is your reputation: you may be very rich, but if you lose your good name you'll never be happy. The thought will always lurk at the back of your mind that people don't trust you.
>
> *(Branson, 2002, p. 100)*

Similarly, a priority of putting other people's interests first also came from his parents.

> At home Mum... always generated work for us.... Whenever we were within Mum's orbit we had to be busy. If we tried to escape by saying that we had something else to do, we were firmly told we were selfish. As a result we grew up with a clear priority of putting other people first.
> *(Branson, 2002, pp. 24–25)*

There was "a great sense of teamwork within our family" (Branson, 2002, p. 24).

These all are very useful lessons for educators who wish to develop wisdom-related skills and wisdom-based performance in today's children–tomorrow's innovators.

Intuition

Great innovators themselves call it "gut instincts," "feelings," "senses," and so on. It is amazing how many important business decisions they made based on their intuition. The extent to which outstanding innovators rely on their intuition in virtually every aspect of business activity is fascinating, as well as the multiple functions of their intuition and the accuracy (and wisdom) of intuition-based decisions.

It looks like all renowned innovators included in this study are distinguished by highly developed intuition. Richard Branson asserts

> I make up my mind about whether a business proposal excites me within about 30 seconds of looking at it. I rely far more on *gut instinct* than researching huge amount of statistics.... I distrust numbers, which I feel can be twisted to prove anything.
> *(2002, p. 216; italics added)*

Also, his "*instinctive understanding* of what your customer wants"[1] is nothing else but his unique intuition.

Richard Branson used his intuition in many business deals. Look, for example, at how *Virgin Money* was born: "*Instinctively*, I felt that the world of financial services was shrouded in mystery and rip-offs and that there must be room for *Virgin* to offer a jargon free alternative with no hidden catches" (Branson, 2002, p. 598; italics added). On another occasion, with respect to a completely different area of management: "*Air New Zeland* made an offer to buy *Virgin Blue* for US$250 million shortly before 11 September. My *instinct* told me that the company was worth more than that, but it wasn't an ungenerous offer" (Branson, 2002, pp. 537–538; italics added). Shortly after *Virgin Blue* became the second largest airline in Australia.

Steve Jobs' feeling of being right and sense for new products expressed in an unalloyed confidence was considered his defining characteristic.[2] Michael Dell is also distinguished by highly developed intuition. Recalling the days before the registration of his company, he wrote, "at age eighteen ... *I definitely felt* that I was diving into something pretty major ... *I felt* that this was absolutely the right time" (Dell, 1999, p. 11). His "instinct" is nothing more than his unique intuition: "*I believe* opportunity is part *instinct*..." "We had *the sense* that we were doing something different, that we were part of something special," noticed Michael Dell about the first months of *Dell Computer Inc.* (1999, p. 29).

Akio Morita is another great example. When he—a co-founder and then the Chairman of *Sony*— intuitively sensed a great market for Walkman and decided to develop it, he faced strong resistance (for a detailed description of this example see Shavinina's chapter on the fundamentals of innovation education, this volume). Morita eventually succeeded and wrote later that it required "human thought, spontaneous intuition, and a lot of courage" (Morita, 1987, p. 83).

Such great cases of practical intuition in outstanding innovators should be included in innovation education aimed to develop students' innovative abilities.

Managerial talent

If we look at the most important manifestation of managerial talent among great innovators, then we can see that this is their unique ability to hire for talent. They always try to find the best talent possible. As Mike Lazaridis, the founder, president, and co-CEO of *Research in Motion* put it, "Always hire people who are smarter than you ... Do not worry about your job. Find people who can do it better than you." Similarly, Robert Noyce insisted that *Intel* was recruiting high achievers to work in the company (Salerno, 1980, p. 124).

This study shows that famous innovators are excellent in managing people. For example, Richard Branson is an exceptional manager in this regard. For instance, he considers his employees as the number one priority. He believes that staff should come first; even if it means making less money. From Richard's point of view, that is the right decision to make. An impressive example of his concern for the staff dates to the early 1990s, when he offered jobs to all his employees at *Virgin Atlantic* after announcing the selling of *Virgin Records*.

Richard Branson's approach to employee selection focuses on finding the right fit between the job and the person based on their unique talents and not necessarily on their qualifications. That is, he hires for talent and does not pay attention to education, experience, and determination as the conventional management wisdom prescribes. A good example is when Richard invited his childhood friend to join the *Student* magazine instead of going to university.

> Nik agreed to delay going to university and come to *Student's* aid. With Nik arrival, *Student* was put back on the rail.... Nik used our Coutts accounts properly... He started writing checks and then checking the stubs off against the bank statements.
>
> *(Branson, 2002, pp. 66–67)*

Nik had just graduated from the high school and did not have any education or experience in finance. However, Richard Branson was able to recognize the potential accounting talent in his friend.

Likewise, Richard was good in identifying talents of other colleagues. Thus, when establishing *Virgin Records*, he noticed that Nik was perfect in accounting, Simon was excellent in choosing potential songs and in predicting what will be sold in the record business every season, and Ken knew how to deal with people and contract negotiations. "While Nik managed the costs of both the mail-order business and the *Virgin Records* shops, Simon began to define the mail order list as well as the *Virgin Records* shops themselves by choosing which records to stock" (Branson, 2002, pp. 106–107). "*Virgin* success, both in mail order and in record shops lay in Simon's skill at buying records" (p. 108).

Real talents and interests of employees are the most important considerations for Richard Branson when he appoints someone to a senior managerial position. It does not matter whether that person has a formal education or not; his or her actual talents are essential. For example, Richard has been entirely relying on Simon Draper's musical talent in selecting new records for *Virgin Records*. Ken Berry started to work as a clerk in *Virgin*, but had an obvious managerial talent, and he eventually ended up negotiating contracts with star singers such as Janet Jackson and others.

Richard Branson always gives an opportunity to his employees to manifest themselves to their full extent or re-invent themselves (e.g., to display their talents in new areas of business). Thus,

Simon Burke, who used to work as a clerk in *Virgin*, applied for the CEO position of *Virgin Shops*, got Richard's blessing, and was successful at this position. The CEO of *Virgin Money* decided to launch *Virgin Vines* and Branson completely supported him in his new venture.

Richard sets expectations by defining the right outcomes and lets his employees figure out the required steps. He believes that *Virgin* staff are entrepreneurs in their own right. Granting autonomy at work empowers *Virgin* employees to reach high levels of performance. For example, in early years of *Virgin Atlantic* one of the stewardesses approached Richard with an idea of creating *Virgin Brides*. She was planning to get married and noticed that there is no one place that could provide all the necessary services for brides including clothes, wedding ceremony, honeymoon travel, and so on. Richard liked this idea and she was immediately appointed to the CEO position of *Virgin Brides*. He always tries to find the right fit between the employees' interests or talents and the area of business activity.

With respect to setting up expectations, Richard Branson gives a complete freedom to his managers. As soon as a new *Virgin* company is created based on an original idea, he keeps 51% or 52% of its shares leaving the rest to the generator of the idea who very often becomes a senior manager (e.g., CEO or president) in charge of the company. Branson does not interfere with their management decisions. The overall performance of each company is the main evidence based on which he can evaluate the managerial success of his CEOs or presidents. He thus manages by remote control and trusts in his employees' ability to do their best from the very beginning of any venture.

Richard Branson motivates his employees by showing sincere appreciation and offering—when it is possible—generous rewards for their efforts. In 1993, he split the compensation—£500,000—received from *British Airways* among the employees of *Virgin Atlantic*.

> Back at Holland Park the party started. I decided to share the £500,000 damages which had been given to me among all the *Virgin Atlantic* staff, since they all had to suffer from the pressure that *British Airways* had put us under, in the form of reduced salaries and cuts in their bonuses.
>
> *(Branson, 2002, p. 488)*

The bottom-line, hence, is that outstanding innovators are usually excellent managers (with some exceptions; for example, Steve Jobs during the first years of *Apple* was not a good manager (see Bennis & Biederman, 1997)).

Entrepreneurial giftedness

The given research reveals that all great innovators are, in fact, gifted entrepreneurs. This is because the nature of individual innovation and entrepreneurial giftedness largely coincide (for more details on the subject see Shavinina's chapter on the fundamentals of innovation education, this volume). It is not, therefore, surprising that innovators with longstanding records of breakthrough innovations and gifted entrepreneurs in childhood share the same characteristics such as: love to generate and implement real-life projects with at least a minimal financial reward; love doing real business plans with predicted financial outcomes; work passionately and hard on executing their plans; wish to do "real" things that bring money and try to do whatever possible to cut unnecessary steps; perseverance to succeed; optimism and "change the world" attitude; early exposure to challenges; competitiveness, excellence, and perfection; independence in thoughts and actions; and a rule-breaking attitude (see Shavinina, 2008; as well as the chapter on entrepreneurial giftedness by Shavinina, this volume). Many of these characteristics explain why prominent innovators are courageous people.

Excellence

Excellence refers to doing ordinary things extraordinarily well (John Gardner). With respect to outstanding innovators with longstanding records of breakthrough innovations excellence means world class products, processes, or services. They do not settle for second best. Innovators are convinced that only companies with great products survive (e.g., *Apple*'s iPhone). For instance, Steve Jobs was well known for his drive for excellence and even perfection. He always demanded from his employees to develop only the most perfect products. This is why *Apple* has a lot of innovations on the market.

Likewise, Michael Dell recalled that a month before opening his company

> I knew in my heart that I was on to a great business opportunity... I knew what I wanted to do: build better computers than IBM, offer great value and service to the customer by selling direct, and become number one in the industry.
>
> *(1999, p. 11)*

Excellence is also in the heart of Michael Dell's entrepreneurial motivation. When 18-years-old Michael almost left university and his father asked him what he was planning to do with his life, he replied: "I want to compete with IBM! I want to build better computers than IBM" (Dell, 1999, p. 10).

Innovators' belief in exceptionally high standards of performance is behind their excellence. They always try to do their best. "Anything I do in life I want to do well and not half-heartedly. I feel I am doing my best in *Student*," wrote sixteen-year-old Richard in a letter home (Branson, 2002, p. 46).

All of the above-discussed qualities of innovators drive their exceptional persistence and determination: "It's interesting to note that many people told us the direct model would fail in virtually every country ... *Believe* in what you're doing," says Michael Dell (1999, p. 29). Similarly, if somebody says that something cannot be done, Richard Branson tries to prove the opposite. "Life insurance? Everyone snorted when they heard the idea. People hate life insurance ... Exactly, I said. It's got potential" (Branson, 2002, p. 497).

Today's children will definitely benefit from these lessons in excellence and determination of outstanding innovators.

Courage

This study shows that renowned innovators were and are very courageous individuals (Shavinina, 2008, 2011). Courage is compulsory for innovators. They are not afraid to go ahead when they are trying to implement creative ideas into practice and, therefore, bring great innovations to the marketplace. Usually, such innovations have not existed before and nobody can predict the response from markets. Herbert Kelleher was not afraid to start up *Southwest Airlines*, an entirely innovative company in the airline industry. He successfully went through all the multiple obstacles created by competitors. The same is true in the case of Richard Branson and *Virgin Airlines*. The story of how he and *Virgin* employees overcame numerous dirty tricks made by *British Airways* is an impressive account of the innovator–entrepreneur's courage (Branson, 2002).

The rule is that markets do not exist for radical innovations such as iPod or iPhone. Innovators must create them. In order to succeed, they should be courageous enough to persuade everyone around them that the market for this and this specific product, process, or service will indeed exist. The Sony Walkman (see Shavinina's chapter on the fundamentals of innovation education, this volume) is an excellent example of Akio Morita's courage. It is interesting to note that innovation suffers in bad times (e.g., recession) due to fear. Courage is, consequently, mandatory for innovators!

It is discouraging that children are not taught to be courageous. This is a real omission in general education. Innovation education should play an important role in developing courage in future generations of innovators.

Joint functioning and compensatory mechanisms

Joint functioning of highly developed creativity, applied wisdom, practical intuition, managerial talent, entrepreneurial giftedness, courage, and excellence was not studied in innovators before. This research found that all these elements are always in play when it comes to innovators: they all work together at all times. Some of them may prevail over the others (and sometimes it is difficult to say with 100% certitude whether, for example, Herbert Kelleher made this decision because of his creativity or wisdom or intuition), but they all are there. However, it is clear that this is a rare and unique combination. Wisdom is an especially rare ability and not everyone has it. If one of those seven talents is not well developed, people and companies are in trouble and at risk. Also, this explains why innovation does not happen very often. It explains as well why there is a high mortality rate amongst new start-up companies. The right combination of those seven unique talents becomes critical in times of recession as the recent economic crisis demonstrated (see the section on wisdom above).

Compensatory mechanisms are also important in this regard. The essence of compensatory mechanisms is straightforward: if you are not, say, exceptionally excellent or wise in some aspects of your business activity, then you must compensate for that. To put it even more briefly: you must become excellent or wise in compensating for your lack of excellence or wisdom in a particular ability required to effectively achieve success. In other words, creativity, wisdom, intuition, entrepreneurial giftedness, managerial talent, excellence, and courage can compensate for one another. If, say, Robert Noyce's wisdom did not point him down the right path in that particular moment, his managerial talent came into play. But the truth is that all the components function together. Broadly speaking, innovators know what they do not know and how to compensate for what they do not know. This is an important aspect of their metacognition. For instance, many would say that Steve Jobs was just a lucky boy who founded a company in the garage of his parents. However, those, who knew the co-founder of Apple, said, "Steve Jobs was not just a lucky kid. *He knew what he didn't know, and sought people who did.* That was a very mature strategy" (Rogers & Larsen, 1984, p. 278, italics added). To be more accurate, that was a very mature *compensatory* strategy.

Here we begin to broach a very important problem. Innovators definitely do not possess all the traits of talented, gifted, and creative people, which were identified by researchers (Shavinina, 1995). They do not worry about this, because they know when, where, and how to compensate for their weaknesses and rely on their strengths. The compensatory mechanisms can thus manifest themselves in conscious relying on strengths and avoiding weaknesses. For example, an integral element of *Intel*'s long-range strategic planning is the analysis of its strengths and weaknesses, as well as the avoidance of certain areas of business. As Robert Noyce emphasized, "We build on strength and try to stay out of competition where we're weak" (Salerno, 1980, p. 129).

For the most part, compensatory mechanisms of great innovators manifest themselves in their ability to find and hire talented employees who could compensate for their lack of knowledge or experience in something. For instance, Simon Draper's love of and taste for music compensated for Richard Branson's lack of knowledge in this area and were behind the success of *Virgin Records*, childhood friend Nik Powell was good in finance and thus compensated for Richard's lack of interest in this field, and so on.

The general imperative is, hence, clear: if we are going to properly develop tomorrow's innovators who will succeed in innovation, then we have to make sure that they will be creative, wise, intuitive, excellent, and entrepreneurial in all facets of their professional activity, as well as possess a particular managerial talent, or compensate for a lack of some of these talents.

Conclusion

This chapter presented the preliminary findings from the study of the phenomenon of individual innovation in the case of outstanding innovators with longstanding records of breakthrough innovations. It described a collective picture of prominent innovators, who possess a unique ability both to generate great ideas and to implement them into practice in the form of new products, services, and processes by putting into place all the necessary organizational, human, and "environmental" structures. This is what the phenomenon of individual innovation is all about. The findings demonstrate that this phenomenon is characterized by a rare combination of highly developed creative abilities, applied wisdom, practical intuition, managerial talent, entrepreneurial giftedness, excellence, courage, which are based on a unique objective vision, function jointly, and compensate for one another. Each of these elements was briefly considered in a special section of the chapter. Practical examples from prominent innovators were used. If teachers and parents want to be successful in fostering innovative abilities of today's children who will make future innovations happen, then they have to develop their creativity, wisdom, intuition, managerial talent, entrepreneurial giftedness, excellence, unique objective vision, compensatory mechanisms, and their joint functioning to the fullest extent, as well as to encourage them to be courageous individuals. These are the main lessons from exceptional innovators with longstanding records of breakthrough innovations.

Acknowledgments

The study reported herein was supported under the Support for Innovative Projects program (Grant AN-129135) of the *Fonds québécois de la recherche sur la société et la culture* (FQRSC). The findings and opinions expressed in this chapter do not reflect the positions or policies of the FQRSC. I am very grateful to Larry Vandervert for his comments on the first draft of the chapter.

Notes

1 Richard Branson is convinced that

> an innovative business is one which lives and breathes "outside the box." It is not just about ideas. It is a combination of good ideas, motivated staff and an instinctive understanding of what your customer wants, and then combining these elements to achieve outstanding results.
>
> *(quoted in Clegg, 1999, p. 96)*

2 Conger (1995) noted that visionary leaders' ability to foresee future events is an intuitive process.

> When he co-founded *Apple* in 1976, Mr. Jobs bet that there would be a mass market for computers. And there was. The launch of the Macintosh in 1984 was predicated on the notion that giving computers a graphical interface, controlled with a mouse (then a real novelty), would broaden their appeal. He was right again ... Mr. Jobs's decision in 1999 to launch a range of iMac computers in different colours was also derided, but proved popular enough to turn *Apple's* fortunes around. Another bold move came in 2003. With the launch of the iTunes Music Store, Mr. Jobs dared to suggest that there might be a way to get people to pay to download music from the Internet rather than steal it. Once again, his nose for a new market proved accurate: Apple now sells millions of songs every month.
>
> *(The Economist, 5 February 2004)*

References

Branson, R. (2002). *Losing My Virginity: The Autobiography*. London: Virgin Books.
Bennis, W., & Biederman, P. (1997). *Organizing Genius: The Secrets of Creative Collaboration*. Cambridge, MA: Perseus Books.
Charest, J. (2006). *Québec's Innovation Strategy*. Québec City: Office of the Prime Minister of Québec.
Clegg, B. (1999). *Creativity and Innovation for Managers*. Oxford, UK: Butterworth.
Conger, J. A. (1995). Creativity and visionary leadership. In C. M. Ford & D. A. Gioia (Eds.), *Creative Action in Organizations* (pp. 53–59). Thousand Oaks, CA: Sage Publications.
Dell, M. (1999). *Direct From Dell*. New York: Harper Business.
Kholodnaya, M. A. (1990). Is there intelligence as a psychological reality? *Voprosu psichologii, 5*, 121–128.
Krames, J. A. (2003). *What the Best CEOs Know*. New York: McGraw-Hill.
Morita, A. (1987). *Made in Japan*. London: Collins.
Rogers, E., & Larsen, J. (1984). *Silicon Valley Fever: Growth of High-Technology Culture*. New York: Basic Books.
Salerno, L. (1980). Creativity by the numbers: An interview with Robert N. Noyce. *Harvard Business Review, 58*(3), 122–132.
Shavinina, L. V. (1995). The personality trait approach in the psychology of giftedness. *European Journal for High Ability, 6*(1), 27–37.
Shavinina, L. V. (2003). *The International Handbook on Innovation*. Oxford, UK: Elsevier.
Shavinina, L. V. (2007). Comment l'innovation peut-elle accroitre la performance organisationnelle? In L. Chaput (Ed.), *Modèles Contemporains en Gestion* (pp. 167–197). Le Delta: Presses de l'Université du Québec.
Shavinina, L. V. (2008). Early signs of entrepreneurial giftedness. *Gifted and Talented International, 23*(2), 3–17.
Shavinina, L. V. (2009). A unique type of representation is the essence of giftedness: Towards a cognitive-developmental theory. In L. V. Shavinina (Ed.), *International Handbook on Giftedness* (pp. 231–257). Dordrecht, Netherlands: Springer Science.
Shavinina, L. V. (2011). Discovering a unique talent: On the nature of individual innovation leadership. *Talent Development and Excellence, 3*(2), 165–185.
Shavinina, L. V. (2012a). How to develop innovators? Innovation education for the gifted. *Gifted Education International, 28*(3), 1–15.
Shavinina, L. V. (2012b). The impact of the Apollo Project on creativity and innovation management at Sony: The implications for project management. *Journal of Global Business Administration, 4*(2), 85–93.
Simonton, D. K. (1999). Significant samples: The psychological study of eminent individuals. *Psychological Methods, 4*(4), 425–451.

37

The role of entrepreneurs' career solidarity toward innovation

An irreplaceable relationship in career capital pyramid

Masaru Yamashita and Jin-ichiro Yamada

AOYAMA GAKUIN UNIVERSITY, JAPAN, AND OSAKA CITY UNIVERSITY, JAPAN

Summary: Recent studies on entrepreneurial activities and innovation have paid more attention to entrepreneurs as a team rather than the role of a single outstanding individual. This chapter addresses the importance of a relational aspect, which we term "career solidarity", observed not only in the innovation process but also in the development process of high performance teams. Career solidarity requires the accumulation of career capital, which is composed of the assets of Knowing-whom, Knowing-why, and Knowing-how. By applying the lens of career solidarity, a best practice case from the Japanese film industry is explained. We propose that this irreplaceable relationship strongly influences entrepreneurial creativity through career development and the establishment of innovation dynamics.

Key words: career solidarity, irreplaceable relationship, career capital pyramid, knowing-whom, knowing-why, knowing-how, entrepreneur, innovation, network.

Introduction

This chapter focuses on the fact that, in recent accounts, entrepreneurship is more successful following team work rather than individual work. This concept, called "new combinations of already existing elements", was coined by Schumpeter (1971), and can also be applied in an entrepreneurship role. If the elements were easily available via simply purchasing, an entrepreneur would be able to achieve his/her plans alone. However, when the elements are not easily purchased (e.g., specialized knowledge or skills), the entrepreneur must create new combinations through new partnerships. This is the reason working as an (entrepreneurial) team is required for entrepreneurship.

After acquiring the necessary resources (elements), which beforehand were very difficult to acquire before but are now easily accessible, what will the team do? Will they dissolve their entrepreneurial team, seeking to become individual entrepreneurs? Is easier access to resources merely an advantage for entrepreneurial teams?

In this chapter, we advocate that truly successful and significant entrepreneurial teams have not only access to resources but, so to speak, their own irreplaceability, making their replacement by other combinations impossible. Thus, instead of each newly combined element being modulated, each element, once newly combined, has an integrated value and knowledge that no longer can be degraded. In this chapter, we call these teams with integrated nature "career solidarity" (not the modular combined ones).

In this chapter we will, first, introduce and organize previous studies, discussing anew the superiority of entrepreneurial teams. Our focus here is on how entrepreneurial teams with career solidarity generate innovative ideas through their force as a team, not by the intentions (ideas) of individual entrepreneurs leading to successful results. Second, to consider how such career solidarity actually emerges, we use the long-term perspective method of the entrepreneur's career. Generally, entrepreneurs are considered to go through their careers boundaryless. As for the characteristics of workers in the creative industries, they tend to form their teams throughout their careers by depending on the individual's skills. Third, as a result of integrating previous studies, we present the framework of our analysis called the Career Capital Pyramid Model (CCPM). This CCPM can comprehensively explain individual entrepreneurs going through boundaryless careers, devoted employees climbing up the career ladder within a single company, and individuals forming their career solidarities. Fourth, we explain the integrated nature and the formation process of career solidarity by using the CCPM framework. Fifth, we describe, using the founder of a successful film production company in Japan as an example, the concept of CCPM and career solidarity to gain an in-depth understanding of the founder. In conclusion, we offer suggestions on the implications for future entrepreneurship using our frameworks and the concepts explained herein.

Entrepreneurial teams and their formation

One of the great myths of entrepreneurship has been the lone hero type, against a backdrop of economic, social, and other environmental forces prior to their achievement of innovation. However, recent entrepreneurial activity studies have paid more attention to entrepreneurial teams rather than a single outstanding individual (Ruef, Aldrich, & Carter, 2003). There is a good body of empirical studies on the relationship between the scale of entrepreneurial teams and corporate achievements, revealing that businesses established by teams are more successful than those started-up by individuals (Chandler, Honig, & Wiklund, 2005). There is also a study that attempts to define entrepreneurial activities in terms of whether or not some sort of organization has been newly created (Gartner, 1988), further evidence of the importance of team formation.

Although earlier studies clarified that, as innovation leaders, entrepreneurs need a team or partnership, they did not look to the process between the formation of a partnership and the result. In particular, they paid little attention to the view that innovation is induced by diverse interaction processes between several individuals rather than the strong leadership of a single entrepreneur seizing a market opportunity and starting a business. As for the social and relational view of entrepreneur team formation, it characterizes the phenomenon as a result of interpersonal interaction in the business creation process and innovation. This underlying theory is "similarity-attraction", which constitutes stable and consistent results of entrepreneurial efforts (Byrne, 1971; Jones, 1996; Morry, 2005).

Each individual obtains the resources and knowledge as the composite elements for innovation while the individual plays their career's roles. Recently, the Architectural Approach has been mentioned to explain such connections between people and the relationships between organizations and individuals.[1] Various resources required for innovation are obtained one-by-one modularly and integrated later by entrepreneurs (Galunic & Eisenhardt, 2001). Meanwhile, preceding resource integration, studies have only pointed out the importance of the entrepreneurial teams' configuration; a key cog in the combinations for innovations, such as the importance of obtaining knowledge and information in the field of new business and the significance of having personal connections

(Shane, 2004). However, little attention is paid to the fact that connections between people enable them to work together like management resource modules are not easily formed. Before continuing with the intention for entrepreneurial activities and the formation of business plans, we need to further discuss how an individual's career journey leads to the start of a new business and ends up forming a new entrepreneurial team.

Entrepreneurs' career and network

Since Bygrave (1989) and Gartner (1988, 1999) stressed the need to focus on the process itself to select the various individual's independence and the start of their own businesses, Shaver and Scott (1991), Dyer (1994), and Shaver (2003) have instigated refinements of models concerning entrepreneurial career formation. Just as an entrepreneurial team cannot be organized overnight, individuals must be considered from the perspective of their mid- to long-term vocational careers. However, entrepreneurial activities have rarely been stressed in the usual career researches. Career theory initially relied on developmental psychology's stage theory, therefore assuming that individuals formed a career in stages within a company.

However, with the transformation of social conditions around us, individuals forming a career agnostic in a particular company are actually growing in numbers worldwide (Moralee, 1998; Pink, 2001). What we seek now is to determine the circumstances of the network the individual entrepreneur's progress their careers whilst proactively making decisions.

Due to the impact of recent network theory, as an alternative to the internal corporate career, the debate regarding the boundaryless career has developed to capture a career tailored to an era of change. These studies are based on new concepts, such as the boundaryless career (Arthur, 1994; Arthur & Rousseau, 1996) and the protean career (Hall, 1976; Hall & Mirvis, 1996), and attempt to explain the reality of careers not limited within a particular company. With the trend for such studies, those individuals who choose independence by starting new businesses are proof of the existence of career agnostics within a company and, therefore, are finally open as a topic for discussion.

However, the problem of evaluation is even more serious in boundaryless careers since they are unable to go through an evaluation system within a single corporate entity. In such situations, the human network, which is project based, has become an alternative to the evaluation system for companies. Jones (1996), whose subject of study was the boundaryless career of filmmakers, identified four career stages common to successful filmmakers: (1) the career beginning, (2) construction, (3) direction, and (4) maintenance. In her study, she discovered more talented filmmakers had greater densities of the network throughout the career stages. While the human network here serves to evaluate the capability for those filmmakers, it also provides the next work (project) commensurate with the evaluation by sharing the information among the network. Alternatively, those individuals who want a good evaluation learn the rules of the special systems within their industry. It must be essential for them to present their own career capitals in preparation for the opportunities of projects they specifically want to be involved in (Jones, 2002). Career capital is described in more detail in the next section. Through such mechanisms, their reputation is born in the industries and communities beyond companies (the individual's career capital is evaluated) and a creative career gradually forms (DeFillippi & Arthur, 2002).

Similarly, entrepreneurs are considered to form their teams via a process that obtains important career capitals during these career stages. Alvarez and Svejenova (2005) presented the "united career" concept to explain the process of formation, development, and dispersal, regarding the interpersonal relationships of an organization's top managers. They examined partnerships, such as Whitehead and Weinberg of Goldman Sachs, or Ibuka and Morita, founders of Sony, indicating the utility of considering such partnerships as a single career block instead of each individual's career

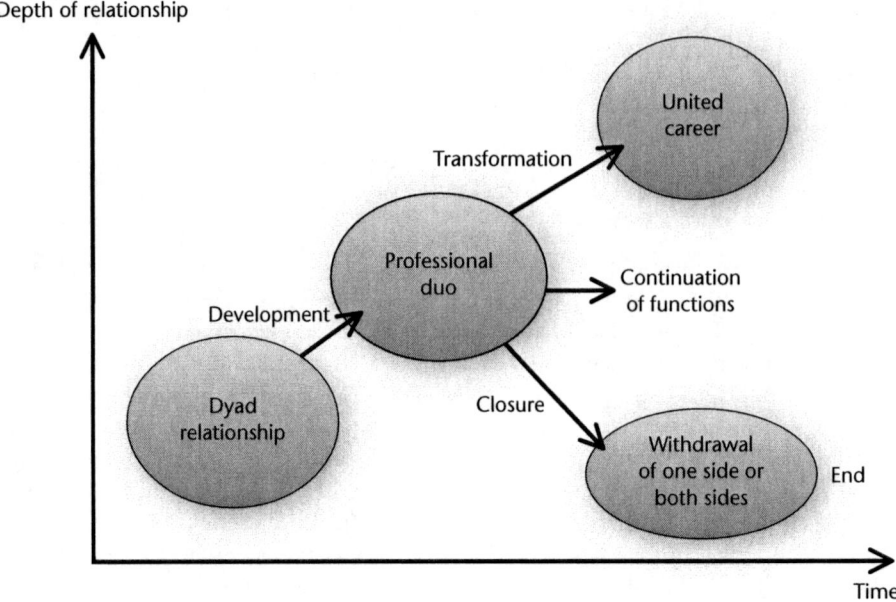

Figure 37.1 Formation process of the united career, diagrammatizing Alvarez and Svejenova (2005)'s relationship types (source: Yamashita & Yamada, 2010).

as one unit. Mere "dyad relationships" of professional duties develop into "professional duo" relationships as individuals collaborate many times. "Professional duo" indicates the combination of moderately high degrees of integration in a business relationship between two parties. Dependencies between them enable complementary strengths through their organic links (Zaleznik, 1965). Alvarez and Svejenova (2005) propose three possible relationship types: keeping the function constant, ending in withdrawal, or development into a "united career".

A stage model of partnerships is important when explaining an entrepreneurial team's formation process. However, Alvarez and Svejenova (2005) only provide suggestive ideas, not a concrete directions, about what individuals need to obtain to promote themselves to the next stage of entrepreneurial relationship development.[2] Thus, the concept of career capital is also necessary in building the integrated framework. Furthermore, studies suggest there are two types of entrepreneurial teams; teams easily explained in the development of stages by the clear division of their labors as shown in the Architectural Approach about product development and human resources (teams that are integrated human resource modules of management resources), and teams that are not sufficiently explained by the first type. Alvarez and Svejenova (2005) only discussed the latter type. The formation patterns of entrepreneurial careers and the nature of each team in the integrated manners have rarely been explained.

Career capital pyramid model and three patterns of career formation

Here, we present the career capital pyramid model and three patterns of career formation as integrated frameworks explaining entrepreneurs and their teams' careers. The career capital pyramid model integrates the concept of career capital advocated by Arthur (1994), and stage models Jones (1996), and Alvarez and Svejenova (2005), created. The three patterns of career formation are classified by considering a new concept called "irreplaceability".

Career capital pyramid model

Career capital[3] can be roughly classified into three categories; Knowing-whom, Knowing-why, and Knowing-how (Arthur, 1994; Arthur, Inkson, & Pringle, 1999; Arthur, DeFillippi, & Jones, 2001). Knowing-whom recognizes who one has connections to; literally personal connections and the human network. Knowing-why recognizes what an individual wants to do with his/her career, including finding value and identity. Knowing-how recognizes how or what one should do to become skillful, by finding a way to improve their skills and techniques.[4] Arthur *et al.* (1999) never specifically addressed these three career capitals in hierarchical systems, but we can better interpret them as a hierarchical framework by integrating the career stage model of Jones (1996) and Alvarez and Svejenova (2005).

In Jones' model (1996), building the human network (Knowing-whom) is a challenge for creative filmmakers early in their careers (career stages 1 and 2; beginning and construction). Only those who accomplish this challenge reach the next stage (direction) where they face another challenge: building their identities and values (Knowing-why). Similarly, only successful challengers proceed to the final stage (maintenance) confronting the last challenge, which is obtaining the knowledge and skills to realize their values. In the model by Alvarez and Svejenova (2005), two parties merely have a dyad relationship, like working together, during the first stages following formation of entrepreneurial teams. However, upon obtaining the Knowing-whom, their relationship transforms into the professional duo; a cooperative relationship within their organic links. If they agree upon Knowing-why, this professional duo relationship develops into a "united career" and their unique identities are shared.[5]

In the career capital pyramid model described above, individuals obtain Knowing-whom, then Knowing-why, and finally Knowing-how. According to this model, individuals cannot develop and obtain Knowing-why immediately alone. First they must obtain Knowing-whom, and then, through human networks, they can obtain Knowing-why. Likewise, Knowing-why is necessary knowledge for the actualization of their values (Knowing-how).

Three patterns of career formation

The career capital pyramid model explains not only the formation process of entrepreneurial teams, but also the career development process for individual entrepreneurs. Career formation can roughly

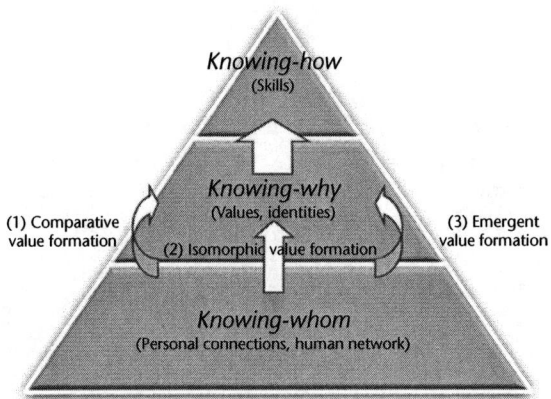

Figure 37.2 Career capital pyramid model (source: Yamashita & Yamada, 2010).

be classified into three patterns: (1) comparative career formation, (2) isomorphic career formation, and (3) emergent career formation.

The reason career formation is classified into three patterns is due to differences in how individuals obtain Knowing-why. People in comparative career formation tend to create their own unique values through a strong awareness of the value differences in their human network, built in the Knowing-whom stage. People in isomorphic career formation tend to obtain Knowing-why by accepting values already shared in the human network. People in emergent career formation tend to bring out their own unique values spontaneously during communications, especially with significant members of their human network.

Furthermore, these three career formation patterns can be defined by the concept of the human network's "irreplaceability". Irreplaceability indicates the degree of impossibility when obtaining the same Knowing-why from different persons in the human network.[6] Since the human network is only an objective to compare differences of values in comparative career formation, there is not much difference in the Knowing-why, which is the final destination of whomever this happens to be. From this point of view, irreplaceability is low in comparative career formation. In isomorphic career formation, people belong to a particular community because they accept the values of their human network. However, the network's members needn't be a specific person as long as they are in the same community; therefore, irreplaceability is considered moderate. In emergent career formation, identical Knowing-why cannot be obtained without a specific person (since they create their own values during communications with specific people). Irreplaceability for emergent career formation is extremely high.

Comparative career formation

It is considered that many individuals who aspire to boundaryless careers build comparative career formations (Arthur, 1994; Arthur & Rousseau, 1996). They do not accept a particular organization's values, instead forming unique values while traversing a variety of organizations. Conversely, because they do not depend on any specific human network, they must receive resources from various people to realize their own values. Therefore, the skills that they need for obtaining Knowing-how are often generic skills that facilitate communications with various people. Teams tend to be modular with low irreplaceability when forming the team; therefore there is always the possibility of forming a new team. The entrepreneurs or entrepreneurial teams that generally and typically come to mind are in this type of career formation. They are usually less constrained by the hierarchy of career capitals since their irreplaceability is low.

Isomorphic career formation

This explains career progression by accepting the values of particular organizations and is similar to the "legitimate peripheral participation" approach (Lave & Wenger, 1991). The Knowing-why obtained is usually traditional, sometimes with little transformation; therefore, newness with regards to individual's own unique values is scarce. The curriculum for "Knowing-how" is well maintained for efficient dissemination within the organization; therefore, people in this formation obtain Knowing-how relatively easily. However, the skills and knowledge are less generic; they have specialized to realize the unique organization's values. Therefore, one tends to remain within and pursue the isomorphic career formation once begun.

Table 37.1 Three patterns of career formation

	Comparative career formation	Isomorphic career formation	Emergent career formation
Knowing-how	Generic skills	Special skills	Unique skills
Knowing-why	(Newness)	(Newness)	(Newness)
	High	Low	High
	(Source)	(Source)	(Source)
	Newly developed in comparison among people communicated with	Socialized from organizations where people communicated with belong to	Cooperatively developed by emerging from people communicated with
	(Irreplaceability)	(Irreplaceability)	(Irreplaceability)
	Low	Moderate	Extremely high
Knowing-whom	Extensive acquaintances beyong organization's boundaries	Colleagues of reference group within organization	Significant others within organization

Source: Yamashita & Yamada, 2010.

Emergent career formation

The "united career" (Alvarez & Svejenova, 2005) is considered an emergent career formation consisting of a dyad relationship with extremely high irreplaceability. The Knowing-why here is fostered and shared between significant others in the dyad relationship. Resulting from collaborative work, their values spontaneously emerged and form a relatively clear step towards realization of "Knowing-how". However, the team's specialized Knowing-how is particular. In order for those in such a small team to obtain the Knowing-how, they require reasonable assessments from the people surrounding them.

These three career formation patterns have very different natures, but are not necessarily exclusive. Whether pursuing comparative career formation or emergent career formation, some degree of nature of isomorphic career formation will exist as long as the individuals belong to a particular organization. It is, however, unlikely that an individual will have all three aspects equally, most are committed to just one.

Isomorphic career formation seems to have fewer links to entrepreneurs as far as their careers are concerned since people in isomorphic career formation must inherit existing values. In other words, people in this formation are not suited to entrepreneurship. Comparative and emergent career formation, which present their own unique values, are considered more appropriate when explaining the careers of entrepreneurs or entrepreneurial teams. Thus pursuit of these two career formations may lead to entrepreneurship.

Career solidarity generated from irreplaceable relationships

In this chapter, we assume that results produced by entrepreneurial teams are more successful than those from individual entrepreneurs. As explained above, entrepreneurs' careers are classified into two formations: comparative and emergent. While the former is suited to explaining the success of individual entrepreneurs who establish intrinsic values (Knowing-why) which lead them to innovation alone, the latter is more suited to explaining the successful results of entrepreneurial teams which entwine the values of specific people of high irreplaceability. However, we would like to say that our interest leans more towards emergent career formation in line with the assumptions mentioned above.

Here, we shall think again about irreplaceability, an essential feature of emergent career formation. To examine and recapture irreplaceability's importance we must delve into the relationships of social structures behind organizations' divisional structures (roles and explicit contractual relationships). In the field of new knowledge and value creation, the difficult to substitute social relationships are the source, making it difficult to imitate while producing results of high originality and creativity. The irreplaceability within this kind of dyad relationship results in major differentiation compared to other common combinations. It may occasionally be constrained by convention but in many cases leads to a competitive advantage.

In terms of business administration studies, when we think about the relationship for the division of labor, labor division is usually considered in terms of improving productivity through the specialization of tasks and development of personal skills. This is a result of designing to producing greater quantities of better quality products with the same effort. That is, it has been assumed that the division of labor only brings the specialization of occupational abilities and power (Fayol, 1916). This comes from the viewpoint that people are replaceable resources like modifiable modules.

To discuss the relationship between entrepreneurs on interdependent terms with high irreplaceability (occasionally seen in the past among successful entrepreneurial teams) Methodological Relationalism is espoused, instead of the view of organizations based on traditional economic division of labor. Using this, we would like to redefine the phenomenon of career development and deployment for creative individuals (like entrepreneurs) who rely on a highly irreplaceable relationship, raising anew the concept of career solidarity.

Career solidarity as addressed here is the concept of a united career developed by Alvarez and Svejenova (2005). They never discussed irreplaceability or Methodological Relationalism, however they focused on the size of successful innovation results produced by not-easily substitutable dyad relationships. However, they never systematically described how the united career differed from other perspectives on careers, nor discussed in detail what kinds of elements develop dyad relationships into the next stage. Therefore, the united career was difficult to use as a framework for analyzing individuals' careers. Based on the discussions so far, we would like to redefine the concept of career solidarity as follows in this chapter: career solidarity is the foundation for a social relation that is difficult to substitute and whose actors who are significant to each other can co-create and share their values, forming their creative careers. The development process of career solidarity is organized and shown in Figure 37.3.

As Alvarez and Svejenova (2005) and the framework of the career capital pyramid show, in the level 1 dyad relationship (the dyad relationship of the occupations), the actors have their roles linked via personal connections. This means that level 1 is the stage described by "whom do you know (Knowing-whom) not what do you know" and it is important to recognize significant others. After opportunities that lead the actors to recognize significant others for their career capital development, the dyad relationship proceeds to level 2, the foundation for the professional duo relation, during certain projects or organizational situations. Here, some actors engage in the search for their occupational identities and values at a deeper level with these significant others. Career solidarity is established here if the parties form emergent values (Knowing-why) together, enabling the dyad relationship to reach level 3 (unmatured career solidarity). The united career's development process as explained by Alvarez and Svejenova (2005) finishes at this point. However, career solidarity matures as the parties obtain further individual unique skills and practical experience, progressing to level 4. This further stage makes it possible for them to survive creatively in their careers (matured careers solidarity).

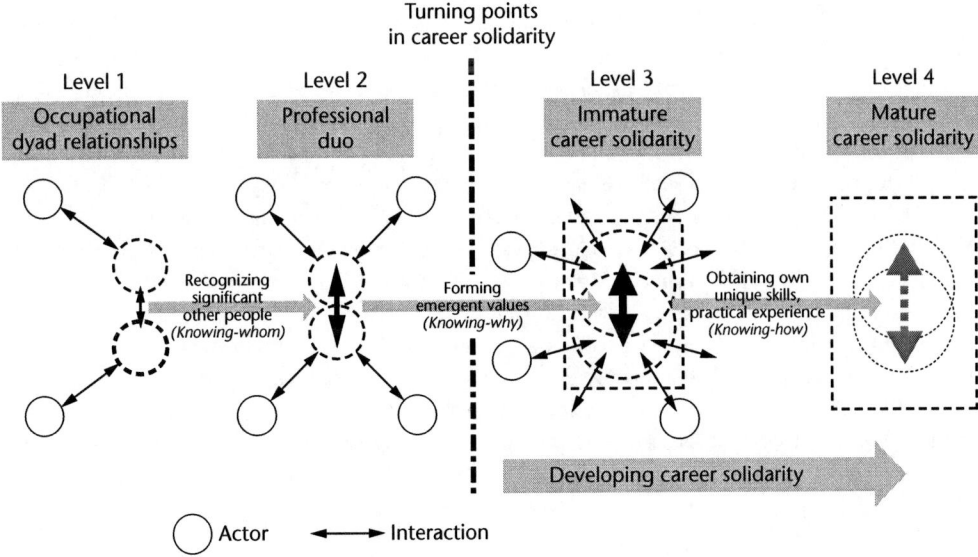

Figure 37.3 A new analytical framework: the formation process of career solidarity (source: Yamashita & Yamada, 2010).

Case study

Here, as a career solidarity example, we introduce a successful team led by Shoji Masui, who established a small film company in Japan. First, we briefly describe Masui's career, then add an explanation of his case within the career capital pyramid framework and the concept of career solidarity (emergent career formation).

The career of Shoji Masui

Masui started working for a small film company, Daiei, in the 1980s following graduation from university. While following the company's policies, he accumulated experience as a producer of adult films and TV drama productions and gained opportunities to meet many people. Among these, he built a strong rapport with the director, Masayuki Suo. When Suo wrote a comedy movie screenplay about a temple priest, Masui asked the company for a production budget of over 100 million yen to produce it. Daiei agreed since it was a small budget. The movie's quality was good but was not widely distributed and therefore didn't gain any popular recognition. Next, Masui produced another small budget movie with Itsumichi Isomura, Suo's protégé. Thus, Masui gained experience producing general movies for Daiei. In 1992, Masui produced a comical movie about a Japanese university's sumo club, with screenplay by Suo. This movie was distributed by a leading film distributor, Toho, thus reaching a much larger audience enabling recognition of Suo's talent such that he won film awards in Japan that year.

Masui realized the limitations of his team's production activities arose from belonging to Daiei. He thus left Daiei in 1993 and established a production company called Altamira Pictures with Suo and Isomura. The new company's first screenplay, written by Suo, was titled "Shall we dance?" As Altamira Pictures had insufficient funds at that time, Daiei contributed the production funding. It was estimated that 500 million yen was necessary based on past experience (a standard amount for production expenses on a movie published at the national level in Japan), but Daiei paid only 400 million yen for the production budget and 30 million yen for the project fees. To cover costs, Masui asked Suo for minor changes to the script, attempting to maintain the production within budget. This movie, again distributed by Toho, enabled Masui to appoint the advertising salesperson he had worked with previously, establishing a respectful relationship. Therefore, he expected the promotional work to follow Altamira Pictures' viewpoint. This was important as Suo intended "Shall we dance?" to be a romance film while Daiei and Toho were considering promoting it as a comedy. The advertising salesperson promoted the movie as a romance film in response to Masui's expectation despite pressure from Toho. As a result, many middle-aged audiences, not usual theater-goers, attended and "Shall we dance?" became a blockbuster in Japan in 1996.

However, the production budget of 400 million yen was not insufficient resulting in the project fees of 30 million yen (saved as working capital for the company) being spent. While the movie became a big hit, with Daiei making large profits, Altamira Pictures stayed in the red. Masui had met with Fuji Television broadcasting station's representative when looking for an equity partner. Due to the relationship of respect between Altamira Pictures and this representative, Fuji Television agreed to collaborate with Altamira Pictures as an equity partner. Since then, a number of films have been made using this collaborative system, enabling Masui to successfully bring not only Suo, but also Isomura, to the public. Altamira Pictures is now a leading Japanese independent film production company. However, they still making movies only with close peers.

Description of the case

The occupational career of Masui began at Daiei. He can be said to have pursued an isomorphic career because, despite not knowing the basics of film making, he initiated a network in the production field, taking advantage of the opportunity when producing an adult film for the company. He was able to obtain a lot of knowledge and experience there. This illustrates that people choose isomorphic career formation not because of their originality, but as a result of accepting that existing ideas and knowledge are necessary at the starting point of their occupational careers. In fact, many entrepreneurs in Japan start their careers as an employee of a large company (Takahashi, 2005).

The meeting of Masui and Suo, a significant other in the community of producing adult films, transformed his career into an emergent career formation. Their relationship was bound through understanding that they would perform better in collaboration as opposed to working with a variety of others. This stage signifies that they had already obtained the career capital of Knowing-whom and their career solidarity had reached level 2.

It is not clear in this example how they obtained Knowing-why, that is, how they created their own values. However, although Masui left Daiei, becoming independent, and managed his own team during the promotion of "Shall we dance?" he tried to maintain Masui and Suo's values despite conflicts with sponsors and distributors. It may be meaningless to pursue the Knowing-why since it differs between people. What is important here is whether the specific team members created shared values, to be protected throughout their careers. It may be assumed that the establishment of Altamira Pictures or production of "Shall we dance?" indicates obtaining their Knowing-why and a career solidarity of level 3.

If only the unique values of the team are established it will not be successful. It is necessary to find the means (Knowing-how) to actualize the values. By appointing a familiar advertising salesperson, they attempted to protect their work, possibly indicating Knowing-how. However, Altamira Pictures went over budget, a long-term issue, as a result of trying to enhance the quality of their work (protect their values). This led to financial struggle and jeopardized their values. Ultimately a solution was found via a collaborating equity business partner who was legally contracted. This partner understood the values of their team; the most important Knowing-how they obtained. Generally, while Knowing-whom (obtained in level 2) and Knowing-why (level 3) are built inside a team, Knowing-how (level 4) seems to be related to influences from outside the team. Masui and his team reached level 4 by obtaining an external partner which enabled them to realize their values while continuing their career solidarity.

Conclusion and implications

With the assumption that entrepreneurs receive more successful innovation results when acting in a team, we have introduced and discussed a careers perspective that has not been discussed significantly until now to foster such entrepreneurs. We presented the concept of the career capital pyramid as a framework accounting for the careers of creative individuals, including entrepreneurs, as well as the concept of career solidarity or a strong team in the career capital pyramid that leads irreplaceable dyad relationships to create their own unique values. Based on these discussions, we would like to make suggestions regarding entrepreneurship.

First, according to the career capital pyramid framework, individuals who work creatively as entrepreneurs need to pursue either comparative or emergent career formation. However, since collaborative experience is necessary for both formations and obtaining at least a basic knowledge is required in any business, they must still undergo a preparatory training period. As shown in the case study, isomorphic career formation through working for a large company during early career periods might be a typical preparatory training method. A pitfall for individuals who wish to become individual entrepreneurs may be the acquisition of superficial occupational skills too quickly, while underplaying the experience of isomorphic career formation in these early career periods. This kind of attitude neglects personal connections within the communities and companies they belong to, inhibiting acquisition of

Knowing-whom, the career capital foundation. As a result, they cannot receive the values of their communities, let alone create their own unique values.

In contrast, individuals less aware of being individual entrepreneurs tend to drag out the isomorphic career formation, losing the chance to exercise creative activities. Prolonged isomorphic career formation makes them experts in the communities they belong to, enabling them access to power and influence but they are not able to challenge anything new. The first challenge for the career of an entrepreneur is not merely to form a wide human network in the community/company one belongs to, but to seek the Knowing-whom from outside the community/company, or to only seek significant others by narrowing the search range.

Next, in this chapter we asserted that forming a team is more advantageous for an entrepreneur than acting alone. However, here we would like to add that emergent career formation is more desirable than comparative career formation (so to speak, building career solidarity is desirable), if he/she wants to form an entrepreneurial team and expect to be very successful. What he/she has to be careful of here when building career solidarity is to decide first "whom to do it with" (Knowing-whom), not "what to do" (Knowing-why). "What to do" will be spontaneously produced later, during collaboration with the partner. However, it does not mean that all they need to be partners is closeness to each other. The framework of career solidarity shows that the dyad relationship in level 2 already turns into the professional duo, and the professional duo, because their perspectives are very similar, assumes that they work together much more smoothly than in pairs with others. Whether being able to find a partner who satisfies this assumption is the big challenge when an entrepreneur builds his/her career solidarity.

On the other hand, individual entrepreneurs who pursue comparative career formation usually decide "what to do" earlier than "whom to do it with". For this type of entrepreneur, "who to do it with" is not regarded as Knowing-whom but as Knowing-how. Their own unique values have already been established and, therefore, the only reason they need a team is to actualize their values. In this case, anybody can be their team member as long as they have the knowledge and resources required, in other words, the members are sufficiently replaceable. The typical image of entrepreneurs or entrepreneurial teams we envision is most likely this type, but the future of this type of entrepreneurial team often seems difficult. Individual entrepreneurs give each team member a specific role and expect them to perform it well; however, they often come to speak and act beyond their original roles since there is no clear job description for each member. Such attitudes sometimes urge modifications or changes to the values themselves. Since the individual entrepreneurs who pursue comparative career formations often adhere too stringently to values determined by them, they then tend to disband their teams and switch partners. Why would they do such a thing? It is because their teams are replaceable. However, it is very rare that an individual's intention leads to innovation without being affected by the surroundings (Yamada & Yamashita, 2006). Such events are possible only by a handful of geniuses and it is not realistic to assume any team can mimic this. Therefore, individuals who have pursued comparative career formations need to respect any new values that emerge naturally in their teams without obsessing over their own values. This requires them to stop regarding their partners as replaceable and start pursuing an emergent career formation, in other words, their career stance needs to transform to one of career solidarity.

This chapter introduced a perspective with career solidarity and the career capital pyramid model as a new direction in entrepreneurs' career development. Its structure consists of three routes to accumulate career capital as discussed above. This chapter highlighted that career solidarity is highly related to innovation and creativity in entrepreneurs' success. If innovation educators are concerned with quality and impact of entrepreneurs' teams, this new perspective based on entrepreneurs' career solidarity and career capital pyramid model will help to examine it and distinguish from the extant dominant view of the modular-combination type of entrepreneurs' team.

Notes

1 Lepak and Snell (1999) advocated the theory of human resource architecture in order for people to confront the idea that all personnel (human resources) are not necessarily strategically important. This theory takes a skeptical stance against many studies that have highlighted the value all general employees give to the competitive advantage of their companies.
2 Alvarez and Svejenova (2002) discussed the relationships between entrepreneurs' creative visions and the traditional isomorphic pressure. As a measure of avoiding institutional pressure from the environment, individual artists create symbiotic careers (partnerships) with their protective stakeholders, e.g., producers, establishing their own companies. Through these actions, they create a nested structure that corresponds to the pressure and draws the audience and critics to their side. Finally, they are able to exert their creativity in the context of wide industrial communities, achieving innovation.
3 Lave and Wenger (1991) explains their theory of legitimate peripheral participation as being similar to the hierarchy of career capital; the values (Knowing-why) are accumulated in the human connections (Knowing-whom), and next will be the acquisition of the skills and techniques (Knowing-how). However, each element in the theory has never been considered separately. The theory has never paid attention to the flow of the process for the acquisition of the values (Knowing-why) nor to the creation and practice of the new innovative know-how. Therefore, the theory differs from our point of view in terms of the patterns of entrepreneurial career formations that we explain in this chapter.
4 Skills in this context include the intellectual proficiency mentioned by Koike (1988). Intellectual proficiency formation indicates development of the ability to deal with changes and abnormalities. This ability is cultivated by the internal promotion and long-term selection found in Japanese long-term employment.
5 Alvarez and Svejenova (2005) (Figure 37.1) stopped at the point where the relationship of a united career could end. However, as Jones (2002) mentions, there should be another stage sustaining the relationship. Knowing-how must be sought after this stage.
6 Relationships with high irreplaceability can be regarded as the source for a deeper level by delving into the relationships of social structures behind the organization's divisional structures (roles and explicit contractual relationships). In the field of innovation, the concrete social relationship is the source thus making imitation difficult while producing highly original and creative results. Module-like roles and knowledge that can easily be procured on the markets make it difficult for dyad relationship to innovate according to the concept, "preciousness is the irreplaceability of individuals".

References

Alvarez, J. L., & Svejenova, S. (2002). Symbiotic careers in movie making: Pedro and Agustín Almodóvar. In M. Peiperl, M. Arthur, R. Goffee & N. Anand (Eds.), *Career Creativity: Explorations in the Remaking of Work* (pp. 183–208). Oxford, UK: Oxford University Press.
Alvarez, J. L., & Svejenova, S. (2005). *Sharing Executive Power: Roles and Relationships at the Top*. Cambridge, UK: Cambridge University Press.
Arthur, M. B. (1994). The boundaryless career: A new perspective for organizational inquiry. *Journal of Organizational Behavior, 15*(4), 295–306.
Arthur, M. B., DeFillippi, R. J., & Jones, C. (2001). Project-based learning as the interplay of career and company non-financial capital. *Management Learning, 32*(1), 99–117.
Arthur, M. B., Inkson, K., & Pringle, J. K. (1999). *The New Careers: Individual Action and Economic Change*, Thousand Oaks, CA: Sage.
Arthur, M. B., & Rousseau D. M. (1996). The boundaryless career as a new employment principle. In M. B. Arthur & D. M. Rousseau (Eds.), *The Boundaryless Career: A New Employment Principle for a New Organizational Era* (pp. 3–22). Oxford, UK: Oxford University Press.
Bygrave, W. D. (1989). The entrepreneurship paradigm (1): A philosophical look at its research methodologies. *Entrepreneurship Theory and Practice, 14*(2), 7–26.
Bygrave, W. D., & Zacharakis, A. (2008). *Entrepreneurship*. New York: John Wiley & Sons.
Byrne, D. (1971). *The Attraction Paradigm*. New York: Academic Press.
Chandler, G. N., Honig, B., & Wiklund, J. (2005). Antecedents, moderators, and performance consequences of membership change in new venture teams. *Journal of Business Venturing, 20*(5), 705–725.
DeFillippi, R., & Arthur, M. (2002). Career creativity to industry influence: A blueprint for the knowledge economy? In M. Peiperl, M. Arthur, R. Goffee & N. Anand (Eds.), *Career Creativity: Explorations in the Remaking of Work* (pp. 298–313). Oxford, UK: Oxford University Press.

Dyer, W. G. Jr (1994). Toward a theory of entrepreneurial careers. *Entrepreneurship Theory and Practice, 19*(2), 7–21.

Fayol, H. (1916). *Administration Industrielle et Generale, Dunod*. Saint-Etiennne, Paris: Siege de la Societe.

Gartner, W. B. (1988). Who is an entrepreneur? Is the wrong question. *American Journal of Small Business, 12*(4), 11–32.

Gartner, W. B. (1989). Some suggestions for research on entrepreneurial traits and characteristics. *Entrepreneurship Theory and Practice, 12*(4), 27–37.

Galunic, D. C., & Eisenhardt, K. M. (2001). Architectural innovation and modular corporate forms. *Academy of Management Journal, 44*(6), 1229–1249.

Hall, D. T. (1976). *Careers in Organizations*. Pactific Palisades, CA: Goodyear Publishing.

Hall, D. T., & Mirvis, P. H. (1996). The new protean career: Psychological success and the path with a heart. In D. T. Hall (Ed.), *The Career is Dead: Long Live the Career: A Relational Approach to Careers* (pp. 15–45). San Francisco, CA: Jossey-Bass.

Jones, C. (1996). Careers in project networks: The case of the film industry. In M. B. Arthur & D. M. Rousseau (Eds.), *The Boundaryless Career: A New Employment Principal For New Organizational Era* (pp. 58–75). Oxford, UK: Oxford University Press.

Jones, C. (2002). Signaling expertise: How signals shape careers in creative industries. In M. Peiperl, M. Arthur, R. Goffee & N. Anand (Eds.), *Career Creativity: Explorations in the Remaking of Work* (pp. 209–228). Oxford, UK: Oxford University Press.

Koike, K. (1988). *Understanding Industrial Relations in Modern Japan*. London: Macmillan.

Lave, J., & Wenger, E. (1991). *Situated Learning: Legitimate Peripheral Participation*. Cambridge, UK: Cambridge University Press.

Lepak, D. P., & Snell, S. A. (1999). The human resource architecture: Toward a theory of human capital allocation and development. *Academy of Management Review, 24*(1), 31–48.

Moralee, L. (1998). Self-employment in the 1990s. *Labour Market Trends, 106*(3), 121–130.

Morry, M. M. (2005). Relationship satisfaction as a predictor of similarity ratings: A test of the attraction–similarity hypothesis. *Journal of Social and Personal Relationships, 22*(4), 561–584.

Pink, D. H. (2001). *Free Agent Nation: The Future of Working for Yourself*. New York: Warner Books.

Ruef, M., Aldrich, H., & Carter, N. (2003). The structure of founding teams: Homophily, strong ties and isolation among US entrepreneurs. *American Sociological Review, 68*(2), 195–222.

Schumpeter, J. (1971). The fundamental phenomenon of economic development. In P. Kilby (Ed.), *Entrepreneurship and Economic Development* (pp. 43–70). New York: The Free Press.

Shane, S. (2004). *Academic Entrepreneurship: University Spinoffs and Wealth Creation*. Cheltenham, UK: Edward Elgar.

Shaver, K. G. (2003). The social psychology of entrepreneurial behaviour. In Z. J. Acs & D. B. Audretsch (Eds.), *The Handbook of Entrepreneurship Research* (pp. 331–357). London: Kluwer Law International.

Shaver, K. G., & Scott, L. R. (1991). Person, process, choice: The psychology of new venture creation. *Entrepreneurship Theory and Practice, 16*(2), 23–42.

Takahashi, N. (2005). Profile of entrepreneurs. In K. Kutsuna & T. Yasuda (Eds.), *Start-up Business in Japan* (pp. 1–25). Tokyo: Hakuto-Shobo (*Kaigyosha no purofiru, Nihon no shinki kaigyo kigyo*, in Japanese).

Yamada, J., & Yamashita, M. (2006). Entrepreneurs' intentions and partnership towards innovation: Evidence from the Japanese film industry. *Creativity and Innovation Management, 15*(3), 258–267.

Yamashita, M., & Yamada, J. (2010). *A Sense of Solidarity that Guides Careers of Producers: Strategic Collaborative Organization of Creative Individuals in the Japanese Film Industry*, Tokyo, Japan: Hakuto Shobo (*Purodyusa no kyaria rentai: Eiga sangyo ni okeru sozoteki kojin no sosikika senryaku*, in Japanese).

Zaleznik, A. (1965). Interpersonal relations in organizations. In J. G. March (Ed.), *Handbook of Organizations* (pp. 574–613). Chicago, IL: Rand McNally.

38

Modeling the firm

Constructing an integrated entrepreneurship course for undergraduate engineers

Pius Baschera, Fredrik Hacklin, Georg von Krogh, and Boris Battistini

ETH ZURICH, SWITZERLAND

Summary: In traditional engineering curricula, management education typically takes place on complementary, sporadic, and unintegrated teaching "islands." However, as the reality of industry becomes increasingly complex and boundary spanning, today's engineers are required to take on integrative and interdisciplinary coordination and management roles early on in their career. We argue that contemporary teaching programs need to respond to this trend by equipping student engineers with a more holistic and integrated view of managerial concepts, enabling them to act as innovators and entrepreneurs in their field. In this chapter, we start by elaborating on the increasing managerial challenges of innovation and entrepreneurship from the perspective of engineering students. We then present an example of an integrative course, built along the dimensions of a firm's business model, and discuss our experiences of devising and teaching the course at ETH Zurich.

Key words: Entrepreneurship education, integrative and interdisciplinary entrepreneurship course, simulation, innovation.

Introduction

Innovative and entrepreneurial behavior is increasingly needed, not only for starting new businesses, but also for renewing and improving existing ones. While much of society's innovative and entrepreneurial potential is attributed to the engineers of tomorrow, today's student engineers are frequently in the dark when it comes to gaining the skills required. Yet, fostering and encouraging entrepreneurship among students is critical for sustained national competitiveness (e.g., von Krogh, 2011).

In order to respond to this need, universities' engineering curricula have started to offer a growing number of courses in innovation and entrepreneurship. While these are generally perceived as very successful at offering complementary skills to engineers' technical curricula, they often remain isolated initiatives without being properly integrated, either within different entrepreneurial courses, or within the rest of the curriculum. In this chapter, we argue that education in innovation and entrepreneurship needs to be better integrated into the engineering curriculum, rather than presented as standalone courses rolled out toward the end of degree courses. We begin by discussing the increasing managerial challenges of innovation and entrepreneurship from the perspective of engineering students. We then present an example of an integrative course, built along the dimensions of a firm's business model, and discuss our experiences of devising and teaching the course at ETH Zurich.

The rise of entrepreneurial courses

In the development of courses, programs, and curricula for educating university students on innovation, it is crucial to revisit the definition of the term. While scientific discovery is important, innovation is not simply invention; it refers to the entire process from research, through opportunity recognition, to commercial application and business growth (Schumpeter, 1939). Therefore, as a whole, the process of innovation requires the expertise of a variety of professionals—the scientists or engineers responsible for the invention, corporate experts who evaluate commercialization models, and lawyers involved in protecting intellectual property (Thursby, Fuller, & Thursby, 2009). Hence, innovation, by its very nature, represents an interdisciplinary construct that needs to be taken into account in the education of university students. This has been acknowledged by a broad variety of studies on education within the field of technology entrepreneurship. These studies emphasize the need for equipping students with broad and integrated skill-sets related to various areas of innovation (e.g., Kingon, Thomas, Markham, Aiman-Smit, & Debo, 2001; Thursby, 2005; Barr, Baker, Markham, & Kingon, 2009). Many of them focus on graduate-level education (PhD, MBA, and other professional degrees), where the curriculum allows more freedom to span the boundaries between different disciplines, and between traditional coursework and team-based project learning (Thursby et al., 2009). This has sparked a broad wave of entrepreneurship courses in educational institutions, and in fact is not an entirely new phenomenon; the first entrepreneurship course in the USA was given at Harvard Business School by Myles Mace as early as in 1947 (Katz, 2003). Since then, however, entrepreneurship education has evolved, showing high growth rates in terms of numbers of courses offered and faculty hired during the past three decades (see e.g., Katz, 1994, 2003; Hills, 1988).

In reality, not all students have the luxury of applying for specialized degree programs in innovation and entrepreneurship, or of waiting until graduate level to have access to interdisciplinary course offerings. This is particularly the case for engineering students, who, depending on the national education system, need to invest three to four years of their education in gaining the technical in-depth expertise required to master an engineering discipline. At the same time, as successful technological ventures frequently emerge as spin-offs from engineering programs, one can argue that there is an obvious need for motivating engineering students to acquire entrepreneurial skills. As a result, several faculties have recognized the importance of providing entrepreneurship education to engineers, and initiated various educational programs (Wang & Kleppe, 2001). Nevertheless, given the predefined requirements of traditional engineering programs, students have been somewhat constrained when it comes to the amount of time available for complementary and interdisciplinary course offerings. As a result, entrepreneurship-related courses are often given as supplementary or optional rather than compulsory courses. As early as 1987, McMullan and Long (1987) pointed to the emerging challenges for entrepreneurship education in the decade to follow, emphasizing the risk of entrepreneurship being seen strictly as an add-on to education in management or engineering. They suggested that entrepreneurship education should be more strategically organized, enabling it to be embedded in the overall degree program, and a more substantial part of students' learning experience. In line with this view, Haase and Lautenschläger (2007) suggest that entrepreneurship education should go beyond solely teaching knowledge on business creation and focus instead on experiencing entrepreneurship. However, this raises the question of how this experience should be facilitated in a meaningful and actionable way, and how "experiencing entrepreneurship" can be integrated into established programs. Given this background, there is a need for a new educational framework that will allow integration to take place (Thursby et al., 2009).

A framework for an integrative entrepreneurship course

In previous management-related educational programs targeted at engineers, the approach of simulating a "company" has often been a key element in allowing students to experience the dynamics of markets and organizations (e.g., the University of Nevada, Reno, project, discussed by Wang & Kleppe, 2001). Building on this, the faculty at the Department of Management, Technology, and Economics at ETH Zurich decided to launch an integrative course offering, allowing engineering undergraduates to experience the dynamics of innovation and entrepreneurship within the context of a firm. Instead of offering complementary entrepreneurship-oriented courses toward the end of a postgraduate program, the purpose of the "Discovering Entrepreneurship" course is to give them a head start during their undergraduate studies. The course is offered in different undergraduate programs throughout the university, and aims to provide a platform for building up initial entrepreneurial skills; it also attracts students to the Master's program in Management, Technology, and Economics. "Discovering Entrepreneurship" is not about starting new ventures; it's about discovering entrepreneurial opportunities in existing organizations. One major teaching goal of the course is to weaken the prevalent implicit association between entrepreneurship and start-up firms, and to foster awareness of entrepreneurial careers within established corporations.

Unlike one-off, standalone courses in entrepreneurship, the "Discovering Entrepreneurship" course at ETH Zurich has two major integrative characteristics.

First, it is not run by a single teacher, but by a broad mix of faculty drawn from the Department of Management, Technology, and Economics at ETH Zurich. This means there is not only alignment of interests between all department chairs involved, but, much more importantly, students are also introduced to an interdisciplinary view of entrepreneurship, encompassing multiple perspectives from different areas of research. Specifically, 11 members of the faculty are involved, working in technology and innovation management, corporate sustainability, human resource management, operations and supply chain management, quality management and business excellence, financial management and accounting, marketing and sales, risk management, macroeconomics, management information systems, corporate strategy, and corporate renewal.

Second, a simple, yet integrative framework aligns this highly heterogeneous set of subject areas into one course. Course structure follows a corporate business model (Figure 38.1), which is based on an excellence model proposed by the European Foundation for Quality Management (EFQM).[1] The EFQM model represents a framework for an organizational management system, and is traditionally used to increase the competitiveness of organizations.

Although this model is primarily used as a basis for quality programs within an organization, its application to the "Discovering Entrepreneurship" course framework ensured comprehensive and exhaustive coverage of different firm-internal perspectives on entrepreneurship. Each element of the framework within one course module is taught by a professor with expertise in that area; this adds depth to the breadth of the framework. Furthermore, the element of continuous improvement in the EFQM model translates well to the teaching methodology of the course (see Figure 38.1). This is reinforced by establishing cross-references between different lectures, by discussing relationships with other modules and implications for them, and by continually emphasizing the interaction between all elements of the framework.

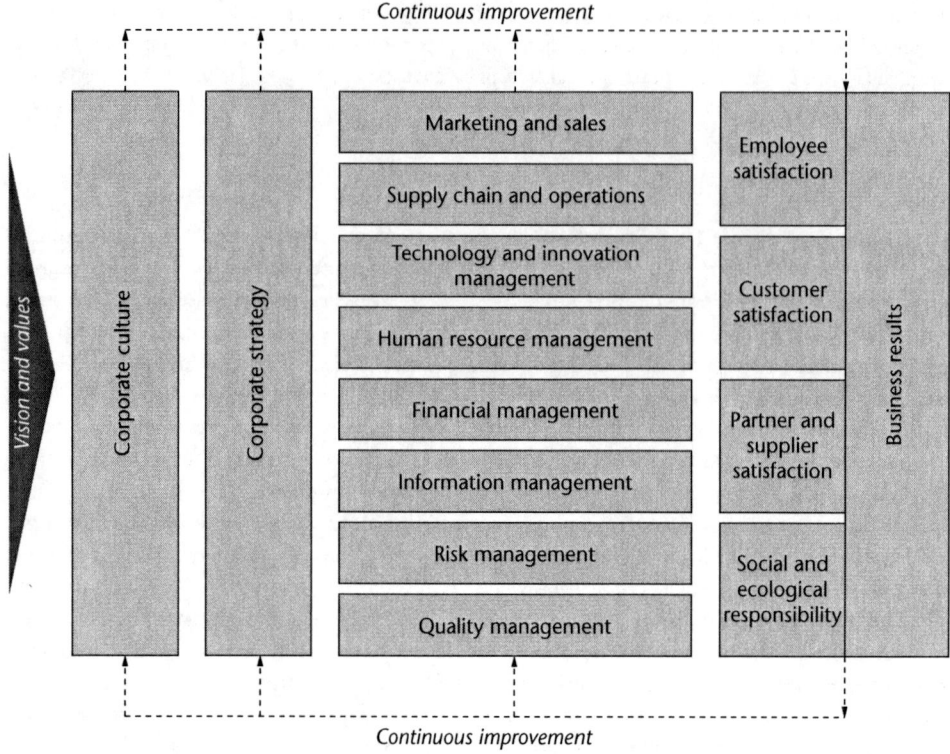

Figure 38.1 The "Discovering Entrepreneurship" course structure.

Experiencing entrepreneurial dynamics

Entrepreneurial environments, within the context of established organizations and markets or elsewhere, are often described as complex agents of change. They call into question well-established certainties, challenge the obvious, and transform ways of working (e.g., Kuratko, Hornsby, Naffziger, & Montagno, 1993). To understand and manage frequent and complex transformations, organizations cannot merely rely upon planning; they need to engage in continuous and systematic learning activities (De Geus, 1988). Traditionally, decision-making under uncertainty has made extensive use of advanced planning and forecasting methodologies, such as scenario planning (Bood & Postma, 1997; Coates, 2000):

> Strategy is the art of making choices—investing both for current and future success. To understand these choices clearly, organizations should identify a business idea and test it in substantially different scenarios. This process can help an organization to develop a business idea that will serve it well as the future evolves.
>
> *(van der Heijden, 2005, p. 351)*

Building on this paradigm, teaching in engineering faculties has usually focused on transferring specific skills to students, for example, planning and forecasting based on learning from examples, cases, and best practices. However, the intensified competitiveness in global markets has stimulated new needs and generated new challenges for corporate decision-makers. In response to new

An integrated entrepreneurship course

demands, organizations are increasingly applying simulation techniques to the development of the skills and competences required to manage and analyze effectively the future business environment (Herman, Forst, & Kurz, 2009; Oriesek & Schwarz, 2008; Schwarz, 2009).

Against this background, we need to determine and develop the skills and tools that will enable engineers to succeed in such environments. This requires more specific teaching techniques to help us develop the entrepreneurial behavior necessary to become an effective team member and leader.

We believe that, if integrated into a holistic teaching framework, business simulations used in close iteration with case studies provide a window into the future of organizations and the management capabilities needed to contribute effectively to value creation in an intensive work environment. This provides a valuable learning platform to integrate the development of skills and competences for entrepreneurial decision-making into existing engineering and scientific curricula.

Business wargaming[2] is the simulation of a dynamic business situation, based on role-play and involving a series of teams, each assigned to assume the identity of an entity with a stake in the scenario. A typical wargame involves several "sessions" representing different time periods—months, quarters, or years. Alternatively, a "session" might represent different phases in the life of a product launch, a plan to make an acquisition, to win a major order, or some other venture (Kurtz, 2003). Figure 38.2 illustrates major characteristics of a business wargame.

There are some set steps in business wargaming. First, to simulate real-life group dynamics, heterogeneous teams are formed from people with different academic and socio-cultural backgrounds. Team members are coached to develop effective cooperative behaviors that will enable them to engage in creative problem-solving and search for successful solutions through mediation. Second, team members have to assign specific tasks and responsibilities within the group, as well as decide how to coordinate their actions and decisions. Finally, the simulation provides accurate data based on the decisions taken. Team members analyze the results, compare them to their own benchmarks, the results of competing teams, and similar cases discussed in class, and make decisions accordingly. In class each week, an invited corporate executive presents a real-life case featuring problems and decisions similar to those that characterize the current episode of the simulation. This systematic feedback loop generates a learning process in which comparison, analysis, and discussion of the results provides the basis for informed decision-making (Kurtz, 2003).

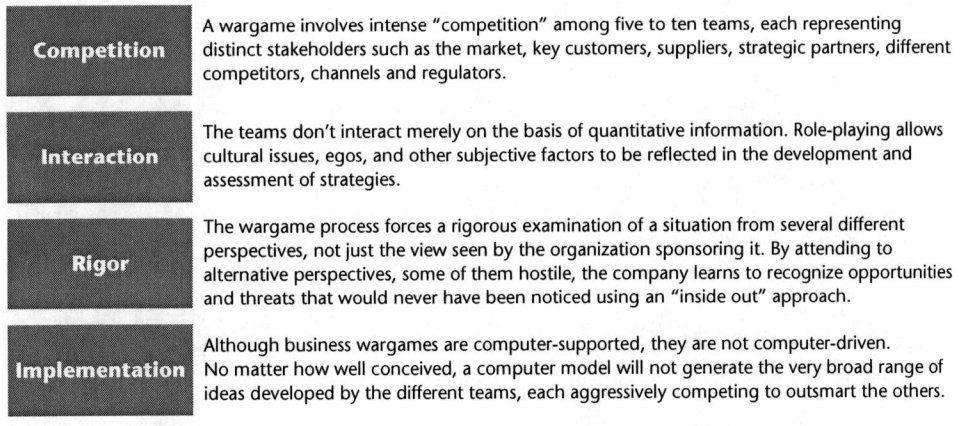

Figure 38.2 Characteristics of a business wargame (source: based on Kurtz, 2003).

Discussion and conclusions

In this chapter, we discuss challenges related to education in innovation and entrepreneurship that, due to the interdisciplinary nature of the problems involved, require more comprehensive and integrative responses in the design of curricula and programs. Many courses in this area are given as standalone, uncoordinated offerings that are not integrated with existing degree programs and cause fragmentation—and confusion—in the educational market. Based on our experiences, we suggest three ways in which educational institutions could increase the efficiency and effectiveness of teaching innovative and entrepreneurial skills. First, existing educational initiatives need to be coordinated and integrated, reducing overlap and redundancy and increasing impact. Second, we urge universities to begin educating an entrepreneurial mindset in the early phases of an undergraduate career, instead of offering elective courses toward the end of a degree program. Third, the interdisciplinary challenges inherent in innovation and entrepreneurship can be addressed by teaching input from a diverse set of faculty members, and a balance between in-class teaching and simulation or case-based learning. We demonstrate how these initiatives are applied in a description of the "Discovering Entrepreneurship" course at ETH Zurich.

Notes

1 See also www.efqm.org.
2 Oriesek and Schwarz (2008) note that "the term 'wargame' is the translation of the German *Kriegsspiel*. [...] In the business environment some discomfort exists with the terms 'war' and 'game' ... [W]argames therefore have also been described as dynamic strategic simulations or simply strategy simulations."

References

Barr, S. H., Baker, T., Markham, S. K., & Kingon, A. I. (2009). Bridging the valley of death: Lessons learned from fourteen years of commercialization of technology education. *Academy of Management Learning and Education.* 8(3), 370–388.
Bood, R., & Postma, T. (1997). Strategic learning with scenarios. *European Management Journal.* 15(6), 633.
Coates, J. F. (2000). Scenario planning. *Technological Forecasting and Social Change.* 65, 115–123.
De Geus, A. (1988). Planning as learning. *Harvard Business Review.* March–April, 70.
Haase, H., & Lautenschläger, A. (2007). The "teachability dilemma" of entrepreneurship. *International Entrepreneurship and Management Journal.* 7(2), 145–162.
Herman, M., Forst, M., & Kurz, R. (2009). *Wargaming for Leaders: Strategic Decision Making from the Battlefield to the Boardroom.* New York: McGraw-Hill.
Hills, G. E. (1988). Variations in university entrepreneurship education: An empirical study of an evolving field. *Journal of Business Venturing.* 3(2), 109–122.
Katz, J. A. (1994). Growth of endowments, chairs, and programs in entrepreneurship on the college campus. In F. Hoy, T. G. Monroy & J. Reichert (Eds.), *The Art and Science of Entrepreneurship Education*, vol. 1 (pp. 127–149), Cleveland, OH: Baldwin-Wallace College.
Katz, J. A. (2003). The chronology and intellectual trajectory of American entrepreneurship education 1876–1999. *Journal of Business Venturing.* 18(2), 283–300.
Kingon, A. I., Thomas, R., Markham, S. K., Aiman-Smith, L., & Debo, R. (2001). An integrated approach to teaching high technology entrepreneurship at the graduate level. *Proceedings of the 2001 American Society for Engineering Education Annual Conference and Exposition*, Albuquerque, NM: American Society for Engineering Education.
Kuratko, D. F., Hornsby, J. S., Naffziger, D. W., & Montagno, R. V. (1993). Implementing entrepreneurial thinking in established organizations. *S.A.M. Advanced Management Journal.* 58(1), 28–39.
Kurtz, J. (2003). "Business wargaming": Simulations guide crucial strategy decisions. *Strategy and Leadership.* 31(6), 12–21.
McMullan, W. E., & Long, W. A. (1987). Entrepreneurship education in the nineties. *Journal of Business Venturing.* 2(3), 261–275.

Oriesek, D. F., & Schwarz, J. O. (2008). *Business Wargaming: Securing Corporate Value*. Aldershot, UK: Gower Publishing.

Schumpeter, J. (1939). *Business Cycles: A Theoretical Historical and Statistical Analysis of the Capitalist Process*. New York: McGraw-Hill.

Schwarz, J. O. (2009). Business wargaming: Developing foresight within a strategic simulation. *Technology Analysis and Strategic Management. 21*(3), 291–305.

Thursby, M. C. (2005). Introducing technology entrepreneurship to graduate education: An integrative approach. In E. D. Libecap (Ed.), *University Entrepreneurship and Technology Transfer: Process, Design, and Intellectual Property* (Advances in the Study of Entrepreneurship, Innovation, and Economic Growth, *16*, pp. 211–240). Bingley, UK: Emerald Group Publishing.

Thursby, M. C., Fuller, A. W., & Thursby, J. (2009). An integrated approach to education professionals for careers in innovation. *Academy of Management Learning and Education. 8*(3), 389–405.

van der Heijden, K. (2005). *Scenarios: The Art of Strategic Conversation*. Chichester, UK: John Wiley & Sons, Ltd.

von Krogh, G. (2011). Der Forschungsstandort Schweiz steht auf dem Spiel ("Switzerland as a research location is at stake"), *NZZ am Sonntag*, November 12, 2011, 50, 19.

Wang, E. L., & Kleppe, J. A. (2001). Teaching invention, innovation, and entrepreneurship in engineering. *Journal of Engineering Education. 90*(4), 565–570.

39
Igniting the spark
Utilization of positive emotions in developing radical innovators

Birgitta Sandberg

UNIVERSITY OF TURKU, FINLAND

Summary: Radical innovators are assumed to play a central role in economic development and we therefore need to pay more attention to their education. Cases of radical innovation are exceptional, characterized by significant uncertainties, mixed emotions, and rapid changes. This presents a considerable challenge to the prevalent education and requires the development of novel – perhaps even radical – forms of teaching. Even though the role of the affective dimensions in learning is acknowledged, the influence of emotions in higher education is still a relatively unexplored field. The aim in this chapter is to determine how positive emotions can be utilized in university education targeted at developing radical innovators. The discussion focuses first on the role of education in this endeavor, and then moves on to the utilization of positive emotions in university education in general. Finally, three positive emotions – joy, enthusiasm, and pride – are considered in some detail, and their use in the development of radical innovators is assessed.

Key words: Radical innovators, innovation education, emotions, enthusiasm, joy, pride.

The education of radical innovators

Radical innovations are those that are new both to the firm and to the market. They provide the foundation on which future generations of products or services are built, and are thus acknowledged as critical to the long-term survival of many firms (McDermott & O'Connor, 2002). Developing and marketing these innovations are very demanding tasks, and many of the developmental challenges are related to technological uncertainty. However, overcoming the technological challenges is not enough to turn an invention into an innovation, in other words to make it succeed commercially. Commercializing a radical invention also requires coping with considerable market uncertainty (McDermott & O'Connor, 2002; Veryzer, 1998).

The terminology related to radical innovations is rather confusing (for a review of the concepts, see Sandberg, 2008). The definition adopted here reflects both customer and technological perspectives, thereby taking into account the unique challenges involved in both development and commercialization (cf. Ali, 1994). Hence, a radical innovation is defined as a new product or service that requires considerable change in customer behavior, is perceived to offer substantially enhanced benefits, and is also technologically new (cf. Veryzer, 1998).

Consequently, radical innovators are defined as individuals, teams, or firms capable of creating radical innovations. Even though the paramount role of individuals, often referred to as innovation champions, is acknowledged in the literature (e.g., Howell, Shea, & Higgins, 2005), radical innovation tends to require the combining of knowledge from various fields (Schmickl & Kieser, 2008; Subramaniam & Youndt, 2005). In fact, the complexity of radical innovations often leads to the adoption of an interdisciplinary developmental approach, and to close co-operation with individuals from different backgrounds (Johansson, 2004).

Flexibility, in other words the ability of managers to experiment and move quickly from one project to another, seems to be a prerequisite for the creation of radical innovations (Koberg, Detienne, & Heppard, 2003). Furthermore, rapidly evolving technologies stimulate the need for individuals who are motivated and have the ability to engage in life-long learning (Herrmann, Tomczak, & Befurt, 2006). It is acknowledged that the educational background of certain individuals plays an important role in the innovation activities of an organization in that it reflects their knowledge, skills, and cognitive base: Hambrick and Mason (1984) argue, for example, that managers with a high level of formal education tend to be more innovation-prone. Even if it is hard to find ways of systematically boosting the creation of radical innovations on account of their complexity, this is no reason to neglect trying. Rather, as O'Connor and Ayers (2005, p. 23) state, "it is in companies' and society's best interest to figure out how". In fact, a review of the existing literature indicates that education could foster many of the skills required for the creation of radical innovations, including those related to the coordination of diverse knowledge streams (Schmickl & Kieser, 2008), the evaluation, elaboration, and development of raw ideas into bigger concepts, conceptualization, opportunity discovery, market learning, market creation, coping with uncertainties and risks, managing teams and high-growth business (O'Connor & Ayers, 2005), networking, and information sharing (Subramaniam & Youndt, 2005). These skills are hard to teach, however, and it is argued here that, given the assumed influence of emotions on the judgments people make, the material recalled from the memory, creativeness, and inductive and deductive reasoning (George, 2000), they may well have a place in the challenging task of educating radical innovators.

Emotions are an inherent component of social behavior (Forgas & George, 2001; Tran, 1998), and it would therefore be worthwhile finding out how they can be utilized in the development of radical innovators. Moreover, despite their central role in cognitive processes and behavior, they have been largely ignored in scientific research on organizations (Muchinsky, 2000; Tran, 1998). Organizations traditionally attempt to control emotional expressions, given that emotions are generally regarded as irrational and therefore undesirable (Pescosolido, 2002), and as inhibiting effective decision-making (Albrow, 1992). It is only in recent decades, when interest in concepts such as "emotional intelligence" has been growing (e.g., Goleman, 1995), that researchers have started to call for more studies on the role of emotions in organizations.

Emotions are particularly likely to influence judgments when decision makers are faced with a complex task requiring extensive and constructive information processing. Their role is further accentuated in situations involving ambiguity and uncertainty, when new information needs to be assimilated, and when decision makers have to make accurate judgments and good decisions (Forgas & George, 2001). This could describe the situation of radical innovators. They are often faced with abundant information characterized by uncertainty and ambiguity, from which they need to make decisions that chart the course of their organizations (Sandberg, 2008). In other words, they are dealing with complex information characterized by high uncertainty and the need to be accurate. The implication is that the role of emotions in their behavior is considerable.

The explicit inclusion of emotions in the education of radical innovators could increase participants' awareness of their own emotion-management skills and also give them the tools to better manage the emotions of others. It is acknowledged that this kind of emotional intelligence contributes to effective leadership in organizations (George, 2000) – and it is especially needed in firms creating radical innovations. Knowing how to communicate emotions appropriately is an advantage in terms of understanding the organizational culture (Rafaeli & Sutton, 1989), communicating effectively in cross-cultural settings (Eid & Diener, 2001), and, eventually, achieving career success (Staw, Sutton, & Pelled, 1994). The discussion now moves on to the deliberate utilization of emotions in university education.

University education and positive emotions

Ever since the days of Plato education has been regarded as an emotional matter. Emotions are inherent in group-work situations, for example, in relationships inside the classroom, in deadlines, and in performance evaluation. Many recent studies have shown that various emotions influence both cognitive information processing and thinking quality. Many students and teachers have strong education-related emotions, and this affects their behavior and performance. Emotions are also known to affect motivation, memory, information processing, and problem solving, thus it is argued that those responsible for planning the aims, content, and forms of education should take emotional matters into account (Dirkx, 2008; Griffiths, 1984; Isen, Daubman, & Nowicki, 1987; Meyer & Turner, 2002).

One characteristic of emotions is that they tend to be contagious: the emotions of individuals combined create the emotional climate of the organization, which in turn affects individual emotions (Bond, 2009). Tran (1998) suggests that the emotional climate has an impact on learning in that individuals who are emotionally upset are not able to give their attention, remember, learn, or make decisions clearly. Indeed, according to Dirkx (2008, p. 9), "helping learners understand and make sense of these emotion-laden experiences within the context of the curriculum represents one of the most important and most challenging tasks for adult educators".

Although there has been growing interest in the role of emotions in education in general in recent decades (Linnenbrink, 2006), there is only limited knowledge in the context of higher education (Beard, Clegg, & Smith, 2007; Dirkx, 2008). Given that university students experience a vast array of emotions related not only to educational encounters but also to their personal lives and social relations (Beard et al., 2007), this chapter concentrates only on academic emotions, in other words "emotions that are directly linked to academic learning, classroom interaction, and achievement" (Pekrun, Goetz, Titz, & Perry, 2002, p. 92). The focus is thus not only on students' achievement emotions but also on emotions related to instruction and the process of studying, for example (Pekrun et al., 2002).

According to Dirkx (2008), emotion-laden experiences in adult learning may well be related to disagreements over values or interests, learning tasks and evaluation, or curricular content. Disagreements over values or interests (e.g., on how to conduct group work) tend to arise due to the diversity of students, color the relations among them, and also affect their learning. Learning tasks and evaluation evoke a variety of emotions ranging from joy and relief to anxiety and fear. In terms of curricular content, powerful emotions may arise, for instance, if the examples or cases involved relate to the emotion-laden memories of the students. Dirkx (2008) also mentions that particular subjects (e.g., math) may evoke considerable anxiety among some adult learners.

Emotions experienced by students range from the negative to the positive. Positive emotions are often seen as energizing and as promoting learning, whereas negative emotions are considered distracting. This is simplistic, however, in that negative emotions such as anger or mild anxiety may also energize learning efforts (Bruinsma, 2004; Dirkx, 2008). Given that most of the extant research on learning targets negative emotions, particularly anxiety, there is still a rather limited view of the role of positive emotions (Fredrickson, 1998). The focus in this chapter, therefore, is on the role of positive emotions in learning.

There is research evidence that positive emotions often help individuals to make more unusual cognitive associations and thus to be more creative. They are also likely to broaden the scope of individual actions by prompting the more unusual instead of the typical (Fredrickson, 1998; Isen et al., 1987; Kaufmann, 2003). Fredrickson (1998) argues that high-energy positive emotions, such as joy, create the urge to engage in action and thus build both physical and intellectual resources. Furthermore, shared experiences of positive emotions help in the creation and sustenance of social relationships. In sum, positive emotions motivate people to engage with their environment and to take part in different activities (Fredrickson, 2001). As Fredrickson (2001, p. 218) notes, "positive emotions are worth cultivating, not just as end states in themselves but also as a means to achieving psychological growth and improved well-being over time".

In the context of students' learning processes, positive emotions affect motivation, learning occurrence, and achievement (cf. Pekrun et al., 2002; Schutz & DeCuir, 2002). First, in the academic context they may trigger or sustain motivation. Emotions influence both intrinsic (i.e. learning because it is enjoyable) and extrinsic (i.e. learning in order to attain outcomes) motivation (Pekrun et al., 2002). Second, these emotions facilitate the use of flexible and creative learning strategies and are likely to enhance self-regulated learning (Pekrun et al., 2002), thereby influencing learning occurrence. Learning occurrence could be described as what takes place when the stimulus situation together with the contents of the memory affects the learner in such a way that his or her performance changes (Gagne, 1985). Third, it is assumed that through their effect on motivation and learning occurrence, positive academic emotions increase achievement outcomes (e.g., grades and exam scores). Thus, emotions seem to influence the learning process in various ways (see Figure 39.1). However, it is worth noting that the causation is likely to be reciprocal: experiences

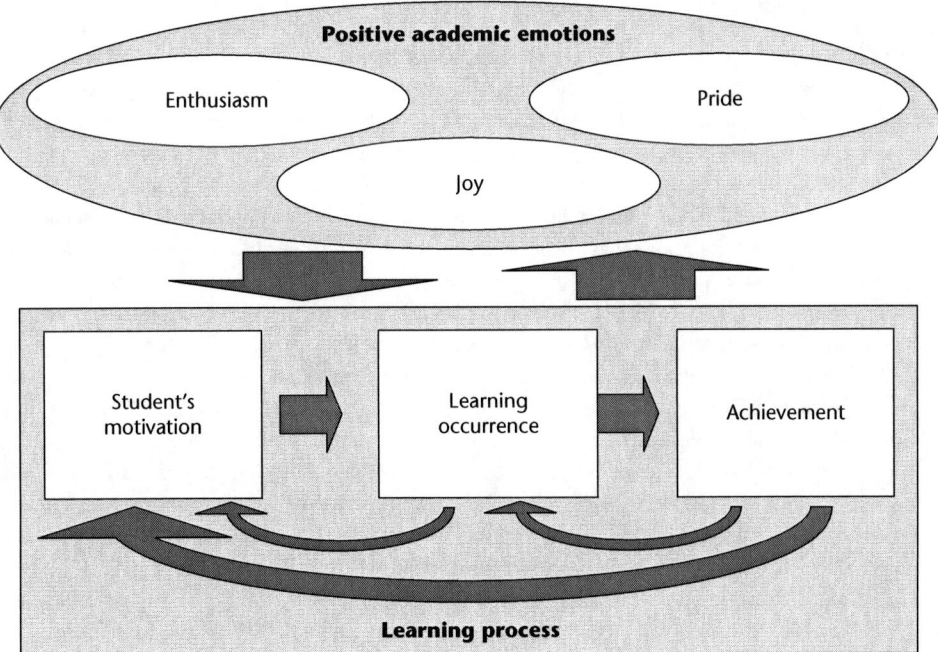

Figure 39.1 The relationship between positive academic emotions and the learning process.

during the learning process also influence academic emotions (Pekrun et al., 2002). It is therefore possible that positive emotions and learning processes together form a virtuous cycle of enhanced learning and stronger positive emotions.

There have been various attempts to list and classify emotions. Pride, joy, and enthusiasm (interest) are among the pleasant ones listed in Tran (1998) and Pekrun et al. (2002). Given the apparent emphasis on these emotions in processes of radical innovation development (cf. Sandberg, 2008) they are discussed in the following sections. The suggestions for applying the ideas in courses on radical innovations in particular are based on the author's own experience of teaching radical-innovation management and of coordinating the Global Innovation Management Master's Degree Programme. In order to concretize the issues the utilization of emotions is discussed with reference to a Business Development Laboratory course. The author has not taught the course, and the description is based on an interview with the coordinator and numerous informal discussions the author has had over the years with students and teachers involved.

The concept of the Business Development Laboratory was developed in 2007 in cooperation between Turku School of Economics and the University of Turku, which were later merged. A local science park was also involved as a partner. In 2009 the Finnish Association of Business School Graduates gave the course an award for innovative teaching. The aim of the course is to promote research-based business activities. Students from Turku School of Economics and the Faculty of Law at the University of Turku work in the Business Development Laboratory together with natural-science or technology researchers in order to commercialize their inventions. In four months each interdisciplinary group of students prepares a common business plan for the invention. The course begins with an intensive business-planning module that covers the basic concepts and tools. Personal mentoring by an experienced entrepreneur or industry specialist is available to the groups throughout the course (Hautala, Malinen, Orava, & Puhakainen, 2009).

Fostering enthusiasm

The experience of enthusiasm is highly subjective and transcendental, and thus extremely difficult to describe in words. It is often characterized as a strong feeling about something (e.g., *Oxford Advanced Learner's Dictionary of Current English*, 1989; *Webster's Ninth New Collegiate Dictionary*, 1988). Glassman and McAfee (1990, p. 4) describe it as "being excited or highly aroused", or having an "ardent zeal" for something. *Collins Cobuild English Language Dictionary* (1987, p. 471) gives a more detailed definition: "great eagerness to be involved in particular activity, because it is something you like and enjoy or that you think is important". This definition is used in this chapter because it is more extensive and encompasses both feelings and action, i.e. "being involved". Marcus and Mackuen (1993, p. 673) also acknowledge the behavioral aspect of enthusiasm: "enthusiasts throw themselves into the cause".

Patrick, Hisley, and Kempler (2000, p. 217) characterize enthusiastic students as alert, energized, engaged, and "seemingly curious and eager to learn". They state further that these students want to be in the classroom and they are excited about learning. Student enthusiasm seems to facilitate the execution of challenging and novel activities. Indeed, enthusiastic students are willing to commit to their work, and they invest in learning capabilities that produce personal results (e.g., make tasks easier to accomplish and generate feelings of self-worth) (Gläser-Zikuda, Fuß, Laukenmann, Metz, & Randler, 2005; cf. Senge, Kleiner, Roberts, Ross, Roth, & Smith, 1999; Walker, 2002).

Enthusiasm is also contagious (Sandberg, 2007; Senge et al., 1999). There is ample evidence (Patrick et al., 2000; Pekrun et al., 2002; Ramsden, 1979; Streeter, 1986) that teacher enthusiasm tends to spread to students, improving their attitudes toward reading and making them more interested, curious, energetic, and excited about learning. Although, to a degree, the level of enthusiasm is beyond their control, it has been shown that teachers can be trained to become more enthusiastic (Streeter, 1986), and that those who adopt a teaching style that is natural to them and who are able to choose freely the topics they consider most important for their students are likely to be more enthusiastic (Baum, 2002).

According to Lewis (2009), there are two vital aspects that should be taken into account in the fostering of enthusiasm: the meaningfulness of the action, and progress. In the case of *meaningfulness* this would mean compatibility between the learning objectives of a course and the students' own personal goals, and a feeling among the students that the ways of learning promoted in the course are valuable. Radical innovation as a course subject tends to be demanding in that it goes against many of the forms of thinking emphasized in other business courses. For example, the old wisdom of satisfying the needs of customers learned in marketing may not hold when the firm is creating a totally new market with its breakthrough product, and traditional business models may not be applicable to completely new business forms. Consequently, the teacher needs to put much effort into proposing and fostering different kinds of thinking – and at the same time should not underestimate the value of what students have learned earlier in other business courses. Constructive and continuous feedback will support students' in their *progress* toward their goals (cf. Lewis, 2009). A major challenge in teaching radical innovations is to constantly encourage novel solutions and new ways of dealing with assignments, while at the same time demanding mastery of the course material.

Considerable effort is put into selecting the most motivated and capable students for the Business Development Laboratory. Students apply for a place as they would apply for a job. Some of them are interviewed, and recommendations from professors are also taken into account. One of the selection criteria is motivation. The participants are close to finalizing their studies and are eager to put into practice what they have learned over the years. Working on a real case evokes natural enthusiasm. This is evident in the strong devotion in terms of both time and effort that the student groups show in preparing the business plans. Continuous feedback from the mentors inspires the groups to devote even more energy to the tasks. Students tend to have rather narrow views about the possible business models when they start to prepare their business plans (Hautala et al., 2009). Thus, one of the mentors' tasks is to encourage non-traditional thinking and inspire the students to develop new kinds of revenue logic. Mentors and teachers tend to be very enthusiastic, and see the course as an avenue on which to meet young inspired students.

Bringing joy into education

Joy is defined as a feeling of great pleasure, happiness, or contentment (*Collins Cobuild English Language Dictionary*, 1987; *Oxford Advanced Learner's Dictionary of Current English*, 1989; *Webster's Ninth New Collegiate Dictionary*, 1988). It is a high-arousal positive emotion that creates the urge to play and be playful. Play takes many forms, and intellectual, imaginative play "involves exploration, invention, and just plain fooling around" (Fredrickson, 1998, p. 305). Although often aimless, it seems to create the urge to push the limits and be creative, thereby building an individual's intellectual and social skills (Ellsworth & Smith, 1988; Fredrickson, 1998).

According to Csikszentmihalyi (1991, p. 2), "if educators invested a fraction of the energy on stimulating the students' enjoyment of learning that they now spend in trying to transmit information we could achieve much better results". He goes on to suggest that when students are intrinsically motivated to learn they may have flow experiences during their learning, in other words they find it so enjoyable it that it is like being carried away by a current, or being in a flow. Fostering motivation and carefully planning course contents stimulate the flow. Pedagogical goals should be articulated as meaningful challenges that fit students' own goals, desires, and skills (Csikszentmihalyi,

1991; Csikszentmihalyi & Wong, 1991). This means that teachers should know their students well, thus implying the need for smaller class sizes so the teacher can devote more time to individual students. This, however, is becoming more and more unrealistic given the increasing emphasis on output at universities.

There is surprisingly little literature focusing explicitly on enjoyment in learning among university students, and what there is mostly relates to distance learning. The question remains whether or not it is possible to teach serious science at the university level and at the same time to make it a joyful experience – or perhaps even fun – for the students. If it is possible, how does it happen? One widely used method is through simulations and games, which according to Ruben (1999) bring playfulness and joy into the classrooms and allow for both emotional as well as cognitive learning. However, current business games are designed to present the realities faced by typical firms making typical decisions, and thus do not grasp the realities faced by firms making things differently and creating radical innovations.

Students in the Business Development Laboratory work with real business ideas and inventions. This fosters motivation, but also a degree of seriousness: the students have signed confidentiality agreements, they are stuck in the group with students from different backgrounds, and they have promised to deliver a business plan on a due date. Group spirit is enhanced in a special, somewhat informal opening session that takes place outside the university. Given the big differences in dynamics, the coordinator has to follow the progress of each group and, if necessary, step in and help the students to resolve any conflicts between the members. Some groups clearly enjoy the course so much that the students continue to meet up afterwards.

Promoting pride in the education

Pride is defined as the feeling of satisfaction or pleasure one gets from doing something well, and as the feeling of being superior to others (*Collins Cobuild English Language Dictionary*, 1987; *Oxford Advanced Learner's Dictionary of Current English*, 1989; *Webster's Ninth New Collegiate Dictionary*, 1988). The first definition, authentic pride, in other words pride in one's accomplishments (Tracy & Robins, 2007), is applied in this chapter because it can be linked to students' accomplishments, whereas hubristic pride, in other words overgeneralized hubris at one's superiority (Tracy & Robins, 2007), is a personality trait and therefore beyond its scope.

Tracy and Robins (2007) suggest that pride may be the most important human emotion given the motivation of social behavior: it sparks the desire to achieve and to attain power. According to Williams and DeSteno (2009), one of its primary functions is to motivate hedonically costly short-term efforts aimed at acquiring skills that increase one's status and add value to one's social group in the longer term. Hence, pride motivates increased perseverance with socially valued tasks, such as learning and education.

Experiencing pride after task accomplishment fosters improved performance in subsequent tasks (Herrald & Tomaka, 2002). Good results alone are not sufficient to generate it, however, and in order to experience pride in their achievements students need to intrinsically value the knowledge and skills they acquire (Pekrun et al., 2002). Thus, in a learning context pride is experienced in relation to students' goals: it intensifies if they feel that they are successful in their attempts to reach their goals and that they are in control of their accomplishments (i.e. are not being unfairly treated by the teacher) (Schutz & DeCuir, 2002). Consequently, the way in which students' achievements are evaluated is critical. Biggs (2005) argues that students should be involved in the assessment process, both in establishing the criteria and in the evaluation itself. Self-assessment and peer assessment are being used more and more in courses on innovation management. The challenge in courses concentrating on radical innovations lies not only in giving credit for mastery of the course material but also rewarding novel and – perhaps even radical – solutions.

An evaluation group consisting mainly of business professionals evaluates the business plans drawn up in the Business Development Laboratory. Each group receives both oral comments and written feedback. The course coordinator gives the final grade. Even though self and peer assessment may be used in the future, the advantage of the current practice lies in the fact that the course coordinator is better able to reward the groups that have put in the most effort, and also to take into account the differences between the assignments: the task may be relatively easy in some groups, whereas in others it may involve solving considerable problems related to commercialization. Hence, in the laboratory and also through the evaluation students learn that there is no ready commercialization blueprint, and that flexibility and opportunity exploitation are crucial. The best plan is announced at the cocktail party given at the end of the course. Many of the groups have gone on to enter the Venture Cap Business Plan Competition over the years, and have won several prizes and received honorable mentions (Hautala et al., 2009).

Conclusions

The process of developing radical innovations could be described as an emotional rollercoaster ride. Those involved need to deal with rapidly changing intensive emotions. The motivation for this chapter was the idea that the explicit exploitation of emotions in the education of radical innovators is likely to increase students' awareness of their own emotion-management skills, and also to give them tools to better manage the emotions of others. Thus, deliberately taking emotions into consideration in their education is likely to encourage the students to build up their emotional intelligence. It is essential for those involved in developing the education of future radical innovators to understand how emotions work and influence learning. Teachers who know how to evoke and enhance positive emotions are more likely to foster student engagement and learning. Furthermore, if students themselves become better aware of their emotions and can recognize how they influence their behavior, and if they have the language to think through and describe them, they should develop a better understanding of emotion management in working life.

Even though emotions are critical in terms of promoting student learning, they have been neglected, especially in the context of university education. This chapter concentrates on three positive emotions – enthusiasm, joy, and pride – which are emphasized in the processes of learning and radical innovation. Different ways of fostering these emotions in education focused on radical innovation are discussed. A theme that appears to be common to the three emotions is the fit between student goals and pedagogical goals, without which there would seem to be little room for genuine enthusiasm, joy, or pride in learning. Radical innovations require novel solutions and atypical decisions. This makes the education of radical innovators fun, but also challenging in that conventional business cases and simulations are not appropriate. One solution is to allow the students to work on real, ongoing cases, as happens in the Business Development Laboratory.

This chapter has limitations due to its exploratory nature. First, it does not take into account the trait affect, in other words the generalized tendency of an individual to have a high or low propensity to be enthusiastic, joyful, or proud (cf. Watson & Walker, 1996). This was excluded because the genetic dispositions of students are beyond the control of educators (cf. (Pekrun et al., 2002). Personality traits ought to be taken into account in further research on the subject, however. Second, cultural differences are not discussed, even though they are likely to create additional challenges for increasingly internationalizing universities. It should be remembered that the expression and suppression of emotions varies between cultures, and the desirability of different emotions is not constant. For instance, pride is considered appropriate in the United States, whereas many people in more collectivist countries such as China still believe that it is not acceptable to express it (Eid & Diener, 2001).

In spite of the limitations, it is hoped that this preliminary discussion on the utilization of positive emotions in the education of radical innovators will tempt curriculum planners to explicitly consider the role of emotions in the courses. It is also encouraging that there are now concrete tools available for developing students' emotional intelligence. A recent example of this is the training material freely provided by the Leonardo Programme of the European Union (Training Innovation, 2011).

References

Albrow, M. (1992). Sine ira et studio: or do organizations have feelings? *Organization Studies*, *13*(3), 313–329.
Ali, A. (1994). Pioneering versus incremental innovation: Review and research propositions. *Journal of Product Innovation Management*, *11*(1), 46–61.
Baum, L. (2002). Enthusiasm in teaching. *Political Science and Politics*, *35*(1), 87–90.
Beard, C., Clegg, S., & Smith, K. (2007). Acknowledging the affective in higher education. *British Educational Research Journal*, *33*(2), 235–252.
Biggs, J. (2005). *Teaching for Quality Learning at University*. Maidenhead: Open University Press/McGraw-Hill Education.
Bond, M. (2009. January 3). Three degrees of contagion. *New Scientist*, (2689), 24–27.
Bruinsma, M. (2004). Motivation, cognitive processing and achievement in higher education. *Learning and Instruction*, *14*(6), 549–568.
Collins Cobuild English Language Dictionary (1987). London: Collins.
Csikszentmihalyi, M. (1991). Thoughts about education. In D. Dickinson (Ed.), *Creating the Future: Perspectives on Educational Change*. Retrieved from www-bcf.usc.edu/~genzuk/Thoughts_About_Education_Mihaly_Csikszentmihalyi.pdf.
Csikszentmihalyi, M., & Wong, M. M.-H. (1991). The situational and personal correlates of happiness: A cross-national comparison. In F. Strack, M. Argyle, & N. Schwarz (Eds.), *Subjective Well-Being: An Interdisciplinary Perspective* (pp. 193–212). Oxford: Pergamon Press.
Dirkx, J. M. (2008). The meaning and role of emotions in adult learning. *New Directions for Adult and Continuing Education*, 2008 (120), 7–18.
Eid, M., & Diener, E. (2001). Norms for experiencing emotions in different cultures: Inter- and intranational differences. *Journal of Personality and Social Psychology*, *81*(5), 869–885.
Ellsworth, P. C., & Smith, C. A. (1988). Shades of joy: Patterns of appraisal differentiating pleasant emotions. *Cognition and Emotion*, *2*(4), 301–331.
Forgas, J. P., & George, J. M. (2001). Affective influences on judgements and behavior in organizations: An information processing perspective. *Organizational Behavior and Human Decision Processes*, *86*(1), 3–34.
Fredrickson, B. L. (1998). What good are positive emotions? *Review of General Psychology*, *2*(3), 300–319.
Fredrickson, B. L. (2001). The role of positive emotions in positive psychology: The broaden-and-build theory of positive emotions. *American Psychologist*, *56*(3), 218–226.
Gagne, R. M. (1985). *The Conditions of Learning*. New York: Holt, Rinehart and Winston.
George, J. M. (2000). Emotions and leadership: The role of emotional intelligence. *Human Relations*, *53*(8), 1027–1055.
Glassman, M., & McAfee, R. B. (1990). Enthusiasm: The missing link in leadership. *S.A.M. Advanced Management Journal*, *55*(3), 4–6, 29.
Gläser-Zikuda, M., Fuß, S., Laukenmann, M., Metz, K., & Randler, C. (2005). Promoting students' emotions and achievement: Instructional design and evaluation of the ECOLE-approach. *Learning and Instruction*, *15*(5), 481–495.
Goleman, D. (1995). *Emotional Intelligence*. New York: Bantam Books.
Griffiths, M. (1984). Emotions and education, *Journal of Philosophy of Education*, *18*(2), 223–231.
Hambrick, D. C., & Mason, P. A. (1984). Upper echelons: The organization as a reflection of its top managers. *Academy of Management Review*, *9*(2), 193–206.
Hautala, V., Malinen, P., Orava, M., & Puhakainen, J. (2009). Developing the regional innovation system: Business Development Laboratory as a promoter of science-based new ventures. *Research in Economics and Business: Central and Eastern Europe*, *27*(1), 18–34.

Herrald, M. M., & Tomaka, J. (2002). Patterns of emotion-specific appraisal, coping, and cardiovascular reactivity during an ongoing emotional episode. *Journal of Personality and Social Psychology, 83*(2), 434–450.

Herrmann, A., Tomczak, T., & Befurt, R. (2006). Determinants of radical product innovations. *European Journal of Innovation Management, 9*(1), 20–43.

Howell, J. M., Shea, C. M., & Higgins, C. A. (2005). Champions of product innovations: Defining, developing, and validating a measure of champion behavior. *Journal of Business Venturing, 20*(5), 641–661.

Isen, A. M., Daubman, K. A., & Nowicki, G. P. (1987). Positive affect facilitates creative problem solving. *Journal of Personality and Social Psychology, 52*(6), 1122–1131.

Johansson, F. (2004). *The Medici Effect: Breakthrough Insights at the Intersection of Ideas, Concepts, and Cultures.* Boston: Harvard Business School Press.

Kaufmann, G. (2003). The effect of mood on creativity in the innovation process. In L. V. Shavinina (Ed.), *The International Handbook on Innovation* (pp. 191–203), Oxford: Elsevier Science.

Koberg, C. S., Detienne, D. R., & Heppard, K. A. (2003). An empirical test of environmental, organizational, and process factors affecting incremental and radical innovation. *Journal of High Technology Management Research, 14*(1), 21–45.

Lewis, C. (2009, January). Help people perform with passion. *Training Journal,* 39–43.

Linnenbrink, E. A. (2006). Emotion research in education: Theoretical and methodological perspectives on the integration of affect, motivation, and cognition. *Educational Psychology Review, 18*(4), 307–314.

Marcus, G. E., & Mackuen, M. B. (1993). Anxiety, enthusiasm, and the vote: The emotional underpinnings of learning and involvement during presidential campaigns. *American Political Science Review, 87*(3), 672–685.

McDermott, C., & O'Connor, G. C. (2002). Managing radical innovation: An overview of emergent strategy issues. *Journal of Product Innovation Management, 19*(6), 424–438.

Meyer, D. K., & Turner, J. C. (2002). Discovering emotion in classroom motivation research. *Educational Psychologist, 37*(2), 107–114.

Muchinsky, P. M. (2000). Emotions in the workplace: The neglect of organizational behaviour. *Journal of Organizational Behavior, 21*(7), 801–805.

O'Connor, G. C., & Ayers, A. D. (2005). Building a radical innovation competency. *Research Technology Management, 48*(1), 23–31.

Oxford Advanced Learner's Dictionary of Current English (1989). Oxford: Oxford University Press.

Patrick, B. C., Hisley, J., & Kempler, T. (2000). "What's everybody so excited about?" The effects of teacher enthusiasm on student intrinsic motivation and vitality. *Journal of Experimental Education, 68*(3), 217–236.

Pekrun, R., Goetz, T., Titz, W., & Perry R. P. (2002). Academic emotions in students' self-regulated learning and achievement: A program of qualitative and quantitative research. *Educational Psychologist, 37*(2), 91–105.

Pescosolido, A. T. (2002). Emergent leaders as managers of group emotion. *Leadership Quarterly, 13*(5), 583–599.

Rafaeli, A., & Sutton, R. I. (1989). The expression of emotions in organizational life. *Research in Organizational Behavior,* 11, 1–42.

Ramsden, P. (1979). Student learning and perceptions of the academic environment. *Higher Education, 8*(4), 411–427.

Ruben, B. D. (1999). Simulations, games, and experience-based learning: The quest for a new paradigm for teaching and learning. *Simulation Gaming, 30*(4), 498–505.

Sandberg, B. (2007). Enthusiasm in the development of radical innovations. *Creativity and Innovation Management, 16*(3), 265–273.

Sandberg, B. (2008). *Managing and Marketing Radical Innovations: Marketing New Technology.* London: Routledge.

Schmickl, C., & Kieser A. (2008). How much do specialists have to learn from each other when they jointly develop radical product innovations? *Research Policy, 37*(3), 473–491.

Schutz, P. A., & DeCuir, J. T. (2002). Inquiry on emotions in education. *Educational Psychologist, 37*(2), 125–134.

Senge, P., Kleiner, A., Roberts, C., Ross, R., Roth, G., & Smith, B. (1999). *The Dance of Change: The Challenges of Sustaining Momentum in Learning Organizations.* Sydney: Random House.

Staw, B. M., Sutton, R. I., & Pelled, L. H. (1994). Employee positive emotion and favourable outcomes at the workplace. *Organization Science, 5*(1), 51–71.

Streeter, B. B. (1986). The effects of training experienced teachers in enthusiasm on students' attitudes toward reading. *Reading Psychology, 7*(4), 249–259.

Subramaniam, M., & Youndt, M. A. (2005). The influence of intellectual capital on the types of innovative capabilities. *Academy of Management Journal, 48*(3), 450–463.

Tracy, J. L., & Robins, R. W. (2007). Emerging insights into the nature and function of pride. *Current Directions in Psychological Science, 16*(3), 147–150.

Training Innovation – Development of tools to help the teachers to display, in their learners, key informal competences linked to innovation. (2011). A DVD developed in a project "Training Innovation: Tools to valorize, develop and mobilize the competences linked to innovation and acquired through informal learning". Leonardo da Vinci, Multilateral Projects, Transfer of innovation. Project number: ES/09/LLP-Ldv/TOI149050. European Union.

Tran, V. (1998). The role of the emotional climate in learning organizations. *Learning Organization, 5*(2), 99–103.

Veryzer, Jr., R. W. (1998). Discontinuous innovation and the new product development process. *Journal of Product Innovation Management, 15*(4), 304–321.

Walker, D. H. T. (2002). Enthusiasm, commitment and project alliancing: An Australian experience. *Construction Innovation, 2*(1), 15–31.

Watson, D., & Walker, L. (1996). The long-term stability and predictive validity of trait measures of affect. *Journal of Personality and Social Psychology, 70*(3), 567–577.

Webster's Ninth New Collegiate Dictionary (1988). Springfield: Merriam-Webster.

Williams, L. A., & DeSteno, D. (2009). Pride: Adaptive social emotion or seventh sin? *Psychological Science, 20*(3), 284–288.

40

Introducing the phenomenon of the "abortion" of new ideas and describing the impact of "saved" ideas and thus implemented innovations on the economy in the case of distinguished innovators

Larisa V. Shavinina

UNIVERSITÉ DU QUÉBEC EN OUTAOUAIS, CANADA

Summary: The traditional practice of economy and innovation science focuses on the implemented innovations (such as, for example, iPod and iPhone) and counts the revenues received from them. However, neglecting the fact of the "killed" ideas and, consequently, the value of lost innovations is a huge blank spot in research. When people intentionally or unintentionally abandon their ideas without a wish to develop them any further and eventually implement them into practice, they thus abort potential innovations. This is what the phenomenon of the abortion of ideas is all about. The chapter concentrates on the well-known cases when individuals resisted abandoning their creative ideas and finally implemented them. These cases shed light on what today's children, adolescents, and adults can learn from outstanding innovators in order to be able to save, develop, and implement their ideas into practice in the form of new products, processes, and services.

Key words: Phenomenon of the abortion of ideas, "killed" ideas, lost innovations, economy, distinguished innovators, innovation education.

Introduction

> Never give up! Never, never, never, never, never, never!
>
> *(Sir Winston Churchill)*

Some people are concerned with abortion killing potential human beings. However, nobody appears to concern themselves with the "abortion" of new ideas resulting in the killing of potential scientific, technological, and societal innovations. Being implemented into practice in the form of innovative products, processes, or services, creative ideas thus lead to enhanced economic growth and competitiveness by increasing employment and prosperity for all. Consequently, the "abortion" of ideas is dangerous for any society. All means should, therefore, be used to develop and implement ideas into practice.

During many years of work on the bestselling *International Handbook on Innovation* (Shavinina, 2003), I discovered that no one teaches people how to implement their ideas into practice in the form of innovative products, processes, and services. It was also discovered that today nobody evaluates the potential impact of "killed" ideas and, therefore, unborn or lost innovations on the economy. It has never been studied. The world can recover from an economic recession and avoid future economic downturns only through innovations and every effort should hence be made to prevent the "abortion" of ideas and to ensure their implementation into practice. After the invited presentation on how to develop the next generation of innovators in science, technology, engineering, and mathematics (STEM) disciplines at the US National Science Board, with a subsequent report to Congress and President Obama in August 2009, I came to believe that the time is right for initiating this new research direction. Innovation education has an important role to play in teaching everyone how to implement his or her ideas into practice in the form of new products, processes, and services.

This chapter will describe the renowned cases of distinguished innovators who resisted abandoning their great ideas and eventually implemented them. Akio Morita, Fred Smith, Richard Branson, Herbert Kelher, Bill Gates, Michael Dell, and Jeff Bezos, just to mention a few, are among such outstanding innovators, which will be considered in the chapter. They are characterized by a unique ability to both generate great ideas and to implement them into practice. These cases have significant implications for innovation education, namely: they shed light on what can today's children, adolescents, and adults learn from eminent innovators in order to be able to save, develop, and implement their ideas thus making innovations happen.

Defining the phenomenon of the "abortion" of new ideas

While the traditional practice of economic and innovation sciences focuses on implemented scientific and technological innovations (such as, for example, iPod or iPhone, just to mention two) and the counting of revenues received from them, I found that neglecting "killed" ideas and, consequently, the value of lost innovations represents a huge gap in research. When people intentionally or unintentionally abandon their ideas without any desire to further develop them and to eventually implement them into practice, they also abort potential innovations. This is the phenomenon of the "abortion" of new ideas.

There are many well-known cases of individuals resisting abandoning their creative ideas and finally successfully implementing them. Akio Morita, a co-founder and the former Chairman of *Sony*,[1] is an excellent example. He insisted on the idea to develop the "Walkman" when everyone else at *Sony* resisted his idea including senior executives. They based their decisions on the results of marketing research, which had shown that nobody wanted to buy this future product. Relying on his intuition and courage,[2] Akio Morita insisted. "Everybody gave me a hard time. It seemed as though nobody liked the idea" (Morita, 1987, p. 79). He could have easily given up, but he did not. When the resistance from senior management reached its height, Akio Morita threatened to leave his position as the Chairman if *Sony* did not sell 100,000 Walkman in the first half a year. When the Walkman was eventually developed, it became *Sony*'s best-selling product. Moreover, in the next 12 years of its production *Sony* pioneered a hyper strategy of innovation: the company developed four platforms and 122 incremental innovations! Nobody else in the whole industry ever matched this achievement, which resulted in profit and increased prosperity at *Sony*.

Fred Smith is another good example. While attending Yale University, he wrote a paper for an economics class, outlining an overnight courier service delivery in a computer information age. He received a C for this paper. However, Fred Smith insisted on his idea of creating such a service delivery despite of the professor's disapproval. Eventually, he founded *FedEx*, the first overnight express delivery company in the world, and the largest in the United States, as well as becoming its chairman, president, and CEO. Today, *FedEx* has more than 290,000 employees and is consistently ranked among the world's most admired companies. For instance, *FedEx* has been featured on *Fortune* magazine's "100 Best Companies to Work For" every year since 1998 and was ranked No. 91 on the 2010 list.

The ever-optimistic Richard Branson was reportedly depressed on his 40th birthday because he was not able to develop *Virgin Atlantic* as fast as he wanted (Branson, 2002). If he had abandoned his ideas, *Virgin* might never have happened and the 60,000 people the company now employs might not have had jobs (Branson, 2011c). Not every of his creative ideas was a great success, but he never gave up and persisted in implementing them into practice (Branson, 2008).

Herbert Kelleher did not give up either when competitors tried to keep the just-founded *Southwest Airlines* grounded. He overcame a year's worth of legal challenges from them and thus kept his idea implemented. *Southwest* has been an innovative enterprise just from its inception by using a ten minute turnaround, airhostesses in hot pants, and free bottles of alcohol with every ticket. The company has never had an in-flight fatality. With its more than 37,000 employees, *Southwest* is consistently named among the top five Most Admired Corporations in America in *Fortune* magazine's annual poll. *Fortune* has also called Herbert Kelleher perhaps the best CEO in America.

Michael Dell is another excellent example who resisted abandoning his great idea. If he had followed his father's advice and his brother's steps and continued university studies, *Dell Computer Inc.* would not exist and 46,000 people would not have jobs. In contrast to his parents' expectations, he left university and decided to "compete with IBM" by founding his own company and building better computers than IBM (Dell, 1999).

However, how many similarly remarkable ideas were abandoned and potential innovations lost? No one tried to estimate it. This chapter thus opens a new research direction with significant scientific potential and important practical implications.

Explaining the phenomenon of the "abortion" of new ideas

Although innovation is exceptionally important for the economic development of any society, it should be acknowledged that innovation does not happen often. Why do people abandon their ideas? One of the possible explanations consists in the fact that there exist many constraints or factors inhibiting innovation. The concept of *innovation gap* is a critical one. It means that people have a lot of creative ideas, but they are not able to implement them into practice due to various reasons. There are multiple barriers to innovation.

A barrier to innovation is any factor that influences negatively the innovation process. Researchers found that the existence of barriers in innovation is the rule rather than the exception. In most cases market, government, societal, organizational, and business procedures work against both successful development and use of innovative products, processes, and/or services (Hadjimanolis, 2003). In contrast, facilitators are factors with a positive influence on the development of innovation. Barriers and facilitators are related. Research shows that many barriers exist due to lack of facilitators (Shavinina, 2003).

Richard Branson has been demonstrating many times how to overcome numerous obstacles on a way to implementing creative ideas into practice. For instance, in 1970 a new law had just been passed that allowed people to sell records at discounted prices, and Branson was among the first to take advantage. He thus launched his second major venture: a mail-order record business. Like his *Student* magazine before, Richard's new company was a great success. Sales skyrocketed, and

Branson scrambled to find employees to keep up with the tremendous order load. When a postal strike crushed the mail-order endeavor, the ever-creative Richard Branson responded by opening a small, discount record shop on Oxford Street in London that was a hit as well. A chain of *Virgin Record* stores was the next.

Early setbacks, such as the postal strike, were representative of the great obstacles that Branson would be forced to overcome in the UK anti-business climate of the 1970s and even 1980s. Indeed, during the 1970s the United Kingdom was mired in economic malaise. Tax rates on unearned income were as high as 98%, and labor strikes such as the one that nearly destroyed *Virgin* were the norm. Furthermore, a general disdain for entrepreneurs and "new money" permeated the business and social environment, making it more difficult for would-be capitalists to get their ideas off the ground. A mid-1980s survey, for example, showed that 29% of the executives in the United Kingdom viewed business owners as having the lowest status in the country, while only 13% thought they had the highest status (www.fundinguniverse.com)

Nevertheless, Britain's political, social, and economic environments were perfect for Richard Branson. A rebel by nature, he loved a good challenge from early childhood and enjoyed bucking convention since school years. These characteristics were most conspicuously evidenced by the name that he chose for his company. He used Virgin to signify his lack of knowledge about the businesses into which he entered. While business convention demanded that entrepreneurs have experience in the ventures they began, Branson elected to enter businesses that interested him, regardless of his background; he would ask questions and invent his own route to success. Having no preconceived ideas about an industry, he was able to identify unnecessary hurdles that his competitors took for granted, as well as to recognize hidden opportunities and go ahead. Many of Richard's subsequent ventures testify to this.

For example, in 1984 he came across another industry that interested him and about which he knew relatively little: the airline industry. Critics effectively laughed off Branson's idea to begin providing long-haul air service between London and New York. Nonetheless, he purchased a Boeing 747 and began flying people back and forth between the UK and USA, offering improved service and unique features. *Virgin Atlantic Airways* wowed observers by posting a profit in its second year. "It's not so divorced from the music business," Richard Branson pointed out in the November 14, 1988, *Forbes*, noting that "if people are traveling for ten hours, they want to be entertained." Entertainment was, indeed, an important element of *Virgin Atlantic*'s success during the 1980s and early 1990s. Passengers were entertained with videos and, in some cases, live performances. One can conclude that creativity helps innovators to overcome the existing obstacles and implement their ideas into practice instead of abandoning them.

Richard Branson is convinced that

> obstacles and challenges are healthy for everyone, not just entrepreneurs. They force you to think outside the box, so to speak—to be creative. The challenge is to follow through on a great idea. I think if [you've] got a great idea, you need to just give it a try.
>
> (Branson, 2009b)

And he lives up to this principle. Thus, after a failed around-the-world balloon trip, Branson said of his experience, "It has been like hitting up against a solid brick wall. All day and all night long, we battled to get through it" (Branson, 2009a). This battle is a familiar one to Richard Branson, who has seen his share of failed business ventures. But, in typical Branson fashion, he rebounds from his failures with the same youthful energy he had the very first day he created *Virgin Records*. His passion to generate original ideas and implement them into practice by creating new companies cannot be quelled by any barrier no matter how large. "He's not driven like other people. He's driven to do stuff," the *Virgin* executive Tom Alexander said. "The money is the byproduct. If it makes money, well, then great, because then he can go off and do more stuff. Doing nothing is not an option. If you've ever been on holiday with him, it's hard work" (Alexander, 2012).

Therefore, this section sheds light on why and under which circumstances people "kill" their ideas. It happens because there are multiple barriers to innovation.

Never give up! The key internal obstacles to innovation

The above-described external obstacles to innovation constitute one part of the explanation of the phenomenon of the "abortion" of new ideas. However, this is not the whole story. Another major group of barriers is related to personal factors such as a lack of courage, persistence, a wish to do the impossible, and all those distinguished characteristics of innovators and entrepreneurs discussed in Chapters 16 and 36 of this volume, respectively. These qualities are highly developed in great innovators who never abandon their new and original ideas.

The case in point is Richard Branson again and the early years of *Virgin Atlantic*. The airline became his main focus during the 1990s. In July 1991 he reached his key goal of expanding service to London's Heathrow Airport. This achievement was a signal victory in Branson's bitter struggle with *British Airways*, which had sought to block *Virgin Atlantic*'s growth through political influence and underhanded tactics. Among the latter was the establishment of an espionage unit to spy on Richard Branson and harass *Virgin* customers in person and by telephone. Lord King, the *British Airways* chairman, spread rumors that his competitor was about to go bankrupt. Virgin's chronic cash flow problems lent credence to these stories. In 1992 Branson made the painful decision to sell *Virgin Music Group* to *Thorn-EMI* for approximately $1 billion to keep *Virgin Atlantic* aloft. This sale enabled him to upgrade the airline with such luxuries as seat-back video screens, full-sized sleeper seats, in-flight massages and manicures, and free ground transportation by limousine. Nevertheless, the ongoing battle with *British Airways* continued. The British press wondered whether Richard Branson had finally taken on a battle he could not win. An editorial in the *Sunday Telegraph* wondered whether Branson was "too old to rock 'n' roll, too young to fly" (March 15, 1992). Richard fought back, casting himself as an upstart David against a greedy Goliath. He continued to accuse *British Airways* of unethical tactics, prompting Lord King to question Branson's truthfulness publicly. Branson sued *British Airways* for libel in December 1992, and *British Airways* offered the highest uncontested libel payment (£610,000) in British history. Richard Branson shared the settlement with the *Virgin Atlantic* staff. The court victory marked a turning point for his airline. By the end of the 1990s it had become the third-largest European carrier and the most profitable company in the *Virgin* group.

What would have happened if Richard Branson had given up in his struggle with *British Airways*? Many thousands of people *Virgin Atlantic* now employs might not have had jobs (Branson, 2011b) and the consumers from around the world might not have flown one of the best airlines. That would be a huge loss for the UK and the global economy.

Bill Gates also did not give up, although he was extremely depressed by a lack of his company's success in its very early years. At one point, Gates wanted to sell the rights to his BASIC software language for just $6,500 because his products weren't selling well. Had he done that, *Microsoft* may never have come to exist (Wallace & Erickson, 1992). Up to now *Microsoft* created 92,000 jobs worldwide and had been recognized as the number one global workplace by the Great Places to Work Institute in 2011.

Akio Morita did not give up either when he could not find American distributors for *Sony* products in the 1960s—the same line of products that later included *Sony*'s ubiquitous and highly profitable Walkman.

As discussed above, Herbert Kelleher did not give up in his battle with competitors and, as a result, there is one of the excellent airlines. In 2010 *Southwest Airlines* flew 86 million passengers, more than any other airline within the United States. It operates more than 3,300 flights a day and, as of January 2012, the company has scheduled service to 97 destinations in 42 American states.

The personality-related factors are thus an important group of character traits, which help innovators to overcome the multiple barriers they face on the way to implementing their original ideas into practice in the form of new products, processes, and services.

Ignore naysayers!

One of the attitudes that help great innovators not to give up is their ignorance of naysayers. For instance, when Michael Dell decided to expand his young computer company internationally, everyone told him he was out of his mind. So, he did what any innovator would: he went ahead with it anyway.

Dell's first international expansion was to the United Kingdom in 1987 and the business was profitable from its very first days. Now *Dell U.K.* is almost a $2-billion-dollar-a-year company. Michael Dell recalled later that all but one of the 22 reporters at the press conference announcing the expansion predicted failure. They said it was a bad idea, that the direct business model was an American invention that would not work in other countries. Even *Dell* employees believed it was silly.

So, what does *Dell*'s success with international expansion say about innovators? In his own words, the lesson is "believe in what you're doing. If you've got an idea that's really powerful, you've just got to ignore the people who tell you it won't work, and hire people who embrace your vision" (Dell, 1999, p. 4).

Richard Branson echoes this approach. According to him, every true entrepreneur has to love solving complex problems and wish to do what others believe is impossible.

> A successful entrepreneur likes to prove that people around him are not right. 90% of businessmen know that when a great idea comes to their mind and they discuss it with friends and colleagues; nobody ever supports them. Almost everyone will put arguments forward against the implementation of that idea. Good entrepreneurs are those who accept this as a challenge and want to demonstrate to all skeptics that they are wrong.
>
> *(Branson, 2012)*

And he proved it on numerous occasions. Thus, even in bad times (e.g., in 2001) he continued to dream up new ventures. One project, *Virginstudent.com*, was a youth-oriented Website that recalled Branson's *Student* days. In interviews he dismissed talk that the *Virgin* brand had become overextended. "That's been said for about 30 years," he told the *Sunday Telegraph* (Branson, 2011a). It is amazing that such talks do not stop him.

The ability to never give up is, therefore, a distinguishing characteristic of outstanding innovators.

"It's okay to risk making mistakes, but it's not okay to be fearful"

Jeff Bezos, the founder of *Amazon.com* is another impressive example, who, as Vandervert put it, seems to have the "almost did not do it, but did" magic (Vandervert, 2012). Although *Amazon.com* is an amazing success story today, during its accelerated growth the company had equally amazing problems.

Sales were staggering from the beginning, but covering business expenses was costly. Whatever money the company made was already spent. Some shareholders grew impatient with Bezos's promises of profit ... Amazon.com was spending more than they were earning. Or, as Bezos put it, "*Amazon.com* was actually profitable in December 1995 ... for, oh, about one hour" (Sherman, 2001, p. 62).

People, of course, were wondering if there was something wrong with the way *Amazon* was being run. How could the company have so many customers, they asked, yet not show a long-lasting profit? Definitely, Amazon's customers seemed to be happy with what Amazon was offering. But the "problem" was that Bezos was continuing his practice of turning most of the company's earnings back into Amazon to pay for improvements rather than showing a profit (Sherman, 2001). The future of the company was thus uncertain. There was a possibility it would not survive (Garty, 2003).

Skeptics, critics, and doubters fully expected that brick-and-mortar retailers like *Barnes & Noble* or *Borders* would soon shoulder the young start-up out of the online book market. Others said the company was burning through its cash too quickly. It was facing lawsuits from competitors and partners. *Walmart* claimed that *Amazon.com* was hiring away employees, while Toys "R" Us sued to end its partnership with *Amazon.com*. Many were sure the company would fail (Robinson, 2010).

However, Bezos did not back down. "We want to build something the world has never seen" (Ryan, 2005, p. 73) and "we are going to be unprofitable for a long time. And that is our strategy," Bezos told *Inc.* magazine in 1997. He and his chief financial officer, Joy Covey, made it clear to investors that *Amazon.com* was different. "The Company believes," stated the plan they presented, "that it will incur substantial operating losses for the foreseeable future, and that the rate at which such losses will be incurred will increase significantly from current levels" (Ryan, 2005, p. 54). That is, they warned the investment worlds that *Amazon.com* was not a profitable operation even though it intended to be one someday. In September 1999, *Fortune* magazine credited Joy for "convincing Wall Street that a profitless company was worth $22 billion" (Ryan, 2005, p. 77). This overall outcome sums up the value of one of Bezos's basic beliefs: it's okay to risk making mistakes, but it's not okay to be fearful.

Jeff did not give up. The doom-and-gloom predictions turned out to be wrong. *Amazon.com* earned its first full-year profit in 2003 and, by 2008, the company's revenue had reached $4 billion. *Amazon.com* succeeded in large part because of Bezos's vision that mostly consisted in quickly embracing e-commerce innovations that improved its customer experience. Such standard operating procedures as one-click shopping, e-mail verification of orders, and customer product reviews were not on the radar until *Amazon.com* adopted them. In order to implement his vision, Jeff had used whatever was earned from sales to expand the business. This is why in 1999 alone he purchased nine smaller Internet companies, opened seven new online stores, and built five huge warehouse distribution centers. He preferred to focus on investing in new and wider markets, which totaled 13 million customers, rather than declaring profits which Wall Street would applaud (Ryan, 2005). Bezos proved the powerful intuitions behind his strategy were correct, and he won. Today, *Amazon.com* hires 65,600 employees.

The impact of "saved" ideas and thus implemented innovations on the economy

If one looks at the end results of those ideas, which great innovators implemented into practice by founding or co-founding new companies, one can see a tremendous impact on the economy. Together, Akio Morita, Fred Smith, Richard Branson, Herbert Kelher, Michael Dell, Bill Gates, and Jeff Bezos created 758,800 jobs worldwide. It is quite remarkable both for national and global economies. Plus, taxes paid from the billions of dollars in annual revenues. Besides that, their companies became the leaders in their respective areas of business thus introducing and defining the rules of the game in those industries.

Moreover, innovators not only created many new jobs; they created the best places to work. It reflects itself, for instance, in a great number of résumés that their companies receive every year. It demonstrates how much people want to work in those organizations. Thus, as it follows from the company's website, in 2009 *Southwest Airlines* received 90,043 résumés but hired a mere 831 people. In 2010, it received 143,143 résumés but hired only 2,188 people, making it harder to get a job at *Southwest* than to get into a prestigious Ivy League college.

Therefore, the impact of innovators' implemented ideas on the economy is great.

Conclusion

This chapter introduced and described the phenomenon of the "abortion" of new ideas. It offered one of the possible explanations regarding why people abandon their ideas. It also roughly estimated the impact of "saved" ideas and, therefore, implemented innovations on the economy in the case of prominent innovators known for their ability to both generate creative ideas and to put them into practice in the form of new products, processes, and/or services. The presented case studies of Akio Morita, Fred Smith, Richard Branson, Herbert Kelleher, Michael Dell, and Bill Gates demonstrated how they did not give up in the face of multiple obstacles and implemented their ideas.

Further research is definitely needed. It should be related to the evaluation of the potential value of innovators' abandoned or aborted ideas (they all had such ideas in addition to the implemented ones) by estimating the possible profit from them and then counting the potential impact of "killed" ideas and, consequently, lost innovations on the economy. This is new, groundbreaking research with a great potential to advance knowledge in innovation science, education, economy, entrepreneurship, business, and public policy.

The "abortion" of ideas is dangerous for any society. Therefore, every effort should be made in order to teach children, adolescents, and adults how to save, develop, and implement their ideas into practice. This is a vital goal of innovation education. By saving many potential innovations, we will thus fuel the global innovation-based economy.

Acknowledgments

The work presented herein was supported under the Support for Innovative Projects program (Grant AN-143064) of the *Fonds québécois de la recherche sur la société et la culture* (FQRSC). The findings and opinions expressed in this chapter do not reflect the positions or policies of the FQRSC. I am grateful to Larry Vandervert for his helpful comments on early drafts of this chapter.

Notes

1 Together with Mr. Ibuka, Akio Morita implemented his idea of a company producing highly innovative products and thus founded *Sony*, which currently consists of 168,200 employees worldwide.
2 This case—as an example of innovators' courage—is also briefly discussed in Chapter 3, *The fundamentals of innovation education*, included in this volume.

References

Alexander, T. (2012). Interview. *Forbes*. Downloaded from www.referenceforbusiness.com/biography/A-E/Branson-Richard-1950.html#b on January 17, 2012.

Branson, R. (2002). *Losing My Virginity: The Autobiography*. London: Virgin Books.

Branson, R. (2008). *Screw It, Let's Do It*. London: Virgin Books.

Branson, R. (2009a). *How Branson Achieved Success*. Downloaded from www.youngentrepreneur.com/forum/f37-famous-entrepreneurs/richard-branson-25278.html on September 17, 2009.

Branson, R. (2009b). Virgin entrepreneur. *Success Magazine*. Downloaded from www.successmagazine.com/virgin-entrepreneur/PARAMS/article/712 on September 17, 2012.

Branson, R. (2011a). Interview. *Sunday Telegraph* (February 25). Downloaded from www.referenceforbusiness.com/biography/A-E/Branson-Richard-1950.html#b on January 17, 2012.

Branson, R. (2011b, November 15). Interview. *Guardian*. Downloaded from www.guardian.co.uk/society/2011/nov/15/richard-branson-champions-employment-ex-offenders on January 17, 2012.

Branson, R. (2011c). *Screw Business as Usual*. New York: Portfolio/Penguin.

Branson, R. (2012). Interview. Downloaded from http://infobank.by/831/Default.aspx on January 20, 2102.

Dell, M. (1999). *Direct from Dell*. New York: HarperBusiness.

Garty, J. (2003). *Jeff Bezos: Business Genius of Amazon.com*. Berkeley Heights, NJ: Enslow Publishers.

Hadjimanolis, A. (2003). The barriers approach to innovation. In L. V. Shavinina (Ed.), *The International Handbook on Innovation* (pp. 559–573). Oxford, UK: Elsevier Science.

Morita, A. (1987). *Made in Japan*. London: Collins.

Robinson, T. (2010). *Jeff Bezos: Amazon.com Architect*. Edina, MN: ABDO Publishing Company.

Ryan, B. (2005). *Jeff Bezos: Business Executive and Founder of Amazon.com*. New York: Ferguson.

Shavinina, L. V. (Ed.) (2003). *The International Handbook on Innovation*. Oxford, UK: Elsevier.

Sherman, J. (2001). *Jeff Bezos: King of Amazon*. New York: 17th Street Productions.

Vandervert, L. (2012). Personal e-mail communication, February 17.

Wallace, J., & Erickson, J. (1992). *Hard Drive: Bill Gates and the Making of the Microsoft Empire*. New York: HarperCollins Publishers.

Part XI
Policy implications, institutional, and government efforts in innovation education

41

Innovation education through science, technology, engineering and math (STEM) subjects

The UK experience

Frank Banks

THE OPEN UNIVERSITY, UK

Summary: This chapter draws on 30 years of projects and initiatives across England and the other three states of the United Kingdom intended to increase the relevance of the curriculum to life outside the school, to promote creativity and enterprise and to foster innovation through 'minds-on' as well as 'hands-on' teaching strategies. Through an analysis of both successful approaches and a discussion of 'lessons to be learned', there is a consideration of the Technical and Vocational Educational Initiative (TVEI) in the 1980s, the introduction in the 1990s of the manufacture of innovative products through the new subject 'Design and Technology' for all students aged 5–16 years and, recently, the collaboration with other STEM subjects. There is much to celebrate in promoting and facilitating innovative Product Design for students of all ages across UK, but there have been many obstacles to overcome at national and local levels too. This chapter explores both.

Key words: Design, Technology, Science, Mathematics, STEM, TVEI

Introduction

Many teachers approaching retirement in the United Kingdom began their teaching career in the mid 1970s. It was, in terms of teaching, then a completely different world from the one in which they teach today. As one would expect, there have been changes in the access to educational technology; to electronic whiteboards and easy access to the information on the internet. But more significantly, there has been a sea-change in the external control of the curriculum and the monitoring of student attainment through national standardised testing. There has been a move from each school – or even a class teacher – designing their own curriculum, to a nationwide detailed prescription by government agencies of what should be taught. What is meant by 'national' has changed too. The United Kingdom (UK) is composed of the four nations of England, Northern Ireland, Scotland and Wales, and over the last 30 years the different states have taken increasing direction and control of their own education systems. Scotland always has been markedly different in the way it has organised its education from the rest of the UK, but Wales and Northern Ireland have established their own national government assemblies and now stress different aspects of the curriculum for their own needs as nation-states and, in particular, how aspects of the cultural and creative subjects, including Design, could promote the skills, knowledge and attitudes for future innovative citizens.

In 1976, James Callaghan, the Prime Minister, made a speech at Ruskin College, Oxford, in which he called for a 'Great Debate' in education. There had been a series of pamphlets published by opposition right-wing politicians and their supporters severely criticising the then standard of teaching and the nature of what was taught in government schools. Subjects such as Peace Studies were on the curriculum of many London schools, for example, infuriating the right-wing press, and there was a general feeling that teachers as a profession were themselves rather too much in control of the nation's future. The UK economy was moving from one based on large-scale manufacturing (sometime by state-controlled industries) and the need for many generally low-skilled jobs, to one requiring a much higher skilled workforce for high value-added industries and a dependence on financial services.

In short, it was considered that the school curriculum was inappropriate, particularly in the way it dealt with vocational subjects and Callaghan wished the government to regain the initiative in the debate. Fifteen years after the Ruskin 'Great Debate' speech Callaghan confided to me that his civil servants in the Department of Education were very unhappy that he had decided to make a speech on education at all. They advised that it was not the business of politicians to interfere in what schools should be teaching their students and that he would be accused of suggesting political indoctrination at worst and improper government interference at best. With the benefit of hindsight it is amazing how naive and complacent were the views of those civil servants. Relying on the 'back-wash' effect of the school-leaving public examination system to dictate what was taught in schools meant that there was an emphasis on cerebral academic subjects at the expense of the creative and vocational, and generally very little was taught that promoted innovation and enterprise. Some deviations from the academic thrust of the curriculum were possible through what was known as Mode 3 Certificate in Education courses, where teachers designed both the curriculum and its assessment themselves in their school. For example, courses in Engineering were taught in some schools, but there was no parity of esteem of these vocationally oriented areas of the curriculum with the more academic, and rather abstract, subjects such as Physics.

In 1979 there was a very big change in attitude to public services in the United Kingdom with the election of Margaret Thatcher. James Callaghan's premiership had ended with a 'Winter of Discontent' and a series of wide-scale industrial and public service strikes resulting in rubbish piling up uncollected in the streets and even the dead not being buried in a timely and respectful fashion. Thatcher's view was that the Unions were too powerful and their ability to strike needed curbing. Further, there should be a new era of private enterprise and that the attitude to the concerns and needs of business should be revised in the public's mind. The promotion of innovation and enterprise should be encouraged. The government proceeded to raise considerable sums of money through selling off shares by 'privatising' the public-owned companies and investing in ways to change perceptions of who could run businesses. In Thatcher's view, Britain would once again become not as Hitler once sneered 'a nation of shopkeepers' exactly, but certainly a 'nation of new entrepreneurs'.

TVEI

The Conservative government of the mid 1980s was concerned that school leavers were very ill-equipped for the workplace. To take the lead in rectifying this, in 1983 the government introduced the Technical and Vocational Education Initiative (TVEI). It was funded by the Department of Industry rather than the Department of Education and by the time of its eventual demise in 1997 almost GBP 1 billion had been spent (Yeomans, 2002). There were two broad aims of TVEI; first to align the school curriculum more closely to the 'needs' of industry and commerce and rectify some of the knowledge, skill and particularly the 'attitude deficits' of school leavers. The argument from politicians is simple and linear; that there is a direct link between better vocation preparation and better individual contribution to the workplace, and consequently greater economic growth. This view that the UK needs to 'get vocational education right' still continues today and over the years there have been many criticisms of the UK's vocational education system in comparison with that of much of the rest of Europe (Prais, 1995; Smithers, 1993). In the mid 1980s, the view that positive attitudes to commerce, industry and innovation were not properly promoted were considered to be due to a politically left-leaning teaching force who wished to educate their students broadly and who shunned any suggestion of a curriculum that might become too narrowly occupationalist.

The second aim was a view that learning should be more active and practical. This, in contrast, was embraced warmly by teachers and such a broad notion of vocational education – where new knowledge is required for the workplace along with the development of new interpersonal skills through team work and collaborative learning – was widely accepted and welcomed by the teaching profession. As a move to more active learning approaches, and new 'constructivist' ideas about learning were gaining wide approval in the teaching profession (Driver & Oldham, 1986), the considerable funding available through TVEI was often described by teachers themselves as being 'subverted'. Although some teachers might have baulked at teaching students to become 'merely factory fodder', they were more than content to adopt teaching techniques that promoted critical thinking and teach through links to real life and a practical purpose.

Two examples illustrate this change in emphasis in schools through TVEI funding and the promotion of innovation and enterprise. Many local government authorities had a TVEI officer who would provide small-scale seed funding to those schools who suggested new curriculum ideas which promoted innovative thinking by the students. In Powys in mid-Wales, for instance, students were asked to work in teams to design and make novel garden furniture such as 'plant boxes' and picnic tables which would be attractive for sale both to private users and robust enough for public use in parks. The students proposed a range of possible designs and negotiated what was wanted with potential customers and local government officials and, as a small collective, manufactured, advertised and sold their new products. At the other end of the scale was the TVEI funded national (UK-wide) promotion of 'Technology' as a new subject in the curriculum. Drawing together different curriculum subject areas such as craft and design with the physical sciences, a new subject was created that enabled students to design innovative products which required as part of their learning, in a practically based way, the real-world use of some scientific concepts. It was an 'applied science' approach to technology education which is still in place in Northern Ireland today.

Teacher preparation for this new practical subject was also funded through TVEI and involved teacher professional development using a mobile classroom/workroom containing appropriate equipment, and supported by a dedicated staff of teacher trainers. The training was very successful and the following lessons can be learned about how to prepare teachers for a new curriculum that promotes problem solving and encourages innovation in students:

- The syllabus was nation-wide and linked to a new examination.
- The mobile classroom staff offered a twenty-day programme but broken into four-week slots with a month between visits to enable suggestions on the training for new classroom practice could be carried out in school before the following session.
- Mobile classroom master-trainer staff were trained centrally to provide a high-quality and systematic experience.

- The mobile classrooms were equipped centrally to the same standard and through bulk-buying to increase cost-effectiveness – equipment included construction kits, electronic and pneumatic components, computers and the means to link the computers to control external devices.

This new subject combined the practical hands-on experience of the craft department in schools with the more cerebral (but often abstract) experience of the physical sciences. It was 'minds-on' as well as 'hands-on' and took advantage of new equipment such as system electronic boards which enabled students to invent their own electronic devices as they used the functional system blocks (INPUT-PROCESS-OUTPUT) to physically 'jig-saw' together their ideas to solve everyday problems and create new electronic products such as a liquid level detector for the blind, or an automatic window-opener for a greenhouse.

But such cross-subject cooperation was not to last. In 1988 the Department of Education was tasked with producing a national curriculum for all students for the years of compulsory schooling in England and Wales; from the ages of five to 16. Northern Ireland followed quickly, but Scotland preferred to produce a set of suggested 'guidelines' rather than legal documents. In England, Wales and Northern Ireland, in contrast, the new national curriculum was enforceable by law in government schools. Science was introduced first and the extent of the prescribed curriculum was such that it excluded any time for collaboration with other subject areas. It was as if the Department of Education did not know that the Department of Industry funded TVEI programme with its cross-curricular sympathies even existed. All of the collaborative work on a more integrated 14–19 curriculum was largely lost as a rather 'traditional' return to separate academic-based subjects won the day. Two separate subjects were created science and 'technology', soon changed to 'design and technology'. These along with mathematics now form the STEM subjects in school: Science, Technology, Engineering and Mathematics. As the success of a school is largely judged on its performance in so-called 'league tables' of examination results, there followed a period of inter-subject wrangling, each subject stressing the difference between their purposes and the importance of how they each contribute to the students' capabilities in later life.

Despite the school politics that has surrounded the teaching of the two subject areas of 'Science' and 'Technology' for many years, which stresses the differences, there are indeed some clear commonalities. Both subjects opened up the academic breadth of the group of students who would take the subject. For example, science moved from being a specialist subject just for those going on to study it at higher levels to embrace a 'science for citizenship' focus for all. Design and Technology with its emphasis on manual dexterity had to appeal to the academically 'able' as well as the traditional group of 'non-academic' students; indeed both subjects make much of 'hands-on' learning; both try to explicitly link school tasks to useful learning for everyday life and the needs of the work-place; along with mathematics the STEM subjects claim to promote problem solving and other 'processes'.

What do the STEM subjects contribute to innovation education?

The following statements are from the current national curriculum in England published in 2007 (my emphasis) and taken from the Department for Education website:

The importance of science
The study of science fires students' curiosity about phenomena in the world around them and offers opportunities to find explanations. It engages learners at many levels, linking direct practical experience with scientific ideas. Experimentation and modelling are used to develop and evaluate explanations, **encouraging critical and creative thought**. Students learn how

knowledge and understanding in science are rooted in evidence. They discover how scientific ideas contribute to technological change – affecting industry, business and medicine and improving quality of life. They trace the development of science worldwide and recognise its cultural significance. **They learn to question and discuss issues that may affect their own lives, the directions of societies and the future of the world.**

The importance of design and technology
In design and technology students combine practical and technological skills with creative thinking **to design and make products and systems that meet human needs**. They learn to use current technologies and consider the impact of future technological developments. They learn to think creatively and intervene to improve the quality of life, solving problems as individuals and members of a team.

Working in stimulating contexts that provide a range of opportunities and draw on the local ethos, community and wider world, students identify needs and opportunities. They respond with ideas, products and systems, challenging expectations where appropriate. They combine practical and intellectual skills with an understanding of aesthetic, technical, cultural, health, social, emotional, economic, industrial and environmental issues. As they do so, they evaluate present and past design and technology, and its uses and effects. Through design and technology students develop confidence in using practical skills and become discriminating users of products. **They apply their creative thinking and learn to innovate.**

The importance of mathematics
Mathematical thinking is important for all members of a modern society as a habit of mind for its **use in the workplace, business and finance; and for personal decision-making**. Mathematics is fundamental to national prosperity in providing tools for understanding science, engineering, technology and economics. It is essential in public decision-making and for participation in the knowledge economy.

Mathematics equips students with uniquely powerful ways to describe, analyse and change the world. It can stimulate moments of pleasure and wonder for all students when they solve a problem for the first time, discover a more elegant solution, or notice hidden connections. Students who are functional in mathematics and financially capable are able to think independently in applied and abstract ways, and can reason, solve problems and assess risk.

Mathematics is a creative discipline. The language of mathematics is international. The subject transcends cultural boundaries and its importance is universally recognised. Mathematics has developed over time as a means of solving problems and also for its own sake.

(DfE, 2007)

These three statements lay out what has been the culmination of a change process in the STEM school curricula in England over the last 25 years, namely the rationale for the designation of these separate subjects as required areas of study during the compulsory school years. In 2005 the requirement for all students to study Design and Technology (D&T) was restricted, and D&T is struggling to remain an obligatory subject between the ages of 5–14 years. A review of the national curriculum in 2011 recommended D&T should be removed completely from the national curriculum, meaning it should not be specified in law, although it should still be taught in schools. Learning of Science and Mathematics, however, is reinforced as a requirement for *all* students up to 16 years of age.

What lessons can be drawn about making these subjects compulsory for all students? What is the curriculum rationale behind the 'importance statements' listed above, particularly in relation to promoting innovation in students? What decisions were taken about what all should be able to 'know, understand and do' as a result of studying STEM in the school curriculum and what was communicated to teachers, students and their parents?

Science led the way in 1988 and continues to do so. The nature of a science curriculum for all students – a science for citizenship or 'scientific literacy' – was being debated which is still highlighted now with an emphasis on 'How Science Works' in the current curriculum. In particular, the importance of learning 'facts' in science was questioned and a case was made that the *processes* of the scientific method were much more important for all students. The government policy document Science 5–16 (DES, 1985), that pre-dated the national curriculum, did not merely define what should be taught in terms of content such as 'electricity' or 'plants', but instead emphasised the importance of a process approach. Indeed, science curriculum innovation in the mid 1980s saw a large number of new courses such as 'Warwick Process Science' and 'Science in Process' for secondary schools. These focused, not on science concepts, but rather on processes such as observation, interpretation and classification – aspects critical to 'the scientific method'. The idea is that when all the science facts have been forgotten, what remain are the useful processes that promote critical thinking and facilitate innovation.

There emerged a generally common consensus that science might be more accessible to all students if it emphasised skills applicable to other areas of life both inside and outside school. The attention to 'doing' science – raising questions that could be answered by an investigation – became the cornerstone of the developing science curriculum in elementary schools. For example, the question 'What is the best carrier bag?' would be turned into an investigation question such as 'Which carrier bag carries the greatest weight?' in what was considered a problem-solving approach. To answer such a question, 'independent and dependent' variables were identified. Thus the importance of the procedural knowledge of science was developed.

There was something of a backlash to the 'process is all that is important' line and the debate became heated (see Millar & Driver, 1987; Millar, 1988; Screen, 1988; Wellington, 1988, 1989; Woolnough, 1988). Some argued that, for example, 'observation' in isolation and done just for the sake of it was pointless – one had to apply the process of observation within the context of the understanding of science concepts. This was also accompanied by research-led initiatives emphasising constructivist learning ideas, resulting in a concern for student conceptual development and the associated pedagogy. The science curriculum as the acquisition of 'facts' was, however, very deep-rooted. The national curriculum for Science was first published in 1988 and, although it had an area devoted to process issues, was largely a re-emphasis on teaching 'content' or conceptual knowledge. The balance had shifted again away from procedural knowledge, reinforced at all levels by national testing which, despite the rhetoric of a concern for assessing understanding, rather emphasised memorisation and did not include an assessed practical element for younger students. In the rapid revisions of the science curriculum over the last 20 years, the push has been to cut back on the extent of content in the curriculum but the premier position of scientific method, so to the fore in the early 1980s, would never be regained. Throughout that period however we have the shift in concern for the balance of procedural and conceptual knowledge, a theme which is reflected too, although in different ways, in both the subjects of technology and mathematics.

Design and Technology is a relative newcomer to the curriculum for all students from 5 to 16 years. The compulsory national curriculum was introduced in 1990 and focused on Technology (as it was first called) as a process concerning design. It had four attainment targets:

- Attainment target 1 – identifying needs and opportunities;
- Attainment target 2 – generating a design;
- Attainment target 3 – planning and making;
- Attainment target 4 – evaluating,

This process-based curriculum was difficult to implement for both secondary and primary schools. Primary teachers were unused to considering formal aspects of designing, although craft activities had long been a feature of primary school life. It was also suggested that, at secondary level, a wide range of teachers become involved to cover technological processes in material areas such as food and textiles, and aspects of business studies, as well as the more traditional materials of wood, metal and plastics. Few secondary teachers had practical experience of design in the way they had high skills for craft work.

After only two years of the new curriculum, the Engineering Council in England produced a damning report which declared that 'Technology in the National Curriculum is a mess' (Smithers & Robinson, 1992, p. 1). Their main criticism was that by defining technology solely through a process approach meant that almost all problem-solving activity such as producing a play or writing a book could be considered as 'technology'. 'Defined on problem-solving alone, most activities become technology – writing this report, conducting a scientific experiment, finding one's way to a railway station. What is needed is some statement of technology's domain' (Smithers & Robinson, 1992, p. 3). The report made recommendations as to what should be considered the subject domain of technology and what should not, and for a better balance between process and content. It also tried to untangle the 'vocational' and 'basic skills' labels that some had attached to the new compulsory subject, and it advocated a consideration of the 'literature' of technology; looking at and learning from the products and artefacts that already exist which can inform designing and making. Subsequent developments tried to address these concerns. In 1995 a new version of the curriculum for England and Wales gave a clearer steer to what D&T was, and the main activities that should be employed:

> Students should be given opportunities to develop their design & technology capability through:
> Assignments in which they design and make products, focussing on different contexts and materials and making use of:
> Resistant materials;
> Compliant materials and/or food (DMAs – Design and Make Assignments).
> Focused practical tasks (FPTs) in which they develop and practise particular skills and knowledge;
> Activities in which they Investigate, Disassemble and Evaluate familiar products and Applications (IDEAS).
>
> *(DFE/WO, 1995, p. 6)*

This methodology strongly reflected the pedagogic model promoted by Nuffield Design and Technology (Barlex *et al.*, 1994). There was a reduction to two attainment targets that had looked for progress in each part of a process, 'Designing' and 'Making' and, eventually, as even this separation was considered as unhelpful, to just one attainment target 'Design and Making'. Although there was a better balance including *cultural understanding, creativity and critical evaluation*, the emphasis was still based around the design and making process, with students being expected to achieve the following at the highest level (my emphasis):

> Responding creatively to briefs, they are discriminating in their selection and use of information sources to support their work. They interpret and apply knowledge and understanding creatively in new design contexts and communicate ideas in new or unexpected ways. They use understanding of others' designing in innovative ways. They work with tools, equipment, materials, ingredients and components to a high degree of precision. **They make products that are reliable and robust and that fully meet the quality requirements given in the design proposal**. They reflect critically and effectively throughout designing and making processes.
>
> *(DES, 1990, p. 23)*

Technology (now D&T), although a subject required to be taught to all students under the 1990 national curriculum, was always under attack from those who could not see the justification for that position. The reasons for the animosity range from those who would put 'science and technology' together as one curriculum domain (especially at primary school level) to those who, more prosaically, just considered the subject too expensive to teach to all pupils in terms of tools, equipment and materials. Education for Innovation is expensive. The response from the D&T lobby was to argue that the subject was important as it prepared 'students to participate in tomorrow's rapidly changing technologies' and so much was done to introduce new technologies such as Computer Aided Design and Manufacture (CAD/CAM) mainly at secondary school level as a tool for the designing and making processes (Banks & Owen-Jackson, 2007).

Mathematics teachers have never had to convince government ministers that it is a core subject that all students should study. It has always been seen as an essential element in a curriculum for all. However a consensus as to *what* the content of the mathematics curriculum should be was not so straightforward. Just as the nature of the science curriculum was extensively reviewed in the 1960s, starting in the USA as 'New Math' so was mathematics. New mathematics was promoted extensively in the UK by the Schools Mathematics Project (SMP). With an emphasis on concepts such as set theory, functions, number bases other than 10, matrices and algebraic inequalities the notion was that such a foundation would enable children to understand theorems later. This was very much a view pointed up still in the above importance statement that 'Mathematics has developed over time as a means of solving problems and also for its own sake'. However, much criticism of this approach to mathematics teaching came both from parents who complained that the concepts were too remote from the child's everyday experience and from teachers who did not fully understand many of the new concepts that they were required to teach. There were more principled objections too as some argued that mathematics is cumulative, and it is not possible to grasp new mathematical ideas unless one has grasped older ideas first.

Mathematics under the national curriculum in England gave much more common structure as a compulsory subject and, in 2009, received some attention as part of a key national strategy into five strands of progression

1. Mathematical processes and applications
2. Number
3. Algebra
4. Geometry and measures
5. Statistics.

The sub-categories in Statistics are particularly interesting in a STEM context:

- 5.1 Specifying a problem, planning and collecting data
- 5.2 Processing and representing data
- 5.3 Interpreting and discussing results
- 5.4 Probability

and have a certain similarity with the 'design or technology process' (DfE, accessed 2012).

So, in science, technology, and mathematics there has been debate over the last 25 years as to the balance that should exist in the specified curriculum between procedural knowledge and conceptual knowledge. However, these debates have largely been within each subject community, independent of each other, and tend to emphasise the inevitable differences between the goals of each subject rather than the common ground the subjects occupy. What can be learned generally, and what can the different STEM subjects learn from each other to promote student creativity that leads to innovation?

A key lesson to be learned by the rapid revisions of the specified curriculum of science, D&T and mathematics in England over the last 25 years is that it is very difficult to impose a curriculum on to teachers be it from central government or from within a school structure (Banks et al., 2004). A top-down method of seeking to describe the curriculum in close detail without working with teachers – and those involved in pre-service and in-service teacher education – to develop a common understanding of purpose, leads to a mismatch between a teacher's own view about their subject and what is prescribed to be taught. Teachers have a personal view about what their subject is about and, although they wish their students to do well in externally set examinations, when the specified curriculum moves independently of these deeply held views teachers feel obliged to 'teach to the tests' but lose some of the fire and passion for their subject. It is therefore imperative that the tests accurately reflect the intentions of the curriculum designers.

The importance of *teaching* problem solving to promote innovation

From our research, and that of Bob McCormick and other colleagues at The Open University, there is considerable evidence that problem solving – a key aspect of all STEM subjects that leads to innovative practices – is often conducted in a sort of 'ritual' way in school classrooms (see Banks, 2009; McCormick & Davidson, 1996; McCormick et al., 1994). In D&T, for example, this ritual is the way the problem solving process of designing is enacted by some teachers, with a limiting experience for students. This was one of the results of the imposed national curriculum.

Let us follow a small case study of a teacher with 12–13-year-old students working on an electronic badge project based on a 'face' with LEDs for eyes these cases are drawn from (Banks & McCormick, 2006). The teacher deliberately did not emphasise the design process; it was not one of his main aims, and he seemed to view designing as a logical approach rather than as a process that involved sub-processes to be taught and learnt. He said:

> although I'd like them to understand and use the design process and I think it's quite a nice framework for them to fit things on to, I don't think there's a great need to be dogmatic about it and say you must learn it... the nature of projects leads them through the design process despite the teacher's bit, going through it with them in front of the class.

He appeared to see the 'logical approach' as a 'way of working', and in that sense the sub-processes were of little significance to him. For him the design process was very much in the background, not just in this project but in general: 'I'm relying rather a lot on a subconscious level of going through things. Some of them won't do it, some will'.

In D&T lessons the particular view that a teacher takes of the design process affects the way tasks are structured, the kinds of interventions that are made by the teacher and the assessment of students' work. Not all of these will be consistent either with each other, or with the view espoused by a teacher, but collectively they will have a profound effect on the students' perceptions and activities. But, whatever view is taken of designing, there is a tendency to see it as an algorithm to be applied in a variety of situations.

The teacher involved in the *electronic badge project* began it with the 'Situation' being presented:

> A theme park has opened in [place] and it wants to advertise itself. It plans to sell cheap lapel badges based on cartoon characters in the park. To make these badges more interesting, a basic electronic circuit will make something happen on the badge.

This was set within the general title of 'Festivals', but the links to the 'Situation' were not discussed, and from then on no further reference was made to festivals. The teacher continued in the session by asking the students to define the 'Design brief' and draw up a spider diagram of 'Considerations' (a specification), tasks which all the students seemed familiar with. He did not, however, elaborate on the 'Situation' or the 'Design brief', or invite students to discuss them in the context of the planned project.

The three students we followed (we'll call them Bill, Tanvir and Rose) produced different design briefs that illustrated how the 'Situation' was interpreted by them. Bill and Tanvir interpreted it as a 'button is pressed to light up the eyes', whereas Rose makes no such inference: 'to design and make a clock badge'. Their initial ideas of their personal 'briefs' lingered and influenced future tasks; for example, Rose continued to talk about a 'clock face' for several lessons and abandoned the idea only when she realised that the electronics would not be like that of a watch. She also imagined that the battery would resemble that in a watch and was almost incredulous when the teacher showed a comparatively large conventional dry 9-volt battery that she (rightly) considered too heavy for a lapel badge. The teacher's discussion with Rose about this issue indicated that unlike Rose, he had not entered into the 'Situation' and 'Design brief' in a meaningful way, but only ritualistically – his ultimate answer to the problem was to 'have a strong pin for the badge', a response Rose felt very dissatisfied with!

Next the teacher gave several tasks relating to drawing the faces for the badge, which implicitly reflected the sub-processes of 'generating ideas', 'developing a chosen idea' and 'planning the making'. However, this was again done in a ritualistic way as the following indicates. At the end of the first session students were asked, for homework, to create four cartoon faces as potential designs for the badge. No parameters were given other than that all four should fit into the design sheet and that students should be 'creative'. As with the 'Situation', 'Design brief' and 'Considerations', this step of producing four designs appeared to be a standard one and, again, was accepted without question by the students. However, in the next session students were asked to re-draw the faces so that they touch the sides of a fixed drawn square (70×70 mm). The reason for this was not made clear until a later session. Evidence from the students' folders indicates that students had to modify their designs in order to fit these new demands. For example, Rose had originally drawn a thin 'carrot' character, which she had to distort to make it fat enough for it to touch the sides of the square. The fact that the creation of several designs is sometimes perceived by students to be merely a ritual is seen in Rose's comments to the teacher implying she had in fact already made a final choice while she is still completing the four 'possible outcomes' drawings.

In looking at STEM teaching in the classroom we discovered some of the strategies that students actually adopted in response to the various ways the teachers viewed and enacted the problem solving process. These strategies certainly do not resemble the 'algorithms' or 'ways of problem solving' that are so often taught. The first strategy is what we characterised as *problem solving as dealing with classroom culture*. This occurs when students try to 'work out' the rules the teacher sets in the classroom, and play to those rules. Examples of students seeking this culture out is contrasted in the experience of two girls (Kathy and Alice) producing a mobile. Alice wanted to do something that clinks when the wind blows, and so had an idea of using metal. So, given a restricted choice of material, she chose to cut thick mild steel in the form of disks about two inches diameter. Because she played the rules of the classroom, Alice ended up with very sore hands, and took a long time; her endeavour resulted in a very inappropriate way of creating the effect she wanted. (But she did learn quite a lot about mild steel, as it turned out!)

Kathy had designed a moon and planets going around it, and wanted some kind of glinting material. When presented with the choice of material, Kathy in contrast to Alice looked elsewhere and saw some aluminium (not available to the class) and asked to use this. The teacher agreed, and she cut this easily with tin snips. Kathy took this approach many times throughout the project. She broke the rules of the classroom, knowing what she could and couldn't get away with. She experienced different kinds of issues and problems from Alice, but she was avoiding many technological problems that Alice faced.

The second strategy is *problem solving as giving and finding a solution*, illustrated in a project involving a moisture sensor. The teacher in this study defined the task in terms of making a box in which to put the electronics (the transistor circuit, the bulb or the little speaker, switch, etc.). This had to be appropriate to the situation of detecting moisture or lack of it. He taught them to cut the material (styrene) in straight lines with a steel ruler and a knife because when he said 'box', he had in mind a rectangular box. He also gave them a jig so that they could put the two edges together at right angles and run the solvent along to stick the two together. But some students wanted curved shaped boxes, which gave some of them at least three emergent problems. First, they had to cut a curved shape, and students asked each other and the teacher how to cut the shape as the steel ruler method wouldn't work (the solution was to cut it slowly). Second, a curved profile on one part of the box required one side to bend to follow the profile, but the styrene they were given was too thick. The students asked the teacher who simply gave them a thinner gauge of styrene, without any discussion. Third, the students did not know how to support or hold the thinner styrene in place to apply the solvent, and so again asked the teacher. This time the teacher had to think and was obviously solving the problem himself, but again he gave the *results* of his thinking as a ready-made solution to the students and did not involve them in his problem solving process. All they received was the solution without being involved in the problem solving. This continually being 'given solutions' becomes a culture of the classroom at the expense of a 'problem solving' culture.

In contrast, we found a teacher in an elementary school, who worked with younger children (ten and 11 year-olds), who was able to create this *problem solving culture* through interactions with her students. When students came up with problems, the teacher asked questions about their problem, or posed alternative solutions (because sometimes students cannot cope with the questions or provide solutions). Students were given more than one solution, because the teacher was trying to engage students in the problem and the problem solving process. Such a teacher has to set up a completely different culture in the classroom. It takes longer, and it is harder to do, but it is crucial to foster productive problem solving which leads to innovation.

The final strategy is the *student collaboration model*, and that happens in a variety of ways (see Hennessy and Murphy (1999) for the literature on collaborative activity and Murphy *et al.* (1996 and 2004) for an analysis of examples of collaboration). One way is through cooperation. In D&T in England students are usually set individual projects, so they may be working alongside each other on a table or a bench, and they can cooperate because they are doing similar things; they are not identical, but similar enough to help each other and share tasks. The second form of collaboration involved students in dividing up the task: 'You do this bit, I'll do that bit. You're good at that and I'm good at this'. Some of the learning is lost in this approach. But at least it is a way of collaborating, because they have to put the two bits together at some stage, and that has an element of good collaborative problem solving. The final form of collaboration occurs when students have a shared task, and they can talk about it. This means the design of the task must *require* the students to collaborate. Designed correctly tasks should require solutions to a problem to be considered by all students through discussion and decision making.

These four strategies of problem solving in the Design and Technology classroom differ from the way problem solving is depicted in the national curriculum, and the way STEM educators normally think about it. Without sensitivity to students' experience of problem solving the enacted curriculum will not have the required impact imagined by the teacher.

Summing up

What lessons can be drawn from a consideration of STEM education in the UK over the last 25 years?

- *It is very difficult to control the intended learning of students by an elaborate specification in law of what students should know.*

A curriculum specified as a legal document is open to challenge in the court if it is not carried out in schools. If teachers themselves are not part of the discussion on what STEM in school should be, they will 'teach to the test' to cover themselves leading to teaching strategies that have, for example, elements of 'ritual'. There will be a clash between their personal view of their subject and that specified by the State, and classroom practice will go through a period of extremes until some commonly shared beliefs of what constitutes 'good' teaching emerge.

- *In an effort to direct the learning outcomes for all students and make the tasks manageable in the classroom, teachers tend to closely direct the activity of students which can stifle the creativity that leads to innovation.*

Through constraints of time and resources, teachers transfer their subject into a form of 'School Knowledge' and students play the game of discovering what that is. Some students never quite understand the rules of the game and the relevance of the subject becomes lost to them; others pick up incidental aspects because teachers have either not made clear what is salient or their classroom culture produces effects at odds with their rhetoric.

- *The way that students engage in problem solving in D&T and in Science and in Mathematics depends on the view of designing and of investigating held by the teacher.*
- *Shared in-service training of teachers in STEM is highly successful. D&T teachers have much to teach Science teachers on the handling of processes and the Science teachers much to teach Technology teachers about the problems associated with acquiring conceptual knowledge. Mathematics teachers can help both with data handling and can learn about making a subject relevant to all.*

Our observation of the classrooms discussed above, however, would be that without organised collaborative teacher development, good practice in STEM classrooms is not shared well across schools and between schools. As new equipment such as ICT produces yet more teaching opportunities, we need to find out about their impact on the curriculum experienced by students inside and outside schools. In bridging the gap between the creative teaching of the production of innovative products, STEM offers some very exciting opportunities.

References

Banks, F. (2009). Research on teaching and learning in technology education. In A. Jones & M. J. de Vries (Eds.), *International Handbook of Research and Development in Technology Education* (pp. 373–390). Rotterdam, Netherlands: Sense Publishers.

Banks, F., Barlex, D., Jarvinen, E.-M., O'Sullivan, G., Owen-Jackson, G., & Rutland, M. (2004). DEPTH – Developing Professional Thinking for Technology Teachers: An International Study. *International Journal of Technology and Design Education. 14*(2), 141–157.

Banks, F., & McCormick, R. (2006). A case study of the relationship between science and technology: England 1984–2004. In M. J. de Vries & R. Custer (Eds.), *International Handbook of Technology Education* (pp. 285–312). Rotterdam, Netherlands: Sense Publishers.

Banks, F., & Owen-Jackson, G. (2007). The role of making in design and technology. In D. Barlex (Ed.), *Design & Technology for the Next Generation* (pp. 186–197). Whitchurch, UK: TEP & Cliffe and Company.

Barlex, D., Black, P., & Harrison, G. (1994). *Nuffield Design and Technology: INSET Guide*. Harlow, UK: Longman.

Driver, R., & Oldham, V. (1986). A constructivist approach to curriculum development in science. *Studies in Science Education, 13*(1), 105–122.

Department of Education and Science [DES] (1985). *Science 5–16: A Statement of Policy*. London: HMSO.

Department of Education and Science [DES] (1990). *Technology in the National Curriculum*. London: HMSO.

Department for Education/Welsh Office [DfE/WO] (1995). *Design and Technology in the National Curriculum*. London: HMSO.

Department for Education [DfE] (2007). *The National Curriculum*. Downloaded from www.education.gov.uk/schools/teachingandlearning/curriculum/secondary on 14 February 2012.

Hennessy, S., & Murphy, P. F. (1999). The potential for collaborative problem solving in D&T. *International Journal of Technology and Design Education, 9*(1), 1–6.

McCormick, R., & Davidson, M. (1996). Problem solving and the tyranny of product outcomes. *Journal of Design and Technology Education, 1*(3), 230–241.

McCormick, R., Murphy, P., & Davidson, M. (1994). Design and technology as revelation and ritual. In J. S. Smith (Ed.), *IDATER 94: International Conference on Design and Technology Educational Research and Curriculum Development* (pp. 38–42). Loughborough, UK: University of Loughborough.

Millar, R., & Driver, R. (1987). Beyond processes. *Studies in Science Education, 14*(1), 33–62.

Millar, R. (1988). The pursuit of the impossible. *Physics Education, 23*(3), 156–159.

Murphy, P., Lunn, S. A., McCormick, R., Davidson, M., & Jones, H. (2004). *EiS Final Evaluation Report. Evaluation of the Promotion of Electronics in Schools Regional Pilot: Final Report of the Evaluation*. Milton Keynes, UK: Open University.

Murphy, P., Scanlon, E., & Issroff, K. with Hodgson, B., & Whitelegg, E. (1996). Group work in primary science: Emerging issues for learning and teaching. In K. Schnack (Ed.), *Studies in Educational Theory and Curriculum*, vol. 14. Copenhagen, Denmark: Danish School of Educational Studies.

Prais, S. (1995). *Productivity, Education and Training: An International Perspective*. Cambridge, UK: Cambridge University Press.

Screen, P. (1988). A case for a process approach: The Warwick experience. *Physics Education. 23*(3), 146–149.

Smithers, A. (1993). *All Our Futures: Britain's Education Revolution*. London: Channel Four Television.

Smithers, A., & Robinson, P. (1992). *Technology in the National Curriculum: Getting it Right*. London: Engineering Council.

Wellington, J. (1988). Process and content in physics education. *Physics Education, 23*(3), 150–155.

Wellington, J. (Ed.) (1989). *Skills and Processes in Science Education: A Critical Analysis*. London: Routledge.

Woolnough, B. (1988). Whither Process in Science Teaching? *Physics Education, 23*(3), 139–140.

Yeomans, D. (2002). *Constructing Vocational Education: From TVEI to GNVQ*. Leeds UK: Post 14 Research Group, University of Leeds.

42
Policy on knowledge exchange, innovation and entrepreneurship

Alice Frost

HIGHER EDUCATION FUNDING COUNCIL FOR ENGLAND, UK

Summary: This chapter examines policies in the United Kingdom to connect higher education with innovation and enterprise in the economy and society, through support for a third mission of 'knowledge exchange'. It provides the historical and academic research underpinnings to this third mission, as well as locating these in international practice. It explores how the third mission leads towards an embedded environment in universities for innovation education, and draws some tentative conclusions on achievement to date and prescriptions for next steps.

Key words: Higher education, knowledge exchange, innovation, entrepreneurship, technology transfer.

Introduction

The central problem of this handbook is the development of the innovator: the creative individual who can put new insights into practice and, hence, spark, transform or make step change in, innovation in products and services, or broader economic and social developments.

While this central problem can be easily highlighted, the topic is actually about a great range of humans, human activities and situations of innovation that are not so easily characterised in simple terms or in a uniform model. One perspective is to focus on the heroic and creative individual – the great business entrepreneur or technologist; but there is also an argument that some of this creativity, particularly from the entrepreneur, is innate and not formed, or is formed in the fairly disordered 'university of life'. However, another perspective is that our economies and societies – and indeed knowledge and innovation themselves – are now so complex that they develop through mass endeavours and teams of knowledge-creators and innovators, who build incrementally upon a stock of human knowledge and experience (something like the World Wide Web).

In the UK, innovation has been a developing policy pre-occupation for some decades. The UK Government's Annual Innovation Report, BIS/NESTA (2010), calculated that innovation had accounted for 63% of annual labour productivity growth in this country since 2000. Early policy and research focused on commercialisation and the role of science and technology, particularly in manufacturing, as critical elements to innovation. The scope of innovation policy in the UK has though changed over time, covering a wider range of industries, including services, and the public and charitable sectors. Similarly, understanding of the mechanisms of innovation has changed, with a wider focus than research and development, now covering intangible investments. What is meant by 'innovators' and 'innovation education' needs to be set within this frame of much broader definitions and approaches to innovation.

This chapter then provides a national policy perspective on the framework and infrastructural conditions for the creation of innovators and innovation education, specifically focused on the role of higher education (HE). The chapter is set in the context of an argument that higher education has a distinctive and indeed unique role to play in the development of innovators, due to a range of factors related to the purpose of HE:

- An historic role in the development of thinkers and leaders, and increasingly of the technological specialists and professionals needed in a more knowledge-based economy and society.
- A prime, indeed possibly largely unique, responsibility for the development and expansion of the stock of knowledge in society; as well as responsibilities for archiving and preserving the knowledge of the past, and reconsidering its relevance to innovation in the present and future
- A major contribution to the development of methods of analysing data and evidence, as well as standards, to create new opportunities for innovation (new quantitative or qualitative tools, or use of computing).
- The primary lead in society for defining and developing the knowledge, skills and capabilities needed in highly and very highly skilled human capital (and hence their contributions to innovation), through the design of the curriculum. Research from NESTA (the National Endowment for Science, Technology and the Arts – an independent agency in the UK that focusses on innovation) in the Annual Innovation Index notes that graduates are particularly important in innovation; as an illustration, innovation active businesses have double the graduates of non-innovation active firms.
- An increasing responsibility to provide the appropriate environment for the transition between education and work of a significant proportion of the population in any country (around 40% of young people in the UK attend higher education (see HEFCE, 2005, while this may be much greater in other developed countries, such as South Korea with 80–90% participation). Increasingly this includes a focus on the development of employability skills within the HE curriculum and in practical experience, including support for enterprise and social enterprise. This focus on employability, including direct means to develop skills and experience in students, has increased in England due to the change to a greater student contribution to the funding of higher education (see for example BIS, 2011).
- A major role in the updating and refreshment of knowledge and skills in professional and technical experts.
- An increasing role in the provision of new knowledge and advice on inventive application of existing knowledge in new fields and sectors, through the range of knowledge exchange activities conducted in higher education, in conjunction with businesses and public and third sector organisations. This has always been an element in the work of universities due to their public benefit responsibilities to make their knowledge widely available. It has become a more important task as economies have become more knowledge based; and hence governments look increasingly to universities to contribute to the solution of global challenges, including now, economic recovery and growth. This role may become even more significant in future due to the scale, difficulty and cost involved in new technological development, leading to companies out-sourcing more research and development to universities and the adoption of open innovation models.
- The development of entrepreneurial skills in academics and students, with a view to enhancing knowledge exchange capabilities and to making direct economic and social contributions, such as formation of spin-out companies or social enterprises.

- A collective responsibility from the HE community to forming part of national innovation eco-systems – such as described in UK HM Treasury (2007), which includes responsibilities to conduct innovation, research and knowledge exchange activity and to raise absorptive capacity in the economy (the latter alongside others such as further education).
- A responsibility to contribute to regional and local economic development – with universities acting as 'anchors' (explorations on the anchor role in for example, The Work Foundation, 2010) that will remain rooted in a place and have the capability to support place-based innovation.

It could then be argued that the contribution of universities to innovation broadly – and more narrowly to the development of the innovator – flows from the conjunction of all the factors described above. This chapter initially describes the overall innovation approach in the UK. The chapter then examines policy and funding for knowledge exchange, how this includes enterprise education and how knowledge exchange is embedded within the processes of a university, to create an environment for innovation education. Finally, the chapter provides details of underpinning academic literature, and relevant international development. The chapter concludes with a presentation of UK achievements to date, and draws some very tentative conclusions on issues relative to the further development of innovation education.

The innovation context

In the 20th century, the role of university research in technological development became more important. There are some differences though between continental Europe and the Anglo-Saxon models, with free-standing research and innovation institutes (such as the Fraunhofer and Heimholz institutes in Germany) in the former, and universities playing a pre-eminent part in the latter (for example, described in Rathenau Instituut, 2009). US universities such as the Massachusetts Institute of Technology and Stanford University became particularly important players in technological development, in part due to impetus from Second World War defence investment (described in Clark, 1998). The UK with a similar higher education and research model to the US, imported policies towards a stronger contribution from universities to technology and innovation – technology transfer – in the 1970s. These policies followed a linear or 'push' model, with universities pushing out technologies into industry through Intellectual Property (IP) methods, such as licensing and spin-outs (the creation of new companies based on novel ideas). The technology transfer model succeeded an 'open science' model whereby academics' roles were primarily to publish findings, with handling of IP and innovation solely the responsibility of industry (described in Capart & Sandelin, 2004).

Academic understanding of the process of innovation developed over the 1980s, and into the 1990s (for example, see work of the Science Policy Research Unit, University of Sussex – such as Gibbons et al., 1994). The Government and other policy-makers in the UK rejected the linear model of innovation – the push model – towards new policies focused on more complex interactions between universities and business and other users (what is now termed 'knowledge exchange'). The UK Government's Office of Science and Technology (1993) White Paper, 'Realising our Potential', was one reflection of this new understanding of the innovation process. The linear model was rejected for a range of reasons:

- An understanding that a range of market – pull – related reasons affected which technologies could be taken up and commercialised. Producing technologies out of universities did not mean that these would or could be taken up by business, with the potential then for a high failure rate.

- Increased evidence that the transfer of technology involves a transfer of tacit or uncodified knowledge, and hence a company might not be able to use IP without developing these more complex understandings. An emphasis was then placed on people flow, as a means to transfer these more complex tacit knowledge domains.
- The recognition that absorptive capacity in business and other users was an important part of the equation of what could or could not be commercialised.

The Labour government in the UK in the 2000s took particular interest in the links between science/research and innovation, for example, in the HM Treasury (2004) *Science and Innovation Investment Framework*, which set out policy for a decade – 2004–14. Knowledge exchange between HE and business and other users was recognised as an important component of this innovation framework.

Over the 2000s, there have also been rapid developments in understanding of innovation systems and consequent policy changes. The innovation research organisation, NESTA, in the UK, conducted important research on the characteristics of innovation and requirements for innovation policy. As an example, NESTA highlighted the concept of Hidden Innovation in NESTA (2007) – this reflected that the UK, as with many developed economies, has a significant services sector. NESTA noted that innovation did occur in the services sector, but that its characteristics were different from those of manufacturing, the latter often associated with IP and industrial Research and Development. NESTA working with the relevant UK Government department, BIS (the Department for Business, Innovation and Skills), has produced an Annual Innovation Report and Index, BIS/NESTA (2010), that includes wider measures of innovation, such as investment in intangibles, including training and software, design and marketing and new business models. This wider approach is particularly important in the UK with a slightly higher proportion of economic activity in services than major competitors, such as Germany and Japan.

The UK approach described above has increasingly followed the Finnish model, introduced in the 1990s by the then Finnish President Aho, of an innovation eco-system (see for example Finland Ministry of Trade and Industry, 2003). The eco-system approach seeks to determine the roles of the range of actors and infrastructure contributors (such as universities and research institutes) relevant to innovation, and to set out frameworks and incentives overseen by government to encourage participants to work most effectively together for the national good.

The UK is presently working on an Innovation and Research Strategy to move these policies even further forward.

Knowledge exchange policy in the UK

As part of these trends in innovation described above, at the end of the 1990s, the Higher Education Funding Council for England (HEFCE) led development of a new concept of a *third stream or third mission* to higher education, described in HEFCE (1999), which was specifically focussed on the interactions between universities and users of knowledge of all kinds (business, public sector, charities and community bodies) – what is now termed in the UK, knowledge exchange. Given the inter-connected problems of technological innovation and absorptive capacity, the English third stream focussed on activities related to both research and teaching, and hence Continuing Professional Development (CPD) is as much a part of knowledge exchange in the UK as IP. The primary focus has been direct connections of all sorts between HE and the economy and society, which ultimately support innovation and impact. The UK approach to knowledge exchange in HE has fitted with the breadth of the policies on innovation in the UK described above.

An initial important step in the development of the policy on higher education's contribution to innovation was the creation of a survey to measure the state and volume of knowledge exchange activities, or interactions. The HEFCE Higher Education Business and Community Interaction (HE-BCI) Survey (2001) was first conducted in 2001 and has been run annually since, and is now fully embedded in HE processes in the UK. The Survey provides information on a range of infrastructure indicators suggesting the extent of capacity and capability to contribute to innovation, but also on income measures of various aspects of knowledge exchange (such as contract research, consultancy or CPD or IP licensing income), from all types of users. (Although the HE-BCI Survey is probably the most advanced measurement system in the world for describing HE's contribution to innovation, it is recognised that it yet covers only a small subset of all interactions (see for example, academic studies of interactions, which include 'soft measures' such as attending events and universities as 'public spaces' and sources of open innovation, which cannot yet be metricised, such as PACEC/CBR, 2009).)

Funding for knowledge exchange was initiated by HEFCE in 1999. Funding evolved over the 2000s as the Higher Education Innovation Fund (HEIF), increasing from £78 million for HEIF round 1 (2001–04) to £150 million per annum at present. This reflects the very strong performance of the HE sector in seizing the opportunities to contribute to innovation. Income for knowledge exchange into HE in the UK measured in the HE-BCI has increased 34% from 2003–04 to more than £3 billion in the latest survey 2009–10. HEFCE's investment in knowledge exchange has given the nation a very promising return: for every £1 invested through HEIF, universities in England have produced a return of at least £5 (see PACEC/CBR, 2009).

HEFCE's vision from the outset was that knowledge exchange would ultimately be a cause led by universities themselves, defining their own roles in the innovation eco-system. This reflects that it was not expected that all HE institutions would have the same roles in knowledge exchange and innovation – and that these would differ by missions, and would respond to the great range of user needs. This was realised through a transition from project-based funding for various aspects of knowledge exchange (including proof of concept support, business mentors and now social entrepreneurship) to a formula-based system. Under the formula system now, universities gain an allocation through their performance in a range of income metrics (see HEFCE, 2011), drawn largely from the HE-BCI Survey. Universities are free though to use funds as they wish and to innovate and explore new partnerships. HEFCE exercises a light touch scrutiny of developments through collecting and analysing institutional strategies for knowledge exchange. These strategies also provide an invaluable source of national intelligence on the state of HE contributions to innovation in England (strategies for 2001–111 HEIF round 4 are discussed in HEFCE, 2008).

Enterprise education

As noted above, the UK has sought to take a very broad and inclusive view of knowledge exchange in the country – both IP and people related (including graduate) contributions to innovation. Enterprise in terms of the creation of companies based on new ideas in research – 'spin-outs' – has always been at the core of technology transfer, and hence has formed part of our approach to knowledge exchange. In the HE-BCI Survey though, data collected have always covered start-ups, which are new companies that may be formed based on expertise or new business models, not IP, including academic and student start-ups. There has also been increasing interest recently in social enterprises, as a way to tackle major societal challenges in innovative and enterprising ways, and HEFCE has supported a major Social Entrepreneurship programme in HE in England, working with a leading social enterprise body, UnLtd (HEFCE/UnLtd, 2011).

The range of ways in which knowledge exchange supports entrepreneurship across academics and students (described in HEFCE HE-BCI, 2001) includes:

- the provision of physical spaces for new business start-ups, such as hatcheries, incubators and science parks;
- entrepreneurs in residence and other role models and mentors (including opportunities from links with alumni and mature and part-time learners who may bring wider experience back into the university);
- enterprise and entrepreneurship non-accredited courses, student enterprise societies and business plan competitions;
- contacts with venture capital, business angels and the like;
- access to sources of funding, such as European Union structural funds programmes, charities and alumni contributions;
- links with local and regional economic partnerships and development agencies, providing opportunities to explore the local market conditions for entrepreneurship and access business support, advice and resources;
- international networks of entrepreneurs and enterprise societies and clubs;
- the environment created by close interactions between universities and academics and businesses, providing a site to link innovation and entrepreneurship contributions. A university business school may also play a part here.

As in many countries, the UK has had particular focus recently on enterprise education (UK Government, 2011), with a view to increasing business activity after a protracted economic downturn. These developments work across different levels of education – schools, further and higher education – with a view to unlocking potential of all in society. While higher education participation in these wider developments is important to change overall cultural attitudes to entrepreneurship, higher education has a very distinctive role to play in enterprise policy, particularly related to the links to innovation. To begin with, higher education is focussed on adults, even if many are young and yet to embark on the world of work. This means that there are many opportunities *to educate and to practice* entrepreneurship at the same time in higher education. Second, higher education study in many disciplines provides an initial quarry of ideas and expertise for business start-up. This is most obvious in the business subject area, but also runs throughout a range of other disciplines, such as art and design and engineering. Similarly some disciplines lend them to social enterprise start-ups, such as health, social work, development or social sciences more broadly. For these reasons, the UK has approached entrepreneurship as a component part of knowledge exchange, and hence of HE's contribution to innovation.

Embedding within a university

This chapter has so far described the policy introduced in the UK on knowledge exchange, which provides an important environment for innovation education. But how does this policy become realised on the ground within a university?

It is important to stress that universities and HE colleges in the UK, as in most countries, are very diverse, varying from small specialist colleges of art, medicine or agriculture, to large multi-disciplinary research-based universities, as well as a range of more teaching and professionally focussed universities of varying sizes, discipline mixes – and all with different contributions to innovation, economic and social and global to local. Contributions to innovation from different HE institutions are affected by the research and teaching mix, the disciplinary mix, but also the general character of the institution and the strategic direction set by institutional leadership and senior management – but above all by the nature of innovating partners.

Figure 42.1 (which may have some resonance with the academic work of Burton Clark and Shattock described later in this chapter) is taken from a research working paper supported by

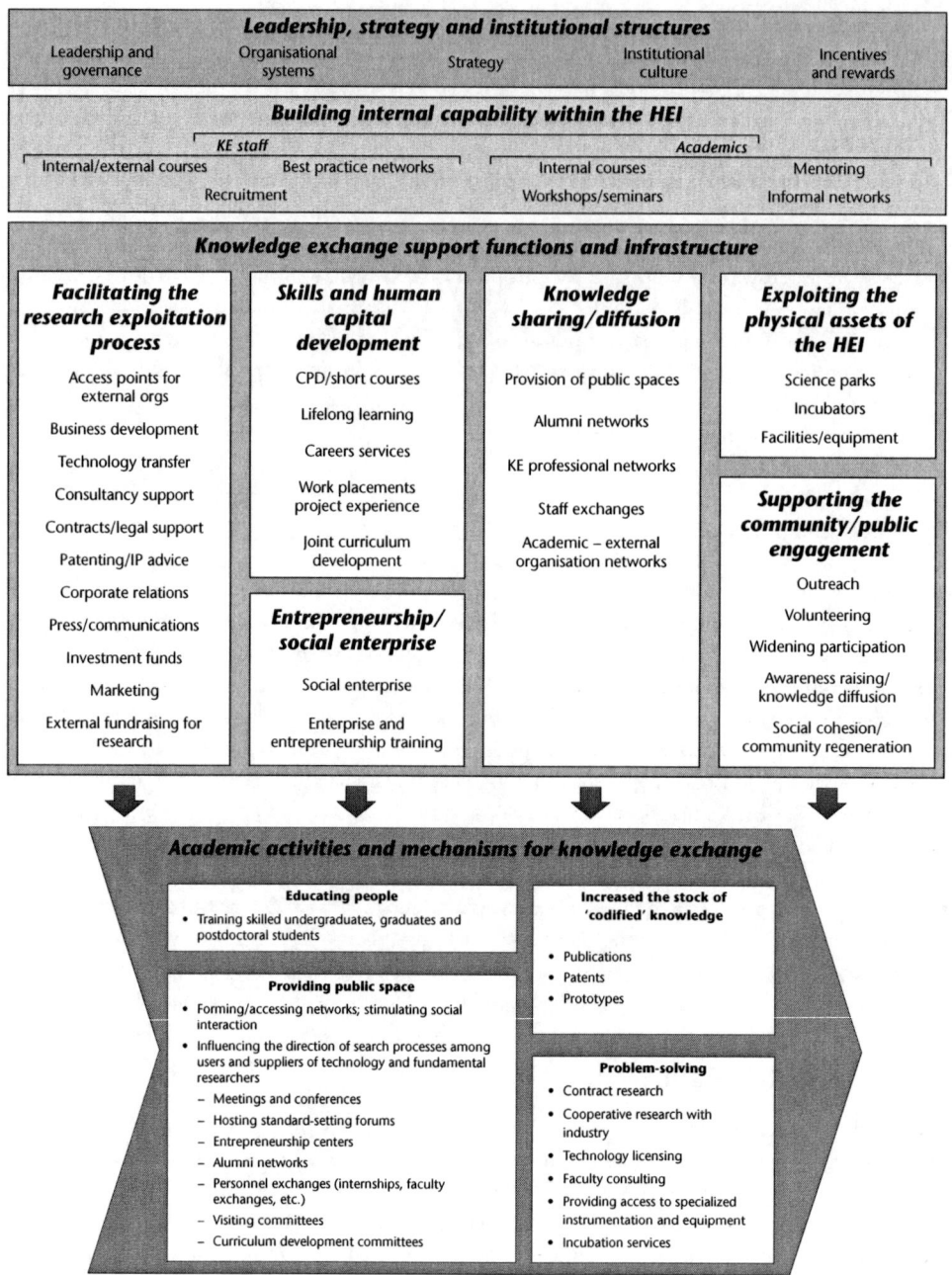

Figure 42.1 PACEC/CBR – embedding knowledge exchange: conceptual framework.

HEFCE as part of a series from the University of Cambridge Centre for Business Research and PACEC (Public and Corporate Economic Consultants) (CBR/PACEC (2011)) to advance understanding of knowledge exchange strategy and practice in the UK. It is a stylised picture of what goes on within a university to embed knowledge exchange, and demonstrates the wide range of elements that need to be in place to enable a university to realise fully its potential in knowledge exchange, including providing an appropriate environment for entrepreneurship and innovation education. The specific elements and their configuration are context specific and can vary significantly by institution, depending not least on their internal capabilities and strengths, the nature of demand they face, as well as the influence of legacies and the prevailing culture amongst academics to engage in knowledge exchange.

It is interesting that approaches to embedding knowledge exchange in the mission of HE in this country have not focussed on the contribution of university business schools. The focus in the UK has been more on how individual disciplines interact with the wider range of users relevant to them, and the forms of entrepreneurship this inspires, rather than the generic business school dimension. This conforms somewhat to American models; for example, at Stanford University where entrepreneurship was initially embedded within the engineering school. Many universities now do involve business schools in, for example, entrepreneurship development – across academic staff and students – although this is not the only model.

Underpinning academic literature

The developments in higher education policy described above, and particularly the adoption of the third – knowledge exchange – mission, are underpinned and critiqued by a number of academic theories. An important strand of these academic conceptualisations concerns *mutation theories*, discussed, for example, in Frost and Blackwell (2006). These theories assert that HE is being fundamentally transformed wholly or mainly by its engagement with the economy and society and its role in innovation eco-systems, and seek to identify and explain the factors and processes involved. Two theories are particularly relevant to analysis of policies on innovation: Gibbons et al. (1994) on the alleged shift from 'Mode 1' and 'Mode 2' knowledge production; and Etzkowitz (2002) (and in other works – Etkowitz & Leydesdorff, 2000; Etkowitz et al., 2000) on the 'triple helix' of government, industry, university interactions.

Gibbons and colleagues characterise traditional discipline-based research, deriving its focus, status and relevance from peers as Mode 1 of knowledge production. Knowledge production is internally referenced and separate from use and application in the economy and society. Mode 2 knowledge production is 'new', focused on real world problems and use, involves actors from different parts of academe and is 'transdisciplinary' in nature – the subjects it tackles and the outcomes it produces crosses or sits upon disciplinary boundaries. They cannot be easily categorised as 'from' any particular discipline. This process is applicable to all parts of the academy and is increasingly supplanting Mode 1 processes. Institutions need urgently to adapt and change their structures and organisational cultures in order to respond to this shift to Mode 2 knowledge (and policy developments for knowledge exchange or third mission in higher education are relevant to this shift).

Etzkowitz is generally credited with developing the notion of the so-called 'triple helix' of university–industry–government relations. This proposes overlapping boundaries between the three circles with boundaries becoming increasingly blurred in a complex system of interactions (consonant with the eco-systems approaches to innovation described elsewhere in this chapter). 'Scientists' pursue their work with a view not only to theoretical and methodological advancement but also commercial potential. The entrepreneurial university, a hybrid based on a third mission of knowledge exchange, is created. Patenting, licensing, spin-off companies, business incubators and science parks are all signifiers of this trend. In the early work of Etzkowitz this is an irresistible and unavoidable process and the University of the Future will be a business incubator.

Another strand to academic enquiry relevant to this chapter considers drivers of HE institutional transformation. If the mutation theories described above have accurately identified powerful transforming drivers one would expect these to be prominent in the findings of research into institutional transformation. Most relevant is the work of Clark (1998) in the US, who identifies five pathways to 'organisational transformation'. These are:

- A strengthened steering core. This is about the quality and capacity of central leadership combining managerial knowledge and capacity.
- Enhanced development periphery through for example a series of units attached to discipline-based departments which are externally focused and undertake applied and commissioned research. They are project and problem oriented, flexible and so on but founded on the fundamental research of the home department.
- Discretionary funding base. These organisations need economic flexibility and multiple sources of funding including the ability to generate capital investment. This includes negotiating more effectively with the state, being competent in competitive bids and finding appropriate niches in the higher education marketplace.
- The academic heartland. The core academic heartland needs to be motivated to be innovative and entrepreneurial. This means that their revenue generating activity must be experienced as positive both in academic and financial terms.
- The entrepreneurial belief. To establish a new entrepreneurial culture it is important to identify an idea about institutional change that is related to important traditional activities. This idea can become a trademark label which stimulates feelings of belonging and creates identity amongst staff and students.

Recent academic literature has explicitly linked Burton Clark's original conception of the entrepreneurial university with the third mission developments and innovation described in this chapter, such as, for example, Shattock (2009).

International developments

Although this chapter is focussed around the experience in the UK, some of these types of changes can be seen going on around the world. The OECD (Organisation for Economic Co-operation and Development) held a major event on the effects of the economic downturn on HE in September 2010 called 'Higher Education in a World Changed Utterly' (OECD Conference Proceedings, 2012). The event highlighted a twin dilemma for higher education posed by the recession: the decreases in public funding of higher education likely to ensue as countries seek to reduce budget deficits; but the increased expectation of higher education and research as sources for innovation to help economies and societies escape from the economic downturn.

As already noted in this chapter, the United States of America is always looked to as a leading edge player in HE and particularly with regard to links with business and contributions to innovation and entrepreneurship. Two universities in particular tend to be picked out as exemplifying the best of American practice in these areas – Stanford and MIT, though there is a range of other distinguished players in an HE sector in the US that is very large and very diverse. These two universities are highlighted for the deep and wide-ranging links that they have with industry, including collaborative R&D, licensing activity, alumni donations and 'people flow' of all forms (see for example detailed descriptions of these universities in Clark, 1998). They are also notably linked with a major world technological development in Information and Communications Technology (ICT) that dominates innovation and the economic growth landscape. It is important to keep in mind the unique nature of these cases, linked to this single technology, which could not be adopted elsewhere. That said, there is a lot to be learnt from the American experience in this area, particularly on discipline-based entrepreneurship (the Stanford model).

In developing a new policy for innovation and knowledge exchange in the UK, it has been important to benchmark activity and performance against good practice elsewhere, and hence America is an inevitable focus for our efforts on overseas comparisons. Due to differences of culture and practice, and hence definitions, there is a tendency to focus on international comparisons narrowly in the field of technology transfer, focusing on licensing and spin-out activity (see HEFCE HE-BCI, 2001). A study by the technology transfer professional body Unico in the UK made links with similar organisations in America, AUTM (the Association of University Technology Managers), and Canada, to look at comparisons (Library House, 2008). As a result of this work, the USA has taken increasing interest in work in the UK, and in a report in October 2010, the US National Academies of Science commended practice in the UK and recommended that the USA adopt similar practices, including conducting surveys similar in nature to HE-BCI in the UK in order to illustrate the breadth of the HE contribution to innovation.

To understand the approaches adopted in America, CBR/PACEC (2010) conducted a study of US–UK comparisons in knowledge exchange. The report concluded that practice of technology transfer and knowledge exchange was similar in both countries, and both countries were looking to each other for new ideas. It concluded, however, that American universities had gone further in civic and community dimensions of knowledge exchange, and indeed entrepreneurship is a stronger thrust. Closer working with the USA in future may provide opportunities to align better the scope and definitions used in this area, including comparable metric developments. This may also provide more insights into the strengths and weaknesses of different approaches to developing the environment within a university that can support entrepreneurship and innovation education.

There have been recent developments in both the OECD and the European Union that point towards a stronger focus on knowledge exchange, and the overall issue of universities as sites for innovation, enterprise and hence innovation education. In 2009 under the Swedish Presidency of the EU, a major focus was placed on the 'knowledge triangle' – the links to be made between HE, research and innovation, with a view to increasing the competitiveness of the EU economy and development of society. The knowledge triangle is fairly embedded in the UK due to strong links between HE and research (with fewer independent research institutes than on the continent), but also from the strong focus on knowledge exchange in HE in UK policy over the last 20 or so years. However, arrangements in continental Europe are more complex, and abilities to join up the triangle are weaker due to a range of factors – such as differences in location of research and applied work outside universities, and a stronger statist ethos in universities with lower autonomy and entrepreneurialism. There is also a perception that levels of entrepreneurialism are lower in Europe than elsewhere (the European paradox – as in OECD, 2008). A strong focus in the EU has been placed then upon entrepreneurship (variously defined), including in new initiatives such as the European Institute of Technology.

Practices and definitions in knowledge exchange and innovation vary across Europe (as well as university and research arrangements) making consistent metrics development that can inform international comparisons very difficult. There remains a tendency still in Europe to have relatively strong divisions between research and innovation agendas on one hand, and education, skills and training on the other. This makes comparisons with the more integrated UK approach to knowledge exchange and wider scope to innovation difficult – and may also then be a hamper to development of an infrastructure and approach to innovation education. New work is underway in many European countries to devise approaches to knowledge exchange and new measurement systems (such as 'knowledge valorisation' in Netherlands). There has also been work under the European Commission on knowledge exchange metrics (EC, 2011).

Conclusions

A major opportunity from the approach taken in the UK to knowledge exchange – crossing domains of technology transfer, HE–industry links and entrepreneurship – is the potential it provides to identify, develop and educate innovators. The enterprise agenda contributes the potential to identify a market or business model, the appetite to take risk and the capabilities to turn ideas into real world opportunities. The agenda of links between HE and business contributes the potential to identify innovation and market opportunity from new discipline knowledge and the connections, sources of expertise and role models to get started.

In the past, there was a very large gulf between the street savvy entrepreneur who was the central focus of enterprise, and the engineering boffin spinning out some arcane technology. Time has moved on, and these stereotypes are very far from the present broader range of entrepreneurs and innovators that are needed to inhabit a sustainable and balanced economy and vibrant society. Indeed, it may be that in some areas, academic innovators may be the most likely entrepreneurs. A survey of engineering and physical sciences academics in this country for EPSRC (2010) highlighted that UK academics appeared to be highly entrepreneurial. A significant proportion of academics were directly involved in developing new ventures, and respondents to the survey were almost five times more likely to be involved in entrepreneurial efforts than general members of the UK population.

How can the progress of the developments in the UK set out in this chapter be described and measured? The most recent HE-BCI Survey for 2009–10 describes some of the latest achievements in the UK in entrepreneurship:

- Over 2,000 graduates established new enterprises in the period to capitalise on the knowledge and experience gained while studying.
- In 2009–10, 273 new businesses were set up based on the leading edge research carried out by UK universities making the total number of active spin-offs 1,340; these companies employed around 17,000 people and turned over nearly £1.8 billion during the year.
- UK universities formed one new company per £23 million of research funding during 2009–10. This far exceeds the record of US universities (one new company per £56 million).

This chapter finishes with some reflections and prescriptions for how the agenda to improve innovation education might go forward in the future:

- It may be necessary to break down further the conceptions held in many heads of a divide between theory and practice, and between academic knowledge and the 'real world' (Modes 1 and 2). There is increasing interest in experiential learning in HE as an effective learning tool for students. HE is particularly well placed to provide a situation for entrepreneurship and innovation education, as it combines relevant knowledge sets and expertise with the opportunities of an adult learning community on the cusp of the transition to the world of work.
- While research commercialisation and its contribution to innovation is better understood in policy circles, it may be necessary to deepen understanding and promote the opportunities to develop links between teaching and learning, knowledge exchange and innovation. A feature of higher education is the development of research skills as part of the curriculum; and increasingly it may be expected that practice or enterprise skills should be similarly a distinctive feature of the higher education sector. This area may make an important contribution to

innovation education in the future. The role of business schools in these developments, and their role in innovation education, may also become more apparent.

- It will be important to promote the widest range of innovation opportunities appropriate to different disciplines, across the full spectrum of subjects, so that multi-disciplinary developments are enabled which are often most relevant to innovation. This provides the basis for an entrepreneurial university that can provide an appropriate environment for innovation education. It will also be important to seek out all sources of potential for innovation, relevant to the mix of a developed economy now – across products and service sectors. And it will be necessary to raise awareness generally of the innovation and enterprise potential in all types of HE activity – the potential of graduates and postgraduates and post-doctoral researchers, as well as licensing of IP and collaborative work with industry.
- It will be important to promote the role models of academic entrepreneurs and innovators – with a view to changing perceptions of what universities can contribute. Most vitally, it will be essential to change perceptions of graduates about what roles they can play in the economy (as entrepreneurs and innovators), what are sources of value in those roles (such as research knowledge and research skills) and how the university can be a long-term source of innovation support and inspiration throughout their lives.
- And all need to challenge out-of-date stereotypes of entrepreneurs, scientists, technologists and engineers and creatives which dog a more connected and inclusive approach to innovation and hence innovation education, through wider public engagement about the nature of research and HE and its criticality in real innovation and economic growth in the economy and society.

References

BIS (2011). *Higher Education: Students at the Heart of the System*. London: Stationery Office.
BIS/NESTA (2010). *Annual Innovation Report 2010*. Downloaded from www.nesta.org.uk/publications/assets/features/annual_innovation_report on 11 October 2011.
Capart, J., & Sandelin, J. (2004). *Models of, and Missions for, Transfer Offices from Public Research Organizations*. Downloaded from http://otl.stanford.edu/documents/JSMissionsModelsPaper-1.pdf on 11 October 2011.
CBR/PACEC (2010). *The HE Knowledge Exchange System in US*. Cambridge: PACEC publication. Available www.pacec.co.uk/index.php/publications/easytablerecord/5-publications/204.
Clark, B. (1998). *Creating Entrepreneurial Universities: Organisational Pathways of Transformation*. Oxford: IAU Press and Pergamon.
European Commission (2011). *A Composite Indicator for Knowledge Transfer: Report from the European Commission's Expert Group on Knowledge Transfer Indicators*. Brussels: EC.
EPSRC (2010). *The Republic of Engagement: Exploring UK Academic Attitudes to Collaborating with Industry and Entrepreneurship*. Swindon: Engineering and Physical Sciences Research Council.
Etzkowitz, H. (2002). Incubation of incubators: Innovation as a triple helix of university–industry–government networks. *Science and Public Policy*, 29(2) 115–128.
Etzkowitz, H., & Leydesdorff, L. (2000). The dynamics of innovation: From national systems and mode 2 to a triple helix of university–industry–government relations. *Research Policy*, 29(2) 109–123.
Etzkowitz, H., Schuler, J., & Gulbrandsen, M. (2000). The evolution of the entrepreneurial university. In M. Jacob & T. Hellstrom (Eds.), *The Future of Knowledge Production in the Academy*. Buckingham: SRHE/OUP.
Finland Ministry of Trade and Industry (2003). *Evaluation of the Finnish Innovation Support System*. Downloaded from http://julkaisurekisteri.ktm.fi/ktm_jur/ktmjur.nsf/All/172616819C0174ECC2256D2B003CA685/$file/ju5teoeng.pdf and http://julkaisurekisteri.ktm.fi/ktm_jur/ktmjur.nsf/All/172616819C0174ECC22 56D2B003CA685/$file/ju5teoeng.pdf on 11 October 2011.
Frost, A., & Blackwell, R. (2006). Issues in embedding the 'third stream' in HE. Society for Research into Higher Education Annual Conference Proceedings.
Gibbons, M., Limoges, C., Nowotny, H., Schwartzman, S., Scott, P., & Trow, M. (1994). *The New Production of Knowledge: The Dynamics of Science and Research in Contemporary Societies*. London: Sage.

HEFCE (1999). *HE Reach-out to Business and the Community*. Bristol: HEFCE publication. HEFCE 99/40.
HEFCE (2005). *Young Participation in Higher Education*. Bristol: HEFCE publication. HEFCE 2005/03.
HEFCE (2008). *Higher Education Innovation Fund Round Four Institutional Strategies: Overview and Commentary*. Bristol: HEFCE publication. HEFCE 2008/35.
HEFCE (2011). *Higher Education Innovation Funding 2011–12 to 2014–15: Policy, Final Allocations and Request for Institutional Strategies*. Bristol: HEFCE publication. HEFCE 2011/16.
HEFCE HE-BCI (2001). The survey is described, with links to the series of survey reports: downloaded at www.hefce.ac.uk/econsoc/buscom/measure/ on 11 October 2011.
HEFCE/Unltd (2011). Further details of the Social Entrepreneurship Awards Scheme; downloaded from www.hefce.ac.uk/econsoc/buscom/socent/ on 11 October 2011.
HM Treasury (2004). *Science and Innovation Investment Framework 2004–2014*. London: Stationery Office.
HM Treasury (2006). *Science and Innovation Investment Framework 2004–2014: Next steps*. London: Stationery Office.
HM Treasury (2007). *The Race to the Top: A Review of Government's Science and Innovation Policies'. Lord Sainsbury of Turville*. London: Stationery Office.
Library House (2008). *Metrics for the Evaluation of Knowledge Transfer Activities at Universities*. Cambridge: Praxis-Unico publication.
NESTA (2007). Hidden Innovation: how innovation happens in six 'low innovation' sectors. Downloaded from www.nesta.org.uk/publications/reports/assets/features/hidden_innovation on 11 October 2011.
OECD (2008). *Tertiary Education for the Knowledge Society: OECD Thematic Review of Tertiary Education: Synthesis Report, Vol. 2*. Paris: OECD.
OECD (2010). Conference proceedings and reports downloaded from www.oecd.org/site/0,3407,en_21571 361_43541789_1_1_1_1_1,00.html on 11 October 2012.
Office of Science and Technology (1993). *Realising our Potential: A Strategy for Science, Engineering and Technology*. London: Stationery Office.
PACEC/CBR (2009). *Evaluation of the Effectiveness and Role of HEFCE/OSI Third Stream Funding: Report to HEFCE by PACEC and the Centre for Business Research, University of Cambridge*. Bristol: HEFCE publication. HEFCE 2009/15.
PACEC (2011). *Understanding the Knowledge Exchange Infrastructure in the English HE Sector*. Cambridge: PACEC publication. Available: www.pacec.co.uk/index.php/publications/easytablerecord/5-publications/200.
Rathenau Instituut (October 2009). *Science Systems Compared: A First Description of Governance Innovations in Six Science Systems*. Amsterdam: Rathenau Instituut.
Shattock, M. (Ed.) (2009). *Entrepreneurialism in Universities and the Knowledge Economy: Diversification and Organizational Change in European Higher Education*. New York: SRHE/OUP.
The Work Foundation (2010). *Anchoring Growth: The Role of 'Anchor Institutions' in the Regeneration of UK Cities*. London: NESTA publication.
UK Government (2011). Government-led campaign Start-Up Britain. Downloaded from www.startupbritain.org/ on 11 October 2011.
US National Academies of Science (2010). *Managing University IP in the Public Interest*. Washington, DC: National Academies Press.

43

The worldwide interest in developing innovators

The case of the Center for Talented Youth (United States) and PERMATApintar (Malaysia)

Julian Jones and Noriah Mohd. Ishak

JOHNS HOPKINS UNIVERSITY, USA, AND UNIVERSITI KEBANGSAAN MALAYSIA, MALAYSIA

Summary: The Johns Hopkins University Center for Talented Youth (CTY) identifies and nurtures high ability pre-university youth. Although it has received considerable academic recognition worldwide over its 32-year history, international attention has broadened in the last four years to include CTY's economic development potential. This interest appears to have been encouraged by research that connects innovation with a nation's successful shift into knowledge-intensive industries. Malaysia was the first major adapter of the CTY approach to integrate education of highly gifted youth into plans for economic transformation. This chapter discusses Malaysian government and university interest in deepening the pool of youthful genius to encourage economic growth, generally, and the knowledge economy, in particular. It also describes how CTY helped establish the Malaysian version of CTY at the Universiti Kebangsaan Malaysia (UKM). The success to date of the Malaysian model called PERMATApintar ("Gifted Gems") is considered as well.

Key words: Brain race, creativity, innovation, innovation education, globalization of education, Center for Talented Youth (CTY), PERMATApintar

Introduction

When Hans Morgenthau published *Politics Among Nations* in 1948, "innovation" was not among the leading measures of national success, but natural resources were at the top of his "Elements of National Power". Sixty-plus years later innovation had become the driver of the new knowledge economy, and writers like Robert Reich, Richard Florida and John Kao heralded it as a key metric of national economic success. Kao's 2007 book, *Innovation Nation*, laid out this thinking in detail, and his March 2009 *Harvard Business Review* article, "Tapping the World's Innovation Hot Spots", made sure the idea was well understood in every government ministry around the world: innovation is the key to moving from the old manufacturing base to the new, high-value-added knowledge economy. As a *Wall Street Journal* editorial stated in August 2011, entrepreneurs have become "the heroes of the economy".

After Kao's *Innovation Nation* (2007), the worldwide search for brainpower and the innovation it could unleash began in earnest. One part of that search focused on establishing great universities from the ground up, like King Abdullah University of Science and Technology (KAUST) in Saudi Arabia. Another part sought to strengthen existing ones such as National University of Singapore or the C9, the Chinese Ivy League. And a third approach imported leading universities to new homes, as Yale in Singapore and New York University (NYU) in Abu Dhabi. The globalization of top research universities is well described in Ben Wildavsky's *The Great Brain Race: How Global Universities are Reshaping the World* (Wildavsky, 2010).

A problem with importing highly regarded universities or creating ones that aspire to top international rank is finding local student talent who could be admitted, and, once in, would thrive in a new teaching/learning environment aimed at fostering innovation. Abu Dhabi may have felt disappointment at seeing only a few Emirati students admitted to NYU Abu Dhabi in its opening semester, fall 2010 (Redden, 2010). The admissions issue was not students' lack of brainpower but their uneven preparation, the challenge of the "SAT" (the "SAT Reasoning Test" is a standardized test used for university admissions in the United States), and inexperience in being interviewed. Then, too, Confucian and other regional traditions sometimes emphasized memorization and recitation at the expense of the critical thinking skills U.S. admissions officers seek. A Johns Hopkins University Center for Talented Youth team observed firsthand the admissions challenge bright students face in the United Arab Emirates. Some 50 top-performing middle school students took CTY's School and College Ability Test in January 2011. Its mathematics section required judgment beyond a keen memory for formulas to solve problems, leaving some students looking for "the" answer and confused. A few years in the future these same extraordinary students may have difficulty gaining admission to world-class universities.

In part because of such testing issues, the brain race began moving to younger ages in order to identify and nurture bright pre-university students and prepare them for admission to top American, upgraded local or newly established universities. This shift became apparent to staff at CTY as early as 2008 when the number of international visitors to its advanced programs for high ability youth increased from an average of four per year earlier in the decade to ten in 2008 and 26 in 2011. The international visitors grew steadily more focused as the decade wore on, looking for help in replicating the CTY far from its U.S. roots. Interest centered on the practicalities of identifying and nurturing local genius who, it was expected, would move to top universities and lead economic transformation in the next generation. This practical interest introduced by visiting international groups had never been explicit at CTY in the United States.

The CTY model

The Center for Talented Youth developed from research in the 1970s led by a Johns Hopkins University psychometrician, Dr. Julian Stanley. He had been approached for advice by parents of bright middle school students in Baltimore whose sons had exhausted all mathematics courses available in the local public schools and who sought additional challenge for their children. On an inspired whim, Stanley asked the students to take the "SAT-M" (mathematics portion of the standardized test for college admission in the United States), and, to his amazement, they scored well enough at age 13 to gain admission to Johns Hopkins University. This chance encounter led to more research into high ability 7th and 8th grade students and eventually to a regional talent search and a supplemental summer program that challenged, excited and nurtured their extraordinary ability. The Center for Talented Youth opened its doors in 1979.

Dr. Stanley traveled the world for the next decades presenting at gifted education conferences and meeting with faculty colleagues to discuss his brainchild, but the idea of establishing local

CTYs took hold only in Ireland, which opened an identical model in 1992 with the help of a handsome grant from an international philanthropist. For the most part, though, world interest in CTY remained academic, limited to university research.

In the meantime, CTY in the United States expanded from a few hundred participants at one summer location in 1980 to over 9,000 students at 24 U.S. summer sites in 2011. The locations in 2011 ranged from Johns Hopkins and Princeton on the East Coast to Berkeley and Stanford on the West Coast. The Center added CTY*Online* in the 1990s which is now its fastest growing division. Specialized units conduct research, support the profoundly gifted and provide diagnostic and counseling services to bright students and their parents, rounding out the CTY picture.

The "intake valve" at the Center remains its talent search designed by Dr. Stanley based on student scores on above-grade-level tests such as the "SAT" as well as CTY's own School and College Ability Test (SCAT) and Spatial Test Battery (STB). Qualifying CTY students score in the top 3% of the student population in mathematical or verbal reasoning ability. Once admitted through testing, the various CTY programs and services become available to students and parents. All are tailored to the students' extraordinary cognitive abilities.

Malaysia's search for innovators

Growing international interest in the CTY model of identifying and nurturing high talent encouraged the opening of a CTY international program unit in 2008. Among its first partners was Malaysia, and in many ways the Johns Hopkins CTY–Malaysian collaboration set the pattern for other versions of CTY established in locations from Egypt to Hong Kong. The next pages look at why and how Malaysia's PERMATApintar ("Gifted Gems") holiday period program developed, with what success it has met and how CTY assisted.

Connecting the education of Malaysia's most gifted pre-university students to the country's economic growth is a recent phenomenon. Malaysia, like the United States, has experienced an ambivalent relationship with gifted education. This stretches back almost to Malaysia's 1957 independence from Britain. A widespread perception from the beginning was that gifted children, however defined, were likely to achieve academic and life success on their own. Government need not include them in the modified education it provides other children who have special needs.

Nevertheless, Malaysia's Ministry of Education made sporadic efforts to introduce special programs for gifted and talented students starting in the 1960s. Among these were Express Classes (a form of acceleration) established in 1962 and the Level One Assessment System created in 1996 (Malaysian Education Statistics, 2006). Both failed to provide for the educational needs of gifted students because of weaknesses in planning and leadership and lack of funding.

One event that began to focus attention on gifted students was former Prime Minister Dr. Mahathir Mohamad's Vision 2020 first presented in 1991 (Sixth Malaysia Plan, 1991). Its goal was to move Malaysia to the status of developed country, defined as meeting World Bank standards of full industrialization and high income. There were only 19 such countries at the time. As the 2020 date has come closer, government studies of how to reach the Vision 2020 goals influenced Malaysia's successive five-year plans to become more specific on what education would have to contribute. The five-year Malaysian plans also began to change as the effects of rapid globalization and the IT revolution became apparent.

For example, developing the potential of Malaysian gifted and talented children was a focus in the ninth Five Year Plan (RMK-9), prepared by the Malaysian Inter-Agency Planning Group (IAPG) to cover the years 2006–2010 (Malaysian IAPG Report, 2008). In 2010, Malaysian Prime Minister Dato' Sri Najib Tun Razak raised the issue in his budget speech asserting that Vision 2020 would require a new economic model based on innovation, creativity and high-value-added activities, key points made in Kao's previously mentioned *Innovation Nation* (2007). To remain relevant in the new competitive global market, he continued, Malaysia must seek innovative and creative transformation in all sectors including education. "To help students think outside the box", he asserted, "innovation and creativity should be inculcated in the country's learning system ... and efforts to educate individuals to [their] potentials must be stepped up" (Malaysian Budget Report, 2010).

The Prime Minister designated 2010 as the year of creativity and innovation, spelling out that "creativity and innovation are among the strategic thrusts ... transforming Malaysia to a developed nation status by 2020" (SME Innovation Showcase, 2010). He added that transformation in education and emphasis on talent development among Malaysians, especially the gifted and talented, would be a crucial part of reaching Vision 2020.

Along with this official recognition of the importance of gifted students, Datin Sri Rosmah Mansor, Malaysia's First Lady, showed a similar concern for gifted and talented students. Her interest grew after she met a Malaysian boy with extraordinary mathematical prowess. Oddly similar to the story of Julian Stanley's transformation after discovering a few children of high ability in the 1970s, Datin Sri formed a committee to meet the special educational needs of the Malaysian youth and others like him. She visited many gifted centers around the world to examine first hand what they were doing and to seek best practices for Malaysia. After these visits and much internal discussion, the Center for Talented Youth was selected by Malaysia as the best model. The First Lady came to the Johns Hopkins University campus in Baltimore and learned of CTY's long history with students of high ability and observed the range and quality of its programs. Also impressive to the Malaysian visitors were the number of students worldwide engaged with CTY and the adaptability of its model to other countries.

CTY and PERMATApintar

In 2008, the Universiti Kebangsaan Malaysia (UKM), the National University of Malaysia, established Pusat PERMATApintar Negara, the National Gifted Center. Among its first tasks was to plan, execute and manage the PERMATApintar (Gifted Gems) project, establishing a supplemental education camp for the country's brightest students on a model adapted from CTY. The National Gifted Center soon developed its own testing known as PERMATApintar UKM Test1 and PERMATApintar UKM Test2. Both tests have now been used by more than one million students from Malaysia, and another 500 students from other countries who are invited to test for the Malaysian school holiday program, its version of CTY's summer programs, described below.

After some months of negotiation and deliberation, the CTY and UKM teams came to an agreement signed with great fanfare in Kuala Lumpur in April 2009. The agreement provides for CTY advisory services to UKM, training of UKM faculty and staff in the United States and Malaysia, and transfer of CTY intellectual property to UKM. In addition, PERMATApintar students would participate in CTY summer programs in the United States, CTY*Online* courses and in the Cogito.org website for those interested in mathematics and science. CTY staff worked closely with UKM for the whole of 2009 to guide UKM in its first venture into a supplementary school holiday program for high ability students. The opening program in November/December 2009 was a success. The second year of the agreement was equally successful and in December 2011 Pusat PERMATApintar Negara, completed its third school holiday camp with the help of CTY.

UKM staff training is taking place over a period of three to five years. For every contracted year, UKM sends five to seven instructors to CTY Johns Hopkins to be mentored in various approaches and courses. The training is conducted by CTY staff as part of CTY's International Educators Institute. It unfolds in two parts spread over a period of five weeks. Participants receive intensive exposure on issues pertaining to gifted students and their education for the first two weeks. Over the next three weeks they shadow an instructor at a chosen CTY summer site. Upon returning home, these Malaysian participants help select and instruct other staff members who will be teaching the various courses offered during the Malaysian holiday-period camp. Malaysian staff may adapt the CTY syllabus to fit the Malaysian students' learning needs and to ensure that materials used can be found locally.

As of late 2011, Pusat PERMATApintar Negara offers 16 courses, 13 of which are CTY courses, and the remaining are courses developed by UKM's own research institute. The courses engage students from ages nine to 15. The Malaysian Center follows closely the model used by CTY and has adopted policies and regulations used by CTY in the United States. For example, CTY staff helped Malaysia's Center develop the CTY-type living-learning community in its holiday camp, promoting healthy learning environments for all students. This is important given the unique characteristics of gifted students, and their sense of competitiveness which, if not properly channeled, can lead to negative behaviors such as bullying. Instead, respect for others' opinions are encouraged and demonstrated by instructors and staff. Pusat PERMATApintar Negara also adopted a similar administrative structure to the one used by CTY in the United States. The Malaysian holiday period camp is located on the UKM campus and managed by a site director with the help of an academic dean and a residential dean, much in the same manner as U.S. CTY locations. During the three-week academic camp, the Malaysian Center employs about 130 temporary staff to teach and help manage some 500 students' daily academic and out-of-class activities. All temporary staff are trained prior to the camp, and the training is normally conducted by senior staff of Pusat PERMATApintar Negara who have attended the CTY International Educators Institute. A representative from CTY participates in the training as well.

The pedagogical approach used is also worth mentioning. CTY emphasizes differentiated learning. Pre- and post-testing is widely used in its classroom to gauge the students' initial level of understanding of the topic and their progress at the end of three weeks of intensive study and individualized instruction. Students are also given challenging materials to work on during the academic session. In the first year of the Malaysian adaptation of CTY, anxiety ran high among the teaching staff who were unfamiliar with the curriculum. Some academic staff were also skeptical of the students' ability to cope with the above-grade-level curriculum. One staff member wrote

> Malaysian students are not familiar with a higher syllabus [CTY's] that is not comparable to the national syllabus, and therefore, the syllabus offered must follow closely to the national syllabus... or else the camp will be useless because it will not help the students get better marks in their exams when they go back to their respective schools.

Nonetheless, upon completing the first three-week program in Malaysia, the academic staff members were surprised to see that the students were able to follow the new syllabus with ease. The instructors reported common themes: "the students are more talkative ... in a positive sense, they think before they say anything, and they are able to reason",

> when they first came, they would just absorb... I can see that they were memorizing all formulas. Now they can analyze, synthesize and even evaluate... so they consider the pros and cons... before they make any decision on how to solve the problems,

and "I am surprised that they scored well in the post-test. All of them scored above 80%. That is high for little students who are taking this course [CTY's "Be A Scientist" course]."

PERMATApintar's success in Malaysia

The instructors' reactions matched the students' perceptions. Based on a questionnaire distributed to the students to examine changes in their thinking ability, it was found that the students were able to develop metacognitive skills that equal the three highest levels (analysis, synthesis and evaluation) of Bloom's Taxonomy, a well-established classification of levels of intellectual behavior important in learning (Bloom, 1956). The students were also given open-ended questions to examine differences in thinking patterns after attending the PERMATApintar program. Responses given by the students were very insightful: "I never thought about what I am thinking, but now I ask myself why I think like that", "Now my thinking is different. I see similarities and differences. I compare and contrast before I decide" and "I used to think that numbers can start from 1 to infinity. Now I am thinking what if numbers have an end, with no absolute zero." Analysis was also conducted on the pre- and post-test given by the instructors during the school holiday camp. The data analyzed to compare the pre- and post-test scores for each course shows that for the last two years (2009 and 2010) students' understanding on topics discussed in each course improved by at least 70%. Differences in pre- and post-test scores are found to be significant (all $p < 0.05$) across all courses, particularly in the areas of science and mathematics.

The success of CTY-UKM camp supports a number of conclusions. First, the screening tests used to identify the top Malaysian students were effective. Second, the challenging syllabuses kept the students focused. Third, the differentiated teaching technique used helped the instructors provide the right approach for each student. Further, the teaching materials proved effective.

Another measure of success was how PERMATApintar students did at CTY in the United States. The Malaysian Center sent students to the CTY summer program, and these students participated in some of the most difficult courses offered by CTY. All of them completed the U.S. CTY summer program successfully. One of the course instructors at CTY who taught Number Theory praised the Malaysian student who participated in his course. He wrote that "every time [the Malaysian student] … asks a question, I have to open my book". This particular student had attended the Malaysian version of CTY consecutively for two years. The CTY-UKM program appears to have prepared him well for the academic adventure at the U.S. CTY summer program.

Although CTY's role is limited to advisory services on the Malaysian school holiday camp, CTY extended its help to developing a high school program for top Malaysian gifted and talented students that has been part of the larger gifted education plan adopted by the government of Malaysia to reach Vision 2020. A residential high school, built on the success of Malaysia's version of CTY, was launched in January 2011 and is now functioning at the Pusat PERMATApintar Negara, UKM campus.

The Center for Talented Youth and Pusat PERMATApintar Negara are considering other ways they may collaborate. For example, they would like to conduct research in the areas of comparative gifted education, contributing to what Lynn H. Fox describes elsewhere in this volume as the "emerging innovation education discipline". That will require, among other steps, a longitudinal study of both CTY USA and Malaysian PERMATApintar students.

Conclusion

What seems clear at this point is that the CTY model of talent identification and nurture is readily transferrable to a supportive Asian setting and enthusiastically welcomed by Malaysian students and parents. In fact, the Malaysians greatly extended their talent search throughout the country, even encompassing isolated river and forest settlements, well beyond the largely urban and suburban

U.S. model's reach. They also took an innovative step of their own, much admired by CTY staff, opening a high school designed for top students in the PERMATApintar holiday-period program. This program and its new high school are clearly serving Malaysia's national interest in reaching the lofty Vision 2020 goals by encouraging creativity and innovative thinking among high-ability students. As Prime Minister Najib put it: "knowledge is the main instigator of global economic growth and ... our ability to encourage innovation [is] the starting point in our goal of national development" (Najib, 2010).

References

Bloom, B. (1956). *Taxonomy of Educational Objectives Book One: Cognitive Domain*, 2nd ed. (1984). Boston: Addison Wesley.
Economic planning unit (2006). *Ninth Malaysian Plan (RMK-9) 2006–2010*. Economic planning unit, Prime Minister's Department and the Finance Ministry of Malaysia.
Economic Planning Unit (1996). *Seventh Malaysia Plan 1996–2000*. Economic planning unit, Prime Minister's Department and the Finance Ministry of Malaysia.
Economic planning unit (1991). *Sixth Malaysia Plan 1991–1995*. Economic planning unit, Prime Minister's Department and the Finance Ministry of Malaysia.
Kao, J. (2007). *Innovation Nation: How America Is Losing Its Innovation Edge, Why It Matters, and What We Can Do to Get It Back*. New York: Free Press.
Kao, J. (2009). Tapping the world's innovation hot spots. *Harvard Business Review*, March.
Malaysian Budget Report (2010). Ministry of Finance, Malaysia.
Malaysian Education Statistics (2006). Educational planning and research division (EPRD) report.
Malaysian inter-agency planning group (IAPG) Report (2008). Malaysia.
Morgenthau, H. (1948). *Politics Among Nations: The Struggle for Power and Peace*. New York: Knopf.
Najib T. R. (2010). Najib's answers: Solving yesterday's problems, today's conflicts and tomorrow's challenges. M. Ismail, K. A. Jushoh, & M. Muhammad (Eds.). Kuala Lumpur: Institut Terjemahan and MPH Group.
Redden, E. (2010). The world's honors college? *Inside Higher Ed*, June.
Small and medium enterprise corporation Malaysia (SME) Innovation Showcase, 2010.
Vision 2020 (2008, January 14). Malaysia, 1991–2020. Economic planning unit.
Wildavsky, B. (2010). *The Great Brain Race: How Global Universities are Reshaping the World*. Princeton, NJ: Princeton University Press.

44

How does Singapore foster the development of innovators?

Chwee Geok Quek and Liang See Tan

MINISTRY OF EDUCATION, SINGAPORE NANYANG TECHNOLOGICAL UNIVERSITY, SINGAPORE

Summary: To meet the challenges of global economic development and international competition, Singapore needs people who are willing to try new and untested routes without fear of failure; people who are prepared to be flexible, and are resilient and able to respond nimbly to complex changes, and to inspire teams and organisations to take leaps of innovation. The future workforce must have the ability to think creatively and with originality to generate ideas and solve future, as yet unseen, problems. This chapter will provide an overview of the initiatives introduced by the Ministry of Education in the K-16 system to create an innovation culture, and develop a spirit of innovation and enterprise. Descriptions of key programmes for students, as well as innovations introduced by schools and tertiary institutions will be included.

Key words: Innovation, entrepreneurship, talent development, education reform.

Introduction

The 21st century is fast-paced, risk-filled and complex. Technological advancements have shrunk the distance between continents and economic development has created interdependence among countries such that economies across borders are now inextricably linked in ways that they had never been before. Even as these trends have brought opportunities for international collaboration, competition among countries for talents which is the core of innovation has accelerated exponentially. Countries that fail to anticipate, adapt, evolve and innovate are facing relentless challenges to survive (Canton, 2007). Innovation is imperative for survival, and progress. Being a free economy system, innovation is one of the key thrusts in driving Singapore from an efficiency-driven to an innovation-driven knowledge-based economy in the new millennium (Tan & Phang, 2004).

Contrary to the belief that innovation happens only when the government adopts a "hands off" stance and only needs to give space to innovators, this chapter presents Singapore as a unique case where innovation can be said to be "government led", and that takes place through a virtuous circle of deliberate educational reform, in tandem with economic restructuring which in turn is supported by educational policies. In Singapore, educational policies have been key in facilitating and shaping the development of the innovative environment. Singapore's experiences and efforts to enhance innovative capacity of young children tend to take place while they are in school, where seeds are sowed and the foundation is laid. This approach appears to be supported by Gergen (2009) who contends that we need to "innovate our way out of crises", and that students must be equipped with skills and mindsets to deal with the "new normal". The salient aspects of educational policies which aim at developing innovators through diverse educational pathways will be discussed.

Defining innovation: the Singapore way

Innovation differs from creativity in that it is a process that transforms creative ideas into practical products, services or processes that command greater commercial or societal value. Innovation refers to an inclination to improve through modification, involves conceptualising skills and is embedded in the notion of creativity (Taylor, 1959). Creativity is the "infinite source of innovation" (EC, 2008) and innovation can be understood as the application and implementation of creativity (Craft, 2005). The concept of innovation has been used in different fields. Very often, in applied, technological innovations in the manufacturing sector, creative industries and business organisations[1] or work teams,[2] innovation has been understood as introduction or implementation of ideas, processes, procedures and products[3] for modification and improvement (OECD, 2005; Craft, 2005; West & Richards, 1999). The term "innovation" has even been used to refer to the creative process and work (Sternberg & Lubart, 1999).

Singapore views innovative workers as the powerhouse of sustainable economic growth. While creativity can be exploratory, combinational and transformational (Boden, 2001), Singapore's pursuit of innovation in schools can be said to be incremental in approach. The literature suggests the manifestation of creativity and innovation requires the interplay of the following processes: personal traits (Russ, 1996), emotional or affective processes (Russ, 1996), cognitive abilities (Torrance, 1974; Runco, 1990; Russ, 1996; Amabile, 1998), expertise (Amabile, 1998; Runco, 1990; Esquivel, 1995), imagination (Craft, 2005) and evaluation (Runco, 1990), motivation (Amabile, 1998) and fostering cultural environment (Florida, 2002; Albert & Runco, 1990). Policy makers have increasingly realised the importance of promoting creativity and innovation through fostering cultural and environmental factors. In Singapore, the national innovation policies in the education and economic sectors are intertwined, and aligned to the goal of developing Singapore into a cosmopolitan urban city state where talents congregate and are supported by advanced technology in an environment that is tolerant of failure (Florida, 2002).

Innovation as a tool for nation building

Lack of natural resources for survival

Singapore is a small island country with virtually no natural resources. As such, it is critical for Singapore to leverage on human capital and intellectual capacities as resources and find creative ways to optimize limited resources such as space and strategic position. The experiences of Singapore's struggle for survival and development can be encapsulated by Porter's (1998) four-phase model of national competitive development: factor-driven phase, investment-driven phase, innovation-driven phase and wealth-driven phase. For instance, historically, being strategically located in the centre of Southeast Asia, Singapore was an entrepôt that promoted free trade in the region and internationally. After Independence in 1959, Singapore upgraded its entrepôt port to that of world's gateway hub and many multi-national corporations (MNCs) were brought in to jump-start the industrialization process in the first 20 years after Independence. In the face of intense competition, Singapore realised that to succeed in the future,

> it [would] not be enough to have First World infrastructure. It [would] not be enough to have the rule of law, honest civil servants and consistent reliable regulatory frameworks. It

[would] not be enough to pick potential winners or sectors which the government thinks [would] succeed. It [would] not be enough to improve productivity or to be "cheaper, better or faster". All of the above is necessary but not sufficient for success. If we do all of the above, we will make steady progress, but risk being overtaken by global competition. The only way to generate exponential growth is for innovation and enterprise to flourish.

(Balakrishnan, 2009)

Educational policy has always been and continues to be a prime conduit for promoting economic development since Singapore gained its Independence in 1959 (Tan & Gopinathan, 2000). In order to increase the nation's propensity to innovate, and "to develop entrepreneurial talent, the national school curriculum has been revised to promote innovation and enterprise, and to nurture problem-solving and independent learning abilities".

Building a first class workforce to support an innovation-driven economy

Like many open economies, Singapore experiences cyclical economic turbulence and international competition. In order to withstand the impact of economic turmoil and enable her people to compete globally, building human capital is imperative for Singapore to thrive. Schools, being a social agency, have become the pillar to equip future generations of citizens with a global mindset. This was declared officially to the teaching fraternity by Rear Admiral Teo Chee Hean (2001), then Minister for Education, who encouraged schools to

shift away from the old industrial models on which schools were based – the model of the factory, processing raw materials into a product that met standardized quantity and quality requirements – well-oiled, with everything controlled hierarchically from the top. Schools have to become learning organizations, open to new ideas and drawing upon a wider range of learning resources in their communities and beyond.

Educators are tasked to prepare students "to be enterprising and creative thinkers" in order to develop an "innovative workforce". The need to promote innovation has brought to the fore the need to shift our educational focus from rote learning to creative thinking skills where young minds are primed "to experiment, to make mistakes, to learn and to innovate, in order to be leaders in their own fields".

Although Singapore has consistent high international rankings on the Trends in International Mathematics and Science Study (TIMSS) and the Organisation for Economic Co-operation and Development's (OECD) Programme for International Student Assessment (PISA), the government is not complacent about its current educational achievements. The government is aware of the innovation-driven future the young have to face. In order to prepare young minds beyond examination skills, the Ministry of Education (MOE) has put in place educational infrastructures which require students to learn and "do things differently and with verve and imagination, not by replicating what has been done before" (Tharman, 2005a).

It is evident that Singapore's efforts to nurture innovators through educational policies are intricately tied to the nation's strategic plans to ensure continued economic competitiveness. The speeches of political leaders show that this is a national agenda – to change mindsets and attitudes, and to move away from what used to work, but may now not be sufficient for Singapore to thrive in the economy of the future. The MOE whose mission is to "Mould the Future of the Nation" undoubtedly would have a major role to play in this attempt to build a new culture of innovation, and to do this through educational reforms to inject innovation and enterprise into the education system. The next section provides a description of the educational policies that have been introduced in the past two decades to promote a conducive school environment to nurture innovators.

Promoting innovative culture through educational policies

Several educational policies with the specific goal of nurturing innovators in Singapore have been implemented by the MOE to build an innovative culture among the youth from primary through tertiary levels. Policies such as the Thinking Schools Learning Nation (TSLN), IT Master plans, Innovation and Enterprise (I&E), Teach Less Learn More (TLLM), Project Work (PW) and Integrated Programmes (IP), have been instrumental in shaping the structures, processes and resources to nurture innovations. Since emphasis on basic science and knowledge is seen as the basis for future innovations (MTI, 2006), there are also new educational initiatives to promote interest and talent development in science and technology.

Thinking Schools Learning Nation

In the midst of the 1997 Asian financial crisis, the Singapore government made strategic shifts to ensure Singapore can cope with fierce borderless competition and at the same time enable it to recover faster than others from economic downturns. Recognising that it is the "people's imagination, their ability to seek out new technologies and ideas, and to apply them in everything they do [that] will be the key source of economic growth" (Goh, 1997), the government invested heavily in providing a world class education for students of all abilities. Mr Goh Chok Tong, then Prime Minister of Singapore, hosted the First International Thinking Conference and advocated "Thinking Schools Learning Nation" (TSLN) in 1997. TSLN aimed to shift from content-focused curriculum to a process-oriented one which emphasises critical and creative thinking in learning, and to develop a lifelong passion for learning. In Goh's vision, Thinking Schools would be "crucibles for questioning and searching, within and outside the classroom, to forge this passion for learning" (Goh, 1997). Curriculum and assessments were reviewed to develop critical and creative thinking abilities that were needed to succeed in the evolving new economic landscape. In the same year, TSLN was adopted as the vision statement for the Ministry of Education. To date, it continues to be the over-arching descriptor of the transformation of the education system, encompassing changes in all spheres of education. These changes articulate how MOE would strive towards the Desired Outcomes of Education (DOEs).

Leveraging on technology to promote innovation: the IT masterplans

Technology is one of the bases for creative industrialisation (Florida, 2002). Technological advancements have changed many aspects of our life: learning, working and communicating, as well as our pattern of consumption. Since 1997, the Ministry of Education has implemented three IT Master Plans (The Edumall 2.0). Each master plan is a five-year cycle with specific goals and aligned outcomes. The IT Master Plans aim to increase innovative processes in education, enhance administrative and management infrastructure, promote thinking skills, create opportunities for new curriculum and alternative mode of assessment, and to modify communication and collaborating patterns in teaching and learning. Each cycle is more pervasive than the previous one in building innovative culture. The sustained implementation of IT Master Plans has benefited students as learning environments are being transformed to facilitate the development of critical competencies and dispositions to succeed in a knowledge-based economy. The Master Plans encourage pedagogical innovations by providing a network of educational laboratories where innovations can be prototyped and tested, and subsequently shared among teachers. Schools have access to the latest technologies, and school leaders and teachers receive training to increase the depth of technological, pedagogical and content knowledge in the classroom. To enable students to be future innovators, and producers of knowledge, and to thrive in the 21st century world, the ministry effectively leverages on technology to make students comfortable with new technologies so that they will be able to exploit these new technologies to venture beyond their current boundaries, and hopefully be able to open up new frontiers of knowledge. Supporting the Ministry of Education in this endeavour is the Info-Communication Development Authority of Singapore (IDA) which aligns its flagship programme EdVantage to MOE's IT Master Plans so that it can deploy infocomm strategically to provide a learner-centric, collaborative learning environment within and beyond the classroom, and to create a diverse and vibrant school landscape in the use of infocomm technologies (ICT).

The three core programs are futureSchools@Singapore, Experimentation@Schools and Infocomm@All Schools. These programmes harness ICT in education through innovative pedagogies and flexible learning environments to help schools achieve higher levels of student engagement through an ICT-integrated lifestyle. Students are equipped with the essential skills to be effective workers and citizens in the globalised, digital workplace of the future.

Encouraging project work in schools

Singapore had done well in international studies like TIMSS and PISA, but there continues to be criticism that the system has not been very effective in nurturing students in creative thinking, and problem-finding. Detractors have often attributed this lack to an "over-centralised" system and curriculum, and over-emphasis on national examinations. Cognisant of these shortcomings, the MOE took steps to address them. In 1999, Project Work (PW) was introduced as a component of the curriculum to foster "qualities such as curiosity, creativity and enterprise, [and to] nurture critical skills for the information age, cultivate habits of self-directed inquiry and encourage students to explore the interrelationships and interconnectedness of subject-specific knowledge" (MOE, 2000). To signify its importance, the grade for PW was taken into consideration for applications for admission to the local universities from 2004. The use of PW as an assessment mode signals the shift in emphasis from assessment of end products to assessment of process and skills. This deliberate shift is to encourage students to explore and construct knowledge and to expose students to more authentic and varied performance tasks.

Integrated programmes (IP) and establishment of specialised schools

For students who are clearly university-bound, the MOE went a step further and, in 2004, took another bold measure to de-emphasise examinations by establishing the six-year Integrated Programmes, which allow students to skip a major national examination at Grade 10 – the General Cambridge Examination "Ordinary" Level (GCE "O" Level). The seamless transition from secondary to high school would allow students more room to explore, experiment and develop broader skills. Selected schools were designated IP schools, and granted the autonomy to design their own school-based curriculum instead of following the national curriculum (as long as they continue to adhere to policies such as bilingual education and national education). IP schools could also offer the GCE "A" Level Examination, just like all other Junior Colleges, or opt for alternative examination such as the International Baccalaureate. The removal of a high stakes exam at Grade 10 has enabled IP schools to become wellsprings of educational innovations. The IP schools have developed school-based curriculum that focus on enriched content, and develop thinking dispositions to nurture inquisitive learners who are prepared to solve unseen complex problems. Effort is made to ensure space is given to students to develop passion for what they do, and realise their unique talents and abilities. A ubiquitous feature of the IP curriculum is Research Studies. Students are taught research skills, and expected to embark on research work. It is also common to find IP schools offering "sabbaticals" or "gap semesters" for students – these are "planned absences" from formal lessons in school where students can decide what learning they want to undertake, and how. Such initiatives are in line with the desired outcomes of IP to develop critical and creative thinkers and self-directed learners – all essential attributes of innovators. For instance, one group of IP students worked on isolating bioactive compounds from maggots used in wound treatment. Their project was awarded Gold at a science competition "for their creative approach and systematic study of compounds". Lim (2011) opined that "some of these projects may be done in the school laboratory and due to limited access to specialized equipment, students would have to come up with very innovative solutions to obtain data from their experiments". With more space and opportunities to translate their passions and ideas into action, students will develop new mindsets, see new possibilities and hopefully see their roles in the continually evolving new economic landscape of the 21st century.

In line with the move to provide multiple pathways for students of different abilities and talents to achieve their potential, new Specialized Independent Schools have also been established. For example, specialized schools such as the School of Science and Technology, the National University of Singapore High School of Math and Science and the School of the Arts focus on specific areas of talents in technology, mathematics, science and performing arts, and create learning processes and environments to nurture innovators/creative producers in the different domain areas. Lim (2011) described the case of a mathematics project initiated entirely by a student out of a keen interest and deep understanding in a specific domain in the discipline. This student combined his interest in origami and passion in mathematics to develop an innovative computer algorithm to fold complex three-dimensional structures. Although such schools cater to only about 10% of each cohort of secondary/high school students, the impact of their programmes in developing researchers and innovators of the future cannot be underestimated.

Teach Less Learn More (TLLM)

The announcement of TLLM (Tharman, 2004a) at the national level has invited teachers and stakeholders to reexamine educational assumptions, beliefs and goals; and to redesign pedagogies in order to offer quality educational experiences to diverse learners beyond examination grades. TLLM is aimed at reducing content-laden curriculum in order to minimise regurgitation of course materials; promote problem solving and critical life skills. As explained by Prime Minister Lee Hsien Loong (2006), TLLM is "about spurring independent thinking and learning, and about encouraging students to follow their passions". Under the "top down support for ground-up initiatives" approach, provision of additional manpower within school and white space within curriculum time for customisation of school-based curriculum, the devolution of pedagogic authority from the Ministry to the schools signalled the valuing of educational innovation within the system.

Promoting the spirit of innovation and enterprise

One of the Ministry of Education's Desired Outcomes of Education states that "Students should be innovative – have the spirit of continual improvement, lifelong habit of learning, and an enterprising spirit in undertakings". To help Singapore youth develop a new mindset and attitudes, and equip them with relevant skills to meet the challenges of the 21st century, a strategic plan, Innovation and Enterprise (I&E) was launched in 2004. Mr Tharman Shanmugaratnam (2004b), then Minister for Education eloquently described the spirit of I&E thus:

> At the heart of what we are trying to achieve in I&E is not a new set of activities or programs, but a set of mental attitudes amongst our young, a new culture or outlook on life. We want to nurture in them the mental traits that will serve them well in a future full of challenge and opportunity – a robust spirit of inquiry, a willingness to take untried paths, and a certain ruggedness of character. These are the intangible factors that will make the difference for Singapore in the future.

The core set of life skills and attitudes that the Ministry wants students (and teachers) to develop include:

- a spirit of inquiry and thinking originally;
- a willingness to do something differently, even if there is a risk of failure;
- a ruggedness of character, the ability to bounce back and try again;
- a willingness to stand in a team, lead a team and fight as a team;
- a sense of "giving back" to the community.

Singapore as a hub for entrepreneurs beyond the schools

Singapore has created many other infrastructure, platforms and opportunities to harness innovation at the national level across ministries, and the Ministry of Education works synergistically with these other agencies. To signify the shift towards an innovation-driven economy, and Singapore's intention to promote creativity to sustain growth, Singapore positions itself as a hub for entrepreneurs[4] by establishing several organisations such as the Standards, Productivity and Innovation Board (SPRING Singapore) and Action Community for Entrepreneurship (ACE, www.ace.sg). SPRING Singapore is a statutory board under the Ministry of Trade and Industry. SPRING Singapore helps Singapore enterprise grow by working with partners to help enterprises in financing, capability and management development, technology and innovation, and accessing new markets. Although SPRING Singapore helps to seed and nurture innovative start-ups and nurture a pro-enterprise environment mainly by supporting young entrepreneurs and small and medium enterprise, it also has in place a comprehensive structured entrepreneurship learning programme for students in schools (Grades 7 to 12) as well as students in post-secondary institutions (polytechnics) and vocational institutions (Institutes of Technical Education). The Young Entrepreneurs Scheme for Schools or YES! Schools, for short, launched in 2008 has very specific learning objectives like "Inculcate entrepreneurial mindset and skills in youth" and "nurturing aspiring youths to be entrepreneurs" and lists very specific learning outcomes for students of different grade levels. Schools applying to SPRING Singapore for grants will be evaluated, and two of the criteria emphasise innovation and enterprise: "The proposal should highlight the innovative approach of the project in entrepreneurship learning" and "should cover a comprehensive programme on the entrepreneurship learning activities to cultivate a mindset for enterprise among the students and/or teachers". These educational learning outcomes are developed with the help of the Action Community for Entrepreneurship (ACE) so that schools and their training vendors will have a guide to expose students progressively to entrepreneurship according to the age and/or aptitude of students. Thus it can be seen how the Ministry of Education works with other ministries to develop entrepreneurial talent, to promote innovation and enterprise, and to nurture problem-solving and independent learning abilities. In order to nurture a future generation of innovative and entrepreneurial Singaporeans to build the next wave of Singapore-built global enterprises, the government aims to ensure that the education system is conducive to entrepreneurship learning, and entrepreneurial youth are given the opportunity to learn. Schools who receive funding (ranging from $10,000 to $100,000 per institution, according to type of institution) through YES! Schools can tweak the curriculum to create opportunities for students to have "hands-on" entrepreneurship learning opportunities. Lest students erroneously equate innovation to business or money making, they also receive reminders to tap opportunities schools provide to help them discover their interests and passions, and when pursuing their interests, not to think about the money, but to think about the value that they can add with their innovation.

Tertiary level initiatives

Tertiary institutions have also introduced new initiatives and new foci areas to support Singapore's transformation into a knowledge-based and innovation-driven economy. They realise the need to move beyond their traditional academic missions and adopt an entrepreneurial orientation, infusing an entrepreneurial spirit and cultivating a global outlook among their students. The National University of Singapore (NUS), for example, established NUS Enterprise to provide an enterprise dimension to teaching and research involving the University's students, staff and alumni. To foster the development of an entrepreneurial culture in the NUS community through teaching, training, internship and the nurturing of start-up enterprises, NUS created new divisions such as the NUS Overseas Colleges (NOC) and the NUS Entrepreneurship Centre (NEC). For example, the NOC seek to develop entrepreneurial NUS students with a global mindset by offering selected students an experiential education programme to immerse themselves in the activities of high-tech start-ups located in leading entrepreneurial and academic hubs around the world. The NEC organises events and activities to promote interest in entrepreneurship within the NUS community, and nurtures entrepreneurial start-ups by NUS professors, students and alumni (NUS, n.d.)

The Nanyang Technological University (NTU) set up the Nanyang Innovation and Enterprise Office (NIEO) to support NTU innovators in the commercialisation of intellectual property and to facilitate the transformation of the university's groundbreaking research to market-ready products and services. NIEO also manages the Innovation Centre, which provides aspiring entrepreneurs a place to jump-start their businesses, and provides budding entrepreneurs access to NTU's state of the art facilities as well as the technical know-how and expertise of NTU faculty and staff (NTU, n.d).

To develop and support academic entrepreneurship at the institutes of higher learning, the National Framework for Innovation and Enterprise (NFIE) was established. The NFIE receives funding from the National Research Foundation (NRF) which was set up in 2006 to strengthen Singapore's Research and Development capabilities, encourage greater innovation and nurture the growth of technology-based enterprises in Singapore. The NRF is a department in the Prime Minister's Office, and its functions include the implementation of national research, innovation and enterprise strategies approved by the Research, Innovation and Enterprise Council (RIEC), chaired by the Prime Minister, and the allocation of funds to programmes that meet NRF's strategic objectives. A sum of S$360 million over five years (2008–2012) is allocated to fund the initiatives under NFIE.

These efforts to create the spirit of autonomy and entrepreneurship illustrate the importance of the interface between strong academic grounding and knowledge (e.g., finance, marketing) of management of an enterprise.

Programmes to promote innovation

The next section gives examples of educational programmes that have been implemented to nurture students' spirit of innovation, develop their attitude of exploration and encourage them to think unconventionally, and creatively.

There are a few programmes that specifically focus on nurturing innovators. Among them are the Innovation Programme (IvP) organised by the Ministry of Education, the Tan Kah Kee Young Inventors' Award organised by a philanthropic organisation, the Tan Kah Kee Foundation, and the Singapore Science and Engineering Fair organised by the MOE, the Agency for Science, Technology and Research (A*STAR) and the Science Centre Singapore (SCS).

The Innovation Program (IvP)

The IvP provides a platform for students in primary and secondary schools to be involved in innovation. It gives them a unique experiential learning opportunity to develop their problem-solving and inventive skills, and to give full play to their creativity and imagination to experience the excitement and joy of creating things. Pupils are also encouraged to commercialise their inventions in the spirit of entrepreneurship. The MOE's partners include Institutes of Higher Learning, the Innovators and Entrepreneurs Association, the Intellectual Property Office of Singapore as well as one or two commercial organisations.

Students selected must have an interest in the innovation process, have an open and curious mind, and are willing to learn. They will work on self-initiated projects under the supervision of teacher-mentors in school and expert mentors from the tertiary institutions over a nine-month period. Expert-mentors meet with the pupils during three draft sessions to provide consultation to the pupils. Participants also get to attend a one-day IvP Seminar comprising workshops and talks by professionals experienced in the invention process. This Seminar also provides the opportunity for teacher-mentors and expert-mentors to do a final evaluation of the pupils' products as well as serve as a platform for all schools to present their best projects to all the participants. The IvP culminates in the Young Innovators' Fair during which outstanding projects are showcased at the IvP Show and a poster exhibition. The scientific and technological problem-solving process that students engage in over a prolonged period of time has led a few students to generate innovative solutions in real life.

Tan Kah Kee Young Inventors' Award

Asian Nobel Laureate Professor C.N. Yang, who is also a Tan Kah Kee Foundation adviser and mentor, mooted the Tan Kah Kee Young Inventors' Award (TKKYIA) in 1986 as he observed that Asian students were good at learning but lacked originality in thinking. The award which has since become an annual event seeks to stimulate innovative ideas among the young and to promote scientific and technological research in Singapore. TKKYIA has strengthened over time as it collaborates with many scientific organisations in Singapore, including the Agency for Science, Technology and Research (A*STAR), known as the National Science and Technology Board (NSTB) then and the Defence Science and Technology Agency (DSTA) and DSO National Laboratories and Science Centre Singapore (SCS). Since 1995, the award has been extended to participants beyond school age through the adding of the Open section and Defence Science section in the competition. In 2002, the Foundation launched the same competition in Shanghai and extended collaboration in the region among youth and scientific communities. TKKYIA has nurtured imaginative and innovative thinking among the youth and contributed to nurturing talent by challenging youth to constantly think of new and innovative ways of solving problems in their learning and in life. The Award has contributed to heightened interest in innovation and invention amongst students. To prepare students for this competition, science clubs and mentorship programmes have sprung up in schools.

The Singapore Science and Engineering Fair

The Singapore Science and Engineering Fair (SSEF), a national competition is affiliated to the prestigious Intel International Science and Engineering Fair (Intel ISEF), which is regarded as the Olympics of science competitions. SSEF is open to all secondary and pre-university students between 15 and 21 years of age. Participants are required to submit research projects from the Science Research Programme (SRP), Technology and Engineering Research Programme (TERP), Science Mentorship Programme (SMP) or projects done at school or cluster level. Projects cover all areas of science and engineering. Many of these scientific papers are the products of school-based mentorship programmes supported by tertiary educational institutions.

Students exhibit their projects at the fair, and are assessed by a number of judges from local universities, polytechnics as well as research institutes. Winners will represent Singapore to compete at the Intel ISEF. The Intel ISEF is held annually in the USA and attracts over 1,300 students from 48 states in America and 40 nations to compete for scholarships of over US$2 million, tuition grants, scientific equipment and scientific fieldtrips.

Excellence through Continuous Enterprise and Learning (ExCEL) fest

The Ministry of Education has been organising an ExCEL (which stands for Excellence through Continuous Enterprise and Learning) Fest since 1997. The ExCEL movement seeks to provide quality education through innovation and continuous improvement, and the ExCEL Fest is an annual event that celebrates and shares innovative practices in schools.

To encourage teachers and schools to adopt and adapt innovative practices, the MOE has a generous and systematic way of funding projects. The MOE Innovation Fund introduced in 2000 encourages the testing of new ideas, and the sharing of workable ideas among schools. Many of the projects described in the next section have been shared at past MOE ExCEL Fests.

School initiatives

The implementation of Innovation and Enterprise (I&E) has taken many forms in schools. Many primary schools build the innovative culture by seeding innovative ideas among the young primary pupils. For instance, in Blangah Rise Primary School, teachers made Mathematics lessons exciting and fun for the children. They adapted a driving simulation software and packaged it with a control joystick to create a lesson package called "Drive Around the World". Students could drive from their school to a specific destination, and then work out the links between the amount of time taken and the average speed and the final distance travelled. Another example is the Science department of Radin Mas Primary which organises an annual "Young Innovators' Day" to promote creativity and innovation and use it as a platform for sharing and learning for the pupils across levels. Teachers brainstorm and work on common guidelines on the approach to innovations. The teacher-facilitator in each class may either give a theme, for example, "Household Appliances" for pupils to work on or pose a problem for them to solve. Pupils can work individually, in pairs or in groups of three or four with the teachers as advisers. Pupils are to come up with their inventions or innovations, and the best in each class is selected for presentation and display in the school hall during the "Young Innovators' Day". The innovation is judged according to these five criteria: application of a scientific concept, demonstration of element of creativity, display of practicality of innovation, good layout and presentation.

Schools leverage on technology to explore novel ways to engage students. Cognisant of the popularity of online game platforms among students, teachers from Yusof Ishak Secondary School developed innovative lesson packages integrating such games. For instance, the Nintendo Wii was integrated in the teaching of biology. In the topic on the human body, students play the game, "Trauma Centre". Students take on the role of trainee surgeons and conduct research into the functions of different organs. Teams of students even perform virtual surgery and discuss their understanding of the workings of the human body.

General Paper[5] teachers at Innova Junior College came up with an innovative idea to draw shy and introverted students into discussions. Together with researchers from the Learning Sciences Laboratory at the National Institute of Education, Nanyang Technological University, they developed a curriculum which incorporated an online digital world known as SecondLife. By tapping on their students' interest in, and adeptness at IT, the teachers have enabled students to role-play as avatars to engage in online discussions. Teachers are encouraged that this novel approach has enhanced students' interest, and this approach was subsequently extended to other subjects as well.

I&E has also motivated administrators to relook at school structure and policies. At Tanjong Katong Girls' School (TKGS), I&E is understood as a self-renewing, ongoing review of school processes and outcomes to create good quality learning experiences for students' holistic growth. The school believes in continuously aligning school policies, and experimenting with new initiatives in response to changes in the educational landscape, the evolving demands of a knowledge-based economy and the complexity of a globalised and interdependent world. Bold changes introduced by the school include the Integrated Character Development Programme, the creation of new School Management Committee positions to support the changes of initiatives, the abolition of mid-year examinations, the implementation of a more flexible time-tabling structure and the Integrated Aesthetics Programme (IAP) for Lower Secondary students.

With the IT infrastructure, teachers conduct in-depth discussions with their students and plan more group-oriented work. Quality enrichment programmes developed with external partners and expertise can now be scheduled within curriculum time to benefit an entire cohort of students. Teachers build their capacities to use new pedagogical approaches to conduct engaging lessons. Teachers are involved in school-wide academic programme revision discussions and provide invaluable input on the possibility of reducing curriculum time for some subjects whilst retaining the rigour of academic processes. Curriculum time freed up is used for enrichment activities for students to pursue their interests. TKGS teachers also tap on the School Innovation Fund to come up with innovative pedagogies and practices. For instance, a Biology teacher successfully devised an innovative teaching package that enables students to carry out an experiment and monitor the observations at home, while a Mother Tongue (language) teacher initiated the use of blogs to develop and enhance aural and oral skills in the classroom. The Integrated Aesthetics Programme (IAP) is innovative in the way it weaves the students' learning in Art, Design and Technology, Home Economics, General Music or the Music Elective Programme as students go about working on a thematic and authentic project. There is integration between Humanities and Science on certain topics related to ecology and between Humanities and Mother Tongue on areas of historical interest. Integration of skills on argumentative writing and debating skills is also consciously done between the Languages and the Humanities Departments.

The School is entrepreneurial in developing partnerships with government and private agencies such as the National Environmental Agency and DHL to work on a school-wide recycling project;

Institute of Molecular and Cell Biology to mentor students in Life Sciences research; and the National Heritage Board to involve students in archaeological excavation and research. The Mother Tongue Department leverages on TKGS' relationship with partner schools in Xi'an and QiQihar of the Peoples' Republic of China to organise exchange programmes for students.

To promote a vibrant thinking culture in Jurong Junior College, I&E culture takes the form of JJindogu adapted from the word Chindogu which means "Unusual Tool". Chindogu, an art of invention that originated from Japan in the mid 1980s by Kenji Kawakami, is fundamentally designed to solve daily life problems. Certain aspects of Chindogu that resonate with I&E are adapted. The school celebrates both new inventions and improvements to existing tools and solutions for personal needs and also work- and pedagogy-related issues. Students are encouraged to suspend judgment of ideas, be pragmatic and consider innovation as an essential part of daily living. To encourage aspiring inventors and to promote an innovation culture in school, sharing of innovative ideas is done during curriculum time slots, relevant resources are acquired for the library, inter-class competitions are organised and interactive notice-boards are set up for idea-sharing and dissemination of event updates. Staff and students use college-produced JJindogu postcards to document their inventions. A committee is convened to evaluate the ideas before the students build their prototypes. Upon submission of the prototypes, inventors are reimbursed for their purchases of materials and presented with token coffee vouchers. To maintain sustainability of the JJindogu initiative, all students are also given time during civics lessons monthly to reflect on problems they encounter and engage in JJindogu.

I&E can also take the form of subject-specific enriched learning opportunity. An example of this is "Science Research Excellence" at Anderson Junior College. In line with the I&E initiative, various science interest groups – the Physics Enthusiasts Group, Chemistry Challenge Circle, Life Science Group and Project Eureka, collaborate and organise a Science Innovation Trail for all JC1 arts and science students to appreciate science from different perspectives. Before the Trail, the teachers of the various science interest groups design a variety of thought-provoking hands-on activities involving applications of scientific concepts. The Innovation Trail gives students a chance to approach problems from different perspectives and learn science through a discovery process in a fun and stress-free environment. Furthermore, it encourages students to move on and learn as a team, regardless of their science background. Such learning teams encourage students to brainstorm, respect each other's strengths and weaknesses, without compromising their learning opportunities.

Concluding remarks

In its 2010 report, the Economic Strategies Committee (ESC) envisioned Singapore as a distinctive global city — open and diverse, the best place to grow and reach out to a rising Asia. The committee recommended that Singapore make skills, innovation and productivity the basis for economic growth. There is no doubt that MOE will have to play a key role in preparing the students to help Singapore realise this vision. Education has to equip the young with a creative mindset, and "nurture a spirit of inquiry, of daring to do, and of joyfulness" for "[i]t is only when young Singaporeans live this spirit of innovation ... that we can shape a future for Singapore in an innovation-driven world" (Tharman, 2005b).

Notes

1 Organisation for Economic Co-operation and Development (2005) defined innovation as the "implementation of a new or significantly improved product (good or service), or process, a new marketing method, or a new organisational method in business practices, workplace organisation or external relations".

2 Innovation has also been defined as the "intentional introduction and application within a job, work team, or organisation of ideas, processes, products, or procedures that are new to that job, work team or organisation and that are designed to benefit the job, work team or organisation" (West & Richards, 1999).
3 Craft (2005) sees innovation as the "implementation of new ideas to create something of value, proven through its uptake in marketplace. An innovation can be seen as a new idea being launched on the market for the first time".
4 A Minister in charge of Entrepreneurship was appointed in 2003.
5 General Paper is a course of study at the "A" Level Exam for high school students. It seeks to develop maturity of thought, independent thinking and proficient use of knowledge. Source: www.seab.gov.sg/aLevel/2012Syllabus/8806_2012.pdf.

References

Albert, R. S., & Runco, M. A. (1990). *Theories of Creativity*. Newbury Park, London: Sage Publications.
Amabile, T. M. (1998). How to kill creativity. *Harvard Business Review*, 76(5), 76–87.
Balakrishnan, V. (2009). *Innovation and enterprise critical for growth, says Singapore minister*. Retrieved on 19 October 2011 from www.smbworldasia.com/en/content/innovation-and-enterprise-critical-growth-says-singapore-minister.
Boden, M. (2001). Creativity and knowledge. In A. Craft, B. Jeffrey & M. Leibling (Eds.), *Creativity in Education* (pp. 95–102). London: Continuum.
Canton, J. (2007). *The Extreme Future: The Top Trends that Will Reshape the World in the Next 20 Years*. New York: Penguin Group.
Craft, A. (2005). *Creativity in Schools: Tensions and Dilemmas*. London: Routledge.
EC (2008). Lifelong learning for creativity and innovation: A background paper. Slovenian EU Council Presidency, Retrieved on 31 December 2011 from www.sac.smm.lt/images/12%20Vertimas%20SAC%20Creativity%20and%20innovation%20-%20SI%20Presidency%20paper%20anglu%20k.pdf.
Economic Strategies Committee Report (2010). Retrieved on 25 October 2011 from www.ecdl.org/media/Singapore Economic Committe_2010.pdf.
Esquivel, G. B. (1995). Teacher behaviours that foster creativity. *Educational Psychology Review*, 7(2), 185–202.
Florida, R. L. (2002). *The Rise of the Creative Class: And How it's Transforming Work, Leisure, Community and Everyday Life*. New York: Basic Books.
Gergen, D. (2009). Speech. Professor of Public Service Harvard Kennedy School of Government.
Goh, A. L. S. (2005). Towards an innovation-driven economy through industrial policy-making: An evolutionary analysis of Singapore. *Innovation Journal: The Public Sector Innovation Journal*, 10(3), 1–33.
Goh, C. T. (1997). Shaping our future: Thinking schools learning nation. Speech by PM Goh Chok Tong at the opening of the seventh International Conference on Thinking, 2 June 1997. Retrieved on 1 November 2011 from www.moe.gov.sg/media/speeches/1997/020697.htm.
Infocomm Development Authority (IDA). Available from www.ida.gov.sg/Sector%20Development/20060413170949.aspx.
Innovation and Enterprise. Retrieved on 1 November 2011 from www3.moe.edu.sg/bluesky/ine.htm.
Lee, H. L. (2006). Speech by Prime Minister Lee Hsien Loong at the Teachers' Day Rally on 31 August 2006, The Max Pavilion, Singapore Expo. Retrieved on 1 November 2011 from www.moe.gov.sg/media/speeches/2006/sp20060831.htm.
Lim, T. M. (2011). School science research: A path to nurture talents in innovation. *Innovation*, 10(2), 10–12.
Ministry of Education (2000). *Mission with a passion: Making a difference, MOE yearbook 1999–2000*. Retrieved on 1 November from www3.moe.edu.sg/bluesky/tllm.htm.
Ministry of Trade and Industry (2006). *Science and technology 2010 plan*. Retrieved on 1 November 2011 from http://app.mti.gov.sg/default.asp?id=148&articleID=2461.
Nanyang Technological University (n.d.). Retrieved on 3 November 2011 from www.ntu.edu.sg/nieo/Pages/default.aspx.
National Research Foundation (n.d.). Retrieved on 3 November 2011 from www.nrf.gov.sg/nrf/otherProgrammes.aspx?id=1206.
National University of Singapore (n.d.). Retrieved on 3 November 2011 from www.nus.edu.sg/enterprise/.
Ng, P. T., & Tan, C. (2006). From school to economy: Innovation and enterprise in Singapore. *Innovation Journal: The Public Sector Innovation Journal*, 11(3), 1–12.
OECD. (2005). *Oslo Manual: Proposed Guidelines for Collecting and Interpreting Technological Innovation Data*, 3rd ed. Luxembourg: OECD Publishing.

Porter, M. E. (1998). *The Competitive Advantage of Nations*. London: Macmillan Press.

Runco, M. A. (1990). The divergent thinking of young children: Implications of the research. *Gifted Child Today*, *13*(4), 37–39.

Russ, S. (1996). Development of creative process in children. *New Directions for Child Development*, *72*, 31–42.

SPRING Singapore. Retrieved on 1 November 2011 from www.spring.gov.sg/qualitystandards/be/bea/pages/innovation-excellence-award.aspx.

Sternberg, R. J., & Lubart, T. I. (1999). The concept of creativity: Prospects and paradigms. In R. J. Sternberg (Ed.), *Handbook of Creativity* (pp. 3–15). Cambridge: Cambridge University Press.

Tan, J., & Gopinathan, S. (2000). Education reform in Singapore: Towards greater creativity and innovation? *NIRA Review*, 5–10.

Tan, K. S., & Phang, S. Y. (2004). From efficiency-driven to innovation-driven economic growth: Perspectives from Singapore. *Research Collection School of Economic*, Paper 788.

Taylor, I. A. (1959). The nature of creative process. In P. Smith (Ed.), *Creativity: An Examination of the Creative Process* (pp. 54–61) (A report on the 3rd communications conference of the Art Directors Club of New York). New York: Hasting House.

Teo, C. H. (2001, September). Schools for a new Singapore. Keynote Address by RADM Teo Chee Hean, Minister for Education at MOE Work Plan Seminar, 26 September. Retrieved on 1 November 2011 from www.moe.gov.sg/media/speeches/2001/sp26092001.htm.

Tharman, S. (2004a). Speech by Tharman Shanmugaratnam, Minister for Education, at the MOE Work Plan Seminar on 29 September. Retrieved on 28 October 2011 from www.moe.gov.sg/media/speeches/2004/sp20040929.htm.

Tharman, S. (2004b). The teacher as edupreneur: Exploring new frontiers. Speech by Tharman Shanmugaratnam, Acting Minister for Education at the Teachers' Conference on 8 June. Retrieved on 31 October 2011 from www.moe.gov.sg/media/speeches/2004/sp20040608.htm.

Tharman, S. (2005a). Speech by Tharman Shanmugaratnam, Minister for Education, at the MOE Work Plan Seminar on 22 September. Retrieved on 28 October 2011 from www.moe.gov.sg/media/speeches/2005/sp20050922.htm.

Tharman, S. (2005b). Speech by Tharman Shanmugaratnam, Minister for Education, at the MOE EXCEL FEST 2005, on 1 July. Retrieved on 28 October 2011 from www.moe.gov.sg/media/speeches/2005/sp20050701a.htm.

Torrance, E. P. (1974). *Torrance Tests of Creative Thinking*. Lexington, MA: Personnel Press.

West, M. A., & Richards, T. (1999). Innovation. In M. A. Runco & S. R. Pritzker (Eds.), *Encyclopedia of Creativity* (pp. 45–56). San Diego, CA, London: Academic.

Part XII
Conclusions

45

Overall perspectives on the future promise (and forward thrusts) of innovation education

Larry R. Vandervert

AMERICAN NONLINEAR SYSTEMS, USA

Summary: A review of the many insightful contributions of the chapters in this handbook led to a framework of overarching questions which provides perspectives on current and future innovation education. These larger questions include: (1) *Why is innovation education necessary in the first place?* (2) *Does innovation education actually work?* and (3) *How do the innovative abilities of the gifted individual interpenetrate with the social frameworks of innovative teams?* Answers to these questions draw upon both (1) the long history of human innovation and (2) the innovation concepts, methods, and programs discussed in the chapters in this handbook.

Key words: Innovation education, Morrill Act, natural-born innovators, entrepreneurial drive, Richard Branson, Sigmund Freud.

Why is innovation education necessary in the first place? The larger pre-historical and historical pictures

If, as has been widely supported by anthropological and archaeological evidence, human cultures evolved on the basis of strong human *innovative* adaptability (Ambrose, 2001; Maslin & Trauth, 2009; Potts, 1998; Ridley, 2010; Tooby & DeVore, 1987; Vandervert & Vandervert-Weathers, Chapter 6, this volume), why would we have to bother with the development of something like innovation *education*? According to the above-mentioned anthropological and archaeological evidence, aren't *all* humans natural-born innovators? Wouldn't all individual humans and all groups of humans naturally strive to be the fiercest of innovative competitors? Why do we need innovation education in the first place?

Humans are natural-born innovators: the origins of the entrepreneurial drive

Around 50,000 years ago, a creative and "innovative explosion" is noted in the archaeological record (Ambrose, 2001; Pfeiffer, 1982; Ridley, 2010; van der Leewu, 2010). Those were times when hundreds of thousands of years of the gradual and progressively accelerating natural selection of cognitive and manual abilities that produced stone-tool technology, language, and the arts (Vandervert, in press) were culminating in producing an abundance of innovations:

[40,000–5,000 years ago, there suddenly appeared] ground, polished, drilled, and perforated bone, ivory, antler, shell and stone, shaped into projectiles, harpoons, buttons, awls, needles, and ornaments.... Traces of more perishable materials, including string and woven fibers that may have been made into nets, ropes, bags, and clothing are also well documented. These innovations are among many that signify modern human behavior, including art, ornamentation, symbolism, ritual burial, sophisticated architecture, land use planning, resource exploitation, and strategic social alliances.

(Ambrose, 2001, p. 1752)

This artifact record, indeed, indicates a roughly 10,000-year-long period of relatively rapid innovation. And, on the basis of hundreds of thousands of years of natural selection related to regular advances in stone tool technology combined with archaeological accounts such as that described above by Ambrose, humans do appear to have evolved to be "natural-born innovators", both as individuals and as members of concentrated groups. Ridley (2010) suggested that periods of rapid innovation such as that described above by Ambrose resulted whenever large concentrations of people began to occur. Ridley convincingly argued that the upsurges of innovation were based upon (1) the evolved propensity of *individuals* to innovate in specialized areas, and (2) the *sharing or trading* of these specialized innovations in the formulation of ever more innovations, the two, combined, thus constituting a positive feedback loop of innovation.

Moreover, this evolved propensity toward innovation by the individual within the feedback context provided by the innovative resources of others *within an overall economy* is illustrated in Shavinina's (Chapters 16 and 36, this volume) studies of the entrepreneurial giftedness in childhood (the childhoods of Richard Branson, Warren Buffett, Bill Gates, and other prominent entrepreneurs). That is, Shavinina's arguments of the fact of an unquenchable *entrepreneurial drive* in some children is a strong indication that entrepreneurial giftedness has evolutionary origins, and Ridley's (2010) model of the evolutionary history innovation may explain precisely why this is so.

Ironically, however, the accumulation of the products of innovation began to get in the way of further innovation

However, by 10,000–12,000 years ago these accumulations of new innovations and the detailed processes that produced them began to give rise to likewise accumulations of technology, knowledge, and beliefs (for the most part, the early precursors of science, technology, engineering, and mathematics [STEM]) (Vandervert & Vandervert-Weathers, Chapter 6, this volume). These accumulations of knowledge had to be mastered by each new generation of children if the early, developing agricultural villages were to survive and grow. Because people then had to begin to spend their childhoods almost exclusively learning *pre-existing, standardized* knowledge about the ways things were known to work, they began to think *less and less* about the way things *could* work and about the way they themselves might change things through innovation.

The accumulation of fixed ways of doing things tends to block innovation

When we learn *anything* about accumulated knowledge of *any* sort it tends to *fix* how those things are thought about. This cognitive phenomenon is known as *functional fixedness* (Chrysikou, 2006; Chrysikou & Weisberg, 2005; Duncker, 1945; German & Barrett, 2005). For example, once humans learned to make stone tools certain ways, future innovations in stone tools were to a degree blocked by the very fact of the accepted, standardized way of making tools (German & Barrett, 2005). Each accepted accumulation of technology, story-telling, and art tended to block innovative thought. Thus as early as 10,000 years ago, people were beginning to spend a great deal of time being taught to *conform* their thinking to what was *known*, and this, ironically, began to cause an overall problem for further innovation. As will be shown below, education in China currently represents this very situation.

Today, the older, evolved innovative propensities or sensitivities, which exploded 40,00–50,000 years ago, remain widespread in the population. When high levels of these sensitivities occur in idiosyncratic "innovation supportive" conditions (idiosyncratic innovation education), geniuses like Albert Einstein can be the result (Shavinina, Chapter 7, this volume). Vandervert (2009a, 2009b) argued that as natural, inborn sensitivities toward innovation met with large accumulations of rule-governed knowledge, child prodigies naturally emerged. Because rule-governed knowledge (for example, that of the STEM disciplines) has underlying regularities, and because it is relatively clear as to what is to be mastered, rule-governed knowledge domains like mathematics or classical music tend to appear most often as the forte of gifted children, including child prodigies (Winner, 1996, p. 5). But, of course, not all innovative geniuses and gifted children show their unique strengths in highly rule-governed STEM-related areas. For example, some children as young as age five are gifted writers of child-oriented novels (Edmonds & Noel, 2003; Noel & Edmunds, 2007). Noonan (Chapter 17, this volume) analyzes the novelistic innovative genius of J. K. Rowling (of the Harry Potter series) to illustrate supportive conditions of innovation education *within* traditional "conformity-based" education that can encourage innovation in all children. J. K. Rowling definitely showed the high sensitivities and the rage to master of an innovative genius in her strong desire to be a writer from age five or six (www.accio-quote.org/articles/1998/autobiography.html).

An extreme emphasis on exam-focused education can block creativity and innovation

In some cultural circumstances today, an emphasis on educating children primarily to conform to the accumulation of what is known has led to cultural crises. It is well known in China, for example, that there is such an extreme emphasis on rote memorization in *exam-focused* education that, when compared to other nations, creativity and innovation are largely blocked (Fryer, 2009; Geller, 2011; Jiang, 2011; Tatlow, 2010). Chinese educational scholars ask "the Qian Xuesen question", which originated from Mr. Qian, an American-educated, Chinese rocket scientist, who had asked, "Why does China produce so many clever people, but so few [creative] geniuses?" (Tatlow, 2010)—where are the Bill Gates, the Google guys, or Steve Jobs?

Geller (2011, p. 1) provides insights as to how innovation in China is being blocked by its emphasis on conformity:

> Although China has massive numbers of engineers coming out of its universities, most don't have strong "soft" skills. They're very good at the projects they are assigned to, but don't necessarily think about the next problem coming down the pike. The Chinese education system is still very focused on rote memorization and examinations. Moreover, intense pressure on scientists to advance has resulted in a great deal of plagiarism and data theft.
>
> We've witnessed a massive expansion of entrepreneurship, particularly among young Chinese. But the focus of most Chinese startups is on incremental and business process innovation, or what's called C2C ["copy to China"]: taking a U.S. model and then applying it to the Chinese market. It's where the money is being made. Government intervention also distorts the market. Because the government is so focused on reducing dependence on foreign technology, they say, "Here is the cutting edge. We think it's this Intel processor or this database software. We want you to copy that." This government direction encourages startups to

reverse engineer, rather than to focus on science-based product innovation. Across the board, we see heavy-handed government intervention. There's still a great deal of deference to political authority and to seniority, which is not supportive of individual initiative.

To be sure, education in China is changing, but it will take many years to make significant changes in its extreme emphasis on exam-focused education. Perhaps innovation educators in China will be looking to models like the collaboration of the Johns Hopkins Center for Talented Youth with the Universiti Kebangsaan Malaysia (Jones & Ishak, Chapter 43, this volume) to find their future Bill Gates and Google guys. Further, Chinese educational scholars might undertake overarching analysis of the dynamics of innovation as outlined in Xu (Chapter 10, this volume).

Two vastly different streams of education

There appear to be two vastly different streams of education in the world today. One stream is the standardized, exam-focused education consisting of "reading, writing, and arithmetic" that is given to children around the world. Standardized education is generally aimed at educating children to conform to the needs of society, so that they can become its future functioning and producing members. The other stream of education is *innovation education* which is the subject matter of this handbook. *Innovation education is now needed throughout the world to off-set the effects of standardized, exam-focused education.*

Does innovation education actually work?

As the many different contributions to this handbook attest there are various approaches to innovation education, each of which contains arguments supporting their effectiveness. But do efforts toward educating children (and adults, see Shavinina, Chapter 3, this volume) to become innovators actually work in producing new crops of innovators, new ideas, new ways of doing things, new products, and new increases in profitability? Why should we ask such a question? The answer is quite simple. Jaruzelski, Loehr, and Holman (2011, p. 1) of the worldwide Booz and Company recently reported:

> We continue to emphasize the key finding that our Global Innovation 1000 study of the world's biggest spenders on research and development has reaffirmed in each of the past seven years: *There is no statistically significant relationship between financial performance and innovation spending, in terms of either total R&D dollars or R&D as a percentage of revenues* [italics added].

Research and development (R&D) is a company's investment in the future, the future robustness of its innovation culture, its future capacity to put new products and new applications of existing products on the market; R&D is therefore a company's innovation education program. Simply throwing money at traditional R&D innovation appears not work!

So, given these findings on the effectiveness of R&D spending, how do we test whether innovation education actually works? In the last section of this chapter the innovation-blocking effects of *functional-fixedness* (Duncker, 1945) were described. So, the first question is, can functional fixedness be reduced through training and education? Experimental studies have shown that, yes, functional fixedness can be reduced through training (Chrysikou, 2006). But how can such "fixedness-busting" be achieved outside the experimental lab? Jaruzelski et al. (2011) found that what *does* work in the R&D real world is *not* the amount of money involved, but (1) whether or not the company has an innovation "culture" and (2) that R&D is firmly connected with the company's strategic goals. These findings set the stage to show that *innovation education actually does work* both for adults and across ever-new generations of children and, in fact can change the course of a nation.

The Morrill Act of 1862: the establishment of a culture of innovation (and innovation education) that extended across generations

One remarkable national program started in the United States 150 years ago has shown that when a *culture* of innovation education exists and innovation education is connected with *strategic goals*, it can dramatically change a nation! The Morrill Act of 1862 is described here, because it informs us about the variables of innovation education that produce the greatest results. This long-developing program, while not often recognized as an innovation program or even as a "program", produced more innovation in the United States and its territories that any other innovation effort—it produced far more innovation than NASA's moon-landing, its space shuttle program, and its journeys to Mars combined. The "program" began in 1862 with the Morrill Act. The Morrill act provided huge tracts of land to fund colleges and universities in *each* of the states of the United States and in *each* of its territories for the purpose of establishing the following:

> at least one college where the leading object shall be, without excluding other scientific and classical studies and including military tactics, to teach such branches of learning as are related to *agriculture and the mechanic arts… in order to promote the liberal and practical education of the industrial classes in the several pursuits and professions in life* [italics added].
>
> (Section 4: www.csrees.usda.gov/about/offices/legis/morrill.html)

Innovation is about producing new and useful ideas and products. The Morrill Act was an act aimed directly at innovation; its purpose was to extend higher education into the new areas of agricultural and mechanical arts specifically for the industrial classes (the "working classes"). If something equivalent to the Morrill Act were created today, we would likely refer to it as a massive, nation-changing program of innovation education. Whole bodies of agricultural and mechanical knowledge had to be gathered and re-formulated into higher education curricula suitable for consumption and practical use of a huge and wholly new population of students entering higher education. The Morrill Act brought into being a *culture of innovation education* for the industrial classes that was organized around the agricultural and mechanical arts. The Morrill Act demonstrates a simple principle: science and technology can be hooked up with innovation (see Jones and Buntting, Chapter 30, this volume for associated problems) simply by placing them in larger cultural initiatives that are aimed directly at improving some sector of a nation's economy, be it transportation, fisheries, space exploration, or, of course, agriculture.

Twenty-five years after the United States Congress passed the Morrill Act, its innovative purposes were augmented by the Hatch Act of 1887. The Hatch Act of 1887 provided funding for the establishment of agricultural experiment stations to be added to the colleges and universities set up by the Morrill Act. The purpose of the agricultural experiment stations was specifically to generate original research and research verification that would acquire and diffuse new practical and useful agricultural information. Then, continuing in the innovative traditions of the Morrill Act and the Hatch Act of 1887, the Smith–Lever Act (1914) was added by the United States Congress to make the entire "innovation education" program complete. The Smith–Lever Act provided additional funding for the colleges and universities of the Morrill Act to establish agricultural extension services for the following purposes:

the development of practical applications of research knowledge and giving of instruction and practical demonstrations of existing or improved practices or technologies in agriculture, home economics, and rural energy, and subjects relating thereto to persons not attending or resident in said colleges [those created in each state and territory by the Morrill Act] in the several communities, and imparting information on said subjects through demonstrations, publications, and otherwise and for the necessary printing and distribution of information in connection with the foregoing.

(Section 2: www.higher-ed.org/resources/smith.htm)

The Hatch Act and the Smith–Lever Act thus completed the development of a "culture" of innovation among agriculturalists at American colleges and universities. This culture of innovation extended out into the American farming and rural communities, the consumers of its products, *and it taught each new generation to be constantly innovative in their approach to agriculture*. In this way, the Hatch and Smith–Lever acts served to align the actual innovative strategies of college and university agricultural programs with the original overall intentions of the Morrill Act, namely, *to bring into being a culture of innovation education for the industrial classes that for the first time in America was organized around the agricultural and mechanical arts*.

The Morrill Act was not about conformity-based education in American colleges and universities, in the main it was about innovation education. The national Acts that followed the path of the Morrill Act brought the *intent* of this nationwide program of innovation education to all the people and to unparalleled agricultural wealth. The Morrill Act and its subsequent follow-up Acts have shown that *innovation education works*, and they have shown how it works across generations of innovative agriculturalists.

The specific message for innovation education, starting from the original wording of the Morrill Act to the research stations created by the Hatch Act and on to the Smith–Lever cooperative education programs that extended the first two acts out into the consumer community, is, as Jaruzelski et al. (2011) above found: *innovation strategies produce the greatest innovation when they are closely aligned with the larger, overall corporate, national, or school strategies*. Since innovation *education*, too, would be most successful when couched in larger corporate, national, or school strategies, placing the United Kingdom's Technical and Vocational Educational Initiative (TVEI) within, for example, a larger strategy of transportation renewal, or agricultural renewal in the United Kingdom would likely have led to the more positive results which Banks (Chapter 41, this volume) had hoped for.

How are individual innovative geniuses related to teams of innovators working within social-institutional frameworks?

The question of the relative contributions of individual innovative geniuses (for example, Thomas Edison, Albert Einstein, or Richard Branson) on the one hand, and of teams of innovators working within social-institutional frameworks on the other is often addressed (see the following chapters in this volume: Banks, Chapter 41; Frost, Chapter 42; Jones & Ishak, Chapter 43; Quek & Tan, Chapter 44). How are the processes that create individual innovative geniuses related to those processes at work within social-institutional frameworks? Should innovation education for individual innovators be the same as that for teams of innovators working, for example, in a corporate setting, or should it be different in certain ways?

The story of the motives that drive innovation both in individuals and in groups at large was captured by one of the world's greatest thinkers on human thought and behavior, Sigmund Freud. Freud (1930/1961, pp. 37–38) proposed that control over the environment is the driving motive (both in individuals and in groups) behind the innovative rise of the technological aspects of culture:

With every tool man is perfecting his own organs, whether motor or sensory, or is removing the limits to their functioning. Motor power places gigantic forces at his disposal, which like his muscles, he can employ in any direction; thanks to ships and aircraft neither water nor air can hinder his movements; by means of spectacles he corrects defects of the lens of his own eye; by means of the telescope he sees into the far distance; and by means of the microscope he overcomes the limits of visibility set by the structure of the retina. In the photographic camera he has created an instrument which retains the fleeting visual impressions, just as the gramophone disk retains the equally fleeting auditory ones; both are at bottom materializations of the power he possesses of recollection, his memory. With the help of the telephone he can hear at distances which would be respected as unattainable even in a fairy tale. Writing was in its origin the voice of an absent person.

According to Freud, the whole point of innovation is to make culture/civilization increasingly workable by controlling the forces of nature. And, because the individual and the group share the same motor and sensory functions in this control of nature they inextricably share the advantages of innovation: *The extension ("perfecting") of the powers of the motor and sensory capabilities by an individual human, extends the same collective motor and sensory powers of the group, and extending the powers of the motor and sensory capabilities by the group extends the same motor and sensory powers of the individual.* For Freud, the motive toward innovation in the individual and in the group is thus totally interpenetrated and interdependent—the motive in both cases emanates from the same innate motor and sensory sources, and through innovation they are both headed in the same direction.[1] In the pages following the above quote, Freud went on to make deep connections between the human motive to control the environment (to innovate) and the rise of religion and the arts. Essentially, he concluded that through their accelerating control of environmental forces, humans were becoming "prosthetic gods". Freud might well have said that *both* the individual and the group were beginning to live the life of the gods they, in more primitive times, once feared and used as models of perfection to guide innovative attempts at their *own* perfection.

Individual radical innovators: what really drives them toward entrepreneurial giftedness?

Freud made the above observations on the basis of technological advancements through the late 1920s, over ninety years ago. Since then, individual innovators and groups of innovators have produced "tools" (as Freud called them) like radar, television, computers, cell-phones, space technology that has flown to the Moon, the Global Positioning System (GPS), and a space plane (like Richard Branson's Virgin Galactic [www.virgingalactic.com/]) which can fly people to high altitudes to view the earth from, as perhaps Freud would have put it, a god's point of view. Certainly, in the case of Richard Branson's Virgin Galactic space plane we have an example a "radical innovation" accomplished by a "radical innovator", as described by Sandberg (Chapter 39, this volume): "a radical innovation is defined as a new product or service that requires considerable change in customer behavior, is perceived to offer substantially enhanced benefits, and is also technologically new". As Sandberg further points out, such radical innovations have economic and cultural impacts which extend far into the future and inspire further innovation. Visionary projects like Branson's require constant innovation through the breaking down of traditional disciplinary boundaries across STEM disciplines as outlined by Feldon, Hurst, Rates and Elliott (Chapter 25, this volume), and such visionary projects require innovation education techniques which foster fundamental "concept changes" as outlined by Vosniadou and Kampylis (Chapter 5, this volume).

The trajectory leading to individual innovation

But what really drives gifted entrepreneurs like Richard Branson? What catalyzes the drive? Shavinina (Chapter 16, this volume) describes the strong influence of a "crystallizing experience" that set financial guru Warren Buffett in motion. When Buffett was age ten he met Sidney Weinberg, the most famous investor in New York at the time. Weinberg's personal attention and kindness to Buffett replayed in Buffett's mind for the rest of his life. Buffett's crystallizing childhood experience shows the powerful interplay between the sensitivities of entrepreneurial giftedness in the individual and, as the social psychologist would put it, *the actual, imagined or implied presence of others*. But why do crystallizing moments such as these have such a powerful, resonating effect? Vandervert (2009a) argued that these special *social-psychological* experiences are replayed over and over in the sensitive child's working memory, and *because* they are replayed, the brain's cerebellum sets up as a positive feedback loop with the rest of the brain which compounds the emerging crystallizing experience that is driving thought and behavior. This positive feedback loop often culminates in the feeling of the type of insight that is associated with creativity or innovation where a new idea or perspective suddenly "pops" into view (Vandervert, Schimpf, & Liu, 2007), and thus "feels" like a crystallizing moment of some extraordinary significance. The "aha" moments of creativity and the crystallizing moments which guide the course of one's life appear to have a common source. This cerebellum–cerebral cortex positive feedback loop, Vandervert believes, is a fundamental part of the "psychological carrier" of innovation suggested by Shavinina (Chapters 12 and 16, this volume) to lead to an individual innovator's crystallizing moment such as that of Warren Buffett cited above.

However, teams of innovators working within social-institutional framework tend to stay "on top of things"

Freud felt that the higher mental motives of humans bound them together in religion, aesthetics, and order. The individual's motives or innovations must take into account the motives of the group, that is, what people need or want. For example, Edison's overall goal, in his own words, was "to invent useful things every man, woman, and child in the world wants ... at a price they could afford to pay" (Josephson, 1992, p. 314)—there is no need to invent things other people don't need or want. *Teams of innovators must do the same thing.* Corporations that innovate in accordance with constant advice from their consumers are far more successful (Jaruzelski et al., 2011). So, the individual innovative genius and groups of innovators working within social-institutional frameworks interpenetrate one another on the basis of the common motives that both lead to innovations in the first place and that make those innovations useful. Because of the complexities of today's world, including the fast pace of change, groups of innovators working within social-institutional settings and paying close attention to trends are likely to be more in contact with the needs and desires of consumers, and therefore produce more widely successful innovations.

Hybridization: an institutional team of innovators uses a prototyping factory to produce more individual innovators

At the same time, however, it now appears that more individuals will be getting a new kind of *hybrid* "innovation education", and thus more individuals will be able to do the work of the once-rare individual innovative genius. This hybridization of innovation education promises to be successful, *because the individual works within a facilitative team structure.* Heffernan (2012) describes how sophisticated prototyping machinery and software are becoming inexpensive and easier to use and can therefore be used to produce more innovation. Heffernan illustrates how TechShop, taking advantage of this cheap abundance of prototyping machinery and software, operates prototyping facilities open to the public of any age or any interest—each location has robust innovation education programs. She goes on to describe how Ford Global Technology, a division of Ford Motor Company, negotiated to set up one of TechShop's prototyping locations (there are currently five TechShop locations in the United States) in Detroit, Michigan, to inspire technological innovations. This hybridization of who does what in the whole range of issues in the marriages of company, university, and the Internet in the preparation and delivery of innovation amplifies the lessons of several chapters in this volume, for example, Wallace (Chapter 31), Salyers and Carter (Chapter 32), Thorsteinsson (Chapter 33), Fish and Kim (Chapter 34), Keinz and Prügl (Chapter 35). Within a year, the Ford team's innovative output was up 17%. Even at that, TechShop might take advantage of Delcourt and Renzulli's (Chapter 9, this volume) concluding "Tips to Foster Innovative Behaviors". Putting sophisticated, rapid prototyping capacities into the hands (and minds) of *individual* members of TechShop social-institutional innovation teams creates a *hybridization* consisting of individual innovative genius and innovative teams that seems to represent the best of both individual and social-institutional creative worlds. Ritchie and Tomas (Chapter 27, this volume) support this hybridization or interpenetration through teaching writing that is based on the fact that innovation always involves the interpenetration of the individual and social contexts; *innovation always occurs, often to some small unexpected degree, within the context of a community of other thinkers, past or present.*

Vandervert and Vandervert-Weathers (Chapter 6, this volume) discuss in some detail how prototyping is the fundamental mechanism of innovation, whether in individuals or within teams working in social contexts. They describe a prototype as an early model or sample envisioned and "built" for the purposes of testing the workability of a new concept or product. This prototyping mechanism is what is being actualized for more and more individual innovators working within the prototyping framework at TechShop (and at Ford Global Technology).

An interesting, and I think valid, connection can be made between how prototyping leads to innovation for the individual at TechShop and how prototyping leads to innovation in the developmental histories of prominent innovators. Shavinina (Chapters 16 and 36, this volume) described the development histories of six well-known entrepreneurially gifted innovators, namely, Richard Branson, Warren Buffett, Steven Case, Michael Dell, Bill Gates, and Sam Walton. She described how each of these developmental stories is the story of personal propensities and conditions that drove and supported staying with developing themes of entrepreneurial behavior until their fruition. When one looks at the common patterns of the overall developmental trajectories of these prominent innovators, a *macro*-trajectory of the essence the *micro*-trajectory of the prototyping that occurs in every instance of individual innovation reveals itself. All six of these gifted entrepreneurs produced early business models (actually, prototypes of their lives to be), so that they could test and re-test through constant revisions the workability of a new concept or a new product. This strong pattern of envisioning a business model, then testing and re-testing of their own vision is how these personalities build and actualize themselves. But, as Shavinina (Chapter 16, this volume) clearly reported, this "personality" is always formulated within the context of others.

Summary

The original three questions posed at the beginning of this chapter were formulated as follows:

1 *Why is innovation education necessary in the first place?*
2 *Does innovation education actually work?*

3   *How do the innovative abilities of the gifted individual interpenetrate with the social frameworks of innovative teams?*

A brief summary of the answers to these three questions that were provided in this chapter are presented below.

Why is innovation education necessary in the first place?

It was argued that innovation education has become a necessity in today's world by illustrating that while humans evolved to have a strong, inborn drive to innovate, ironically, the accumulations of knowledge produced by that innovation led to societies where standardized, conformity-based education became necessary to their survival and growth. China's strong emphasis on exam-focused education was discussed as a contemporary example of this situation. *Innovation is necessary throughout the world today to off-set standardized, conformity-based education and in order to help re-capture the evolved potential of all humans to innovate.* That is the future of innovation education.

Does innovation education actually work?

Since it has been found that there is no relationship between financial performance in business and industry and spending on research and development (R&D) (which is the company's investment in innovation for the future), whether or not innovation education actually works can be a legitimate and perplexing question for innovation education. To get to the heart of this issue the great successes of the innovation education inspired by America's Morrill Act (1862) and the resulting huge increase in across-the-board farming production was examined. The overall conclusion of this examination, backed up by recent massive studies of innovation dynamics, and repeated here, was that: *innovation strategies produce the greatest innovation when they are closely aligned with the larger, overall corporate, national, or school strategies.* The message of this section was that: *Yes, innovation education actually works.* And, if innovation education is aligned with larger institutional strategies, it can be shown to involve the innovative capacities of more people and to induce dramatically more productive outcomes for school systems, corporations, and even nations. This, too, is the future of innovation education.

How do the innovative abilities of the gifted individual interpenetrate with the social frameworks of innovative teams?

Re-visiting Sigmund Freud's view on how innovations extend or amplify human motor and sensory capabilities, it was pointed out that: *Because the individual and the group inextricably share the evolutionary advantages of all motor and sensory functions, extending ("perfecting") the powers of the motor and sensory capabilities of a human, extends the same collective motor and sensory powers of the group, and extending the powers of the motor and sensory capabilities of the group extends the same motor and sensory powers of the individual.*

Individual radical innovative geniuses like Albert Einstein, Thomas Edison, and Richard Branson can, to use the vernacular, blow the top off culture as it is currently known and practiced. However, teams of innovators working within social-institutional frameworks tend (repeat, *tend*) to stay more on top on trends. At the same time, it was shown how newer hybrid models of innovation education that, like those of the prototyping facilities at ShopTech, involve individual innovators with facilitative teams can both keep up with changing trends and quickly produce more individual innovators.

Conclusion

Starting with Freud's comments on the human motive for innovation, and running through the relative innovative contributions of individual innovative geniuses on the one hand and the contributions of teams working within social-institutional frameworks on the other, the lesson of this section was: innovation always involves the interpenetration of the individual and social contexts; *innovation always occurs within the context of a community of other thinkers, past, present, and future.*

Note

1 It is surprising that Freud did not mention *motion pictures* by which we compose innovative sequences of visual-spatial images and sounds which can be shared with everyone. Moreover, these external visual-spatial/sound patterns, because they are constructed to parallel problem-solving and imagination running in our heads, can show others how to solve problems and to create new ideas. Motion pictures impart power to the individual and the group, because they solve problems and spread culture through a sort of "effortless thought", "effortless entertainment", and "effortless information-gathering" (i.e., the "news" as seen in motion pictures or on television). Elaborating on Freud, it is not far-fetched that some radical, innovative genius like Richard Branson will someday combine motion picture formats (now in digital form on CDs) with the computerized mixing of visual-spatial imagery patterns modeled after those produced in the human brain's cerebellum (Imamizu, Higuchi, Toda, & Kawato, 2007; Vandervert, in press) so that the combined system automatically proposes useful innovations.

References

Ambrose, S. (2001, March 2). Paleolithic technology and human evolution. *Science, 291*, 1748–1753.
Chrysikou, E. G. (2006). When shoes become hammers: Goal-derived categorization training enhances problem solving performance. *Journal of Experimental Psychology: Learning, Memory, and Cognition, 32*, 935–942.
Chrysikou, E. G., & Weisberg, R. W. (2005). Following the wrong footsteps: Fixation effects on pictorial examples in a design problem-solving task. *Journal of Experimental Psychology: Learning, Memory and Cognition, 31*, 1134–1148.
Duncker, K. (1945). On problem solving. *Psychological Monographs, 58* (Whole No. 270).
Edmunds, A., & Noel, P. (2003). Literary precocity: An exceptional case among exceptional cases. *Roeper Review, 25*, 185–194.
Freud, S. (1961). *Civilization and its Discontents.* New York: W.W. Norton (original work published 1930).
Fryer, W. (2009). Creativity and innovation in Chinese society and schools. www.speedofcreativity.org/2009/11/03/creativity-and-innovation-in-chinese-society-and-schools/.
Geller, L. (2011). The innovation advantage. www.strategy-business.com/article/00069.
German, T., & Barrett, H. C. (2005). Functional fixedness in a technologically sparse culture. *Psychological Science, 16*, 1–5.
Heffernan, M. (2012). How to boost creativity. www.cbsnews.com/8301-505125_162-57374516/how-to-boost-creativity/.
Imamizu, H., Higuchi, S., Toda, A., & Kawato, M. (2007). Reorganization of brain activity for multiple internal models after short but intensive training. *Cortex, 43*, 338–349.
Jaruzelski, B., Loehr, J., & Holman, R. (2011). The global innovation 1000: Why culture is key. www.strategy-business.com/article/11404.
Jiang, X. (2011). Resistance futile in Chinese classroom. http://the-diplomat.com/china-power/2011/06/14/resistance-futile-in-chinese-class/#respond.
Josephson, M. (1992). *Edison: A Biography.* New York: John Wiley and Sons, Inc. www.nps.gov/nr/twhp/wwwlps/lessons/25edison/25edison.htm.
Maslin, M., & Trauth, M. (2009). Plio-Pleistocene East African pulsed climate variability and its influence on early human evolution. In F. Grine, R. Leakey, & J. Fleagle (Eds.), *The First Humans: Origins of the Genus Homo* (pp. 151–158). Netherlands: Springer Science+Business Media BV.
Noel, K., & Edmunds, A. (2007). Constructing a synthetic-analytic framework for precocious writing. *Roeper Review, 29*, 125–131.
Pfeiffer, J. (1982). *The Creative Explosion: An Inquiry into the Origins of Art and Religion.* New York: Harper and Row.

Potts, R. (1998). Variability selection in hominid evolution. *Evolutionary Anthropology, 7*, 81–96.
Ridley, M. (2010). *The Rational Optimist: How Prosperity Evolves.* New York: Harper-Collins.
Tatlow, D. (2010). Letter from China: Education as a path to conformity. www.nytimes.com/2010/01/27/world/asia/27iht-letter.html?pagewanted=all.
Tooby, J., & DeVore, I. (1987). The reconstruction of hominid behavioral evolution through strategic modeling. In W. G. Kinzey (Ed.), *The Evolution of Human Behavior: Primate Models* (183–237). Albany, NY: State University of New York Press.
Van der Leeuw, S. (2010). The archaeology of innovation: Lessons for our times. http://athensdialogues.chs.harvard.edu/cgi-bin/WebObjects/athensdialogues.woa/wa/dist?dis=83.
Vandervert, L. (2009a). The appearance of the child prodigy 10,000 years ago: An evolutionary and developmental explanation. *Journal of Mind and Behavior, 30*, 15–32.
Vandervert, L. (2009b). Working memory, the cognitive functions of the cerebellum and the child prodigy. In L. Shavinina (Ed.), *International Handbook on Giftedness* (pp. 295–316). New York: Springer.
Vandervert, L. (2011). The evolution of language: The cerebro-cerebellar blending of visual-spatial working memory with vocalizations. *Journal of Mind and Behavior, 32*, 317–332.
Vandervert, L. (in press). How the blending of cerebellar internal models can explain the evolution of thought and language. *Cerebellum.*
Vandervert, L., Schimpf, P., & Liu, H. (2007). How working memory and the cognitive functions of the cerebellum collaborate to produce creativity and innovation. *Creativity Research Journal, 19*, 1–18.
Winner, E. (1996). *Gifted children: Myths and Realities.* New York: Basic Books.

Index

Abbott, W. 343
"abortion" of new ideas phenomenon 545–6; defining 546–7; explaining 547–9
above-average ability 56, 129–31, 132, 195, 205, 223, 304
abstractness of titles 156–7; decrease in 158–9
accelerated development 95, 103–4
accumulated knowledge 608–9
achievement motivation 245–6, 252–3
Action Community for Entrepreneurship (ACE), Singapore 596–7
Action-based Problem Solving (AbPS) 219, 220, 222
active learning strategies 275, 396–7, 437–8, 559
activist education 409–12
Adam, B. 21
adaptive creativity 153, 154–7, 159
adaptive expertise 58, 61, 63, 366
adult education 46
aesthetic preferences 338
affective factors: radical innovators 534–42; science learning 375–81
agricultural education 611–12
Alexander, T. 549
Alferov, Zhores I. 267
Allchin, D. 411
Alperovitz, G. 405
Altimira Pictures 521–3
Alvarez, J.L. 515–16, 517, 520, 521
Amabile, T. 53, 55, 63, 117–18, 284, 591
Ambrose, S. 83, 84, 89
analogical deduction 75, 363–5, 367
Anderson Junior College, Singapore 601
Anderson, D. 296
anonymity 433–5
Apple 112, 508, 509, 510
applied mathematics 342–55
applied science 420, 422, 559–60
apprenticeship 412
Aptitude-Treatment Interaction (ATI) 61
Architectural Approach 514, 516
architecture 330, 331; interdisciplinarity as means of innovation in 338–9

Arthur, M. 515, 516, 517, 518
arts: creativity as domain of 162; *see also* mathematics and the arts
Asch, S. 435–6
assessment imperative 62–3
assessment: authenticity of 426–7; new approach to 171–9; theory of individual innovation 169–71
assignments, innovative schools course 291–2
Association to Advance Collegiate Schools of Business (AACSB) 475
astronomy 129, 136, 343, 348, 354, 381
asynchronous development 94, 103, 443, 444
Attention-Deficit/Hyperactivity Disorder (AD/HD) 160
Austin Discovery School 295
authentic problem-solving 365–6
autonomy 119, 124, 125, 253, 254, 508
avatars 87, 432, 434, 437, 464, 600
Axelrod, P. 407
Ayers, A.D. 535

Babiak, P. 409
bacterial theory of ulcers 70–1
Bakan, J. 408, 409
Balakrishnan, V. 592
Bandura, A. 24, 344
Banting, Fred 37
Barber, B.R. 408
Bauman, Z. 273
Baumol, W.J. 216, 217
Beghetto, R.A. 52, 62, 63
Bell, P. 374–5
Bell, R.L. 367, 411
belonging 253–4
Bem, D. 437
Bencze, J.L. 409, 410, 412
Bennis, W. 501, 508
Bereiter, C. 57–8, 60–1, 63, 125
Bernstein, B. 277, 278, 285
Berry, J.S. 324
Beswick, K. 338
Beyers, R.N. 72

Index

Bianco, B.H. 194
bias 194, 436–7
Biederman, P. 501, 508
Biggs, J. 540
Bill, B. 331
biographies 93, 228
biology, integrated approach *see* computer based learning (CBL) systems
BioStories project: background 387–8; design and evolution 388–91; implications for conceptualising innovation 391–3
Biotechnology Learning Hub 424–5
Blanagh Rise Primary School, Singapore 599
Blanton, P. 288
blended learning 438, 443, 444, 446, 448, 450, 451, 456, 465, 467
Bloom, B. 588
books: early access to 98, 101–2, 260–3; *see also* textbooks
boundaryless careers 514, 515, 518
brain-imaging 82–5
brain race: Center for Talented Youth model 584–5; Malaysia's search for innovators 585–6; PERMATApintar and Centre for Talented Youth 586–7; PERMATApintar's success in Malaysia 588
brainstorming 143–5
Branson, Richard 34, 38–9, 40, 501–11; early development 230–1; impact of early development 236–9
breakthrough innovations 501–11
Brewer, W.F. 70, 74
Brody, L.E. 188
Bronfenbrenner, U. 277
Brooks, N. 405
Broussau, G. 317
Brown, A. 35
Brown, E. 61, 202
Brown, R. 233, 331
Bruder, P. 348
Bruner, J.S. 22, 23, 57, 246
Buffet, Warren 614; early development 231–2; impact of early development 236–9
"Bug Reports and Two Cools" exercises 477
Bullen, M. 447
Buntting, C. 422, 423, 424
Burke, C. 275, 283
Business Development Laboratory courses 538–41
business pitch development 476–7
business plans 32, 79, 147, 236, 237–8, 475, 479, 538, 541
business startups 478, 574–5, 609–10
business wargaming 531
business–science partnerships 406–7

Callaghan, James 558
Camp Invention: accomplishments 87–8; activities 85–6; future goals 88–9; INNOVATE program 87; schedule 85–6
Can Do Society 478–9
Canada *see* university online education
Canisius College *see* entrepreneurship collegiate academic program
Canisius Entrepreneurs' Organization (CEO) 478, 479
capacity building imperative 61–2
Career Capital Pyramid Model (CCPM) 514, 515–16, 517, 521, 523, 524
career solidarity: case study 521–3; entrepreneurial teams and their formation 514–15; entrepreneurs' career and network 515–16; generated from irreplaceable relationships 520–1; implications for entrepreneurship 523–4; patterns of career formation 516–20
career opportunities 479
Carey, S. 69
Carlson, Arvid 261
Carter, L. 409, 449
case studies, research on 72–3
case study lessons 460–6
Case, Steven: early development 233; impact of early development 236–9
casual learning 374
Ceci, S.J. 154, 364
cell theory 399
Center for Gifted Education, College of William and Mary 123, 195
Center for Mathematical Talent, New York University 310–11
Center for Talented Youth (CTY), John Hopkins University 187, 188, 584–8; *Online* program 438–9, 585, 586; and PERMATApintar 586–8; School and College Ability Test 584
Cerini, B. 377, 381
chance hypothesis 361–2
charter schools 293–6
Chen, Z. 366
Cherry, K. 344
Chi, M.T. 42, 363, 367
childhood play 159–60
China: blocked innovation 609–10; creative education 146, 148
Cho, S.H. 156
Chu, Steven 262, 268
Clandinin, D.J. 298
Clark, B. 572, 578
classroom change 277
classroom culture, problem solving as 566–7
classroom-as-laboratory 304–6
classrooms, innovation education in 278–81
Clayburn Cross, A. 363
Clement, J. 75
coaching skills 247, 250–1
Cognitive Abilities Test (CogAT) 205
cognitive conflict 74–5

Index

cognitive experience 99, 103–4, 170–1, 175, 178, 179, 318–20, 326, 327
cognitive sensitivity 94, 102–4
cognitive style 176, 178, 316
Cohen-Tannoudji, Claude 261
Cohen, Stanley 261
Coleman Faculty Entrepreneurship Fellows Program 482
Coleman Foundation 474, 478, 480, 481, 482
collaborative practices 385–93
collaborative problem solving 567–8
Collins, A.M. 58, 61
Collins, B. 331
commercialisation 376, 497, 570, 580–1, 597
commitment 117–18, 129–30, 131–2, 137–8
Common Core Standards Initiative (2011) 219–20
common good, socioscientific innovation for 409–12
communication skills 134, 203, 479, 485
communitarianism 411
community of practice, linking classrooms with 423–5, 427
Community Problem Solving (CmPS) 217, 222, 223
comparative career formation 518, 520, 523, 524
compartmentalism 343–4
compensatory mechanisms 510–11
competence 63, 217, 251–2, 316, 319, 320, 322–3
competitions 18, 204, 401, 421, 478, 481, 59; as "locomotive" for innovation 311–12; Renzulli Academy 209–12
competitiveness 32, 33, 210, 217, 341, 359–60, 527, 545–6
Computer Aided Design (CAD) 457, 464, 467, 564
computer based learning (CBL) systems: methodological apparatus 397; procedural technology 397–8; technology of developing practical skills to use acquired knowledge 398–9; technology of ecologizing education and creating foundations for nature-conservation activity 401–2; technology of individualized knowledge-testing and self-checking 400–1; technology of individualized orientation to learning 400; technology of promoting healthy lifestyle 399–400; technology of providing equal access to knowledge 400
computer-mediated communication (CMC) 433–4
computer-supported collaborative work (CSCW) 437–8
concentration 98, 100, 103, 175–6
conceptual change: definitional issues 69–70; relevance for innovation 70–1; research on innovation education 71–3; science education context 73–6
conceptual knowledge 364–6, 562, 565, 568
"Conceptual Synthesis" test 173, 174, 175, 176
Connecticut Invention Convention 204, 211
Connecticut Mastery Test (CMT) 204–5, 212–13
Connelly, F.M. 298
conscientization 411

consumerism 404–5, 406–10
context dimension of creativity 57–8, 60–3
continuing professional development (CPD) 559–60, 573
continuous innovation 148, 529–30
controlled pedagogy 278, 280, 284
convergent thinking 153–7, 190
Corbin, J. 463
Cornell, Eric A. 163–4, 265, 267
corporate innovation, integration of users 488–9
cost externalization 408–9
courage 41, 436–7, 509–10
Cowie, B. 426, 427
Cox, C. 216
Craft, A. 58, 275, 591
Cramond, B. 115, 218
Crane, V. 373
creative contents, HICEMTs 45
Creative Education Foundation (CEF) 143–4
creative education: environment around venture and entrepreneurial education 142–3; new creative education 146–8; traditional creative education 143–6
creative leadership types/skill sets 192
creative needs 55, 161–3
creative problem-solving (CPS) 55–7, 119–21, 190; definitions of education for innovation 215–17; Future Problem Solving Program International 217–24
creative productivity: barriers 138–9; catalysts 136–8; enrichment triad model 132–6; identification of gifted students 132; and three-ring conception of giftedness 129–32; tips to foster innovative behaviors 139
creative relevant skills (CRS) 19, 24–7, 117–18
creative teaching 289–90
creative thinking 119–21, 144–8, 561; combining with critical thinking 123–4; testing 154–7
creativity 129–30, 132, 275, 316, 320; changes in 158–9; critical components to development 483–6; decrease in 158; definition of 111–12; elements of 246–8; identifying and cultivating 60; measurement of changes 154–7; models promoting 117–19; naturalizing 57–9
creativity assessments 162
creativity crisis, possible reasons for 159–64
creativity enhancement, taxonomy of 62–3
creators: contemporary examples 112; distinctions with innovators and implications for school-based learning 113–14; historical examples 112–14
CREST (Creativity Education in Science and Technology) 421
Creswell, J.W. 463
Crèvecoeur, H.S.J. 162
critical thinking 53, 61, 114, 117, 444, 208, 373, 559, 561, 562; combining with creative thinking 111, 123–4; and innovation 121–3

621

Croft, D. 70–1
Cross, N. 363
"crystallizing experience" 103, 209, 238, 614
Csikszentmihalyi, M. 55, 58, 62, 117, 119, 135, 246, 247, 337, 539–40
culturally and linguistically different (CLD) students 192, 195–6, 202
Cummins, Emily 376–7, 379, 381
curiosity 55, 103, 96, 101, 102, 155, 157, 161
curriculum challenges 32, 247, 249–50
curriculum change 276–7
curriculum differentiation 191–2
curriculum imperative 60
curriculum initiatives: Canisius College 476–80; Renzulli Academy 206–12; science education 385–6
Curriculum-based Future Problem Solving (FPSPI) 218, 220
customer-active paradigm 487–9

Dahl, Roald 136
Dai, D.Y. 53, 59, 60, 61, 112, 125
Dalke, A. 337
Daly, L. 405
Danilov, S.A. 401
Danilova, O.V. 397, 398, 401
Darwin, Charles 112–14
De Bono, E. 145, 155–6
De Simone, D. 216
deadline management 30, 40
Deci, E.L. 251, 254
Delcourt, M.A.B. 115, 135, 137–8
Dell Computer Inc. 40, 234, 237, 506, 547
Dell, Michael 34, 506, 509, 547, 550; early development 233–4; impact of early development 236–9
democratic schools 295–6
Denham, S.A. 194
Denton, H. 72, 458, 459, 460, 464, 465
DeRoche, E.F. 347
design education 337–9, 559–60; contribution of innovation education 561–4; problem solving approach 565–8
design-a-school project 297
DeSteno, D. 540
developing school setting 282, 284–5
developmental capacities 99–104
developmental influences, new targets of 43–4
Dewey, J. 373, 291
diagnostic-prescriptive instruction 188
Dierking, L.D. 372, 374, 377
Diener, E. 536, 542
distance learning 189, 430, 437, 443, 449, 462, 466, 480, 540; *see also* online learning
distinguished innovators: defining phenomenon of "abortion" of new ideas 546–7; economic impact of "saved" ideas and implemented innovations 551–2; explaining phenomenon of "abortion" of new ideas 547–9; ignoring sceptics 550; key internal obstacles to innovation 549–50; risk of mistakes and fear 550–1
divergent thinking 55, 56, 144–5, 154–7, 159, 210
diversity issues 192–6, 436
dodecahedrons 335–6
Dodier, N. 278
domain-specific knowledge 42, 56, 117–18, 363
dormant school setting 282
Dörrie, H. 311
Dukas, H. 97, 98, 100, 101
Dunbar, K. 362, 364
"dyad" relationships 516, 517, 520, 521, 523, 524
Dyer, W.G. 515
Dymaxion Maps 334–5
dynamic mathematics curriculum: classroom-as-laboratory 304–6; competitions as "locomotive" for innovation 311–12; difficulties and problems 313; discussion 309–10; eliciting information in unexpected places 310–11; game theory and isomorphism 308–9; group theory and symmetry 306–7; interactive journals 312–13; linear algebra 308
dysynchronous development 94, 103

e-learning *see* online learning
e-portfolios 427
Eames, M. 425
early childhood and adolescent education: micro-social factors 259–60; Nobel laureates 36–7; role of parents 260–7; special teachers 267–70
eco-system approach 572–4, 577
"Ecological Forecast of Future Developments of the Earth" test 175, 176
ecologizing education, technology of 401–2
economic impact of innovation 551–2
economization 405–6
Edison, Thomas 80–1, 83, 87, 88–9, 614
Educate America Act (1994) 161
educational and methodical set (EMS) 318, 321–3
educational changes 275–7
educational policies, promoting innovative culture through 593–6
educational technologies 397–402
educational trends 222–3
"edutainment" 464
Eid, M. 536, 542
Einstein, Albert 37, 43, 174; impact of sensitive periods on development of 100–2; overlapping sensitive periods and resulting unique cognitive experience 103–4; sensitive periods 96–9; understanding exceptionality of development of 102–3
Ekman, P. 390
elaboration 156–7, 159; decrease in 158
Elder, L. 122–3

electronic badge project 565–7
emancipatory pedagogy 278, 280, 281, 284
emergent career formation 518, 520, 521, 523, 524
emerging curriculum areas (ECAs) and settings: changing education 275–7; discussion 283–5; findings 278–83; research methodology 277–8
Emmer, M. 330, 331
emotional sensitivity 102
emotions 534–42
employee selection 507–8
enclosed school setting 282, 284–5
engagement in learning 247, 249–50, 254–5; curriculum innovation 385–6; generating ideas with others 386–7; *see also* BioStories project
engineering: experiencing entrepreneurial dynamics 530–1; framework for integrative entrepreneurship course 529–30; problem-solving as solution search 362–3; rise of entrepreneurial courses 528–9
Ennis Model of Critical Thinking 121–2
enterprise education 274, 574–5
enthusiasm, fostering 537–9
entrepreneurial abilities, great entrepreneurs 33–5
entrepreneurial collegiate academic program: assurance of learning results 484; curriculum 476–80; developing creativity and innovation at collegiate level 483–6; extending entrepreneurship and innovation to other academic areas 480–3; extending entrepreneurial learning outside the classroom 478–80; history of 474–5; learning goals 475; literature review 474
entrepreneurial courses, rise of 528–9
entrepreneurial drive 84–5; origins of 607–8; promotion of 595–6
entrepreneurial dynamics, experiencing 530–1
entrepreneurial giftedness: and breakthrough innovations 508; case studies 229–36; as cornerstone of innovation education 31–2; as curriculum friendly prototyping skills 80–1; discussion 236–9; early manifestations of 32–3; literature on 229
entrepreneurial leadership 476–83
entrepreneurial teams and their formation 514–15
Entrepreneurs on Campus, Canisius College 479, 482
entrepreneurs, Singapore as hub for 596–7
equal access to knowledge, technology of 400
Erickson, J. 33, 235, 549
Ericsson, K.A. 82, 270, 362, 366
ETH Zurich 529–30
ethical behavior 505–6
ethical maturation 460
ethnicity issues 195–6, 202, 293
Etzkowitz, H. 577
EU, knowledge exchange 579
Euclidean geometry 97, 101, 331–2
European Foundation for Quality Management (EFQM) 529–30

exam-focused education 342, 343, 565, 609–10
excellence 33, 39, 46, 268, 509
Excellence through Continuous Enterprise and Learning (ExCEL) Fest 599
executive abilities 30, 34, 35–6, 71
"expanding on lessons from Hollywood" exercise 477
experiential learning 375–81, 460–1, 476–7
experimental design strategies 363, 365–6
expert validation 246
explanatory models 75–6
extracognitive abilities 34–5, 36, 170–1, 178–9

face-to-face learning environments 434–6, 437, 448, 449–51
facilitation of creativity for ideation (FCI) process 19, 20, 24–7
faculty buy-in, university innovation 452
Falk, J.H. 372, 374, 377
family values 259, 260–7
family–career conflict 194
"famous five" exercise 249–50
fear as obstacle to innovation 550–1
feedback 138, 367, 539, 541
Feldhusen, J.F. 60, 63
Fenical, Bill 377
Fensham, P. 342, 372, 426
Feynman, Richard 3
"Fifteen Game" 308–9
film industry 515, 517, 521–3
Finke, R.A. 55, 59
Fish, L.A. 477
Flake, M.A. 161
flexible learning: in action 44–5; definition of 444–5; Nipissing University 445–9; nursing education 449–51; reflections and recommendations 451–2
Florida, R.L. 591
fluency 156–7, 159; decrease in 158–9
Fölsing, A. 98, 102
for-profit innovations 406–9
Ford, D.Y. 195, 202
formal learning: compared to formal learning 372–5, 377–80, 381; lessons from informal learning 382–3
"Formulation of Problems" test 174–5, 176
Foucault, M. 406
Foundation for Critical Thinking 123
Fox, L.H. 194
fractal art 336–7
framing 277, 278–81, 284, 363, 366
Fredricks, J. 252, 254
Fredrickson, B.L. 537, 539
Freeman, J. 193
Freud, Sigmund 612–13
Frick, R. 251, 252, 254
Friesen, W.V. 390

Frumkina, R.M. 179
Fuller, Buckmeister 331, 334–5
Fuller, S. 405, 411
functional fixedness 608–9, 610–11
Future Problem Solving Program International (FPSPI): competitive components 217–18; and current trends in education 222–3; development of innovation skills 218–21; development of innovative dispositions and meeting affective needs 221–2; effectiveness 223–4
Futureintech 424
futures thinking 420, 425–6, 427

Gabbard, D.A. 405, 407
Gagné, F. 247, 248, 255
Galton, Sir Francis 112–14
game theory and isomorphism 308–9
Gandal, M. 161
Gardner, H. 238
Gardner, P.L. 407, 422
Gates, Bill 33, 112; early development 234–5; impact of early development 236–9
Gelfman, E.G. 317, 321, 326
gender issues 192–4, 195, 196
Gendron, G. 233, 238
general ability 130–1
general exploratory activities 133–4, 135–6
genius hypothesis 361–2
Gentner, D.R. 75
Gentry, M. 161, 248, 249
geometry: and architectural design 338–9; Einstein's interest in 97–8, 101; and fractal art 336–7; global projections 331–6; and visualization 331
Gergen, D. 590
Getzels, J. 54, 55, 119
Gibbons, M. 572, 577
gifted education: advances contributing to innovation education 53–7; advances in innovation education contributing to 57–9; creativity research 114–15; diversity issues 192–6; links with innovation education 52–3; mathematics curriculum 304–13; programs focused on talent identification and facilitation 186–9; prospects of interaction with innovation education 59; research on 71–2; shortcomings of 31; synthesis across identification and facilitation models 189–90; systemic reform approach 190–2; synthesis across systemic models 192
giftedness: identification of 132–6; research on 115–17; see also entrepreneurial giftedness; three-ring conception of giftedness
gifts 247, 248–55
Giles, C. 296, 288
Glaser, B.G. 456, 461, 467
Glass, C. 338–9
Gleick, J. 336
global education market 448

Global Issues Problem Solving (GIPS) 217, 219–20, 222–3
global projections 331–6
globalization 222–3, 409, 583–9, 590
Goh, B.E. 63
Golden Key schools 296
governance structure, innovative schools 296
Grayson, D. 276, 277
Grosvenor, I. 275
grounded theory 456, 463, 467
group decision support software (GDSS) 435
group dynamics 435–6
group theory and symmetry 306–7
"Grouping Dots" test 179
Growther, J.G. 98, 101
Guilford, J.P. 53, 54, 155, 459
Gunnarsdottir, R. 19–20, 458, 465
Gunstone, R.F. 385

Haase, H. 528
Haensly, P.A. 136–7
Halpern, D.F. 364, 367
Halverson, R. 58, 61
Hambrick, D.C. 535
Hare, R.D. 409
Hargreaves, A. 288, 296
harmful innovation 409
Harpaz, Y. 221
Harry Potter and the Sorcerer's Stone (1997) 247–55
Hatano, G. 68, 74
Hatch Act (1887) 611–12
Hautala, V. 538, 539, 541
healthy lifestyle, technology of promotion 399–400
Hébert, T.P. 134
hidden innovative abilities 177
Hiebert, J. 288
high expectations transitions 210–12
high intellectual and creative educational multimedia technologies (HICEMTs) 71; adaptation to individual's psychological organization 44; activation of fundamental cognitive mechanisms 43; general psychological basis 42; new targets of educational and developmental influences 43–5; numbers of 45–6; psycho-edutainment 44–5; specific characteristics 45
high potential /low income (HP/LI) students 202, 204–5, 206, 212–13
high-stakes testing 161, 162, 288, 294, 342, 421
Higher Education Funding Council for England (HEFCE) 573–4, 576
higher education: embedding knowledge exchange within 575–7; enterprise education 574–5; institutional transformation 577–8; international developments in knowledge exchange 578–9; knowledge exchange and innovation 573–4; roles and responsibilities 571–2; technological development role 572–3; see also universities

higher order thinking 60–1, 208, 437–8
historical perspective on innovation 607–10
Hitchings, George H. 262
Hmelo-Silver, C.E. 364, 489
Hodson, D. 411, 412
Hoffman, B. 97, 98, 100, 101, 450
Holtzman, L. 291
Holyoak, K. 61, 364
Hounsfield, Godfrey N. 266
Howe, M.J.A. 259, 260
Hulse, Russell A. 265
human capital formation 592–3
Human Genome Project 361
Hunkin, Tim 378, 380
Hunt, Tim 268
Hwang, S.W. 387, 392
hybrid education 196, 577, 614–16
hybridized writing 388

ICARE model 449–51, 452
Iceland, innovation education 17–28, 289; *see also* emerging curriculum areas (ECAs); virtual reality learning environment (VRLE)
icosahedrons 331–6
"Ideal Computer" test 172–3, 174, 175, 176, 177
idealized representations 407–9, 411
ideation 72, 458–9, 467; and innovation 459; research findings 463–4; role in building innovativeness through general education 460; process 19, 20, 24–7; supporting with VRLE 460–2
imagery training 72
Implicit Association Test (IAT) 338
immersive education 85, 431, 432, 437
impression management 434, 436–7
Inagaki, K. 68, 74
individual innovation 33, 40, 385–6, 391; basis of 175; phenomenon of 501–11; theory 169–71; trajectory leading to 614
individual innovative geniuses, relation to innovative teams 612–15, 616
individualized knowledge-testing and self-checking, technology of 400–1
individualized learning 437, 438; technology of 400
industry–school links 423–5, 427
infantilization 408
Info-Communication Development Authority of Singapore 594
informal learning; in life of innovators 375–81; in science 373–5
information and communications technology (ICT) 428, 431–4, 578; *see also* user innovation
initiative 20, 27, 316, 320, 460
INNOCREX 46
Innova Junior College, Singapore 600
INNOVATE program 87
Innovation and Enterprise (I&E) plan, Singapore 596, 599–600

innovation culture 590, 592–6, 611–12
innovation education: agenda to improve 580–1; benefits of 148; in context 433–4; defining phenomenon of 20–2; definition of 215–17; effectiveness of 610–12, 616; finding the balance 19; key concepts 19–20; necessity for 607–10, 616; practical description 18–19; research on 71–3; sequence of achievements for 84–5; structure of 31–2; theoretical language 22–4
innovation gap 35–6
innovation products, accumulation of 608
Innovation Program (IvP), Singapore 598
innovation science, basics of 40
innovation trends, UK 572–3
innovation: blocks on 608–10; conceptualising 391–3; critical components in developing 483–6; definitions of 111–12, 444, 591; elements of 246–8; generating 361–2; internal obstacles to 549–50; models promoting 117–19; programmes to promote 598–601; research on 115–17
innovative abilities assessment: new approach to 171–9; theory of individual innovation 169–71
innovative behaviors, tips to foster 139
innovative capacity: evolution of 82–3; STEM instructional strategies 365–7
innovative creativity 153, 154–7
innovative culture, promotion of 593–6
innovative dispositions, development of 221–2
innovative potential testing 177
innovative schools course: course description 290–2; creative teaching and innovation education 289–90; student reflections and comments 297–9; supporting reflections and connections 296–7; themes from school visits 293–6
innovator teams, relation to individual innovative geniuses 612–15, 616
innovators: contemporary examples 112; distinction with creators and implications for school-based learning 113–14; historical examples 112–14; informal learning 375–81; young examples 128–9, 136
inquiry-based learning 294, 296–7, 360–7, 373, 382, 412
institutional change 577–8
instructional analogies 75
integrated and interdisciplinary education: experiencing entrepreneurial dynamics 530–1; framework for 529–30; rise of entrepreneurial courses 528–9
Intel International Science and Engineering Fair (ISEF) 135, 599
intellectual contents, HICEMTs 45
intellectual development: educational texts as means of 320–6; enrichment of mental experience as psychological basis for development of intelligence 318–20
intellectual property (IP) 493, 352, 489, 572–5, 578

Index

intelligence: changes in 154; and creativity 153–4
intentional (emotional–evaluative) experience 318–20, 326, 328
Interest-A-Lyzer 134
internal obstacles to innovation 549–50
International Biology Olympiads (IBOs) 401
intra-government cooperation, Singapore 596–7
intrinsic motivation 117–18, 161, 162, 251, 537, 539–40
intuition 34–5, 506–7
IQ tests 54, 153–4, 156, 172, 174, 175–6, 187, 236
irreplaceable relationships 516, 518, 520–1
Isaacson, W. 100, 101, 112
Isaaksen, W. 57, 63, 112, 119
isomorphic career formation 518, 520, 522, 523–4
IT masterplans, Singapore 593–4

Jackson, P.W. 54
Japan, entrepreneurial education 142–3, 147–8
Jarvis, J.M. 249–50
Java applets 71
Javits programs 113, 114
Jeffrey, B. 275
Jobs, Steve 112, 501, 506, 508, 509, 510
Johansson, F. 436, 535
joint functioning 510
Jónasdóttir, S.R. 274–81
Jones, A. 420, 421, 422, 423, 424, 425–7
Jones, C. 515, 516, 517
Josephson, M. 81
joy in education 537–8, 539–40
"junior version" innovations, support for 61–2
Jurong Junior College, Singapore 601

Kagan, J. 178
Kaleidoscope project 196
Kamen, Dean 216
Kao, J. 583, 586
Kaplan, S.N. 206, 208
Kaufman, J.C. 52, 62, 63
Kegan, R. 254
Keinz, P. 489, 492
Kelleher, Herbert 504–5, 509, 510
Ketterle, Wolfgang 260, 264, 267–8
Khan Academy 82–3
Kholodnaya, M.A. 45, 94, 104, 171, 172, 173, 174, 177, 178, 179, 180, 317, 318, 319, 321, 326, 502
Kim, K.H. 153–4, 156, 158, 162, 247, 249
King, Martin Luther Jr. 293
Kirton, M.J. 154–5
Klahr, D. 362, 363, 366
Klein, E.J. 291
Kleppe, J.A. 528, 529
Knol, K. 74
knowing-how 517, 518, 520, 523, 524
knowing-whom 517, 518, 521, 523, 524
knowing-why 517, 518, 520, 523, 524

knowledge base 27, 42, 111, 135, 162, 260, 361, 422
knowledge economies 116, 407–8, 571, 590, 593–4, 597
knowledge exchange: embedding within universities 575–7; enterprise education 574–5; innovation context 572–3; international developments 578–9; UK policy 573–4; underpinning academic literature 577–8
Knowledge is Power (KIPP) Program 294
knowledge transfer for solving new problems 364–5
Koen, B.V. 360
Krames, J.A. 504, 505
Krathwohl, D.R. 120
Krimsky, S. 406
Kroemer, Herbert 261
Krüger, M. 338
Kuhn, T. 69, 113, 217

Lakhani, K.R. 492
language education 480; Renzulli Academy 207–8
Larsen, J. 510
lateral thinking 155–7
Latour, B. 405, 408
Lautenschläger, A. 528
Lave, J. 62, 518
lead-user method 489–93
learning communities 276, 288, 296, 297, 580, 587
learning preferences 444, 447–8
Lee, Y.-J. 385, 391–2
Lehrer, R. 61, 366
Leites, N.S. 94, 95–6, 99
Leonard, A. 409
Lewis, C. 539
lifelong learning 59–60, 117, 535
Limon, M. 74
Lintott, Chris 380
Little, C. 112, 114
Long, W.A. 528
Lorenz, K. 258
love of learning 245–6, 251–2
low-income students 202, 204–5, 206, 212–13, 293–4
Lowenstein, R. 232
Luftig, R.L. 337

"m-learning" 432
McClintock, Barbara 37
McCormick, R. 565
Macdonald, A. 274, 275, 277, 278, 279, 280
McGiffert, I. 161
Mackuen, M.B. 538
McLaren, P. 411, 412
McLean, K. 375
McMullan, W.E. 528
McQuaig, L. 405
macro-social factors 35

McTighe, J. 191
magic squares game 304–11
magnetic fields 96, 100–1, 102, 104
Malaysia, search for innovators 585–8
managed learning environment (MLE) 460–6
managerial challenges, engineering 528–9
managerial talent 507–8; lessons from great managers 39
Mandelbrot, Benoit 336, 337
Manhattan Country School 293
map making 331–3
Marconi, Guglielmo 37
Marcus, G.E. 538
market research 424, 483, 488, 492
Marks, R. 334, 335
Marland, S.P. 54
Marshall, Barry 379
Mason, P.A. 535
Mastandrea, S. 338
Masui, Shoji 522–3
"Matching Familiar Figures" (MFF) test 178
mathematically gifted youth 186–8, 193
mathematics and the arts: fostering for mathematical innovation 337–8; fractal art and geometry 336–7; global projections 331–6; interdisciplinarity as means of innovation 338–9; visualization and geometry 331
mathematics education: contribution to innovation education 561, 564–5; Renzulli Academy 206; see also dynamic mathematics curriculum
"Mathematics, Psychology, Intelligence" education project (MPI) 315, 321–6
mathematics problems 342–55
mathematics textbooks: as means for intellectual development 320–3; psycho-didactic requirements 323–6; psycho-didactic typology 326, 327–8
mathematics–science gap, bridging 341–5
Mather, John C. 263, 265
Means, B. 434, 439
mechanical education 611–12
mechanical toys 96, 98–9, 100, 103–4
media, learning from 374, 380–1, 431
mediation 23–4, 27–8
Medvid, M. 38
Meier, D. 291
mental models 75, 82–3, 84, 89
Mercator, Gerardus 334
Meridian School 295
metacognition 30, 34, 73–5, 125, 170–1, 178–9, 318–20, 326, 328, 364–5; in action 35–6
methodological apparatus, computer based learning systems 397
methodological relativism 520–1
methodological space 362–3, 366
Metz, K.E. 361, 365
Micklus, C. & S. 210

micro-social factors 35, 229–30, 238–9, 259–60
Microsoft 235, 237, 549
misconceptions 69, 70, 74, 345
mistakes, risk of 550–1
model-based reasoning 61, 68, 75–6
Moller, A.C. 251, 254
Moon, S.M. 115
Moore, Patrick 380–1
moral imagination 246, 249
moral responsibility 38–9, 505–6
Moreland, J. 423, 426
Morita, Akio 41, 502, 503, 506, 510, 546
Morrill Act (1862) 611–12
motivation 34, 117–18, 131–2, 137–8, 202–4, 205, 210–12, 252–3, 539–40
motivational pathways 251–4
multimedia technologies 41–6
mutation theories 577–8
MyLinkFace 480
Myhrvold, N. 235

Nantz, K. 343
Nanyang Technological University (NTU), Singapore 597
NASA eClips 350–2
NASA IMAGE satellite education program 345
NASA press releases and mission statements: brief history 345–7; extracting mathematics from 348–50; impact and evaluation of *SpaceMath@NASA* 352–5; and mathematics behind science 342–5; newspapers in education 347–8; video programs 350–2
nation building 591–3
National Curriculum: Iceland 18, 27, 273–4, 289, 457, 460; UK 560–5, 568
National Endowment for Science, Technology and the Arts (NESTA), UK 571, 573
National Framework for Innovation and Enterprise (NFIE), Singapore 597
National History Day Competition 208–9, 211–12
National Science Board 37, 114–15
National University of Singapore (NUS) 597
natural resource deficit, Singapore 591–2
nature-conservation activity foundations, technology of 401–2
Neill, A.S. 283, 291
neoliberal influenced science and technology 405–10
Nersessian, N.J. 75
networks/networking 432, 436, 515–16, 517
"new combinations of already existing elements" 513–14
new creative education 146–9
New York City mathematics team 311–12
New York University (NYU) in Abu Dhabi 584
New Zealand, technology education 421, 423–5
Newell, A. 55, 362

Index

Newspapers in Education (NIE) programs 347–8
NextBizSolutions 480
Nipissing University: Centre for Flexible Teaching and Learning 448–9, 452; overview 445–6; rationale for flexible learning 446–8
No Child Left Behind (NCLB) Act (2001) 161, 341–2
Nobel Laureates in Science: lessons learned from 36–8; micro-social factors in development 259–60; parents role in development 260–7; special teachers role in development 267–70
nonconformity 435–6
Nossal, Gustav 378
Nurse, Sir Paul 269
nursing education: contextual similarities between schools of nursing 451; ICARE model 450–1; need for innovative approaches 449–50; research results 451–2
Nüsslein-Volhard, C. 260–1, 268

O'Connor, G.C. 534, 535
Odyssey of the Mind 210
Olson, D.R. 172
online discussion groups 435
online learning: case study context 444–5; characteristics of environments 432–4; and educational innovation in context 443–4; evolution of environments 431–2; flexible learning at Nipissing University 445–9; and innovation education 434–8; nursing education 449–51; reflections and recommendations 451–2
"open character", innovation tests 174–5
open-ended learning 56, 87–8, 169, 195, 218, 289, 294, 372–3, 380, 382
optimism 32, 508
originality 156–7, 159; decrease in 158
Osborn, A. 190
Osborne, J. 373, 388

Palade, George E. 263
parallel curriculum model (PCM) 191
parents: role in development 203, 204, 260–7, 376; value of education 260–3
passion 245–6, 247, 249
Pasteur, L. 362
Patrick, B.C. 538
Patton, M.Q. 463
Paul, R. 122–3
pedagogical imperative 60–3
pedagogy 18–19, 458–9, 466–7, modes 277, 278–81; discussion 283–5; innovative schools 294–5; 587; interaction with school settings 283; and new technology 432
Pekrun, R. 536, 537, 538, 540, 541
Perkins, D.N. 61–2, 118, 221, 364
PERMATApintar 585–8
perseverance 32, 96, 100, 238, 382–3

personality dimension of creativity 54–7, 60–3
personality traits 20, 21, 32–3, 138, 155–7, 159, 238, 322
Pfeiffer, J. 83, 89
Phillips, William D. 262, 264–5, 269
philosophy, Einstein's interest in 98, 102
physical distance, online environments 433–4
Piechowski, M.M. 94
Piirto, J. 193–4
Polak, F.L. 175
Polanyi, M. 58, 60
polyfunctional psychological systems 317–18
polyhedrons 333–5
polymathy 33, 38, 330
Ponomarev, E.A. 37, 44, 432
Porter, M.E. 591
positive emotions 536–41
practical skills, technology of developing 398–9
preconceptions 69
preservice teachers: course description 290–2; creative teaching and innovation education 289–90; student reflections and comments 297–9; supporting reflections and connections 296–7; themes from school visits 293–6
pride in education 537–8, 540–1
problem-based learning (PBL) 72, 120–1, 190–1, 222, 294; user innovation 489–97
problem-solving approach: importance of teaching 562–3, 565–8; science and engineering 362–7; strategies 137–8; see also creative problem–solving
procedural knowledge 73–4, 322, 365–6, 562–3, 565
procedural technology 397–9
process dimension of creativity 55–7, 60–3
"professional duo" relationships 516, 517, 524
professional occupations, parents 263–4
Programme for International Student Assessment (PISA) 342
progressive pedagogy 278, 280, 284
Project M^3, Mentoring Mathematical Minds 206
Project Rainbow 195–6
project selection 137
Project Work (PW) policy, Singapore 594
project-based learning 137–8, 190–1
prominent innovators: case studies 229–36; discussion 236–9; literature on 229
propulsion theory of creativity 156
prospective assessments 177
prototypical design structures 338
prototyping design factories 614–15
prototyping: and Camp Invention 85–9; entrepreneurial giftedness as prototyping skills 80–1; reasons and methods of innovativeness of 82–5
Prügl, R. 489, 492
psycho-didactic approach 316–18
psycho-didactic requirements, educational tests 323–6

psycho-didactic typology, educational texts 326, 327–8
psycho-educational multimedia technologies (PMTs) 41–6
psycho-edutainment 44–6
psychological assessment of innovative abilities 171–9
psychological basis: HICEMTs 42; individual innovation 169–71; intellectual development 318–20
psychological functions tests 175–6
psychological mental context 172–4
psychological organization, adaptation to 44
Purdue Three-Stage Model of enrichment 56, 115
Pythagorean Theorem 97

"Quadratic equations" 324–6
Quantum (journal) 312–13

Rabbitt, P. 176
Rabi, Isidor 262–3
radical innovators: bringing joy into education 539–40; drive toward entrepreneurial giftedness 613; education of 534–6; fostering enthusiasm 538–9; promoting pride in education 540–1; university education and positive emotions 536–8
Radin Mas Primary School, Singapore 599–600
Rapaport Academy 293, 294
reader-oriented theory 321
reading: parental encouragement of 260–3; Renzulli Academy 206–8
real world issues 62, 72, 111–12, 115, 117, 120–3, 135–6, 236, 460, 477–8, 480, 485
reasoning elements model 122–3
reasoning strategies, STEM disciplines 363–4
reflectivity-impulsivity cognitive style 176, 178
Reines, Frederick 269
Reis, S.M. 57, 62, 194, 202, 203, 204, 206–7, 208, 209, 210, 248, 250
Reiser, A. 97, 101
religious feeling 96–7, 101
Rennie, L.J. 373–4, 420–1
Renzulli Academy 201–2: curriculum 206–12; innovation and creativity 212; notable accomplishments 212–13; talent identification 204–5; philosophy 202–4
Renzulli Learning System (RLS) 62, 134
Renzulli Talent Pool Identification Model 204–5
Renzulli, J.S. 52–9, 62, 129, 132, 134, 202–4, 206, 210
research and development (R&D) 610–11
Research, Innovation and Enterprise Council (RIEC), Singapore 597
resistance to premature closure 156–7, 159; decrease in 159
resources, effective use of 137
retrospective assessments 177

Rinearson, P. 235
Riordan, M. 291
risk-taking 434–5
Ritchhart, R. 289, 291
Ritchie, S.M. 387, 389, 390, 393
Roberts, Richard J. 262, 266, 269
Robinson, P. 563
Rogan, J. 276, 277
Rogers, E. 459, 510
role models 229, 239, 259–60, 378–80, 382
roles, classification of 277, 278–81
Root-Bernstein, R. 38, 115, 330, 338
Rosenblatt, J.S. 95
Rosmah Mansor, Datin Sri 586
Roth, W.-M. 385–7, 390, 391–2, 393, 407, 408
Rowling, J.K. 245–6, 247–55
Roychoudhury, A. 386
Runco, M. 62, 63, 591
Russ, S. 159, 591

Sagan, Carl 381
Salerno, L. 507, 510
Sandberg, B. 535, 538
"saved" ideas, impact on economy 551–2
Sawyer, R.K. 53, 57, 58–9, 62, 71–2
Scardamalia, M. 57–8, 60–1, 63
Scenario Writing (SW) 218, 223
sceptics, ignoring 550
Schauble, L. 61, 366
Schneider, W. 42, 137, 252
Scholastic Achievement Test (SAT) 187, 188, 193, 196, 584, 585
school initiatives, Singapore 599–601
school level change 276–7
school missions 293–6
school settings: discussion 283–5; innovation education in 281–3; interaction with pedagogy 283
school structure 296
school visits 290–1, 292–6
Schools Mathematics Project (SMP), UK 564
Schoolwide Enrichment Model in Reading (SEM-R) 206–7
Schoolwide Enrichment Triad Model (SEM) 55–7, 188–90, 202–4, 205, 250, 321–3; enrichment clusters 209; forms of implementation 135–6; Type 1 enrichment 133–4, 135–6; Type 2 enrichment 134, 135–6; Type 3 enrichment 135–6
Schrage, M. 81
Schragen, J. 363
Schroeder, A. 231, 232, 238
Schumpeter, J. 513, 528
science education: BioStories Project 387–93; contribution to innovation education 560–1, 562; curriculum initiatives 385–6; generating ideas with others 386–7; informal learning 373–5;

science education *continued*
    nature of and links to innovation education 420–2; problem-solving as solution search 362–3; Renzulli Academy 208; teaching from conceptual change point of view 73–6
science and technology and societies and environments (STSE) education 411–12
science fairs 129, 135, 136, 138–9, 210, 212–13, 598
science–mathematics gap, bridging 341–5
scientific creativity 361–2
scientific experimentation, home beginnings 264–7
scientific innovators-geniuses: development psychology and high ability studies on sensitive periods 93–6; explaining emergence of 99–104; implications for innovation education 104–5; methodological issues 93; sensitive periods in individual development of Albert Einstein 96–9; *see also* Nobel Laureates in Science
search field definition 490–2
search specification, identification of 495–6
Seeling, C. 98, 101
Seeratan, K. 104, 169–70, 178–9
self-confidence 193, 436–7
self-directed learning 295–6, 489, 595
self-evaluation 124, 400–1, 540–1
self-explanation 367
self-regulation 135, 207–8, 210–13, 316, 320, 406
self-talk 377–8
semiotics 407–10
"sensing" 436, 502
sensitive periods: chains of 102–3; as development foundation of scientific innovator–genius 99–100; findings from development psychology and high ability studies 93–6; implications for innovation education 104–5; in individual development of Albert Einstein 96–9, 100–3; overlapping periods and resulting unique cognitive experience 103–4
Shantz, D. 290
Shavinina, L.V. 30–6, 40–1, 70–1, 94–5, 102–4, 169–80, 227, 229–33, 330, 386, 432, 500–3, 508–10, 546, 547
Sherman, J. 234, 235, 551
Shulman, L.S. 247, 284
Shute, V.J. 63, 367
Silverman, L. 95, 176
Simon, H.A. 55, 229, 362, 363
Simon, M. 317
Simonton, D.K. 216–17, 212, 221, 228, 361, 362, 365, 500
simulation 529, 531
Singapore Science and Engineering Fair (SSEF) 599
Singaporean talent development: defining innovation the Singaporean way 591; innovation as tool for nation building 591–3; programmes to promote innovation 598–601; promoting innovative culture through educational policies 593–6; Singapore as hub for entrepreneurs beyond schools 596–7
Sky at Night (BBC) 380–1
small business management course 475, 477–8
Smith, C.L. 75
Smith, Hamilton O. 263, 265, 269
Smith, M.K. 304
Smith-Lever Act (1914) 611–12
social activism 409–12
social context, science learning 374–81
social ecology 276–7, 284, 285
social justice 293–4
social matrix 21–2
social needs, meeting 221–2
social norms, willingness to defy 435–6
Social Sciences and Humanities Research Council (SSHRC) of Canada 259
social sensitivity 102
social studies, Renzulli Academy 208–9
sociocultural learning theories 385–93
socioscientific innovation: for the common good 409–12; in neoliberal times 405–9
solution giving and finding, problem solving as 567
solution problem space 363, 366
Sony Walkman 41, 502, 503, 506, 509–10, 546
Southwestern Airlines 504–5, 509
SpaceMath@NASA 345–6, 348–52; evaluation of 354–5; impact of 352–4
Space Science Problem of the Week 345
special schools 189
special teachers 267–70
specialization 38, 131, 203, 238, 337, 363, 520, 608
Specialized Independent Schools, Singapore 594–5
spin-outs 571, 572, 574–5, 579
Sriraman, B. 330
stage theory 515, 516
Standards, Productivity and Innovation Board (SPRING Singapore) 596
Stanley, J.C. 186, 584–5
STEM (science, technology, engineering and mathematics) education 29–30, 84, 185–6, 189–90, 192–3, 194, 422; call for 115–17; common modes of reasoning 363–4; contribution to innovation education 560–5, 568; instructional strategies to enhance capacity for innovation 365–7; knowledge transfer for solving new problems 364–5; lessons from UK 568; NASA support for 355; problem solving approach in 362–3, 565–8; relevant distinctions between disciplines 360; scientific creativity 361–2
Stenhouse, L. 304
Stepwise curriculum framework 410–12
stereotypes 194, 436–7
Sternberg, R.J. 113, 118, 125, 156, 172, 176, 192, 195–6, 246, 434, 435, 591
Stigler, J.W. 288
Stocklmayer, S.M. 373, 374, 375, 377, 380

stone-tool evolution 83, 84, 608
Stormer, Horst L. 261, 267
Strauss, A.L. 456, 463, 467
structured learning environments 373–4
STS (science-technology-society) approaches 422
student democracy 295–6
student market, responding to 447–8
Students in Free Enterprise (SIFE) 482
Study of Exceptional Talent (SET) project 187, 188
Study of Mathematically Precocious Youth (SMPY) 188–9, 193, 196; research 188; talent identification 186–7; variety of interventions 187–8
subject-centred approach 316–18
Subramaniam, M. 535
successful intelligence 192; and WICS 195–6
Sullivan, P. 434, 437
supported independent reading (SIR) 207
sustainability, innovative schools 296
Suzuki, David 376
Svejenova, S. 515–16, 517, 520, 521
synchronous learning environments 443
systemic reform approach 190–2

talent development: challenging curriculum 249–50; creativity and innovation 246–8; defining innovation the Singapore way 591; expert teachers 248–9; innovation as tool for nation building 591–3; motivational pathways and pedagogy 251–4; programmes to promote innovation 598–601; promoting innovative culture through educational policies 593–6; Singapore as hub for entrepreneurs beyond schools 596–7; skilled coaching 250–1; student engagement in learning 254–5
talent identification/facilitation; programs focusing on 186–90; Renzulli Academy 204–5; three-ring conception of giftedness 129–32
Tan Kah Kee Young Inventors' Award (TKKYIA) 598–9
Tanjong Katong Girls' School, Singapore 600–1
Tapscott, D. 442, 447
task commitment 55–6, 117–18, 129–30, 131–2, 137–9, 203, 205, 209
Teach Less Learn More policy, Singapore 595
teachers: expertise 247, 248–9; imposition of curriculum 565; influence on high-ability development 267–70; professional development 559–60, 573; role of 275; specialization 203; subject knowledge 425–6; subject subculture 422–3, 427; *see also* preservice teachers
Techlink 423
Technical and Vocational Educational Initiative (TVEI), UK 558–60, 612
technical competence leveraging method 489, 493–7
technological readiness 444
technology application areas, identification/analysis of 496–7

technology as curriculum area 422–3
technology education: authentic assessment 426–7; contribution to innovation education 561–4; fostering innovation education in 422–6; nature of and links to innovation education 420–2; problem solving approach 565–8; UK 559–60
technology leveraging, Singapore 593–4
technology transfer 572–80
technology-push strategies 493–4, 572–3
TechShop 615
Teo, C.H. 592
Terman, L. 54
tertiary level initiatives, Singapore 597
Teutsch, A. 235
Texas Instruments 503
textbooks: shortcomings of 373; space science 344–5, 350, 351; *see also* mathematics textbooks
Thagard, P. 61, 69, 70–1
theoretical language 22–4
theoretical space 362–3, 366
Thiessen, B. 330, 337, 338
Thinking Schools Learning Nation (TSLN) policy, Singapore 593
thinking skills 72, 221–2
Thorsteinsson, G. 72, 457, 458–9, 460, 464, 465
Three Ring Conception of Giftedness 56, 115, 129–32, 205
Thursby, M.C. 528, 529
time constraints, innovation tests 176–7, 179
Tobin, K. 386, 393
Tomas, L. 390
Tomlinson, C.A. 61, 191
Torrance Tests of Creating Thinking (TTCT) 57, 154–7
Torrance, E.P. 54–5, 57, 58–9, 115, 160–1, 175, 591
Tracy, J.L. 540
traditional creative education 143–6
"trajectory analysis"/"trajectory assessment" 168, 169
Tran, V. 536, 538
transmissive pedagogy 278, 280, 284
Traytak, D.I. 396
Treagust, D.F. 385
Treffinger, D.J. 56, 57, 63, 121, 190, 222, 223, 224
Tribus, M. 217
"triple helix" university-industry-government relations 577
Tsui, Daniel C. 267
"two-eyed theory of teaching" 255

UK: enterprise education 574–5; innovation trends 572–3; knowledge exchange policy 573–4; lessons from STEM education 568; STEM subjects contribution to innovation education 560–5; teaching problem solving to promote innovation 565–8; Technical and Vocational Educational Initiative (TVEI) 558–60

underachievers 36, 99, 162–3, 230, 258, 501
unique vision 34, 59, 169, 171, 177, 179, 316, 320, 501–2
"united career" 515–16, 517, 520, 521
Universiti Kebangsaan Malaysia (UKM) 586–7
university admission tests 584, 585, 586–7
university culture 452
University of Northern British Columbia 450–1
University of San Diego 450
university online education: case studies 445–51; case study context 444–5; in context of educational innovation 443–4; reflections and recommendations 451–2
university-based scientists 406–7
urban environments 192, 195–6, 201–2, 212–13
US: culture of innovation 611–12; knowledge exchange comparisons with UK 578–9
user innovation: common user approaches how to teach them 489; as new paradigm in innovation 487–9; problem-based learning courses 489–97; success factors 497

Vallentin, A. 97, 102
values 194, 457
van der Heijden, K. 530
VanTassel-Baska, J. 61, 112, 114, 162, 191, 196, 202, 205
Vandervert, L.R. 46, 80, 82–3, 88, 169, 180, 550
Vane, John R. 265–6
Veltman, Martinus J.G. 269
Venetian merchants 163–4
Venville, G. 411
vertical thinking 155–7
video games 160
Vienna University of Economics and Business: "New Business Development" course 493–7; "Sources of Innovation" course 489–93
Virgin group 229–30, 238, 501, 502, 505, 506, 507–8, 509, 510–11
virtual reality learning environment (VRLE) 432; discussion 464–6; focus and aims of research project 462; general findings 463–4; ideation and its role in building innovativeness through general education 460; innovation and ideation 459; for innovation education 457–8; pedagogy of innovation education 458–9; research methodology development 463; upgrading pedagogical model 466–7; using to support innovation education 460–2
Vision 2020 585–6
visual representations 179
visualization and geometry 331
vocational education 558–60
Vockell, E.L. 347
von Hippel, E. 488, 489, 490, 492

Vosniadou, S. 70, 74, 75
Vygotsky, L.S. 22–4, 26–7, 61, 63, 95, 159, 177, 276, 296, 458, 464

Waldorf schools 294–5, 296, 299
Wallace, J. 33, 235, 549
Wallen, N.E. 137
Walton, Sam: early development 235–6; impact of early development 236–9
Wang, E.L. 528, 529
"water and aqueous solutions" lessons 398, 400, 401
Wellington, J. 373, 388
Wenger, E. 62, 412, 518
Wharton, C.M. 365
Whitrow, G.J. 96, 97, 102
WICS (Wisdom, Intelligence and Creativity Synthesized) 192, 195–6
Wieman, Carl E. 262, 268
Wildausky, B. 584
Williams, L.A. 540
Williams, W.M. 154
wisdom 38–9, 504–6
Wiser, M. 75
wizards 248–55
Women's Business Center, Canisius College 478, 480
women/girls 35, 192–5
wonder 96, 101, 102
Wong, R.M.F. 367
Woods, P. 275
World Wide Day of Science 379
Wozniak, Steve 112
Writing, Renzulli Academy 206–8

Xu, F. 143, 144, 145, 146

Yalow, Rosalyn 261, 263, 267
Yamada, J. 516, 517, 519, 521, 524
Yamashita, M. 516, 517, 519, 521, 524
Yeomans, D. 559
Yin, R.K. 278
Youndt, M.A. 535
Young Entrepreneurs Scheme for Schools (YES! Schools), Singapore 596
Young Innovators' Fair, Singapore 598
Young Inventors Competition (YIC) 18
Young, R.A. 376
Yusof Ishak Secondary School, Singapore 600

Zaleznik, A. 516
Zaporozhets, A.V. 95
zeitgeist hypothesis 361–2
Zhang, J. 58
Zimmerman, B.J. 135
Zone of Proximal Development 23–4, 26–7, 466